# eWORK AND eBUSINESS IN ARCHITECTURE, ENGINEERING AND CONSTRUCTION

PROCEEDINGS OF THE 12TH EUROPEAN CONFERENCE ON PRODUCT AND PROCESS MODELLING (ECPPM 2018), COPENHAGEN, DENMARK, 12–14 SEPTEMBER 2018

# eWork and eBusiness in Architecture, Engineering and Construction

*Editors*

**Jan Karlshøj**
*Department of Civil Engineering, Technical University of Denmark, Lyngby, Denmark*

**Raimar Scherer**
*University of Technology, Dresden, Germany*

CRC Press is an imprint of the
Taylor & Francis Group, an **informa** business

A BALKEMA BOOK

CRC Press/Balkema is an imprint of the Taylor & Francis Group, an informa business

© 2018 Taylor & Francis Group, London, UK

Typeset by V Publishing Solutions Pvt Ltd., Chennai, India

All rights reserved. No part of this publication or the information contained herein may be reproduced, stored in a retrieval system, or transmitted in any form or by any means, electronic, mechanical, by photocopying, recording or otherwise, without written prior permission from the publisher.

Although all care is taken to ensure integrity and the quality of this publication and the information herein, no responsibility is assumed by the publishers nor the author for any damage to the property or persons as a result of operation or use of this publication and/or the information contained herein.

Published by: CRC Press/Balkema
　　　　　　　Schipholweg 107C, 2316 XC Leiden, The Netherlands
　　　　　　　e-mail: Pub.NL@taylorandfrancis.com
　　　　　　　www.crcpress.com – www.taylorandfrancis.com

ISBN: 978-1-138-58413-6 (Hbk)
ISBN: 978-0-429-50621-5 (eBook)

*eWork and eBusiness in Architecture, Engineering and Construction – Karlshøj & Scherer (Eds)*
© 2018 Taylor & Francis Group, London, ISBN 978-1-138-58413-6

# Table of contents

Preface ix

Organization xi

### BIM implementation and deployment

Experiences from Norway on implementing BIM in existing bachelor engineering curriculum 3
*E. Hjelseth*

Understand the value of knowledge management in a virtual asset management environment 13
*C. Mirarchi, L. Pinti, M. Munir, S. Bonelli, A. Brizzolari & A. Kiviniemi*

Integrated information management for the FM: Building information modelling and database integration for the Italian Public Administration 21
*L. Pinti, S. Bonelli, A. Brizzolari, C. Mirarchi, M.C. Dejaco & A. Kiviniemi*

### Building performance simulation and nD modelling

Use-case analysis for assessing the role of Building Information Modeling in energy efficiency 31
*A. Alhamami, I. Petri & Y. Rezgui*

A new design for gas dehydration units 39
*D. Asgari*

Fast track BIM integration for structural fire design of steel elements 43
*L. Beltrani, L. Giuliani & J. Karlshøj*

BIM tools for structural analysis in the Wenchuan earthquake aftermath 51
*G. Cerè, Y. Rezgui & W. Zhao*

A tool for IFC building energy performance simulation suitability checking 57
*G.N. Lilis, G. Giannakis, K. Katsigarakis & D.V. Rovas*

Implications of operational, zoning-related, and climatic model input assumptions for the results of building energy simulation 65
*G. Pilati, G. Pernigotto, A. Gasparella, T. Farhang & A. Mahdavi*

BIM-to-BEPS conversion tool for automatic generation of building energy models 73
*M. Regidor, M. Andrés & S. Álvarez*

Energy-saving potential of large space public buildings based on BIM: A case study of the building in high-speed railway station 79
*N. Wang, J.L. Wang, C.H. Liu & L. Liu*

Multi-objective optimization design approach for green building retrofit based on BIM and BPS in the cold region of China 87
*H. Yang & C. Liu*

### Collaboration methods

IPD/BIM collaboration requirements on oil, gas and petrochemical projects 97
*A.H. Fakhimi, J.M. Sardroud, S.R. Ghoreishi & S. Azhar*

| | |
|---|---|
| Towards level 3 BIM process maps with IFC & XES process mining<br>S. Kouhestani & M. Nik-Bakht | 103 |
| An agenda for implementing semi-immersive virtual reality in design meetings involving clients and end-users<br>S. Mastrolembo Ventura & F. Castronovo | 113 |
| Improving the integration between BIMs and agent-based simulations: The Swarm Building Modeling – SBM<br>G. Novembri, F.L. Rossini & A. Fioravanti | 123 |
| Holistic methodology to understand the complexity of collaboration in the construction process<br>S.F. Sujan, A. Kiviniemi, S.W. Jones, J.M. Wheatcroft, E. Hjelseth, B. Mwiya, O. Alhava & A. Haavisto | 127 |

## *Integrated modelling of the built environment, incl. smart cities*

| | |
|---|---|
| Building/city information model for simulation and data management<br>T. Delval, A. Jolibois, S. Carré, S. Aguinaga, A. Mailhac, A. Brachet, J. Soula & S. Deom | 137 |
| Defining quality metrics for photogrammetric construction site monitoring<br>F. Eickeler, A. Borrmann & A. Mistre | 147 |
| HBIM acceptance among carpenters working with heritage buildings<br>S.A. Namork & C. Nordahl-Rolfsen | 155 |
| Towards the application of the BIMgrid framework for aged bridge behavior identification<br>M. Polter & R.J. Scherer | 163 |
| Collaborative platform based on standard services for the semi-automated generation of the 3D city model on the cloud<br>I. Prieto, J.L. Izkara, A. Mediavilla, J. Arambarri & A. Arroyo | 169 |
| Visualizing earthwork and information on a linear infrastructure project using BIM 4D<br>L. Schneider Jakobsen, J. Lodewijks, P.N. Gade & E. Kjems | 177 |

## *Interoperability and standardization of data structures and platforms*

| | |
|---|---|
| Multi-LOD model for describing uncertainty and checking requirements in different design stages<br>J. Abualdenien & A. Borrmann | 187 |
| NovaDM: Towards a formal, unified Renovation Domain Model for the generation of holistic renovation scenarios<br>A. Kamari, C. Schultz & P.H. Kirkegaard | 197 |
| Software library for path planning in complex construction environments<br>K. Kazakov, S. Morozov, V. Semenov & V. Zolotov | 207 |
| Delivering COBie with ProNIC—compliance and implementation<br>P. Mêda, J. Moreira & H. Sousa | 215 |
| Automatic development of Building Automation Control Network (BACN) using IFC4-based BIM models<br>R. Sanz, S. Álvarez, C. Valmaseda & D.V. Rovas | 223 |
| Design-to-design exchange of bridge models using IFC: A case study with Revit and Allplan<br>M. Trzeciak & A. Borrmann | 231 |

## *IoT, sensor and industrialized production*

| | |
|---|---|
| Modeling construction equipment in 4D simulation<br>R. Amrollahibuki & A. Hammad | 243 |

| | |
|---|---|
| IoT 2.0 for BIM, Bluetooth technology applied to BIM facility management oriented<br>*S. Giangiacomi & R. Seferi* | 251 |
| A specialized information schema for production planning and control of road construction<br>*E. Haronian & R. Sacks* | 257 |
| Formwork detection in UAV pictures of construction sites<br>*K. Jahr, A. Braun & A. Borrmann* | 265 |
| Fostering prefabrication in construction projects—case MEP in Finland<br>*R.H. Lavikka, K. Chauhan, A. Peltokorpi & O. Seppänen* | 273 |
| RenoBIM: Collaboration platform based on open BIM workflows for energy renovation of buildings using timber prefabricated products<br>*A. Mediavilla, X. Arenaza, V. Sánchez, Y. Sebesi & P. Philipps* | 281 |

*Modelling of design, construction, operation, maintenance management processes*

| | |
|---|---|
| BIM model methods for suppliers in the building process<br>*A. Barbero, M.D. Giudice & F. Manzone* | 291 |
| Implementation framework for BIM-based risk management<br>*I. Björnsson, M. Molnár & A. Ekholm* | 297 |
| BIM-based model checking in a business process management environment<br>*P.N. Gade, R. Hansen & K. Svidt* | 305 |
| Building Information Modelling (BIM) value realisation framework for asset owners<br>*M. Munir, A. Kiviniemi & S. Jones* | 313 |
| BIM solutions for construction lifecycle: A myth or a tangible future?<br>*E. Papadonikolaki, M. Leon & A.M. Mahamadu* | 321 |
| Schema-based workflows and inter-scalar search interfaces for building design<br>*P. Poinet, M. Tamke, M.R. Thomsen, F. Scheurer & A. Fisher* | 329 |
| 4D BIM model adaptation based on construction progress monitoring<br>*K. Sigalov & M. König* | 337 |

*Ontology, semantic web and linked data*

| | |
|---|---|
| A novel workflow to combine BIM and linked data for existing buildings<br>*M. Bonduel, M. Vergauwen, R. Klein, M.H. Rasmussen & P. Pauwels* | 347 |
| Linking sensory data to BIM by extending IFC—case study of fire evacuation<br>*R. Eftekharirad, M. Nik-Bakht & A. Hammad* | 355 |
| Semantic BIM reasoner for the verification of IFC Models<br>*M. Fahad, N. Bus & B. Fies* | 361 |
| Modular concatenation of reference damage patterns<br>*A. Hamdan & R.J. Scherer* | 369 |
| A graph-based approach for management and linking of BIM models with further AEC domain models<br>*A. Ismail & R.J. Scherer* | 377 |
| A building performance indicator ontology<br>*A. Mahdavi & M. Taheri* | 385 |
| From patterns to evidence: Enhancing sustainable building design with pattern recognition and information retrieval approaches<br>*E. Petrova, K. Svidt, R.L. Jensen & P. Pauwels* | 391 |
| Managing space requirements of new buildings using linked building data technologies<br>*M.H. Rasmussen, C.A. Hviid, J. Karlshøj & M. Bonduel* | 399 |

Linked building data for modular building information modelling of a smart home  407
*G.F. Schneider, M.H. Rasmussen, P. Bonsma, J. Oraskari & P. Pauwels*

Ontology and data formats for the structured exchange of occupancy related building information  415
*M. Taheri & A. Mahdavi*

Graph representations and methods for querying, examination, and analysis of IFC data  421
*H. Tauscher & J. Crawford*

Integration of an ontology with IFC for efficient knowledge discovery in the construction domain  429
*Z.S. Usman, J.H.M. Tah, F.H. Abanda & C. Nche*

The RIMcomb research project: Towards the application of building information modeling in Railway Equipment Engineering  439
*S. Vilgertshofer, D. Stoitchkov, S. Esser, A. Borrmann, S. Muhič & T. Winkelbauer*

SolConPro: Describing multi-functional building products using semantic web technologies  447
*A. Wagner, L.K. Möller, C. Leifgen & U. Rüppel*

*Regulatory and legal aspects*

Semantic topological querying for compliance checking  459
*N. Bus, F. Muhammad, B. Fies & A. Roxin*

BIM-based compliance audit requirements for building consent processing  465
*J. Dimyadi & R. Amor*

Contract obligations and award criteria in public tenders for the case study of ANAS BIM implementation  473
*F. Semeraro, N. Rapetti & A. Osello*

Author index  479

# Preface

Dear Reader,

The Architectural, Engineering, Construction, Owner and Operator (AECOO) industry serves an important role in modern society as the built environment is a key component in creating growth and prosperity, but at the same time construction and operation of the built environment consume an enormous amount of natural and financial resources at the global level. The industry generates jobs for millions of people and represents their livelihood, but it is also known for challenging working conditions that can even have fatal consequences for workers at construction sites. The industry is helping to solve the increasing need for housing, production facilities, offices, roads, railways, public facilities, like hospitals, and is appreciated for this, but is also criticized for a lack of quality, failures, delays and cost overruns.

Information and Communication Technology has been used for more than three decades in many companies and organisations in the AECOO industry, but despite the widespread use of ICT, the industry often still struggles with the above-mentioned challenges. For more than two decades, the *European Conference on Product and Process Modelling* (ECPPM) has been held every two years, with a focus on how, from a scientific point of view, the AECOO industry can design, construct and operate the built environment better by applying ICT, such as by changing processes based on and benefitting from structured data and methods for modelling products.

Scientific knowledge covers a variety of issues related to product and process modelling including, but not limited to, Building Information Modelling (BIM) implementation and deployment, performance simulation, integrated modelling of the built environment, information and knowledge management, ontologies, the semantic web, linked data, the Internet of Things (IoT) and interoperability, standardisation, collaboration methods and legal aspects.

The following sections discuss most of the above-mentioned important components of product and process modelling, as presented at the *Twelfth European Conference on Product and Process Modelling (ECPPM2018),* held in Copenhagen, Denmark (12–14 Sep., 2018). ECPPM is the flagship conference event of the European Association of Product and Process Modelling (EAPPM). The conference aims to provide an international forum for the exchange of scientific information and knowledge-sharing on state-of-the-art research efforts and on contemporary product and process modelling issues, covering a large spectrum of topics pertaining to ICT deployment instances in AEC/FM, attracting high-quality research papers and providing a platform for the cross fertilization of new ideas and know-how.

The work presented and included in the conference proceedings constitutes cutting-edge research, scientific and applied knowledge and case-studies, which should all be of great interest to both researchers and practitioners. In particular, the conference offers the European and the international community of product and process modelling professionals with a number of high-quality papers related to interoperability and especially Industry Foundation Classes (IFC) and linked data. The proceedings include work from researchers from a total of 20 countries from Europe and oversees.

J. Karlshøj, PhD
*Conference Host and Chair*

# Organization

STEERING COMMITTEE

*Chairperson*

Raimar J. Scherer, *Technische Universitat Dresden, Germany*

*Vice Chairpersons*

Ziga Turk, *University of Ljubljana, Slovenia*
Symeon Christodoulou, *University of Cyprus, Cyprus*
Ardeshir Mandavi, *Vienna University of Technology, Austria*

*Members*

Robert Amor, *University of Auckland, New Zealand*
Ezio Arlati, *Politecnico di Milano, Italy*
Jakob Beetz, *Eindhoven University of Technology, The Netherlands*
Adam Borkowski, *Institute of Fundamental Technological Research, Polish Academy of Sciences, Poland*
Jan Cervenka, *Cervenka Consulting, Czech Republic*
Attila Dikbas, *Istanbul Technical University, Turkey*
Ricardo Gongalves, *New University of Lisbon, UNINOVA, Portugal*
Gudni Gudnason, *Innovation Centre, Iceland*
Noemi Jimenez Redondo, *CEMOSA, Spain*
Jan Karlshøj, *Technical University of Denmark, Denmark*
Tuomas Laine, *Granlund, Finland*
Karsten Menzel, *University College Cork, Ireland*
Sergio Munoz, *AIDICO, Institute Technologia de la Construction, Spain*
Pieter Pauwels, *Ghent University, Belgium*
Byron Protopsaltis, *Sofistik Hellas, Greece*
Svetla Radeva, *College of Telecommunications and Post, Sofia, Bulgaria*
Yacine Rezgui, *Cardiff University, UK*
Dimitrios Rovas, *Technical University of Crete, Greece*
Vitaly Semenov, *Institute for System Programming RAS, Russia*
Ales Siroky, *Nemetschek, Slovakia*
Ian Smith, *EPFL—Ecole Polytechnique Federale de Lausanne, Switzerland*
Rasso Steinmann, *Institute for Applied Building Informatics, University of Munich, Germany*
Väino Tarandi, *KTH—Royal Institute of Technology, Sweden*
Alain Zarli, *CSTB, France*

*Retired Members*

Bo-Christer Bjork, *Swedish School of Economics and Business Administration, Finland*
Per Christiansson, *Per Christiansson Ingenjors Byrd HB, Sweden*
Anders Ekholm, *Lund University, Sweden*
Gudfried Augenbroe, *Georgia Institute of Technology, USA*
Matti Hannus, *VTT Technical Research Centre of Finland, Finland*
Ulrich Walder, *Graz University of Technology, Austria*

## INTERNATIONAL SCIENTIFIC COMMITTEE

Robert Amor, *University of Auckland, New Zealand*
Chimay Anumba, *Pennsylvania State University, USA*
Ezio Arlati, *Politecnico di Milano, Italy*
Godfried Augenbroe, *Georgia Institute of Technology, USA*
Håvard Bell, *Catenda, Norway*
Michel Bohms, *TNO, The Netherlands*
André Borrmann, *Technische Universität München, Germany*
Tomo Cerovsek, *University of Ljubljana, Slovenia*
Jan Cervenka, *Cervenka Consulting, Czech Republic*
Edwin Dado, *Nederlandse Defensie Academie, The Netherlands*
Nashwan N. Dawood, *Centre for Construction Innovation and Research, University of Teesside, UK*
Attila Dikbas, *Istanbul Technical University, Turkey*
Robin Drogemuller, *Queensland University of Technology UT CSIRO, Australia*
Anders Ekholm, *Lund University, Sweden*
Bruno Fies, *CSTB, France*
Martin Fischer, *Center for Integrated Facility Engineering, Stanford University, USA*
Thomas Froese, *University of British Columbia, Canada*
Gudni Gudnason, *Innovation Centre, Iceland*
Tarek Hassan, *Loughborough University, UK*
Eilif Hjelseth, *Oslo Met, Norway*
Wolfgang Huhnt, *Technische Universität Berlin, Germany*
Ricardo Jardim-Goncalves, *Universidade Nova de Lisboa, Portugal*
Jan Karlshøj, *Technical University of Denmark, Denmark*
Peter Katranuschkov, *Technische Universitaet Dresden, Germany*
Abdul Samad (Sami) Kazi, *VTT Technical Research Centre of Finland, Finland*
Arto Kiviniemi, *University of Liverpool, UK*
Bob Martens, *Vienna University of Technology, Austria*
Karsten Menzel, *University College Cork, Ireland*
Sergio Munoz, *AIDICO, Instituto Technologia de la Construcción, Spain*
Svetla Radeva, *College of Telecommunications and Post, Sofia, Bulgaria*
Iñaki Angulo Redondo, *TECNALIA, ICT Division, European Software Institute, Spain*
Yacine Rezgui, *Cardiff University, UK*
Uwe Rueppel, *Technical University of Darmstadt, Germany*
Vitaly Semenov, *Institute for System Programming RAS, Russia*
Miroslaw J. Skibniewski, *University of Maryland, USA*
Ian Smith, *EPFL—Ecole Polytechnique Fdrale de Lausanne, Switzerland*
Rasso Steinmann, *Institute for Applied Building Informatics. University of Munich, Germany*
Väino Tarandi, *KTH—Royal Institute of Technology, Sweden*
Walid Tizani, *University of Nottingham, UK*
Hakan Yaman, *Istanbul Technical University, Turkey*
Pedro Nuno Mêda Magalhães, *Porto University, Portugal*

## LOCAL ORGANIZING COMMITTEE

Jan Karlshøj, (Chair), *Technical University of Denmark, Denmark*
Line Leth Christiansen, *Technical University of Denmark, Denmark*
Melena Schjøth, *DIScongress, Denmark*

*BIM implementation and deployment*

# Experiences from Norway on implementing BIM in existing bachelor engineering curriculum

E. Hjelseth
*Oslo Metropolitan University, Oslo, Norway*

ABSTRACT: This study explore experiences from ongoing implementation of BIM in existing bachelor engineering courses at Oslo Metropolitan University in Norway. This is done by a combination of semi-structured interview, net-based survey of management, lectures and students at the department. The findings are analysed by use of the Multi-motive Information Systems Continuance (MISC) model, which focus on hedonic, extrinsic and intrinsic motivation. This study do not confirm the traditional view that young students are positive and old teachers are negative to BIM, or that use of BIM will increase by itself when I become more mature. The most important aspects for increased implantation of BIM is to create a dynamic learning environment who support and combine all three types of motivation by having an intentional attitude to learning objectives, assessment criteria and context and relevance of competence. This approach can support implantation of BIM in all professional course in an engineering study as an integrated part of the learning outcome by focusing on "Use of BIM to learn Construction".

## 1 INTRODUCTION TO THE CASE STUDY

Student love working with Building Information Model (BIM) based tools, and the Architects, Engineers, Contractors/Facility management (AEC/FM) industry demand BIM competency. However, implementation in higher architect and engineering education has been rather slow. This is a paradox since his type of professional education has tradition for embedding the best industry practice into their curriculum. Lectures in higher education (HE) are also involved in research, and by this used to pay attention to latest trends and solution within their profession. Continuously development of own competency is embedded in the lectures way of working. It should therefore not be lack of awareness that explain lack of BIM in the curriculum in HE.

The challenge is therefore how to give the engineering students BIM-based competency, when the curriculum is packed with important professional content? A simple solution is of course to let the students work with authoring tools like Revit from Autodesk, ArchiCAD from Graphisoft, Vectorworks from Nemetschek or MicroStation from Bentley Systems in project work. This is a very practical approach to give the students skills in use of software, and a way to support teamwork. This approach imply that BIM is mostly used in courses where students work with design related projects in team. Training in use of BIM tools (software) is normally included in the course structure. The consequence is that BIM is restricted to a small number of courses using BIM tools, while all other courses continue in the "old way".

This approach understand BIM as just software skills. This stands therefore in contradiction to both the need in the AEC/FM industry for increased BIM competency for future engineers and the impact of BIM as catalyst for new ways for working and collaborating to enable better solutions.

The research question is: *What can be done to increase the implementation of BIM to support learning of construction engineering?*

This study use the ongoing implementation at the construction engineering bachelor study program at Department of Civil Engineering and Energy Technology at Oslo Metropolitan University (OsloMet) in Norway as case. The intent to develop a *"BIM-string"*, see Table 1 for overview, which embeds BIM as an integrated part of all engineering course in in the entire study program.

## 2 WHAT DOES THE LITERATURE SAYS

What is the status of BIM in HE? Is this problems already solved? A study by Badrinath et al. (2016) identified 70 academic BIM education publications. Of these, half were published in 2015, 71% of which were conference papers. Case studies and experiences were the dominant type of publication in this type of studies.

There are challenges regarding BIM in terms of establishing (1) a common understanding of what BIM really is and (2) how to determine whether, to what extent, and for what purpose BIM is introduced in HE. The first challenge has been experienced by other scholars; for instance, in a study about BIM teaching strategies by Barison and Santos (2010, p. 1), the authors stated: "*it is still unclear how BIM should be taught as most experiences are very recent.*" The NATSPEC survey (Ronney, 2015), however, did suggest increased interest in and a focus on BIM in a number of countries. According to Rooney (2014, p. 1): "*It would appear that the majority of BIM education available to date focuses on training in the use of particular BIM software packages, particularly seen as a lot of training for professionals appears to be provided by the software vendors. Training for both graduates and professionals in openBIM concepts, BIM management and working in collaborative BIM environments appears to be still in its infancy*".

The Ph.D. thesis by Hjelseth (2015) introduce a dynamic understanding of BIM which combines the Model/Modelling/Managements do be applicable by focus on program/processes/person/as illustrated in Figure 1.

The second challenge is based to the above referenced observations indicate that the dominant view of BIM is related to the use of software. HE is by nature theory focused, not on practical skills in the use of particular BIM software. Measuring the status of BIM in education must therefore include a better understanding of what BIM really is. Studies by Becker et al. (2011), Salman (2014), and Rooney (2014, 2016) demonstrate that many educational institutions across the globe investigates how to incorporate BIM in HE.

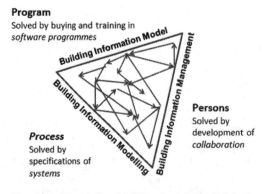

Figure 1. The trinity of BIM understanding (Hjelseth, 2015).

Peterson (et al., 2011) give an example of how teaching construction project management with BIM support. This more integrated perspective is quite different from e.g. use of BIM software in project management to develop a 5-D schedule like Synchro based on import from authoring tool like Revit.

A study in UK by Underwood and Ayoade (2015) illustrate the situation in HE by following quote: "*Despite an overwhelming level of support for the importance of BIM related accreditation criteria of courses in academic institutions, the level of conviction for actual change is however debatable*".

## 3 THEORETICAL LENS

Answering the research question can be answered just by asking direct question to the head of course. However, this will not give a reasoned answer. Introducing BIM in curriculum, and keep it as part require multiple actions—motivated by multiple aspects. This study introduce the "*Multi-Motive Information Systems Continuance Model (MISC)*" by Lowry, Gaskin & Mood (2015) to explore this situation. The "*Intention to continue*" focuses on three type of motivation: "*hedonic, intrinsic and extrinsic*". All these three are included in each of the following perspectives: "*Expectations*", "*Disconfirmation*" and "*Performance*" with respectively relation to "*Attitude*", "*Satisfaction*" and "*Attitude*". MISC can be arranged as quantitative study supported by regression analysis. This study is more explorative and we do therefore use a qualitative approach based on thematic analyses classify and analyse the three types of motivation in the MISC-framework.

## 4 BIM-STRING DIDCACTICS

Implementing BIM in an overloaded curriculum need an adapted approach. Figure 2. illustrate an integrated approach proposed by Hjelseth (2017b). This integrated concept is based on The Technological Pedagogical Content Knowledge (TPACK) framework by Koehler (et al., 2014).

The concept need to be applicable in the real context. Hjelseth (2017b) presents examples on how BIM tools and processes supports specification of relevant information (IDM) for use in e.g. structural calculation, use of model checking and information take off to give facts for cost calculations, or the develop an BIM execution plan to manage the information flow during the life cycle of the building project. Further details is included in the case description.

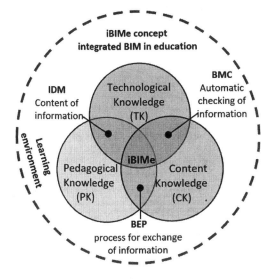

Figure 2. The BIM methods related to the TPACK framework (Hjelseth, 2017b).

Table 1. Overview of courses in the BIM-string.

| Semester: Course, ECTS | Software | Process |
|---|---|---|
| 1: Introduction to building professions, 10 | J Revit | bSDM |
| 2: Building technology, 10 | J Revit | bSP |
| 3: Building materials and concrete design, 10 | J Robot | bSDD |
| 4: Technology management, 10 | J Process modeller | bSP, BEP |
| 5: Design of Steel- and timber structures S), 10 | S Robot | bSDM |
| 5: Land Use and Transport Planning P), 10 | P Novapoint | bSDM |
| 5: Elective BIM course, 10 #) | J Revit, Nova-point, Solibri | All |
| 6: The Building Process, 10 | J Process modeller | bSP, BEP |
| 6: Bachelor thesis, 20 | J All | All |

ECTS, 30 = 100% work load of one semester. J: Joint for both study specialisations S: Study with specialisation in structural engineering P: Study with specialisation in infrastructure planning #: Proposal, not approved by the department management bSDM – buildingSMART DataModel (IFC) – interoperability bSP – buildingSMART Process (IDM) – Collaboration bSDD – building S–. DataDictionary (IFD) – Product/Material data BEP – BIM Execution Plan – Collaboration/Management All – combination of all above + others (OsloMet – Engineering study, 2018).

Implementation is supported by the "BIM-group", three lectures who support practical implementation. They offer the course coordinator—and the students services like developing exercises, having introduction BIM lectures, and discussions/advices on how to implement. Implementation has been discussed with all relevant course coordinators, and solutions are developed for 1st and 2nd semester.

## 5 CASE DESCRIPTION

### 5.1 Overview of the BIM-sting

This study is based on the ongoing implementation in the bachelor study in construction engineering at OsloMet in Norway. This is a traditional engineering study in compliance with the national framework.

Table 1 illustrate professional engineering course relevant for embedding BIM as integrated part of the curriculum and defined learning objectives.

The curriculum is, as for these types studies, overbooked with professional content; lectures, exercises, projects and exams. Use of software like Revit and Robot has been used in students' projects for a long time. Systematic implementation called the "BIM-string" started with first year students (FYS) autumn 2017. Use of BIM software in the string is not an add-on is each course, but a tool to support the learning objectives in each course. However, use of BIM and has a long and unsystematic history: We have therefor include last year students (LYS) as part of the study.

### 5.2 Status of implementation

The systematic implementation of the BIM-string (BIM in BE) started autumn 2017 BIM as use or Revit software is embedded in 1st semester course in *"Introduction to Building Professions"*. In first semester, all students work in teams of five, have different roles; architect, structural engineer, building engineer, contractor etc. The teachers has written a course book. This include a chapter (25 pages) about "Information management", which is place before the major part about Revit training (135 pages). This illustrate that the focus on BIM is not just to learn Revit to design a building, but to be aware of the different need for information in each roles troughs the entire life cycle of the building.

In 2nd semester does the course *"Building Technology"* following up the BIM sting by focus on material properties—and how this information can be entered, processed, presented and distributed in BIM. This is done by a separate lecture, an exercise manual and hand-in of a small report. An interesting perspective is that the course coordinator has projects in the industry: He is rather critical to the quality of BIM models used as support for production and document building details. However, this is actually a good foundation for the need for improving the quality of what the BIM software "deliver". This include both need for relevant details (Level of Development) and content of product documentation (Level of Information). The role model is the *"Critical engineer"*, which require knowledge about good professional (technical) – and document this by use of BIM tools.

The *"BIM-string"* is continued in 3rd semester in the *"Building Materials and Concrete Design"* course. This will continue the approach from 2nd semester with increased focus on quality of profession facts/information. The 4th semester will in the *"Technology management"* course give priority to processes and new way of working. In last year, 5th and 6th semester, of the bachelor study will the professional courses intent to combine BIM as *"Product- and process modelling"*. An elaborative course for specialisation in BIM has been proposed by the BIM research group, but have so far not been supported by the management. In the 6th semester can the *"The Building Process"* course play the role as integrator of BIM. The students can of course, in limited or higher extent, choose to include BIM in their bachelor thesis.

## 6 METHODS

This study done by a combination of a net-based survey to all students in first and last year of the bachelor study in construction engineering. The course coordinators (responsible teachers) for the engineering courses listed in Table 1 participated in semi-structured interview. The findings was analysed by use of the MISC framework.

The net-based survey included all (approx. 150) first year students (FYS). We got feedback from 43 students (30%). The have experiencer from two course in the "BIM-sting". The survey included all (approx. 120) last year students (LYS), where we got feedback for 47 student (40%). The have experience from use of BIM based software (like Revit, Robot, Solibri, Novapoint) in variable degree in several engineering courses, see also Table 1. The answers is rounded to nearest 5% to avoid too strict interpretation of small differences on limited answering basis.

This study included also semi-structured interviews with course coordinators (responsible teachers) for the engineering courses listed in Table 1. The teachers at the department was well informed about the potential of BIM in the industry. All teachers was informed and aware of the industry focus on BIM in general. However, most teachers did not find BIM relevant for his or her own courses. The general experience was that BIM was already included in the *"Introduction to Building Professions"* course in 1st semester, and limited need for continuing with BIM.

The management group at the department shared the same attitude as the teachers. They was aware of the high interest for BIM in the industry, but do not want to give recommendation to implement. They delegated this decision has by each teacher individually. The other attitude was that BIM most include extra cost, or change in lecturing coordination.

## 7 RESULTS

### 7.1 Interpretation of feedback from students and teachers

This study is use MISC as framework for structuring the results into: hedonic, intrinsic and extrinsic motivation. Analysis based on MISC has *"Intention to use"* as final outcome. In our respect is related to in which extent integration of BIM into existing curriculum has support to be continued. The results was grouped into *"Students"* and *"Teacher"* finding, which then is extracted into a joint result.

### 7.2 Feedback from the net-based survey to the students

The first section of the questionnaire was related to Digitization/use of BIM in the construction industry. On the question" *To what extent do you think digitalization/BIM will change the construction industry for the next 3–5 years?"* chosen 15% of first year students (FYS) and 50% of last year students (LYS) the: *"In very high extent"* as answering option.

Regarding the feedback on the question *"To what extent do you think digitalization/BIM will change your way of working when you get out of work?"* reported 10% of (FYS) and 40% of (LYS) *"In very high extent"*.

On the question *"To what extent do you think digitalization/BIM will change your way of working when you get out of work?"* reported 15% of first year students (FYS) and 60% of last year students (LYS) *"In very high extent"*. The feedback from LYS is reflecting the interest in the industry. Approx. 2 of 3 had been to interview for job. Of these answered 3 of 4 that they had been asked about digital competency. One LYS commented in the net-based survey that he did not get the job due to limited BIM competency. More student commented that the industry was focused on BIM, and that OsloMet should increase focus on BIM in the curriculum.

On the other side: 3 of 5 student did also ask the industry about their plans for digitalisation when they was at job interview.

### 7.3 Experiences with software training at OsloMet

To what extent do you think that what you learned about Revit will be useful in engineering studies?

There was a high correlation feedback between the expectation from FYS, and the experience form LYS, Approx. 15% reported "In very high extent".

However, on the two last answering options: (Limited and little extent) had no answers from FYS, while 15% of LYS used these answering options.

There was limited comments from FYS. These commented that teamwork was relatively time consuming compared to working individual.

The comment from LYS was as expected more extensive, and the questionnaire for LYS was there extended with additional questions: "*Focus in the course was on generating a 3D model in BIM, and not in exploiting the information in BIM*". Another commented following: "We have also not learned about Solibri", while another said that "*BIM should be regarded as a process*". One student comparted experiences with other courses: "*In math, we are encouraged to program in Matlab, that's good. But then we have fed and revised the construction of ALL subjects for three years. Then I get a small extent in the scale of use*".

### 7.4 Factors contributing to motivation

The feedback on question: "What have been the most motivating working with BIM/Digital model?" is presented in Table 2 below.

The answers on "*Fun to create yourself*" can be interpreted as intrinsic motivation is the dominating. On the answering option "*Easier to get good results*" was there some difference between FYS and LYS. 8%. The difference indicate extrinsic motivation is relatively low and become increased during the sty when awareness of industry demand become clearer. On "Less job when working in group" the answering difference indicate clear differences in hedonic motivation. The grade itself as motivation decrease during the study.

The questionnaire followed up with an open text answer on the question: "*What do you think are the main reasons you want to work more digital/BIM?*" We got comments was only from LYS:

Here is a comment from a student with craftsman experience. "*Has worked a lot of paper drawings as craftsman in the past: A massive advantage with BIM is that you get a lot easier on building/items. It is also much easier to convey ideas and thoughts in 3D than 2D*". This comment highlight the role of statements form the industry to (extrinsic) motivation.

Following comment focus on the importance in lack of intrinsic motivation. "*Seems it was a joke when it came to something, but generally not motivating to work because we do not learn that very well. Along the way, we have received very little info about how much digital tools are actually used, and it's hard to see the importance of learning this*".

Other comments mentioned the importance from the industry: "*Think it will be standard to use in the industry.*"/"*Will acquire knowledge in order to do a better job.*"/"*Think it will make the industry more efficient and help make it more environmentally friendly.*"/"*All big companies use it, it should be more focused on learning how to use it in the study so that you are more prepared for working life.*"/"*BIM is the future, it is needed at work.*"/"*Learn how they do this out in the industry.*"/"*Learn to facilitate the work is done with this type of tools.*"/"*Work more with the information in the BIM and use this for cost estimates e.g. LCC, FDV, etc.*"/"*This is the present and future of the construction industry.*"

This imply that intrinsic motivation is dominating by both FYS and LYS. The importance of extrinsic motivation is mostly connected to impact of this type of competence can have for getting a job. This correspond with an understating of a good engineering education is not only good grades, but competency which is relevant for the industry. BIM/digitalisation can act as a marker for OsloMet as the future oriented engineering education study.

### 7.5 Factors contributing to demotivation

The feedback on question: "What have been the most de-motivating working with BIM/Digital model?" is presented in Table 3 below.

The intention with this question was to identify if elements that reduced the motivation and priority to increased use or BIM/digitalisation in both learning, teaching and assessment process.

The questionnaire followed up with an open text answer about the question: "*What do you think are

Table 2. Overview factors motivating for BIM use.

| Answering option | LYS | FYS |
|---|---|---|
| Fun to create yourself: | 70% | 65% |
| Easier to get good results | 0% | 25% |
| Less job when working in group: | 15% | 0% |
| Do not know/else: | 10% | 10% |

Table 3. Overview factors demotivating for BIM use.

| Answering option | LYS | FYS |
|---|---|---|
| Nothing: | 5% | 20% |
| Difficult to use Revit: | 30% | 10% |
| Much job in relation to learning: | 15% | 25% |
| Much job in relation to grade: | 40% | 30% |
| Do not know/else: | 10% | 20% |

*the main reasons you want to work more digital/ BIM?"* We the comments was also only from LYS:

> *"It takes some time to get into. I think it took some time before you saw the results. It was also a bit demotivating every time you got an error, because since it is at a beginner level, you have no skills other than following the course/template/recipe."*
>
> *"We have used BIM as a tool in project assignments, but this has not been a work requirement, so it does not matter to the grade."*
>
> *"Lack of continuity. Learned a lot of first semester in constructional introduction. A little used in the teaching after that, and a half-hearted attempt to implement it in the plumbing field."*
>
> *"Feeling this question is not in place. All businesses focus on BIM, this is the industry students are going into. When people sow about using BIM, it's because they are generally tired of the studies and are going to find something to blame on. It's not hard to use when you follow and make compulsory. There is much work in relation to the learning (intuitive), much work in relation to character. What can be demotivating is for the Educational Introduction that one must "play" all the roles instead of having one who represents the line, cover all the fields of responsibility, while focusing on BIM. Then there can be a lot and people can be demotivated."*

### 7.6 Increased use in the engineering study

The experience from LYS is an important indicator of the potential for increased digitalisation in existing curriculum (FYS is part or the new "BIM-string" and was therefore not asked).

Over 85% of the LYS chosen the answering options "To a very large or large extent" on the question: "To what extent do you think it is possible to take more use of professional software in the teaching of engineering education?". This positive interest for something relatively unknown indicate that the motivation probably is based on intrinsic motivation for learning more or learn in a different was. In addition can increased awareness of the demand in the industry for this competence trigger extrinsic motivation. Some selected feedback in the open text field was: "Increased learning by user of BIM tools."/"Use it in several courses, not just the first year as we experienced."/"Important to use the programs over time so you remember how to use it."/"It is time consuming to get into these programs, and this should not be at the expense of important engineering such as constructional, steel and wood constructions, etc."/"The basic knowledge in engineering MUST be good: Shit in, shit out."/"The tasks in the 1st semester can be done with more focus on digital work than paper work. Drawings have previously been handed in largely on paper—switched to digital reinforcement drawings, etc. Exercises in other subjects may be done slightly less regarding handwritten, but with more focus on how software can be used to solve. One might also use programs in a larger part of the classroom, to show, for example, how different forces work on constructions."/"In the structural engineering courses we have dimensioned, we could use Robot and Revit more accurately and dimensioned in the programs, and supplemented to hand calculation the see the information use in the formula and standards."/"Learning a lot, but not always as easy and understanding everything that the software does. For example, for curvature for roads by use of Novapoint Trimble software."/"This will be needed in all jobs in the future."/"One feels more ready for work".

### 7.7 Students PC situation

The survey include a question about to see if the teaching in increased degree could be based on students own PCs, see Table 4.

An interesting observations that FYS have more Gaming PC than LYS. This indicate that they invest more in PC, and use it for more advanced tasks with high quality visualisation. Another observation it that LYS have very high extent of Mac. This is interesting, since Mac cost more that PC, and do normally create problems for most BIM- / engineering software. This indicate that the students do not experience in use of professional software, and therefore can choose Mac as preferred laptop for "office" work.

Asking the student about willingness to invest e.g. 1.500 € in a gaming laptop. Approx. 30% of FYS and 45% LYS was positive to invest. Many students also claimed their own PC/Mac was good

Table 4. Type of computer for the students.

| Type of computer | FYS | LYS |
|---|---|---|
| Laptop PC | 75% | 40% |
| Portable gaming PC | 15% | 10% |
| Portable mac | 10% | 50% |
| Does not have own computer | 0% | 0% |
| Desktop gaming PC | 30% | 10% |

enough. In the open text was FYS was negative, while some of the LYS argument for the benefit for investing in high performance computer equipment. This feedback indicate that one can go for flexible solution for software training on students PC – if the students are informed about the requirements – and that the PCs actually become used regularly in the courses. However, this is not the case today.

7.8 *Teachers point of view*

The teachers are all very well aware of the focus of BIM in the industry, and are general positive to include BIM in the study program. However, not in mine course, it is already too packed. The main focus of the teachers is to give good courses in their own subject. This imply focus is on teaching the students as much as possible, and to introduce advanced methods. This "silo" situation no unique of the case study. This situation has been the main motivation for inventing a new integrated approach to implement BIM in existing curriculum.

The course coordinator faces two challenges when introducing BIM in an existing professional course: What to do in a professional aspect, and how to do it in a pedagogical aspect. This imply that project based courses are more likely candidate than lecture based curses. The approach in the "BIM-sting" is to support leaning of existing learning objectives in the curriculum by use of BIM-based tools and new was of learning and collaborating.

However, letting the course coordinator answering this question alone is maybe not polite. The BIM-group does in this respect play an important role to support practical implementation adapted to each course. This is a learning processes for all, based on dialogue and establishing a joint understanding. When the course coordinator see that this is also supported (demanded) by the students, this contributes to increase their extrinsic motivation for course evaluation, and maybe more important, trigger intrinsic motivation for felling development in own lecturing. This is not "proven", but course coordinators and lectures wants to increase integration of BIM support in their courses.

7.9 *Support from the department management*

The dominating view in the department management is that BIM is interesting and relevant. Changing existing curriculum with existing staff is in general a hard challenge. It is the course coordinator whish has the mandate to change an approved course curriculum. (The changes must in next step be approved by the Board of education).

The "BIM-string" is supported as a way to motivate course coordinators to include BIM and digital way of working in their lecturing. However, the department management has not stated in meeting or other statements encouraged including the BIM-sting is important, or asked for report of ongoing progress. Implementing is supported if this does not include increased costs, can be done with existing structure, the responsible teacher give full support (no encouragement to change), and very important or if existing part of curriculum do not become reduced. The priority is given to the "hard" engineering topics like structural calculations, and difficult topics they have to learn at school. BIM is a "soft topic", which the students will learn when the go into job.

However, digitalisation has in last years been given priority at strategic level by the university management. When the budget situation for current year is improved, the department management is supporting investment in a "BigRoom" with interactive white board (smart board screens) and VR-equipment.

This imply that funding of investment in equipment and study facilities is supported, but changes in didactic and curriculum is not given attention. This attitude has allowed the BIM-group at the department to act dynamically, based on intrinsic motivation, to implement the "BIM-sting".

## 8 DISCUSSIONS

8.1 *Motivation and management of change*

This study identify that both students and teachers are driven by different types of motivation in different contexts and learning environments. Awareness of the impact of which types of motivations different learning environments support is most critical.

This study have not identified simple relations the like young students are positive and old teachers are negative to BIM as arguments for level of BIM use. The study have not identified relations like BIM will be used more in HE when it becomes more common in the AEC industry the industry.

There has been hard to find similar studies in use of BIM in HE. Most examples of studies is based on experience by use of BIM-based software to design as specified solution. The outcome has been assed on the quality of the designed solution, which in most cases is based on visualisation, and not the content of information in the BIM. Most studies of BIM is in HE has focus on use of BIM software to design. This design can be rather advanced, and is then often performed by teams. It has therefor been hard to find relevant studies to include in this discussion.

A study by Lassen et al. (2017) do also confirm the importance on intrinsic motivation explaining why student invested much time in the pass/fail assessed project task in the first semester course "*Introduction to Building Professions*" at OsloMet.

### 8.2 Relevance of digitalisation for HE

Numerous presentations state clearly that BIM is an abbreviation for building information modelling, and where modelling indicates that BIM is about processes, it is a new way of working and collaborating. These types of presentations have received recognizing nods and applause. "*BIM is a process that's enabled by technology. You can't buy it in a box, it's not a software solution, and it's something you can't do in isolation*" (Mordue et al. 2015, p. 349).

A study by Hjelseth (2017a) show that there is wide variation in understanding of what BIM is, and where the dominating underrating is use of software tools (programs), not focus on processes when becoming introduced to BIM/digitalisation, as illustrated in Figure 3.

However, learning software, and especially maintain this competency is not the scope of HE. The feedback from first year's students did not give significant support for introduction of BIM to support focus on processes. The attitude was more related to use BIM to produce something that can be delivered. The students are in general very positive to use BIM based tool, but there is challenge that it take a relative high amount of study hours compared to the impact on the grade of the course. This approach do also contribute with limited learning outcome in first phase as illustrated in Figure 4.

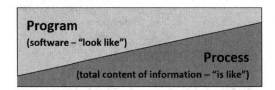

Figure 3. Shift of focus in BIM understanding during the engineering study.

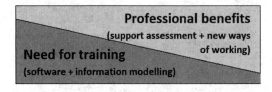

Figure 4. Return of investment in software training in engineering curriculum.

### 8.3 Need for deeper understanding

Norway do not have a defined and clearly expresses BIM strategy to support investment in competency. The BIM strategy at OsloMet can be regarded to aim at Level 3 in the UK wedge for BIM strategy (McPartland, 2018). However, the origin of OsloMet strategy was not to aim for a pre-defined level, but to support engineering education by introduction of technology that support increased understanding of working as engineer.

There is a difference between having success with BIM/digitalisation in a single dedicated course, versus integration into a curriculum with focus on engineering competency.

## 9 CONCLUSION REMARKS

Implementing BIM in the curriculum in HE is a complex task with multiple aspects when one give priority to learning of professional competency and not only use of software to solve a project task. The presented case from OsloMet to not give a fixed answer, but outline an applicable concept for integration of BIM in professional engineering courses. Focus g given to BIM as process for providing, processing and presenting information (facts) to support professional tasks as engineer (student).

This study identify the importance for combing learning activities (project task, exercises, lectures, software training, industry contribution etc. which trigger multiple motivations; hedonic, extrinsic and intrinsic, as an important factor for continuously implementation and improvements through the entire study program.

The integrated approach can support implantation of BIM in all professional course in engineering studies when it becomes part of the learning outcome by "*Use of BIM to learn Construction*".

## REFERENCES

Barison, M.B., & Santos, E.T. (2010). BIM teaching strategies: An overview of the current approaches. Proceedings of the International Conference on Computing in Civil and Building Engineering (ICCCBE 2010). Paper 289, p. 577–564. Nottingham, UK: Nottingham University Press. Retrieved from http://www.engineering.nottingham.ac.uk/icccbe/proceedings/pdf/pf289.pdf.

Becker, T.C., Jaselskis, E.J., & Mcdermott, C.P. (April, 2011). Implications of construction industry trends on the educational requirements for future construction professionals. Proceedings of the 47th ASC Annual International Conference. Omaha, Nebraska, USA.

CIB-IDDS. (2015). Integrated Design and Delivery Solutions (IDDS), International Council for Research and Innovation in Building and Construction. International Council for Building. Retrieved from http://www.cibworld.nl/site/programme/priority_themes/integrated_design_and_delivery_solutions.html.

Hjelseth, E. (2015). Foundations for BIM-based model checking systems, Transforming regulations into computable rules in BIM-based model checking systems. PhD Thesis: 2015:54, Department of Mathematical Sciences and Technology, Norwegian University of Life Sciences, Nor-way, ISSN: 1894–6402, ISBN: 978-82-575-1294-1.

Hjelseth, E. (2017a). BIM understanding and activities. Presented at the 2nd International Conference on Building Information Modelling (BIM) in Design, Construction and Operations, 10th – 12th May 2017, Alicante, Spain. http://www.wessex.ac.uk/conferences/2017/bim-2017.

Hjelseth, E. (2017b). Building Information Modeling (BIM) in Higher Education Based on Pedagogical Concepts and Standardised Methods. The International Journal of 3-D Information Modeling (IJ3DIM), IGI Global, ISSN: 2156-1710, ISSN: 2156-1702, DOI: 10.4018/IJ3DIM http://www.igi-global.com/journal/international-journal-information-modeling-ij3dim/41967.

Koehler, M.J., Mishra, P., Kereluik, K., Shin, T.S., & Graham, C.R. (2014). The technological pedagogical content knowledge framework. In J.M. Specter, M.D. Merrill, J. Elen, & M.J. Bishop (Eds.), Handbook of research on edu-cational communications and technology (pp. 101–111). New York: Springer. doi:10.1007/978-1-4614-3185-5_9.

Lassen, A.K., Hjelseth, E. & Tollnes, T. (2017). Enhancing learning outcomes by introducing BIM in civil engineering studies—experiences from a university college in Norway. developed. ISSN: 1743-7601 (paper format), ISSN: 1743-61X (online), DOI: 10.2495/SDP-V13-N1-62-72.

Lowry, Paul Benjamin and Gaskin, James and Moody, Gregory D, (2015) Proposing the Multi-Motive Information Systems Continuance Model (MISC) to Better Explain End-User System Evaluations and Continuance Intentions. Journal of the Association for Information Systems, vol. 16(7), pp. 515–579. Available at SSRN: https://ssrn.com/abstract=2534937.

McPartland, R. (2018). BIM Levels explained, National Building Specification, RIBA Enterprises Ltd, https://www.thenbs.com/knowledge/bim-levels-explained.

Mordue, S., Swaddle, P. & Philp, D. (2015). BIM for Dummies, p. 349, Building Information Modeling For Dummies, ISBN: 978-1-119-06005-5, John Wiley & Sons, Inc.

NATSPEC_Documents/BIM_Education_Global_2016_Update_Report_V3.0.pdf.

OsloMet—Engineering study (2018). Information about the bachelor program in engineering—structural and infrastructure planning, (Norwegian), http://www.hioa.no/Studier-og-kurs/TKD/Bachelor/Byggingenioer.

Peterson, F., Hartmann, T., Fruchter, R., Fischer, M. (2011). Teaching construction project management with BIM support: Experience and lessons learned, Automation in Construction 20, p. 115–125, doi:10.1016/j.autcon.2010.09.009.

Rooney, K. (2014). BIM Education—Global—Summary Report. NATSPEC Construction Information. Sydney, Australia. Retrieved from http://bim2.natspec.org/images/NATSPEC_Documents/BIM_Education_Paper__Final.pdf.

Rooney, K. (2016). BIM Education—Update Report. NATSPEC Construction Information. Sydney, Australia. Retrieved from http://bim.natspec.org/images/.

Salman, H. (2014). Preparing architectural technology students for BIM 2016 mandate. Proceedings of the 4th International Congress of Architectural Technology (pp. 142–158), UK.

Underwood, J, & Ayoade, O. (2015). Current Position and Associated Challenges of BIM education in UK Higher Education. Retrieved from https://media.thebimhub.com/user_uploads/baf_bim_education_report_2015.pdf.

# Understand the value of knowledge management in a virtual asset management environment

C. Mirarchi & L. Pinti
*Department of Architecture, Built Environment and Construction Engineering, Politecnico di Milano, Milano, Italy*

M. Munir
*School of Engineering, University of Liverpool, Liverpool, UK*

S. Bonelli & A. Brizzolari
*Department of Architecture, Built Environment and Construction Engineering, Politecnico di Milano, Milano, Italy*

A. Kiviniemi
*School of Architecture, University of Liverpool, Liverpool, UK*

ABSTRACT: The research provides a practical case study to demonstrate how clients can derive value from digital information technologies and technics devoted to the management of data and information with reference to Facility Management (FM) and Asset Management (AM) activities. During the study, public assets of around 450 buildings in a municipality were analysed over a two-year time-frame. The public body acts both as client and as asset manager, facilitating the study of interactions between the two phases that is usually hindered due to the fragmentation of the subjects involved. This study demonstrates in the first instance the value of an effective information management process and secondly possible impacts and value areas of KM in building information modelling-based FM and AM. The case study presented can help clients understanding the value of information and knowledge management technologies and techniques in current asset interventions and/or in the development of future projects.

## 1 INTRODUCTION

The use of Building Information Modelling (BIM) in Facilities Management (FM) and Asset Management (AM) is transforming the way assets are operated and managed (Love et al. 2014). As such, BIM offers the opportunity to utilise object based intelligent models for FM and AM tasks. The inclusion of geometric and non-geometric information based on a shared semantic structure paves the way for optimised information management processes resulting in less errors, greater consistency, clarity, accuracy, and clear responsibility of authorship. BIM can be viewed as a mean for facilitating Knowledge Management (KM) activities including acquisition, extraction, storage, sharing and update of knowledge (Deshpande et al. 2014). However, research efforts on the role of KM in BIM-based FM are lacking (Charlesraj 2014) and even more serious are the efforts towards AM. With the utilisation of BIM as a central knowledge resource, there is the potential of improving information exchange and better information management within owner-operator organisations. Nevertheless, one of the problems of the Architectural, Engineering, Construction and Owner-operator (AECO) industry is the lack of learning from experiences of the use and operations of existing assets (Jensen 2009). Similarly, this deficiency has been tackled by some organisations in the industry through the development of lessons learnt databases, communities of practice, project closeout interviews and other informal techniques (Rezgui et al. 2010). The recent focus of the AECO industry on BIM and facility data brings forth the opportunity for KM in FM and AM.

The issues revealed in the diffusion of innovative technologies and techniques in FM and AM can be related to the difficulties in demonstrating the business value of information and knowledge management processes and technologies. This research provides a practical case study to individuate where clients can look for the identification of value in the use of digital technologies devoted to the management of data and information. It is recognised in case studies the more appropriate investigation to evaluate the business benefit of information systems (Bakis et al. 2006).

In first place, the research highlights issues related to a non-organised information process. Furthermore, results show how the combination of different data sources generated during operation and maintenance phases can provide useful insights for the definition of future projects requirements. Thus, showing the relation between the characteristic of a building and/or of its components and the maintenance costs. Starting from the optimisation of information management processes, the study envisages the use of facility information for knowledge generation and its applicability in decision-making processes. This study demonstrates in the first instance the value of an effective information management process and secondly possible impacts and value areas of KM in BIM-based FM and AM. The case study presented can help clients in understanding the value of information and knowledge management technologies and techniques in current asset interventions and/or in the development of future projects.

The rest of the paper is organised as follows. Section 2 introduces the background including the identification of the value embedded in information. Section 3 presents the proposed case study including the methods used for its development. Section 4 and 5 contain a discussion about the obtained results and conclusions.

## 2 BACKGROUND

BIM on its own can offer a useful and personalised graphical visualisation of the contents of a database. However, even if representation of the data can provide useful information to individual users it cannot be seen as knowledge transfer. Knowledge needs the identification of patterns behind the information and the human action to gain power in the process (Kamaruzzaman et al. 2016). Information needs to be merged, combined and analysed including domain specific a-priori knowledge to find patterns and extract knowledge (Fayyad et al. 1996). Nevertheless, measuring the value of intangible assets like KM is difficult (Kaplan & Norton 2004) and an incorrect perception can undermine its effective introduction. Hence, a clear vision on the value of information and the possible implication of KM in AM is required.

### 2.1 Knowledge Management (KM) in Asset Management (AM)

KM is defined as any process that enhances organisational learning and performance through the process of creating, acquiring, capturing, sharing and using knowledge (Scarborough et al. 1999). Similarly, Jennex (2005) defines KM as the process of utilising organisational experience as a knowledge base and the selective application of those experiences to current and future decisions with the sole aim of improving organisational effectiveness. There are several benefits that can be gained from the knowledge accumulated in AM activities. This is because, an organisation's competitive advantage depends on what it knows, how it uses that knowledge, and how fast it can learn something new (Charlesraj 2014). Asset managers may require the integration of various types of information such as work orders and maintenance records across different parts of the business for decisions on asset interventions in order to apply knowledge-based decisions (Motawa & Almarshad 2013).

On the other hand, Rainer & Turban (2009) define knowledge as *'data and/or information that have been organized and processed to convey understanding, experience, accumulated learning and expertise as they apply to a current problem'*. Similarly, Alavi & Leidner (2001) suggest that knowledge becomes information when it can be interpreted by individuals. Furthermore, they go on to describe that knowledge can become information if it can be expressed in words, graphic or other representations.

### 2.2 Value of information

Senn (1990) defines information as data presented in a form that is meaningful to the recipient. Therefore, information management processes and techniques have become important aspects of AM. Asset managers utilise hardware and software as mechanisms to create and maintain information within their organisation. These technologies and processes are means of delivering information, whilst information is the asset that can be used to gain strategic advantage by the organisation (Moody & Walsh 1999).

In trying to identify the value of information, Moody & Walsh (1999) proposed seven laws of information, these are:

- information is (infinitely) shareable; the value of information increases with use
- information is perishable
- the value of information increases with accuracy
- the value of information increases when combined with other information
- more is not necessarily better
- information is not depletable.

Similarly Burk & Horton (1998) identified nine similarities between information and assets, these are:

- information is acquired at a definite measurable cost
- information possesses a definite value, which may be quantified and treated as an accountable asset
- information consumption can be quantified
- cost accounting techniques can be applied to improve the control of costs associated to information
- information has identifiable and measurable characteristics
- information has a clear life-cycle
- information may be processed and refined
- substitutes for any specified item or collection of information is available and may be quantified as more or less expensive
- choices are available to management in making trade-offs between different grades, types and prices for information.

Miller (1996) suggests that in order to information to be valuable it has to have these qualities, they are: relevance, accuracy, timeliness, completeness, coherence, format, accessibility, compatibility, security, and validity. On the other hand, there is the view of information from two perspectives of value, which are: philosophical and practical (Repo 1986, Repo 1989) value. These two perspectives can be used to highlight that value depends on the users' perception. Engelsman (2007) reviewed six approaches to valuing information in the normative literature, they are: valuation in risk perspective (Poore 2000); Historical cost valuation (Moody & Walsh 1999); usage over time valuation (Chen 2005); utility value of information (Glazer 1993); and valuation of knowledge assets (Wilkinns et al. 1997). Furthermore, Engelsman (2007), proposed a four-step framework to valuing information, they are; identify the information asset; determine the audience for valuation; determine context and value information.

Asset managers have tried to create systems that enable them to utilise their knowledge-base through the delivery of quality data but end up having results that are missing required information or embedded with significant amounts of meaningless data (Brous et al. 2015). The problem for organisations is not the production of data but capturing, interpreting and managing them for future decisions. Information is at its lowest value when the organisation does not know it exists. Unused information can be termed as a liability for asset owners because they extracts no value from it and the organisation continuously incurs cost for storage and maintenance (Moody & Walsh 1999). The value of data increases when an organisation is able to integrate its systems and collect data from various aspects of its business in order to make simulations and analysis that will aid decision making and improve the execution of future processes.

Despite gaining recognition as an asset in its own right, information has so far resisted quantitative measurement because it has no real value until it is utilised (Moody & Walsh 1999). As a result, there is no consensus on how to measure the value of information.

### 2.3 Asset intervention

Asset intervention which is also referred to as building maintenance *'is the combination of all technical and administrative actions, including supervision actions, intended to retain an item in, or restore it to, a state in which it can perform a required function'* (BS 3811 1984). Motawa & Almarshad (2013) classified maintenance into two main categories, preventive and corrective. Similarly, BS 3811 (1984) divides maintenance into two: planned and unplanned maintenance. It further presented other classifications of planned maintenance which are preventive, corrective, condition-based and scheduled maintenance. Furthermore, Chanter & Swallow (2007) distinguish maintenance activities into planned and unplanned maintenance. They also described maintenance which is based on operational decisions as shown in Figure 1.

#### 2.3.1 Planned Maintenance (PM)

PM involves maintenance activities carried out by an organisation according to a predetermined plan (BS 3811 1984). This may also refer to preven-

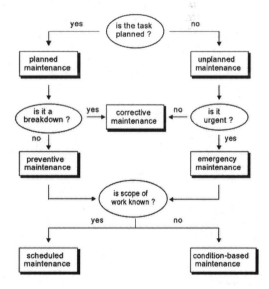

Figure 1. Types of maintenance (Chanter and Swallow, 2007).

tive, scheduled or condition-based maintenance (Chanter & Swallow 2007). These activities are planned to prevent failure and to extend the service life of assets. The technique of PM was developed to address the traditional maintenance approach of only fixing broken assets (Munchiri et al. 2017).

Asset owners benefit from this approach because PM extends asset life, reduces down-time, improves safety, and reduces the need for premature capital investments.

### 2.3.2 *Unplanned maintenance*

Unplanned maintenance involves ad-hoc maintenance activities carried out by an organisation with no predetermined plan (BS 3811 1984). This is a reactive strategy where assets are only fixed when they are broken (Munchiri et al. 2017). This asset intervention strategy is most suitable for inexpensive elements and those that are easy to replace.

## 3 CASE STUDY

### 3.1 *Methods*

During the study, public assets of around 450 buildings has been analysed in a municipality over a two-year time frame. The public body acts both as client and asset manager facilitating the study of possible interactions between the two phases that is usually hindered due to the fragmentation of the subjects involved. These aspects facilitated the analysis of possible repercussions of the proposed study on the definition of future requirements for new projects and/or renovation activities.

Archival analysis of facility information based on asset interventions were conducted for the purpose of this study. A first step of data cleaning based on data quality techniques (Hernández & Stolfo 1998) has been developed. In particular, facility information was non-homogeneous and distributed on several distinct sources (files and/or databases). Hence, during the data cleaning activities, the information has been aggregated and uniformed to allow future analysis. Moreover, a high number of empty cells have been registered with consequent problems related to the correct interpretation of the missing information (Zaniolo 1984).

Starting from the cleaned data, two main kinds of analysis have been developed. On the one hand, the information has been represented in meaningful graphical form to provide a better comprehension of possible relations. This first step can be interpreted as an improvement in the information management process with the consequent increase of the information availability and accessibility.

On the other hand, the information has been analysed using statistical techniques such as linear regression to identify unknown patterns behind the data and thus provide knowledge to the facility manger. This second step of analysis can be viewed as a further one in comparison to the information management because the effective use of the information provides unseen patterns moving the focus from information to knowledge generation.

While information visualisation (first level) is mainly devoted to the analysis of the entire dataset aggregating the information, the application of linear regression analysis (second level) needs the study of data related to single buildings. Since the data cleaning activity highlighted the impossibility to define precise data on all the single buildings, only a sub-set of the asset has been included in the second level analysis. To avoid possible deviations in the results, the areas characterised by a high concentration of historical buildings have been omitted.

Figure 2 represents the research path followed in the development of the proposed case study. The statistical analysis proposed takes into account the value of data integration that will be discussed in detail in section 4.

### 3.2 *First level of analysis: information management*

As described in section 3.1, the first level of analysis is focused on the representation of the data contained in the cleaned data set. This first phase can be used to demonstrate the potential benefit of better information management processes that can facilitate the availability, accessibility and usability of data through information visualisation means. In particular, three main areas have been analysed, namely total maintenance costs associated to specific work types, frequency of maintenance intervention associated to specific work types and the order of works characteristics in terms of work optimisation.

In this section, the main results associated to the proposed case study in terms of data visualisation are described, while a detailed analysis with reference to value association will be proposed in section 4.

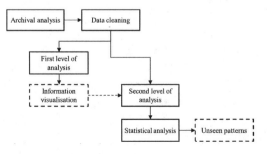

Figure 2. Research path.

Figure 3 shows the total cost of maintenance on the entire asset divided according to the work types registered in the selected time span. This visualisation can help in identifying the main areas of costs in maintenance processes guiding the facility manager in monitoring activities. For example, in the proposed visualisation it is clear that there is great impact of minor maintenance on roofs, followed by minor maintenance of smith works and wood works.

Figure 4 shows the frequency of maintenance intervention in relation to the same work types analysed in Figure 3. Following the same principle presented for Figure 3, the identification of the work types that requires a high frequency in maintenance intervention can guide the facility manager in identifying possible issues and activate corrective actions. In this case, minor maintenance of smith works and wood works have the highest impact.

Combining the two analyses can help identifying possible issues related to minor maintenance of smith works and wood works due to their high impact in terms of both costs and frequency.

The third part of the analysis included in the first level is focused on the study of the orders of works. Work orders identify the means through which the facility manager requires maintenance works to the company in charge. In particular, 584 orders have been included in the analysis studying both the number of works comprised in one order and the total cost of each order. Starting from the global representation of the analysis that reports a maximum cost per order equal to 18000 euros and a maximum number of works in one order equal to 18, Figure 5 shows a focus of the graph in the area possible issues can be identified. In fact, the definition of work orders with a small cost and a small number of associated works can reveal possible inefficiencies helping the facility manager in the identification of possible corrective actions in the process.

Starting from this principle, in Figure 5 3 main clusters have been identified to better understand the characteristics of the maintenance activities. The first cluster (dark grey area) identifies all the work orders with a total cost minor or equal to 600 euros and with a total number of works minor or equal to two. The second cluster (middle grey area) identifies all the work orders with a total cost minor or equal to 800 euros and with a total number of works minor or equal to 3. The

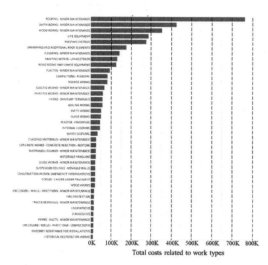

Figure 3. Total costs (in thousands of euros) related to work types.

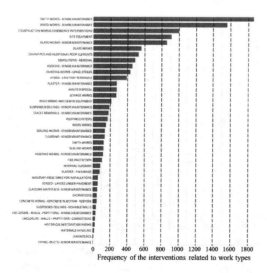

Figure 4. Frequency of the intervention related to work types.

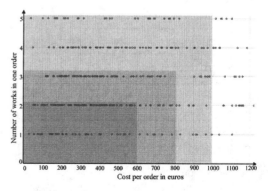

Figure 5. Order of works during the year and related cost.

third cluster (light grey area) identifies all the work orders with a total cost minor or equal to 1000 euros and with a total number of works minor or equal to 5. The number of work orders included in each cluster are following listed (percentage impact on the total in brackets):

- First cluster 173 (29,72%).
- Second cluster 284 (48,63%).
- Third cluster 403 (69,01%).

The high number of work orders included in these clusters highlights the prevalent presence of unplanned maintenance. The detailed study of the work types included in the work orders in the different clusters can help the facility manager in the definition of possible action devoted to the optimisation of future interventions.

The results proposed in this section demonstrated how the integration of data visualisation means in information management processes can provide highly accessible information with a consequent increase in the information value as discussed in section 4.

### 3.3 Second level of analysis: knowledge creation

In an organisation, the budget devoted to maintenance activities on buildings can be defined on annual basis according to relevant parameters related to the buildings such as the surface or the number of daily users. In the proposed case study, all the buildings included in the statistical analysis were schools. In school buildings parameters can be the number of students or the dimension of the school in square or cubic meters. However, the definition of a maintenance cost per year per square meter is difficult and do not consider the characteristics of the school, bringing to possible over or underestimation of the costs.

Starting from the data organised and cleaned, it is possible to identify relations among parameters of the school data to obtain better forecasts for future maintenance. An example is the use of a linear regression model to identify unknown relations between the characteristics of the schools and their maintenance costs.

Due to the short time frame considered in this study, i.e. two year' time frame, some results yielded insignificant relationships, for example the relation between year of construction of the building and costs. Nevertheless, the number of floors of the schools shows a significant relation with the costs of maintenance. Table 1 proposes the results of the analysis where the constant term represent the cost of maintenance per year per cubic meter.

Results show how the number of floors is indirectly related to the maintenance cost. In particular, starting from a constant cost of 3.057 euros per

Table 1. Results of the linear regression model.

| Terms of the model | Results |
|---|---|
| Constant Term | 3.057 (0.00)* |
| Number of floors | −0.746 (0.00)* |
| Adjusted $R^2$ | 0.3606 |
| Joint F-test | 54.01 |
| Number of observations | 95 |

*Significant test reported in brackets.

cubic meter, a school with one floor would require 2.311 euros per cubic meter per year, while a school with two floors would require 1.565 euros per cubic meter per year.

These insights can be also used for the definition of requirements for future schools. For example, the proposed analysis highlights how the number of floors in a school diminishes the costs of maintenance. Hence, future projects for new schools can require buildings with more floors to optimise the maintenance costs.

## 4 DISCUSSION

The results obtained through the case study proposed in section 3, can be used to identify and clarify the value of information and knowledge in AM and FM practices supported by digital information technologies and techniques.

Starting from the data cleaning activities, it can be noted that several buildings have been removed from the analysis due to a high number of unknown information and/or not coherent information. Furthermore, among the buildings maintained in the analysis, several empty cells have been individuated. Empty cells can represent a not trivial issue due to the impossibility in defining the reasons behind the empty value. Consequently, the related information is lost. Furthermore, with reference to geometrical information, the presence of documents that contain the information in a not computable format (e.g. 2D drawings in pdf format) reduces the number of information that can be included in the analysis. For example, the analysis proposed in section 2.3, shows the possibility to provide useful information to the administration. However, due to the limited number of geometrical information available in machine readable format, the analysis is bounded to a poor number of parameters. In the future, the use of digital models related to the asset will provide precise geometrical information that can be used in the improvement of the proposed framework to guarantee better performance and more precise information. Hence, BIM can be viewed as a facilitator for the development of data analytics application on buildings data, thanks to the increased availability of machine readable information both in terms of geometrical and non-geometrical data.

These first considerations highlight a valuable point in the identification of the value behind the introduction of digital based information management processes. On the one hand, it can be noted a direct improvement in the accessibility and usability of the data with the consequent reduction in the related time and costs. On the other hand, the increased availability of the data paves the way for the development of statistical applications as proposed in section 3.3.

Relating these evidences with the parameters connected to information value presented in section 2.2 it can be stated that the case study demonstrated a value increase improving the possible uses of information and the quality and consequently the accuracy of information.

The availability of structured data related to different aspects of the maintenance process (e.g. costs, building characteristics and maintenance activities) can be used to generate new knowledge as proposed in section 3.3. This aspect acts directly on the value of information associated to the combination of sources. However, the main benefit can be related to the increased comprehension of the dynamics behind the maintenance process thanks to the extraction of unseen knowledge from an extended data set. In fact, the introduction of information management processes can produce defined benefits as discussed in the previous paragraphs. However, these benefits do not express the complete potential behind the increased information availability, that is the creation of new knowledge that can be used to change and improve the entire maintenance process.

This has been already discussed as the analysis proposed in the second level can produce unseen knowledge. However, it can be noted in the literature a discrepancy between the represented knowledge and the knowledge perceived by the user according to his or her past knowledge. In this study the passage between the knowledge creation to the knowledge conversion and integration in human mind is not addressed and can be identified as a future line of research.

Finally, the example proposed in section 3.3 demonstrates how the possible impacts derived from the use of information is not limited to the maintenance process but can be extended to other activities such as the design of new buildings. The information can be used to identify possible relation between maintenance costs and building characteristics or to study the characteristics of elements that demonstrated low performances during the building use. Hence, this information can constitute the basis for the development of information-based project requirements towards the definition of computable project requirements. This last area is gaining increasing interest in the scientific field and it represents a critical point for the introduction of BIM in structured clients such as public administrations.

## 5 CONCLUSION

The understanding of the value (in terms of possible positive effects) of the use of BIM or in general digital techniques and technologies for better information and knowledge management is a critical issue and can hinder the adoption of BIM in FM and AM. This paper presents a practical case study that can contribute in the understanding of the value of effective information and knowledge management in FM and AM to push the industry and the research in the exploration of new uses for data and information. In particular, through the study of the data produced during maintenance activities on public assets of 450 buildings in a two year' time span, has presented possible applications that can help in the identification of the positive effects derived from a structured data organisation and an effective use of this data. The results obtained in the case study, have been compared to key value indicators associated to information value to highlight the impacts of digital technologies and techniques in the maintenance process. Furthermore, the paper propose a perspective interpretation that looks on possible future applications that the increasing production of machine-readable data can produce. In particular, the introduction of statistical means and more in general of data analytics applications on maintenance. This has demonstrated the possibility to provide useful information and consequently produce important change in actual processes in the logic of learning from the past through the data.

The study highlights several issues with reference to the availability and accessibility of the data. These issues hindered the extended exploration of possible data analytics applications. Hence, future case studies can explore the integration of digital models and/or of computable information related to the characteristics of buildings and of their elements (e.g. geometry and main features) to integrate the proposed analysis with further evidences that can extend the presented uses.

## REFERNCES

Alavi, M. & Leidner, D.E. 2001. Knowledge management and knowledge management systems: Conceptual foundations and research issues. *MIS Quarterly*, 25(1): 107–136.

Bakis, N., Kagioglou, M. & Aouad, G. 2006. Evaluating the business benefits of information systems, *3rd International SCRI Symposium, Salford Centre for Research and Innovation, University of Salford*: 280–294.

Brous, P., Herder, P. & Janssen, M. 2015. Towards modelling data infrastructures in the asset management domain, *Procedia Computer Science.* 61(1): 274–280.

BS 3811 1984. Glossary of maintenance management terms in terotechnology. London: British Standard Institution.

Burk, C.F. & Horton, F.W. 1998. *INFOMAP: a completeguide to discovering corporate information resources.* Englewood Cliffs, NJ: Prentice Hall.

Chanter, B. & Swallow, P. 2007. *Building Maintenance Management.* 2nd ed. Oxford; Malden, MA: Blackwell.

Charlesraj, V.P.C. 2014. Knowledge-based Building Information Modeling (K-BIM) for Facilities Management, in *Proceedings of the 31st International Symposium on Automation and Robotics in Construction and Mining (ISARC 2014).*

Chen, Y. 2005. Information valuation for Information, in *ICAC 2005. Proceedings. Second International Conference on on Autonomic Computing, June 2005.*

Deshpande, A., Azhar, S. & Amireddy, S. 2014. A framework for a BIM-based knowledge management system, *Procedia Engineering* 85: 113–122.

Engelsman, W. 2007. Information assets and their value, in *Proceedings of the 6th Twente student conference on IT.* Enschede, Netherlands: University of Twente.

Fayyad, U., Piatetsky-Shapiro, G. & Smyth, P. 1996. From Data Mining to Knowledge Discovery in Databases, *AI Magazine* 17(3): 37–54.

Glazer, R. 1993. Measuring the Value of Information: The Information Intensive Organization, *IBM Systems Journal* 32(1).

Hernández, M.A. & Stolfo, S.J. 1998. Real-world Data is Dirty: Data Cleansing and The Merge/Purge Problem, *Data Mining and Knowledge Discovery* 2(1): 9–37.

Jennex, M.E. 2005. What is Knowledge Management?, *International Journal of Knowledge Management* 1(4): 1–4.

Jensen, P.A. 2009. Design Integration of Facilities Management, *Architectural Engineering and Design Management* 5(3) 124–135.

Kamaruzzaman, S.N., Zawawi, A.M.A., Shafie, M.O. & Noor, A.M. 2016. Assessing the readiness of facilities management organizations in implementing knowledge mangament systems, *Journal of Facilities Management*, 14(1): 69–83.

Kaplan, R.S. & Norton, D.P. 2004. Measuring the Strategic Readiness of Intagible Assets, *Harvard Business Review* 82(2): 52–63.

Love, P.E.D., Matthews, J., Simpson, I., Hill, A. & Olatunji, O.A. 2014. A benefits realization management building information modeling framework for asset owners, *Automation in Construction.* 37: 1–10.

Miller, H. 1996. The multiple dimensions of information quality, *Information Systems Managemen* 37(1): 1–10.

Moody, D. & Walsh, P. 1999. Measuring The Value Of Information: An Asset Valuation Approach, in *Seventh European Conference on Information Systems (ECIS'99).* Frederiksberg, Denmark: 1–17

Motawa, I. & Almarshad, A. 2013. A knowledge-based BIM system for building maintenance, *Automation in Construction* 29: 173–182.

Munchiri, A.K., Ikua, B.W., Munchiri, P.N. & Irungu, P.K. 2017. Development of a theoretical framework for evaluating maintenance practices, *International Journal of System Assurance Engineering Management* 8(1): 198–207.

Poore, R.S. 2000. Valuing information assets for security risk management, *Information Systems Security* 9(4): 17–23.

Rainer, R.K. & Turban, E. 2009. *Introduction to information systems.* 2nd ed. Hoboken, NJ: John Wiley & Sons.

Repo, A.J. 1986. The dual approach to the value of information—an appraisal of use and exchange values, *Information processing & management* 22(5): 373–383.

Repo, A.J. 1989. The Value of Information: Approaches in Economics, Accounting, and Management Science, *Journal of the American society for information* 40(2): 68–85.

Rezgui, Y., Hopfe, C.J. & Vorakulpipat, C. 2010. Generations of knowledge management in the architecture, engineering and construction industry: An evolutionary perspective, *Advanced Engineering Informatics* 24(2): 219–228.

Scarborough, H., Swan, J. & Preston, J. 1999. Knowledge Management: Literature Review, *London: Institute of Personnel and Development.*

Senn, J.A. 1990. *Information Systems in Management.* 4th ed. Belmont, California, USA: Wadsworth Publishing.

Wilkinns, J., Wegen, B.V. & Hoog, D.R. 1997. Understanding and Valuing Knowledge Assets, *Expert systems with applications* 13(1): 55–72.

Zaniolo, C. 1984. Database relations with null values, *Journal of Computer and System Sciences* 28(1): 142–166.

# Integrated information management for the FM: Building information modelling and database integration for the Italian Public Administration

L. Pinti, S. Bonelli, A. Brizzolari, C. Mirarchi & M.C. Dejaco
*Department of Architecture, Built Environment and Construction Engineering, Politecnico di Milano, Milano, Italy*

A. Kiviniemi
*School of Architecture, University of Liverpool, Liverpool, UK*

ABSTRACT: The focus of the research is the integration of Building Information Modelling (BIM) with dynamic database in order to provide a powerful instrument to control the variables of Facility Management, Operation and Maintenance (FM-O&M) field. The paper underlines the results of a theoretical and practical study about BIM and FM integration in the Public Administration field, combining the main benefits of information storage offered by database application with the advantages of Building Information Modelling. The analysis is performed on a building portfolio of 450 public assets and develops a replicable procedure that implements innovative tools and methods to manage the variety of processes within the built environment.

## 1 INTRODUCTION

Building Information Modelling is a process of representation, which creates and maintains multidimensional data-rich views throughout a project lifecycle to support communication (sharing data); collaboration (acting on shared data); simulation (using data for prediction); and optimization (using feedback to improve design, documentation and delivery) (Laiserin, 2012).

Therefore, when we refer to Building Information Modelling (BIM) we are talking about a process that allows you to generate, collect and manage all the data about the building throughout its life cycle: from concept and design to the stage of demolition or reuse (Succar et al., 2007; Vanlande, Nicolle and Cruz, 2008; Gu and London, 2010).

The life cycle of a building can be summarised according to some main phases: design, construction and facility management (Ahmed et al., 2017).

In the traditional methodology, all these phases are often disconnected with each other because they involve different professionals (Love et al., 2013; Edirisinghe, Kalutara and London, 2016). Even within the same phase there are many professionals involved and they all tend to produce documentation only related to their own interests. This is one of the reasons for the fragmentation of data during the building process (Ahmed et al., 2017).

In addition, there is usually a clear separation between the construction phase and the facility management phase (Kubler, Buda and Fr, 2016). In fact, the models used in the design and construction stages are not interoperable with the existing Computer Aided Facility Management (CAFM) systems used by the Facility Manager (Vanlande, Nicolle and Cruz, 2008; Akcamete, Akinci and Garrett, 2010).

Today BIM is widely used in the initial phases of the building life cycle (conception, planning and construction) while in the post-construction phases the methodology still encounters difficulty to be integrated (Eadie et al., 2013; Edirisinghe, Kalutara and London, 2016).

Effectively, BIM has transformed the planning and construction processes, simplifying the multidisciplinary coordination and integrating the 3D design with the project documentation (Enoma, 2005).

Design, planning and construction phases affect only a minor part of the building life in terms of time and cost, compared to the facility management (Liu et al., 1994; Clayton, Johnson and Song, 1999; Edirisinghe, Kalutara and London, 2016).

In fact, the operation and maintenance phase of a project is the main trigger that increases the total life cycle cost, since it is the phase with the higher duration, we can say that the incorrect management of Facility Management, Operation and

Maintenance (FM/O&M) implies a wide economic impact.

Moreover, the actual wrong way of managing in-formation since the early stages of the process, and the lack of regulated procedures and standards among the operators involved, negatively influence the exchange and quality of data.

The problems presented are even more obvious when we look at the Public Administration (PA). In fact, in addition to the obstacles described above, the PA must deal with a limited budget and with a large number of professionals with very heterogeneous skills.

The paper underlines the possibility of BIM and FM integration in the Public Administration and highlights the main barriers to BIM for FM adoption, the problems with the current practice, and the most considerable benefits of this merging (Migilinskas et al., 2013).

## 2 BACKGROUND

The demand of BIM is increasing recently, as much international organizations and governments are taking the initiative to promote BIM in all industry life cycle. BIM has been used most extensively in design and construction and it is possible to affirm that adoption and use for FM purposes is a complex issue.

When facility managers and maintenance operators have to consider the extensive documentations of information needed for their activities, it is clear that finding efficient ways to gather, access and update this information, is very important.

In BIM a digital representation of the building process can be used to facilitate the exchange and interoperability of information in digital format (Eastman et al., 2011).

Therefore, Information is the most emerging necessity for FM, since it requires tremendous amount of data for efficient O&M of the facilities (Wang et al., 2013).

### 2.1 Maintenance as a strategy

Concerning maintenance matters, it is possible to say that Italy is traditionally marked by an old historical building stock, which has suffered over the years the indifference of the operators and of public administrations. The vulnerability of our assets, today affected by catastrophic events, drives the need of coordinated maintenance actions and regulated frameworks. Therefore, after years of negligence, we feel the need for maintenance and starts to consider it also from the legislative point of view introducing a regulatory framework.

For all these reasons, today we have to deal with buildings that, for the most part, are not designed to be maintained. Thus, we need to build a maintenance knowledge, considering maintenance as an innovative strategy.

Differently from the past, maintenance is no more considered a reparation task after failure or a low-profile activity but is a way to preserve resources extending the building life cycle. In fact, the process of dismantling a building can be postponed or the substitution of degraded elements can avoid the economic and environmental costs of a new phase.

In order to extend the building life cycle, it is necessary to adopt since the design phases criteria to choose technical elements and materials with higher durability and reliability, to minimise the probability to have unexpected breakdowns during the useful life, applying a constant preventive maintenance approach.

The maintenance management has the assignment to preserve the building system in a context that is characterised by high variability and several transformations. For example, the performance required, the needs, the regulations, external events, are not constant in time. In this context, the maintenance management assumes the role of mediator in rearranging a system which is subject to all these factors.

The fundamental point of the maintenance management practice is represented by the availability of information, so by an efficient informative system. In fact, the area of information is the link between the two other fields within the maintenance management which are the design area and the operational area. So, the correct implementation of an integrated information system for maintenance allows the connection among design, construction, technical and economic data which are useful for defining the most suitable management strategy.

Hence, the Facility Manager must be able to keep information available, editable and updated, working on centralized archives that enable to make forecasts and budget estimation, optimizing the operational efficiency.

### 2.2 Information traceability and the value of data

One of the most important element in managing the built environment is linked to the relationship between action and knowledge. Only with a deep knowledge of the real estate assets, it is possible to enhance, manage and make buildings adequate to the users' needs, reaching quality goals following efficiency and costs saving principles.

Therefore, all the actors that intervene in the management operations and that have to control the building functionality, have to establish their activities on a detailed analysis, which is able to capture all the strategic objectives, costs and technical regulation of the asset.

The acquisition of a complete set of data represents the fundamental condition which permits a correct development of the maintenance services. The data which are necessary are linked to the object itself, so to the building and its parts, and also to the policies and management objectives. These kinds of data constitute the informative system of the building.

Further, facility managers encounter the issue of quality and timely-access to information for O&M activities, and the need for more efficient methods to manage that (Sabol, 2013).

Moreover, the flow of information throughout the building life cycle is crucial and should be considered a primary objective for each construction endeavour from its beginning to the current operation stage. So, the data management in order to be efficient, must be based on a systematic collection, analysis, flow of information across the multidisciplinary environment throughout the life cycle of the building asset.

In the usual current practice, O&M costs are often overlooked at the design phase by owners and project stakeholders although they could amount to over half of the total building life cycle costs (Becerik-Gerber *et al.*, 2012).

In addition, the transfer of such data to FM systems is a cost and time-consuming process, resulting in prolonged periods before optimal building performance can be reached, because optimisation decisions concerning maintenance, energy or safety are delayed at the FM stage (Sabol, 2008).

As it is highlighted above, data and information handover to the FM phase is often left until the completion of the construction phase and information is typically delivered in manual or non-digital formats. With this late delivery of unstructured data, it becomes very challenging for owners and facility managers to assess whether the information they need is included in the handover data.

For all these reasons, finding methods to organize the shape and structure of data for FM & OM purposes, since the preliminary stages of the life cycle should be a primary topic.

## 3 CASE STUDY

### 3.1 *Research approach to the current situation*

The FM area is characterized by high complexity of data and tangled actors' relations. In particular, there is the need to integrate the main benefits of information storage, offered by database applications, with the advantages of BIM software in terms of visualization, localisation and immediate awareness in virtual environment. The research provides a method to effectively apply and cover the items stated above, analysing the portfolio of a Public Administration which consists of around 450 buildings.

During the analysis, the conventional methodology is firstly applied and the current tools used by the PA for the building management are investigated. This first step allows to identify the input and the output data that are needed by the PA to carry out the standard maintenance practice and management activities, and to take stock of the functioning of the current system. The aim of the research is in fact not to offer a new tool to the final user, which is different from the one currently in use, but to provide a better-organized method to collect, store, find and share data quickly in the most efficient way.

The problem faced by many technicians, actually, is exactly the time needed to find the correct information among available data folders. Moreover, it is often necessary to acquire and look for new data to update the information to the actual building condition. All these aspects become more critical when a large number of professionals has to deal with new tools and this fact automatically implies further waste of time to appropriate train the personnel. For these reasons, the developed applicative tries to retrace the systems currently used, introducing an innovative logic approach behind the management and maintenance mechanisms.

### 3.2 *Data analysis and re-elaboration*

The analysis of all the available data and information of the buildings is useful to fully understand the processes behind the current database and specially to capture which is the amount of data and information required. Then in this way, data useful for the management software or pointless information can be identified. Furthermore, concerning the absence of specific data, it is possible to underline the lack of all the information related to the technical elements like the location of the intervention, the technical characteristics of the item and so on, forcing on-site inspections, which greatly delay the completion date of the intervention. Finally, the database is managed autonomously and not in a sharable way by the different technicians and this implies difficulties in exchanging knowledge. Therefore, the main problems observed during the analysis of the database are the difficulties in finding and reading the data, the inconsistency because of information inserted in multiple fields depending on the building analysed, the long time spent to find or to type the desired data and lastly the non-contextualisation of the intervention.

After this kind of analysis, the database is cleaned up removing the non-necessary information. At this point, the data elaboration shows that the system is essentially based on a macro vision, which is mostly focused on understanding the overall monetary outflows instead of governing problems and matters specific for the individual building. The database collects information of more than four hundred buildings in a fragmented and lacking way.

Frequently public administrations divide their buildings in zones according to their location within the municipality, and this means that there are different technicians in charged to manage more than one building at the same time.

Each technician can enter only that information related to his/her own area and so manage a limited number of information. This can represent both an advantage and a weakness. The former because each technician can access only the information related to his/her area of competence; the latter because this limited vision does not allow the insertion of data in a coordinated and uniform way, facilitating the incoherence of data between different zones.

Moreover, technicians have to manage not only the database but also a huge amount of files and documents that, if handled in an autonomous and not coordinated way, can lead to data dispersion.

In the Table 1, we can look at the high number of columns that each technician has to fill for each building; this explains the reasons why most of the time these fields are not fully compiled or they are totally empty. In general this behaviour is due to the lack of time and will of the technician to complete all these documents in restricted times. Moreover, the presence of empty fields leads to wondering whether the data is already present in other documents or the data is unknown.

After these considerations, the database has been cleaned from those empty fields and from those rows that present values lower or equal to zero, and the final result is a file that consists of the 50% of less data.

The possibility to re-elaborate those data and to merge them allows to better understand which are the zones that have a higher impact on the maintenance expenditure (Figure 1).

Adding a mean value to these results we can investigate also which are the reasons why the other zones are characterized by lower maintenance expenses w.r.t the previous ones.

The research highlights how correct data organisation and a coherent information exchange and management of all the information, allow the effective use of data and provide means for the control and forecast of maintenance costs.

The second phase of the research is about the development of a tool able to integrate database and innovative 3D parametric models in order to manage the overall procedure and to fill the gap between the first phases of the process and the last FM-O&M steps. We decide to exploit the functionalities of parametric modelling software that permit to export in a database format all the information associated to a building information model.

These databases recognize all the elements inserted in the model and allow to attach additional information.

If the model is appropriately characterized by a classification plan-based breakdown structure, it is possible to establish the link between the technical elements of the model and the attributes of the database corresponding to those elements. This process enables data synchronization and a constant model-database update.

### 3.3 *Structured data framework*

The need of a structured data framework is a prerequisite for an efficient data management. As it is

Table 1. Total cells of zone database.

| Zone | Rows | Columns | Total cells |
|---|---|---|---|
| Zone 1 | 2.514 | 52 | 130.728 |
| Zone 2 | 2.075 | 52 | 107.900 |
| Zone 3 | 1.686 | 52 | 87.672 |
| Zone 4 | 2.228 | 52 | 115.856 |
| Zone 5 | 1.202 | 52 | 62.504 |
| Zone 6 | 1.398 | 52 | 72.696 |
| Zone 7 | 1.796 | 52 | 93.392 |
| Zone 8 | 2.234 | 52 | 116.168 |
| Zone 9 | 1.979 | 52 | 102.908 |
| Total cells | | | 889.824 |

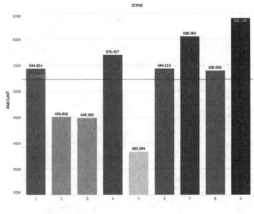

Figure 1. Maintenance total amount per zone.

underlined before, the model characterization can be supported by a breakdown structure. This is the key tool to govern the overall process because it gives a proper framework and a clear codification, which is strategic for the course development. The WBS is the tool capable of linking technical elements, which are the expression of design choices and the detail of interventions/works related to the construction phase of the process and later to the management and maintenance phase. In particular, for O&M purposes the implementation of a Space Breakdown Structure is required to enrich the database and to localize the building items. In particular the localization label takes advantage of the Omniclass table 13 classifying elements according to "Class of functional space" (class of spaces with the same function) and "Elementary space" (single zone within the same class of functional space). Then each element presents an additional code related to the building block, floor and room number.

The database organized in this way has the potentialities of accessing information and data collected in the same place, permits to query those data and derive from them information useful to manage buildings interventions.

In order to have a single database, an application is created. In this way, using multiple interfaces, it is possible to collect data and produce different documents each time these are required.

### 3.4 Integration between database and 3D model

As highlighted before, the focus of the work is the integration between database and 3D parametric models. Concerning the database the process adopts the Relational Database Management System (RDBMS) that allows data creation and elaboration from a DBA Administrator. The Relational Database Model enables a collection of structured data in columns and rows through SQL Language; it combines the formal mathematical theory with usual application with the aim to manipulate and access data.

In the case study development, the parametric application is able to export the information from the model to the database. Then, after changes in the data tables, it is possible to import the updates back to the model.

The database generated from the 3D model is a set of complex relations which are represented in Figure 2.

As it is possible to deduce from the Figure 2, it is not easy to understand all these links. For this reason, one of the first point of the study has been to test the operating principle of the mechanism, exporting and re-importing from an empty 3D model so that networks can appear clearer.

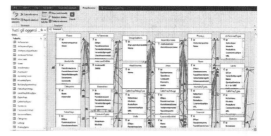

Figure 2. Schema of the relationships between the database and the building information model.

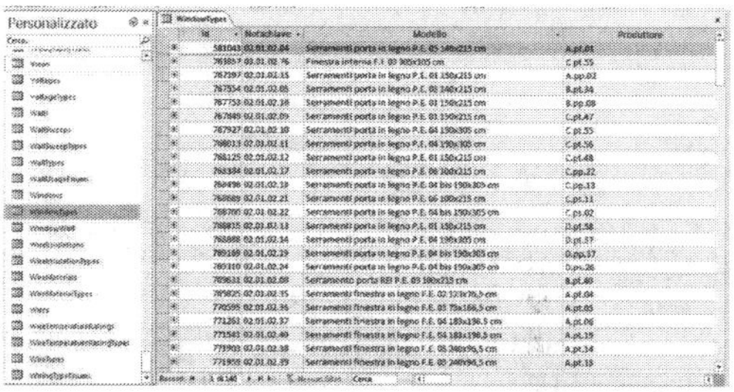

Figure 3. Scheme of database wall types table from the building information model.

The database file presents all the tables corresponding to the 3D model elements and can also work with additional shared parameters that appear as new fields in the respective table.

One of the main advantages of the file exported from the parametric 3D application is the possibility to customize the relational database dashboard in order to group tables of data directly coming from the 3D model and separate them by the other elaborated objects. In this way, the user can easily access data and understand which are its personal tables, forms and queries. These can be linked to those from the model, using internal database software relationships (primary keys); in this way, shared parameters with associated data within the model, can be extracted and can automatically fill the connected user table. Therefore, the act of inserting data can be automated, avoiding typing mistakes in filling the database.

Another relevant feature is the generation of a report that shows the import details. This is a powerful tool to immediately understand if the export of changed and modified data brings to a successful conclusion.

### 3.5 Model characterisation

Based on what we have claimed above, we develop and carry out the following process, to test the efficiency and validity of the methodology.

After the definition of the database (WBS, SBS) according to the reasons explained before, we examine the way of attributing information to the model. Thus for each individual space (elementary space), we link all the technical elements related to that precise area focusing mainly on the finishing layers, doors, windows, external and partition walls.

The model behaves as an information container and gathers all the useful data related to this phase of the process.

In this way, the model perfectly fits its real conditions and includes information about materials and types of technical elements that are useful for the future phases. At this moment, the school model must be implemented with additional data which are strategic for executing and planning maintenance. So, we decide to introduce specific parameters that can facilitate the facility managers, owners and maintenance operators work.

Moreover, we analyse which is the best way to characterize the model with shared parameters and we understand which are the default fields that we can use to associate attributes to types and instances; we examine the default relational database tables created through the export to plan our procedure.

The choice of what kind of data to introduce in the application, is based on the study of the owner database and therefore of the information they store and require on building elements over time. So, to improve their system, we find that the implementation of certain data can represent a key tool to cover their main deficiencies.

3.6 *Application*

The primary objective of this stage is the development of an application to manage in an automatic and integrated way the data system. Therefore, the main challenge is to connect the model information with the PA database needs.

The result of the export from the parametric 3D model to the database software is a file split in tables that correspond to the modelled elements. In particular, there are tables that show all the single elements like walls, windows, floors and others that grouped them according to their typology. In those tables, we find also the shared parameters set in the 3D model with the associated information.

At this point we define this application considering the three main actors. In fact, we provide forms that can track the communication between the Users of each building, the Public Administration and the Technicians that perform the maintenance actions.

Therefore, we set a first login window divided for each entity involved, in this way they can access their area without interferences and managing their own data.

Therefore through this system, the User can point out a problem that affect a space, an element, like for example a door or a window, sending an advisory to the Public Administration. It can simply fill a panel indicating which is the element that presents an anomaly, where it is located, which is the type of failure that can be documented by an attached picture.

On the other hand, the Public Administration after logging in, can select the district in which the asset is located and choose the building in the areas list, after that it can see all the data related to the building and also to the model.

In the Public Administration section a collection of the use manual, maintenance manual and maintenance schedule is available for each element and can be updated in case of changes or substitutions (Figure 4).

In the same section the PA can access the building ID document in which all the relevant data about the current condition of the asset are registered.

Here, the Public Administration can also see all the requests sent by the User, the history of the executed interventions and it can also communicate with the operators to point out the need for an intervention.

In the end the Technicians can log-in, browse the work order requests received by the Public Administration and then, after the maintenance execution can forward its own feedback to the PA attaching a picture of the work executed, compiling the fields with technical comments, and indicating the amount with its discounted value that has to be paid by the Public Administration.

It is clear that the proposed application can be implemented with additional sections, modules and tables according to the needs of the actors involved.

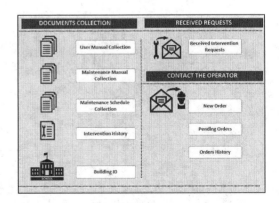

Figure 4. PA interface.

## 4 CONCLUSIONS

This paper presents the results of a theoretical and practical research about BIM and FM integration, as the natural evolution of a methodology that needs to follow innovation and to implement new processes and tools.

The paper demonstrates how managers, owners and public authorities can achieve significant advantages using this framework and these processes. In addition, the paper highlights how BIM can support team collaboration and reduces errors, time and costs, leading to a more efficient and reliable management.

This research work highlights the need of a methodology able to involve all the actors that interact in the construction area, manage the exchange of information among them and control the whole building life cycle.

Moreover, a link between the first design phase of the process and the later stages related to operation and maintenance, is required.

Therefore, this research draws a process to improve the maintenance management practice, integrating Building Information Modelling and Database and it provides new tools to optimize the FM overall procedure.

The application on the tangible case study of the Public Administration portfolio shows that a connection is possible. In fact, as a result, we succeed in obtaining the association between model attributes and database information, collecting and exchanging key data.

In particular, the information is efficiently tracked and stored within the system, typing fields and consequently typing mistakes are limited.

The same steps, which are here customized on the Public Administration data and assets, can be easily applied to other contexts.

The overall process, starting from the work and space breakdown structure association, to the following parameters definition and then to the relational database connection, could be summarized in a matrix that can generally fit similar procedures on the built environment.

So, the final achieved goal has been to combine the main benefits of information storage, offered by database applications, with the advantages of BIM software in terms of visualization, localisation and immediate awareness in virtual environments with the aim to produce a coherent data framework.

This work is mainly focused on the activities and dialogue of/among the actors that characterise the Public Administration daily activity; so, forms and data in the sketched application concentrate on stored orders, received requests, maintenance documents collection and history of interventions, but for a further development, the relational database system can be implemented with new tables and forms, new shared parameters within the software, needed for other kind of purposes.

Concerning the information, the Public Administration provides in the first stage, it is possible to say that the re-shaped buildings data that we have organised in a more efficient and coherent way, gives the chance to make some reasoning on maintenance expenses and frequency for building category.

The implementation of this new procedure can decrease all the costs components linked to the information loss and can influence the outflows related to the collection of specific data. Moreover the new structured framework of data is able to point out the main problems/costs of the whole management. Thanks to the new tables and diagrams, data can be easily analysed, so that consecutive actions and strategic policies can be reshaped consciously. In fact, in case of some peaks, a future option can be that of reconsider the choices of certain technical elements which still generating a negative impact on costs. On the other hand, through an inverse approach, it is possible to deduce from the design phase, the estimation of operation and maintenance costs for the registered elements.

In the end, the Data engineering method for maintenance management, demonstrates that Building information modelling and database integration can be powerful for the process optimization. This kind of innovative tools based on collaboration, interoperability, data exchange over the whole life cycle are progressively becoming the key for the construction field, Public Administrations and operators to success in the future.

## REFERENCES

Ahmed, V. *et al.* (2017) 'The Future of Big Data in facilities management: opportunities and challenges', *Facilities*, 35(13/14), pp. 725–745.

Akcamete, A., Akinci, B. and Garrett, J.H. (2010) 'Potential utilization of building information models for planning maintenance activities', in TIZANI, W. (ed.) *Proceedings of the International Conference on Computing in Civil and Building Engineering*. Nottingham, UK, pp. 151–157.

Becerik-Gerber, B. *et al.* (2012) 'Application Areas and Data Requirements for BIM-Enabled Facilities Management', *Journal of Construction Engineering and Management*, 138(3), pp. 431–442.

Clayton, M.J., Johnson, R.E. and Song, Y. (1999) 'Operations documents: Addressing the information needs of facility managers', *Durability of Building Materials and Components Proceedings*, 8(14), pp. 2441–2451.

Eadie, R. *et al.* (2013) 'BIM implementation throughout the UK construction project lifecycle: An analysis', *Automation in Construction*, 36, pp. 145–151.

Eastman, C. et al. (2011) *BIM handbook : a guide to building information modeling for owners, managers, designers, engineers, and contractors.* Second Ed. Hoboken, New Jersey: John Wiley & Sons.

Edirisinghe, R., Kalutara, P. and London, K. (2016) 'An investigation of factors affecting BIM adoption in Facility Management: an institutional case in Australia', in *The RICS annual construction and building research conference (COBRA 2016)*. Toronto, Canada.

Enoma, A. (2005) 'The Role of Facilities Management at the Design Stage', in Khosrowshahi, F. (ed.) *21st Annual ARCOM Conference*, pp. 421–30.

Gu, N. and London, K. (2010) 'Understanding and facilitating BIM adoption in the AEC industry', *Automation in Construction*. Elsevier, 19(8), pp. 988–999.

Laiserin, J. (2012) 'Building with words', *Architectural Research Quarterly*, 16(03), pp. 195–196.

Liu, L.Y. et al. (1994) 'Capturing as-built project information for facility management', in Khozeimeh, K. (ed.) *Computing in Civil Engineering (New York)*. Washington, DC, USA: Publ by ASCE, pp. 614–621.

Love, P.E.D. et al. (2013) 'From justification to evaluation: Building Information Modeling for asset owners', *Automation in Construction*, 35, pp. 208–216.

Migilinskas, D. et al. (2013) 'The Benefits, Obstacles and Problems of Practical Bim Implementation', in *Procedia Engineering*. Elsevier, pp. 767–774.

Sabol, L. (2008) 'Building Information Modeling & Facility Management', *Design + Construction Strategies*, p. 13.

Sabol, L. (2013) 'BIM Technology for FM', in *BIM for Facility Managers*, pp. 17–45.

Succar, B. et al. (2007) 'A Proposed Framework to Investigate Building Information Modelling through Knowledge Elicitation and Visual Models', in *2007 Conference of the Australasian Universities Building Education Association*. Melbourne: AUBEA, pp. 308–325.

Vanlande, R., Nicolle, C. and Cruz, C. (2008) 'IFC and building lifecycle management', *Automation in Construction*. Elsevier, 18(1), pp. 70–78.

Wang, Y. et al. (2013) 'Engagement of Facilities Management in Design Stage through BIM: Framework and a Case Study', *Advances in Civil Engineering*. Hindawi, 2013, pp. 1–8.

*Building performance simulation and nD modelling*

# Use-case analysis for assessing the role of Building Information Modeling in energy efficiency

A. Alhamami, I. Petri & Y. Rezgui
*BRE Institute of Sustainable Engineering, Cardiff University, UK*

ABSTRACT: Global warming has drastically increased the pressure to reduce energy use in buildings as reflected by a stringent regulatory landscape. The construction industry is expected to adopt new methods and strategies to address such requirements primarily focusing on reducing energy demand, improving process efficiency and reducing carbon emissions. The realization of these emerging requirements has been constrained by the highly fragmented nature of the industry, often portrayed as involving a culture of "adversarial relationships", "risk avoidance", exacerbated by a "linear workflow". Recurring problems include low process efficiency, delays and construction waste.

Building Information Modelling (BIM) provides a unique opportunity to enhance building energy efficiency and open new pathways towards a more digitalised industry and society. BIM is foreseen as a mean to waste and emissions reduction, performance gap minimization, in-use energy enhancements, and total lifecycle assessment. It also targets the whole supply chain related to design, construction as well as management and use of facility, at the different qualifications levels (including blue collar workers).

This paper provides an evaluation of the key criterions and strategies that can promote BIM as an efficacious tool for facilitating energy efficiency. We explore how different use-cases variables influence the impact of BIM on energy efficiency and provide useful insights on how building lifecycle, discipline and buildings types can influence different BIM energy efficiency scenarios.

*Keywords*: Building information modeling, energy efficiency, life-cycle, requirements, skills, construction

## 1 INTRODUCTION

With the pressure to reduce energy use in buildings, the European Commission has defined a clear 2020 target to reduce by 20% the energy consumption and the $CO_2$ emissions and increase by 20% the share of renewable energies (European Commission 2005). Reducing energy consumption and carbon emission have become important objectives that have been translated into stringent regulations and policies at the European and National levels. For instance, the recast of the Energy Performance of Buildings Directive (2010/31/EU) imposes stringent energy efficiency requirements for new and retrofitted buildings.

Studies and researching communities are seeking for solutions into paving the way to a fundamental step change in delivering systematic, measurable and effective energy efficient buildings through BIM training with a view to effectively address European energy and carbon reduction targets. Actors in the engineering domain need to evolve into a well-trained world leading generation of decision makers, practitioners, and blue collars in BIM for energy efficiency and establishing a world-leading platform for BIM for energy efficiency training nurtured by an established community of interest (Thomson and Miner 2010). Studies and associated benchmarks exist at Europe-wide BIM trainings across the building value chain (including lifecycle and supply chain), highlighting energy efficiency linkages, as well as qualification targets, delivery channels, skills, accreditation mechanisms, while highlighting training gaps and enhancement potential. This will include: (a) better determination of future capability needs; (b) clear routes of entry and clear career progression pathways; (c) clear, standard means of recognising competence; (d) exploring the scope to make apprenticeships more flexible; (e) an industry review of the current skills and capability delivery mechanisms; (f) review of approaches to career planning, training and development with a commitment to rationalise.

The technology and corresponding implementations of Building Information Modeling (BIM) have been in continuous growth over the years trying to facilitate a more effective multi-disciplinary collaborations with a total lifecycle and supply chain integration perspective (Petri et al. 2014). BIM is the process of generating and managing data and information about built environment during its entire life cycle from concept design to decommissioning (Figure 1). The technology of BIM has led to significant transformative power into AEC/FM domain (Architecture, Engineering and Construction/Facility Management) during the last decade in terms of its fundamental life cycle and supply chain integration and digital collaboration (Eadie et al. 2013). Such BIM technology enacts and exposes methods and key-aspects that can revolutionize the construction industry, which is forecasted to reach over $11 trillion global yearly spending by 2020 (Cummings and Blanford 2013).

A number of researching studies are aiming at harmonizing energy related BIM qualifications and skills frameworks available across Europe with a view of reaching a global consensus through a BIM for energy efficiency External Expert Advisory Board (EEAB). The focus is on setting up a mutual recognition scheme of qualifications and certifications among different Member States supported by an effective strategy to ensure that qualification and training schemes are sustained after the end of the study.

This paper investigates what are the key aspects that can support the implementation of BIM for energy efficiency with associated BIM training by undertaking an in-depth analysis and gaps identification of skills and competencies involved. Consultations and interviews have been used as a method to collect requirements and a portfolio of use-case has been created to understand existing BIM practices and determine existing limitations and gaps in BIM training. The research presented in this paper is part of the EU H2020 BIMEET project delivering training and education for BIM implementation for energy efficiency.

In Section 2, we will present background on BIM research. Section 3 describes the overall methodology and system requirements. Section 4 presents the evaluation process alongside with the outcome of the research from Section 5. The conclusions are presented in Section 6.

## 2 RELATED WORK

There are several technological developments in the engineering sector that can facilitate dramatic reductions in building energy domain can be achieved. In the new European Union regulations, there is a special directive for promoting energy performance in buildings, taking into account cost-effectiveness and local conditions and requirements (energy consumption in buildings is highly influenced by local climates and cultures) (Petri et al. 2017).

The global construction market is forecast to grow by over 70% by 2025 (Global Construction Perspectives and Oxford Economics 2015). Several countries have already set-up the target to achieve sizeable objectives, such as the UK Construction agenda: (a) 33% reduction in both the initial cost of construction and the whole life cost of assets; (b) 50% reduction in the overall time from inception to completion for new build and refurbished assets; (c) 50% reduction in greenhouse gas emissions in the built environment; (d) 50% reduction in the trade gap between total exports and total imports for construction products and materials (Magnier and Haghighat 2010; Rezvan et al. 2013).

Industrial sensocommunities and the engineering section are now searching for key technical solutions to reduce energy demand, improve process efficiency and reduce carbon emissions. Such industry is also traditionally highly fragmented and often portrayed as involving a culture of "adversarial relationships", "risk avoidance", exacerbated by a "linear workflow", which often leads to low efficiency, delays and construction waste (Rezgui 2011). The process of designing, re-purposing, constructing and operating a building or facility involves not only the traditional disciplines, but also many new professions in areas such as energy and environment (Rezgui 2011); also there is an increasing alignment of interest between those who design and construct a facility and those who subsequently occupy and manage it, and that demands dedicated skills and competencies to address multi-objective sustainability (including energy) requirements (Bryde et al. 2013).

Digitalization of the built environment is targeted in the vision for Horizon2020 by setting this sector firmly on a path towards competitiveness and sustainable growth; the European Commission's modern industrial policy recognizes the strategic importance of the construction industry, as witnessed by the Public Private Partnership Energy

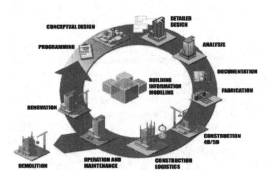

Figure 1. BIM uses across building lifecycle: it presents the entire life cycle of projects from concept design to decommissioning.

Efficient Buildings launched under the Recovery Plan in FP7 and now supported in H2020.

The construction industry in Europe has a wide range of training and education providers with an equally diverse set of training courses. It is essential to improve the collaboration process by using novel virtual methods for creating communities of professional and to address in-depth, quantity and quality of educated and trained professionals in the built environment that can support an effective BIM agenda across Europe (European Comission 2016).

In fact, a number of training and education offerings concentrated on quite a narrow band of the industry; main courses focus on design and construction and not on briefing or planning and the impact of BIM to improving the operations of assets. Also, training courses largely target technical users rather blue-collar workers or management teams and strategic roles in organisations. In addition, BIM education and training is focused on Buildings, and rarely Infrastructure (HM Government 2015).

Education and engagement with BIM practices may therefore be seen as both an individual and collective phenomenon reflected in the team and project-based nature of the construction industry. This is in line with related literature drawn from environmental psychology which highlights the influences of past behaviour, knowledge, experiences, feelings, social networks, and institutional trust on individual attitudes and behaviour towards environmental issues.

## 3 METHODOLOGY

In this paper we aim at collecting quantitative and qualitative evidence on the gaps and required skills for improving BIM practices for energy efficiency. The methodology we adopt is presented in sections 3.1 and 3.2.

### 3.1 *General methodology*

As part of our research we conduct extensive consultations approached as a twofold strategy:

- a user engagement instrument in the form of an online collaborative platform to support with the requirement capture activity of the project while maximizing users' engagement by the creation of a community of practice around the theme of **BIM for energy efficiency**,
- an online web Europe-wide BIM use-case collection template and questionnaire (November 2017– February 2018) from which 38 best practice use-cases have been collected, (3) experts panel consultations in Europe comprising 1 workshop (c. **40** participants in total), (4) a series of 15 semi-structured interviews with key industry representatives (December 2017 – February 2018), and (5) other focus meetings with project partners.

To conduct such consultation studies we have created and exposed an open community of users that share resources and experiences related to BIM energy training supported by **energy-bim.com**. The objectives of the consultations were to determine best practices, regulation awareness and gaps in BIM for energy efficiency domain and to determine a set of training requirements. The subsequent combined consultations explored stakeholders' knowledge, understanding, and behaviours, and helped identify key barriers to BIM applicability for energy efficiency. The identified barriers were discussed and debated from a variety of socio-technical perspectives. A total of **40** experts took part in the consultations (workshop), including: construction companies and practitioners, advisory groups, professional organisations, consultants, policy makers and education and training bodies.

The results of the use-cases and interview analysis are presented in Section 4. We have undertaken a set of actions, as part of the methodology process (see Figure 2) and focused on: (i) carrying out the study consultation while maximizing continuous engagement with our Expert panel and Community of Practice, (ii) use partners, expert panel members, and community of practice members to register on study portal to provide authoritative sources of information, (iii) develop a framework to categorise all retained use cases using two dimensions, i.e. lifecycle (from Briefing to Recycling) and supply chain (i.e. Architects, Structural engineers, to blue collars), and (vi) Develop a template to report selected use cases, implemented directly on the study portal. This has culminated a community exposure by publishing the study use-cases widely inviting people to register if they want to access study materials.

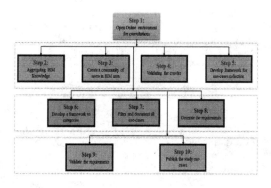

Figure 2. General requirements methodology.

## 3.2 Researching sampling techniques

The methodology has been focused on community knowledge extraction involving project consortia partners, expert panel members and skilled BIM experts. Such experts have been involved in validating the use-cases collection template and questionnaire elaboration. Based on the use-case collection template, the consortia partners have been asked to provide five relevant use-cases from their country of origin in order to cover a wider European BIM perspective. Use-cases have been collected from Greece, Finland, France and UK followed by analysis and requirements elicitation. Using this wide community of experts, interviews and consultations have been conducted as a mean to validate the findings in the assessment of the use-cases and leading to a more comprehensive BIM training set of requirements. One workshop for consulting the BIM community on the existing BIM practices, areas of improvement in BIM trainings and education for energy efficiency has been organised in Brussels. Brainstorming sessions with experts have been organised as part of the workshop, in order to understand existing gaps in the field of BIM for energy efficiency and to aggregate new best practices use-cases.

### 3.2.1 Searching authoritative URIs

To support in the process of use-case collection and BIM knowledge aggregation, partners and experts have been asked to contribute and register a list of authoritative URI sources. These have been registered within the energy-bim.com platform, indexed for crawling and BIM knowledge has been aggregated. Such sources have been integrated in the search service aiming at facilitating users of BIM to extract best practices, regulations and to support with requirements definition and training. As part of the energy-bim.com platform a specialised crawler service has been implemented to help with BIM knowledge harvesting from the provided URIs and to create a BIM knowledge repository for a community of users. A human based process has been utilised to validate these relevant sources and searching URIs based on specialised keywords. These have been validated by experts in the field of BIM and supported by the consortium partners. Such keywords include: BIM, energy efficiency, best practice, case study, training and education.

### 3.2.2 Searching education indexed engines

To support with the process of requirements elicitation, we have conducted searching in educational indexed engines such as Scopus and google scholar based on which requirements have been determined and additional use-cases practices have been identified and included in the study use-cases repository.

We undertook a broad critical review of the academic literature, international standards, legislation, and key economic and political events surround-

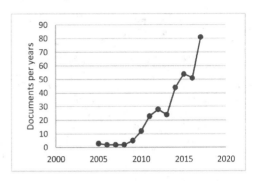

Figure 3. Popularity of BIM for Energy Efficiency research over time as number of relevant Scopus articles per year.

ing BIM, training and education, energy systems and their management. The body of literature was then broken down into chronological and thematic groupings. Following the observation of new challenges and opportunities arising imminently from a mismatch in these projections, key concepts were identified to address these from related fields and novel management paradigms. The rest of this section details the scope of the review and initial observations of the subject domain.

It was apparent that as an emerging field, 'building information modelling for energy efficiency' encompasses many other fields, mandating a well-considered scope. We therefore disregarded papers which only focused on national or building level energy management, or which only considered the design phase of energy systems. We also placed an emphasis on recent publications due to the accelerating change in technologies and focused on BIM training for energy efficiency. Based on this, a trend of increasing popularity in the field was observed since circa 2005, as depicted in Figure 3. The sources were filtered to those deemed most relevant and influential, to a final bibliography of circa 250 references.

## 4 EVALUATION

The evaluation process has been focused on in-depth analysis of best practice use-cases and gaps and skills identification in the field of BIM for energy. The results of the requirement capture process are based on 6 months of work on collecting data and sources, as facilitated by the partners collaboration process. This has culminated with identification of a set of requirements collected for the training process (presented in Section 5).

### 4.1 Use-case collection

With the use-cases collection and analysis we aim at determining how BIM can support energy-efficient design, construction and building maintenance in many ways. In principal, BIM can boost and ease energy-efficient buildings on the basis of better data exchange and communication flows, and in practice for example by accelerating energy simulations and searching for beneficial solutions, supporting end users' involvement, requirement setting and commissioning, and by providing an opportunity for systematic maintenance management. Amidst the positive impacts brought about by BIM, AEC/FM industry can leverage BIM for greater energy efficiency in new designs as well as in retrofit and renovation projects. This study demonstrates the key aspects to address in delivering BIM collaboration for energy-efficient building by collecting and providing use cases with associated analysis. Table 1 shows two examples of use-cases where life cycle applicability is aligned with eight work stages of RIBA plan of work 2013.

## 4.2 Use-case type analysis

We have collected 38 use-cases from users in European countries and applied analysis for gaps identification and requirement capture. The results reported in this section present the distribution based on criterions such as: discipline, building type, impact, lifecycle stage.

### 4.2.1 Use-cases type analysis

There are three types of use cases in this evaluation as: 1) Research & Development, 2) Real world application and 3) BIM Guideline. As per the analysis, it can be observed that Research & Development covers a number of 17 use cases, and Real-world application has 13 use cases and BIM guideline has only 1 use-case (at the time of writing this paper, additional ones are expected in this category) (see Figure 4).

### 4.2.2 Target discipline analysis

The portfolio of use-cases is structured based on the target discipline. Figure 5 presents the

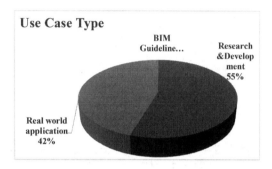

Figure 4. Use-case type analysis of use BIM for energy efficiency.

Table 1. BIM based best practice use-cases.

| Variables/Use-cases | Use case 1 | Use case 2 |
|---|---|---|
| Title | Reduce the gap between predicted and actual energy consumption in buildings: Knoholem project | Shopping center using around half the energy of a typical development |
| Use case type | Research & Development | Real-world application |
| Target discipline | Facility Management | Architectural design/Structural engineering/HVAC engineering/Electrical engineering/Builders/Construction companies/Building managers |
| Target building type | Public | Commercial |
| Lifecycle applicability | In Use | Preparation and brief, concept design, developed design, technical design, construction, In Use |
| Brief description | This study presents a novel BIM-based approach with the objective to reduce the gap between predicted and actual energy consumption in buildings during their operation stage(Yuce and Rezgui 2017). | The project is a large shopping center and commercial development in Pori, southwestern Finland. The development was designed to LEED Gold and has won a global BIM award for its innovative use of modeling during design and construction (Skanska 2014). |
| Impacts | The use of BIM has helped achieve a reduction of 25% energy compared to baseline figures. | BIM was effectively used in a project where 50% energy savings were achieved compared with Finnish Code and 50% savings in water consumption compared with conventional retail development in Finland. Also measured energy production of geothermal heat pumps and gains of free energy for heating and cooling have exceeded expectations. |

distribution of use-cases based on the target discipline. Architecture design and Facility management discipline projects use BIM more frequently whereas structure engineer and mechanical engineer projects utilise BIM in a lower percentage. In the analysis we have used different target disciplines such as architecture design, facility management, structure engineer, mechanical engineer, and other. Architecture designers are targeted by 29% of the collected use-case studies, facility management by 25% whereas the structure and mechanical engineers are targeted by 16% and 14%, respectively.

### 4.2.3 Building type analysis

In this part we assess the use-cases based on the type of building project where BIM has been utilized. As reported in Figure 6, the majority of projects are for public buildings whereas domestic, commercial and industrial buildings seem less popular in adopting BIM. From the set of building types that we have used in our evaluation, the most popular are public buildings whereas domestic building, commercial building, and industrial building have lower percentage.

As reported in Figure 6, 65% of these use cases have applied BIM in public building, 17.5% in domestic building, and the rest of them in commercial and industrial buildings.

### 4.2.4 Lifecycle stage analysis

For the analysis, we have used RIBA stage life-cycles and this part aims at determining associated life-cycle stages of each BIM best practice use-case. Figure 7 shows that, 56% from the recorded projects use BIM for energy efficiency in the design stages in lifecycle of the project, whereas in-use stage identifies 13% in the lifecycle of the projects.

### 4.2.5 Project type analysis

In this part we investigate how the set of use-cases that have adopted BIM, classifies in relation to the project type variable.

From the analysis reported in Figure 8, it can be observed that a majority of use-cases utilise BIM for existing and new buildings, whereas extension and renovation projects tend to not adopt BIM. In percentage, 84% of project types are existing and new build projects and the rest of the project types are renovation and extension projects.

### 4.2.6 Target discipline and impacts

The first variable used for the analysis is the target discipline which we compare with the impacts to find the corresponding association between the target discipline and the impacts of use cases. Figure 9 shows that the majority of use cases that implement BIM for energy efficiency are associated with the facility management discipline. However, there are a number of use-cases that implement BIM for energy

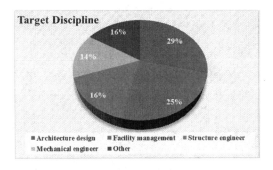

Figure 5. Target discipline analysis of use BIM for energy efficiency.

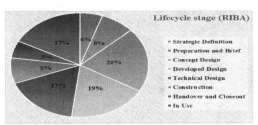

Figure 7. Lifecycle stages analysis of use BIM for energy efficiency.

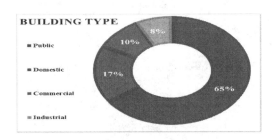

Figure 6. Building type analysis of use BIM for energy efficiency.

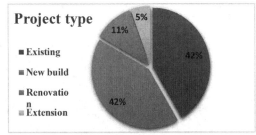

Figure 8. Project type analysis of use BIM for energy efficiency.

efficiency methodology for multiple disciplines with great impacts on energy and water savings.

To this day, BIM has been implemented more and with more powerful results for some building types. Especially certain cases of retail and office buildings provide good examples how BIM has supported demanding requirement management, simulations and searching solutions for ambitious energy targets. For instance, availability and use of BIM data aid towards 25% of energy reduction in facility management (use case 1). Likewise, BIM has been effectively used in a Shopping Center (use case 2) using around half the energy of a typical development, results associated with commercial buildings report about 50% energy saving and 50% saving in water consumption.

In other hand, using RIBA Plan of Work for lifecycle applicability we can observe also associations between lifecycles and BIM impact on energy efficiency. It reflects increasing requirements for sustainability and BIM and it allows simple, project-specific plans to be created. The RIBA Plan of Work organizes the design process into different stages including briefing, designing, constructing, maintaining, operating and using building. According to these stages,various ways of use and levels of impact can be identified for the use of BIM for energy efficiency.

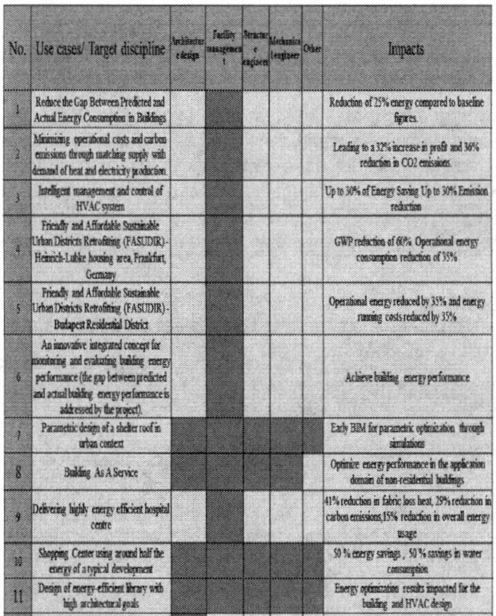

| No. | Use cases/ Target discipline | Architecture design | Facility management | Structure engineer | Mechanical engineer | Other | Impacts |
|---|---|---|---|---|---|---|---|
| 1 | Reduce the Gap Between Predicted and Actual Energy Consumption in Buildings | | | | | | Reduction of 25% energy compared to baseline figures. |
| 2 | Minimizing operational costs and carbon emissions through matching supply with demand of heat and electricity production | | | | | | Leading to a 32% increase in profit and 36% reduction in CO2 emissions. |
| 3 | Intelligent management and control of HVAC system | | | | | | Up to 30% of Energy Saving Up to 30% Emission reduction |
| 4 | Friendly and Affordable Sustainable Urban Districts Retrofitting (FASUDIR)- Heinrich-Lubke housing area, Frankfurt, Germany | | | | | | GWP reduction of 60% Operational energy consumption reduction of 35% |
| 5 | Friendly and Affordable Sustainable Urban Districts Retrofitting (FASUDIR)- Budapest Residential District | | | | | | Operational energy reduced by 35% and energy running costs reduced by 35% |
| 6 | An innovative integrated concept for monitoring and evaluating building energy performance (the gap between predicted and actual building energy performance is addressed by the project). | | | | | | Achieve building energy performance |
| 7 | Parametric design of a shelter roof in urban context | | | | | | Early BIM for parametric optimization through simulations |
| 8 | Building As A Service | | | | | | Optimize energy performance in the application domain of non-residential buildings |
| 9 | Delivering highly energy efficient hospital centre | | | | | | 41% reduction in fabric loss heat, 29% reduction in carbon emissions,15% reduction in overall energy usage |
| 10 | Shopping Center using around half the energy of a typical development | | | | | | 50 % energy savings , 50 % savings in water consumption |
| 11 | Design of energy-efficient library with high architectural goals | | | | | | Energy optimization results impacted for the building and HVAC design |
| 12 | Use of Optimization tool to compare hundreds of concepts energy efficiency before actual design | | | | | | Use of Optimization tool has the potential to save money and time while directing to more optimal energy efficiency solutions. |

Figure 9. Relevance between target discipline and the impacts: This figure shows how the impact evolves with different disciplines.

## 5 VALIDATION

In this paper we have critically reviewed and investigated the current BIM practices landscape and gathered the requirements for developing a BIM training scheme to address current industry collaboration problems on projects. The aim of this solution is to facilitate and to govern the collaboration processes of construction teams taking into account construction practitioners' requirements. Further, a number of requirements for developing a training scheme have been identified and classified into two main categories: (a) socio-organisational and legal requirements and (b) technical requirements. In addition to contributing to the growing body of BIM adoption and collaboration knowledge, this evaluation underlines the importance of BIM training laying out as the foundation for future research and development in this area.

This section identifies a set of general and specific requirements for developing BIM skills, competencies and training with particular emphasis on energy efficiency as informed by the use-cases analysis. This subsection provides the list of gaps as identified by the use-case analysis and validated by the interviews conducted. Table 2 presents the gaps/requirements that have been identified by our research.

Table 2. Use cases analysis identified gaps.

| No. | Parameters | Requirements and training |
|---|---|---|
| 1 | Use case type | Users need training in understanding and applying BIM Guideline see (Fig. 4). |
| 2 | Building type | Training is required for enhancing skills and competencies for using BIM for industrial and commercial buildingssee (Fig. 6). |
| 3 | Project type | Training is required for expanding BIM applicability for renovation and extension projects see (Fig. 8). |
| 4 | Target discipline | Training is required for education on BIM methodology towards mechanical and structure engineers see (Fig. 5). |
| 5 | Lifecycle stages | Training is needed to address other RIBA stages lifecycles such as Strategic Definition, Preparation and Brief, Construction, and Handover and Closeout see (Fig. 7). |
| 6 | Impacts on discipline | Increase BIM applicability and impact for architecture and design, structural engineers, mechanical engineers see (Fig. 9). |

## 6 CONCLUSIONS

In this paper we present an in-depth BIM analysis process for determining gaps and new strategies in delivering BIM training for energy efficiency.

As part of the methodology, we aggregate content and best practice use-cases from the field of BIM engaging with key stakeholder communities in order to help identify and then screen/analyse past and ongoing projects related to energy efficiency involving aspects of BIM. The analysis provided in the evaluation part of the research show how BIM is implemented and approached in the engineering community and also assesses the role of BIM in achieving energy efficiency in buildings across the whole value chain. Based on an industry consultation and with a portfolio of 38 best practices use-cases from the field of BIM for energy efficiency, we have conducted in-depth analysis in order to understand which are the gaps in BIM for energy efficiency and propose training and possible areas of improvement. The portfolio of use-cases are collected, stored and maintained on a web platform (www.energy-bim.com) and exposed to potential users across Europe. The evaluation process included several criterions such as stage and discipline, highlighting stakeholder targets ranging from blue-collar workers to decision makers.

## ACKNOWLEDGMENTS

This work is part of EU H2020 BIMEET project: "BIM-based EU-wide Standardized Qualification Framework for achieving Energy Efficiency Training", grant reference: 753994.

## REFERENCES

Bryde, D. et al. 2013. The project benefits of Building Information Modelling (BIM). *International Journal of Project Management* 31(7), pp. 971–980. Available at: https://www.sciencedirect.com/science/article/pii/S0263786312001779?via%3Dihub [Accessed: 26 January 2018].

Cummings, D. and Blanford, K. 2013. *Global Construction Outlook: Executive Outlook*. Available at: https://www.ihs.com/pdf/IHS_Global_Construction_Exec-Summary_Feb2014_140852110913052132.pdf [Accessed: 20 December 2017].

Eadie, R. et al. 2013. BIM implementation throughout the UK construction project lifecycle: An analysis. *Automation in Construction* 36, pp. 145–151. Available at: https://www.sciencedirect.com/science/article/pii/S0926580513001507#bb0070 [Accessed: 25 January 2018].

European Comission 2016. The European construction sector., pp. 1–16.

European Commission 2005. *Challenging and Changing Europe's Built Environment A vision for a sustainable and competitive construction sector by 2030*. Available at: https://www.certh.gr/dat/79DC02 A3/file.pdf [Accessed: 20 December 2017].

Global Construction Perspectives and Oxford Economics 2015. *Global Construction 2030 A global forecast for the construction industry to 2030*. Available at: https://www.pwc.com/gx/en/engineering-construction/pdf/global-construction-summit-2030-enr.pdf [Accessed: 20 December 2017].

HM Government 2015. 3-Digital Built Britain Level 3 Building Information Modelling—Strategic Plan. *UK Government* (February), pp. 1–47.

Magnier, L. and Haghighat, F. 2010. Multiobjective optimization of building design using TRNSYS simulations, genetic algorithm, and Artificial Neural Network. *Building and Environment* 45(3), pp. 739–746. Available at: https://www.sciencedirect.com/science/article/pii/S0360132309002091 [Accessed: 10 May 2018].

Petri, I. et al. 2014. Engaging construction stakeholders with sustainability through a knowledge harvesting platform. *Computers in Industry* 65(3), pp. 449–469. Available at: https://www.sciencedirect.com/science/article/pii/S0166361514000244?via%3Dihub [Accessed: 26 January 2018].

Petri, I. et al. 2017. Optimizing Energy Efficiency in Operating Built Environment Assets through Building Information Modeling: A Case Study. *Energies* 10(8), p. 1167. Available at: http://www.mdpi.com/1996-1073/10/8/1167 [Accessed: 26 January 2018].

Rezgui, Y. 2011. *Harvesting and Managing Knowledge in Construction: From Theoretical Foundations to Business Applications*. Routledge.

Rezvan, A.T. et al. 2013. Optimization of distributed generation capacities in buildings under uncertainty in load demand. *Energy and Buildings* 57, pp. 58–64. Available at: https://www.sciencedirect.com/science/article/pii/S0378778812005476 [Accessed: 10 May 2018].

Skanska 2014. Case studies | www.skanska.co.uk. Available at: https://www.skanska.co.uk/about-skanska/sustainability/health-and-safety/case-studies/ [Accessed: 25 April 2018].

Thomson, D.B. and Miner, R.G. 2010. Building Information Modeling—BIM: Contractual Risks are Changing with Technology. Available at: https://s3.amazonaws.com/academia.edu.documents/7562887/ge-2006_09-building information modeling.pdf?AWSAccessKeyId=AKIAIWOWYYGZ2Y53UL3A&Expires=1516-926400&Signature=tiOvYHsY3b4Ncpj4qZiqyRsgwCU%3D&response-content-disposition=inline%3Bf [Accessed: 25 January 2018].

Yuce, B. and Rezgui, Y. 2017. An ANN-GA Semantic Rule-Based System to Reduce the Gap Between Predicted and Actual Energy Consumption in Buildings. *IEEE Transactions on Automation Science and Engineering* 14(3), pp. 1351–1363. Available at: http://ieeexplore.ieee.org/document/7317804/ [Accessed: 26 January 2018].

# A new design for gas dehydration units

D. Asgari
*Chemical Engineering, SGPC*

ABSTRACT: A major concern of new civilized world is Air pollution that has a serious problem on human health and the environment. Carbon dioxide, a greenhouse gas, is the main pollutant that is warming Earth. Take a variety of measures were agreed by the most people and governments that curb the emissions of carbon dioxide and other greenhouse gases. The Khangiran sour gas refinery is the biggest refinery plant in Northern East of Iran. In this refinery, four drying towers were used for natural gas dehydration. As the gas flows through one tower containing the solid desiccant, water and hydrocarbons are adsorbed onto the surface of the material. After adsorbing for 80 minutes, the material becomes fully saturated with water and hydrocarbons. While two tower are on adsorption time, the other two towers are being regenerated by the application of heat through a small portion (about 10 percent) of entering gas and subsequent cooling. In this process, the furnace temperature is set to 300°C. The regeneration time is 30 minutes less than the need adsorption time for towers therefore large amounts of energy were wasted until the next period. In order to reduce fuel consumption, a Temperature Transmitter (TT) was installed where the heating stream goes out from the towers. This plan is suitable when the bed is regenerated, the furnace output temperature is reduced to the appropriate value. In this research, furnace fuel consumption and carbon dioxide ($CO_2$) generation versus temperature transmitter installation were investigated. Then heat optimization was achieved by the mentioned process simulation by AspenTech softwares. The simulation results shows that by this project reduction in fuel consumption was (about 833.7 MSCM per year). Moreover, The results show that silica gel life increases with this design. Furthermore, it was shown that this scheme is economically viable and caused reduction of 1663.9 metric ton of $CO_2$ emission per year.

## 1 INTRODUCTION

A Green House Gas (GHG) is any gas in the atmosphere which absorbs heat, and thereby keeps the planet's atmosphere. Over the last 150 human activities make the earth temperature warmer than it otherwise would be [1]. Greenhouse effect Increases with increasing greenhouse gases in the atmosphere which is lead to global warming and consequently climate change [2].

Industry is one of the main part of greenhouse gas emissions that is primarily send greenhouse gases to the atmosphere by burning fossil fuels for energy. Increasing gas prices since 2005 lead to growing concerns about the scarcity of oil and gas resources and attract attention for fuel saving issues. In industrial use, fuel consumption In addition to have a significant percentage of the total processing cost, it can cause greenhouse gases emissions [3] Energy efficiency, fuel switching, combined heat and power, use of renewable energy, and the more efficient use and recycling of materials are ways for greenhouse gas reduction to the atmosphere from the industrial sector. A variety of different strategies are available for process and equipment designers to improve industrial heat transfer.

Zhang et al. [4] investigated on the effects of process and operating parameters on the performance of carbon dioxide vacuum swing adsorption ($CO_2$VSA) processes for $CO_2$ capture from gas, especially as it affects power consumption. they demonstrated that the $CO_2$VSA process has good recovery when operating with 40°C feed gas provided relatively deep vacuum is used. Habib et al. [5] studied numerically on the problem of NOx pollution using a model furnace of an industrial boiler utilizing fuel gas. They worked with various operating conditions and showed that as the combustion air temperature increases, furnace temperature increases and the thermal NO concentration increases sharply.

Khangiran sour gas refinery was originally founded in late 1970 decade and commissioned in early 80s. The original plant consisted of three gas treating units (GTU) refining around 30 MMSCMD Mozdouran reservoir sour gas at the peak capacity. The refinery production rate was increased to 50 MMSCMD at the beginning of new millennium by constructing two additional GTUs and supplying gas to six northern provinces [6]. All GTUs have a dehydration unit with solid desiccants which removes hydrocarbons and water that is associated with natural gases in vapor form. The potential for corrosion, hydrate formation and freezing in the pipeline are reduced by gas dehydration [7]. In this plant, a process cycle with four adsorption tower perform the task of dehydration and also a furnace with three burner which provides heat of silica gel regeneration. In this plant, a portion (approximately 10%) of the entering wet gas is sent to furnace and is heated to temperature of 300°C for regeneration purpose. Due to heat losses, the gas regeneration temperature at the point that is entered the bed considered as 273°C. In this condition, 80 minute bed regeneration time is considered.

This paper examine on the effect of installation a temperature transmitter at towers outlet stream in khangiran plant. This paper study about fuel consumption and environment protection from GHGs emission due to temperature transmitter installation. Also annual benefits from this project were then calculated. The mentioned Gas Dehydration unit was simulated using simulation software and the results are presented.

## 2 PROCESS DESCRIPTION

Fig. 1 shows the flow diagram for a gas dehydration unit in AspenTech softwares. According to the Figure 1 and the assumptions that is needed to simulate the dehydration plant is, two bed is on the adsorption cycle that removes water and hydrocarbons from natural gas stream, while one bed is in state of heating to vaporize and drive off the adsorbed moisture from the surface of the silica gels and the other is in state of Cooling to achieve its normal operating temperature prior to starting the drying cycle. Each cycle take place in 80 minutes and The towers were changed each 80 minute. The unit also consists of a furnace, which uses fuel gas, to supply the necessary heat for regeneration. The gas compositions for this simulation are reported in Fig. 2.

There are two different type of silica gels in beds, H and W. Top layers of the bed (protective layer, type w) directly adsorbs the water vapor. At the bottom section, as the cycle proceeds, Dry hydrocarbons gas components (pentan, hexan plus) are adsorbed by H type silica gels with crossing the hydrocarbon gas through the bed and replacing the lighter components with heavier components. Fig. 3 shows the formation layers of towers. Dehydration of natural gas with sufficient and convenient regeneration time are important for successful and economical operation. In this study, a portion (approximately 10%) of the entering wet gas are used and fixed for regeneration purposes. performed by using.

This gas is heated up to 300°C by sending it through a furnace, that uses fuel gas to supply heat of regeneration and then is sent to the tower that is in heating state.

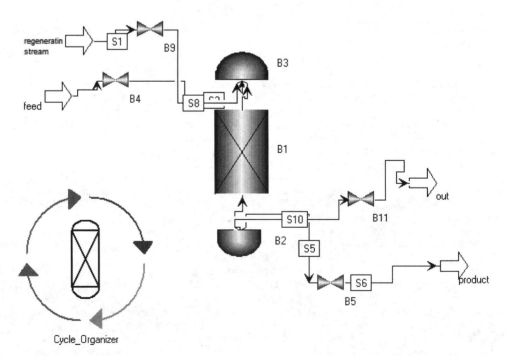

Figure 1. Khangiran gas dehydration plant in simulator software.

|  | Value | Units | Spec | Description |
|---|---|---|---|---|
| Y_Fwd(*) |  |  |  |  |
| Y_Fwd("C2H6") | 0/0063 | kmol/kmol | Fixed | Composition in forward direction |
| Y_Fwd("C3H8") | 1/e-003 | kmol/kmol | Fixed | Composition in forward direction |
| Y_Fwd("C4H10") | 5/e-004 | kmol/kmol | Fixed | Composition in forward direction |
| Y_Fwd("CARBO-01") | 1/e-004 | kmol/kmol | Fixed | Composition in forward direction |
| Y_Fwd("METHA-01") | 0/99 | kmol/kmol | Fixed | Composition in forward direction |
| Y_Fwd("N-HEX-01") | 0/0011 | kmol/kmol | Fixed | Composition in forward direction |
| Y_Fwd("N-PEN-01") | 4/e-004 | kmol/kmol | Fixed | Composition in forward direction |
| Y_Fwd("WATER") | 9/e-004 | kmol/kmol | Fixed | Composition in forward direction |
| P | 71/9 | bar | Fixed | Boundary pressure |

Figure 2. Gas compositions in khangiran gas adsorption plant.

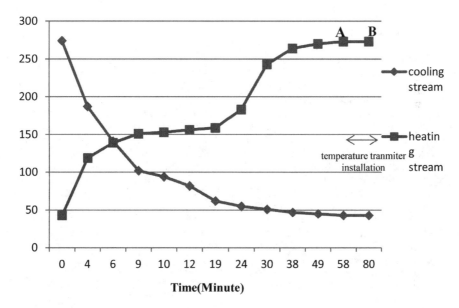

Figure 3. Temperature profile during heating cycle.

In heating state, bed Saturation decreases with time as regeneration cycle proceeds and. when the bed is completely de-saturated, the outlet temperature increases rapidly until it reaches the temperature value of the hot gas at the inlet. Fig. 3 graphically shows the temperature change in heating and cooling bed with respect to time. A is a time which the temperature transmitter is ready for contact reduction in furnace fuel consumption and point B represents a time which heating cycle is finished and cooling cycle is started.

From Figure 3, AB is an indication that the water and hydrocarbon desorption from the silica gels has been completely finished because the hot gas that is exited from the bed almost equals to the one that is entered the bed [8]. In khangiran plant, in order to reduce fuel consumption in furnace, after the temperature of tower's outlet heating stream reached to 273°C, (A point) a transmitter put in circuit and orders to reduce the input fuel to the furnace. This idea lead to reduction in furnace's fuel consumption and GHGs generation. Therefore main purpose in this study is to evaluate the heat savings and reduce air pollution due to temperature transmitter (is a instrument that controls the furnace output temperature) installation.

3 CASE STUDY

The simulation results of temperature transmitter installation idea are checked and are reviewed below to show the reliability of this idea. A usual

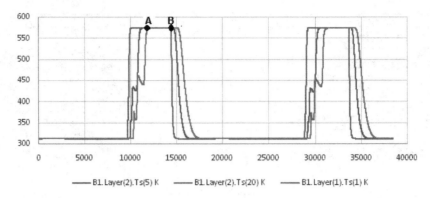

Figure 4. Temperature changes with time at three points of the bed.

regeneration flowrate is 35000 SCM per hour with pressure of 71.9 kg/cm$^2$. To be consistent in the case study it will be assumed that flow rates and compositions of cooling and heating gas stream are constant and does not change with time, as well as it's pressure profile. Also all losses is ignored.

Figure 4 shows the Temperature changes with time at three points of the bed in two continuous cycle that is taken from the output of ASPENAD-SIM software. Ts(1) and Ts(20) respectively related to bed' bottom and the top of the bed in one cycle. In accordance to the Figure 4, AB is a line which the water and hydrocarbon desorption from the silica gels has been completely finished therefore, the desorption of water and hydrocarbon from tower is not required so when the tower's exiting temperature reaches to 273°C, TT decrease the temperature of the furnace to 200°C.

In the following, fuel consumption reduction were calculated due to temperature transmitter installation.

Energy saving due to TT installation was about 31.8% lower than the usual orbit on the last 22 minute in each cycle. So reduction in furnace fuel consumption was 126.9 SCM as expected. According to the initial state, 1663.9 metric ton reduction (per year) in carbon dioxide emission makes the environment less polluted. Electric energy saving in regeneration gas cooling fans is about 766.35 GJ per year which means that a fan can be used (instead of two fans). It should be noted that it is not possible to optimize further than U*A = 400 kj/°C because of cross temperature.

4 CONCLUSIONS

In this paper, installation a temperature transmitter have been examined for heat optimization in gas treating plant. The results show that by this installation, furnace fuel consumption savings is about $ 1881 per year and savings in regeneration gas cooling fans is about $ 12387. (total saving cost is about $ 14268 per year). GHGs emissions reduction was more than 1663.9 metric ton per year that is make environment less polluted. It should be noted that, it is not possible to reduce the temperature of the furnace by more than 200°C because of the problem of the next bed regeneration and also furnace tubes may be damaged due to the high temperature stress. Consequently, it is recommended that designing and installation of a temperature transmitter is useful for all dehydration units.

REFERENCES

[1] Change, I.P.O.C., *Climate change 2007: The physical science basis.* Agenda, 2007. **6**(07): p. 333.
[2] Allison, I., *The science of climate change: questions and answers.* Australian Academy of Science, 2015 (February): pp. 1–44.
[3] Ishii, T., C. Zhang, and S. Sugiyama, *Numerical simulations of highly preheated air combustion in an industrial furnace.* Journal of energy resources technology, 1998. **120**(4): pp. 276–284.
[4] Zhang, J., P.A. Webley, and P. Xiao, *Effect of process parameters on power requirements of vacuum swing adsorption technology for CO2 capture from flue gas.* Energy Conversion and Management, 2008. **49**(2): pp. 346–356.
[5] Habib, M., M. Elshafei, and M. Dajani, *Influence of combustion parameters on NOx production in an industrial boiler.* computers & fluids, 2008. **37**(1): pp. 12–23.
[6] Shahsavand, A. and A. Garmroodi, *Simulation of Khangiran gas treating units for various cooling scenarios.* Journal of Natural Gas Science and Engineering, 2010. **2**(6): pp. 277–283.
[7] Gandhidasan, P., A.A. Al-Farayedhi, and A.A. Al-Mubarak, *Dehydration of natural gas using solid desiccants.* Energy, 2001. **26**(9): pp. 855–868.
[8] Akpabio, E. and V. Aimikhe, *Dynamics of Solid Bed Dehydration in a Niger Delta Natural Gas Liquids Plant.* International Journal of Engineering and Technology, 2012. **2**(12).

# Fast track BIM integration for structural fire design of steel elements

L. Beltrani, L. Giuliani & J. Karlshøj
*Technical University of Denmark, Kongens Lyngby, Copenhagen, Denmark*

ABSTRACT: By allowing a greater inter-connection between design phases, BIM software is an essential tool for decreasing costs and delivery time of building projects, as well as increasing productivity and quality. However, a major hindering in a widespread use of BIM consists in the difficulty of efficiently exporting selected building information to third-parties tool for other engineering applications. This work aims at developing a dynamo script for an efficient information exchange between BIM and a sectional analysis tool for fire design called SteFi. The script resorts an algorithm capable of identifying most common steel profiles and exporting both the geometrical and mechanical properties of the element into a file that is then passed to SteFi, The results of the structural analysis are found in an output file and imported back into Revit. The developed script works both by direct selection of elements in the Revit model and from the IFC file.

## 1 INTRODUCTION

Building Information Modeling (BIM) is one of the most promising recent development in the Architecture, Engineering and Construction (AEC) industry (Y.-S. Jeong, 2009). Being able of storing and connecting building information and making it accessible for all the stakeholders involved in the project (Oogink, 2015). The interoperability between stakeholders is a key factor in decreasing project costs and delivery time, as well as increasing productivity and quality. Although BIM may allow achieving such goals, there are no widely used solutions enabling an efficient utilization of building data generated (Törmä, 2015) for other engineering applications, such as structural and geotechnical design, installation for indoor climate or fire safety.

Industry Foundation Classes (IFC) is the complete and fully stable open international standard for exchanging BIM data developed by building SMART (buildingSMART, 2018). It describes models exported from different BIM software with the same conceptual schema and with well-defined conceptual relations (Törmä, 2015).

Although all major BIM software can import/export the building geometry into IFC file, the geometrical data can hardly be directly imported without modification into another software, such as, e.g., Computational Fluid Dynamic (CFD) programs for fire analysis or Finite Element Method (FEM) software for structural analysis. This is due to the fact that most software companies have only implemented support for the Coordination Model View Definition. This view only supports the basic spatial structure of a building, geometric description of the entities and common properties (buildingSMART, 2018). Furthermore, only a few third-party software can interpret the full content of the IFC file. This means that, in most of the cases, a separate model must be created from scratch in another software.

For instance, structural engineers will typically create a separate model in an FE software. Even in case of simple buildings, where a sectional analysis is sufficient for designing the structural element, geometrical data and mechanical properties of each element must be found in the Revit model, and then manually exported in a sectional analysis tool. Afterwards, the BIM model must be manually updated by changing the profiles of the elements found in need of structural changes.

As long as data need to be re-entered and retyped into separate programs (Neuhold, 2015), there will be no real integration with other software for building design. In these conditions, working with BIM will, in fact, mean spending more and not less time for design and the use of BIM will not spread, as it could. Therefore, it is of outmost importance to develop tools capable of providing a proper integration between BIM and the specific software used in different areas of building design.

### 1.1 *State of art*

During the past years, many tests have highlighted numerous limitations in exchanging both geometric shape and other semantically meaningful information (Y.-S. Jeong, 2009) between the BIM model and technical software. Since, improving the exchange of data means improving the productivity and the quality of the works, researchers and companies are attracted to this topic.

Many companies are making their own structural software programs compatible with BIM models. Some of them have already developed plug-ins, which synchronize the BIM model with their own sectional analysis software, but many problems related to the import/export process are still not solved. The large amount of manual work to correct the model is often cumbersome and it is often easier to build a new model from scratch in the new software.

A way that has been used to solve this integration problem is to create a specific BIM library able to translate data during the import/export process. In the field of building energy, a physical BIM library for building thermal energy simulation called ModelicaBIM has been developed by J.B. Kim (2015). Through this custom library, every component is automatically translated in the correspondent component of a software for thermal analysis called Dymola.

An alternative way of exchanging data is the use of an algorithmic design platform. Dynamo, for example, is a widely used open-source software platform for computational design. As W. Wahbeh (2017) highlighted, the connection between Revit and Dynamo allows the interaction through two-way data exchange, which includes geometry and parameters.

Additionally, an MSc project at Technical University of Denmark (DTU) showed that it is possible to correct the Revit model by using Dynamo (Hejnfelt, 2016). The work shows that aligning the analytical model to the grid and the levels in Revit allows the direct utilization of the latter for the FEM analysis. The developed script has been also used to apply loads to the model.

While some basic modules have been developed for FEM integration, nothing has been done for simpler design tools (e.g., sectional analysis tools). However, these are widely used for performing sectional verifications of single elements, especially at the early design stage.

## 1.2 Early design stage

In the construction industry, the 70–80% of construction costs are determined by designers' decisions in the early design stage (Jungsik Choi, 2015). The cost for making changes to the project exponentially increases during the time the project is developed. In order to have a feedback on the quality of the decision that has been taken, it is convenient to perform simulations at the early design stage (Julien Nembrini, 2014).

R. Kumanayake (2017) developed a tool for making decisions in the early design stage on building materials since they have a high impact on the building sustainability. The study also highlights how the correct decision also influences the quality of the final product.

Another design area that would greatly benefit from a better integration in the early design phases of the building is structural fire safety. This is especially true for steel buildings, where the need of passive insulation significantly affects the costs and the maintenance of the building. The steel profiles are typically dimensioned and optimized at the ultimate and service limit state respectively. Structural fire safety verifications are performed as one of the last design steps when a change of profiles would pose too many difficulties. As a consequence, a bigger amount of insulation is added to the elements that what would be needed, if fire design would have been accounted for from the beginning, thus nullifying the economical savings expected by the profile optimization made at earlier stages. This issue can be easily exemplified by the widespread use highly insulated I-shape cross-section profile. I-shape cross-section possesses the highest elastic and plastic modulus of resistance in comparison with other cross-section having the same steel area (Madsen S., 2016). However, this profile has one of the highest section factors (ratio of exposed perimeter and steel area) and therefore most rapid heating of the elements in case of fire.

Finally, the automatic extraction of data from the BIM model highly reduces the time spent on making structural analyses. This allows the designers to perform a new analysis whenever the design changes and to have a quick feedback on the quality of the decision. Therefore, using a process that automatically extracts information from the BIM model allows to reduce costs and to improve the quality of the design.

## 1.3 Objective and method

The work presented in this paper focuses on the development of an automated procedure to improve the communication between BIM software and sectional analysis tools, thus reducing the work time for structural design and helping to make decisions at the early design stage.

The developed method is a two-way process capable of (i) exporting data from the BIM model to a third-software party, and (ii) importing back the new geometry as well as the data obtained after the external analysis. A sectional analysis tool for structural fire design of steel elements recently developed at DTU is taken as reference for the integration.

Figure 1. Automatized data sharing between BIM software and sectional analysis tool.

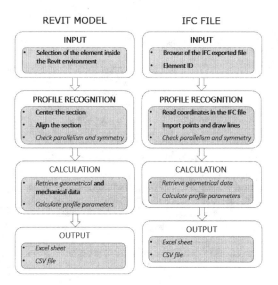

Figure 2. Comparison of the two processes for extracting data developed in the project.

The exported information is automatically read by the sectional analysis tool and the analysis output (i.e., the amount and type of insulation needed for ensuring the required fire resistance) is imported back as new parameters of the structural element and new physical elements inside the BIM model.

This paper describes the two processes developed for exchanging data between the BIM platform and the sectional analysis tools. The main difference between them is the input given to the script. The first input is designed to work in combination with the Revit model (Chapter 2), while the second input uses the exported IFC file (Chapter 3).

## 2 INTEGRATION PROCEDURES

### 2.1 Used software and tools

The chosen BIM software is Autodesk Revit, which is the most used BIM platform in Denmark. The model contains a rich database where the information regarding many disciplines can be stored. Revit uses a strictly hierarchical organization. The "Family" is the building block of the Revit model and it belongs to a specific category. Families have multiple properties, called parameters. A specific configuration of these parameters of a family is called "Type". Any component belonging to a specific type and placed in the Revit environment with unique properties is called "Instance".

Dynamo is a visual programming software developed by Autodesk. Through a friendly visual interface, Dynamo constructs logic routines to smooth and automate workflows (Autodesk, 2018). The strict connection between Revit and Dynamo provides the Revit user with the possibility of improving its experience and exceeding different limits of the main program. Dynamo can be used as an add-in of Revit but also as a stand-alone platform (Dynamo Studio version). In the Dynamo workplace, it is possible to connect nodes with wires, by specifying the logical flow of the resulting visual program. Each node performs an operation, which may be the simple storage of a number, or the more complex creation or query of a geometry (Jezyk, 2018). The Dynamo library contains nodes grouped by category with the default installation, but the user has the possibility to extend this basic functionality with custom nodes and additional packages.

The sectional analysis tool for structural fire design considered for the integration is a software developed by M. Andersen and T. Dyhr (2018) in the framework of an MSc project at DTU. The software is called SteFi and is a verification tool for steel beams and columns that includes both standard and parametric fire curves as design fire as well as different material degradation models.

The integration of Revit with SteFi is based on two processes developed by using Dynamo. The first process exports parameters of the element to be designed in a .xlsx format or .csv format file that is readable by SteFi. The second process reads the SteFi's output and imports the new information about the element fire resistance and insulation into the Revit model.

### 2.2 Export

A sectional analysis tool needs, as input, the geometry of the section, the material properties, the definition of the boundary conditions and the load distribution. Since the integration has been developed for SteFi, which is a sectional analysis tool for fire design, the thermal parameters of the material are also used.

All this information is available in the Revit model and can be extracted by using a customized script. The main challenge in extracting data from Revit is that the parameters are not always called the same way, depending on the standards followed (i.e., European, American, Imperial). This means that extracting the parameters values by using their names is not possible. However, the geometrical parameters can be retrieved from the geometry. Accordingly, they can be calculated for every section by meshing the shape or describing it as a series of functions. Since SteFi performs the verification and the design of steel elements, the following paragraph focuses on the retrieval of the geometrical parameters of standard steel profiles.

### 2.2.1 *Geometrical parameters*

By selecting the element to be designed from the Revit model, it is possible to obtain the geometry of the profile as a series of curves, which describes the cross-section. The script uses the coordinates of the starting and the ending points of each curve to calculate the meaningful parameters for the specific profile. This procedure is different for every profile because each of them is represented by different parameters. For example, an I-Profile is represented by its height and width, the thickness of the flange and the web, as well as the radius of the connection between the flange and the web. A CHS, instead, is represented by the external radius and the thickness of the profile.

Using this data is possible to calculate any properties derived from the geometry through mathematical rules. Therefore, the script also calculates perimeter, area, the moment of inertia for the strong and the weak axis, elastic and plastic modulus of resistance for the strong and the weak axis. Using the length of the element, our script calculates also the volume and the external surface of the element. The ratio between these two values is called "section factor" and is a driving parameter in the heating of steel and therefore in the structural fire design.

### 2.2.2 *Profile recognition*

As mentioned above, the procedures to retrieve geometrical parameters are different for each profile. For this reason, a process capable of recognizing the profile from the geometry has been developed by using Dynamo.

The recognition is based on the check of parallelisms and symmetries with respect to the two axes. For doing this, the section must be placed on the XY plane, centered at the origin and aligned to the axes. The process consists of three custom nodes in sequence: "Center Section", "Align Section" and "Recognize Profile".

By selecting the element from the Revit model, it is possible to export the element as a solid into the Dynamo environment at the spatial position

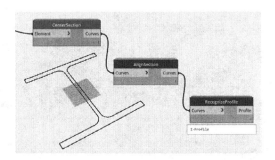

Figure 3. Dynamo visual programming for centering, aligning and recognizing the profile.

and with the same orientation, it has in the Revit model.

"Center Section" places the cross-section of the element on the XY plane and places the center of gravity of the section at the origin of the axes. The center of gravity of the section is already available in the Revit model, as is one of the positions where the analytical model assigned to the structural element can lay. Since the analytical model can be placed in another position, this custom node temporary moves it to the center of gravity and resets it on the initial position after the center is found. The script places the element orthogonally to the XY plane and puts the center of gravity at the origin of the axes. From the solid, our script takes only the surface that intersects the XY plane and gets the curves of the section from the surface.

"Align Section" rotates the geometry around the origin of the axes for aligning one of its sides to the Y-axis. The script gets the angle between the chosen side and the Y-axis and it rotates the section of that angle. Since the angle is always obtained as a positive value, it is not known whether is correct to rotate the section clockwise or counter clockwise. For this reason, after the section is rotated, the script checks the angle again and if it is not null, the section will be rotated to the opposite direction.

"Recognize Profile" divides the curves of the section into lines and arcs and it counts the number of each. Afterwards, it checks the parallelism of the lines and the symmetry of the section with respect to the two axes. This process checks whether the shape could be represented by a parametric form and be recognized as a standard profile. For example, a section that has six lines parallel both to the Y-axis and to the X-axis and it is symmetrical with respect to the two axes can be univocally identified as I-Profile.

To recognize profiles, it is necessary to add more conditions. For example, the C-Profile and the T-Profile have the same number of lines and arcs, the same number of parallel lines to the two axes, and both are symmetric with respect to only one axis. The implemented additional condition is based on the fact that both profiles have four points with the same Y coordinate. The algorithm checks whether the Y coordinate of these four points corresponds to the maximum/minimum Y value. In this case, the output is a C-Profile, differently a T-Profile.

### 2.2.3 *Mechanical and thermal properties*

By selecting the element in the Revit model, Dynamo can retrieve the material and its properties, which are basically mechanical and thermal.

Both of them may be relevant for structural fire design, but since the integration is based on the functionality of SteFi, only the mechanical properties have been taken from the list of properties and exported to the output file. The tools, indeed, is not able to import the thermal properties of the material up to now. Therefore, the exported parameters are: density, yield strength, young modulus, and shear modulus. However, it is possible to also export thermal properties such as thermal expansion coefficient, thermal conductivity, and specific heat.

### 2.2.4 *Load and boundary conditions*

In Revit, the boundary conditions are defined in the analytical model of the element. Among the parameters assigned to it, it is possible to find "Start release" and "End release" if the element is a beam and "Base release" and "Top release" if the element is a column.

These boundary conditions can be defined as "Fixed", "Pinned" or "User-defined" to use a custom constraint. The fixed condition refers to a moment resistant connection to the adjacent object such as a beam or column, a wall, or the ground. Hence, it would be incorrect to export the boundary conditions assigned in Revit to a structural tool performing analysis of single elements. In order to calculate the rotational stiffness of the connection, a structural analysis of the whole building should be performed, which would be out of scope for the integration with simple element analysis tools like the one considered in this paper.

The same applies to the export of mechanical loads acting on the elements. Since area loads are often applied on slabs in the Revit model, a structural analysis of the whole building would be required to derive line and point loads acting on beams and columns respectively, when moment resisting connections are employed in the structure.

For this reason, the developed script does not include the boundary conditions and the mechanical load among the information to be exported. Therefore, this information must be defined by the user of the sectional analysis tool.

However, the script calculates the self-weight of the element by using its volume and the density of the material provided in Revit.

In order to perform a structural verification against fire, thermal loads should also be considered. However, in structural fire design, the elements are generally assumed to be exposed to the ISO834 fire curve (generally referred to as standard fire), which is a nominal, monotonically increasing temperature-time curve used for a limited amount of time, depending on the resistance class prescribed for the element. Alternatively, a parametric fire exposure can be considered, where the fire curve depends on the property of the fire compartment (fuel load, ventilation, material properties of the enclosure). While the standard fire is independent on the Revit model, information on the compartment could be retrieved in Revit to define the parametric fire. However, since fire compartments are hardly identified in Revit models, it has been chosen not to include thermal loads in the export file and let the user define them in the sectional analysis tool.

### 2.2.5 *Output file*

The last part of the Dynamo script merges all the calculated parameters and other meaningful information such as the element ID and the element category and sorts them, making them ready

| | A | B | C |
|---|---|---|---|
| 1 | Element ID | 282134 | $ |
| 2 | Category | Structural Framing | $ |
| 3 | Profile | I-profile | $ |
| 4 | b | 300 | mm |
| 5 | h | 340 | mm |
| 6 | tw | 12 | mm |
| 7 | tf | 21,5 | mm |
| 8 | r | 27 | mm |
| 9 | Length | 2700 | mm |
| 10 | Perimeter | 1809,646003 | mm |
| 11 | Area | 17089,77896 | mm^2 |
| 12 | External surface | 4886044,209 | mm^2 |
| 13 | Volume | 46142403,18 | mm^3 |
| 14 | Wely | 2156258,792 | mm^3 |
| 15 | Wply | 2408105,361 | mm^3 |
| 16 | Iy | 366563994,6 | mm^4 |
| 17 | Welz | 645995,8966 | mm^3 |
| 18 | Wplz | 985720,66 | mm^3 |
| 19 | Iz | 96899384,5 | mm^4 |
| 20 | Material | Metal | $ |
| 21 | Unit weight | 77 | KN/m^3 |
| 22 | Yield strength | 345 | MPa |
| 23 | Young modulus | 200007 | MPa |
| 24 | Shear modulus | 77523 | MPa |
| 25 | Self weight | 3,552965045 | KN |

Figure 4. Outputted excel sheet with exported parameters of the section.

for the export phase. Having data always allocated with the same structure in the output file is very important because any software using this export file must be able to find the information it needs.

The export phase creates two output files, a .xlsx format and a .csv format file. These files have been structured to have three columns for storing data. The property name is placed in column one, its value in column two and the unit of measurement in column three. Since different profiles have different parameters, each of them has its own script to name it and to add the units of measurement.

The file name and the directory where to store it can be chosen by the user. However, if the user doesn't define them, the file will be placed on the desktop with a default name shaped by the element type and the element ID (e.g., I-Profile 282134). If a file with the same name already exists in that directory, it will be overwritten by the script.

### 2.3 *Import*

The geometrical parameters previously calculated by the algorithm can be directly imported back into the Revit model by using a simple Dynamo script. Adding these data to the element parameters inside the model (e.g., moment of inertia), may be useful for eventual future different calculations. In fact, when the information is stored in the model, it can be taken by any software that interacts with Revit for performing calculations.

To complete the integration, it is needed a script capable of reading the output of the sectional analysis tool by inserting the new information into the Revit model and updating it.

The script has been developed based on the SteFi's output, therefore the process must be changed for the integration with other software.

#### 2.3.1 *Data storage*

SteFi exports information related to the design fires, the load capacity of the element as well as the solicitant load and the information about the designed insulation: type, material and thickness.

Browsing the file path of the output of the sectional analysis tool, the script finds the related element in the Revit model by reading the element ID inside the file.

The new parameters do not exist inside the Revit model unless another similar importation has previously been done. Therefore, it is generally necessary to create those parameters before setting their values. However, the script checks whether they already exist to avoid the possibility of creating them twice.

SteFi exports different parameters depending on whether the designed element is a beam or a column. Therefore, different parameters are

| Fire Protection | |
|---|---|
| Solicitant load in fire | 250.00 kN/m |
| Load-bearing capacity | 867.68 kN/m |
| Fire type | DS Parametric fire |
| Resistance time | 70.000 min |
| Opening factor | 0,04 [m^(1/2)] |
| Thermal inertia | 1160 [J/(m^2*s^(1/2)*K)] |
| Fuel load | 200 [MJ/m^2] |
| Insulation type | Hollow encasement |
| Insulation material | Mineral wool slabs |
| Insulation thickness | 5.0 mm |
| Thermal resistance | 0,026 [m² *C/W] |

Figure 5. Analysis results imported into the Revit instance.

created in the Revit instance depending on the element category.

Both the parameter name and value are taken from the output of the sectional analysis tool. While the parameter type (e.g., Period, Force, Length) is added by the algorithm. If the parameter type is defined, Revit can import the new parameter as a number and with the correct unit of measurement. In the other case, the unit of measurement is read from the SteFi's output and imported into Revit as a string, together with the parameter value.

#### 2.3.2 *Insulation modeling*

A step forward than adding information regarding the insulation to the element is to model the physical element in Revit.

Many insulation materials are already available in the Revit library, included gypsum board and mineral wool, which are used by SteFi for designing the insulation. If a custom insulation material is used for the design, it is possible to create a new material and to add thermal and physical properties to it by using Dynamo.

Since the coordinates of the external perimeter points of the designed element are known, both the geometry of the box and the contour encasement can be created. From the defined geometry and material, it is possible to create a new family type representing the insulation. Afterwards, the new insulation can be placed in the Revit model around the designed element using the spatial coordinates retrieved from the element ID, which is stored in the SteFi's output.

While modeling the shape of the box encasement is meaningful for the entire design, insulation such as spayed mineral wool or fire retardant paint does not require more than a few millimeters of space and is not even visible in the Revit model unless the "Visual Style" allows to visualize the material color. However, modeling them is useful for storing information regarding the insulation without creating new parameters on the structural element, which would be in an inappropriate place.

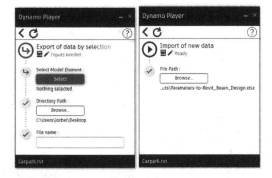

Figure 6. Dynamo player interface for the developed processes.

## 2.4 Summary and user interface

The developed BIM integration for sectional analysis tools is able to handle and to export parameters for I-Profile (HE and IPE), IPN-Profile, RHS, CHS, T-Profile, C-Profile, UPN-Profile, L-Profile, Z-Profile, Rectangular and Circular profiles.

Since the selection of the element is referred to the medium accuracy visibility, if the user needs higher accuracy, it is possible to change the visibility settings inside the family of the element to be designed.

While the export can be always used for the extraction of data from the Revit model, the import process has been designed specifically for the output of SteFi. Therefore, it is not possible to use the import in combination with other tools, unless they export the same parameters, stored in the same way as SteFi.

The user of the integration does not need to open the visual programming environment to run the script. Dynamo Player can be used directly from inside Revit. It requires the user to define the inputs that are visualized on the screen. For the export, inputs are the selection of the element, the directory where the file will be stored and the name of the file. For the import, the only input is the browse of the file generated by SteFi. After the inputs are defined, it is possible to directly run the script.

## 3 INTEGRATION WITH IFC

Within BIM, interoperability has been addressed by IFC technologies. The benefit is that since the models created with different BIM software are ultimately based on the same primitives, the content of the model can be interpreted in the same way (Törmä, 2015).

Since from the Revit environment it is possible to export the model as an IFC file, an alternative method for exporting data for the design phase has been developed.

### 3.1 IFC from Revit

In the IFC file exported from Revit, data are limited to the geometrical parameters, since the material properties are not included in the file. This is not a limitation related to the IFC specifications, but due to the fact that Revit has not implemented support for export of structural properties. In the IFC domain, the standard parameters for the material properties do exist, but they are not used in the module developed by Autodesk for exporting the model.

The recognition of sections, instead, it is something that Autodesk implemented in the export script and the profile is stored in the IFC file under the specific profile type (e.g., #IFCISHAPEPROFILEDEF). However, not every profile can be recognized, and, in such a case, it is stored under the voice #IFCARBITRARYCLOSEDPROFILEDEF. The reason is that Autodesk implemented the most common ones, leaving behind the less used profiles, even if the IFC standards contain much more profiles than those.

Developing these aspects means taking full advantage from the potential of the IFC, and this is a challenge for the future works of Autodesk.

### 3.2 Alternative import/export method

While the previous method requires the selection of the element to be designed in the Revit model, this one requires the browsing of the IFC file and the ID of the element to be designed. This means that the user of the structural tool does not need to have the Revit license to perform the structural design. If the user does not have Revit license, should have license for Dynamo Studio, which is more affordable than the former one, to have the possibility of using the software independently.

The IFC file loaded into the Dynamo environment, appears as a series of strings and the script uses the element ID to find the row containing the information regarding the element. From that string, the script retrieves the category of the element and references that lead to find other information such as the profile type and the element length.

When the profile type is recognized (e.g., I-Profile, CHS, Rectangular), the script finds the geometrical parameters that represent the shape and it uses them for the calculation of other parameters as previously described.

Instead, when the profile type is not recognized (e.g., RHS, L-Profile, C-Profile), the script, through references that can be found in the row, finds the coordinates of the points that describe the perimeter of the section. Since the points are given in sequence, it is possible to know how they are connected to each other.

The coordinates of the points are given by the IFC file with the center of gravity of the profile centered on the origin and the profile aligned to the axes. Therefore, it is possible to use the custom node previously described ("Recognize Profile") to recognize the profile. From this point, the method uses the same script as the method shown in Chapter 2.

The import process adds additional information to the IFC file. The Dynamo script appends data as strings, adding new rows in the file. The new data are inserted using the schema definition of the IFC. That means that each new row must contain the reference to the row where the ID of the designed element is stored and to the rows where data related to the new row can be found. For example, the row containing the thermal resistance of the insulation must have the reference to the row containing the information regarding the insulation material.

## 4 CONCLUSIONS

The work presented in this paper focused on the development of a dynamo code that allows an efficient information exchange between BIM and a sectional analysis tool for fire design called SteFi. The script resorts an algorithm capable of identifying most common steel profiles and exporting both the geometrical and mechanical properties of the element into a .csv and a .xlsx file, both readable by SteFi, The results of the structural analysis are found in an output file and imported back into Revit. The use of the script, which works both by direct selection of elements in the Revit model and from the IFC file, is expected to foster the early integration of the fire design issues into other aspects of building design, thus allowing for a reduction of the passive insulation of steel buildings and the connected costs of application and maintenance.

Therefore, the procedures shown in this paper can lead to a considerable reduction of the construction costs and to a valuable increase of the design quality.

## REFERENCES

Autodesk. (10. May 2018). *Explore|Dynamo BIM*. Hentet fra Dynamo BIM: http://dynamobim.org

buildingSMART. (9. May 2018). *IFC Overview summary—Welcome to buildingSMART-Tech.org*. Hentet fra Welcome to buildingSMART-Tech.org: http://www.buildingsmart-tech.org/

buildingSMART. (2. May 2018). *IFC4 Documentation*. Hentet fra www.buildingsmart-tech.org: http://www.buildingsmart-tech.org/ifc/IFC4/final/html/

Hejnfelt, A. (2016). *Structural Design and Analysis through Visual Programming*. Copenhagen: Technical University of Denmark.

Jezyk, M. (26. April 2018). *About|The Dynamo Primer*. Hentet fra The Dynamo Primer: http://dynamoprimer.com

Jong Bum Kim, W. J. (2015). Developing a physical BIMlibrary for building thermal energy simulation. *Automation in Construction*, 16–28.

Julien Nembrini, S. S. (2014). Parametric scripting for early design performance simulation. *Energy and Buildings*, 786–798.

Jungsik Choi, H. K. (2015). Open BIM-based quantity take-off system for schematic estimation of building frame in the early design stage. *Journal of Computational Design and Engineering 2*, 16–25.

Madsen S., L. N. (2016). Topology optimization for simplified structural fire safety. *Engineering Structures 124*, 333–343.

Mikkel Dohm Andersen, T. D. (2018). *Automatic and BIM-Integrated Fire Design of Steel Elements*. Copenhagen: Technical University of Denmark.

Neuhold, E. J. (2015). Interoperability and semantics—An introduction into the past and look at the future. *10th European Conference on Product and Process Modelling (ECPPM 2014)* (s. 7–9). Vienna, Austria: CRC Press.

Oogink, H. (2015). Introducing BIM+: An open platform for building faster and better. *10th European Conference on Product and Process Modelling (ECPPM 2014)* (s. 3–6). Vienna, Austria: CRC Press.

QIN Ling, D. X.-y.-l. (2011). Industry Foundation Classes Based Integration of Architectural Design and Structural Analysis. *J. Shanghai Jiaotong Univ. (Sci.)*, 83–90.

Ramya Kumanayake, H. L. (2017). Development of an automated tool for buildings' sustainability assessment in the early design stage. *Procedia Engineering 196*, 903–910.

State, A. (15. April 2018). *Steel beam tables—properties and dimensions*. Hentet fra www.structural-drafting-net-expert.com: http://www.structural-drafting-net-expert.com/steel-beam.html

Törmä, S. (2015). Web of building data—integrating IFC with the Web of Data. *10th European Conference on Product and Process Modelling (ECPPM 2014)* (s. 141–147). Vienna, Austria: CRC Press.

Wahbeh, W. (2017). Building skins, parametric design tools and BIM platforms. *12th Conference of Advanced Building Skins* (s. 1104–1111). Bern, Switzerland: ResearchGate.

Y.-S. Jeong, C. E. (2009). Benchmark tests for BIM data exchanges of precast concrete. *Automation in Construction*, 469–484.

# BIM tools for structural analysis in the Wenchuan earthquake aftermath

G. Cerè, Y. Rezgui & W. Zhao
*BRE Trust Centre for Sustainable Engineering, Cardiff, UK*

ABSTRACT: The concept of Building Information Modelling (BIM) is increasingly adopted as a means of gathering digital and semantic building information across the life-cycle of constructions. The present work aims at assessing the structural resilience of the existing building stock in face of earthquakes and seismically triggered geo-environmental hazards. For this purpose, an inclusive methodology is proposed in this paper to perform structural analysis on existing buildings relying on the interoperability between Autodesk Revit and Robot Structural Analysis. Case study applications consist in two Reinforced Concrete (RC) frame structures in the Old Beichuan County (China), respectively, i.e., a residential building and a Hotel reception which were damaged by the 2008 Wenchuan Earthquake. Both structures have been architecturally characterized in Revit, followed by a modal analysis on the residential building in Robot.

## 1 INTRODUCTION

Building Information Modelling (BIM) consists of a comprehensive process which has been widely implemented both in the research and industrial environment. The definition of BIM is still a controversial topic and there is no univocal way of universally defining it, but different institutions provide various descriptions (Czmoch & Pękala 2014) all sharing the idea of a virtual cluster of information that can be systematically updated by the different figures involved in a design process. The BIM approach has been evolving through time with the aim of characterizing itself as a holistic representation of a project's information dataset and has been implementing more features, such as time (i.e., BIM 4D), costs (i.e., BIM 5D), sustainability (i.e., BIM 6D) and facility management (i.e., BIM 7D) (Czmoch and Pękala 2014).

Several applications of BIM have been devised and they encompass different domains, such as: design of new buildings, performance-based recovery of existing structures, restoration of protected heritage buildings (Barazzetti et al. 2015). Rezgui and colleagues (2013) highlighted that the adoption of BIM is spread more in the architectural community rather than in the engineering ones. Although recently, BIM has started to be coupled to structural analysis perhaps also thank to the consistent implementation of parametric software (e.g., ArchiCAD, Revit) that enable a high level of data integration. An increasing trend in this domain is proved in the work by Chi and colleagues (Chi et al. 2015), who outlined a drift towards more computational strategies (e.g., optimization) regarding BIM-based structural analysis in buildings. A remarkable application of adoption of BIM in the context of historical buildings can be found in the work by Barazzetti and colleagues (Barazzetti et al. 2015, Banfi et al. 2017), who proposed a methodology involving the collection of point clouds, their conversion into a BIM model and eventually performing a Finite Element Analysis (FEA).

It is worth mentioning that the implementation of BIM should not be confused with the adoption of tools that enable the achievement of BIM standards. In particular, different software generally address specific domain, therefore the overall integration that is required for the achievement of BIM cannot be found in the use of just one tool. This instead lies in how these information can be combined and made intelligible even if consisting in output of different software. This objective has been achieved through the introduction of standardized formats like IFC (i.e., industrial foundation class), a comprehensive and neutral standard that is functional for the Architecture, Engineering and Construction (AEC) domains in order to collect all the specific project-related in one file that can be accessed, updated and shared systematically. Despite that, the adoption of an IFC object format might not be enough for the data to be perfectly intelligible by different actors in the design process. Industrial figures have in fact highlighted (Rezgui et al. 2013) that between IFC-based products data losses can occur, resulting therefore in additional costs (Eastman et al. 2010) and burdensome work aimed at filling the gaps caused by the lack of information.

Nawari (2011) presents an exhaustive schematization of how structural analysis should be embedded in an integrated perspective of design, given its systematic and reciprocal influence on the architectural and MEP domains. Nonetheless, a slight resistance in the implementation of BIM is still registered where the adoption of paper might appear as a more straightforward solution in the short term but being more demanding in terms of update in the long run, such as the case of technical drawings (Rezgui et al. 2013). Another obstacle for a holistic implementation of BIM has been highlighted to be the compliance of regulatory systems (Rezgui et al. 2013), especially because the domains involved are various and hence the related regulatory frameworks.

Particularly in the structural engineering field a strong persistence in the adoption traditional methods is registered, statistically placing itself half-way in terms of frequency of use (Hunt 2013, Kreider et al. 2010). Disaster engineering is outlined in the work by Kreider and colleagues (2010) as the field with the lowest frequency of adoption. To this regard, a higher implementation of the domain should be achieved considering the impact of natural hazards on the existing building stock (Cerè et al. 2017). The same authors also registered significantly positive feedbacks from the designers who decided to implement BIM in their working routine, meaning an encouraging trend for the adoption of a more inclusive strategy of design. The work by Kreider and colleagues (2010) also highlights a strong prevalence in the adoption of BIM by architectural, managerial or management figures while engineers account for about a tenth of this percentage. In light of that and considering the role of structural components in the context of buildings, it is identified a gap in the implementation of BIM tools for structural modelling and analysis.

The current paper first introduces in section 2 the tools adopted and in the following chapter presents a methodology aimed at speeding up the process for building modelling and its analysis. To this regard, section 4 will involve two case-scenarios of RC framed structures in Old Beichuan, in China.

## 2 PRELIMINARIES ON ROBOT AND REVIT

Robot (Autodesk 2010) and Revit (Autodesk n.d.) are produced by the software company Autodesk and they respectively address mostly the structural and architectural domain of the design process. However, Revit is produced in different versions, mainly Architecture, Structure and MEP (i.e., Mechanical, Electrical and Plumbing), each entailing different modelling options. In particular, the first two Revit types reflect their name and have been devised mostly for architectural and structural purposed, allowing differential levels of details for the specific domain, whereas the MEP version targets principally installations and its use it is out of the scope of the current research. Since the refined structural analysis and relative detailing is performed through Robot, Revit Architecture has been adopted to model the building.

Revit is perhaps one of the BIM design tools that are mostly adopted in industry given the ease of implementation of the different building components (e.g., architecture, structure and installations) and the provision of a set of linkable software that enable an easier information-sharing between the figures involved in the design process. It is worth highlighting that Revit, as long as his predecessor Graphisoft ArchiCAD, is a parametric software, meaning that the elements can be systematically modified tweaking the desired variables (e.g., length, pertaining storey, section geometry). Robot instead does not allow this flexibility, as the majority of structural software (e.g., SAP2000, MasterSap) in which the bars are defined by means of the start and end nodes. As far as Robot is concerned, similarly to Revit it consists in a parametric software and hence each element is identified by a series of attributes, such as an ID, elevation, spatial location and dimensions, just to cite some examples. It is provided with a set of the most up-to-date regulatory framework and it is designed to perform linear, nonlinear, static and dynamic analysis.

## 3 METHODOLOGY

### 3.1 *Overview and data collection*

The methodology presented in this section relies on the interoperability amongst Revit and Robot, devising an overall breakdown regarding the typology of data that might be required in case of existing building. In this particular case no retrofitting measures were embodied because of the irretrievable structural condition that would make any recovering attempts impossible.

Figure 1 shows the process that has been adopted in this context and its application will be further discussed in the following section specifically addressing two existing buildings in Old Beichuan, in China. In particular, the first step involved the collection of information during two field trips carried out in December 2016 and July 2017 in several sites of the Sichuan province in China that were affected by the consequences of the 2008 Wenchuan Earthquake.

In particular, one of the most impressive and devastated locations was the city of Old Beichuan, in which the great majority of the building stock was completely destroyed. Preliminary analysis at the time of the field work allowed to hypothesize the occurrence of seismically-induced liquefaction leading to the collapse of the vast majority of the existing building stock. This has been thought to have such a significant impact not just for the

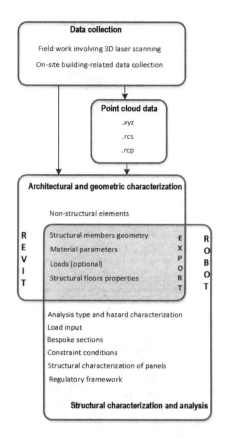

Figure 1. Adopted methodology.

phenomenon magnitude alone, but because it has not been contrasted with a suitable structural design and geotechnical analysis.

Visiting the site allowed to collect a rich photographic inventory which has been then fundamental for the point cloud analysis and the characterization of the different structures. Relevant point cloud data were available for one of the two buildings, whereas the other one has been modelled adopting standard sizes as a reference (e.g., brick dimension).

It goes without saying that the knowledge of the structures, especially while dealing with existing building, is of primary importance and it consists in the first step to accomplish before any type of virtual modelling. Therefore, prior to any modelling activity the designer must have a clear understanding of the structure and its functioning.

### 3.2 Methodology insight

With particular attention to the point cloud data, Revit allows to import them in several formats and in Figure 1 some examples are shown, although. xyz is perhaps amongst the most common ones.

The point cloud is detected by Revit as a single object and it can be used as a reference for the modelling, particularly because it is detected also in the first level of the "Floor Plans" category and hence can be easily adopted for outlining the building's footprint. If cut planes are introduced, the point cloud can be accessed also through the different sections, allowing a better understanding of the structure for a more accurate modelling.

Following that, the actual modelling of the building can be carried out implementing the information collected during the field works. The first action is to place the load-bearing elements, characterizing them in terms of material and section typologies. After that it is possible to progress introducing the slabs and flooring systems. Although Revit provides several choices for flooring typologies, in this particular case it was necessary just to introduce a horizontal element that in Robot would have been then converted in quasi-zero thickness rigid diaphragm.

Non-structural elements such as infill walls or windows can be then added in order to provide a more realistic representation of the building. These last features are not transferred into the Robot model once exported it from Revit, as showed in Figure 1. In fact, just the elements listed in the "export" section are preserved, preventing the repetition of non-necessary tasks and reducing the likelihood of mistakes since the geometry does not have to be defined twice while moving from one software to another.

It should be pointed out that a BIM Revit model consists of a comprehensive combination of both 3D visualization and semantics (Eastman et al. 2010) in order to comply to BIM standards. However, even if Autodesk provides structural versions of Revit, a thorough structural analysis has to be performed by means of a dedicated tools, such as Robot. Therefore Robot is complementary to Revit in providing the merely structural simulation tool and its use becomes essential if achieving a numerical analysis on buildings' behavior is needed.

As outlined above, Figure 1 shows that almost the whole geometric characterization can be carried out in Revit while Robot allows to adjust the parameters that characterize the structural performance and behavior of the different structural members. As an example, the floorings can be modelled in Revit specifying for instance thickness, material, slab typology and potential reinforcement. When the model is then exported to Robot it will result in a panel including all the previous features. Nonetheless, in order to make the element perform and behave as desired it is necessary to adjust the different parameters that pertain the merely structural aspect. As such, slabs will have to be characterized for instance as shells or diaphragms and in the first case it is also required to specify which type of meshing to apply, the size of the finite elements and their shape. On the contrary, in case of a diaphragm it has to be selected whether it is rigid or not and in which direction because this will affect the overall weight distribution on structural members and the behavior of the building in its entirety.

With regard to load application for the purpose of this research it is preferred to input the loads in Robot for consistency reasons, although it could be done in Revit by means of the relative tool. Revit, in fact, allows to define load combinations and while exporting the model it is possible to specify which load case should account for the self-weight. Besides, it is preferred here to separate the architectural side from the structural characterization and therefore all the specifically structural features are left for the tool devised for the purpose (i.e., Robot). Potentially, load combinations could be defined in Revit, but for the reasons above it is not done.

As it will be better explained in section 4, a relevant issue connected to the export of the Revit models concerns bespoke structural sections that are not preserved in the transfer, as it can be observed from Figures 2 and 3. This can be solved adjusting the different elements singularly where the section is available as a parametric object, or defining the new section with the "Section Builder" tool of Robot. Alternatively, where both these solutions are not applicable because of structural limitations inherent in the design (such as the case of the residential building that will be outlined in the following section), the whole element can be converted into a load having the same effect of the non-structural member before its removal.

The characterization of the hazard has to be strictly conducted in Robot. Additionally, the specification of the code adopted for calculation, verification and spectra definition allows the compliance

Figure 2. Residential building in Old Beichuan.

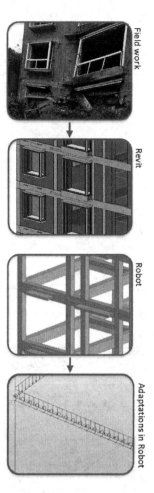

Figure 3. Details of residential building façade.

to regulatory frameworks, which partially covers the issue raised by Rezgui and his colleagues (2013).

It should be noted that once the model has been exported to Robot and analyzed it is neither envisaged nor allowed to reverse this process back to Revit. With respect to reinforcement, it is recommended to input it directly in Robot and potentially afterwards exporting it back into the desired Revit model.

The adoption of an interoperable suite of tools allows to better comprehend the behavior of the buildings in the analyzed context. In particular, the aim consists of modelling the structure according to the situation prior to the disaster, applying the hazard and then assessing if the effect is compatible with the state of the art. Consequently, this approach leads to a more thorough understanding of the causes that lead to the collapse mechanism and subsequently can be meaningful for prevention purposes in the future.

## 4 CASE STUDY

This section addresses the work conducted on two buildings in Old Beichuan (Sichuan Province, China) significantly compromised by the 2008 Wenchuan Earthquake. Both buildings are made of RC frames infilled with different typologies of masonry elements. While the residential building (Fig. 2) infills consist of UNI bricks, the Hotel (Fig. 4) was designed adopting masonry semi-hollow blocks and hence leading to a much lighter structure.

The choice of the building has been driven by different factors. In particular, for the residential building it has been motivated by the clear concurrence of both liquefaction-induced differential settlements and the presence of a clearly identifiable structural skeleton. The incidence of geotechnical-related failure modes has been registered as one of the most significant in the area of Old Beichuan. Conversely, in the case of the hotel the failure mechanism was mainly due to torsional phenomena of the superstructure rather than a lack of capacity at the foundation level.

With regard to the Hotel building and as it can be observed from Figure 1 and Figures 4–7 the process starts with importing the point cloud in Revit and from there achieving a representation of the structure. Conversely to the residential building, the model of Beichuan Hotel has not been yet enriched with all the architectural features but this did not preclude from exporting the main structure into Robot since any additional non-structural element is not anyway relevant for the structural analysis because it is being accounted separately as a load in Robot, unless it has an effect of the main structure.

In fact, elements that have been defined as non-structural in the Revit environment (e.g., infill walls) are not exported in Robot. On the other hand and as mentioned in section 3, floorings' properties are consistently preserved through the export relatively to material, thickness and

Figure 5. Beichuan Hotel point cloud.

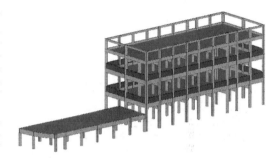

Figure 6. Beichuan Hotel Revit model.

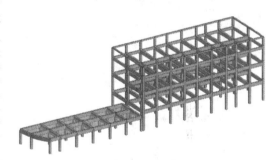

Figure 7. Beichuan Hotel Robot model.

reinforcement. According to the type of spatial dynamic constraints that the designer wants to assign to the structure, it is possible either to keep the "shell" option for the slab characterization, or modelling it as a rigid diaphragm. For the scope of the analysis, this last option has been chosen, therefore meshing is not required for this element.

A loss of information is detected while dealing with bespoke elements such as the L-section cantilever beam that characterize the residential building, as in Figures 2 and 3. Observing the building picture detail in Figure 3 it can be seen that a beam was implemented in the infill wall, disjointed from the main structural skeleton. Robot does not allow to introduce structural members not connected to the frame and to solve this issue linearly distributed moments were applied to the beam to simulate the

Figure 4. Beichuan Hotel reception building.

torsional action caused by the cantilever elements' indirect effect on the beam. Proof of this failure mechanism can be found in the cracks that vertically go from the disjointed element to the underlying beam as in Figure 3, showing a detachment in the masonry infill consistent with the hypothesis of torsional effects on the beam.

In this particular case all the elements that do not share nodes with structural elements are automatically jointed to the closest one, such as in the Robot details illustrated in Figure 3. Similarly, members that in Revit where placed assigning a certain offset, as the cantilever beams above and below the window elements as in Figure 3, are in Robot automatically placed in their original position.

With regard to the residential building only, fixed constraints have been assigned to the ground nodes, in order to account for the uncertainties deriving from the lack of geotechnical data. The final step of the analysis has been characterizing the seismic hazard in Robot in the context of the residential structure. Following to that a spectral analysis has been performed, based on seismometer velocity records downloaded from the IRIS database in order to achieve a realistic representation of the 2008seismic event. Results have then shown that displacements are negligible for this building showing that the earthquake itself did not account for the main failure cause but the seismic-induced liquefaction played a significant role jeopardizing the stability of the superstructure.

## 5 CONCLUSIONS

A methodology has been introduced in this paper for showing potential adoption of BIM tools for structural analysis, contextualizing it in real case scenarios and implementation of point cloud data. The presented methodology allowed to achieve a comprehensive representation of two real structures and to take into account the potential for point cloud data implementation. The interoperability between the two tools analyzed here (i.e., Autodesk Revit and Robot) consists in a key element that allows to speed up the modelling and analysis process, enabling a faster creation of 2D drawings and calculations. As a matter of fact, it is much faster from 3D parametric models to generate 2D plans and sections of building rather than relying just on a 2-dimensional representation of the structure, increasing hence the likelihood of mistakes and leading to cumbersome tasks when having to modifying them.

Both the scenarios cannot disregard the implementation of the information collected during the field works and that allowed the virtual representations to resemble reality as much as possible. Virtual models hence require to be constantly updated and being able to rely on tools designed to fit the purpose of BIM, such as Revit and Robot, allowing to achieve a higher detail of analysis in shorter time frames compared with traditional approaches.

## REFERENCES

Autodesk, Autodesk Revit Getting Started. Available at: https://knowledge.autodesk.com/support/revit-products/getting-started?cg=GettingStarted&p=REVITPRODUCTS&v=2018&sort=score&page=1&knowledgeSource=ProductDocumentation. Accessed May 10, 2018.

Autodesk 2010. Autodesk Robot Structural Analysis. Metric Getting Started Guide.

Banfi, F. et al., (2017). Historic BIM: A new repository for structural health monitoring. *The International Archives of the Photogrammetry, Remote Sensing and Spatial Information Sciences*, XLII-5(W1), pp. 269–274. Available at: http://www.gicarus.polimi.it [Accessed May 4, 2018].

Barazzetti, L. et al. 2015. BIM from laser clouds and finite element analysis: combining structural analysis and geometric complexity. In *The International Archives of Photogrammetry, Remote Sensing and Spatial Information Sciences*. Gottingen: Copernicus GmbH.

Cerè, G., Rezgui, Y. & Zhao, W. 2017. Critical review of existing built environment resilience frameworks: Directions for future research. *International Journal of Disaster Risk Reduction*, 25: 173–189.

Chi, H.-L., Wang, X. & Jiao, Y. 2015. BIM-Enabled Structural Design: Impacts and Future Developments in Structural Modelling, Analysis and Optimisation Processes. *Archives of Computational Methods in Engineering*, 22(1): 135–151.

Czmoch, I. & Pękala, A., (2014). Traditional Design versus BIM Based Design. *Procedia Engineering*, 91: 210–215.

Eastman, C.M. et al. 2010. Exchange Model and Exchange Object Concepts for Implementation of National BIM Standards. *Journal of Computing in Civil Engineering*, 24(1): 25–34.

Hunt, C.A. 2013. *The Benefits of Using Building Information Modeling in Structural Engineering*. Utah State University.

Kreider, R., Messner, J. & Dubler, C. 2010. Determining the Frequency and Impact of Applying BIM for Different Purposes on Projects. In *Innovation in AEC Conference*. The Pennsylvania State University, University Park, PA.

Nawari, N. 2011. Standardization of Structural BIM. In *International Workshop on Computing in Civil Engineering, Miami, 19–22 June 2011*. American Society of Civil Engineers.

Rezgui, Y., Beach, T. & Rana, O. 2013. A governance approach for BIM management across lifecycle and supply chains using mixed-modes of information delivery. *Journal of Civil Engineering and Management*, 19(2): 239–258.

# A tool for IFC building energy performance simulation suitability checking

G.N. Lilis, G. Giannakis & K. Katsigarakis
*Department of Production Engineering and Management, Technical University of Crete, Chania, Greece*

D.V. Rovas
*Institute for Environmental Design and Engineering, University College London, London, UK*

ABSTRACT: Data quality of BIM models is a key determinant in the value that can be extracted out of these data. Yet, despite this importance the discussion of data quality is often relegated to an afterthought. One potential use of BIM model data is the generation of building energy performance simulation models. Within this paper a checking procedure is presented, to ensure that user-supplied BIM models meet threshold data quality criteria and are suitable for the generation of input data files for energy analysis. The checking procedure comprises of three sets of checking operations: consistency, correctness and completeness checks. Consistency checks ensure that the input data are schema compatible; data completeness checks invoke the sequential execution of checking rules to verify the existence of required data; data correctness checks perform more elaborate detection of geometric errors appearing in surfaces, space volumes and clashes between architectural elements, which affect the building energy performance simulation model generation process. The checking procedure has been implemented and tested in two case-study buildings. Although the BIM modelers had been provided with modeling guidelines, multiple inaccuracies and data insufficiencies were still present, highlighting the importance of *a posteriori* process implementation that checks the validity of the model in relation to the purpose of its use.

## 1 INTRODUCTION

BIM models are becoming increasingly available for new but also for existing buildings. Similar to modeling in other domains, there is always the question of what should be modeled and at what level of detail. The level of modeling detail is intricately linked to the intended use of these models for various purposes that might include architectural and building services design, sustainability assessment, and facility management to name but a few. Capturing data requirements is an important first step, followed by the need to ensure that the data provided are of sufficient quality to be useful for the intended purpose. If such purpose is the setting up of Building Energy Performance Simulation (BEPS) models, often required in the context of sustainability discussions, data integration from multiple domains is often required (Cormier et al., 2011; Gudnason et al., 2014; Senave and Boeykens, 2015). To yield any useful outcomes, the input data provided in the first instance should meet certain quality threshold criteria (Gholami et al., 2015). However, due to high-complexity of data required for energy analysis purposes, it is quite common that the provided data fail to meet such criteria (Hitchcock and Wong, 2011; Nasyrov et al., 2014). Quality control in that sense becomes as important as the modeling process itself, while having ways to capture such errors is quite important, given the complexity of the problem. In this paper, data quality checking operations are classified into three categories, which are implemented in a sequential order and include consistency, completeness and correctness checking operations, mirroring checking and ensuring interoperability at three levels: syntactic, semantic and pragmatic.

For consistency checks, the syntactic checking of the input file is performed to ensure that the provided BIM model adheres to a specific schema. For the purposes of this paper BIM files are assumed to follow the IFC4 Add2 schema (ISO16739, 2013). In the completeness, checks the data integrity of the BIM files is investigated, using a number of checking rules and any missing information is reported. These checking rules are implemented using the IFC Model View Definition (MVD) methodology. Recently, several MVDs for BEPS model generation have been proposed (Pinheiro et al., 2018; Alshehri1 et al., 2017). Finally in the correctness checks, the BIM files' information is checked for possible errors. These checking operations should be applied to any BIM input data file in order to guarantee correctness of generated BEPS models.

```
┌─────────────────────────────────────┐
│   Design stage                      │
│   • BIM design guidelines      ✎    │
│   • IFC exporter                    │
└─────────────────────────────────────┘

┌─────────────────────────────────────┐
│   IFC4 file to be checked      ❖    │
└─────────────────────────────────────┘

① ┌───────────────────────────────────┐
  │  Data consistency                 │
  │  • IFC schema compatibility check ☑│
  └───────────────────────────────────┘

② ┌───────────────────────────────────┐
  │  Data completeness                │
  │  • Space boundary rules check     │
  │  • Material rules check         ☑ │
  │  • Space rules check              │
  └───────────────────────────────────┘

③ ┌───────────────────────────────────┐
  │  Data correctness                 │
  │  • Geometric Error Detection    ☑ │
  └───────────────────────────────────┘

┌─────────────────────────────────────┐
│   Checking results report       📄  │
└─────────────────────────────────────┘
```

Figure 1. Overview of the proposed BIM data quality checking process.

These operations are included in a stand-alone BIM for BEPS suitability checking tool, which is the main topic of the present work. As it is illustrated by the diagram of Figure 1, the tool receives IFC4 BIM files, designed followed specific guidelines and exported via a dedicated exporter, passes them trough the three data quality checking stages (consistency, completeness and correctness) and reports the data quality results in its output. The tool is an essential first step in the process of generation input data files for the EnergyPlus simulation environment.

## 2 DATA QUALITY

As described in the introduction, data quality checking operations can be classified into three categories: data consistency, data correctness and data completeness checks as described in the following sections.

### 2.1 Data consistency

Building Information Models, being an object-based digital representation of a building, are an information-rich source for setting up BEPS models. More specifically, in the current version of IFC (IFC4 Add2), which became an ISO standard (ISO16739, 2013), the data are stored in a STEP file, with the schema defined using the EXPRESS data modeling language. The IFC file which is usually generated by the exporter component of a BIM-authoring tool, should be consistent with the EXPRESS schema. This consistency check of the input IFC precedes any other completeness or correctness check.

### 2.2 Data completeness

BEPS model generation requires certain data to be present in the input IFC BIM files and to be "BEPS-complete". To ensure that IFC files contain these data a dedicated IFC exporter, a document containing appropriate BIM design guidelines and a BIM checking tool have been developed and offered to the user of the platform, which are detailed in the following sections.

#### 2.2.1 IFC exporter

Many popular commercial BIM authoring tools e.g. Revit™ (Autodesk, 2018), ArchiCad™ (Graphisoft, 2018) and ALLPLAN™ (Nemetschek Group, 2018), support exportation of IFC files. However, exportation is often not perfect: the exported models can be of poor quality and not directly usable. The Revit™ IFC4 DTV exporter is far from perfect: although, IFC can incorporate information about thermal and optical properties of each building entitys construction material, internal gains (schedules and densities) and inverse relations in case of curtain walls, the current version of the exporter is not capable of exporting this information. Therefore, a dedicated Revit based IFC exporter has been developed which supports these data exportation needs.

#### 2.2.2 BIM design guidelines

BIM models exported by appropriate exporters are not always suitable for BEPS model generation as information might be missing or their data might be incorrectly defined. For example, BIM building geometric data are oftentimes inappropriate for BEMS model generation. For these reason BIM design guidelines have been published (Maile et al., 2013) in order to guide the designers towards building designs suitable for BEPS model generation. Aligned to this direction the tools allows the user to download a pdf document containing appropriate guidelines. This document contains geometric as well as non-geometric rules to export a suitable ("BEPS-complete" and "BEPS-correct") IFC4 file. Snapshots of this document are displayed in Figure 2.

#### 2.2.3 BIM checking tool

In this section checking rules, embedded in the BIM Checking tool, are presented; these rules are applied

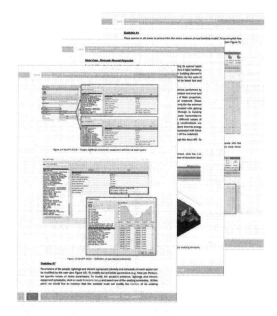

Figure 2. Snapshots of BIM design guidelines.

to check the data availability in the IFC4 file, from a Building Energy Performance Simulation (BEPS) perspective. Three main categories of rules are considered, described thoroughly below: space boundary rules; space rules; and material rules.

2.2.3.1 Space boundary rules

From a BEPS viewpoint, the partition of building construction (interior/exterior walls, floors, roofs) and opening (door, window) surfaces into $2^{nd}$-level space boundary (Bazjanac, 2010) surfaces, is a prerequisite. This process is performed by the Common Boundary Intersection Projection (CBIP) tool. The result of the CBIP tools execution is the enrichment of the IFC file with relevant $2^{nd}$-level space boundary information (IfcRelSpaceBoundary2ndLevel class population).

Before checking the $2^{nd}$-level space boundary content of the IFC file certain data requirements of the CBIP $2^{nd}$-level space boundary calculation and IFC data enrichment process is checked by the BIM Checking tool. More specifically, the existence of the following three IFC objects is examined: (a) one building object; (b) one site object attached to the building (to define the ground boundary conditions of the BEPS); (c) at least one space object (to obtain the space boundary surfaces related to this space volume) (d) at least on building element object (wall, slab, etc); and some unit objects.

The $2^{nd}$-level space boundary content of the enriched IFC file is checked by the BIM checking tool, using the following four checking rules, which are implemented sequentially:

1. The existence of the $2^{nd}$-level space boundaries is checked by examining the presence of the IfcRelSpaceBoundary2ndLevel class instances in the enriched IFC file. If there are no such instances, the following error message is reported: "2nd level space boundaries are missing". If this rule is passed the next rule is examined.
2. For each Ifc Rel Space Boundary 2nd Level populated instance, the presence of its related building element is examined. If related element exists the check proceeds to the next rule, otherwise a "related building element is missing" message, is reported.
3. For each Ifc Rel Space Boundary 2nd Level instance, the related building elements type, the value of the ExeternalOrInternal property, the corresponding space boundary (if the space boundary is INTERNAL) and the parent space boundary (if the space boundary refers to an opening), are examined. If any information is missing error messages are reported according to Table 3.
4. For each Ifc Rel Space Boundary 2nd Level instance the relating space index, is checked. If this index is missing, the following error message is reported: "relating space is missing".
5. Finally, the last rule checks if there is an Ifc Rel Space Boundary 2nd Level instance with ExeternalOrInternal property value "EXTERNAL_EARTH". If no such instance exists, a warming message "ground boundary condition is missing", is reported.

If any missing information is detected appropriate error messages are generated, which are summarized for different space boundary types (internal, external, corresponding and parent) and related building elements (wall, slab, door, window, plate, opening and virtual element) in Table 1.

2.2.3.2 Space rules

Some BEPS programs require information about the space occupants presence, artificial lighting, electrical equipment operation (which act as thermal sources in terms of BEPS) and related operation schedules. These data are supported by the provided IFC Revit™ exporter, which populates a new Pset class, named Pset_Space InternalGainsDesign, which is assigned to each IfcSpace class. The properties of this new class are checked and appropriate error messages are reported according to Table 2.

2.2.3.3 Material rules

BEPS require a number of properties to be assigned to materials of various building constructions. These properties, depending on the material type, are: for opaque materials Conductivity, Density and Specific Heat; for transparent materials U-factor and Solar heat gain coefficient (SHGC).

Table 1. Error messages of space boundary checking rules E: external, I: internal, C: corresponding, P: parent, M: missing.

| IFC class | E/I | C | P | Error message |
|---|---|---|---|---|
| IfcWall | I | M | – | corresponding space boundary is missing for internal wall |
| IfcSlab | I | M | – | corresponding space boundary is missing for internal slab |
| IfcDoor | E | – | M | parent space boundary is missing for external door |
|  | I | – | M | parent space boundary is missing for internal door |
|  | I | M | – | corresponding space boundary is missing for internal door |
| IfcWindow | E | – | M | parent space boundary is missing for external window |
|  | I | – | M | parent space boundary is missing for internal window |
|  | I | M | – | corresponding space boundary is missing for internal window |
| IfcPlate | E | – | M | parent space boundary is missing for plate |
|  | I | – | – | plate cannot have an internal space boundary |
| IfcOpening | E | – | – | external space boundary with IfcOpeningElement as related building element was found |
| IfcVirtual Element | E | – | – | external space boundary with IfcVirtualElement as related building element was found |

Table 2. Error messages of space checking rules.

| Checked property | Error message |
|---|---|
| OccupancySchedule | occupancy schedule is missing |
| LightingSchedule | lighting schedule is missing |
| EquipmentSchedule | equipment schedule is missing |
| HeatGainLighting | heat gain lighting parameter is missing |
| AreaPerOccupant | area per occupant parameter is missing |
| HeatGainPerOccupant | heat gain equipment parameter is missing |
| InfiltrationRate | infiltration rate parameter is missing |

Table 3. Error messages of opaque material property checking.

| Checked property | Error message |
|---|---|
| SpecificHeatCapacity | SpecificHeatCapacity parameter is missing |
| ThermalConductivity | ThermalConductivity parameter is missing |
| MassDensity | MassDensity parameter is missing |

Table 4. Error messages of transparent material property checking.

| Checked property | Error message |
|---|---|
| Visual light transmittance | Visual light transmittance parameter is missing |
| solar heat gain coefficient | solar heat gain coefficient parameter is missing |
| heat transfer coefficient | heat transfer coefficient parameter is missing |
| thermal resistance | thermal resistance parameter is missing |

Two checking rules depending on the material type are applied, which report error messages that are listed in Tables 3 (opaque materials) and 4 (transparent materials).

### 2.3 Data correctness

Even if the BEPS modeler uses an error-free exporter, has followed appropriate design guidelines to develop the BIM file and the exported IFC file has passed all data completeness checks, there are cases where the IFC data might be inaccurate. In these cases manual corrections should be applied. These inaccuracies of the values contained in IFC classes affect the quality of the generated BEPS models. There are various such inaccuracies. For example, an intersection or clash between the geometric description of an internal building volume and the geometric description of a surrounding building construction element such as an internal wall, affect the related space boundary surfaces (surfaces through which the space exchanges heat with its environment), and thus the correctness of the respective BEPS model is affected as well. In the present work only the geometric inaccuracies will be examined, however the data correctness checks can be extended to non-geometric elements, such as the properties of building elements which should obtain numerical values within a predefined range.

There are certain geometric inaccuracies which affect the generation of a building energy performance simulation input data file, grouped in the following three categories: clashes; surface errors; and space incorrect definitions Lilis et al. (2015). These geometric error types are presented in the following sections.

#### 2.3.1 Clashes

Clashes are intersections between the solid geometric representations of architectural building components (walls, slabs, spaces, etc.). Certain clash types affect the generation of BEPS models by altering the space boundary surface topology of the building (Bazjanac, 2010). These clash types refer to intersections of the building internal space volumes with other neighbor solid geometric representations (walls, slabs, etc.), as displayed in the example of a special clash type (containment) in part II of Figure 1. In this example, the wall volume is contained entirely in the space volume resulting to omitting the space boundary surfaces related to this wall. All other (not space-related) intersections affect BEPS model generation only on when the intersection surfaces are attached to pairs of building space volumes or a building space volume and the building site. A Wall Opening—Wall Opening clash example, where the intersection volume is attached to a space pair, causing space boundary duplication, is illustrated in part III of Figure 3.

### 2.3.2 *Surface errors*

Surface errors exist, when one or more surfaces from the boundary surface representation of architectural components (walls, slabs, etc.) are either incorrectly oriented i.e. the direction of the normal vector of the surface calculated using the right-hand rule is reversed (first type of surface error), or are missing from the boundary representation (second type of surface error). Examples of surface errors are displayed in Figure 4. Part I of Figure 4 displays a correctly oriented boundary representation of a building slab. In part II of Figure 4, the first type of a surface error is illustrated.

The boundary representation of the slab used in part I of the same figure, has some surface normals inverted. Finally in part III an example of the

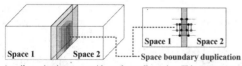

A wall opening intersects with another wall opening. This non-space clash affects BEP simulation model generation because the intersection surfaces are attached to a space volume, resulting to a space boundary duplication.

Figure 3. Clash types and their effect to BEPS model generation.

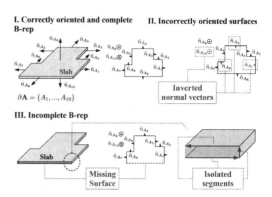

Figure 4. Examples of surface errors: (a) Surface orientation error, (b) Incomplete shell error (missing surface).

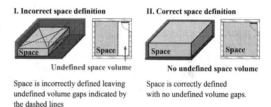

Figure 5. Examples of incorrect and correct space definition.

second type of a surface error is displayed, where the a boundary representation has a missing surface defined by a set of isolated segments (segments which belong to only one boundary surface polygon), on the boundary of this missing surface.

### 2.3.3 *Space definition errors*

Space definition errors occur when an internal building space volume is not surrounded completely by other architectural elements (walls, slabs, openings) and small gaps of undefined space volumes exist between the internal space and surrounding building architectural elements, as presented in part I of Figure 5. In part II of Figure 5 a correctly defined building space is presented, and for comparison an incorrectly defined space is displayed in part I of the same figure.

### 2.3.4 *Geometry error detection tool*

The Geometry Error Detection tool helps the modeler to generate a geometric error-free IFC BIM file by reporting any detected errors in in an XML form and exporting them in obj file format, for visualization purposes.

#### 2.3.4.1 Clash errors

Clash errors are classified depending on the pair of involved entities. Several types such as Wall-Space, Slab-Space and others exist. Such classification is reported in by the GED tool as the following XML instances illustrate.

```
<ClashErrors>
    <Wall-Space>
        <ClashError id="1" , ...">
        <ClashError id="2" . ...">
        ...
    </Wall-Space>
</ClashErrors>
```

As displayed in the following XML instances, a clash error report contains the local building element ids (ID) and IFC global IDs (GID) of the two involved entities and the set of surfaces, which define the intersections among the two involved solid bodies.

```
<ClashError id=1, ID_1=3, GID_1=..., ID_2=4, GID_2=...>
    <clashErrorSurfaces>
        <surface> ... </surface>
        ...
    </clashErrorSurfaces>
    <spaceBoundaryDuplicationSurfaces>
        <surface> ... </surface>
        ...
    </spaceBoundaryDuplicationSurfaces>
</ClashError>
```

Every surface consists of an outer and possible multiple inner contours (holes) as described next:

```
<surface>
    <outerContour>
        ...
    </outerContour>
    <innerContour>
        ...
    </innerContour>
    ...
</surface>
```

Finally the above outer and the inner contours are described by a set of points in three dimensions:

```
<point3D id="1" x="19.0" y="3.2" z="8.9"/>
<point3D id="2" x="19.0" y="3.2" z="6.2"/>
...
```

Containment clash errors are also reported as regular clash errors with the word "containment" replaced by the word "clash" and with only the "spaceBoundaryDuplicationSurfaces" field.

#### 2.3.4.2 Surface errors

Surface errors are linked to a solid body types, which can be either a wall, slab, space, e.t.c., and are reported by the GED tool as the following XML section indicates.

```
<SurfaceErrors>
    <"Solid Body Type">
        <SurfaceError ...>
        ...
        <SurfaceError ...>
    </"Solid Body Type">
</SurfaceErrors>
```

For each individual surface error GED tool reports the local ID of the related building entity, its IFC global ID (GID), the total surface norm of its boundary representation, the number of its boundary surfaces and a set of either isolated segments or inverted surfaces, according to the following XML instance:

```
<SurfaceError id=1, ID=4, GID=...,
    SurfaceNorm="2.3", NumberOfSurfaces="23"/>
    <IsolatedSegments>
        <segment3D> ... </segment3D>
        ...
    </IsolatedSegments>
    <InvertedSurfaces>
        <surface> ... </surface>
        ...
    </InvertedSurfaces>
</SurfaceError>
```

The inverted surfaces are surfaces which have incorrectly inverted normal vector and the isolated segments are line segments which belong to only one boundary representation surface and thus define a hole or a edge misalignment of the solid of the building entity. The inverted surfaces are defined by XML instances as regular surfaces as reported in the case of clash errors described previously, and the isolated segments are defined by the coordinate triplets of the initial ($x_1$, $y_1$, $z_1$) and the final points ($x_2$, $y_2$, $z_2$) of the segment as follows:

`<segment3D id="1", length=... , x1=..., y1, z1, x2, y2, z2/>`

The total surface norm of the boundary representation of the solid body which has a surface error, is the Euclidean norm of the sum of the surface vectors of the boundary representation of the solid body. Any divergence of this surface norm from the zero value could be an indication of a surface error. The magnitude of the divergence of the surface norm from the zero value, is proportional to the severity of this surface error.

#### 2.3.4.3 Space errors

Finally, a space definition error is reported from the GED tool, as displayed in the following XML instance:

```
<SpaceError id="1" SpaceID="3" SpaceGID=...>
    <boundaryRepresentation>
        <surface> ... </surface>
    </boundaryRepresentation>
</SpaceError>
```

As it is presented in previous XML instance, for a space definition error the IFC global id of the internal space is given, together with the boundary surfaces of the space which are not attached to any other building architectural element. The space boundary surface is specified as polygon(s) with outer (and possible inner) 3D point contours.

## 3 EXAMPLES

### 3.1 Case study buildings

The introduced BIM data quality checking tool is validated on two buildings: the Demo4 building which is an imaginary two-story office building, pictured in part I of Figure 6 and the Turina Torre building located in Valladolid in Spain, displayed in part II of Figure 6.

### 3.2 Application of the BIM checking tool on Demo4 building

The BIM Checking tool subcomponent is validated on Demo4 building and some error report message examples are presented next.

- The rule "space boundaries, parent space boundaries, corresponding space boundaries" has the following problems:

   issue related to the MODEL, 2nd level space boundaries are missing.

- The rule "spaces, loads, parameters, schedules" has passed.

- The rule "doors, thermal parameters" has the following problem

   issue related to Single-Flush:30″ × 80″ 2:245315 (1ChqrYs197G8E9uZmTikEW) "visual light transmittance" parameter is missing ...

- The rule "windows, thermal parameters" has the following problems:

   issue related to M_Curtain Wall Awning: M_Curtain Wall Awning:218240 (3RvIiUq4X8 GweJx_3dFm2 s) "visual light transmittance" parameter is missing ...

- The rule "plates, thermal parameters" has the following problems:

   issue related to System Panel:Glazed:214779 (3RvIiUq4X8GweJx_3dFnBD) "visual light transmittance" parameter is missing ...

### 3.3 Application of the GED tool on Turina Torre building

The operation of the Geometric Error Detection tool subcomponent (GED tool) is demonstrated on the Turina building. As it is displayed in the MeshLab screen shots of the obj file exported by the GED tool, which are collected in Figure 7, sev-

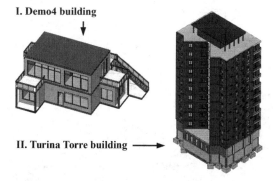

Figure 6. Case study buildings for the introduced BIM quality checking tool.

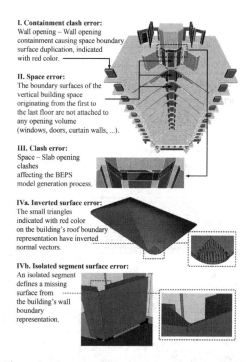

Figure 7. Demonstration of Geometric Error Detection tool's operation on Turina building (MeshLab screenshots of the obj file exported by the GED tool).

eral geometric errors were detected in the IFC file of the Turina Torre building. (space)-(slab opening) clashes (part I), wall opening containment clashes (part III), space errors caused by omitted opening volumes (part II) and surface errors cased either by normal inversion (part IVa) or isolated segments (part IVb), all affecting BEPS model generation, were detected.

## 4 CONCLUSIONS

A stand alone tool which performs BIM for BEPS suitability checking operations on IFC4 files was introduced. These operations are data quality checking operations which include: consistency checks (IFC4 schema compatibility check), completeness checks performed by the BIM Checking tool (space boundary, material and space related data checks) and correctness checks performed by the GED tool (detection of clash, space and surface geometric errors affecting BEPS model generation). To facilitate "BEPS-ready" BIM generation, a dedicated IFC exporter for Revit™ and appropriate BIM design guidelines are also provided to the user of the introduced tool.

The BIM for BEPS data quality checks performed by the introduced tool, should not be limited to building geometry, materials and operation schedules but should extend to other building components such as systems and their properties. Such investigation is a topic of ongoing research work. Additionally other error reporting formats such as the BCF file format (Stangeland, 2013), will be also included in the future.

Although multiple general BIM checking tools are available such as Solibri Model Checker™ (Solibri, 2018) and TEKLA's Model Checker Suite™ (TEKLA, 2014), the novelty of the introduced tool relays on the fact that it contains only the checking operations that will guarantee accurate BEPS model generation and, in this sense, is more specialized. Finally, it is important to mention that without the modeler's experience, even with the best checking tool available, the BIM for BEPS accuracy cannot be guaranteed.

## ACKNOWLEDGEMENTS

Part of the work presented in this paper is based on research conducted within the project "Optimised Energy Efficient Design Platform for Refurbishment at District Level", which has received funding from the European Union Horizon 2020 Framework Programme (H2020/2014-2020) under grant agreement n° 680676.

## REFERENCES

Alshehri1, F., Kenny, P., Pinheiro, S., and James, O. (2017). Model View Definition (MVD) for Thermal Comfort Simulation in Conventional BEPS tools. In *PLEA*, Edinburgh, UK.

Autodesk (2018). Revit Architecture. https://www.autodesk.com/products/revit/architecture.

Bazjanac, V. (2010). Space boundary requirements for modeling of building geometry for energy and other performance simulation. In *CIB W78*, Cairo, Egypt.

Cormier, A., Robert, S., Roger, P., Stephan, L., and Wurtz, E. (2011). Towards a BIM-based service oriented platform: application to building energy performance simulation. In *Building Simulation IBPSA Conference*, pages 2309–2316, Sydney, Australia.

Gholami, E., Kiviniemi, A., and Sharples, S. (2015). Implementing Building Information Modelling (BIM) in Energy-Efficient Domestic Retrofit: Quality Checking of BIM Model. In *CIB W78 conference*.

Graphisoft (2018). Archicad 22 - BIM inside and out. http://www.graphisoft.com/archicad/.

Gudnason, G., Katranuschkov, P., Balaras, C., and Scherer, R. (2014). Framework for Interoperability of Information Resources in the Building Energy Simulation Domain. In *Computing in Civil and Building Engineering*, pages 617–624.

Hitchcock, R.J. and Wong, J. (2011). Transforming IFC architectural view BIMs for energy simulation: 2011. In *Building Simulation conference of IBPSA*.

ISO16739 (2013). Industry Foundation Classes (IFC) for data sharing in the construction and facility management industries. *International Organization for Standardization (ISO)*.

Lilis, G.N., Giannakis, G., and Rovas, D. (2015). Detection and semi-automatic correction of geometric inaccuracies in IFC files. In *Building Simulation IBPSA Conference*, pages 2182–2189, Hyderabad, India.

Maile, T., ODonnell, J., Bazjanac, V., and Rose, C. (2013). BIM-Geometry modelling guidelines for building energy performance simulation. In *Building Simulation IBPSA Conference*, pages 3242–3249.

Nasyrov, V., Strath¨ucker, S., Ritter, F., Borrmann, A., Hua, S., and Lindauer, M. (2014). Building information models as input for building energy performance simulation–the current state of industrial implementations. In *European Conference on Product and Process Modelling*, pages 479–486.

Nemetschek Group (2018). Allplan. https://www.allplan.com/en/.

Pinheiro, S., Wimmer, R., ODonnell, J., Muhic, S., Bazjanac, V., Maile, T., Frisch, J., and van Treeck, C. (2018). MVD based information exchange between BIM and building energy performance simulation. *Automation in Construction*, 90:91–103.

Senave, M. and Boeykens, S. (2015). Link between BIM and energy simulation. *Building Information Modelling (BIM) in Design, Construction and Operations*, 149:341–352.

Solibri (2018). Solibri Model Checker. https://www.solibri.com/products/solibri-model-checker/.

Stangeland, B. (2013). BIM Collaboration Format (BCF). *BuildingSMART International User Group; BuildingSMART: Bakkmoen, Norway*.

TEKLA (2014). Model Checker Suite. https://www.tekla.com/.

# Implications of operational, zoning-related, and climatic model input assumptions for the results of building energy simulation

Gianluca Pilati, Giovanni Pernigotto & Andrea Gasparella
*Free University of Bozen-Bolzano, Italy*

Tahmasebi Farhang & Ardeshir Mahdavi
*Technische Universität Wien, Austria*

ABSTRACT: Computational modelling of building energy performance can provide designers and operators with relevant information toward optimization of design quality and operational performance. However, the efficacy of simulation-supported design strongly depends on the reliability and consistency of model input assumptions, such as those related to thermal zoning resolution and occupants' behavior. Moreover, the actual impact of those input data and modelling choices can be different according, for example, to the climate, the type of HVAC system, and the set-points chosen for ensuring acceptable indoor thermal conditions. The present study deploys an energy model of a prototypical office floor together with a stochastic occupancy model to parametrically explore the implications of thermal zoning, climatic variations, and different control strategies for building energy use.

## 1 INTRODUCTION

Building energy consumption is influenced by several factors, such as the external ambient conditions, the characteristics of envelope and HVAC systems, and the occupant behavior. The latter, in particular, is one of the most challenging aspects to predict and model in building energy simulation, as observed in several studies (Mahdavi & Tahmasebi, 2015; Page et al., 2008; Tahmasebi & Mahdavi, 2016; Yan et al., 2016). Internal gains, HVAC system controls, and ventilation rates are strongly affected by seemingly random patterns of occupants' presence and behaviour.

Moreover, to account for the actual occupants' presence, the built environment must be subdivided into a number of thermal zones that are representative of actual occupancy profiles, distinguishing periods of presence and absence. Consequently, an important role is played also by building thermal zoning (Dogan et al., 2014; Smith et al., 2011). In building energy simulation codes, a zone is generally defined as a part of the building that is assumed to have homogeneous thermal characteristics. Usually, experts and technical standards recommend a minimum number of zones equal to the number of devices (e.g., fans, radiators), not necessarily corresponding to the number of rooms of the building (UNI, 2014; DOE, 2017). Thereby, with detailed zoning in system design and energy models, it is also possible to develop smart HVAC controls that can improve energy savings (Nguyen & Aiello, 2013; Oldewurtel et al., 2013; Yang & Becerik-Gerber, 2014).

In this study, the impact of thermal zoning on building simulation outcome and the benefits of advanced occupancy-responsive HVAC controls are discussed based on a case-study and considering different climatic and thermal comfort scenarios.

## 2 APPROACH

### 2.1 Case study and geometrical configurations

The present study analyses a reference building according to the *Small office-commercial reference building—year 2007 model* for new construction given by the U.S. Department of Energy (DOE, 2007). The DOE case-study is a one-story building but for this research it has been considered as a floor of a multi-story office building.

It includes two different types of thermal zones: a peripheral office zone and a core zone, where facilities such as staircase and elevator are positioned. The focus is on the office zone. As such, the internal office zone boundaries are defined as adiabatic. The reference model has 20 identical windows (width = 1.83 m; h = 1.52 m). The U-Values for external walls and windows are 0.363, and 2.371 $W.m^{-2}.K^{-1}$ respectively (Deru et al., 2011).

The office zone accommodates 24 users (Fig. 1). Each person has his/her own office resulting in 24 different single-occupancy offices.

Since one of the targets of this work is to assess the thermal zoning influence, three different configurations were considered:

1. Basic resolution configuration (model 1): no internal partition wall has been considered, the whole offices zone consists in a single thermal zone (Fig. 2);
2. Medium resolution configuration (model 2): 4 internal partition walls have been assumed as per the original DOE model (Fig. 3);
3. High resolution configuration (model 3): all the 24 offices represent separate thermal zones; the number of windows has been increased to 28 while keeping the same *WWR* (Windows/Wall Ratio) to have, at least, one window per office (Fig. 4);

For each model, annual heating and cooling energy needs were investigated.

### 2.2 Simulation settings

EnergyPlus v.8.8.0 (DOE 2018a) was used to simulate the building dynamic behavior with 15-minute time steps. For the heat transfer through the opaque components, the Conduction Transfer Function methods was chosen while TARP algorithm was adopted for the inside and outside surface convection.

The geometry and zoning were modeled with OpenStudio plug-in for SketchUp (2018). The boundary conditions of all the zones are summarized in Table 1: *Surface BC* is used for internal partitions and it allows heat transfer between the zones, *outdoor BC* is used when a surface is exposed to outside and, consequently, to sun and wind, *adiabatic BC* means that there is not temperature difference at the two sides of a surface (DOE, 2017).

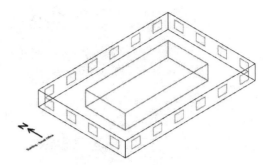

Figure 2. Basic resolution model configuration.

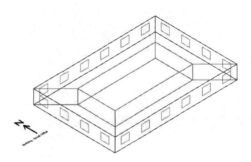

Figure 3. Medium resolution model configuration.

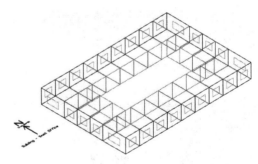

Figure 4. High resolution model configuration.

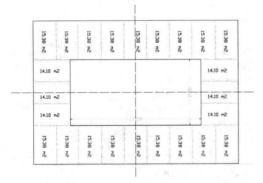

Figure 1. Offices configuration with symmetry axes and offices areas.

Table 1. Surfaces boundaries and exposure.

| Surface | Boundary | Exposure to sun and wind |
|---|---|---|
| Floor | Adiabatic | No |
| Ceiling | Adiabatic | No |
| Ext walls | Outdoor | Yes |
| Int walls | Non-adiabatic | No |
| Core-office separation walls | Adiabatic | No |

There are some minor differences between the three geometry configurations. The 24 internal walls were included only in model 3. In model 2, only the internal surfaces that divide the four offices zones were defined. In model 1, no internal wall was modelled. To consistently represent thermal inertia across the three configurations, in models 1 and 2, EnergyPlus internal mass object was deployed.

### 2.3 *Modelling of occupancy and Internal gains*

Occupancy was modelled based on the number of people in each zone. An internal algorithm is used to determine what fraction of the total heat is sensible and what fraction is latent. Following ASHRAE (2013a), a typical metabolic heat generation of 67 W.m$^{-2}$ (office activity), and a reference human body area of 1.80 m$^2$ were assumed, resulting in 120 W per occupant.

Regarding people's presence in building, the Table *G-G. Office occupancy* in ASHRAE (2013b) was followed. Figure 5 shows schedules for weekdays, Saturdays, and Holidays.

Implementation of schedules in model 3 was derived using a stochastic occupancy model (Page *et al.* 2008). This model uses as input a profile of presence probability and parameter of mobility (defined as the ratio of state change probability to state persistence probability) and returns random non-repeating daily profiles of occupancy states (present: value 1; not present: value 0). The model has been formulated based on the hypothesis that the value of occupancy at each time step depends on the previous occupancy state and the probability of transition from this state to either the same state or its opposite state.

A Matlab® script (2018) that implements the model was written, with a re-sample of input occupancy values from hourly to sub-hourly values and an independent calculation for each office. Output data are Boolean-random schedules for 1 year with time step of 15 min in csv format, with a total of 35040 values. Since an occupancy distribution for each office was to be generated, the code was executed 24 times. Vacations and long absence were explicitly included into the above-mentioned stochastic occupancy model.

Model reliability was tested and average profiles were reproduced using the script (see Figure 6).

For every simulation time step and office, occupancy has a random unique Boolean value in line with the Standard's profile. These profiles imply if a person is inside or outside the office such that—in the former case—potential interaction with HVAC, lighting, and equipment can be considered.

According to EnergyPlus model (Eq. 1), the thermal gain from internal lighting can be divided into four different fractions (Return Air Fraction, *RAF*, Fraction Radiant, *FR*, Fraction Visible, *FV*, and Fraction Convective, *FC*). This depends on the type of lamp and luminaire and on its installation.

$$FC = 1 - (RAF + FR + FV) \quad (1)$$

A value of 10.5 W.m$^{-2}$ was used in all simulations (ASHRAE, 2013a) for closed offices. *RAF, FR, FV* and *FC* were fixed for overhead fluorescent lighting pendant T8. The light uses was defined according to *ASHRAE 90.1–2013* (ASHRAE, 2013b) (Figure 7).

In the simplified models 1 and 2, it is not possible to take into account the actual lighting usage

Table 2. Models 1 and 2 occupancy schemes.

| Model | Zone – | Area m$^2$ | Density m$^2$/person | People # |
|---|---|---|---|---|
| 1 | Office | 361.5 | 15 | 24 |
| 2 | North, South | 113.5 | 15 | 8 |
| 2 | East, West | 67.3 | 15 | 4 |
| 3 | Type 1 | 14.1 | 15 | 1 |
| 3 | Type 2 | 15.4 | 15 | 1 |

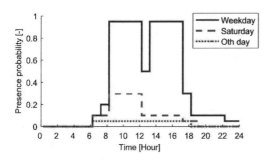

Figure 5. Occupancy schedule based on ASHRAE 90.1 (ASHRAE, 2013b).

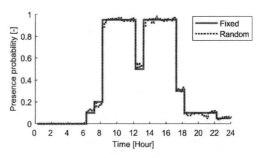

Figure 6. Comparison of the fixed schedule and randomly created profiles for a weekday averaged over the entire year for an office.

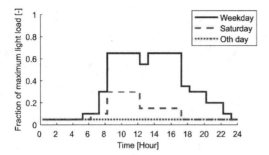

Figure 7. Light schedule based on ASHRAE 90.1 (ASHRAE, 2013b).

Table 3. Light loads comparison.

| Description | Model | Unit | Value |
|---|---|---|---|
| Mean office area | – | m² | 15.06 |
| Light density | – | W.m$^{-2}$ | 10.50 |
| Total light energy | 1, 2 | kWh.m$^{-2}$ | 24.11 |
| Total light energy | 3 | kWh.m$^{-2}$ | 23.96 |

in each single office, whereas in model 3 lighting usage follows people's presence, distinguishing a base load and an occupancy-dependent one. The base load is due to fixed and emergency lights and is equal to 5% of the total lighting gains. The rest load was assumed to be proportional to occupancy level at each time step, following Eq. 2:

$$light_{rnd,i} = occ_{rnd,i} \frac{light_{fix,i} - light_{base,i}}{occ_{fix,i}} \qquad (2)$$

Here $light_{rnd,i}$ is the resulting schedule value, $light_{base,i}$ is the base load schedule value, $occ_{fix,i}$ is the occupancy fixed schedule value, and $occ_{rnd,i}$ is the random occupancy, calculated as previously explained.

In this way, the lighting loads applicable to each occupant were defined as the ratio of occupancy-dependent lighting loads' diversity factors to the presence probability at each time step, both obtained from the fixed schedules used for that occupant (Tahmasebi & Mahdavi, 2017).

A Matlab® script automated this procedure: fixed and random occupancy and fixed lighting schedules were taken as input and for each thermal zone a weighted value of light was calculated at each time step. As output, the code generated 24 csv files with the lighting usage. These could be read as input in EnergyPlus.

Table 3 compares, for the whole building, the total lighting energy demand in the three models.

Office devices that consume electricity (e.g. computers, projectors, portable fans) were assigned a value of 8.61 W.m$^{-2}$ (ASHRAE, 2013a). The implementation was similar to the lighting case, with schedule loads from (ASHRAE, 2013b) in models 1 and 2, as shown in Figure 8.

The schedule values for model 3 were calculated following the method explained for lighting. To verify the accuracy of the aforementioned proportional approach, Table 4 compares the annual equipment energy use of models 1, 2, and 3.

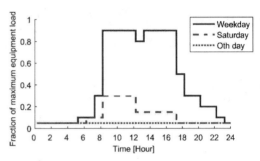

Figure 8. Equipment schedule based on ASHRAE 90.1 (ASHRAE, 2013b).

Table 4. Equipment loads comparison.

| Description | Model | Unit | Value |
|---|---|---|---|
| Mean office area | – | m² | 15.06 |
| Equipment density | – | W.m$^{-2}$ | 8.61 |
| Total equip energy | 1, 2 | kWh.m$^{-2}$ | 25.16 |
| Total equip energy | 3 | kWh.m$^{-2}$ | 25.01 |

## 2.4 HVAC system

The estimate the building's heating and cooling demands, it was assumed to possess an ideal HVAC system with unlimited capacity. The system also provides mechanical ventilation and is controlled based on air temperature and humidity. As shown in Table 5, the constraints for temperature, humidity, and ventilation rate were set based on the three thermal comfort categories defined in EN 15251 (CEN, 2008).

For the unoccupied periods of time, a setback setting $SB$ was defined. Infiltration rate was calculated according ASHRAE (2013a) as in Eq. 3:

$$Q_i = A_L \cdot IDF \qquad (3)$$

Here, $Q_i$ is the calculated air flow rate in l.s$^{-1}$, $A_L$ is the effective building leakage area in cm² and $IDF$ is the Infiltration Driving Force in l.s$^{-1}$·cm$^{-2}$. An average $IDF$ value of 0.069 l.s$^{-1}$·cm$^{-2}$ was set (ASHRAE, 2013a). With regard to the effective building leakage area, Eq. 4 was adopted:

Table 5. Settings for the three thermal comfort categories according to EN 15251 (CEN, 2008).

| Description | Comfort category | | | |
|---|---|---|---|---|
| | A | B | C | SB |
| Heating Temp °C | 21.0 | 20.0 | 19.0 | 16.0 |
| Cooling Temp °C | 25.5 | 26.0 | 27.0 | 30.0 |
| Dehumidification RH% | 50.0 | 60.0 | 70.0 | 70.0 |
| Humidification RH% | 30.0 | 25.0 | 20.0 | 20.0 |
| Ventilation l.s$^{-1}$.pers$^{-1}$ | 10.0 | 7.0 | 4.0 | 3.0 |
| ACH h$^{-1}$ | 0.8 | 0.56 | 0.32 | 0.2 |

$$A_L = A_{es} \cdot A_{ul} \qquad (4)$$

Here, $A_{es}$ is the building exposed surface area (225.2 m$^2$) and $A_{ul}$, the unit leakage area (2.80 cm$^2$.m$^{-2}$) (ASHRAE, 2013a). The infiltration rate was expressed in terms of air changes rate as in Equation 5:

$$ACH = \frac{3.6 \cdot Q_i}{V} \qquad (5)$$

Here $V$ is the whole building volume in m$^3$. A resulting infiltration rate of 0.1 h$^{-1}$ was applied to all perimetral thermal zones.

### 2.5 HVAC mode of operation

Several control strategies were considered in the 3 models as follows:

- HVAC system is always on;
- During weekday, when people are inside offices, HVAC system operates with comfort category constraints according to Table 5:
- During weekday HVAC system is to meet comfort category target values one hour before the arrival of the first office occupant;
- During unoccupied weekday hours, weekdays after 7 pm, and on weekends, HVAC system is to maintain *SB* values.

In model 3, each office is controlled separately and during weekdays between the first arrival and 7 pm, if presence is not detected, the system is to alternatively:

- maintain *SB* values (model 3)
- switch from current category to C category (model 3+switch)

### 2.6 Climate

To give the study a wider coverage in view of climatic conditions, simulations were conducted for three cities, namely Vienna, Bolzano, and Messina based on U.S. DOE weather datasets (DOE 2018b). Vienna has a *Dfb climate* according to the Köppen classification (Peel *et al.*, 2007) (cold climate without dry season and with a warm summer), Bolzano features a *Cfa climate*—temperate climate without dry season and hot summer, whereas Messina shows a *Csa climate*—temperate climate with dry and hot summer.

## 3 RESULTS AND DISCUSSION

A total of 42 alternatives where simulated to capture the variations in the number of thermal zones, comfort categories, occupancy type, and climate (see Figures 9 to 11).

The results suggest that an increase in the number of thermal zones leads to a small increase in estimated energy use. Indeed, if a building is modelled as a single thermal zone, solar gains in the south zone may balance losses in the north zone. Similar effects have been also reported in other studies (Dogan *et al.*, 2014; Smith *et al.*, 2011). It is thus important to remember that the simulation application solves the energy balance equation assuming a perfectly mixed volume of air within each zone. The impact is different considering heating and cooling for different climates. In Vienna, model 3's annual heating energy use is 4% to 6% higher than model 1 for A and C comfort categories respectively. The largest differences in cooling energy use in Bolzano: Depending on comfort category, annual cooling energy need differs 2.8% (A category) and 3.8% (C category) between models 1 and 3.

As to the HVAC systems control, the comparison of models 3 and 3+switch reveals an energy saving potential via the dynamic control of the system based on occupancy presence. This amounts to a saving of up to 12% for heating (see Vienna in Fig. 9) and 3% for cooling (see Messina in Fig. 11).

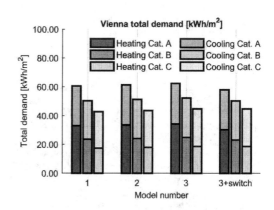

Figure 9. Total energy demand for Vienna.

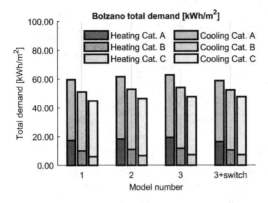

Figure 10. Total energy demand for Bolzano.

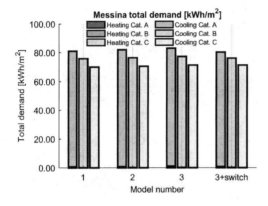

Figure 11. Total energy demand for Messina climate.

## 4 CONCLUSIONS

In this work, assumptions pertaining to people's presence in (office) buildings, thermal zoning resolution, comfort categories, and buildings' location were investigated and discussed in view of their influence on annual heating and cooling energy demand. This was done via a simulation-based case study involving a reference small office building located in three European cities (Vienna, Austria, and Bolzano and Messina, Italy). Three models with increasing level of complexity in thermal zoning were constructed and three alternative control strategies (thermal comfort categories A, B, and C according to EN 15251) were considered. Stochastic occupancy patterns were generated and, for the most complex models, advanced HVAC occupancy-based controls were implemented.

Regarding thermal zoning, the results suggest that:

- Thermal zoning details can have an impact on simulated results of annual energy use and, consequently, on the assessment of building energy performance.
- Other factors being similar, larger number of modelled zones results in slightly higher energy use.
- The impact of thermal zoning depends on is climate and HVAC control regime, with the largest differences in case of C category of thermal comfort.

Regarding HVAC system control options, the results suggest that occupancy-based control strategies can contribute to energy saving.

The study further underlines the importance of the targeted thermal comfort standards for the energy performance of buildings. As compared to model attributes (zoning resolution) and control scenarios, relaxation of mandated thermal comfort conditions was shown to have the most pronounced effect concerning reduction of predicted energy use magnitudes. Needless to say, there are limits to which such relaxation may be applied in practice, as multiple constraints pertaining to building use, productivity considerations, and cultural phenomena would have to be considered.

## ACKNOWLEDGEMENT

This study was funded by University of Trento International mobility scholarship for thesis research and by the project "Klimahouse and Energy Production" in the framework of the programmatic-financial agreement with the Autonomous Province of Bozen-Bolzano of Research Capacity Building.

## REFERENCES

ASHRAE, 2013a. ASHRAE handbook: Fundamentals. Engineering Reference. SI ed. American Society of Heating Refrigerating and Air-Conditioning Engineers. Atlanta, U.S.

ASHRAE, 2013b. ASHRAE 90.1: User's Manual, American Society of Heating Refrigerating and Air-Conditioning Engineers. SI ed. Atlanta, U.S.

Deru, M.; Field, K.; Studer, D.; Benne, K.; Griffith, B.; Torcellini, P. et al. U.S. Department of Energy Commercial Reference Building Models of the National Building Stock, report, February 1, 2011; Golden, Colorado

DOE, 2007. U.S. Department of Energy Reference buildings. https://energy.gov/eere/buildings/new-construction-commercial-reference-buildings.

DOE, 2018a. https://energyplus.net/.

DOE, 2018b. https://energyplus.net/weather/.

Dogan, T., Reinhart, C., & Michalatos, P. (2014). Automated multi-zone building energy model generation for schematic design and urban massing studies. In IBPSA eSim conference, Ottawa, Canada.

European Standard EN. Indoor environmental input parameters for design and assessment of energy performance of buildings addressing indoor air quality, thermal environment, lighting and acoustic. EN 15251 2008 ed.

Mahdavi, A., & Tahmasebi, F. (2015). Predicting people's presence in buildings: An empirically based model performance analysis. Energy and Buildings, 86, 349–355.

Matlab, 2018. https://mathworks.com/products/matlab.html/.

Nguyen, T.A., & Aiello, M. (2013). Energy intelligent buildings based on user activity: A survey. Energy and buildings, 56, 244–257.

Oldewurtel, F., Sturzenegger, D., & Morari, M. (2013). Importance of occupancy information for building climate control. Applied energy, 101, 521–532.

OpenStudio, 2018. https://www.openstudio.net/.

Page, J., Robinson, D., Morel, N., & Scartezzini, J.L. (2008). A generalised stochastic model for the simulation of occupant presence. Energy and buildings, 40(2), 83–98.

Peel, M.C., Finlayson, B.L., & McMahon, T.A. (2007). Updated world map of the Köppen-Geiger climate classification. Hydrology and earth system sciences discussions, 4(2), 439–473.

Smith, L., Bernhardt, K., & Jezyk, M. (2011). Automated energy model creation for conceptual design. In Proceedings of the 2011 Symposium on Simulation for Architecture and Urban Design (pp. 13–20). Society for Computer Simulation International.

Tahmasebi, F., & Mahdavi, A. (2016). An inquiry into the reliability of window operation models in building performance simulation. Building and Environment 105, 343–357.

Tahmasebi, F., Mahdavi, A. 2017. The sensitivity of building performance simulation results to the choice of occupants' presence models: a case study. Journal of Building Performance Simulation, 10:5–6, 625–635.

U.S. Department of Energy. EnergyPlus™ Version 8.8.0 Documentation. Input Output Reference. Build: 7c3bbe4830

UNI. 2014. UNI/TS 11300–1:2014. Prestazioni energetiche degli edifici, Parte 1: Determinazione del fabbisogno di energia termica dell'edificio per la climatizzazione estiva ed invernale. Milan, Italy: UNI.

Yan, D., O'Brien, W., Hong, T., Feng, X., Gunay, H.B., Tahmasebi, F., & Mahdavi, A. (2015). Occupant behavior modeling for building performance simulation: Current state and future challenges. Energy and Buildings, 107, 264–278.

Yang, Z., & Becerik-Gerber, B. (2014). The coupled effects of personalized occupancy profile based HVAC schedules and room reassignment on building energy use. Energy and Buildings, 78, 113–122.

# BIM-to-BEPS conversion tool for automatic generation of building energy models

M. Regidor, M. Andrés & S. Álvarez
*Energy Division, CARTIF Foundation, Valladolid, Spain*

ABSTRACT: Building Energy Performance Simulation (BEPS) software is a powerful tool for the whole building life cycle. However, providing the required input information is often challenging, which has given rise to an increasing research need focused on the automatic generation of BEPS input files from Building Information Modelling (BIM) data. This work presents a first proof of concept of a conversion tool targeting the TRNSYS simulation environment. The tool is capable of automatically extract information from a building model, specified according to the Industry Foundation Classes (IFC) open standard, and consistently creates the required BEPS input files. Data requirements for the IFC-to-TRNSYS interoperability are commented, and finally, the tool is demonstrated through a case study running a yearly thermal load calculation for a simple test building. This contributes to put the TRNSYS modelling potential at disposal of the AEC industry and opens up to very promising research challenges.

## 1 INTRODUCTION

During the last decades the need to reduce the energy demand and increase the efficiency in a wide range of processes of the human activity has become evident. Particularly, buildings account for more than 40% of the total energy consumption at EU level (European Union 2010), thus being a crucial sector where saving-energy actions must focus to help achieve the objectives of the Paris Agreement (United Nations 2015).

Building Energy Performance Simulation (BEPS) constitutes a powerful tool for design and operation phases of the building life cycle. It enables to estimate and compare the performance of different passive and active energy solutions, thus supporting decision-making towards the best final design. However, the configuration of such simulations often involves a time consuming process and require input information that, in most cases, should be entered manually.

To overcome this, a research line has been motivated in recent years based on the great potential of BIM (Building Information Modelling) to feed building energy models in an automatic or semi-automatic process. BIM provides a structured methodology to store, manage and use all relevant information within a construction project, enabling a collaborative and efficient workflow. Data interoperability between BIM and BEPS will facilitate the definition of building geometry as well as the provision of other required input data.

Considering the most extended BIM data model provided by the Industry Foundation Classes (IFC) open standard (buildingSMART 2018), several research works have been recently developed within this field (Bazjanac 2008, 2009, Lilis et al. 2014, Giannakis et al. 2015). Most of them focused on the development of consistent methodologies as well as of first proofs of concept oriented to EnergyPlus and Modelica BEPS applications (Bazjanac 2009, Kim et al. 2015, Remmen et al. 2015, Wetter & Van Treeck 2017).

The aforementioned background highlights the current need and relevance of research focused on the development of conversion tools that bring together the benefits of BIM and BEPS models with the main objective of facilitating the arduous task of data introduction and model creation into energy simulation software.

This work presents the development of a first proof of concept for a BIM-to-BEPS conversion tool, being compatible with TRNSYS 17 energy simulation environment (Klein et al. 2015).

The methodology followed to shape the whole tool is sustained by four main steps: (i) analysis of the IFC data model, (ii) analysis of TRNSYS input requirements, (iii) extraction of information from the BIM model and (iv) generation of BEPS input files.

## 2 ANALYSIS OF DATA REQUIERMENTS

Building energy performance simulation software requires input files containing the features of the construction at issue. Such information can be extracted from BIM models based on open BIM approach, which consists of a collaborative environment of building design and workflows involving open standards.

Due to the different nature and data structures of BEPS information requirements and BIM models, it is essential to map which building information is available and how it must be defined within the simulation input file.

## 2.1 Open BIM data model

The buildingSMART community (formerly, the International Alliance for Interoperability, IAI) adopted the openBIM initiative based on the development of the IFC standard and its corresponding data model. These are under continuous upgrade and improvement, being IFC4 (ISO 2013) the last available version of the standard.

The IFC4 release defines a conceptual data schema according to the EXPRESS data specification language. It comprises different layers of interconnected entities/classes, which are grouped and structured according to their functional characteristics among others.

Although IFC data can be generated and exchanged using other different formats (e.g. XML, ZIP), the '.ifc' textfile format using STEP structure (ISO 2016) is the default one, and it has also been used within the present work.

## 2.2 TRNSYS input files

TRNSYS is a graphically-based software environment for the simulation of transient systems. Particularly, thanks to its high flexibility to extend modelling capabilities and configure innumerable system combinations, it is nowadays one of the most widely-used applications within the energy and construction research community.

However, this versatility is, at the same time, linked to a higher level of complexity and time consumption when it comes to specify the input data for the simulation. This might be one of the reasons why it is not so extended in commercial applications within the Architectural, Engineering & Construction (AEC) industry.

Automation in the generation of TRNSYS input files directly from BIM data will help solve this problem and put a great modelling potential at disposal of the AEC industry at any step of the building life-cycle.

The information needed in order to run a TRNSYS building energy simulation can be classified in different parts according to its nature.

First, building-specific information implies the definition of the thermal zones along with the thermal properties of their bounding elements. It also includes how the thermal zones are interconnected with each other and which type of thermal conditioning is required. This is managed through a single TRNSYS module (Type56) representing a multi-zone building.

One second part includes detailed information on the energy systems, how the constituting elements are joined and the control strategy that is implemented to meet the energy demand imposed by predefined schedules; and finally, some external data are needed to specify boundary conditions, such as weather (solar radiation, temperature, humidity, etc.)

All this information is defined externally to Type56 and managed through a single configuration file together with the basic simulation controls and the processing of calculation results.

In summary, two different files are required to specify all the aforementioned information: data related to the building itself are included in the Type56 .b17 file, while all the rest concerning simulation controls, boundary conditions, externally-modelled elements and results processing is defined in the configuration .dck file.

## 2.3 Data interoperability between IFC and TRNSYS

This section addresses how the non-optional information required by TRNSYS is related with the standard IFC data schema. On account of the complexity of its structure, this work focuses on those requirements for building energy demand calculations, particularly on the .b17 file. Energy systems are kept out of the scope of this work and should be addressed in future developments (see Section 6).

### 2.3.1 Properties

This dataset refers to typical general physical properties. Default values are considered without requiring data extraction from the BIM model.

### 2.3.2 Layers

Each material contained within the BIM model must also exist within the .b17 file, specifying its thermos-physical properties: density, heat capacity and thermal conductivity. This information is contained within the class *IfcMaterial* which makes use of the auxiliary class *IfcPropertySetDefinition* to define each property value.

### 2.3.3 Walls

Structural elements such as walls (*IfcWall*), slabs (*IfcSlab*) or drop ceilings (*IfcCovering*) are included under this label. In essence, these are building elements that bound thermal zones, including doors. The definition must contain the number of material layers and thickness, additionally to some radiation and heat transfer parameters that are missing within the IFC file, although TRNSYS offers predefined values for them. The merger of these building elements and their constitutive materials is given by the class *IfcRelAssociates-Material* which relates each element with a single material or a set of them, making use of abstract supertypes to provide either a material definition (*IfcMaterialDefinition*) or a material usage definition (*IfcMaterialUsageDefinition*).

### 2.3.4 *Windows*

Windows are treated separately from the rest of building elements in TRNSYS. A dedicated library with a pre-loaded set of window types is used to specify window characteristics within Type56.

Then, although it would be possible to create new window types in addition to the predefined ones using BIM parameters, TRNSYS is highly specific on this matter, so the best way to proceed is by using the name of main reference of each window type as the conversion link. This will set a requirement for the BIM modeller, so that window materials (*IfcMaterialDefinition*) are named according to the corresponding reference identification (WINID) from the TRNSYS window libraries.

### 2.3.5 *Orientations*

TRNSYS requires specifying the orientation of each single wall within the model. The orientation calculation can be performed by matrix operations between axis systems, since each element in IFC is provided with placement information (*IfcObjectPlacement*) relative to another one. Thus, for instance, a wall is contained in a building storey, which owns to a building, which is placed in a geo-located site, whose location is referred to the World Coordinate System (WCS) origin. Through some algebraic base changes, it is possible to obtain the final orientation of each element.

### 2.3.6 *Zones*

TRNSYS zones refer to the thermal zone definition input required to perform any building energy simulation. These correspond to the *IfcSpace* entities according to the IFC standard.

Additionally, BEPS software considers that a thermal zone is spatially defined by its physical limits. Then, some space boundaries associated to a type of building element are needed. The IFC standard includes the class *IfcRelSpaceBoundary* for this purpose, giving information about the bounded zone (*IfcSpace*), the total surface area, etc. More specifically, this standard recognises second level space boundaries (*IfcRelSpaceBoundary2ndLevel*) providing more information about the connection between zones and its shared limits as well as its position within the building and physical type.

Second level space boundaries are crucial to enable conduction heat transfer calculations in any building energy simulation. Several efforts have been conducted to define them from BIM geometry data (Bazjanac 2010, Lilis 2017). Nowadays, main BIM authoring tools can export this information, although it is still something not evident that should be considered as one of the primary data requirements to be checked for practical BIM-to-BEPS conversion.

Every other physical feature can be extracted from de BIM model directly or by carrying out some mathematical operations. However, other specifications related to the level of detail and calculation hypotheses of the simulation (such as radiation or geometry modes) will not be found in the BIM model and must be established while creating the file following TRNSYS recommendations.

Thermal zones should also include internal gains such as those emitted by electrical equipment, lighting, people, etc. The definition of spaces within the IFC standard includes these parameters by making use of classes (*IfcRelDefinesByProperties*) which relates property sets (*IfcPropertySet*) with the spaces themselves. Some examples are properties such as *HeatGainPerOccupant*, *HeatGainPerComputer*, etc. The governing schedules can also be present by *IfcTimeSeries*.

### 2.3.7 *Heating and cooling*

Heating and cooling aspects for building thermal load calculations can consist of default typical schedules and temperature set points that do not require data extraction from BIM. In case of more detailed characteristics related to the definition of energy systems, they require going deeper in the IFC standard analysis, which has been kept out of the scope of this conversion tool.

## 3 CONVERSION TOOL

Based on the above-mentioned correspondence of building data, the initial steps of a BIM-to-BEPS conversion tool were developed aiming to produce a first proof of concept. It targets at reducing the existing gap between building modelling and energy simulation, and is composed by two related modules/capabilities: (i) building model visualization and (ii) BEPS input file generation.

Every required operation was deployed in a MATLAB-based algorithm integrated by several functions according to the IFC standard structure.

### 3.1 *Model visualization module*

This module was created as an auxiliary capability to support the progress of the file generation module. It allowed checking the consistency of the data treatment concerning geometry of the different building elements.

Several test BIM models supported the development process, from simple geometries, to progressively incorporate complexity at every step. Figure 1 shows the 3D view of an example model, represented both with an authoring BIM validation tool as well as with the target visualization module.

## 3.2 BEPS input file generation module

The conversion function of the tool itself is deployed through the file generation module. Figure 2 shows the general tool concept in which four different parts can be mentioned separately.

### 3.2.1 IFC data extraction

An algorithm capable of extracting the required information from the IFC file was developed, addressing the programming through MATLAB functionalities based on string processing (MATLAB 2018) from the IFC4-EXPRESS source file.

The followed methodology handles each different building element type by shared auxiliary functions. In this way, materials, location, shape representation, etc. are common issues that can be obtained similarly for every element.

The nature of the auxiliary functions is basically related to two different aspects: (i) geometry and (ii) properties. The first group gathers most of the relevant information for the BEPS model consisting of dimensions and relative positioning of different building elements as well as space representation and boundaries. The second one mainly focuses on material-related information (thermal properties and layer configurations), which are indispensable energy calculations.

### 3.2.2 Intermediate database

Since the required TRNSYS input information is widely dispersed within the IFC model, the extraction algorithm integrates information storage on an intermediate database relying on MATLAB structure data type. It takes the TRNSYS data distribution as a reference with the purpose of facilitating the subsequent file generation.

In addition, it collects all general material properties in a separate field to which detailed material information (layers, wall configurations, etc.) properly refer to ease next steps of the conversion routine.

Figure 3 shows a conceptual scheme of the database structure.

### 3.2.3 Creation of TRNBuild input file (.b17)

The required .b17 input file should contain all building-specific information. The conversion tool embeds a dedicated routine to automatically generate this file from the information on the intermediate database. The file structure (almost identical to the topics addressed in Section 2.3) is strictly followed step by step according to the data format requirements previously mentioned.

The initial sections devoted to the general material definitions and the enumeration of building element types are created from global lists that excludes space-related information. Spaces (and their related building elements and space boundaries) are incorporated to generate the last part of the file, focused on zones/airnodes definition.

It should be remembered that some default input data were considered in this stage without substantial decrease of results quality. This is applied either because such data are (reasonably) not contained in the BIM model (as general physical properties), or because they concerns typical simulation hypotheses. In this sense, processing of working schedules and versatility to provide simulation outputs from the building module (Type56) were not addressed in detail in the current version of the tool.

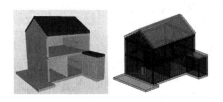

Figure 1. Model visualization example: reference model view (left) and resulting view from the conversion tool (right).

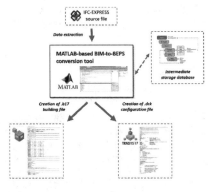

Figure 2. BIM to BEPS conversion tool concept.

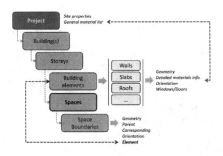

Figure 3. Conceptual map of the intermediate database.

### 3.2.4 Creation of the simulation configuration file (.dck)

A TRNSYS building simulation requires connecting the building module that manages the .b17 file with additional Types that complete indispensable input information and/or allow providing modelling details of additional components (e.g. energy systems). This versatile approach is coordinated by the configuration file (.dck) and comprises the most challenging part in the development of the target BIM-to-BEPS conversion tool, since it should aim a general automatic process to create the required Types and proper connections for a given BIM model.

Such generalization is here limited to the purpose of building thermal load calculation, excluding modelling of energy systems or any other additional component. Then, the conversion tool must be able to create a .dck file with: (i) an initial section devoted to the specification of the main simulation controls and (ii) the information related to three different TRNSYS types:

- Type15 (Weather data processor). Transforms the weather data provided by an external file to make it appropriate for other components. This case study makes use of the *Meteonorm* database focusing on Madrid.
- Type56 (Multizone building). Addresses the thermal behaviour of a building with either one or more thermal zones. It is completely defined in the .b17 file.
- Results generator (e.g. Type65. Online graphical plotter). This kind of elements displays the selected variables to evaluate the system behaviour.

Finally, after the generation of the two required input files, the tool integrated a specific routine capable of calling the TRNSYS simulation engine run the corresponding calculation.

## 4 TOOL DEMONSTRATION CASE STUDY

The developed BIM-to-BEPS conversion tool compatible with TRNSYS simulation software was demonstrated through a simple case study. It successfully completed the automatic creation of a building energy model (geometry and constructive characteristics) as well as the running of a one-year simulation to automatically determine the heating and cooling demands of the targeted test building.

This test building consisted of a simple individual house with rectangular ground plan and gable roof. Two different storeys were considered, defining one space (thermal zone) in each of them. Figure 3 shows its 3D representation as a result of the developed auxiliary visualization tool.

The test building was simulated under continental climate conditions through selection of a weather file corresponding to Madrid (Spain).

Finally, default typical operation schedules and heating and cooling setpoints (21°C and 25°C respectively) were considered within the load calculations.

## 5 RESULTS AND DISCUSSION

The conversion tool proved to perform successfully to run the test case study. Figures 3 and 4 presents the results (graphical MATLAB output) of the model visualization module including the representation of the correctly identified space boundaries.

Moreover, Figure 5 shows the TRNSYS results for the yearly evolution of indoor temperatures and heating and cooling loads corresponding to the 2 defined building spaces.

The obtained results are in accordance with typical reference values based on the same climate and building typology, demonstrating the tool consistency and presenting coherent values throughout the whole year.

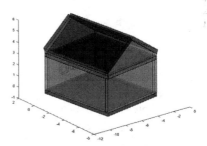

Figure 4. 3D representation of the test building model as result of the model visualization module of the tool.

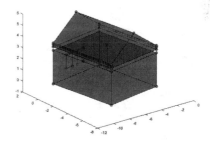

Figure 5. 3D representation of the extracted 2nd level space boundaries for the test building.

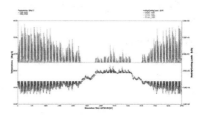

Figure 6. TRNSYS output for the yearly-based building thermal load calculation.

## 6 FUTURE WORK

Automatic generation of BEPS models from BIM data is a current research topic among the international scientific community. This work presented initial steps within a line of work focused on the interoperability with TRNSYS simulation capabilities. However, there is still a great potential ahead derived from this BIM-BEPS relation, which should be addressed in future steps:

- Development of a robust, standardized common data model to ease connection between BIM and BEPS data structures
- Processing of more complex building geometries
- Post-processing of simulation results and calculation of additional indicators
- Integration of energy systems within the conversion workflow
- Development of capabilities in a reverse direction; using BEPS outputs for IFC model enrichment.

## 7 CONCLUSIONS

This work introduces an under development BIM to BEPS tool addressing the interoperability between the input data requirements of building energy simulation software and the information provided by the IFC data schema. To this effect, the analysis on the requisites of TRNSYS input files has been performed and associated them with the information contained within the IFC building model, bringing to light the high conversion potential between them. Some pre-requisites for the proper tool functioning are highlighted, showing possible inconsistencies between both data models. An overview on the conversion tool capabilities is given, and summarized on the following points:

- Building-related information extraction algorithm making use of specifically created IFC-related functions.
- Data structure storage following TRNSYS input data structure.
- Files generation based on the previous data distribution.
- Visualization module oriented to check the correct model information import.

Finally, a simple case study has been used to test the tool, successfully meeting its expectations.

## REFERENCES

Bazjanac, V. 2008. IFC BIM-based methodology for semi-automated building energy performance simulation. In L. Rischmoller (ed.), *CIB W78, Proc. 25th conf., Improving the management of construction projects through IT adoption*, Santiago, Chile: 292–299.

Bazjanac, V. 2009. Implementation of semi-automated energy performance simulation: building geometry. In *CIB W78, Proc. 26th conf. Managing IT in construction*. Instanbul, Turkey: 595–602.

Bazjanac, V. 2010. Space boundary requirements for modelling of building geometry for energy and other performance simulation. *In CIB W78, Proc. 27th conf., Applications of IT in the AEC industry, Cairo, Egypt. 16–18 November 2010.*

buildingSMART. 2018. [online] buildingSMART® International home of openBIM. https://www.buildingsmart.org.

European Union. 2010. Directive 2010/31/EU of the European Parliament and of the Council of 19 May 2010 on the energy performance of buildings (recast). *Official Journal of the European Union.*

Giannakis, G.I.; Lilis, G.N., Garcia, M.A., Kontes, G.D., Valmaseda, C. & Rovas, D.V.; (2015) A methodology to automatically generate geometry inputs for energy performance simulation from IFC BIM models. In *Proceedings of the 14th IBPSA Conference*. Hyderabad, India: 504–511.

International Organization for Standarization (ISO). 2013. ISO 16379:2013 Industry Foundation Classes (IFC) for data sharing in the construction and facility management industries.

International Organization for Standarization (ISO). 2016. ISO 10303-21:2016 Industrial automation systems and integration—Product data representation and exchange—Part 21: Implementation methods: Clear text encoding of the exchange structure.

Kim, J.B., Jeong, W., Clayton, M.J., Haberl, J.S., & Yan, W. 2015. Developing a physical BIM library for building thermal energy simulation. *Automation in construction*, 50: 16–28.

Klein, S.A. et al, 2015, TRNSYS 17: A Transient System Simulation Program, Solar Energy Laboratory, University of Wisconsin, Madison, USA, http://sel.me.wisc.edu/trnsys.

Lilis, G.N., Giannakis, G.I., Kontes G.D & Rovas, D.V. 2014. Semi-automatic thermal simulation model generation from IFC data. In *Proc. of 10th ECPPM*. Vienna, Austria: 503–510.

Lilis, G.N., Giannakis, G.I., & Rovas, D.V. 2017. Automatic generation of second-level space boundary topology from IFC geometry inputs. *Automation in Construction* 76: 108–124.

MATLAB. 2018. [online] Matlab documentation. https://es.mathworks.com/help/matlab/.

Remmen, P., Cao, J., Ebertshäuser, S., Frisch, J., Lauster, M., Maile, T. & Thorade, M. 2015. An open framework for integrated BIM-based building performance simulation using Modelica. In *Proceedings of the 14th IBPSA Conference.* Hyderabad, India: 379–386.

United Nations. 2015. Paris Agreement. United Nations Treaty Collect. 1–27.

Wetter, M. & Van Treeck, C. (ed.) 2017. New generation computational tools for building and community energy systems. International Energy Agency (IEA) EBC Annex 60 Final report. ISBN: 978-0-692-89748-5.

# Energy-saving potential of large space public buildings based on BIM: A case study of the building in high-speed railway station

N. Wang, J.L. Wang & C.H. Liu
*School of Architecture, Tianjin University, Tianjin, China*

L. Liu
*School of Architecture, Tianjin Chengjian University, Tianjin, China*

ABSTRACT: For buildings with large space like high-speed railway stations in China, the energy consumption is much higher than ordinary public buildings. This research tries to explore energy-saving potential of large space public buildings based on Building Information Modelling (BIM) by a case study of the Tianjin West Railway Station. Compared to the original design, the optimized design adopts energy-saving strategies and has a good effect on the energy saving and emissions reduction according to the results. In the case study, BIM provides a technical platform for visualization, high quality, coordination and building performance simulation. As the main BIM tool, Revit and DesignBuilder are used to realize the whole process of 3D modelling and energy simulation. The research has great reference value to the energy-saving design of high-speed railway station and other large space public building in cold regions supported by BIM technology.

## 1 INTRODUCTION

Energy-saving and emission-reduction efforts have become needed accompany with the severe energy crisis and climate change, this situation is very urgent in the developing country especially China. In 2016, buildings consumed about 25% of the total energy consumption in the country's production activities, of which public buildings become the largest energy use item (BECRCTU 2018). In fact, the significant growth in area and energy use intensity of the large space public building are important reasons for energy consumption increase of public buildings in China. With the rapid development of China's high-speed rail transit, large numbers of high-speed railway stations will be constructed and inevitably become the main type for China's railway stations in the future. For the typical public building with large space like in high-speed railway station, the energy use intensity reaches 160–180 $kWh/m^2 \cdot yr$, much higher than that of ordinary public building and house (Hu, R.M. 2006). China's 13th plan Five-Year Plan of national development will vigorously promote energy efficiency of the public building. The energy-saving design of high-speed railway station plays an important role in improving building energy efficiency and reducing emissions of atmospheric primary greenhouse gas $CO_2$ in the construction industry and should be a concern.

Furthermore, Building Information Modelling (BIM) is gradually matured and promoted as a new technology, method and tool in large and complex buildings. It is the creation and use of building information models to support building design and even the entire life cycle. For the high-speed railway station building, BIM provides a visualized platform to incorporate building information integration and sustainability information in the design of buildings, which is beneficial to effectively simplifying the design process and improving building performance (Lu, Y.J. et al. 2017). Through the research on the energy-saving potential of high-speed railway station building, it can provide suggestions for energy-saving design in large space public building, and a method for modelling and building performance simulation based on the BIM.

## 2 RESEARCH METHODS

This paper analyzes energy-saving potential of large space public buildings through the five steps presented below. This section provides a brief overview of the research method for overall understanding. The detailed descriptions of each step are illustrated in the next section. The Tianjin West Railway Station, a typical large space high-speed railway station in China, is selected as a case study building for BIM-based optimization design, energy simulation and carbon emission calculation. The Tianjin West Railway Station that has been built in real life is called original design in this paper, and the new design using energy-saving strategies is called optimized design which makes a contrast with the former.

- Step 1. Original design and energy-saving potential: The situation of Tianjin West Railway Station is introduced and the problems for energy consumption are summarized through the investigation and survey on the data and present condition. The potential for energy saving involves construction scale, space design, function layout, building details and operation mode, which provided a basis for energy-efficient optimization design.
- Step 2. Model development: The original design is modelled using Autodesk Revit, one of the leading BIM tools, and optimized design is also modelled for comparison.
- Step 3. Optimized design and energy-saving strategies: Based on the energy-saving potential analysis in the previous step, relevant energy-saving strategies in five parts are developed to optimized design.
- Step 4. Energy simulation in DesignBuilder: The two building information models are converted to simplified models for energy simulation analysis in DesignBuilder, one of the leading energy simulation software and BIM sustainable analysis software.
- Step 5. Materials data for carbon emission calculation: Materials data from BIM models are used to calculate the life-cycle carbon emissions for comparing the two designs.

## 3 A CASE STUDY APPLICATION: TIANJIN WEST RAILWAY STATION

### 3.1 Original design and energy-saving potential

Tianjin West Railway Station is a famous high-speed railway station located in the Hongqiao District of Tianjin, the cold areas of northern China (Fig. 1). It has modernized new station building, supporting facilities and squares designed by the German gmp-architects and constructed in 2011, with a total construction area of approximately 179,000 $m^2$, of which the station building area is 104,000 $m^2$. The main part is a 2-storey station building with 1-storey basement for passengers' departure. The ground floor consists of entrance hall, and ticket lobby, next to the station platforms. The first floor, which is also called elevated over-crossing waiting floor, mainly comprises waiting room, small shops and other subsidiary rooms.

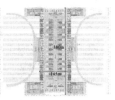

Figure 1. Tianjin West Railway Station and elevated waiting floor diagram.

Figure 2. Interior space of Tianjin West Railway Station.

The waiting room is huge with a clear height of 47 m and an east-west span of 145 m, stretching 380 m from north to south (Fig. 1). Its central is a 39-meter-wide pedestrian corridor, and passengers arrive at the lower platforms through elevators or escalators. The giant arched roof made of white steel framed glass curtain walls is the most prominent external morphological feature (FMeinhard, V.G. et al. 2013).

The authors conducted the survey of station building in Tianjin West Railway Station to investigate what impact its energy use (Fig. 2). Through field survey and analysis, the station building faces the following problems influencing energy consumption: (1) Occupant density of waiting hall is low, reflecting the overlarge construction scale and unreasonable area; (2) Lower vertical space utilization rate and large space excessively leads to high energy consumption; (3) Small shops and other subsidiary rooms on both sides of waiting room lead to poor natural lighting and ventilation conditions in central waiting space because of large east-west span. It needs to be more regulated by lighting and air-conditioning systems; (4) The roof of large-scale glass curtain walls enables too much sunlight to reach the waiting room, which increases the cooling loads in summer; (5) Passenger flows during the hours from 9 pm to 6 am are less, but lighting and air-conditioning systems operate in a normal way. The whole period operation mode results in energy waste. Therefore, the station building has great energy-saving potential in terms of construction scale, space design, function layout, building details and operation mode.

If these design factors can be optimized, building energy consumption will be greatly reduced to achieve energy saving and emission reduction goals.

### 3.2 Model development

Autodesk Revit is selected to be used in this study. Revit is a software of Autodesk Corporation and the most widely used BIM authoring tools currently being applied in the construction industry of China (Song, Q.Y. et al. 2016). According to construction drawings and survey information of original design, the authors extracted major data of design, structure and materials, then created a three-dimensional information model for original design using Revit (Fig. 3). For meeting the following simulation or calculation, the model should include relevant information such as geometric data, physical characteristics, and construction requirements of the building envelope, components, materials, etc. The optimized design was also modelled using Revit in the same way (see Figs. 4–5).

To conduct the energy simulation, the two building models were exported from Revit using gbXML. During the process shown in Figure 6,

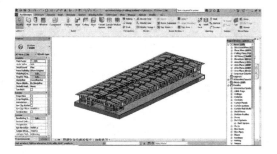

Figure 5. Model of optimized design (first phase and second phase, 2011–2029–2110) in Revit.

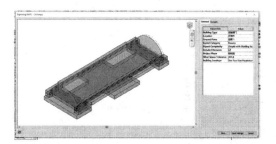

Figure 6. Conversion process to gbXML.

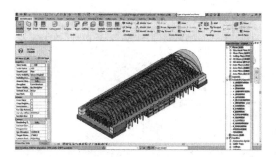

Figure 3. Model of original design in Revit.

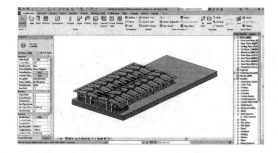

Figure 4. Model of optimized design (first phase, 2011–2029) in Revit.

the building information models are fully converted and can be saved as a gbXML file for energy simulation. Complex Revit model can make simulation time-consuming. In this study, all the models were simplified under the precondition of preserving the primary building space and envelope.

### 3.3 Optimized design and energy-saving strategies

To solve the current problems and tap the energy-saving potential of original design, the authors carried out a research-based design of the Tianjin West Rail
 Station. Following the original site planning and building overall layout, the new optimized design adopting a series of energy-saving design strategies to reduce the energy consumption and carbon emissions.

#### 3.3.1 Construction scale

The actual demand for waiting space is far less than the current construction scale because of low occupant density and short waiting times of passengers in Tianjin West Railway Station. According to the planning of the railway network in the Beijing-Tianjin-Hebei region, the optimized design will be constructed in two phases with the long-term planning of 2030 as a time node (see Fig. 7). The construction of the optimized design in the first phase

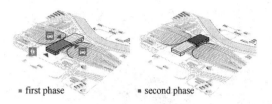

Figure 7. Construction scale of optimized design (first phase and second phase, 2011–2029–2110).

Figure 8. Comparison of space size between the two designs.

will meet the demand for usage from 2011 to 2029, and the area of main functions including waiting room, entrance hall, and subsidiary rooms will be 11,100 m$^2$, which is much smaller than that of the original design. The construction time of the second phase is in 2030, and the expansion area of the main functions will reach 41710 m$^2$ (MOHURD 2012).

### 3.3.2 Space design

From the perspective of energy-saving, station building has great potential in space design. The orientation and function composition of the optimized design are the same as those of the original design. However, the optimized design considers the relationship between building structure and height, and makes multiple structural units support the overall structure of the station building. On the space size, the east-west span and height are reduced to 111 m and 22 m respectively on condition that meeting user requirement and interior comfort (see Fig. 8). The roof of the structural unit is designed to have a top lighting surface, and secondary lighting bands around. The bottom of the structural unit is a ticket check and the upside is not functionally used but expanded to the periphery to reduce the volume of space required for temperature adjustment.

### 3.3.3 Function layout

Two structural units are combined to form a waiting unit. There are five waiting units in the waiting room along the north-south direction (Fig. 9 left). According to different functions for natural lighting and ventilation requirements, the optimized design adjusts the function layout of the main and subsidiary rooms. Waiting room is located on both sides of the central corridor and VIP waiting rooms are on the east. Subsidiary rooms such as shops and toilets are on the west side as buffer space to weaken impact on the waiting room (Fig. 9 right). Because the central corridor is large in span and depth, the glass atrium with green landscape in the central of each two waiting units are designed to improve its lighting and ventilation conditions, regulating local microclimate and collecting rainwater from the roof.

### 3.3.4 Building details

In consideration of cold climate features, some changes have been made in the building details. The lighting surface on roof of the structural unit is divided into south and north parts. Solar panels are installed in the south to achieve shading and solar energy collection, and skylights in the north allow natural lighting and ventilation in the building. For different facades, the south facade makes use of horizontal self-shading to control too much light during summer, and vertical louver shadings are set in the east and west facade because of the low solar angle. These ways can avoid overmuch solar radiation, thereby making the cooling loads in summer on the decrease.

### 3.3.5 Operation mode

For high-speed railway station, flow of people changes constantly; there are more passengers during the day but fewer passengers in the evening. The need for waiting space is lower when people are less. The five waiting units of the optimized design will be controlled by district (Fig. 9 left). The thermal insulation boundary is set between the waiting units and can be turned on or off according to passenger flow and train numbers during the building operation. It's a good operation mode to flexibly reduce the energy consumption of artificial systems.

## 3.4 Energy simulation in design builder

Once the building has been modelled, the next step is to conduct energy simulation demonstrating the fact of energy-saving potential and the advance-

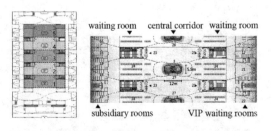

Figure 9. Five waiting units (left) and function layout diagram of optimized design (right).

ment of optimized design. DesignBuilder is chosen for specialist energy simulation and can be used to read a gbXML file. Two Revit models were imported into DesignBuilder software as shown in Figures 10–11, and the final simulation results were acquired after model adjustments and parameter settings (see Table 1). The curve roof of optimized design is built in DesignBuilder for simplification.

Parameter settings such as occupant activity, lighting and HVAC systems are input according to the data from field surveys and construction drawings. The information of envelope construction and openings are from Revit model. The parameters of optimized design and original design are consistent to verify the difference in energy consumption at the building design level. Occupant density is different because of people flow and building area. The HVAC is set as a fan coil unit system and supplied by the city electricity grid. Outside air definition for mechanical ventilation is from the minimum fresh air by per person. The Cop of heating and cooling are 2.202 and 2.5 respectively, but there is no heating and cooling in the basement, staircase, and ticket checking space.

Table 1. Main design parameters of case study in DesignBuilder.

| Case study | Original design | Optimized design |
| --- | --- | --- |
| Density(people/$m^2$) (2011–2029) | 0.1 | 0.4 |
| Density(people/$m^2$) (2030–2110) | 0.2 | 0.34 |
| Interior environment (°C) | 20/26 (winter/summer) | |
| Minimum fresh air (l/s·person) | 2.778 | |
| External wall (W/$m^2$·K) | $U = 0.38$ 70 mm EPS-insulation | |
| Roof (W/$m^2$·K) | $U = 0.48$ | |
| External window, Skylight (W/$m^2$·K) | double low-e glazing (6+18 mm Air +6), $U = 1.62$ | |
| SHGC/LT | 0.29/0.41 | |
| Airtightness (ac/h) | 0.3 | |
| Lighting illuminance (lux) | 200 | |
| Lighting power density (w/$m^2$) | 8 | |

### 3.5 Materials data for carbon emission calculation

This research also focuses the primary greenhouse gas $CO_2$ produced in building life cycle of materials

Figure 10. Simplified model of original design in DesignBuilder.

Figure 11. Simplified model of optimized design (first phase and second phase, 2011–2029–2110) in DesignBuilder.

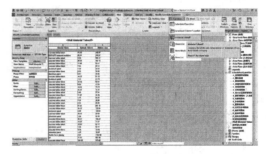

Figure 12. List of material takeoff (take wall for example).

production and transportation, construction, operation, and demolition. Calculation tool of carbon emission is "LCA-Based Carbon Footprint Modelling for Buildings" developed by the research group.

It refers to internationally recognized equations in the 2006 IPCC Guidelines for National Greenhouse Gas Inventories: Greenhouse Gas Emissions = AD × EF, which represent human activity data and emission factor that quantify emissions or removals from unit activity respectively. BIM model carries the information of building materials used. After the models were built, list of materials for the original and optimized design was created including the material name, volume and area. The number of materials used were exported for calculation by the processing based on the created list (Fig. 12).

## 4 RESULTS, ANALYSIS AND DISCUSSIONS

### 4.1 Energy consumption

The interior temperature fluctuations of the waiting room for original design and optimized design were simulated and analyzed, including air temperature, radiation temperature, and sensible temperature. The results show that there is severely overheating in summer and overcooling in winter in the waiting room of the original design. However, it is significantly relieved in the optimized design of this study (Fig. 13).

The energy consumption simulation comprises the three main items of heating, cooling and lighting, and the equivalent electricity method was used to convert between different types of energy. For the total energy use intensity of building operation, the original design is about 160 kWh/m²·a and the first phase of the optimized design is about 100 kWh/m²·a separately; energy-saving ratio reaches 37.5%. On basis of the first phase of the optimized design, the second phase mainly integrates the space of the north entrance hall and elevated waiting floor. The total energy use intensity of two phases of optimized design is approximately 116 kWh/m²·a, which was much less than the energy use intensity of the original design (Fig. 14). For the total energy consumption during the operation period of 100 years, the optimized

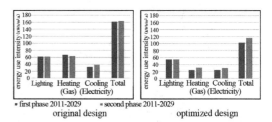

Figure 14. Comparison of the energy use intensity.

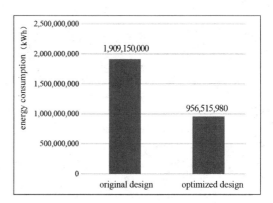

Figure 15. Comparison of the total energy consumption during the operation period of 100 years.

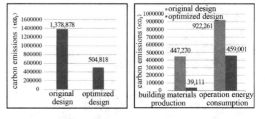

Figure 16. Comparison of the life-cycle carbon emissions (left), and comparison of the carbon emissions between building materials production and operation energy consumption (right).

design is reduced by approaching 50% compared with the original design, and the energy savings amounts to 952,634,020 kWh (Fig. 15).

### 4.2 Life-cycle carbon emissions

For the life-cycle carbon emissions, the optimized design reduces by up to 63.4% compared with the original design and decreases the total $CO_2$ emissions by 874,059 tons (Fig. 16). The results show that the optimized design has the most obvious emission reduction effect in terms of carbon emis-

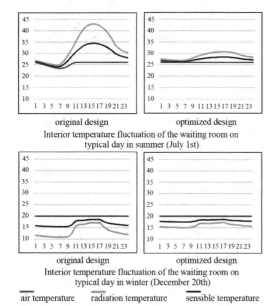

Figure 13. Interior temperature fluctuation of the waiting room in summer and winter.

sions produced by building materials production and operation energy consumption through adopting energy-saving strategies and improving the number of building materials used.

## 5 CONCLUSIONS

Many design factors influence the energy consumption of a large space public building. To some extent, building design optimization can significantly reduce the need for artificial systems making its internal spaces naturally heated and cooled. This research is based on a case study building located in the Tianjin, China. The case study suggest that energy-saving design strategies are adopted for a high-speed railway station. According to the results about energy consumption simulation and carbon emission calculation, we can conclude that the optimized design reduces 50% on the total energy consumption and 63.4% on the life-cycle carbon emissions maximally compared with the original design confirming that above-mentioned energy-saving design strategies are effective. And it also shows that the large space public building has the energy-saving potential on construction scale, space design, function layout, building details and operation mode. At the same time, Building Information Modelling (BIM) is accepted by the research group as an innovative method for integrating many different data related to a building. In this study Revit and DesignBuilder are connected via gbXML to realize the whole process of 3D modelling and energy analysis. It is apparent that the visualization capacity of BIM technology is mainly utilized to improve the quality and speed of modelling, especially for large space public buildings. Furthermore, the BIM model is exported from Revit and computed by DesignBuilder that reuses a subpart of the BIM data as input. The potential for energy saving of large space public building is explored. This study also provides a design method and reference for high-speed railway station and other large space public building in cold regions based on BIM from the perspective of energy saving and emissions reduction.

## ACKNOWLEDGEMENTS

This work was supported by the National Key Research and Development Program of China (2016YFC0700200); The National Natural Science Foundation of China (No.51338006); The Innovation Base Introducing International Talents (Referred as 111 project) (B13011).

## REFERENCES

Abanda, F.H. & Byers. L. 2016. An investigation of the impact of building orientation on energy consumption in a domestic building using emerging BIM (Building Information Modelling). *Energy* 97:517–527.

Azhar, S.2011. Building Information Modelling (BIM): trends, benefits, risks, and challenges for the AEC industry. *Leadership & Management in Engineering* 11 (3):241–252.

Building energy conservation research center, tsinghua university (BECRCTU). 2018. *Research report on China's annual development of building energy conservation*: 4–5. Beijing: China Architecture & Building Press.

Hu, R.M. 2006. *A study on the design of large-space architecture:*162–175. Shanghai: Tongji university.

Lu, Y.J., Wu, Z.L. & Chang R.D. et al. 2017. Building Information Modelling (BIM) for green buildings: A critical review and future directions. *Automation in Construction* 83: 134–148.

Meinhard,V.G., Stefan, S., Stefan R. & Christian G. 2013. The Tianjin West Railway Station. A large scaled, sequential, free space. *Urban Environment Design* 7:74–81.

Ministry of Housing and Urban-Rural Development of the People's Republic of China (MOHURD). 2012. S*tandard for Railway Station Building Design (GB50226-2007)*. Beijing: China Plan Press.

Oduyemi, O. & Okoroh, M. 2016. Building performance modelling for sustainable building design. *International Journal of Sustainable Built Environment* 5 (2):461–469.

Song, Q.Y. & Lu, X.B. 2016. Study on three dimensional energy saving simulation of public buildings based on BIM technology. *Journal of Green Science and Technology* (10): 191–194.

# Multi-objective optimization design approach for green building retrofit based on BIM and BPS in the cold region of China

Hongwei Yang & Conghong Liu
*School of Architecture, Tianjin University, China*

ABSTRACT: This article presents a BIM—and BPS—based optimization design approach employed in the decision-making stage of integrating form, fabric, system and appliance strategies to improve daylighting environment, reduce house-hold energy consumption as well as $CO_2$ emissions. A representative residence built in 1980 is selected as a case study to analysis the building performance improvement by retrofit. Revit is used to build a BIM model, which is imported into both Ecotect—Daysim and Design-Builder to make quantitative analysis of daylighting and energy consumption before and after retrofit. This results will provide the corresponding priorities for measures towards different single-objective. Then, the NSGA-II genetic algorithm is used to complete the multiple-objective optimization compromising $CO_2$ emissions and discomfort hours. This research developed a multi-objective sensitivity tool based on JAVA programming which is applied on the residence case by decision making support for the early design stage.

## 1 INTRODUCTION

According to statistics in 2015, the energy consumption of China's construction operations was 864 million tce (THUBERC, 2017), accounting for 20% of the country's total energy consumption. The "congenitally deficient" statuses have already hidden in existing buildings built in different eras, such as outdated functions, discomfort, retiring equipment and high energy intensity. With worldwide attention, Existing buildings gradually become a key issue for building performance research.

China's total existing building area was about 67.048 billion $m^2$, including 48.879 billion $m^2$ of residential buildings. 71% of these residences were built before 2005 (Yang 2016). With such a huge building stock, relying solely on greening new constructions has far away from achieving the energy efficiency and carbon reduction targets. It's confirmed that deep retrofit would reduce building energy consumption by 50% to 90%. Therefore, it is imperative for green retrofit.

This paper is trying to explore the design approach of green retrofit for residential buildings in the early stage based on BIM and BPS. The following issues will be illustrated: how to clarify the performance optimization level of retrofit design and how to achieve the multi-objective optimization for both emission reduction and comfort enhance?

A literature review revealed that the majority of previous retrofit studies had focused on two categories of research. In one type, the alternative retrofit measures are predefined according to the specific building type, dimensions, function and climatic conditions, which are then verified by simulation. In the other, a series of multi-criteria-based decision-making methods were developed to reach a final decision among implicitly defined solutions. (Yang 2017; Dascalaki et al. 2002; Diakaki et al. 2008). Nowadays, BIM and BPS were widely used in computer aided design for establishing information model and quantifying performance indicators.

## 2 BIM—& BPS—BASED DESIGN APPROACH

At an early design stage of building retrofit, the key problem is to quantize a range of options and identify the relatively effective measures quickly. Different from the new construction, the existing buildings already have a large amount of detailed building information, such as an existed architectural form and material parameters. BIM and BPS can serve as an effective way to predict the performance optimization related to energy reduction, carbon emissions reduction. Figure 1 showed the framework for this research, which is composed of four parts "simulation–based diagnosis", "retrofit design proposal", "performance prediction and verification" and "multi-objective optimization".

### 2.1 *Simulation-based diagnosis*

At the assessment stage, buildings were exhaustively surveyed and was modeled by Autodesk

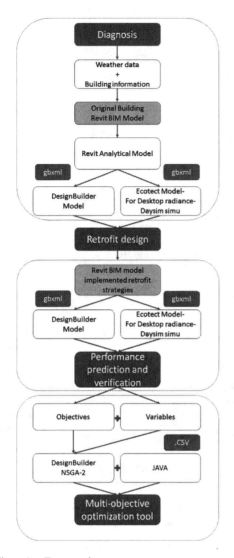

Figure 1. Framework.

Revit with the building information of weather data, dimensions of the building and the fabric including materials, position and area. The first step is a diagnosis of built environment with daylighting simulations to identify the potential occupant discomfort indoors and outdoors. Considering the dynamic lighting environment simulation, the BPS tool Daysim was chosen to calculate Daylighting autonomy *(DA)*, Useful Daylight Illuminance *(UDI)*, as well as Maximum daylighting autonomy *($DA_{max}$)* which can serve as indicators for both lighting uniformity and glare, instead of these static daylighting indicators simulated in Insight360. In order to avoid repeated modeling, Ecotect was used to be the model converter which can both input BIM models of gbXML format to Daysim and present visualization simulation results.

Next, for thermal performance, the 3-D BIM model with location, geometry, built environment and central system operation schedule was built based on the actual situation. The transformation from Revit to DesignBuilder was realized by means of creating an Analytical Model by adding Rooms to the Revit model. Green Building XML (gbXML) data is generated from the Analytical Model. Then, gbXML data is loaded to DesignBuilder for performance analysis. Modelling the built environment could predict the related energy consumption, $CO_2$ emissions, and the performance of the physical environment and support an integrated analytical design process.

2.2 *Retrofit design proposal*

In response to the defects identified by the diagnosis, form redesign, fabric promotion, system optimization and appliance replacement were proposed to improve the performance (Table 1). The form strategy refers to altering the envelope and space by redesign, improving indoor physical environment and architectural appearance. For example, closing open stairwells can reduce heat loss, and a wind-breaking frame can increase outdoor and indoor ventilation. Fabric strategy is an effective passive way to reduce energy demand

Table 1. Retrofit strategies and measures.

| Strategy | Details | Measures |
|---|---|---|
| Form | Improve occupant comfort by architectural redesign | Enclosed staircase, enclosed balcony, atrium, shading, pitched roof, green roof, etc. |
| Fabric | Reduce energy demand and improve the built environment by updating construction | Insulation (external wall, roof, floor, hall), replacing windows |
| System | Switch to an efficient heating and cooling schedule and to supply energy from renewable sources | Sub-metering, room temperature control Solar thermal Solar PV |
| Appliance | Reduce energy consumption of household appliances | Energy-saving lamps |

by optimize building construction, such as adding external wall insulation or roof insulation. System strategy contains renewable energy use, HVAC coefficient of performance upgrade, operation schedule or metering mode alteration. Replacing household white goods is the mainly appliance strategy. Retrofit strategies should be tailored to match the requirements of the retrofit property. In the early design stage, a retrofitted building design with strategies above was promoted.

## 2.3 Performance prediction and feedback

At the decision-making stage, each hypothetical elemental design parameter was simulated in Designbuilder to show the benefits achieved through contrastive analysis. The building performances before and after retrofit were predicted and compared to guide the design optimization. The performance improvements can be identified, such as operating energy savings, $CO_2$ emissions reduction. Different retrofit measures may show contradictions when it comes to different targets. For example, the Sunshade reflecting devices which aimed to optimize daylighting environment, may cause more energy consumption in winter. Thus, the quantitative performance prediction based on BPS is necessary for decision making in the early design stage.

## 2.4 Multi-objective optimization

During the retrofit practice, not all buildings will utilize only one single retrofit measure or have the economic ability to afford the holistic strategies. The research team believes that it is more necessary to clarify the sensitivity of different measures or measure packages, and the interaction of different goals in practice. It can serve as a strategy filter for architects to help make decision. Simultaneously, Conflicts and contradictions were produced when implementing strategies for different targets.

This phase aims to achieve multi-objective optimization to balance carbon emission and indoor comfort. In order to facilitate the use of architects, this research developed a multi-objective performance optimization sensitivity tools for typical case study based on JAVA programming.

## 3 CASE STUDY

A case study is used to illustrate how the building information model and building performance simulation help to make the decision in retrofit and what effects will be achieved by different strategies.

The typical Sijicun residential building was built in 1980s, which is a flat-top five-story apartment containing three units with three families in each unit (Fig. 2). There were two main functional rooms and auxiliary spaces such as kitchens and bedrooms, which leads to poor natural ventilation in each apartment. Non-insulated enclosures and open stairwells and balconies increased heat loss throughout the building.

### 3.1 BIM—based performance modeling and parameter setting

#### 3.1.1 Weather data

Nearly every simulation tool requires the input of a local meteorological file, which can be downloaded from the EnergyPlus official site in an EPW format. The original files used in this research were produced by Tsinghua University and China Meteorological Administration (Yang 2017).

#### 3.1.2 Building model

A building model in Revit has two parts: the geometry of the residence (shown in Fig. 3), and

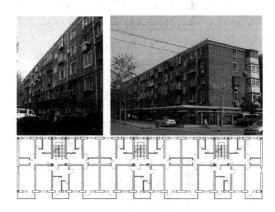

Figure 2. Sijicun residence, Tianjin, China.

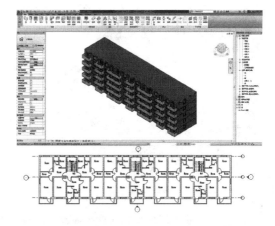

Figure 3. BIM model for Sijicun residence, China.

the other building information including location, construction, structure, core boundary, HVAC System, which were gathered from field surveys and working drawings. For the BIM model using "architectural template", building envelop was modelled by wall, floor, roof, window and other construction elements, which the real construction material and structure layers were set up by modifying type property.

### 3.1.3 Model transformation

Rooms bounded by model elements (such as walls, floors, and ceilings) and separation lines were created and tagged in Revit. As mentioned above, the BIM model (.rvt) was exported to both Ecotect and Designbuilder in gbXML format using "Room/Space Volumes". These geometry and fabric information was contained in the gbXML document. Then, DesignBuilder served as an energy simulation tool, meanwhile Ecotect model was simulated by DAYSIM for daylighting analysis. Data regarding building services includes those associated with heating, cooling, lighting, and domestic hot water (DHW). In addition, heating was switched on or off by sub-metering and room-metering temperature control system) for different spaces to avoid unnecessary energy consumption in non-occupied periods.

### 3.2 Simulation-based diagnosis

The physical environment diagnosis showed in Figure 5 is composed of sun light hour analysis, indoor daylight factors, daylighting autonomy *(DA)*, useful daylight illuminance *(UDI)*. The simulation results indicate that due to relatively low building density and volume fraction, both 2 hours benchmarks of windowful daylight and comfortable air flow conditions outdoors were achieved. The *DFavg* was achieved 4.14%, but the $UDI_{<100}$ was simulated to be 100%, which signified insufficient indoor daylighting, especially at the end of the house. Meanwhile, 59% of all measuring points can reach *DAmax* above 5%, which means severe glare near the window. Hence, the main challenge for retrofitting the physical environment is improving the indoor daylighting and reducing glare near the window.

Table 2 showed the fabric parameters of the building imported from Revit model. The energy consumption for this residence simulated was 139.12 Kwh/m²·a. $CO_2$ emissions was 52.9 kg/m²·a.

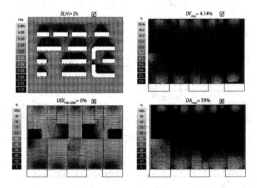

Figure 5. Daylighting simulation for Sijicun residence, China.

Table 2. Thermal performance diagnosis for case.

| Information | Actual building |
| --- | --- |
| Shape coefficient | 0.39 |
| External wall construction | Solid Clay Brick, 360 mm, Solid Clay Brick, 360 mm, U-value = 1.565 W/m²·k |
| Flat roof construction | Concrete, Reinforced, 120 mm, Fly ash ceramics, 30 mm, U-value = 2.634 W/m²·k |
| Window | Single clear (6 mm) windows, U-value = 5.840 W/m²·k |
| Window to wall ratio (S) | 0.34 |
| Window to wall ratio (N) | 0.27 |
| Window to wall ratio (E) | 0.06 |
| Window to wall ratio (W) | 0.06 |
| Energy consumption | 139.12(Kwh/m²·a) |
| $CO_2$ emissions | 52.9(kg/m²·a) |

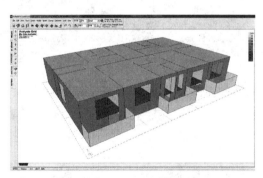

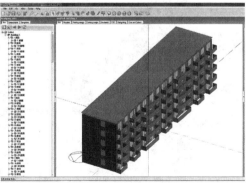

Figure 4. Ecotect and Designbuilder model imported from Revit, China.

According to the simulation results and analysis, the actual building showed great potential for green retrofit by adding sunshade reflector device, the building envelope insulation, adjusting the system efficiency and metering mode, as well as using renewable energy generation.

### 3.3 *Retrofit design proposal*

Following the preliminary lighting physical and thermal diagnosis, 10 different measures were tailored to redesign the building, including enclosed balconies and enclosed staircases with an atrium, wall and roof insulation, window replacement, window shading, flat to pitched roof conversion, space heating sub-metering, solar thermal, solar PV, etc. Figure 6 presented the transformation of the building after retrofit.

### 3.4 *Performance prediction and feedback*

For form strategy, lighting simulations were run to give specific guidance for decision making related to form retrofit (Fig. 7). The enclosed staircase with an atrium can serve as a chimney to promote natural ventilation. According to the prediction, the sloped shading device could effectively block strong sunshine and reduce glare in summer without compromising daylight in winter. As shown in Figure 7, based on the guarantee of the average Daylight factor (2.01%) and *DA* (58%–93%), the sunshade reflector design made the indoor illumination more uniform, $UDI_{100-2000}$ was increased by 5%, and $UDI_{>2000}$ that may cause glare, was decrease by 17%, DAmax also was decreased by 25%. For the fabric retrofit, a layer of 65 mm EPS insulation was added to the external wall, and the existing flat roof was renovated to a pitched roof insulated with 120 mm XPS extruded polystyrene. The simulation results showed that the best fabric

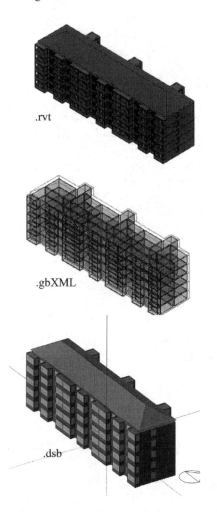

Figure 6. BIM and BPS model for residence after retrofit.

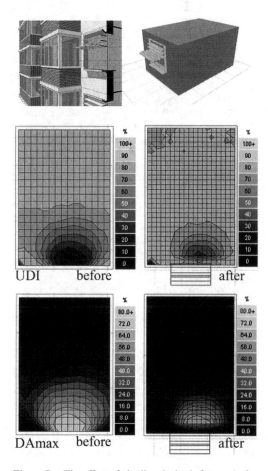

Figure 7. The effect of shading device before and after retrofit.

measure EWI could contribute to a 22.7% energy savings and an 11.7% carbon emissions reduction.

In addition, adjusting the heat metering mode is an effective system retrofit approach. Before the retrofit, the heating was switched on or off for the whole building without treating the rooms with and without occupants differently. By changing the operating schedule from a 24 hour whole flat heating mode to an intermittent heating one adjusted to room occupancy, it could achieve an energy savings of 9.5%. Solar PV showed great potential to $CO_2$ emissions.

Table 3 summarized the prediction results of energy savings, $CO_2$ emissions savings for individual retrofit strategies, and a holistic approach combining all of the measures together.

### 3.5 Multi-objective optimization

#### 3.5.1 Objective functions and decision variables

Occupancy comfort and environmental impact were the two main objectives that contradicted each other during the green retrofit procedure. This research hopes to find a balance between the minimum discomfort hours and the minimum $CO_2$ emissions. Discomfort hours referred to the total time during the year when the indoor humidity ratio and the operating temperature are not within the standard for winter heating and summer cooling. For example, in the period near the heating season or the cooling season, as well as the extreme weather throughout the year, the thermal comfort of the building may be unsatisfactory. This physical quantity can indicate the thermal environment comfort of buildings. The less the discomfort hours, the better the performance. Carbon emissions referred to the carbon emissions from energy consumption in the operation phase, specifically to the $CO_2$ emissions generated by energy-consuming systems such as heating, cooling, lighting, domestic hot water, and electrical equipment. The smaller the $CO_2$ emissions, the smaller the impact on the environment.

The multi-objective optimization in this article was implemented by the "Optimization" module in DesignBuilder. Multi-objective optimization was a simulation-based approach to calculate different design variables and their combinations based on the NSGA-II genetic algorithm to achieve optimal performance. Table 4 summarized the variables and the options for each variable according to the case and China's construction reality.

By setting the "tournament size", "crossover rate" (set to 0.9), "individual mutation probability" (set to 0.2), this simulation tried to calculated every objective indicators generated by different variable combinations. The exhaustive computing solution of all variable options was obtained to clarify the effects of carbon emissions and discomfort hours brought by different retrofit measures, and a part of optimal strategy screening database was provided in Figure 8.

After 180 hours of operation, 3000 performance solutions repeatedly screened by the genetic algorithm were obtained, and the Pareto-optimal two solutions for minimum carbon emissions and minimum discomfort hours were obtained.

Table 3. Energy saving and $CO_2$ reduction of retrofit measures.

| Measures | Details | Energy saving (%) | $CO_2$ reduction (%) |
|---|---|---|---|
| Enclosed staircase with atrium | Reduce shape coefficient, create a chimney effect, develop public communication space. | 3.9 | 2.9 |
| Enclosed balcony | Increase floor area | 1.2 | 1.0 |
| Shading | Window shading on south façade | −0.5 | 0.4 |
| Flat to pitched roof conversion | Concrete, Reinforced, 120 mm, XPS Extruded polystyrene, 120 mm, U-value = 0.265 W/m²·k | 13 | 9 |
| External wall insulation | Solid Clay Brick, 360 mm, EPS Expanded polystyrene, 65 mm, U-value = 0.439 W/m²·k | 22.7 | 11.7 |
| Window | Double-glazed low-E clear windows (3/13 mm), U-value = 1.786 W/m²·k | 10.4 | 6.4 |
| Space heating | Sub-metering and room-metering temperature control | 9.5 | 5.0 |
| Solar thermal | Domestic hot water heating, 135.81 m² solar thermal collector installed on the balconies | 9.7 | 26.9 |
| Solar PV | Solar electrical energy generation, 47.17 kW Photovoltaic panels installed on the roof | 13.6 | 37.8 |
| Electric lighting | LED lighting | 2.5 | 9.6 |
| Holistic approach | Combination of above | 78.1 | 92.6 |

Table 4. Thermal performance diagnosis for case.

| Variables | Options | | | |
|---|---|---|---|---|
| External wall insulation | U = 1.565 | U = 0.45 | U = 0.25 | U = 0.15 |
| Roof insulation | U = 2.634 | U = 1.2 | U = 0.25 | U = 0.15 |
| Window replacement | U = 5.840 | | U = 1.786 | U = 0.8 |
| Heating schedule | Conventional | | Sub-metering | |
| Solar thermal | 0 | | 135.81 m² | |
| Solar PV | 0 | | 47.17 kW | |
| Artificial lighting | Fluorescent lamps | | LED lamp | |

Table 5. Pareto optimal solution.

| Solution | $CO_2$ emissions | DCH |
|---|---|---|
| 1 EWI0.15 + RI0.15 + WR0.8 + sub-metering + Solar thermal + LED lamps+PV | 2739.19(kg/a) 1.17(kg/m²·a) Reduction rate 97.78% | 1973 hr |
| 2 EWI0.15 + RI0.15 + WR0.8 + sub-metering + Solar thermal + Fluorescent lamps + PV | 13426.9(kg/a) 5.75(kg/m²·a) Reduction rate 89.13% | |

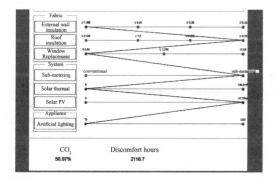

Figure 8. Optimal sulutions (Partial).

Figure 9. Multi-objective optimization tool.

### 3.5.2 Multi-objective optimization tool

According to the calculation of genetic algorithm, a multi-objective optimization prediction tool was developed by JAVA programming, which can present the degree of improvement for both $CO_2$ emission reduction and discomfort hours reduction caused by each combination of fabric, system, and appliance measures. The graphical displays can intuitively express which combination of strategies can be more effectively reduce $CO_2$ emissions, or reduce discomfort hours. It can serve as a sensitivity tool which can offer an architect how to choose optimal solutions, and how to make decision to achieve specific objective, even if he does not have a lot of knowledge about building performance analysis.

When using the tool, moving each slider to reach the variable option point, then the corresponding value of $CO_2$ reduction rate and discomfort hours were generated as shown in Figure 9 to make it easy to choose the appropriate measures according to the needs of users.

## 4 SUMMARY AND DISCUSSION

For the multi-objective analysis in this case, the Pareto optimal solution is calculated by NSGA-II, as shown in the following Table 5. Solution 1 with LED lamps could achieve the highest $CO_2$ reductions, with relatively good indoor comfort. Solution 2 with fluorescent lamps could achieve the best indoor comfort. However, considering the incremental cost and design flexible of the early design stage, multi-objective optimization tool containing more variables and their corresponding results are significantly necessary for designers.

Due to the limitation of time and knowledge of other major, the research only presented a possibility mainly associated with BIM and BPS. The case study described indicated that an interdisciplinary way would be established to conduct retrofit projects. Meanwhile, New design aids is need to be developed to help architects' decision making in the early design stage.

## 5 CONCLUSION

This research presented a methodological retrofit approach of residential building in Tianjin. BIM and BPS were utilized as a flexible method in diagnosis, prediction, and determination. The comparative analysis in this article made tailored retrofit design possible. It was shown that a wide range of available retrofit options need to be selected to specific building stock. The design procedure and multi-optimization tool can be served as propagable methods especially for the building designers who don't have much knowledge in building science. With the help of building information and simulation, a multi-objective optimization retrofit for contemporary China's housing stock should balance environmental impacts, occupant comfort and architectural aesthetics.

Widely interdisciplinary investigation and cooperation are in great demand to aid domestic retrofit practice. For future work, a large number of case studies will be modelled and the results will be validated by monitoring the real buildings. Objective functions and variable values will be expanded, in order to improve the multi-objective optimization tool.

## AKNOWLEDGEMENTS

The research has been supported by National Key R & D Program of China (Grant No. 2016YFC0700200), The Key Program of National Natural Science Foundation of China (Grant No. 51338006), and Innovation Base Introducing International Talents (Referred as 111 project) (Grant No. B13011).

## REFERENCES

Asadi, E. et al. 2012. Multi-objective optimization for building retrofit strategies: A model and an application. *Energy and Buildings*. Elsevier B.V., 44(1): 81–87.

Dascalaki, E. and Santamouris, M. 2002. On the potential of retrofitting scenarios for offices. *Building and Environment*, 37(6): 557–567.

Diakaki, C., Grigoroudis, E. and Kolokotsa, D. 2008. Towards a multi-objective optimization approach for improving energy efficiency in buildings. Energy and Buildings, 40(9): 747–1754.

THUBERC. 2017. *2017 Annual report on china building energy efficiency*. Beijing: China architecture & building press.

Yang H. 2016. Research on design approach and prediction model for green retrofit design based on building performance in the cold region of china. *Doctor Thesis*. School of Architecture, Tianjin University.

Yang, H. et al. 2017. Tailored domestic retrofit decision making towards integrated performance targets in Tianjin, China. *Energy and Buildings*. Elsevier B.V., 140: 480–500.

*Collaboration methods*

# IPD/BIM collaboration requirements on oil, gas and petrochemical projects

A.H. Fakhimi & J. Majrouhi Sardroud
*Department of Civil Engineering, Central Tehran Branch, Islamic Azad University, Tehran, Iran*

S.R. Ghoreishi
*DNV GL—Oil and Gas, Det Norske Veritas, Danmark*

S. Azhar
*McWhorter School of Building Science, Auburn University, AL, USA*

ABSTRACT: The Oil, Gas and Petrochemical (OGP) projects are more complex than Architectural, Engineering and Construction (AEC) projects. Therefore, the use of methods and technologies that can execute the OGP projects at an acceptable level of risk within all the constraints are much needed. Nowadays, benefits of the modern construction management methods and technologies, such as Integrated Project Delivery (IPD) and Building Information Modeling (BIM) in the AEC projects, have been proven and a few OGP companies have used these methods partially. This research study has shown rare evidence for using these modern methods in the OGP projects, while the OGP industry stays at a higher level in terms of knowledge than the AEC industry. So, it is focused to investigate if IPD and BIM can be integrated complementary and synergistically in OGP projects and to draw requirements of this collaboration. To do this, the critical library studies which identify the key parameters of IPD/BIM integration and interviews with OGP experts are combined. Conclusions clarified that IPD and BIM cannot achieve their full capacity without collaboration. Hence the requirements for their collaboration in the OGP projects are investigated. In addition, important shortages related to their collaboration are highlighted.

## 1 INTRODUCTION

The most important goal of the execution of the projects is to achieve the highest quality, within the approved budget and time and to consider all the constraints at an acceptable level of risk. More complexity in parallel with increasing constraints of the projects have intensified the need for modern construction management methods to achieve the goals of the projects (Winch 1998). The Oil, Gas and Petrochemical industry (OGPi) has faced serious challenges in recent years. The decline in oil prices has pushed OGP projects to an uneconomical phase. Therefore, in such a harsh situation the use of methods and technologies that can run the projects of this industry at a lower cost and time, better quality and more safety are highly appreciated (Ehrman 2017). Nowadays, the positive benefits of the use of modern construction management methods and technologies, such as Integrated Project Delivery (IPD) and Building Information Modeling (BIM) in the Architectural, Engineering and Construction (AEC) industry, have been proven. Implementation of these novel methods have improved productivity, efficiency and quality, and reduce cost and time of projects, therefore, their use have become commonplace (Fakhimi et al. 2016 & Bryde et al. 2013). In this regards, a few OGP companies have used these methods in some part of their projects to achieve the mentioned benefits. However, this research has shown that there is rare evidence for using these modern methods in the OGP projects, while the industry stays at a higher level in terms of knowledge and technology relative to the AEC industry (Cheng et al. 2016a).

## 2 GOALS AND METHODOLOGY

The main objective of this paper is to investigate if IPD and BIM can collaborate with each other complementary and synergistically to improve the cost, time and quality of OGP projects and the requirements of their collaboration. To do this, two groups of questions including 8 sub questions have been considered. The first group mainly provides the explanatory knowledge and the second provides the prescriptive knowledge, giving recommendations for how the collaboration can be adopted.

To conduct this research, a critical literature survey was performed in combination with interviews with the experts from various sectors of the OGP industry, including engineering, general contractors, operators, clients, and the owners were performed. The literature was selected from Science Direct, Scopus, Springer, Emerald and university libraries (thesis and dissertation). The literature survey included the definition and the uses of IPD and BIM and, also their tools, benefits, barriers, and challenges, which were explore in AEC and infrastructures industry in three main aspects: process, people and tools. To interview with experts, the results of literature survey were synthesized and categorized. To evaluate and complete the findings of library study, an initial list of questions and the appropriate proposed solutions were prepared and used to interview with OGP experts.

A qualitative semi-structured methodology and brain storm sessions were adopted to respond the research questions. Totally 40 persons from various sector of OGP industry were interviewed averagely for about about 45 minutes per person. The interviewees were selected in accordance with their responsibilities and/or previous experiences. A brain storm meeting was conducted for finalizing the initial list. Finally, another meeting was held for presenting the results of interviews to discuss the findings with all interviewees.

This research is one of the first studies that have employed scientific research for the implementation of the IPD/BIM in the OGP industry.

## 3 IPD/BIM SYNERGISM

### 3.1 *Integrated project delivery*

In 1990, BP in-house team formed an integrated group that combined engineering, subsurface and commercial interests. They called it Alliancing and it continued to emerge with new face called IPD (Sakal 2005). There are many definitions for IPD that they have similarity in essential basis of IPD (Syukran et al. 2016). According to American Institute of Architects, AIA's most recent definition of IPD is defined as "a project delivery method that integrates people, systems, business structures and practices into a process that collaboratively harnesses the talents and insights of all participants to reduce waste and optimize efficiency through all phases of design, fabrication and construction". As shown in Figure 1, IPD approach is built around eleven essential principals that differentiate it from the traditional project delivery (AIA 2014). Despite this different definition, it is clear that IPD

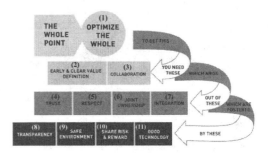

Figure 1. Essential principals of IPD (AIA 2014).

emphasizes five important characteristics that differentiate it from traditional project delivery; (1) relationship, (2) early involvement, (3) collaboration and integration, (4) shared risks and rewards and (5) good technology (Ghassemi et al. 2011, Syukran 2016 & O'Connor 2009).

Many researchers and practitioners have stated that successful implementation of IPD requires compliance with the following:

1. Achieving the benefit that IPD offers requires the participants of the project to follow some key important principles such as mutual trust, mutual reward, early involvement of key participants, early goal definition and leadership (AIA 2014).
2. The IPD should be commenced from the initial design stage and continue to project handover (Ghassemi et al. 2011 & Ilozor et al. 2012).
3. In the early stages of project, sharing the risk and reward between participants should be finalized clearly (Ogunsanmi et al. 2011).

IPD implementation barriers and challenges affect its appropriate adoption and should be prevented and/or solved. The challenges associated with process are mainly legal issues (liability, insurance and risk) and technological barriers and those associated with people are fear and resistance to the changes, and willingness and knowledge of owner (Ghassemi et al. 2011 & Rached et al. 2014).

IPD has been born and grown in the oil industry before it was known to the AEC industry. However, AEC industry has adopted IPD concept and developed it successfully.

### 3.2 *Building information modeling*

As a solution to inefficiencies in the AEC industry, BIM technology has been gradually developed and practically used in the AEC industry in projects started in the mid-2000s and nowadays, BIM is implemented in many countries (Azhar 2012). Over 1000 publications on BIM have attempted to define it in their own terms, so BIM takes a variety of definitions (Barlish et al. 2012). A commonly accepted definition of BIM is: "Building Information Modeling (BIM) is a digital representation of physical and functional characteristics of a facility. A BIM is a shared knowledge resource for information about a facility forming a reliable basis for decisions during its life-cycle; defined as existing from earliest conception to demolition" (NIBS 2007). BIM technology creates a virtual model of a building with quantitative and qualitative characteristics and works through the creation of a database which is compatible with the Industry Foundation Class (IFC) or Construction Operations Building Information Exchange (COBie) compatible that are non-proprietary systems of referencing building and construction industry information. BIM could facilitate the cooperation between the different parties involved in the project and support the project through its lifetime (Rizal 2011). The most important barriers of BIM implementation are cited as: absence of standard BIM contract documents, need for a new business model, need to change information flow management, lack of collaborative work processes, liability and legal issues and registry of communication and information exchange (Azhar 2011, Gu et al. 2010 & Succar 2013).In 2016, a visual guide ( Fig. 2) that introduce main BIM implementation process was launched by the National Building Specification (NBS). It is provided similar to the traditional periodic table of elements and entitled " The Periodic Table of BIM" to ensure a successful BIM implementation (NBS 2016). The periodic table of BIM consists of 10 separate task groups brought together as a step by step roadmap for BIM implementation. Based on the table, BIM implementation starts with defining BIM strategy and goes through process, people and tools to reach the project hand over. Important aspects effecting the removal of barriers to BIM implementation are addressed in this table. The periodic table of BIM as the latest and one of the most complete guide to BIM implementation does not cover some very important issues such as contractual, legal and risk problems. Although this is a common guide that can be used in all industries like the OGPi, there is still no solid evidence for using it in the OGP industry (Cheng et al. 2016b). Therefore, the preparation of BIM to be used in the OGP industry requires a comprehensive research on the adoption of periodic table along with IPD, while the industry is well acquainted with the IPD.

### 3.3 IPD/BIM collaboration and integration

As presented in Figure 1, IPD tries to integrate people, systems, business structures and practices into a collaborative process to reduce waste and optimize efficiency through design, fabrication and construction phases. It is obvious that IPD is concentrated on overall structure of communication between parties via focus on process and people and do not introduce specific tools for implementation. On the other side, as shown in periodic table of BIM, BIM provides the technological territory of information shared between all parties and do not focus on introduction of any framework for cooperation between the different parties in the project. BIM presents a collaborative reliable basis for decisions during projects life-cycle and is concentrated on its favorite implementation.

A deep look at the concepts of IPD and BIM indicates that their overlap as well as the complementary role they can have for each other, are very clear. Better planning and collaboration, integrated and transparent construction process, facilitating/catalyzing the optimum project delivery process, facilitating the use of each effectively for construction projects, increasing the productivity and efficiency and generating more added value to the client, reducing the amount of redundant data, facilitating of the information sharing, improving project relationships, facilitating closer collaboration from early stages of the project serving as an excellent team building tool and accelerating the formation and strategies of projects are main benefits that researchers and practitioners have mentioned for IPD/BIM collaboration (Elmualim et al. 2014; Jones 2014; Hall et al. 2014 & Sarkar 2015). On the other hand, the need for qualified contractors, responsibility, liability and model ownership (legal issues), previous experience of the team as a unit, cultural changes, fair of Changes, interoperability issues and required profound process changes of the involved parties are most reported barriers of their collaboration in AEC industry

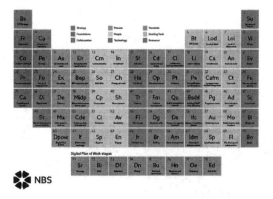

Figure 2. Periodic Table of BIM (NBS 2016).

(Wright 2012; Sarkar 2015; Ilozor et al. 2012 & Hall 2014).

Currently, researchers and practitioners of AEC industry are going to consider integration of IPD and BIM and introduce an integrated framework via their reciprocal synergetic collaborations; so that it could be used as a very good platform customized for the OGP industry.

## 4 DISSCUSSION

Although, a few international oil, gas and petrochemical companies have started using BIM tools and techniques in their projects, no serious evidences for the advent of BIM technology in the OGPi have been found (Cheng et al. 2016b). Since there was found no evidence of using BIM in the OGP industry, it was not possible to integrate or even collaborate with IPD in the oil, gas and petrochemical industry. Conversely, since the OGP industry has been well utilizing IPD techniques in recent decades, it can be used to overcome the challenges of their bilateral collaboration identified in the AEC industry and boost the benefits of IPD/BIM implementation. However, inspired by the experience of using these two novel ideas in the AEC industry, this study aims to provide the requirements for their collaboration and integration in the OGP industry by identification of the strengths and weaknesses of this collaboration. Some articles introducing BIM and IPD as complementary tools stated that: BIM facilitates the integration of information and provides a single platform for storage and retrieval of data, and IPD provides a framework to integrate the shared goals and values of the project participants (Wright 2012). Table 1 is prepared to show IPD principals versus BIM concept based on periodic table of BIM. As Table 1 shows, IPD can embrace all BIM aspects and cover main challenges of BIM implementation including shared` risk and rewards, joint ownership and integration.

It should be noted that the components of the BIM periodic table, same as the periodic table, are separate and a suitable infrastructure must be prepared in order to achieve the proper composition of the elements. Therefore, it can be acknowledged that IPD could act as the best infrastructure for this purpose and could boost BIM benefits and meet the challenges of BIM adoption and implementation.

Successful implementation of IPD in the OGPi over the past decades, the existence of related legal and contractual IPD infrastructure in the OGPi, IPD and BIM overlaps (Table 1) and appreciation of OGPi firms in applying BIM show that IPD could be utilized along with BIM in the OGPi successfully.

On the other hand, the results of the interviews shows that:

1. The processes of information flow between all parties in a project were fully fragmented and each participant followed its own desired procedures. Non-integrities, reworks, misunderstandings, wrong or incomplete orders, delay in all deliverables and claims against other partners were issues needed to spend time and cost to solve it. Challenges often occurring between key participants and the owner had to be met by arranging a powerful team aimed at integrity.
2. All interviewed parties appreciated an integrated platform to cover all the project lifecycle, which was able to solve the mentioned issues.
3. They were very concerned about the misuse of BIM due to the lack of a proper and experienced plan
4. The organizational challenges, new business model, cost and time of implementation, lack of evidence of the financial benefit, better coordination, reduced change orders or RFI's, reliable cost estimates, improved communication within the project team, better documentation, reduced rework, error free drawings, better code compliance, accurate quantity takeoff, increased safety of the site and better logistics were in the face of the interviewees attention.
5. Some aspects of BIM implementation such as better stakeholder services and satisfaction, faster project delivery, early involvement of partners, reduced number of human resources, improved quality, productivity and decision making, reliable cost estimates, rework elimination, redefining staff roles and responsibilities, changing information flow management, lack of collaborative work processes and modeling

Table 1. IPD essential principals versus terms of BIM periodic table.

| IPD essential principals | Terms of BIM periodic table |
| --- | --- |
| 1) Optimize the Whole, not the parts | – |
| 2) Early and Clear Goal Definition | Digital Plan of Work Stages |
| 3) Collaboration | Collaboration |
| 4) Integration | Foundation |
| 5) Joint Ownership | – |
| 6) Respect | People |
| 7) Trust | People |
| 8) Transparency | Process and Technology |
| 9) Safe Environment | Technology and Standards |
| 10) Shared Risk & Reward | – |
| 11) Good Technology | Enabling Tool |

standards, absence of qualified BIM team, interoperability, sufficiently developed specialized software, training issues, integration of information and increasing the safety are the center of their considerations.

6. According to the responses given by most of the interviewees, the main benefits of implementing BIM were unfolded in commissioning/start up and operation stage.

Based on the results of library studies and the results of interviews, it is concluded that no systematic academic and industrial research was done and also no reliable method for implementing BIM or IPD/BIM in OGPi was introduced. The lack of appropriate tools, standards, contractual forms, experienced team and owner interests are main obstacle of IPD/BIM collaboration. So, academic and industrial research is needed to develop a roadmap for implementing BIM and IPD/BIM collaboration in OGPi as the most important perquisite, similar to what the UK government has done for implementing BIM in the UK construction industry (Khosrowshahi et al. 2012 & Fakhimi et al. 2017). Thereafter, a concept map must be provided for the implementation of BIM, in which all concerns are considered and for which a clear path is foreseen.

## 5 CONCLUSION

IPD was started and developed in the OGP industry and entered into the AEC industry to reduce waste of time and cost and to increase its quality and productivity. Conversely, BIM was started and developed in the AEC industry to implement it appropriately, but it has not yet been adopted for the OGP industry. On the other hand the AEC industry moves forward to integrate IPD and BIM to achieve their synergetic collaboration. However, more complexities in the OGP projects in addition to more need for increasing the productivity in the OGP industry, encouraged us to use the successful experiences in this regards.

This study via literature survey and semi structured interview pointed out the capabilities of IPD/BIM collaboration and integration and the necessities of using these novel construction methods in OGP projects. Limited application of BIM in the OGP projects and the lack of academic and industrial research were revealed. The main findings of this study showed the need for preparation of an IPD/BIM adoption roadmap/concept map in the OGP industry in accordance with periodic table of BIM and IPD guidelines and its related concern that should be tackled. The future works may be started by preparing a roadmap and concept map for adopting IPD/BIM in the OGP projects to help with faster implementation and deeper integration of BIM and IPD.

## REFERENCES

American Institute of Architects, C. C. 2014. *INTEGRATED PROJECT DELIVERY—AN UPDATED WORKING DEFINITION.*

Azhar S. 2011. Building Information Modeling (BIM): Trends, Benefits, Risks, and Challenges for the AEC Industry, *Leadership and Management in Engineering* 11: 241–252.

Azhar s., Khalfan M. & Maqsood T. 2012. Building information modelling (BIM): now and beyond, *Australasian Journal of Construction Economics and Building* 12(4): 15–28.

Barlish K. & Sullivan K. 2012. How to measure the benefits of BIM—A case study approach. Arizona State. *Automation in Construction* 24: 149–159.

Bryde D., Broquetas M & Marc Volm J. 2013. The project benefits of Building Information Modelling (BIM). *International Journal of Project Management* 31(7): 971–980.

Cheng J.C.P., Tan Y, Liu X & Wang, X. 2016a. Application of 4D BIM for Evaluating Different Options of Offshore Oil and Gas Platform Decommissioning, *The 16th International Conference on Computing in Civil and Building Engineering*: 1524–1531. Osaka: Japan.

Cheng JC.P., Lu Q. & Deng Y. 2016b. Analytical review and evaluation of civil information modeling, *Automation in Construction* 67: 31–47.

Ehrman M. 2017. The Future of the Canadian Energy Industry in a Low Price Commodity Environment, *LSU Journal of Energy Law and Resources* 5(2): 263–373.

Elmualim A. & Gilder J. 2014. BIM: innovation in design management, influence and challenges of implementation, *Architectural Engineering and Design Management* 10(3): 183–199.

Fakhimi A.H., Majrouhi Sardroud J.& Azhar S. 2016. How can Lean, IPD and BIM Work Together?, *33rd International Symposium on Automation and Robotics in Construction*: 67–75, Auburn:USA

Ghassemi R. & Becerik-Gerber, B. 2011. Transitioning to Integrated Project Delivery: Potential barriers and lessons learned, *LCI Lean Construction Journal, Lean and integrated project delivery special issue*:32–52.

Gu N., London K. 2010. Understanding and facilitating BIM adoption in the AEC industry, *Automation in Construction* 19(8): 988–999.

Hall D., Algiers A. &Lehtinen T. 2014. The role of Integrated Project Delivery Elements in Adoption of Integral Innovation, *Engineering Project Organization Conference,* Colorado:USA.

Ilozor B.D & Kelly, D.J. 2012. Building Information Modeling and Integrated Project Delivery in the Commercial Construction Industry: A Conceptual Study, *Journal of Engineering, Project, and Production Management* 2(1): 23–36.

Jones, B. 2014. Integrated project delivery (IPD) for maximizing design and construction considerations regarding sustainability, *2nd International Conference* on *Sustainable Civil Engineering Structures and Construction Materials:* 528–538. Yogyakarta: Indonesia.

Khosrowshahi F. & Arayici y. 2012. Roadmap for implementation of BIM in the UK construction industry. *Engineering, Construction and Architectural Management* 19(6): 610–635.

National Institute of Building Sciences (NIBS). 2007. *United States National Building Information Model Standard. National Institute of Building Sciences (NIBS), Version 1*.Avalable at http://www.wbdg.org/pdfs/NBIMSv1_p1.pdf. pp 1–2.

NBS. 2016. *National Building Specification.* Retrieved from https://www.thenbs.com/knowledge/periodic-table-of-bim

O'Connor P. 2009. *Integrated Project Delivery: Collaboration Through New Contract Forms.* Minneapolis, Faegre & Benson, LLP:USA.

Ogunsanmi O.S. & Ajayi O. 2011. Risk Clas-sification Model for Design and Build Projects, *Journal of Engineering, Project, and Production Management* 1(1): 46–60.

Rached F., Hraoui Y., Karam A. & Hamzeh f. 2014. Implementation of IPD in the Middle East and its Challenges, *IGLC-22*: 293–304. Oslo: Norway.

Rizal S. 2011. Changing roles of the clients, architects and contractors through BIM, *Engineering, Construction and Architectural Management* 18(2): 176–187.

Sakal M. 2005. Project Alliancing: a relational contracting mecha-nism for dynamic projects, *Lean Construction Journal* 2: 67–79.

Sarkar D. 2015. A framework for development of Lean Integrated Project Delivery Model for infrastructure road projects, *International Journal of Civil and Structural Engineering* 5(3): 267–271.

Succar B. 2013. *Building Information Modelling: conceptual constructs and performance improvement tools*, Thesis for PhD, Universi ty o f Newcastle.

Syukran R, Mohd Nasrun M.N. & Faizatul Akmar, A.N. 2016. Integrated Project Delivery (IPD): A Collaborative Approach to Improve the Construction Industry, *Advanced Science Letter* 22: 1331–1335.

Winch G. 1998. Zephyrs of creative destruction: understanding the management of innovation in construction, *Building Research & Information* 26(5): 268–279.

Wright J. 2012. The Integration of Building Information Modeling and Integrated Project Delivery into the Construction Management Curriculum, *ASEE Annual Conference*.

# Towards level 3 BIM process maps with IFC & XES process mining

S. Kouhestani & M. Nik-Bakht
*Concordia University, Montreal, Quebec, Canada*

ABSTRACT: Building Information Modelling (BIM) corresponds to generation and management of the digital representation of the building products (and processes) by containing building elements in a unique source file. Open BIM, relying on the platform-independent standard Ifc (Industry Foundation Classes), created by BuidlingSMART, is supposed to increase the interoperability in BIM environment. BIM, as a shared work platform in AEC, can be upgraded to act as an Enterprise Resource Management (ERM) system and support data mining for management of design and construction processes. eXtensibile Event Stream (XES) is an eXtensible Markup Language (XML) aims to provide a format for supporting the interchange of event logs. XES-based Event logs normally include some semantics regarding events, called extensions. In the end, process mining on the foundation of BIM (as an ERM system) is used for discovery, monitoring and optimizing BIM processes. The present study aims to facilitate applications of process mining in AEC industry, through BIM.

*Keywords*: BIM processes, data mining, process mining, enterprise resource management

## 1 INTRODUCTION & BACKGROUND

The separation and lack of collaboration and communication between the designer and builder team are known as one of the main challenges in construction projects. This has roots the industry as well as the contractual structure, which is traditionally not collaborative-based, hence urging each party to optimize their individual operation processes regardless of the project performance [1]. Efforts to resolve this issue have resulted in modern project delivery methods (such as Integrated Project Delivery—IPD) and using Building Information Modeling (BIM). Over the past decade, BIM has evolved from a digital product to a "process" and a "culture" in the domain of AEC/FM (Architectural, Engineering, Construction/Facility Management), during the recent years. As a semantic archive of building components, BIM can accommodate a wide range of information, not only from the physical and conceptual elements, but also regarding project phases, dependencies, impacts, actors, etc. These capabilities assist the management of ever-increasing complexity in construction projects by supporting integrated delivery methods (such as IPD).

This has created a new discourse in AEC/FM, called BIM-Management, with roles such as BIM manager or BIM coordinator, and tools such as BIM Execution Plan (BxP). The main role of BxP is providing a protocol for successful BIM adoption in construction projects through defining the expectations (BIM goals and BIM uses); planning for achieving those expectations (BIM processes); and maintaining the timeliness, level of details required, and interoperability for the information flow among the involved parties. BxP templates, such as the CIC *BIM Project Execution Planning guide* – a BuidlingSMART alliance project, introduce two levels of template BIM processes to be designed in projects; level 1, as an overview of the entire project and its BIM uses, and level 2, as a default for each BIM use in the project. These process maps serve as the backbone of BIM implementation and are also critical from contractual and IT infrastructure points of view [2].

On the other hand, the rich source of information collected, archived, and contextualized in BIM provides significant opportunities for distilling data-driven intelligence in projects. Among other benefits, such data analytics can reveal latent patterns of collaboration among project teams and can provide BIM managers assistance with project planning and control. The present research aims to use the data organically generated (and in most cases wasted) during design and construction phases, for measuring the performance of project teams in the implementation of BIM uses. More specifically, this paper reports on using "process mining" to discover and evaluate "as happened" processes (versus "as-planned" ones recorded in the BxP). We metaphorically refer to such processes as "level 3" maps, since they will provide the BIM Manager with a closer, more in-depth, and more case-specific view on the BIM processes executed by the project team. In this regard, "Design Authoring", as a central BIM use in most BIM projects is selected to be studied more closely.

A process can be defined as a sequence of activities having a common objective (process objective). Execution of activities generate "events"; e.g. start and end of an activity are considered two events. Process Mining is the science of discovering the as-is, end-to-end processes from the analysis of recorded data from events collected and stored in "event logs". An event log consists of "traces" (aka cases), each of which may contain one or more events for execution of activities from the beginning through the end. Event logs include instances (rows of data, i.e. events) and variables (columns of data, i.e. "features"). Variables (attributes) of executed events are also stored in event logs and could have numerical (e.g. timestamp) or categorical (ordinal or nominal) nature (e.g. resources associated with events) [3]. Fig. 1 provides a schematic hierarchical view of the event log and the associated concepts introduced. Process mining is, in fact, the analysis of event logs to learn the as happened process models (known as "process discovery"); detect where the as happened is different from as planned (known as "conformance checking"); and re-engineering and optimism business processes (known as "process enhancement").

Despite extensive applications of process mining in different domains; AEC/FM has not taken advantage of this powerful family of techniques, which first and foremost has roots in the lack of availability of event logs (which normally come from the ERMs) in this domain. BIM can potentially archive the digital footprints of processes in design, construction and operation phases of building projects. This can provide an opportunity for upgrading BIM to act as an ERM in the AEC/FM industry. The present paper reports on early steps of an extensive attempt in this regard. The paper reports on acquiring event logs from BIM, and then using them in process mining to discover and analyze as-happened BIM processes. After a brief review on the works is done inside and outside AEC/FM for discovery, modeling, and management of processes in section 2, a tool is introduced in section 3 for logging event logs from analysis of Ifc files. The tool is used in a case study for design authoring processes discovery, details of which are discussed in section 4. Finally, the concluding remarks, as well as the ongoing and future research in this area are highlighted in section 5.

2 WORKS DONE

Process identification, modeling and evaluation has recently attracted attention in AEC/FM ([4], [5], [6], [7] & [8] among others). BIM processes are usually planned either externally by the aid of BIM protocols (in form of BIM frameworks, workflows or process maps [2], [9], [10]); or internally by the project team [4], [11]. Since BIM is value adding, only to the extent that it contributes to the accomplishment of the organization's mission, the process of BIM implementation must be controlled and monitored from an organizational (rather than individual) point of view [12]. Some studies have focused on performance evaluation of BIM processes, taking into consideration different components such as performance, impact, ROI, capabilities, and maturity [13]. While the existing studies mostly focus on outcomes of the BIM processes, bottom-up approaches to evaluate the as-happened processes (when the actors execute activities) have been left understudied. This can be mostly due to the lack of access to detailed information regarding end-to-end processes executed in action. Process mining, however, provides the tools for resolving this challenge and filling the gap.

Successful applications of process mining in process discovery, conformance checking, and process enhancement can be seen in a wide range of industries (including manufacturing, healthcare, Construction, Oil and Gas industry, Silos Management, Delivery Process, Customer Service, and finance [35]). Also, some applications are reported in infrastructure-related workflow management (such as airport management [36] or urban management services [37]). Most of such applications depend on the information exchange standards specifically developed for analyzing event logs.

Extensible Event Stream (XES) [17] is an XML-based standard adopted by IEEE to support the exchange of event logs between ERM systems and data analysis. XES introduces some schema called "extensions", to provide semantics for events attributes in XES-based event logs. For instance, the "Concept extension" defines the name of activities, and "Organizational extension"

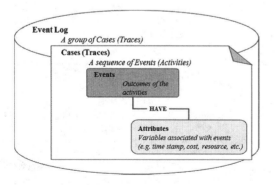

Figure 1. Event logs, cases, event and attributes.

introduces the actors executing and controlling such activities. These extensions can be used to structure the process model in a better format and analyzing them can lead to valuable information about the processes such as the bottlenecks of a process, deficiencies, social network of actors involved, etc. Extensions are, in fact, the key to the openness and extensibility of XES. OpenXES [18], a java-based reference implementation of XES standard, is designed to be implemented by developers in different domains, who aim to extend XES for creating domain-specific solutions [16]. In the recent years, a variety of research-based as well as commercial tools (ProM [17], Disco [18], Celonis [19], myinvenio [20] and Lana [21] among others) have been developed to support the specific purpose of process mining. These tools can process event logs (in XES or CSV format) to discover complex processes, perform compliance checking, and finally optimize the processes. The main feed of the process mining tools is event logs in XES format. An event log conversion from CSV format to XES format can be done by a mapping operation. The mapping elements are the extensions which CSV columns (characteristics of events) can be defined through them. For this purpose, there are already standard extensions, however, there might be some meanings that cannot be defined by them. To do this, there is a possibility to create the desired extension using the OpenXES.

The feasibility of BIM and process mining integration in the construction phase has been recently studied [8]. The study created as-built 4D models by merging several point clouds captured during the construction phase. By comparing the as-built and as-planned 4D models over the time using image processing techniques, the study captured delays as well as ahead-of-schedule activities. It also used process mining engines in combination with BIM-server to analyze the process variants, bottlenecks and the social networks of the project actors. The results of that study showed that the integration between BIM and process mining would not only benefit the ongoing project by detecting the process variations and the as-is social networks but also can help to reuse the project data (which is normally disposed of), and systematically employ them, as lessons learned, in planning future projects. In a case study, the possibility of applying the process mining techniques with maintenance data has been investigated at a hospital [14]. As a result, problematic facility management processes were discovered and optimized. Another case study on the design phase used the data generated from a system engineering tool, called Relatics (which is employed to solve the project complexity). They applied process mining tools to the collected data to examine the possibility of detecting the underlying design process. In spite of performing social network analysis and detecting the key roles in the project, the lack of integration between BIM (as a potential ERM), and data analytics, hence the lack of automation was reported as the major limitation of the work [15].

The recent studies provide a proof of concept for applicability of process mining to the AEC/FM via BIM. However, a successful and meaningful application to harvest the intelligence and achieve the breadth of advantages promised by process mining, will require higher levels of automation. Such an automation has at least two major prerequisites; *(i)* access to event logs in AEC/FM, and *(ii)* availability of information exchange protocols compliant with both BIM needs and XES requirements. The present paper reflects on the former need. We have used Ifc standard to develop a tool for extracting event logs from BIM. This tool and its application are explained in the following section.

## 3 METHODOLOGY

In the proposed case study, at least 4 attributes are required. *(a) Case/Trace ID* – to distinguish different executions of the same process; in our case study, the process would be executed 5 times. The Case ID would be named after designer's role, as *Architect 1 & 2, Structural Engineer 1 & 2 and MEP. (b) Event/Activity ID* – event ID or activity ID are named after the steps in the process. *(c) Timestamp:* the timestamp helps to create the order for events. Moreover, it enables the event log for bottleneck analysis. *(d) Resource:* this attribute can be used for social and organizational network analysis. At a glance, the methodology would look like Fig. 2.

### 3.1 *Design authoring and data preparation*

In this section the required database is produced and saved in a time order to be usable in the next step, which we need to will compare two consecutive Ifc file. To do so, a piece of code is developed in dynamo, which is an API for Revit, to generate Ifc files periodically. In the code, we set the Ifc generation time to be 10 seconds, which presumed to be enough to capture BIM files upon every little change. However, this code may capture redundant BIM files during the idleness. To avoid that, the code is extended to capture Ifc files every 10 seconds, if and only if there is a little change between the new file and previous one. The threshold for this, after trials and errors, is considered "0.0004 * the newer Ifc file size". The name of Ifc files would be their generation time in "MM-DD-YYYY hh-mm-ss PM" format. Finally, we will

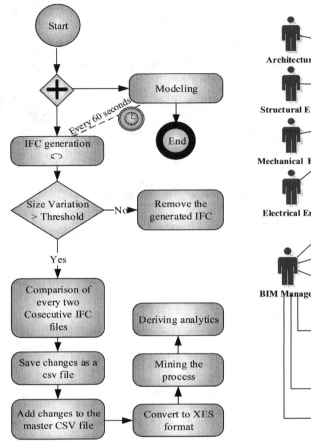
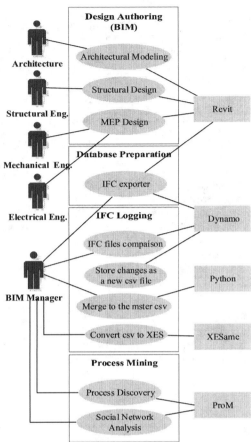

Figure 2a. Activity diagram of BIM & Process Mining integration model.

Figure 2b. Use case diagram of BIM & Process Mining integration model.

have a bunch of Ifc files, from different disciplines, upon every change throughout the design lifecycle.

### 3.2 Ifc logging

In this step, which is the main contribution of our work, a piece of code is written to compare each two consecutive Ifc files in our database, which are produced during the design phase using Ifc generator dynamo code. Upon every change including adding or removing an Ifc element (physical elements like a wall, window, column, etc. and non-physical element such as space, zone, etc.) one activity will be saved with the name of "element added" or "element removed" in CSV format. At any point in time and upon BIM manager request, CSV files will be combined using a piece of dynamo code and stored in another CSV file called "the master CSV event log". Then, the master CSV event log would be mapped via XESame [J.C.A.M. Buijs] to the XES format event log.

### 3.3 Process mining

Then ProM uses the XES format Event log to produce the underlying process happening in the design phase. Finally, the Social Network, the Bottlenecks in the process, and process deviations can be found using the appropriate modules in ProM.

## 4 CASE STUDY—DESIGN AUTHORING PROCESS MINING

To apply and test the system developed, we studied the process of designing a 2-story building using Revit, involving 5 resources including two Architects, two Structural engineers, one MEP engineer. During the design phase, the Ifc generator was running periodically (every 10 seconds), checking if there were changes in the Ifc file size. After completion of the design, every 2 consecutive Ifc files were compared and the differences were saved into a master CSV file. Then, the generated CSV event log, was imported to ProM to produce the XES file using the XESame module. Finally, the process discovery and analytics resulted in finding the social network of the project, the bottlenecks, and process deviations.

### 4.1 The event log

The event log contains 5 different traces (each trace is related to one resource), 840 events (40 unique events). Also, each event contains 5 variables including event name, event ID, time, resource and trace ID. Discovery of the underlying process through the inductive miner module in ProM is depicted in Fig. 3 (the solid blocks are only used for routing purposes and do not have any BPMN semantics [27]). This process model involves every event occurred during the design authoring, regardless of the occurrence frequency of the paths. It is not only hard to read and follow, but also suffers the overfitting issue.

The discovered process is too case-specific and using it for planning purposes will involve the overfitting issue; i.e. the process is not a generalized representation of steps taken in design authoring. Therefore, in order to make the process useful to the BIM Manager, we applied some levels of filtering. We applied the highest level of path filtering (i.e. paths filter of 0.75 in ProM) to remove all the paths which are not happening frequently enough, together with no filtering specifically to the activities (i.e. activity filter as 1.0 in ProM). Nevertheless, four activities were filtered-out due to the path filter applied.

The Petri-Net model of the process after application of the filter is shown in Fig. 4. It must be noticed, that ProM adds dummy activities (rigid blocks in Fig 3) for rerouting of arrows between nodes/activities. We have removed those dummy activities in Fig 4 (for the sake of readability), and that is why we call it a [semi] Petri-net. In addition, it is possible to filter the process based on the actors; in this case, the process model will give insights about the real activities fired by each actor in the design authoring process. Although being insightful, such process models are not shown here, due to the space limitation.

### 4.2 The analytics

Figure 5(a) and (b) show the social networks formed respectively based on "handover of work

Figure 3. The as-happened Petri-net of the process in the case study (overfitting).

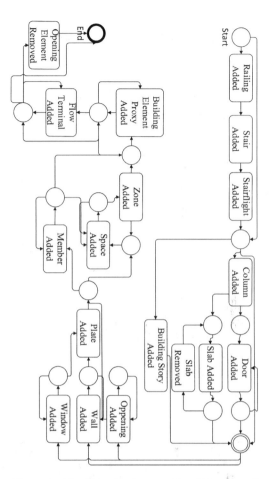

Figure 4. The filtered as-happened [semi] Petri-net of the process of the case study (generalized).

matrix" and "subcontracting". Each color represents one class; the members in each class have more handover of work between each other. The bigger nodes are more strategic and represent the higher "Betweenness" in the social network. Further, the values on edges represent the strength of the connections between each pair of nodes.

The "Handover of Work" figure, shows the resources in 3 different colors and different sizes. To illustrate, Architect 1 and 2 and MEP (represented as green) are collaborating with each other more frequently and forming a strong community. Moreover, Architect 2, Architect 1, and MEP are the strategic roles and construct a core group of the social network.

To illustrate, in the "Subcontracting" figure, structural engineer 1 hires Architect 1 and 2 as subcontractors to execute some of his tasks. Based on this method, all actors are considered as one group; i.e. all actors are extremely dependent on others in the execution of activities so that they cannot be

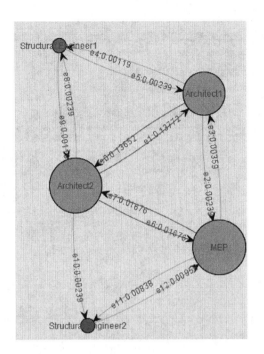

Figure 5a. Social network of the case study-mine for handover of work matrix social network.

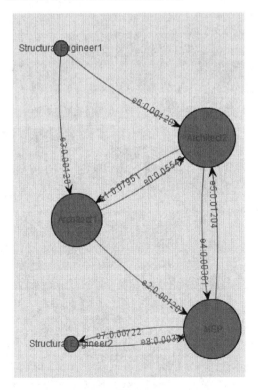

Figure 5b. Social network of the case study-mine for subcontracting social network.

categorized as different communities. Just like the first method, the strategic actors are Architect 2 and 1 and MEP. Furthermore, in both figures, there is no link between the two structural engineers which is a little bit unusual for a design authoring BIM project. Consequently, the reason should be investigated further by the responsible party.

Moreover, Figure 6 depicts an example of deviation in the process model, (the bolded line in red). It exemplifies that the task "Ifc Opening Element Added" was not executed for 4 times. The deviations are due to simplification of the discovered process model; even though, it is beneficial to consider them during the model analysis. After conducting the bottleneck analysis, three activities were found as the most time-consuming, including adding flow terminals and building proxy elements (which is mostly MEP components). The BIM Manager can ask responsible disciplines for further investigation into the causes for the bottlenecks in the as-happened process and fixing the probable issues to improve the performance of the entire process.

### 4.3 Author filtering

There are different types of design specialists involved in the design phase of a construction projects: Architects, and Mechanical/Structural/Electrical Engineers. Previously, authors proposed a method to construct a BIM related XES log by comparing consecutive IFC files captured during the design phase of a project. Then, they discovered as-is process using Inductive Miner algorithm

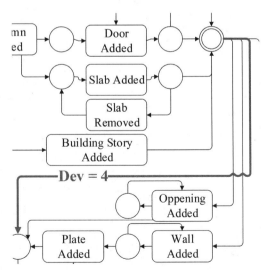

Figure 6. An example of a process deviations existing in process of the case study.

of ProM. Although the idea of having a whole as-is design process is intriguing and might have insights for a BIM manager like design team collaboration, having specific processes for each group of designers (like architects) gives more accurate information on as-is process.

To have different processes for different actors we use "pre-mining filters" to divide the whole design process to Structural, Architect, and MEP processes.

Figure 7, first, shows that MEP team Created Opening elements. Second, the MEP team assign spaces and zones for energy analysis purposes. In the next step, they created Flow terminals, Flow segments. In parallel, the team added mechanical and electrical equipment's which are all categorized as building element proxy; this is a Revit limitation when it comes to mapping MEP elements to IFC schema. Here I explained the main activity sequence, during the MEP modeling.

Figure 8 shows how one group of Structural modeler team first created column then added Slabs. whilst the second team sequentially added rails, stairs and stair flights. Then, this team incorporated in adding the "flow terminal" and "flow segments" which mainly must be within MEP team tasks; it might show a close and healthy communication between the two teams. Also, this collaboration is well explained in the authors' previous paper using social network of the process [cite prev. paper].

In Figure 9, the Architects first added a column and doors and, in parallel, they created Building story; since they start the model from the scratch it makes sense that if they added their first elements, the building story would be created. Then, windows and openings detected from our algorithm; it is also reasonable because when one adds a window an opening will be created in the wall automatically. subsequently, architects added some other elements which all categorized as Ifc Members in our log. This limitation is related to Ifc schema which any load bearing elements other than beam and columns is considered an Ifc Member. Later,

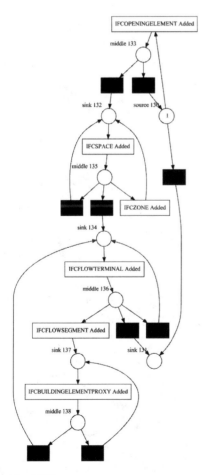

Figure 7.  MEP process.

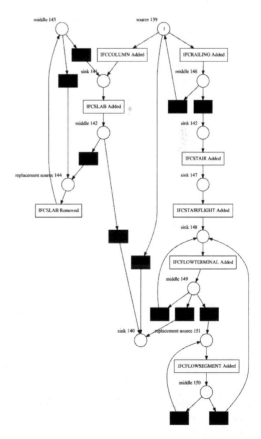

Figure 8.  Structural modelling process.

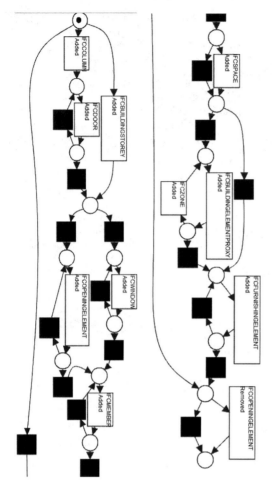

Figure 9. Architectural modelling process (left figure is the first part and the right one is the second part).

the modelers added space to the model. Then, it will be completed by Zones and building element proxies. At the end, furnishing elements are added.

## 5 DISCUSSING THE RESULTS AND CONCLUDING REMARKS

Until now, two levels of BIM process maps have been introduced by BIM protocols, both of which were referring to "as-planned" processes. In this article, we discovered the "as-happened" process for design authoring, which we named level 3 process maps. The methodology of the paper and the tools developed to discover such level 3 maps can be readily applied to other BIM uses. Finally, some analytics have been conducted on the case study, including social network, bottleneck and process deviation analysis.

There are some limitations in the proposed method. First, Revit cannot distinguish among most of the MEP elements and classifies them all as generic Ifc entities "IFCBuildingElement-Proxy". Second, our system must be extended to detect other design activities such as relocation or rotation of elements or adding/removing information to/from them. We are currently working on such extensions. Moreover, reaching the "right" level of filtering to resolve the overfitting issue and provide BIM Managers with simplified-yet-informative process models is the subject of our future studies.

## ACKNOWLEDGMENT

The authors acknowledge the support of the Natural Sciences and Engineering Research Council of Canada (NSERC). This study has been funded by NSERC through RGPIN-2017-06697.

## REFERENCES

BuidlingSMART Alliance. 2015. National BIM standard-United States (NBIMS-US), Version 3. Washington, DC, U.S.: National Institute of Building Science.

Coates, Stephen, Yusuf Aryaici, Lauri Koskela, and K. O'Reilly. 2010. "The key performance indicators of the BIM implementation process." Computing in civil and building engineering. Nottingham.

Golzarpoor, Behrooz, Carl T. Haas, and Derek Rayside. 2016. "Improving process conformance with industry foundation processes (IFP)." Advanced Engineering Informatics 30: 143–156.

Günther, Christian W., and Eric Verbeek. 2014. OpenXES Developer Guide. Eindhoven, Netherlands: Eindhoven University of Technology.

Günther, Christian W., and Eric Verbeek. 2014. OpenXES Standard Definition. Technische Universiteit Eindhoven University of Technology: Eindhoven.

Günther, Christian W., and Eric Verbeek. 2014. XES Standard Definition. Eindhoven: Technische Universiteit Eindhoven University of Technology.

Hola, Bożena. 2015. "Identification and evaluation of processes in a construction enterprise." Archives of Civil and Mechanical Engineering 15: 419–425.

IEEE. 2013. Process mining case studies. http://www.win.tue.nl/ieeetfpm/doku.php?id=shared:process_mining_case_studies.

Kassem, Mohamad, Nahim Iqbal, and Nabil Dawood. 2013. "A practice-oriented BIM framework and workflows." Computing in Civil Engineering 524–532.

Khanzode, Atul, MArtin Fischer, and Dean Reed. 2008. "Benefits and lessons learned of implementing building virtual design and construction (VDC) technologies for coordination of mechanical, electrical and plumbing (MEP) system on a large healthcare project." ITCon 13: 324–342.

Manuel, Alberto. 2012. "Process mining—Ana Aeroportos de Portugal." BPTrends.

myinvenio. n.d. Process Mining Vision. myinvenio.com. Accessed Apr 20, 2018. https://www.my-invenio.com/process-mining-vision/.

Penn State. 2010. BIM project execution planning guide and templates-Version 2.0. State College, PA, USA: CIC Research Group, Pennsylvania State University.

Poirier, Erika A., Sheryl Staub-French, and Daniel Forgues. 2015. "Assessing the performance of the building information modeling (BIM) implementation process within a small specialty contracting enterprise." Canadian Journal of Civil Engineering 42 (10): 766–778.

Rozinat, Anne, Ronny S. Mans, Minseok Song, and Wil M. P. van der Aalst. 2009. "Discovering simulation models." Journal of Information Systems 34 (3): 305–327.

Schaijk, Stijn van. 2016. BIM based process mining: Enabling knowledge reassurance and fact-based problem discovery within AEC industry. Eindhoven: Eindhoven University of Technology.

Slowey, Kim. 2016. "The productivity 'train wreck': why construction struggles to compete with rival industries." ConstructionDIVE. May 19. http://www.constructiondive.com/news/the-productivity-train-wreck-why-construction-struggles-to-compete-with/419450/.

Tsai, Meng-Han, Abdul Matin Md, Shih-Chung Kang, and Shang-Hsien Hsieb. 2014. "Workflow re-engineering of design-build projects using a BIM tool." Journal of the Chinese Institute of Engineering 88–102.

van der Aalst, W.M.P., A Bolt, and S.J. van Zelst. 2017. RapidProM: Mine your processes and not. Eindhoven: CoRR Technical Report abs/1703.03740, arXiv.org e-Print archive. https://arxiv.org/abs/1703.03740.

van der Aalst, Will. 2016. Process mining—Data science in action, second edition. Eindhoven, The Netherlands: Springer.

van Schaijk, Stijn. 2016. Case study: Construction design process mining. Eindhoven: Eindhoven University of Technology.

van Schaijk, Stijn. 2015. Case study: Process Mining with facility management data. Eindhoven: Eindhoven University of Technology.

Celonis. 2018. Apr 20. https://www.celonis.com/.

Fluxicon. 2018. Disco. Apr 20. https://fluxicon.com/disco/.

Lana. 2018. Apr 20. https://lana-labs.com/en/.

ProM. 2018. Apr 20. http://www.promtools.org/doku.php.

USACE. 2010. USACE BIM project execution plan, Version 1.0. Washington DC, U.S.: U.S. Army Corps of Engineers.

# An agenda for implementing semi-immersive virtual reality in design meetings involving clients and end-users

S. Mastrolembo Ventura
*Politecnico di Milano and Construction Technologies Institute of the National Research Council, Milan, Italy*

F. Castronovo
*California State University East Bay, Hayward, USA*

ABSTRACT: An increasing adoption of Virtual Reality (VR) is occurring in the construction industry and it is closely linked to the implementation of Building Information Modelling (BIM). While research in the application of immersive VR in the industry is still growing, a prescriptive procedure for implementing it with various purposes has not been developed yet, especially in the design review process. The paper proposes a standardized agenda for the effective use of semi-immersive multi-user virtual reality when public clients, end-users and designers collaborate in the analysis and communication of a design proposal. A case study was developed to explore VR implementation for a new public building. A qualitative data collection strategy has been applied, collecting data from seven meetings involving twenty-four prospective stakeholders. The steps to follow during the meetings, the media to include in an interactive workspace and the technological aspects and measures to consider are discussed, as well as benefits of VR in user engagement.

## 1 INTRODUCTION

The adoption of Virtual Reality (VR) is growing in the Architecture, Engineering, Construction, and Facility Management (AEC/FM) industry. This growth is closely linked to the ever-increasing implementation of Building Information Modelling (BIM) processes, procedures and technologies in the industry (Castronovo et al., 2013; Liu et al., 2014; Liu, 2017).

Building Information Modelling is a set of management processes and information technologies "enabling multiple stakeholders to collaboratively design, construct and operate a facility" (Succar, 2009). A Building Information Model, as a result of this process, is a shared digital representation and knowledge resource for information (NBIMS-US, 2015), forming a reliable basis for collaborative decision-making processes engaging various stakeholders according to their own role and the stage of the process. Moreover, Eastman et al. (2011) have highlighted how an effective implementation of BIM methods and tools "depends on how well and at what stage the project team works collaboratively".

One application of BIM in design and construction is to perform design reviews with immersive virtual reality technology. Immersive VR has been implemented in the design review process to provide a collaborative environment able to support both communication and analysis of a project (Bullinger et al., 2010; Kumar et al., 2011; Hilfert & König, 2016). Immersive VR, in fact, allows users to experience a full-scale virtual prototype of a proposed design solution while their field of view is covered so that they feel immersed in the reality that is represented (Whyte & Nikolić, 2018). Virtual prototypes are digital representations that can be explored, tested and evaluated before being physically realized (Tutt & Harty, 2013, Kumar et al. 2011).

### 1.1 *Related research*

When experienced in immersive VR, virtual prototypes can support collaborative design review sessions since early design stages, improving the communication among stakeholders and their level of engagement in decision-making processes. A building information model can be the starting point for the creation of a virtual prototype, providing an opportunity for all the project stakeholders to collaborate for an extended review when design issues are detected and resolved systematically (Schaumann et al., 2016).

Moreover, as stated in Heidari et al. (2014), BIM should evolve toward a more User Centred Design (UCD) in order to allow end-users "to experience their daily activity in a virtual environment and understand the space reactions" as well as to support clients in developing a better understanding of a design solution (Shen & Shen, 2010). Immersive VR is an example of the technologies that could support such a BIM use, being effectively implemented in UCD, design review and pre-occupancy evaluation processes when design proposals are presented to various stakeholders in order to collect their feedback (Bullinger, 2010, Heydarian et al. 2015, Liu, 2017). Such a VR-aided review process potentially supports the demand-side (i.e., clients and end-users) (Kumar et al., 2011) reducing the communication gap with stakeholders having a technical expertise (e.g., designers and facility managers) (Liu, 2017). Clients and end-users, in fact, are not usually able to understand technical representations (Boyd et al., 2016), while, by implementing immersive VR, they have the opportunity to experience a design scenario within a virtual prototype. This leads to the enhancement of the users' level of understanding of the proposed solution, which allows for the collection of their feedback and improves the design with the necessary changes according to their needs (Maldovan et al. 2006, Duston et al. 2011). In fact, behaviors and preferences of end-users given different design scenarios can be evaluated (Castronovo et al., 2013, Heydarian et al., 2015) and a pre-occupancy simulation can be performed in order to consider how the space will be used by the end-users during their activities and, at the same time, improving the understanding of the client about the design proposal (Shen & Shen, 2010). For example, the compliance of a proposed design with usability requirements can be validated: Hilfert & König (2016) shown how the access and use performance related to the accessibility for people with disabilities could be demonstrated by simulating their movements within a virtual space.

## 1.2 Research question

While research in the application of immersive VR in the AEC/FM industry is still growing, there are still obstacles to its effective implementation. One common barrier is the lack of understanding of the value that immersive VR can bring to the delivery process (Liu, 2017). Another obstacle is the lack of a prescriptive process to support its implementation in the delivery process. In this paper the authors focus on the latter issue and highlight the need for a framework and procedure for performing design reviews within an immersive VR environment. For that reason, the research question that is investigated in this paper is: "What are the agenda process, steps and measures necessary to conduct clients and end-users design reviews in a semi-immersive virtual reality environment?"

## 1.3 Structure of the paper

The paper is organized as follows. First, the research background and related gap pertinent to the implementation of immersive virtual reality in design review and user involvement processes have been presented in this paragraph as well as the research question. Second, the plan to collect and analyze data is described. Then, the data from the immersive sessions are presented and their implications are discussed. Finally, the paper concludes with a summary of the main results and their contribution to the existing theoretical and practical body of knowledge, introducing research limitations and considerations for future research.

## 2 RESEARCH APPROACH

### 2.1 Research strategy

As main research method, a case study was developed in order to explore the implementation of semi-immersive multi-user virtual reality (i.e., an open footprint single-wall) in the analysis of the design proposal for a new public building (i.e., an educational facility with innovative learning and socializing spaces). A qualitative data collection strategy has been applied. A series of immersive sessions (i.e., meetings during which the design proposal was analyzed by means of immersive VR) have been organized and carried out involving a representative panel of stakeholders (i.e., public clients, designers and end-users). The purpose of these meetings was to collect their feedback on the benefits and limitations of semi-immersive VR in design review and user involvement processes. Moreover, the necessary steps to effectively conduct immersive sessions and obtain reliable results from the participants have been analyzed. In specific terms, operational requirements of functionality and effectiveness of internal spaces have been evaluated during the immersive sessions (Table 1).

A research and data plan organized in three phases was developed, including: (1) technological and logistical preparation, including selection of participants and setting of the immersive virtual environment; (2) data collection; (3) data analysis.

### 2.2 Technological and logistical preparation

The design proposal, which has been used as a testbed during the development of the case study, is related to the BIM-based architectural, structural and MEP detail design of a new educational facility (Figure 1). The first phase of the research and data plan includes the technological and logistical preparation of the immersive sessions. During this phase (1) the immersive virtual environment was prepared, (2) the immersive sessions were planned

Table 1. Design requirements taken into account.

| Operational requirements of functionality and effectiveness* |
| --- |
| 1. Functionality of internal spaces in relation to their use and destination |
| 2. Effectiveness of spaces to ensure learning paths and innovative learning methods |
| 3. Accessibility and safety of circulation paths |
| 4. Interchangeability of space functions: Classroom/ laboratory |
| 5. Flexible aggregation of contiguous spaces to form larger one and vice versa |
| 6. Upgrading of circulation spaces for teaching or aggregative use |

*Guidelines for so-called innovative schools have been published by the Italian Ministry of Education, University and Research (MIUR) in 2013. They focus on flexible learning spaces as required to fit a variety of learning styles and activities with frontal lessons no longer considered the leading model in pedagogy (Giordani et al., 2017).

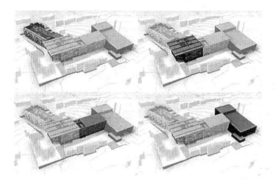

Figure 1. The design proposal includes a primary and a secondary school, a school gym, a canteen and offices (Cominelli et al., 2018).

and (3) a "before-the-experiment" version of the immersive session agenda was proposed.

### 2.2.1 Preparation of the immersive virtual environment

The semi-immersive virtual reality that has been implemented in the case study described in this paper is an open footprint single-wall. Such a technology includes:

– Stereoscopic projector
– Single portable and transportable flat screen
– Visualization package (i.e., Virtalis Visionary Render)
– Graphic card which renders the left and right frames for the projector to show
– Tracking system
– Active glasses that shutter in synchronization with the projector
– Passive polarized glasses, which have been used in group sessions
– VR controllers (i.e., a 3D mouse and a flystick).

The building information model was imported within the virtual environment in the Industry Foundation Classes (IFC) data format in order to visualize both geometrical and non-geometrical attributes during the immersive sessions (Hilfert & König, 2016). On the other hand, the IFC-based representation of the design proposal is simplified in terms of colors, material and lighting settings (Figure 2; Figure 3, left). In order to evaluate if such a representation would have been comprehensible enough to communicate the design proposal to both technical and non-technical stakeholders, an intermediate immersive session was organized during the preparation of the immersive virtual environment, involving an architect with experience in educational facilities and a student representing the stakeholders' category of the end-users (Figure 2, left). They were asked to navigate within that first version of the immersive virtual environment and to evaluate if the design proposal was understandable in terms of internal spaces. Traditional 2D drawings and renders had been prepared to eventually support them during the navigation (Figure 2, right).

During and after the session, the prospective stakeholders provided their feedback. According to the architect, internal spaces should have been more detailed in order to be effectively perceived and evaluated. He was particularly concerned about the necessary increasing need for resources to smoothly implement VR in design processes. At the same time, he recognized the communicative potentials

Figure 2. Intermediate immersive session with an architect expert in innovative schools and a prospective end-user (left). Use of 2D drawings during the immersive sessions to support VR visualization (right).

Figure 3. Visualization of the design proposal imported as IFC file within the VR visualization package (left) and post-processed IFC-based version of the immersive virtual environment (right).

of the tool with a focus on the key role of clients to boost VR implementation as a useful medium to involve end-users and to better manage the interface among the demand- and supply-sides. The young student also found it difficult to evaluate the internal spaces because of the non-realistic colors and a sort of disorientation during the navigation within the immersive virtual environment. Their comments revealed the need for a more detailed representation of the design proposal in order to better communicate it to various stakeholders, including non-technical ones. A second version of the immersive virtual environment was prepared, implementing their feedback in terms of textures and lighting settings (Figure 3, right).

In addition to the VR technology, the workspace where the immersive sessions have been conducted also included traditional 2D drawings and renders. The use of different media during the immersive sessions aimed to support the understanding of the project and to evaluate what type of representation each participant would have preferred to look at during the analysis of the design proposal (i.e., immersive VR, traditional drawings, the joint use of both the media) (Figure 2, right). Moreover, different circulation paths were planned based on the stakeholders involved in order to visualize different aspects of the project with public clients, designers and end-users (i.e., teachers, parents, students, school director).

#### 2.2.2 Planning for the immersive sessions

The immersive sessions have been planned involving a representative panel of prospective stakeholders. They were selected considering the following factors:

- Public clients who have recently faced the need for new educational facilities in their municipalities and, possibly, who have already handled BIM implementation in public procurements.
- Architectural, structural and building services designers, both BIM-aware and analogue ones, possibly with previous experience in the design of innovative schools.
- End-users of a similar building (i.e., school director, teachers, students, parents).

Selected stakeholders have been contacted by e-mail and a call for expression of interest was published online. Finally, seven immersive sessions were organized, involving twenty-four users (Table 2).

#### 2.2.3 "Before-the-experiment" immersive session agenda

A "before-the-experiment" agenda for conducting the immersive sessions was proposed. It was organized in order to (a) explain to the partici-

Table 2. Stakeholders involved in each immersive session.

| Session | Stakeholders involved | Users |
|---|---|---|
| 1 | BIM-aware architect | 1 |
| 2 | BIM-aware architects | 2 |
| 3 | Accessibility expert | 1 |
| 4 | BIM-aware architect, structural engineer and MEP engineer | 3 |
| 5 | Public client and two analogue architects | 3 |
| 6 | Public client and a structural engineer | 3 |
| 7 | End-users | 11 |
|   |   | 24 |

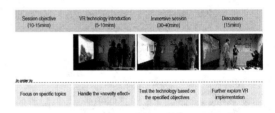

Figure 4. "Before-the-experiment" agenda for immersive sessions.

pants the objective of the immersive sessions, (b) introduce them to immersive VR, reducing the risk for the novelty effect, (c) guide the immersive session within the proposed educational facility and (d) manage a final discussion to further explore aspects related to VR implementation (Figure 4). The four following phases of the agenda and the proposed related timeframes are:

- Define the session objective (10–15 min).
- Introduce users to the immersive VR technology (5–10 min).
- Conduct the immersive session (30–40 min).
- Conduct the discussion (15 min).

### 2.3 Data collection

The method of multiple investigators (Kathleen, 1989) has been adopted, with team members having unique roles during data collection. A team member acted as facilitator and guided the sessions following the macro-phases proposed in the "before-the-experiment" agenda described in 3.3.3. Another team member drove the navigation within the immersive VR by using the 3D mouse and describing the design proposal, whose he was one of the designers. Other two team members, designers as well, took part in the sessions answering questions about design choices. A VR expert managed technical questions about immersive technology. Moreover, the facilitator handled the meeting having a personal interaction with the participants, while other team members were recording notes and observations, keeping a more distant view (Kathleen, 1989). Furthermore, immersive sessions have been audio and video recorded, except when end-users were involved because of the presence of minors.

At the beginning of each session, the objective of the project is stated to the participants: evaluate the implementation of immersive VR to communicate design proposals, involve end-users, manage the client-designer interface and analyze functionality and effectiveness of design proposals in public building projects. The design requirements to consider (Table 1) are reported. In order to avoid the novelty effect that can appear during the first attempt with immersive VR technology, it was decided to let the participants test the tool on a different project (i.e., an automotive virtual prototype). The participants were taught how to (1) navigate within the virtual environment; (2) select and move objects; (3) measure geometric dimensions; (4) visualize non-geometric data embedded in BIM objects, (5) change materials to BIM objects selecting them from a library. The purpose was to avoid that participants would have been too distracted by the technology during the following step. This phase of the session was also used to illustrate why an open footprint single-wall was selected for the development of the case study rather than a full-immersive VR, optioning for emphasizing multi-user interaction instead of a high level of immersion (Castronovo et al., 2013). The performance of the immersive session represents the core phase of the meeting agenda. Starting from a general description of the design proposal, the participants were asked to navigate the virtual prototype following the design circulation paths, simulating the flow that they would have been followed in the real building. The purpose was to let them discover the design proposal and comment it, without intervening directly. The facilitator guided the immersive sessions asking questions only when there was the need to re-focus the participants on the objective of the session. Moreover, this step of the agenda was used to understand if users (1) are able to move by themselves within the immersive virtual environment or if a driver should rather guide their walkthrough; (2) look at the project in a more interactive way than on traditional drawings and what, in their opinion, differs in using the two media; (3) would need further media available in the interactive workspace; (4) are effectively supported by immersive VR in the analysis of operational requirements such as the ones stated in Table 1. After the immersive session, a semi-structured interview was conducted in order to collect further feedback from the participants. During the discussion, a series of open-ended questions has regarded three main aspects: technological, contractual and procedural ones. Only the formers are within the scope of this paper.

### 2.3.1 Virtual reality sickness during immersion

Participants have been introduced to VR technology, as previously described, so to let those who want to navigate by themselves within the virtual environment. In the session involving the accessibility expert (i.e., session 3), who is also a wheelchair user, the navigation was driven by a member of the research team because the user was not able to interact with the technology by himself. In all of the other sessions, users decided to start the navigation using the flystick and the 3D mouse while visualizing the design proposal. However, this navigation mode resulted in more limitations than benefits revealing the need, at a certain point of the session, for a third-person driver who navigates the participants within the virtual prototype of the building following pre-defined paths and while explaining design choices. One of the main reason for that, other than the novelty effect, is related to the virtual reality sickness that affected the immersion during some of the sessions.

Virtual reality sickness affected three sessions out of seven, representing the main issue to data collection. In these cases, after few minutes (i.e. 8 minutes in session 8, 13 minutes in session 3, 15 minutes in session 4) it has been necessary to take a short break and continue the session either with a third-person driving the navigation or turning the stereoscopic visualization off. Usually, the first users feeling the virtual reality sickness were the ones guiding the session (i.e. the one wearing the active glasses) because of the high level of concentration required them to navigate the virtual prototype as if they were in the real building combined with the lack of experience in using 3D mouses and flysticks to move. The observers (i.e. the ones wearing passive glasses) usually felt sick after 30 minutes because of nausea due to observing the movements made by someone else and with glasses that are not synchronized with the tracking system. It is worth to note that even when the stereoscopic visualization had to be turned off, the stakeholders highlighted the benefits of a shared visualization of the project in a 3D and close to full-scale environment and the sessions continued as planned.

## 2.4 Data analysis

Data analysis and data collection often overlapped during the development of the case study. This was to optimize the agenda process from one session to the next. In order to analyze the collected data, a framework has been developed and used to compare the results from each session. The framework allowed researchers to define a general agenda with improvements for the effective implementation of virtual reality in future projects.

For each session, the collected data was organized into the following categories:

- Amendments to the session agenda (i.e., duration in time of each step of the proposed immersive sessions agenda to compare them session by session and evaluate if they were shorter or longer than expected and if further activities should have been introduced in an improved version of the agenda).
- Media that have been used during the session (i.e. traditional drawings, virtual reality, references and examples of similar buildings) and interaction of the users with the immersive technology (e.g. do they prefer someone to guide the session or to move by themselves?).
- Limitations to follow the pre-defined agenda due to virtual reality sickness during immersion (as already described in 3.4.1).
- Technological limitations according to the participants.
- VR benefits in communication with end-users and evaluation of the design proposal

### 2.4.1 Iterations of the session agenda

First, the time spent by the participants in each phase of the proposed immersive session agenda has been considered as well as the need for additional steps.

#### 2.4.1.1 Visualization of the project on traditional drawings

The architect involved in session 1 required to visualize the project on 2D drawings before starting the VR-technology introduction in order to have an idea of the project he was called to discuss about. The visualization of the project on traditional drawings was maintained in the following sessions as an additional phase of the agenda, leaving participants the choice either to use it or to start directly from the VR visualization. Most of the stakeholders involved decided to start visualizing the project on 2D drawings except in one of the two sessions involving public clients (i.e., session 8) and when the multidisciplinary group of BIM-aware designers was involved (i.e., session 5). In fact, they preferred to focus directly on the semi-immersive visualization mode; however, also in these cases at a certain point the need to look at 2D representation became evident in order to provide an overview of the building spaces and reduce the disorientation caused by the navigation within the virtual environment, which was also due to the lack of a mini-map of the building available on the screen. This phase lasted between 5 and 15 minutes depending upon the number of users involved in the session and their level of expertise in reading technical drawings.

End-users looked only at renders and functional layouts, while, 25 minutes were used by a public client who took part in the case study together with two analogue architects (i.e., session 5).

#### 2.4.1.2 Introduction to the meaning of innovative schools and key design requirements to consider

In two sessions it has been necessary to introduce references to normative codes, design guidelines and examples related to innovative building schools in order to clarify some aspects of the design proposal and to guide the participants in the evaluation of the operational requirements stated in Table 1. Such an introduction was implemented with designers who are not expert in educational facilities (i.e., session 3, 5 minutes) and above all with the end-users (i.e., session 6, 15 minutes).

#### 2.4.1.3 The conduction of the immersive session

The immersion took between 30 and 40 minutes, as expected, except in session 8 when two public clients experienced the VR visualization technology for 1 hour and 15 minutes. However, it is important to note that after 40 minutes the stereoscopic visualization was turned off because virtual reality sickness occurred and for the rest of the time the non-immersive mode was used.

#### 2.4.1.4 Discussion

The discussion phase was twice as long as scheduled, from 20 to 40 minutes on the basis of the number of people involved.

### 2.4.2 Need for an interactive workspace

The type of media used by the participants during the sessions was considered. BIM-aware designers, in fact, have shown an interest in VR implementation to perform design review, including clash detection and feedback collection from clients and end-users. Designers (i.e., in particular in sessions 1 and 4) have pointed out how, even if immersive virtual reality was used, it would still be necessary to organize an environment equipped with multiple supports for design communication, including at the same time more conventional representations that could support an overall overview of the design proposal. They also highlighted the need to take note of feedback and detected issues in an interactive way between the VR visualization tool, used as or integrated with a smartboard, and the BIM authoring platform (e.g., by means of the BIM Collaboration Format – BCF).

### 2.4.3 Technological limitations

What has emerged from the analysis of the immersive sessions is that BIM-aware designers are the ones more interested in the type of technology used. An interesting aspect is that all of them have already tried various types of immersive VR, some of them have started to implement if for visualizing building information models, while others have already planned to buy and implement it especially in the low-budget option of VR headsets. For that reason, the open footprint single-wall solution was considered as a limitation in the immersive experience and in term of sense of presence if compared with VR headset and multi-wall semi-immersive technologies: "The impression is that of being in front of a large aquarium" (i.e., session 4). In their opinion, (a) the screen that does not reach the floor alters the users' point of view. The same happens because of the (b) impossibility to set the gravity option while moving within the immersive VR, which is related to the specific adopted visualization package. (d) Some difficulties in the perception of the actual dimension of the spaces exist and (e) a mini-map of the building, including space names, should be included to guide the user during the immersion. Moreover: (f) being the flexibility of the spaces one of the main functional requirements for innovative schools, the use of haptic technology to move pieces of furniture and flexible walls should be tested in order to evaluate in a more interactive way how the space can be reconfigured based on the educational activities to be performed. Furthermore, (g) the real-time management of lighting results for the perception of both natural and artificial lighting should be implemented. Finally, (h) the process of preparation of the VR-compatible version of BIM models should be smoother to implement VR in existing design workflows.

Different feedbacks have come from analogue designers, public clients and end-users who, for the first time, used an immersive VR technology. Analogue designers highlighted the benefits of the tool to communicate the design proposal to stakeholders without technical expertise but, at the same time, they considered as a limitation the wrong perception of the space that sometimes occurs. Moreover, they felt the need for a more detailed representation of materials, lights and furniture as necessary to communicate the design proposal. Public clients and end-users did not take into account the type of technology implemented but they highlighted the strong communication benefits of immersive VR. In general, according to them, a better idea of spatial and functional design aspects that would not have been noticed on 2D drawings is provided. End-users appreciated the possibility to observe the project in full scale and dynamically, moving within the spaces; they started to imagine possible educational activities within the innovative learning environments proposed. Finally, public clients stated they were not interested in the type of technology used but in the benefits that it takes, which they consider to be evident in the analysis of the project going behind normative requirements and evaluating how the building will be used (Figure 5).

Figure 5. According to public clients, the first-person perspective during the walkthrough within the virtual prototype supports usability evaluations going beyond minimum geometrical requirements stated in normative texts.

### 2.4.4 VR-enabled communication and analysis

After the immersion, participants were asked what they believed to be the communicative potential of the tool and how this could support the involvement of end-users. All the stakeholders involved expressed a positive opinion, highlighting the differences with traditional processes and emphasizing, however, the need for some performance measures. Based on their feedback, sessions involving end-users should involve (1) understanding of the project from a functional point of view through 2D static views; (2) illustration of the intended use and operation of the building; (3) guided navigation within the immersive VR. They all agree that leaving freedom of navigation to clients and end-users would entail the risk of a dispersive and unfocused session, which (a) does not consider the key aspects on which to make decisions and, not of minor importance, (b) would go beyond the maximum time for which it was proved that a session can be conducted without causing discomfort and nausea in the people involved due to virtual reality sickness.

Another issue that emerged concerns the creation of expectations in clients and end-users that could be disregarded in the final building. The topic is not actually within the scope of this paper; however, it is useful to report that according to public clients the level of detail in the virtual prototype of a design proposal should be compliant to what is required for each design phase in the public procurement code, paying attention in the use of characterising elements such as furniture and textures within the virtual environment if not already decided. The end-users themselves agreed with this statement adding that, in order to avoid the risk of creating expectations, it would be necessary to clarify at the beginning of the immersive session what has been already decided and what is just a hypothesis and to prefer, is possible, a neutral representation for the latter.

## 3 DISCUSSION AND CONCLUSIONS

### 3.1 Implications

After analyzing, discussing and validating the research findings with the participants, a preliminary standardized "after-the-experiment" immersive session agenda has been developed, including improvements for future similar projects where semi-immersive multi-user virtual reality is implemented.

Phases of the agenda and related timeframes have been modified as follows (Figure 6):

– Define the session objective (10–15 min).
– Introduce users to key design requirements to consider using examples from similar projects and references to normative codes and regulations (optional, 5–15 min.)
– Introduce users to the immersive VR technology, describing the one they are going to use and explaining why this specific type of VR has been selected to support the design meeting. Do not let them try the technology for training purposes during this step (optional, 5–10 min).
– Start recording audio and video, if users agree. Otherwise, some members of the team should be responsible for collecting hand notes.
– Restate the objective of the session, specify the project phase the design proposal refers to and declare the related level of detail, stating what has been already decided and what is just a hypothesis.
– Conduct the immersion without exceeding the 30–40 min in semi-immersive mode in order to prevent virtual reality sickness.
– Take a break to recover from the virtual reality sickness that might have appeared during immersion (5 min)
– Conduct the discussion to collect "after-the-session" feedback related to the scope of the meeting (15 min).

During the immersive session, a member of the research/design team is required to drive the navigation within the immersive virtual environment, following pre-defined paths previously defined based on the type of stakeholders engaged in the meeting and describing design choices. The driver should stop the navigation in interesting spots to analyze in detail, where the participants are let to interact with the single-wall exploiting the immersion and the sense of presence guaranteed to the user wearing active glasses, who can interact with the tracking system observing the full-scale virtual prototype in a first-person perspective. In this way, participants can analyze internal spaces evaluating, for example, how they fit operational purposes. Those steps have to be iteratively repeated based on the number of design aspects to be evaluated.

Moreover, an interactive workspace has to be provided with various media available to support the visualization and analysis of the project on 2D drawings, renders and virtual reality; examples of similar projects may also be available for reference and a smartboard for collecting and tracking comments directly on the BIM model should be provided.

Finally, the facilitator should regularly ask participants if they feel virtual reality sickness during immersion and, if this does happen, the session should take a break or continue in the non-immersive mode. Results, in fact, have shown that more than the degree of immersion, the added value of VR implementation during the sessions was the possibility of looking at a full-scale virtual prototype of the building, levelling the ability to understand and comment on it by both technical and non-technical users improving the way they interact.

### 3.2 Benefits and challenges

The benefits of the proposed preliminary standardized agenda deal with the possibility to follow in future projects a series of steps and performance measures that allow the research/design team to effectively collect feedback while engaging users, obtaining reliable results from VR implementation. Further challenges should be tackled regarding the analysis of the added value of VR in BIM-based design processes based on the adopted project delivery method as well as on specific user engagement processes (e.g., participatory design), which are not within the scope of this paper.

### 3.3 Research limitations and further research

The current study considered only a specific type of public building and operational requirements

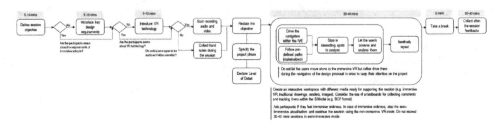

Figure 6. "After-the-experiment" preliminary standardized immersive session agenda.

related to functionality and effectiveness of internal spaces; the findings need to be generalized in other settings. Giving that the growing adoption of VR is related to both public projects and private investments, the above implications and challenges need to be analyzed in additional projects. At the same time, the context of public procurements and related constraints requires an in-depth feasibility study before proposing the adoption of innovative tools such as immersive VR and this paper represents just the starting point of that process. Moreover, an educational facility has been chosen because of the focus on flexibility of spaces related to the activities to be performed that is stressed on governmental guidelines and normative codes about school buildings that, for such a reason, were considered as an optimal test-bed for implementing a collaborative and interactive tool as the semi-immersive multi-user open footprint single wall.

The presented work is a first step towards the development of a prescriptive meeting agenda for end users design reviews in immersive VR environments. This preliminary work will serve for the authors and future researchers to expand the agenda. Therefore, future research would have to focus on finalizing, validating, and generalizing the proposed preliminary immersive session agenda. To generalize the findings and agenda, the authors suggest to conduct additional qualitative research with video recordings of users leveraging immersive technology. Furthermore, to evaluate the validity of the proposed agenda an experimental procedure could be applied to compare the users' perceived value of the technology and the meeting. This could be achieved by comparing an unstructured and structured agenda. To conclude this research is of extreme importance to support the adoption growth of virtual reality in the AEC/FM industry and research.

## ACKNOWLEDGEMENTS

The research project has been sponsored by the PhD scholarship S. Mastrolembo Ventura was awarded by the Construction Technologies Institute of the National Research Council of Italy (ITC-CNR). The visualization system Virtalis Visionary Render has been provided by STR Software and Nuovamacut, part of the TeamSystem Group. The authors would like to acknowledge all the people involved in this research, which has been developed in close cooperation with the University of Brescia (Italy): Prof. Angelo L.C. Ciribini, the researchers Barbara Angi and Lavinia C. Tagliabue and the graduates Carlo Cominelli, Marco Gelfi, Stefano Libretti, who developed the design proposal and set the immersive environment.

## REFERENCES

Boyd, D., Mayouf, M., & Cox, S. 2016. Clients' and Users' Perceptions of BIM: a Study in Phenomenology. Proceedings of the CIB World Building Congress 3:320–331.

Bullinger, H., Bauer, W., Wenzel, G. & Blach, R. 2010. Towards user-centred design (UCD) in architecture based on immersive virtual environments. *Computers in industry* 61(4):372–379.

Castronovo, F., Nikolic, D., Liu, Y. & Messner, J. 2013. An evaluation of immersive virtual reality systems for design reviews. *Proceedings of CONVR 2013*, London, UK.

Cominelli, C., Gelfi, M. & Libretti, S. 2018. *Towards a digital Architecture: Implementing Immersive Virtual Environments to Support Collaborative Decision-Making Processes for Learning Spaces.* M.Sc. dissertation, University of Brescia (Italy).

Dunston, P.S., Arns, L.L., McGlothlin, J.D., Lasker, G.C. & Kushner, A.G. 2011. An immersive virtual reality mock-up for design review of hospital patient rooms. *Collaborative design in virtual environments*: 167–176. Springer Netherlands.

Eastman, C.M., Eastman, C., Teicholz, P., & Sacks, R. 2011. *BIM handbook: A guide to building information modeling for owners, managers, designers, engineers and contractors.* John Wiley & Sons.

Giordani P., Righi A., Mora T.D., Frate M., Peron F. & Romagnoni P. 2017. Energetic and Functional Upgrading of School Buildings. In: Sayigh A. (eds.) *Mediterranean Green Buildings & Renewable Energy.* Springer, Cham.

Heidari Jozam, M., Allameh, E., Vries, de, B., Timmermans, H.J.P. Jessurun, A.J. & Mozaffar, F. 2014. Smart-BIM virtual prototype implementation. *Automation in Construction* 39:134–144.

Heydarian, A., Carneiro, J.P., Gerber, D. & Becerik-Gerber, B. 2015. Immersive virtual environments, understanding the impact of design features and occupant choice upon lighting for building performance. *Building and Environment* 89:217–228.

Hilfert, T. & König, M. 2016. Low-cost virtual reality environment for engineering and construction. *Visualisation in Engineering* 4(2).

Kumar, S., Hedrick, M., Wiacek, C., & Messner, J.I. 2011. Developing an experienced-based design review application for healthcare facilities using a 3D game engine. *Journal of Information Technology in Construction (ITcon)*, 16(6), 85–104.

Liu, Y. 2017. *Evaluating Design Review Meetings and The Use of Virtual Reality for Post-Occupancy Analysis.* PhD dissertation. Pennsylvania State University (USA).

Liu, Y., Lather, J. & Messner, J. 2014. Virtual Reality to Support the Integrated Design Process: A Retrofit Case Study. *Computing in Civil and Building Engineering, American Society of Civil Engineers*: 801–808.

Maldovan, K.D., Messner, J.I. & Faddoul, M. 2006. Framework for reviewing mockups in an immersive environment. *Proceedings of the 6th International Conference on Construction Applications of Virtual Reality.*

National Institute of Building Sciences buildingSMART alliance. 2015. National BIM Standard—United States® Version 3.

Schaumann, D., Pilosof, N.P., Date, K., & Kalay, Y.E. 2016. A study of human behavior simulation in architectural design for healthcare facilities. *Annali dell'Istituto Superiore di Sanità*, 52(1), 24–32.

Shen, W. & Shen, Q. 2011. BIM-based user pre-occupancy evaluation method for supporting the designer-client communication in design stage. *Management and Innovation for a Sustainable Built Environment*.

Succar, B. (2009). Building information modelling framework: A research and delivery foundation for industry stakeholders. *Automation in construction* 18(3):357–375.

Tutt, D. & Harty, C. 2013. Journeys through the CAVE: the use of 3D immersive environments for client engagement practices in hospital design. *Proceedings of the 29th Annual ARCOM Conference*: 111–121.

Whyte, J. & Nicolić, D. 2018. *Virtual reality and the built environment, 2nd Edition*. Routledge.

# Improving the integration between BIMs and agent-based simulations: The Swarm Building Modeling – SBM

Gabriele Novembri & F.L. Rossini
*DICEA, Sapienza, Rome*

A. Fioravanti
*DICEA, Sapienza, Rome*

ABSTRACT: The disruptive factor of the current industrial sectors is the constant spreading of ICT technology into production cycles. Following this trend, several companies improved their productive factors towards the optimization of processes through the digitalization of themselves, also thanks to public and private funds. In the A/B/C sector industry instead, the gap is to face with the traditional design habit to design and building with usual codes, neglecting the use of advanced tools capable not only to automatize representation and computation, but support designers in creative ideas and design optimizations. For these issues, traditional BIM tools are inadequate lacking semantic, reasoning capability and a full dynamic intertwined consequences of design choices. So, a general model of interaction and integration between Building Information Modeling and Agent-Based Modeling is proposed. This model works following the spirit of swarms, in the meaning that every entity of the swarm (i.e. components, procedures context etc.) are linked with others in making different actions aimed at reaching their own goals reach the same result, maintaining a hierarchical order and pursuing a global balance, like is for the building objects in a coherent project.

## 1 INTRODUCTION

The construction industry is, currently, one of the most important sectors of the world economy, considering its impact on pollution, resource consumption and, globally, on 'carbon footprint', although it is managed by outdated ICT systems compared to innovation that spread other sectors such as, for example, the automotive. Till now, in fact, it is estimated that about 30% (Sveikauskas et al, 2016) of the whole resources used in this kind of process, are dissipated due to management inefficiencies, which can be found both in the design phase both in the constructive as well.

This inefficacy is not a new problem, but a constant condition in the construction sector, faced by current developments in digital techniques according to different approaches to the extension of CAD capabilities: from the interactive verification of choices through the use of Augmented, Mediated and Virtual Reality (Park et al, 2013), to the integration between the BIM model and predictive statistical methods, till to the definition of methodologies oriented to verify design solution through simulation (Scherer & Schapke, 2011).

Despite the development of Collaborative Design methodologies, the analysis of results of different lines of research shows the tendency to discretize the design problem in different 'specialist packages', risking losing the sense of the complexity of the building system, and the impacts that design choices can have on the entire building organism. It needed, thus, to approach the design problem with unified models (Jiao et al, 2013), linking BIM to intelligent tools with the sake to provide stakeholder information, knowledge and design intent (Novembri et al, 2017) in a holistic repository, evolutive as the project as well.

The aim of the research is, therefore, to provide Designers of the building process innovative and appropriate models and paradigms, to manage the complexity of buildings, evaluating the outcomes of these choices in a predictive way. Thus, the prototype under development Swarm Building Model - SBM is based on the paradigm of 'Agents Swarm. In conclusion, each agent can receive stimuli from the outside or other agents, reacting according to its behaviour and objectives, and then involving all the other agents of the model: the result is the dynamic adaptation of the entire model to a collectively satisfying swarm behaviour as happens, in nature, with a swarm of birds.

## 2 STATE OF THE ART

### 2.1 *Computer science and architecture: An old feeling/antinomic relationship*

The idea of a collaboration between designer and machine inspired many research centres from the very starting of the "computational-era" (Humbert, 2007). Effectively, since the sixties these capabilities were well known, as Engelbart stated (Engelbart, 1962) "we can begin developing powerful and economically feasible augmentation systems on the basis of what we now know and have. Pursuit of further basic knowledge and improved machines will continue into the unlimited future, and will want to be integrated into the "art" and its improved augmentation systems—but getting started now will provide not only orientation and stimulation for these pursuits, but will give us improved problem-solving effectiveness with which to carry out the pursuits "This resolution was follow-up by Negroponte (Negroponte, 1973), that defined the extended designer like an interactive element immersed in an "ecology of mutual complementation design, augmentations, and substitution".

From the very beginning evidently, actors of the building process have found in computer science a valid ally, able to effectively manage important amounts of data and equip, in an ever more democratic way, the operators of the sector with instruments that are sufficiently complex, with respect to the insidious complexity of the design problem (Kunz & Rittel, 1970). These tools have therefore gradually evolved, depending on the computing power developed by the machines available on the market, allowing at first the possibility to manage geometries and data in separate environments and, from the last decades, modeling information and knowledge in interconnected holistic environments, such as happens in the BIM approach (Loffreda et al, 2013). The open question now is, again, how much could be developed the human-machine interface, and how much machines can imitate humans also via the subjective aspects like taste, judgement and emotions in general terms.

### 2.2 *From geometric representation to the virtual building: BIM as an evolving revolution*

The introduction of BIM systems represented a first important step towards new level of designing, supporting and managing the connected design/building phases and the interaction between different process actors (Fig. 1). Despite the evident limitations (Mettienen & Paavola, 2014) like consistency of data, interoperability real-time sharing

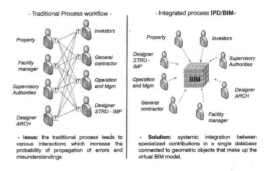

Figure 1. A comparison between traditional and integrated process management approach.

etc., the advantages obtained thanks to these systems now appear to be effective and measurable, even considering the slow and non-homogeneous adoption of this new type of tools. Although there are several valid prototypes of interaction between building design/artificial intelligence (Cambero et al, 2014), the tendency is to focus on the development of these models in specialized field, without going into the holistic vision of the building process.

Actually, it is to be implemented according to methodologies that prefer the collaborative approach in place of a mere sequential integration of specialisms. Beyond these research lines, the integration of BIM and Agents-based Models were developed, maintaining the global vision of the project, towards a synergic collaboration between man/machine and design (Fioravanti, 2008).

## 3 METHODOLOGY

### 3.1 *Learning from nature*

The adopted approach is inspired by the behaviour that some animal species can show by creating numerous and complex groups, able to make surprising geometries based on the simple behaviour of the individual components (Jun, 2012). Some species are indeed able to create groups of subjects that remain compact and coherent, despite the perturbations applied, thanks to the iterative application of very simple rules, because a Swarm represents "the emergence of coherent functional global patterns from the collective behaviours of entities interacting locally with their environment" (Englebrecht, 2005). Despite the apparent simplicity, this approach represents a real revolution (or, in a certain sense, a return on the Cybernetic paradigm (Sutherland, 1975) in the ways in which the behaviour of artificial systems with high levels of interconnection can be simulated. Similarly to what happens in nature for a swarm, the proposed

Figure 2. Bees show a typical swarm behaviour. In the case of the construction of the nest, moreover, they follow 'innate rules' to optimize the work, through mutual synergies, and the optimization of geometry to improve the performance of materials used.

'Swarm Building Model' - SBM is composed of several autonomous agents characterized by specific goal, objectives and rules, working in a collaborative environment (Fioravanti et al, 2017). When the initial balance is altered by internal issues (i.e.: the respect of mutual constraints) or external issues like the context or the functional design development, the 'swarm' will be able to react to these stimuli, creating a new balance that consider the global respect of imposed rules, as happens in natural environment (Fig. 2).

### 3.2 Implementation: Agent swarm acting on the virtual building

In the proposed model, every element included in the building design is an Agent, which is "a computer system that is situated in some environment, and that is capable of autonomous actions in this environment in order to meet its design objectives" (Wooldridge, 2002). So, the need is to address every Agent, characterized by different objectives towards a shared balance, where every Agent reach a suitable level of satisfaction that, at the same time, satisfies the variety of individual objectives required from other agents, setting up a 'Multi-Agent System' – MAS. The global objective, thus, is in the 'swarm' behavior of this MAS, oriented to maintain the coherence into the design choices. The proposed model is composed of the following key elements:

– BIM model, i.e. the digital representation of context and design;
– Knowledge Wrap – KW (Rossini et al., 2017), where are formalized the objectives and rules of agents and, being specific – and evolutive – for each project, 'wraps' precisely the model of the project;
– BIM Event Manager – BEM, that manages updates;

The starting point is the BIM model: it represents every digital element that makes up the context and the consequent design. This model is 'wrapped' in an external repository of knowledge (KW) that encompass the feature of Agents (rules, behaviors, goals etc.) and their correspondence with the instances of BIM model; so, every design change or update, passes prior to the KW, external dynamic-interactive repository. This passage ensures also the avoidance of inconsistency or the propagation of errors into the model: this checking process proceeds for the entire duration of modeling, acting not only on the edited elements but, to ensure the balance of the model, on all the others. The problem, at this point, is to ensure the continuity of model editing and the real update and check of coherence among elements.

In order to avoid these conflict situations, the communication between BIM and KW take place via the BIM Event Manager – BEM. The BEM subscribes some events that the BIM system creates every time a modification occurs in the parametric model, as when an object is created, eliminated or modified. Indeed, events and their subscription are typical mechanisms of the .net framework in which the entire system is developed. Among the events generated by the BIM System, *Idling* plays a major role. The Idling event occurs between user interactions when the BIM system is in a state in which an external system could successfully access the model. Successively, every time that a BIM system signals an Idling state (usually after a stated time), BEM provides modifications in order to apply or search for the necessary information: these are sent directly to the requesting actor by the BEM, without any additional passage (Fig. 3).

A simple example of how this model works has been experimented in verifying the functionality of a corridor in a school. The goal to be reached was the opening of the door from inside the hall towards the corridor; the rule was that the corridor allowed a free passage space of at least 2.00 meters. Therefore, the 'swarm' behavior of the model envisaged the positioning of the doors in such a way as not to occupy, at the time of opening, the space required for the corridor, being positioned in a recess inside the classroom. This hall, therefore losing a part—albeit small—of its surface, automatically recovered the space by changing the position of the partition walls, thus triggering an overall repositioning of the internal partitions of the floor.

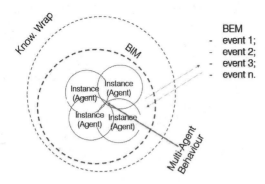

Figure 3. Graphic framework of the proposed model.

## 4 CONCLUSION AND DISCUSSION

The methodology described allows project stakeholders to take the opportunity provided by current computational capability to simplify the complex work of building design and construction using a system able to support process actors in deciding where, when and what to do with awareness of coded risks and the results of simulations performed.

This will be made possible by the capability of intelligent agents equipped with machine-learning capability of ANN to dynamically determine the effective construction sequence needed to improve the quality process rate via the reliable prediction of type and number of work-teams to involve, achieving effective resource allocations.

## REFERENCES, SYMBOLS AND UNITS

Cambeiro F P, Barbeito F P, Castano I G, Bolibar M G and Rodriguez J R. 2014, Integration of Agents in the Construction of a Single-Family House Through use of BIM Technology. Procedia Engineering, 69, p. 584–593.

Engelbart D C. 1962 Augmenting human intellect: a conceptual framework, Ed. Stanford Research Institute: Stanford University, Menlo Park CA, USA.

Englebrecht A P, 2005 Fundamentals of computational swarm intelligence, Wiley.

Fioravanti A, Novembri G, Rossini F L. 2017, Improving Proactive Collaborative Design Through the Integration of BIM and Agent-Based Simulations. In Shock! Sharing of Computable Knowledge, proceedings of the 35th eCAADe International Conference. Rome, 20–22 Sept. 2017. p. 103–108.

Fioravanti A. 2008, An e-Learning Environment to Enhance Quality in Collaborative Design. How to build Intelligent Assistants and 'Filters' Between Them. In: Architecture 'in computro' – Integrating Methods and Techniques, Antwerp, Belgium, 17–20 Sept. 2008 26th eCAADe Conference Proceedings, (ed. M. Muylle), pp. 829–836. Bruxelles: eCAADe.

Humbert M. 2007, Technology and Workforce: comparison between the Information Revolution and the Industrial Revolution. Report-info 210: The information and Services Economy, School of Information, University of California, Berkeley, p. 5–12.

Jiao Y, Wang Y, Zhang S, Li Y, Yang B and Yuan L, 2013, A cloud approach to unified lifecycle data management in architecture, engineering, construction and facilities management: integrating BIMs and SNS. Advanced Engineering Informatics; 27: 173–188.

Jun M., 2012, Research on the Fish Behaviour Simulation based on Swarm Intelligence. In Procedia Engineering; 43: 547–551.

Kunz W and Rittel H. 1970, Issues as elements of information systems. Working paper, Berkeley: Institute of Urban and Regional Development, University of California, Berkeley CA, USA.

Loffreda G, Fioravanti A and Avantaggiato L. 2013, [Architectural] Reasoning over BIM/CAD database. In: Computation and Performance, Delft, The Netherlands, 18–20 Sept. 2013, 31st eCAADe Conference Proceedings, (ed. Stouffs R. and Sariyildiz S.), pp. 495–504. Bruxelles: eCAADe.

Miettinen R and Paavola S, 2014, Beyond the BIM utopia: Approaches to the development and implementation of Building Information Modeling. In Automation in Construction, 43: pp. 84–91.

Negroponte N, 1973, The Architecture Machine: toward a more human environment, Ed. MIT press: MIT, Boston MA, USA.

Novembri G, Rossini F L and Fioravanti A. 2017, Actor-Based modelling of design intentions on BIM systems. In Ciribini, Alaimo, Capone, Daniotti, Dell'Osso, Nicolella, (eds.) Reshaping the construction industry. Ed. Sant'Arcangelo di Romagna: Maggioli, pp. 30–39.

Park C-S, Lee D-Y, Kwon O-S and Wang X. A, 2013, framework for proactive construction defect management using BIM, augmented reality and ontology-based data collection template. Automation in Construction; 33: 61–71.

Rossini, F L, Novembri G and Fioravanti, A. 2017, AS&BIM – A Unified Model of Agent Swarm and BIM to Manage the Complexity of the Building Process. In Future Trajectories of Computation in Design, proceedings of the 17th International Conference CAAD Futures 2017, Istanbul, 12–14 July 2017, p. 321–332.

Scherer R J and Schapke S-E. 2011, A distributed multi-model-based Management Information System for simulation and decision-making on construction projects. Advanced Engineering informatics; 25: 582–599.

Sutherland J W, 1975, System theoretic limits on the cybernetic paradigm. In Behavioral Sciences, 20 (3): 191–200.

Sveikauskas L, Rowe S, Midenberger J, Price J and Young A. 2016, Productivity growth in construction. Journal of Construction Engineering and Management, 142: 10.

Wooldridge M. 2002, An Introduction to MultiAgent Systems, Chichester, England: John Wiley & Sons.

# Holistic methodology to understand the complexity of collaboration in the construction process

S.F. Sujan, A. Kiviniemi, S.W. Jones & J.M. Wheatcroft
*University of Liverpool, Liverpool, UK*

E. Hjelseth
*Oslo Metropolitan University, Oslo, Norway*

B. Mwiya
*University of Zambia, Lusaka, Zambia*

O. Alhava & A. Haavisto
*FIRA, Helsinki, Finland*

ABSTRACT: The modern construction industry is going through changes in the way that it operates, driven by digitalisation. Although digital tools have been developed for over two decades, the adoption has not been as successful as expected. The paper presents a methodology to visualize and understand the complexity of the construction process. A holistic view is presented using the interactions between various thematic structures generated from qualitative data collected from three separate studies. The data was derived from three different approaches and countries in order to increase and understand the generalizability of findings. The analysis in this context departed from understanding how themes found in the three studies interact with data exchange to understand what affects it. Findings confirm the high complexity and capture interactions between various themes and data exchange, bringing a deeper understanding of those interactions; further exemplifying the potential of holistic approaches to understand complexity.

## 1 INTRODUCTION

The Architectural, Engineering, Construction (AEC) industry is known to be inefficient due to its transitory and fragmented nature (Sujan et al., 2017). Positive collaboration between teams at the project level involved in design and construction is critical to ensure project success by creating an environment for high performance teams using technology optimally. Teams in a project are required to collaborate due to the inter-disciplinary nature of deliverables at every stage of construction. Literature shows that there is insufficient research focused on the understanding of people collaborating in the construction industry as compared to the research of tools (Hjelseth, 2017). The authors contend that the understanding of factors that affect collaboration is lacking or highly fragmented in the literature; thereby this study investigates the range of these factors related to data exchange and to devise a comparison between them due to potential for high complexity and interdependence. The holistic context described in the paper is unlike the collaborative aspects investigated in existing literature which are largely studied independently in contradiction to the complex nature of the problem. Therefore, researchers question what impacts practitioner collaboration has at the project level of a construction project.

Human behaviour and collaboration is the product of complex information exchanges at the project level and other external factors that are not well understood (Alperen, 2016). The factors have a high level of complexity due to their inter-related nature; so, as one factor changes this may affect the others depending on the contextual elements of the project (Vandenbroeck et al., 2014).

A construction project involves high levels of social interaction (Cicmil and Marshall, 2005). Research also suggests that there needs to be more social based methods in studying collaboration (Weippert and Kajewski, 2004). Digitalisation in the construction industry is affecting social interaction between practitioners at the project level. Therefore, this study focuses on understanding what practitioners need to consider in relation to data exchange when setting up a construction project in order to create an environment of positive collaboration. Literature suggests that procurement and contracts need to be put in place in a manner to encourage collaboration (Walker and Hampson, 2003), however, the existing literature does not comprehensively document the complex factors that affect people at the project level. The study seeks to understand the various factors that foster a collaborative culture using data collected from three geographically different construction industries.

The study utilises qualitative methods in the form of focus groups and interviews with current practitioners to gain an in-depth understanding. Individual in-depth interviews are used to gain a deeper understanding of the factors described in terms of collaboration. The analytical strategy used thematic analysis, which involves developing codes and themes to sort and rank the qualitative data collected (Braun and Clarke, 2006). Furthermore, the authors assert that national industry specific cultures also affect collaboration at the project level. Therefore, it is significant for industry to understand the generalisability of factors in order to develop an environment where people can collaborate positively and thereby work more efficiently.

The contribution of the paper involves a broad understanding of the human factor in order to help facilitate better use of technology and resources. It should be noted that the examples of interactions presented in this paper are derived to show the initial value of the methodology and model.

## 2 METHODOLOGY

### 2.1 Theoretical underpinning

Various theories such as soft systems thinking and Transdisciplinary Sustainability Science (TSS) provide grounding to approaching a complex problem holistically such as that investigated in this paper. TSS explains the need to study real world problems consisting of both scientific and non-scientific knowledge which requires significant simplification to draw findings (Ruppert-Winkel et al., 2015). In pure organisation science circles, this is considered as an interpretive approach which is argued to be subjective; contextually, due to the need for the researcher's own frame of reference to interpret artefacts and apply semantics to the approach taken by thematic analysis (Deetz, 1996).

In addition to the interpretive approach, a normative approach is also applied; researchers call this law 'like relations' (Deetz, 1996). This was adopted in developing the relationships between the themes described in Section 3.2 which is categorized as shown in Section 3.1 and shown in

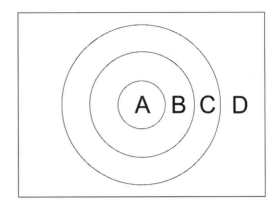

Figure 1. Model structure.

Figure 1. Evidence was found from data to develop the underlying complexity of interactions between themes.

The overall advantages and reasoning of utilising a holistic theoretical departure is:

- Flexibility in the use of parameters allowing for the complex aspects to be more visible. For example, in this context, showing the interactions between the technical, business and social aspects from the perception of the experienced (Vandenbroeck et al., 2014);
- The understanding of collaboration is developed based on studying the effect and working towards the complex underpinning influences unlike in reductionist approaches where only a partial picture of reality can be derived (Checkland, 2000). Simplification cannot assist in understanding the core problems in such a complex interdisciplinary problem (Vandenbroeck et al., 2014);
- Researchers claim that much can be learnt from a little if a holistic approach is applied to a complex problem (Phelan, 2001).

Further, more pragmatically, Taylor (2005) generalised the construction industry as a Project Based Inter-Organisational Network (PBION) with similar social working methods as healthcare, film and defence industries. For example, as can be seen in healthcare, in 1977, a paradigm shift took place in healthcare provision—the biopsychosocial model (Engel, 1977). The model's philosophical underpinning is to encourage holistic thinking of caregivers when providing treatment which requires systematic thinking of biological, psychological and social factors and their complex interactions. Learning from another industry adopting similar methods, it is noted that the AEC industry would also benefit from such holistic approaches in understanding the complexity of the work environment.

Table 1. Method overview.

|  |  | Study | | |
|---|---|---|---|---|
|  |  | Study 1: Zambia | Study 2: Finland | Study 3: Norway |
| Location |  |  |  |  |
| Level of digital technology |  | level 0 to 1 | level 2 to 3 | level 2 to 3 |
| Approach |  | Project to project comparison | End to End, One Firm Perspective | End to End perspective Multiple Firms |
| Duration |  | 30 Days | 5 Days | 5 Days |
| method | Focus groups | Nil | Nil | 5 (14 participants) |
|  | Semi-structured interviews | 15 | 18 | 1 |
|  | Surveys | 22 | Nil | Nil |
| Analysis | Quantitative | Statistical (SPSS) | Not applicable | |
|  | Qualitative |  | Thematic analysis | |
| Participant classification |  | Design Firms (Architects, Structural Engineers, MEP Engineers, Client Representatives | End to End Project Management (Design and Production Managers) | 2 Contractors, 1 Project Management, 1 Public Client, 1 Consultant, 1 Industry Organisation |

## 2.2 Methodology

Table 1 shows an overview of the sample and methodology utilized in the study. Three studies were conducted in three different countries using three different approaches, allowing for increased external validity.

Although approaches varied between the studies, the similarity between the studies was in the focus of semi-structured interviews and focus groups which enabled better understanding of the barriers to collaboration in construction projects; to understand the factors that affect the way people behave and therefore their attitudes towards one another. All semi-structured interviews/focus groups were based on generalized open questions allowing for transfer of control to the interviewee to influence the interview process. Study 1 focused on comparing a highly collaborative project to a less efficient collaborative project and focused on probing the reasons underpinning the efficient collaboration. The quantitative element of the study supported the qualitative claims that one project was more collaborative than the other by applying statistical analysis similar to the perception of the inter-professional collaboration (PINCOM) methodology developed by Ødegård and Strype (2009).

Study 2 utilised the perceptions of design managers and production managers who are heavily involved in the day to day leadership of construction projects and therefore have a holistic view of the end to end construction process.

Study 3 involved an industry wide perspective, as perspectives of participants ranged from the client, designer, industry organisation, management to contracting firms. This was to enable the researcher to understand the differences in the perception of collaboration between the different types of organisations in the industry. Focus groups conducted asked participants what they understood was most critical in influencing collaboration between firms and teams. The participants were probed to question each other's views and were presented with general topics to initiate discussion. The role of the researcher was to manage the discussion rather than control meaning that participants were encouraged to ask why, as articulated as a critical difference by Breen (2006).

Thematic analysis was utilized to analyse qualitative data as used commonly in psychology (Braun and Clarke (2006). The process involved searching for codes in the data and structuring the information under these codes using Nvivo (QRSInternational, 2018), from which overarching themes were defined for a group of codes. The structured data was applied to develop the model structure shown in Figure 1, where each of the categories emerged from the themes. Once the model structure was developed, the interactions between the themes were then mapped using SocnetV2.4 (Socnet, 2018). The approach is similar to that used in social network analysis (SNA) where greater precision was achieved in representing qualitative concepts (Bandyopadhyay et al., 2010). Although used in a different context, the tools developed for SNA were found to be relevant in mapping the complexity of qualitative themes below.

## 3 THE MODEL

### 3.1 Overview

The ideology of the model radiates from the centre which is about human behaviour; attitudes (A) that are brought about as a result of relationship characteristics (B). The controllable (C) and non-controllable factors (D) are selected depending on whether they can be impacted by the teams in the project environment. The justification for utilising A as the centre was by considering the initial question of poor collaboration at the project level. A recurrent statement was 'people's attitudes need changing'. When probed further, most interviewees provided a highly variant response which consisted of multiple factors that contributed to categories B, C or D; the variance was lower in B therefore setting B in connection with A. When generalised, these varying responses were found to involve interactions between various themes, some of which are presented in this paper.

Relationship characteristics (B) consist of trust, openness, respect and comfort which are highly interdependent and affect human behaviour and attitudes, multi-directionally. Evidence for these links is commonly found in psychology defined from an individual's perspective as psychological safety and trust (ref). The interactions between these themes are beyond the scope of this paper, therefore an assumption was made that all the themes in A and B are interdependent due to its complex nature. The main scope of this paper is to understand how the interactions between C and D would affect A directly or via B.

The distinction of controllable and non-controllable is based on the perspective of managing and developing the collaborative environment from the view of a typical project manager. Controllable factors are ones that can be defined and developed by contractual obligations or the way that teams are procured, in other words, factors that can be influenced by the project management or client. Non-controllable factors are all the factors that cannot be defined or controlled in a project operational basis, however still influence the attitudes and behaviour of people at the operational level.

### 3.2 Definition of themes

Themes used to explain the complexity of data exchange in the construction industry found from the three independent studies are based on how the researcher contextually interprets the themes derived from the analytical strategy. The themes are placed into three categories depending on the most suitable position in reference to the model in Figure 1.

- Human Behaviour Themes (A and B)
  - Relationships (R): results from the level of trust (T), openness (O) and respect (R) between teams including the leader.
  - Comfort (CO): psychological ease.
  - Attitudes (A): A way of thinking that affects how decisions are made.
- Controllable Themes (C)
  - Process Management (PM): development of processes regarding how information is shared and how teams inter-operate and solve problems.
  - Liability (LB): the manner in which teams lose profitability.
  - Client Involvement (CI): the day to day involvement of the client in the inter-operation of teams.
  - Contracts (C): a binding legal arrangement enforceable by law.
  - Client Requirements (CR): the needs of the customer.
  - Leadership (L): the way the teams are led and decisions are made.
  - Finance (F): Associated with the aspects of a project that are related to income.
  - Goals (G): the expectations of the project.
- Non-Controllable Themes (D)
  - Organisational Culture (OC): the organisation's visions, values, norms, systems, symbols, language, assumptions, beliefs and habits (Needle, 2010).
  - Local Culture (LC): the norms of the culture in the industry based on a geographical location.
  - Local Policy (LP): laws in society that affect how operation of projects are carried out.
  - Stakeholder Accountability (SA): the responsibility of everyone that is affected by the project.
  - Data Exchange (DE): the transfer of information between teams and individuals which can be either in an informal or formal capacity.
  - Client Knowledge (CK): the client's understanding of the industry.
  - Personality (P): individual differences in characteristics to do with thinking, feeling and behaving (Major et al., 2000).
  - Experience (E): an individual's exposure to the industry.
  - Holistic Understanding (H): the understanding of the whole system that affects how work is done.
  - Change Management (CM): approaches to develop changes in how organisations operate both within themselves and with others.

## 4 RESULTS AND DISCUSSION

### 4.1 Study 1

The data represented in Figure 2 was developed from the comparison of two projects' design teams

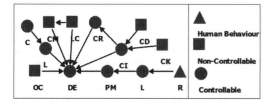

Figure 2. Summary of interactions between themes in Study 1.

where one was found to be more collaborative than the other.

Data exchange (DE) is affected by numerous factors both of controllable and non-controllable nature. DE also has an effect on the themes regarding human behaviour. The approach enabled the researcher to understand the controllable factors closely.

Client involvement (CI) has an effect on how DE occurs both in a formal and informal nature; suggesting that higher CI at the start of the design process, the earlier the design changes start due to changes in client requirements (CR); the vision of the customer becomes clearer and therefore shows that changes need to be made. From the project to project comparison, it was clear that the CI was higher in the more collaborative project which correlated with better client knowledge (CK), as the client team involved end users who understand the industry. Furthermore, the constant involvement of the client team both at formal and informal settings positioned them to be able to make relevant and efficient decisions regarding needs.

Leadership (L) affected DE as the leader was developing the process of DE by defining the channels of communication and exchange. A link was seen between the team relationships (R) and process management (PM) strategy. In the case of the more collaborative project, there was relatively less need to formalise insignificant communications brought about by better relationships due to higher levels of trust as the design teams had good pre-existing relationships. Additionally, an interviewee involved in both projects explained that the nature of the meetings were less heated and more collaborative due to the R between the teams which placed less pressure on the leadership (L). Therefore, creating an open environment where the leader was facilitating discussion rather than controlling conflict.

Liability (LB) in relation to DE was evident as the people in the less collaborative project tended to be more economical with DE as they felt their own mistakes would become more open. Thereby reflecting the contractual obligations (C) made to drive the LB in the project.

Change management (CM) was articulated as participants explained that people do not have the competency in the industry to use the tools. Plus, if they did, the market and the industry's business models are not set up to make the systemic change. A point reiterated by organisational culture (OC) brought about by drastic differences in discipline specific businesses; capacity to utilise tools differs between the teams—for example, a quantity surveyor utilises printed drawings to check the bill of quantities whereas the structural engineer and architects exchange 2D and 3D models, which the L claims is difficult to change.

The Local Culture (LC) in the industry was relevant to DE as information that is shared between teams is perceived as intellectual property by the firms (e.g. in Zambia); this hindered the open sharing of information especially between teams with a new or poor relationship, putting pressure on the L to make the DE more formal.

### 4.2 Study 2

The second study involved data collected from a design and production management firm in Finland, providing a perspective from the leadership of the end to end construction process. Similarly, a summary of the interactions between themes is found in Figure 3.

The way that the contract (C) is developed affects the ability of the leader (L). For example, if a lead firm is not directly contracted to the other teams, the leader may lose leverage; their ability to control the project. The next link is L affecting process management (PM) directly as the leader defines the processes by which DE should occur. Data collected also shows that L affects informal and formal DE as it depends on the style and approach of the leader. The informal DE is affected by the style of the leader, findings suggest that the experience (E) of the L directs the way that the leader leads. Various leadership styles exist, however, according to some participants, the more experienced lead-

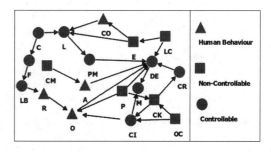

Figure 3. Summary of interactions between themes in Study 2.

ers (E) were more likely to be authoritative whereas the less experienced leaders preferred to focus on facilitating discussion and involving all the teams before making a decision. Furthermore, E is also affected by the local culture (LC) and the way that the LC has been historically changing. As suggested by this model, comfort (CO) is one of the human behaviour themes that interacts with the L theme—as the leader tends to develop a style they find most comfortable from E.

From the data collected, another pathway from C was found to DE. Data suggests that the way that the finance (F) and liability (LB) are defined in the C affect the relationships (R). For example, with a contractual model that makes some teams lose and others win, it constrains the R and therefore the Openness (O). O was directly related to DE, where participants explained that openness about problems or needs hindered DE and PM which placed pressure on the L.

Evidence suggests that the personalities (P) of the people also affect how open they are and thereby affects DE. The interviewed facilitators of the recently introduced big room meetings claim that introverted personalities can sometimes be overpowered by the more extroverted.

Participants also explained how CM has affected individual attitudes (A) historically due to the changes in PM which has been driven by changes in technology (i.e. related to DE).

The organisational culture (OC) of the client firm was evidently affecting client knowledge (CK) and client involvement (CI); participants explain that the high variability of the client's input and client requirements (CR) depend on the nature of the business the client is operating in.

The clarity and definition of the CR was also said to affect the DE, as the need for constant change affects CR which disrupts DE. However, this is counteracted to some extent by positive CI. The definition of the CR and the CI also depends on CK, as a client who has a high CK would ensure there is the suitable amount of involvement in order to ensure that the DE is unaffected by CR. It could also be the case that there is high CI and therefore the CR is always changing. Nevertheless, a client with high CK would find the optimal balance required dependent on the project. Additionally, the CK is said to be dependent on both the P of the representative and the OC of the client where the variance is said to be high. As a result, the common notion is that CK is lacking in the industry.

Participants also explained how the product of positive DE would be motivation (M) of teams. Further, M was also connected to CI which would increase the level of trust and O between the teams as they feel nothing is being hidden. The complexity of M is only partially represented in the figures in this paper (i.e., the focus is on DE) as there are other links in the data. For example, F and L affect M outside the specified scope of the paper—illustrating how the interdependencies of themes demonstrate a highly complex environment.

4.3 *Study 3*

The third study involved an industry wide perspective, therefore providing a more complete idea of the non-controllable factors. The approach also shows the differences in OC between the domains of the industry. Figure 4 shows a summary of interactions between themes.

People tend to lean toward what they are comfortable with (CO) based on experience (E) when it comes to process management (PM) and DE, particularly as the industry remains reliant on having a drawing based output. Participants explained that the value of the information is taken away when a model is turned into drawings. When probed as to why this is the case, one client mentioned that it was due to some members of the supply chain not being engaged in change management (CM). For example, the manufacturers of rebar still expect information (bending schedules) to be provided in the same format as for the last 30 years or more. Further substantiated by the notion that a number of C assume that value is added by producing production drawings.

Evidence suggests that since CM affects the process of DE it thus makes it more difficult to control, again placing pressure on the leadership (L). Additionally, the client requirements (CR) also direct how the processes and DE are defined. Furthermore, contractual obligations (C) with respect to a model rather than drawings are difficult to define; showing that CM is required in the C to better suit the DE. On the other hand, the local culture (LC) in the industry seems to affect the impact of CM on DE. According to practitioners, there appears to be a 'fear of failing' which makes

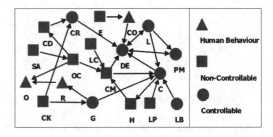

Figure 4. Summary of interactions between themes in Study 3.

change in terms of digitalisation a clear risk perception. Moreover, CM's transfer to project-based DE was said to be affected by the Holistic Understanding (H) of the people—meaning that they need to understand the entire construction process. From the LC, evidence suggests that disciplines are comfortable (C) with thinking about their needs and not motivated to consider needs of the other teams. However, C, such as the Integrated Project Delivery (IPD), have shared goals (G) and liabilities (LB) which force individuals to think more holistically about the processes involved in DE.

Participants in one focus group claimed that based on pilot projects there is a link between DE and the motivation (M) of people, as higher levels of digitalisation reduce the number of repetitive tasks.

The importance of PM in relation to DE is that since the data belongs to different businesses, a robust process needs to be put in place by the L which should be reflected in the C with respect to LB or rewards, in order to ensure that exchanges are efficient.

The client's understanding of the construction process (CK) was also found to develop the G that are shared amongst all the teams; developing relationships (R) that are open (O) and creating an environment where changes can be made as early as possible, avoiding excessive pressure on DE at a later stage. For example, 'Being open can win months', a participant from a contractor claimed. It was seen that the CK affected the C as they are required to decide how the contracts work between the teams; i.e. who the control and power go to.

OC of a firm was also related to O. OC for two of the firms visited were highly hierarchical which meant that the allocation of resources within the business was not done to support CM in DE processes due to the lack of understanding of the management. The workforce signalled that they were not being supported in developing DE and PM. This suggests that the lack of belief of the management created an OC which was not conducive to efficiently manage change (CM) - even though the local policy (LP) that the C followed made such developments a requirement. Additionally, the OC of a client is also found to affect the relationships. For example, participants claimed that public clients tend to be more difficult to work with as compared to private clients. When probed, they explained that the client's representatives from a typical public client would lack in the ability to be decisive (CD) as they would want to go through the 'right channels' to protect themselves the self. Some participants linked this to stakeholder accountability (SA) due to public projects being related to public relations. Therefore, a failure could be led directly back to the decision maker making the decision maker accountable to the people who have stake in the asset.

### 4.4 Key findings

The perspective of the studies presented different interactions between the themes. However, there are similarities in the link diagrams. Some themes that appeared in all three; C, CM, LC, LB, OC, CR, PM, CI, L and CK. This suggests that there is a degree of generalisability of findings between the three different countries' construction industries even though approaches varied.

Although the scope surrounding data exchange was the departing point for analysis of data collected, the methodological approach shows that the construction process is highly complex yet with similarities between countries. Therefore, this demonstrates holistic understanding is required when a change in process or technology occurs.

The following illustrates examples of findings that emerged with the focus of data exchange from the methodology applied:

- Contracts were found to affect data exchange by enforcing liabilities which affect human behaviour and restricts openness of relationships;
- The local culture of the industry was found to affect data exchange both directly and indirectly. Directly in relation to trust between the teams on a project—as teams feel that the information they generate is their intellectual property. Indirectly by historically affecting the experience of the leader and the teams involved—forcing them to reduce discomfort by returning to processes they feel more comfortable with;
- Change management in regards to digitalisation was found to affect the attitudes of the people as there were changes in the processes of data exchange which reduced the need to forcefully collaborate;
- The organisational culture of the client was related to the client's involvement and the client knowledge and thus how client requirements are defined in the project. The common notion that client knowledge is lacking was better understood to be linked to the organisational culture of the client organisation—which was also found to be partially responsible for creating the high variance in client involvement and definition of client requirements.

## 5 CONCLUSIONS

The AEC industry's changing nature due to digitalization requires deep understanding of construction processes. Methodology was developed and applied to create greater holistic understanding of the complexity of the construction process. A model was created to categorise emergent themes from qualitative analysis as a) controllable or b) non-controllable. Controllable factors were those which can be affected by the teams involved in the operation of a project and vice versa for non-controllable.

The focus of the analysis presented in this paper was data exchange due to BIM implementation and other digitalisation-based changes that are common in the modern construction industry. Perspectives presented here show the complexity of the industry. Findings also suggest that holistic understanding of collaboration in the construction process is required to manage change. Changes need to be made by being mindful of human behaviour and potential implications of the same due to both controllable and non-controllable factors. It is recommended that studies similar to those illustrated here are required in order to understand how industry wide changes can be made efficiently.

Finally, it can also be concluded that systemic change with respect to digitalisation in project based inter-organisational networks (PBIONs) should be carried out based on understanding the entire system; comprised of factors that affect one another directly or indirectly.

REFERENCES

Alperen, M.A. 2016. Cognition to Collaboration: User-Centric Approach and Information Behaviour Theories/Models. *Informing Science The International Journal of an Emerging Transdiscipline, Vol 20, Pp 001-020 (2016)*, 001.

Bandyopadhyay, S., Rao, A.R. & Sinha, B.K. 2010. *Models for social networks with statistical applications. [electronic book]*, London: SAGE, 2010.

Braun, V. & Clarke, V. 2006. Using thematic analysis in psychology. *Qualitative Research in Psychology*, 3, 77–101.

Breen, R.L. 2006. A Practical Guide to Focus-Group Research. *Journal of Geography in Higher Education*, 30, 463–475.

Checkland, P. 2000. The emergent properties of SSM in use: a symposium by reflective practitioners. *Systemic Practice and Action Research*, 13, 799–823.

Cicmil, S. & Marshall, D. 2005. Insights into collaboration at the project level: complexity, social interaction and procurement mechanisms. *Building Research & Information*, 33, 523–535.

Deetz, S. 1996. Crossroads—Describing differences in approaches to organization science: Rethinking Burrell and Morgan and their legacy. *Organization science*, 7, 191–207.

Engel, G.L. 1977. The need for a new medical model: a challenge for biomedicine. *Science*, 196, 129–136.

Hjelseth, E. 2017. BIM Understanding and Activities. *WIT Transactions on The Built Environment*, 169, 3–14.

Major, B., Cozzarelli, C., Horowitz, M.J., Colyer, P.J., Fuchs, L.S., Shapiro, E.S., Stoiber, K.C., Malt, U.F., Teo, T. & Winter, D.G. 2000. Encyclopedia of Psychology: 8 Volume Set.

Needle, D. 2010. *Business in context: An introduction to business and its environment*, Cengage Learning EMEA.

Ødegård, A. & Strype, J. 2009. Perceptions of interprofessional collaboration within child mental health care in Norway. *Journal of Interprofessional Care*, 23, 286–296.

Phelan, S.E. 2001. What is complexity science, really? *Emergence, A Journal of Complexity Issues in Organizations and Management*, 3, 120–136.

Qrsinternational. 2018. *Complete your thematic analysis with ease using NVivo* [Online]. Available: https://www.qsrinternational.com/nvivo/enabling-research/thematic-analysis [Accessed].

Ruppert-Winkel, C., Arlinghaus, R., Deppisch, S., Eisenack, K., Gottschlich, D., Hirschl, B., Matzdorf, B., Mölders, T., Padmanabhan, M. & Selbmann, K. 2015. Characteristics, emerging needs, and challenges of transdisciplinary sustainability science: experiences from the German Social-Ecological Research Program. *Ecology and Society*, 20.

Socnet. 2018. *Social Network Visualiser* [Online]. Available: http://socnetv.org/ [Accessed].

Sujan, S., Aksenova, G., Kiviniemi, A. & Jones, S. A Comparative Review of Systemic Innovation in the Construction and Film Industries. eWork and eBusiness in Architecture, Engineering and Construction: ECPPM 2016: Proceedings of the 11th European Conference on Product and Process Modelling (ECPPM 2016), Limassol, Cyprus, 7–9 September 2016, 2017. CRC Press, 89.

Taylor, J. 2005. Three perspectives on innovation in interorganizational networks: Systemic innovation, boundary object change, and the alignment of innovations and networks. *PhD, Stanford University, Stanford, CA*.

Vandenbroeck, P., Dechenne, R., Becher, K., Eyssen, M. & Van Den Heede, K. 2014. Recommendations for the organization of mental health services for children and adolescents in Belgium: use of the soft systems methodology. *Health policy*, 114, 263–268.

Walker, D.H.T. & Hampson, K. 2003. *Procurement strategies: a relationship-based approach*.

Weippert, A. & Kajewski, S.L. 2004. AEC industry culture: a need for change.

*Integrated modelling of the built environment, incl. smart cities*

# Building/city information model for simulation and data management

T. Delval, A. Jolibois, S. Carré, S. Aguinaga, A. Mailhac, A. Brachet & J. Soula
*CSTB (Centre Scientifique et Technique du Bâtiment), Paris, France*

S. Deom
*Paris La Défense, Nanterre, France*

ABSTRACT: This publication is related to a research partnership between the CSTB (Centre Scientifique et Technique du Bâtiment) and the territorial administrative authority of Paris La Défense, regarding multiscale model management and interoperability processes. The project included several steps, starting with the semantization of building data but also information for infrastructures and street furniture. Then, a standard BIM urban model (or City Information Model) has been structured and enriched to address a set of simulations carried out on a pilot urban area. The research challenges were to define data dictionaries, pivotal models and replicable simulation methodologies using the CityGML format and then to integrate them through a multi-scale collaborative format into a database system. In that sense, the objective was to study the potential of BIM as a vector of simulations data for the urban scale.

## 1 SCOPE OF WORKS: MODEL STRUCTURATION AND EXPERIMENTATION

### 1.1 Acquisition process

An acquisition process starting from photogrammetry data has enabled the implementation of a BIM model on the entire territory through an interoperable format adapted to the urban area: the CityGML. The experiment focused on a spatial planning area extracted from the overall model, where urban challenges were expected by local authorities.

Figure 1. Partial view of the generated BIM model in CityGML format including CFD simulation results.

### 1.2 Semantization process

Our process aimed at automatizing the generation of the BIM model, without manual operations made from operators using CAD tools, which would be in any case almost impossible considering the mass of data to be processed on the territory. It consisted in translating topographical plan data into volumes and objects in the CityGML standard format. The achieved semantization allowed to assign a function to each part of the territory, as well as to cluster acquired meshes and forms into infrastructure objects such as roads, railways, green spaces and buildings and to structure them into the CityGML format (see in Figure 1).

During the mesh acquisition process, infrastructure object groups integrated relevant information from the corresponding PostGIS spatial database.

The first sketch, produced by the CSTB, has been stored in database using 3DCityDB solution. In this way, Paris La Défense was able to enrich it semantically (in particular adding generics attributes) and geometrically by usual SQL queries and GIS tools (QGis) before exporting it. The urban model has thus been enriched with structured information that can be requested and manipulated in the same way as the spatial GIS data from which it is derived (using for instance query or filtering), adding 3D volume properties to the BIM semantics.

### 1.3 Implementation and experimentation results

As stated above, the main objective of our research was to define an integrated BIM process (see in Figure 2) taking as an input a section

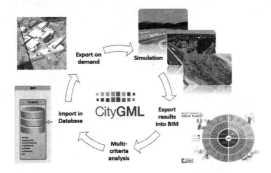

Figure 2. Schematic of the BIM simulation integrated process.

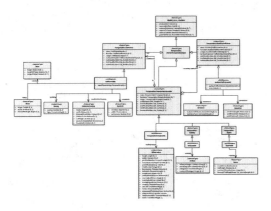

Figure 3. Example of Transportation Complex UML diagram developed for use of simulation.

of the territory, performing a set of simulations and returning the results within the same standard format. This process could allow storage of information (for Data Management through 3DCityDB) and could be used as an input for further simulations.

The first step was to structure and enrich the BIM model to address many simulations from interoperable formats. Our transversal research team—which included experts from several technical domains—carried out wind, acoustic and visual comfort simulations (including shadow and solar reflection simulation) in exterior spaces as well as LCA-based environmental evaluation of buildings. Creating and managing a "simulation-ready" model suited to all performed simulations represented a very challenging task, both from a scientific and technical standpoint. To achieve this, the CityGML format has been extended using extension mechanisms (ADE) as shown in Figure 3. Each extension was built thanks to a tight collaboration between a BIM expert and a simulation expert for each technical domain so as to be able to acquire and structure all required data, and to get the relevant geometrical description, this for each technical domain.

One should point that our purpose is to be able to carry out the results within the OGC consortium to be exploited in future version of the standard.

We now give some details about the way each simulation has been set up, run and integrated back into the CityGML database.

## 2 ACOUSTIC SIMULATION

### 2.1 *Preparing simulation from the BIM model*

As noise exposure is still one of the main concerns for people annoyance and a well-documented cause for adverse health effects (World Health Organization 2011), one of the decided indicator to be calculated was average noise level. Several indicators are typically used in European and French noise regulations, including $L_{day}$ (day-noise indicator), $L_{night}$ (night-noise indicator) and $L_{den}$ (day-evening-night noise indicator). For consistency with the European directive 2002/49/EC, we chose as an example to calculate the $L_{den}$, which is the specified indicator to assess overall annoyance.

To do so, the commercial software MithraSIG v5 (Geomod 2017), codeveloped by CSTB and the company Geomod, has been used. MithraSIG has been developed specifically to simplify as much as possible the integration of GIS data (which can be compared to CityGML LOD0 type data), and was recently extended as part of the European FP7 Holisteec project to directly integrate building objects from CityGML LOD1 files. This however required to develop a connector to automatically generate a LOD1 urban model from a LOD2 model (see in Figure 4), which is what was initially constructed in the BIM model.

To perform noise levels calculations, like all environmental noise software packages, MithraSIG requires several types of data:

Figure 4. Application of the CityGML LOD2 to LOD1 transformation algorithm to buildings objects.

- **Topography** (the ground), here defined as a triangulated complex surface part of a CityGML file
- **Road traffic** (road sections, traffic flow per section, rolling speed and percentage of heavy vehicles), defined as a set of polylines containing traffic flow attributes such as the Average Annual Daily Traffic (AADT) per section, here defined in a GIS (SHP) file
- **Railway traffic** (railway sections, number of pass-bys per section, rolling speed and type of train), defined as a set of polylines for the train tracks, but in which typical number of pass-bys and type of trains were added as attributes.
- **Buildings** (building ground coverages as polygons and building height as an attribute), which usually come from a GIS database, but here directly integrated from the LOD1 BIM model.
- **Urban features**, including **vegetal areas** (polygons, here defined as part of the ground CityGML file) and **noise barriers** (polylines with a height attribute, here defined in a GIS SHP file).

One should point out that for the calculation of the $L_{den}$, one needs to know how traffic conditions vary between the three periods of the day (day: 6h – 18h; evening: 18h – 22h; night: 22h – 6h). Therefore, from average traffic indicators (such as the AADT), a typical distribution of traffic was applied.

Gathering all these data, a complete acoustic simulation model was generated (see Figure 5).

## 2.2 Results computing

MithraSIG can generate several kinds of results, including: horizontal color maps (each color corresponding to a noise level interval), façade color maps, and calculation of a set of predefined points. To generate maps, MithraSIG runs a calculation on a mesh of points generated on the map.

To run the calculation, each road section or train track are divided into equivalent point sources, and ray propagation paths are calculated between each source and each receiver using a ray-tracing algorithm. Then, acoustic propagation effects (reflection, diffraction, geometric attenuation, etc.) for each path are calculated using the standard French method for noise mapping (NMPB'08). Finally, contributions of all paths to a given point are summed to get the level.

The acoustic simulation was first run over an entire neighborhood of Paris La Défense as a horizontal map (at height 1.5 m above the ground) with the following calculation parameters: 20 m receiver mesh size (yielding about 11400 calculation points); 2 reflections per path; 500 m maximum propagation distance; homogenous atmosphere. An example of generated $L_{den}$ noise map in dB(A) is shown in Figure 6.

Figure 6. Generated $L_{den}$ horizontal noise map. dB values vary between below 45 (deep green) to above 75 (deep purple).

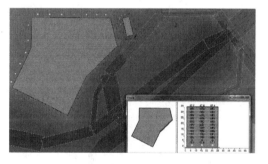

Figure 7. Example of building facade noise level calculation (same color code as above).

Figure 5. View of the MithraSIG acoustic simulation model one all relevant data were integrated.

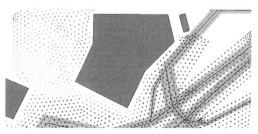

Figure 8. Noise level calculated on a set of predefined set of points (same color code as above).

Figure 9. Integration of a horizontal noise map exported from MithraSIG as a SHP file into the BIM model.

Other simulations were run in order to calculate noise exposure on building facades (see in Figure 7, and on a set of predefined points (see in Figure 8) which was used in coordination with other simulations (see in section 6).

### 2.3 *Integration into the BIM model*

MithraSIG can export its calculation results in many different formats, including GIS type formats (SHP, etc.), PDF for graphical rendering, but also table-like formats (Excel, CSV, etc.) containing a list of objects with their identifiers and attributes values, such as calculated noise level. SHP files can contain polygons defining each colored area.

Since one of the objectives was to re-integrate results into the BIM model, transformations were applied on color maps SHP files to integrate them into the BIM model, both for horizontal maps (see in Figure 9) and for buildings façades.

## 3 SOLAR COMFORT SIMULATIONS

Regarding lighting issues, it was decided to focus the analysis on three subjects:

– **shading analysis** to qualify the impact of a development project on the nearby environment. It is therefore necessary to quantify the importance of solar losses/gains for public or private spaces.
– **importance of solar reflections by glazed facades**, which can induce significant glare risks.
– **public lighting** performance

### 3.1 *Preparing simulation from the BIM model*

Expertise on artificial or natural lighting (sunshine) essentially requires 3D models of building surfaces and the environment as well as the knowledge of the photometric properties of relevant elements: artificial light sources, surface reflection factors, finer descriptions, spectral reflectance, brightness.

Urban BIM models are therefore generally well adapted for this purpose, although it is necessary

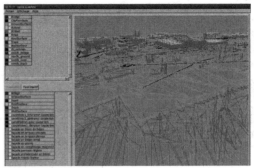

Figure 10. Example of a BIM model imported by Phanie (used lighting simulation software).

to choose the level of detail most suited to the required complexity (see Figure 10).

### 3.2 *Results computing*

For all lighting calculations, the software Phanie, developed by CSTB, has been used. It implements a path tracing method using adaptive records (Ribardière 2011).

The solar shadows analyses are performed on a set of areas of interest. These can be defined based on a list of geo-referenced coordinates or by exploiting various identifiers such as unique IDs of CityGML objects, CityGML types ("road", "wall", ...), or object groups associated with the same material. Then, the most synthetic mode of representation of the results is a map of the number of hours in the shade (or in the sun). This mapping is also available for each time step of the year, which allows for finer analyzes (generation of animations for a typical day for example).

The solar reflections analyses are also performed on a set of areas of interest that can be defined in the same way as for the analysis of solar shadows. The most synthetic mode of representation is a mapping of the number of hours in solar indirect lighting situation via specular elements. This mapping is also available for each time step of the year.

Finally, artificial lighting analyses are also performed on a set of areas of interest that can be defined in the same way as for the analysis of solar shadows. This time, the simplest representation is a map of lighting levels, calculated at a typical human height (see Figure 11).

### 3.3 *Integration into the BIM model*

The calculated photometric quantities are accessible for various analyses and post-treatments (average values, transversal and longitudinal uniformities, etc.), from files with XML formalism.

Figure 11. Map of lighting levels on a set of geo-referenced points.

All this information is geo-localized, or even associated with identifiers of the BIM. They can thus enrich this model of various indicators (comfort, performance) and be represented by tools such as eveBIM or GoogleEarth. The BIM simplifies the definition of the areas of interest and integrates the notion of surface associated with each calculated photometric quantity. This mode of operation is particularly suited to solar studies and those relating to the artificial lighting of urban spaces (excluding roads). The definition of computing grids goes further because it meets normative requirements. For example, it makes it possible to calculate the longitudinal and transverse uniformities of the artificial lighting of the road. Its operation is, however, limited to small areas, which are supposed to be representative of an entire facility. More specific ADEs would be necessary to extend artificial lighting analyses.

## 4 ENVIRONMENTAL SIMULATIONS

### 4.1 Preparing simulation from the BIM model

Two types of environmental simulations have been conducted at district scale: 1) Life-cycle assessment (LCA) and 2) calculation of a biodiversity potential: the biotope area factor (BAF).

Life-cycle assessment (LCA) is a method to assess environmental impacts associated with all the stages of a product's or service's life from raw material extraction through materials processing, manufacturing, distribution, use, repair and maintenance, and disposal or recycling. In the construction sector, LCA has been used since the 1990s to evaluate construction products and building and for few years its application has extended to urban scale, and is more and more used to assess urban precincts and building stocks at large scale (Mastrucci 2017).

In this study, ELODIE, the building LCA software developed by CSTB, has been used (ELODIE 2018). ELODIE is a collaborative tool for quantifying the environmental impacts over the entire life cycle of a building, or an urban object (infrastructure or street furniture).

Performing LCA at building and urban scale requires to collect large amount of information—data—on the foreground system to complete the life cycle inventory (LCI). The integration of BIM to building and urban LCA can reduce efforts during data acquisition, as well as allowing feedback of LCA results into BIM. As mentioned in section 1.3, the CityGML format has been extended using extension mechanisms (ADE) to provide "simulation-ready" model suited to all performed simulations. Regarding environmental simulations, propositions have been made to extend CityGML and ADE Energy standards for exchanging information for LCA simulation at urban scale. Detailed information is given in (Mailhac et al. 2017).

The biotope area factor (BAF) describe the proportion of surfaces favoring the biodiversity into an urban area. In a context of massive extinction, the integration of this indicator is a good way to use the BIM model to take into account the biodiversity conservation.

In the remainder of this section, main data requirements for both LCA and CBS calculation are listed. To perform LCA at urban scale, urban objects within the physical boundaries of the urban project must be identified. This includes in particular:

- **Buildings** (including energy flows related to building consumption, during use);
- **Public infrastructures**: roads, cycle tracks, pedestrian paths, parking, railways; heat network; electricity network; drinking water network; sewage network; telecommunication network
- **Urban furniture**: benches, street lights, bus shelters, fences

Depending on the objective of the LCA study, and the data availability, a screening, simplified or complete/detailed LCA approach can be applied (Wittstock 2012). In this study, the environmental assessment focuses on the elements listed above and consider a 50-year period for the district lifetime. All buildings within the project area are considered to be new construction (i.e. impacts related to their construction, use phase and end of life are taken into account). Data related to each object under study are collected (general information, materials and products used, quantitative data...) and stored in the BIM files.

The BAF expresses the ratio of the ecologically effective surface area to the total land area:

$$BAF = \frac{Ecologically-effective\ surface\ areas}{Total\ land\ area} \quad (1)$$

To calculate the BAF, the different surface's types of the urban area are identified and their areas are calculated thanks to information contained in BIM files. Then, each surface type is weighted according to its ecological value given by the (ADEME 2018), see Table 1.

### 4.2 Results computing

Once foreground information on the district are collected in the BIM files, environmental impacts are calculated using Elodie and information from environmental databases that can be generic; such as Ecoinvent database (Ecoinvent 2018); or specific to the construction sector; such as the INIES database (French national reference database of environmental declarations for products, equipment, and services in the construction sector) (INIES 2017).

LCA results are a set of environmental indicators that reflect the environmental burden of the urban project. The results can be aggregated or presented for each life cycle phase of the project, or for each object.

For convenience reasons, it has been decided to present in this paper results obtained only for the indicator «Global warming potential», in kg eq. $CO_2$ (see and Table 2). However, six other environmental categories have been calculated: hazardous waste disposed, non-hazardous waste disposed, total use of primary energy resources, net use of fresh water, photochemical ozone creation, and total use of non-renewable primary energy resources (primary energy and primary energy resources used as raw materials).

In Table 1, results are presented for each object category and for the entire district considering a 50-year period of analysis. Buildings account for more than 90% of the district's impact on global warming. However, one should remind that 1) results are based on hypotheses taken and 2) main objective of the simulation is to prove the feasibility of an environmental computation using BIM files.

Regarding biodiversity potential of the district, the BAF result is a single score reflecting the ability of the urban project to welcome the biodiversity. The closer the factor is to 1, the more favorable the site is to biodiversity. "Les Groues" district's BAF amount is 0.16 (intermediate results are presented in the Table 2). For this kind of area (central business facilities) a BAF at 0.30 should be reach as recommended in the city of Berlin (Senate Department for the Environment, 2018).

Table 2. Intermediate results for the BAF.

| Surface type | Area in Les Groues (m²) | Weighting factor for each surface type | Ecologically effective surface area (m²) |
|---|---|---|---|
| Sealed | 1.08E+06 | 0 | 0 |
| Partially sealed | 2.36E+05 | 0.3 | 7.07E+04 |
| Semi-open | 1.00E+05 | 0.5 | 5.02E+04 |
| With vegetation: | | | |
| Unconnected to soil (thickness < 80 cm) | 5.02E+04 | 0.5 | 2.51E+04 |
| Unconnected to soil (thickness > 80 cm) | 6.45E+04 | 0.7 | 4.51E+04 |
| Connected to soil | 5.27E+04 | 1 | 5.27E+04 |
| Rainwater infiltration (per m² of roof area) | 0 | 0.2 | 0 |
| Vertical greenery up to 10 m in height | 0 | 0.5 | 0 |
| Greenery on rooftop | 0 | 0.7 | 0 |
| Total ecologically effective surfaces (m²) | | | 2.44E+05 |
| Total surface of the district (m²) | | | 1.60E+06 |
| BAF | | | 0.16 |

Table 1. LCA results on the indicator Global warming potential (kg eq CO2), for the entire district "les Groues".

| | Global warming potential (kg eq. CO2) | % |
|---|---|---|
| Buildings | 1.31E+09 | 91% |
| ○ Construction products and systems | 1.06E+09 | 73% |
| ○ Energy consumption | 2.54E+08 | 18% |
| Public infrastructures | 1.32E+08 | 9% |
| ○ Roads | 1.31E+06 | 0% |
| ○ Heat network | 6.34E+07 | 4% |
| ○ Electricity network | 2.10E+07 | 1% |
| ○ Drinking water network | 3.00E+07 | 2% |
| ○ Sewage network | 1.73E+06 | 0% |
| ○ Telecommunication | 1.50E+07 | 1% |
| Urban Furniture | 1.23E+06 | 0% |
| ○ Bench | 3.81E+04 | 0% |
| ○ Street lights | 9.29E+05 | 0% |
| ○ Bus shelters | 1.07E+04 | 0% |
| ○ Fences | 2.49E+05 | 0% |
| District "Les Groues" | 1.44E+09 | 100% |

## 5 AERAULIC SIMULATION

### 5.1 Preparing simulation from the BIM model

The purpose of the aeraulic simulations is to quantify the pedestrian comfort regarding the wind blowing between the buildings. To do so, it has been chosen to perform CFD (Computational Fluid Dynamics) simulations. They solve the Navier-Stokes equations on a discretized mesh representing the volume of air surrounding the buildings. The quality of the results of the simulations in very dependent of the quality of this mesh. The challenge is then to be able to extract a boundary surface from the CityGML suitable for CFD simulations as shown in the process Figure 12.

This surface has indeed to be watertight and with a Level Of Detail adapted to the flow features at urban scales. As part of this work, an algorithm has been developed in order to simplify the geometry extracted from the CityGML, but a lot of "manual" operations remain to obtain a geometry compliant with CFD simulations. The LOD2 model which has been chosen was a good starting point regarding flow features, however some details are yet to be removed from the geometry to limit the number of cells in the mesh. Finally, the most difficult task is to connect the building geometry with the ground surface to make a single watertight boundary. Once the "cleaned" geometry has been obtained, it must be meshed. As the pedestrian comfort is monitored at 1.5 m from the ground, it is important to have a so-called "boundary layer" mesh which allows a good resolution of the flow near the ground surface (see in Figure 13).

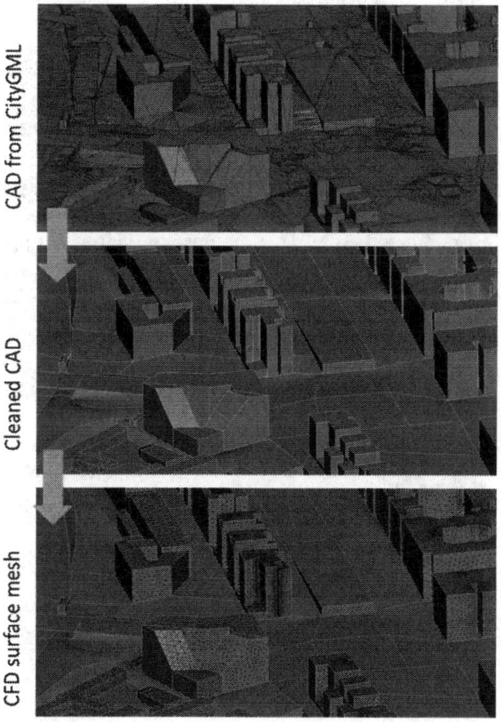

### 5.2 Results computing

Simulations have been carried out using the so called steady RANS (Reynolds Average Navier Stokes) approach. It solves the averaged flow and uses turbulence models which mimic the impact of the unsteady part of the flow on the mean flow. A k-epsilon realizable model is used. The RANS approach proposes the most efficient ratio between confidence in the results and computational time. Simulations are ran using the open source CFD code OpenFOAM.

The flow is solved for 18 wind directions (from 0° to 360° with 20° steps), an example of the wind flows for directions 40° and 240° is shown Figure 14. A boundary layer profile is prescribed at the inlet of the simulations. This profile is relevant of the density of the surrounding urban area according to the Eurocode regulation. Special boundary conditions on the ground (rough walls)

Figure 12. Illustration of the process to generate a correct CFD mesh from the CityGML model.

Figure 13. View of a vertical cut of the CFD mesh with a focus on the boundary layer mesh.

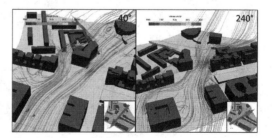

Figure 14. Examples of flow features for two wind directions.

Figure 15. Pedestrian wind comfort map inserted in the BIM model.

Figure 16. Aggregation of simulation results into data points defined as city furniture objects into the CityGML model.

are used in order to maintain this boundary layer profile all over the simulations.

For each wind directions the wind at pedestrian level (1.5 m) is measured and compared to the wind speed prescribed at the inlet of the simulation, which is also relevant of the wind speed which is measured at a local weather station. It is then possible to calculate a "transfer function" between the wind speed measured at the weather station and the wind speed all over the urban area. The wind statistics of the weather station can be then extrapolated to every single point of the simulation.

A quantitative comfort criterion is then defined. It is based on the probability of the wind speed exceeding 3.6 m/s. This "discomfort frequency" shall not exceed a given value depending of the usage of the area. For example, for a restaurant terrace, it shall not exceed 2%, for a strolling area 10%.

### 5.3 Integration into the BIM model

The results of the simulations provide a map of this comfort criterion all over the pedestrian area. This data is then helpful to assess the impact of modifications of the urban area on the pedestrian wind comfort. Those maps are then directly inserted inside the BIM model (see Figure 15). It is also possible to insert some ribbons drawn from the CFD results which picture the flow behavior between the buildings. These ribbons are then helpful to understand the origin of comfort issues and to design corrective devices. For example, in dense areas with high rise buildings, it is not uncommon to see comfort issues induced by streams of wind collected at the top of the tower and sent toward the ground. This can be solved using wind barriers or tall vegetation.

## 6 AGREGATION INTO BIM DATA

### 6.1 Results computing

Although directly displaying simulation results (like noise or pressure maps) allows some form of analysis, the issue of storage and cross-analysis is still a key issue which our work tried to address. The proposed approach was to store all results as BIM data in the same CityGML format as the input data. Multi-disciplinary analysis was then made possible thanks to the development of a generic colorization tool in order to display any type of CityGML object attributes in agreement with attribute values.

To apply this idea, a set of "sensor" points were generated (from a CFD meshing tool) and given as input for all simulations. This set constituted a common basis on which all simulations above were performed (see in sections 2 to 5).

Once all simulations results were generated, the sensor points were integrated into the BIM model. They were structured in the form of City-Furniture objects of implicit geometry, which means that each point geometry was an instanced parallelepiped. The geometry is thus stored only once and then replicated when displayed on all points. Files containing results are thus very light since they only contain point coordinates and values coming from the different simulations (see in Figure 16).

### 6.2 Future works

The next objective of our research is to design a decision support tool for the analysis of public spaces overall quality. The main feature of this tool would be to combine the results of physical simulations, statistical data and socio-urban approaches into a set of key performance indicators.

Besides, our social scientists and the administrative authority in charge of urban strategy are working on innovative user-centric approaches for urban planning and on ways of translating urban quality criteria into BIM standards.

## REFERENCES

ADEME. Le coefficient de biotope par surface (CBS). http://multimedia.ademe.fr/catalogues/CTecosystemes/fiches/outil11p6364.pdf. Accessed 18-04-2018.

« The Ecoinvent Database ». <https://www.ecoinvent.org/>, Accessed 18-04-2018.

« ELODIE Le logiciel de la performance globale des bâtiments ». <http://version3.elodie-cstb.fr/>, Accessed 18-04-2018.

Geomod 2017. Release of MithraSIG v5 software. http://en.geomod.net/gmd-societe-en/gt-accueil/gt-logiciels-2/mithra-suite/mithrasig/

« INIES | Les données environnementales et sanitaires de référence pour le bâtiment ». <http://www.inies.fr/accueil/>, Accessed 27-06-2017.

Mailhac, A., Cor, E., Vesson, M., Rolland, E., Schiopu, N., Schetelat, P., & Lebert, A. (2017). A proposition to extend CityGML and ADE Energy standards for exchanging information for LCA simulation at urban scale. *Presented at the Life Cycle Management Conference,* September 2017, Luxembourg

Mastrucci A., A. Marvuglia, U. Leopold, et E. Benetto, Life Cycle Assessment of building stocks from urban to transnational scales: A review, *Renew. Sustain. Energy Rev.*, vol. 74, p. 316–332, juill. 2017.

Senate Department for the Environment. A green city center—BAF. https://www.berlin.de/senuvk/umwelt/landschaftsplanung/bff/en/anwendungsbereiche.shtml. Accessed 20-04-2018

Ribardière M., Carré S., Bouatouch K. (2011). Adaptive Records for Irradiance Caching. Computer Graphics Forum 30(6):1603–1616. September 2011.

Wittstock B., et al., EEBguide Operational Guidance for Life Cycle Assessment Studies of the Energy Efficient Buildings Initiative, 2012, Available on http://www.eebguide.eu/

World Health Organization 2011. Burden of disease from environmental noise

# Defining quality metrics for photogrammetric construction site monitoring

Felix Eickeler & A. Borrmann
*Chair of Computational Modeling and Simulation, Technische Universität München, Munich, Germany*

Arnaud Mistre
*Centre Scientifique et Technique du Bâtiment, Sophia-Antipolis, France*

ABSTRACT: Point clouds are becoming a quasi-standard as a representation for capturing the existing context, for construction progress monitoring and quality control. While it is possible to create and track sites with a reasonable amount of effort using photogrammetry, different recording strategies and computational power lead to different properties of the point cloud. While the needed specifications are based on the concrete type of analysis and will vary from recording to recording, the overall properties of the reconstruction toolchain are immanent to assess the performance of further processing. Within this paper we will present different criteria for the quality and evaluation of point clouds in respect to construction sites. These indicators together form a benchmark and can be used to evaluate a given toolchain and estimate the properties of a resulting point cloud.

## 1 INTRODUCTION

Capturing construction sites in their as-built state has become a central point in the automation of construction site management and quality control. In the past this was achieved with a high amount of manual labour and untraceable results. Within the digitalization process of multiple disciplines in the construction industry, new opportunities to increase the degree of automation occur. Processed and presented in the right way, the captured data will provide useful information, not only to the main contractor due to a good overview over the progress but also to architects and engineers working in existing context and even insurance companies. The primary step is capturing and documenting the site of interest. Although laser scanned point clouds are of high quality, the expense to create them is rather high. If needed for visualization, colour needs to be added later via images. Capturing images and the photogrammetric reconstruction of point clouds by the combination of Surface from Motion (SfM) and Multiview reconstructions (MVS) are a cheaper and more widely used alternative.

As different automation processes will use the captured data as a basis, the point cloud must suffice specific requirements. Factors as coverage, recording interval, colour schemes, resolution, speed, trueness, precision, observational error and robustness must be taken into consideration. In the workflow of photogrammetric 3D reconstruction, these parameters can be influenced directly, e.g. coverage (our definition in section 4.3) by changing the position of images, or indirectly, depending on the chosen toolchain and reconstruction parameters. Considering 3D reconstruction is a highly computational intensive task and a tradeoff must be made in speed versus quality. For practitioners and early adaptors, this means that using a finished solution may result in insufficient results or unnecessary expenses.

This work will define several indicators and factors that will help the users to evaluate their toolchain. After a brief summary of related research in Section 2, Section 3 will introduce the processes of SfM/MVS that are the algorithm classes used by most applications. In Section 4 we will introduce metrics and deduct how they can be measured before summarizing and suggesting further research topics in Sections 5 and 6.

## 2 RELATED WORK

In recent years, multiple advances for monitoring construction sites via laser scanners or images were developed. Using the base idea of comparing a geometric representation of a construction site and matching it with a captured point cloud (Fischer and Aalami, 1996) lead to several detection algorithms for planed building elements (Scan-vs-

BIM). First approaches went from a simple point to surface with a linear thresholded (<50 mm) recognition of objects (Bosché et al., 2008) over machine learning recognition (Kim 2015) to rasterized point projections onto the geometric model (Rebolj et al., 2017).

While all these publications tend to apply slightly different use cases, the first hurdle for a successful identification is registration of the captured point cloud to the geometric representation. For the relative registration of the geometric model, two base approaches exist in literature: feature-based recognition and the reduction of distances between the alignments (optimization). An Iterative Closes Point (ICP) (Besl and McKay, 1992) showed good results (Bosché, 2010; Masuda et al., 1996) after a successful manual initial alignment. It also worked well in combinational approaches (Huang and You, 2013). A generalized approach for surface to pointcloud was deducted by Segal (D. Holz et al., 2015; Segal et al., 2009). All these approaches need initial alignment or filtering since the ICP is a non-convex method. This problem leads to a global alignment when dealing with noisy construction site captures (Braun et al., 2016; Tuttas et al., 2017).

The SfM/MVS based approach was identified as less accurate but much cheaper (Golparvar-Fard et al., 2011) compared to laser scanning and lead to a discussion of quality (Toth et al., 2013).

To predict the quality of site captures, recent developments showed two different approaches: deducing the quality of the recording based on the result of the identification process (Rebolj et al., 2017) or using pure point cloud related properties and toolchains (Angel Alfredo Martell, 2017; Dyer, 2001; Haala et al., 2013). Following up on these publications, we present simple metrics of quality and emphasize on developing robust independent criteria.

## 3 PROCESS OF RECONSTRUCTION

### 3.1 *Structure from motion*

The structure from motion pipeline provides us with the first step to create a 3D model from taken images. It eliminates the need of a calibrated camera where the extrinsic and intrinsic parameters are fixed and known. In all taken images, points of interest are calculated and their correspondence is determined. Rejecting flawed correspondences, each camera is registered relative to the initial match. With the help of the bundle adjustment, the overall error is reduced significantly as multiple images will be refitted to the current model.

The main goal of the SfM is to generate the initial camera configurations of a scene captured by one or more cameras with multiple images. Since on construction sites, the images are quasi random, it is the first step in retrieving correct 3D information. The correspondences exist as a point cloud deduced for the camera alignment. They already have partial geometric information of the dense reconstruction. The camera alignment is the basis for all common reconstruction pipelines used in construction site monitoring.

### 3.2 *Multi view stereo*

Multi-view stereo algorithms vary significantly in their principles. Seitz categorized the existing methods by six major properties (S. M. Seitz et al., 2006).

1. Scene representation
Voxels, volumes or levelset methods represent the approximate surface, polygon meshes as facets and depth maps as 2D representation.
2. Photo consistency
Two main competitors: Determined by the discretization and projection in scene (reconstruction grid) or image space (pixels of the image).
3. Visibility model
The model verifies, if the view needs to be considered during calculation. This is especially important with larger scenes.
4. Shape prior
During the reconstruction, assumptions for the shapes are imposed e.g. approaches that minimize surfaces.
5. Reconstruction algorithm
Different types are used: calculating the cost of voxels, evolving a surface iteratively, enforced consistency in depth maps and merging them into a 3D scene, fitting a surface to an extracted set.
6. Initialization requirements
Needed initialization may be bounding boxes, fore-/background separation. Image-space algorithms restrict the disparity or depth values.

These properties will later define the quality of different algorithms and in case of 1.) if a point cloud is a suited output. In the next section we will look at current applications available and group them regarding to these fundamental properties.

### 3.3 *Applications*

In the construction industry, only a few selected software solutions are commonly used. Most tools support SfM and MVS and do not need a complementary part. For practitioners the ease of use and the cost can play an important part in their selection of the tool chains. Table 1 lists a selection of applications and their pipelines. When benchmarking different solutions comparing the output of interim results of each reconstruction step helps to identify their limitations. Working with a dataset (construction site 48°08'50.6"N 11°31'33.4"E,

Table 1. Selection of available software solutions for 3D reconstruction from multiple unsorted images. Licenses are commercial (c), non-commercial (nc), and open source solutions. Agisoft (Dmitry Semyonov, 2011), Pix4DMapper (Strecha et al., 2003), ContextCapture (Acute3D, 2018), Visual SFM (CMVS) (Furukawa and Ponce, 2007–2007), Colmap (Schönberger et al., 2016), Sure (Rothermel et al., 2012) openMVS (Demetrescu et al., 2011).

| Application | License | SfM | MVS |
|---|---|---|---|
| Agisoft Photoscan | c | ✓ | depth-map |
| Pix4DMapper | c | ✓ | PDE (whitepaper) |
| ContextCapture | c | ✓ | polygon mesh |
| VisualSFM | nc | ✓ | patch expansion |
| Colmap | o | ✓ | depth-map |
| Sure | nc | ✗ | extracted set |
| openMVS | o | ✓ | depth map |

25th of June 2017, 1087 images) showed that not all software solutions are able to handle the big amount of data from a construction site sufficiently well and that some need a considerable amount of computational resources.

## 4 METRICS

### 4.1 Baseline

For the definition of the quality criteria of a point we will make several assumptions. First, while most of these criteria will also work with all point clouds we will only reference to MVS generated point clouds. Some measurements cannot be performed with the point cloud alone and need a ground truth. For us, there are two possible ways to provide this ground truth. We either generate the point cloud from a synthetically generated set of images (Eickeler and Jahr, 2017; Rebolj et al., 2017) or use actual data (e.g. captured by laser scanning) for the ground truth. With synthetic data generated from a model, it is hard to verify the process of reconstruction due to the assumptions made in the camera model. Opposed to synthetic data, a ground truth from a laser scanner will always have measurement errors and deviations.

Before measuring deviations between captured data and ground truth, the MVS generated model needs to be perfectly aligned. While we can do this with the help of control points, the most suitable alignment (registration) method needs to be determined and the relation to the considered metric must be investigated. This is of particular interest if we take scaling and warping into account. In this regard, we must compare between two different concepts: the alignment via control points with an affine transformation, and the alignment with a ridged transformation to the minimal error. While the first method may induce additional error by distorting the point cloud and fitting the control points to their reference, it may also reduce the error introduced by warping.

Because we cannot assume evenly distributed density, it is not possible to align both point clouds with a point-to-point ICP algorithm and compare the results of the error. Therefore, we will use the generalized ICP approach with point-to-plane matching with a meshed version of the ground truth (or the model itself). Since laser scans are usually much denser than the point cloud generated from an MVS, the introduced error is smaller than the resolution of the tested point cloud (nyquist criteria (Shannon, 1948)).

All criteria will be defined without any relation to the underlying analytical process for Scan-to-BIM, the object recognition. It is our understanding that if we considered these processes, we would bias and tailor the results to any chosen algorithm. Hence, we will only consider metrics that are self-contained within the point cloud, the process of creation or those that we are able to deduce by comparing the reconstruction with the ground truth.

Starting from process parameters that are only partly applicable to all named pipelines in Section 3.2, we will look at the point cloud as an isolated entity and then follow with the comparisons.

### 4.2 Process criteria

Evaluating the process of point cloud creation is meaningful on its own for comparing the resulting quality. Using different configurations and software tools we want to establish some properties for tools first before continuing to investigate point clouds as isolated data structures. This will help to identify possible errors in the image space.

#### 4.2.1 SFM Accuracy

As all pipelines need to find the original camera configuration with SfM, we define the mean error and the variance of the camera positions as our first process criteria. Taking the ridged transformation matrix $H$ where $R$ is a $3 \times 3$ rotation matrix and $t$ the translation vector of the camera.

$$H = \begin{bmatrix} R & t \\ 0^T & 1 \end{bmatrix} \quad (1)$$

With this we define the absolute mean error and the variance as our first two process indicators.

$$\overline{\Delta H} = \frac{1}{n} \sum_{i=1}^{n} | H_{sfm,i} - H_{t,i} | \quad (2)$$

$$\mathcal{V}_H{}^2 = \frac{1}{n}\sum_{i=1}^{n}|\Delta H_i| - \overline{\Delta H} \qquad (3)$$

$H_{sfm}$ is the calculated alignment and $H_t$ the recorded camera positions. For construction sites, the camera positions are not determinable. However for smaller objects, datasets that provide the exact camera positions and orientations exist, (S. M. Seitz et al., 2006).

### 4.2.2 Depth maps, deviation, and noise ratio

Many state-of-the-art reconstruction pipelines use depth maps fusion (see table 1) to generate a point cloud. These maps can easily be compared to the derived depth-map from the ground truth. For this we must create the depth mapping as colour coded view. We use the camera projection matrix from the SfM for this process. A depth map extracted from MVS is shown in Figure 1. After generating the ground truth and normalizing the depth map, the images can be subtracted, and the intensity map evaluated. The variance of this intensity map $\mathcal{V}dm^2$

Figure 3. While higher contours result in higher noise there is also a high frequency interference in the X direction of the depth-map.

is an indicator for the expected accuracy of the point cloud, as it shows the difference before the fusion step. Comparing the Fourier transform in a desired spectrum will compare the spatial resolution of the depth-maps. It is possible to determine the maximal reliable resolution of the resulting point cloud and compare them to the SNR (signal to noise ratio) of the images.

## 4.3 Coverage

The definition of coverage is problematic as we assume knowledge over the following process itself. Normally coverage defines a percentage of the covered ground or volume normed to the overall ground. In current recordings we normally record the image itself, but also the camera parameters and the rough location of where the image was taken and. If (Helge and Nuechter, 2018) recordings are to be taken, this is done by capturing all faces of the building. From the additional data we can determine a rough scale with the differential data of the GPS-Tracker. With this information (and if a higher accuracy is needed, the results from the SfM pipeline) we can deduce the coverage. The process is to fit a closed spline to the projected locations. This spline needs to be offset to the center by the recorded focal length. The coverage can be presented in the length of this offset spline $s_c$ [m] or the inner area $A_c$ [m²].

Figures 1,2. The upper figure shows a depth map from colmap as example. This depth-map has the same resolution as the input images. Figure 2 is the Fourier transform of the same image. A high amount of high frequency spectrum can be seen. The lower image is the result of a low-cut filter. The image set and ground truth was provided by Prof. Nüchter & Helge Lauterbach, Chair of Computer Science VII—Robotics and Telematics, University Würzburg (Helge and Nuechter, 2018).

## 4.4 Resolution and details

### 4.4.1 Point cloud density

The probably most mentioned quality metric in the context of point clouds in construction is the point density. A valid argument considering that most of the progress tracking algorithms use the point density to identify building elements. However, opposed to these concepts, we consider a higher point density only as valuable if the information content is increasing. In our studies we realized that with enough computational power increasing the density is possible but no additional information could be deducted ergo there is no need for a high number of points on a flat surface. We therefore propose a normal weighted density that will increase or decrease with the change of the normal vectors. These will reduce the weighted density in high contour areas and increase the density on flat surfaces without any further information.

$$\rho_w = \frac{1}{i}\sum_{i=0}^{i}\sum_{p,n \in ND} d(p_i, p)\left(c(x) + \frac{\partial n_i}{\partial n}\right) \quad (4)$$

In equation 4, $i$ is the number of points in the point cloud, $p_i$ the current point and $n_i$ complementary normal. $ND$ is a set of $k$ nearest neighbours to $p_i$. Again, $p$ nominates the point and $n$ the normal. The weighting function c is used to control the influence of the normals on the density. This definition also evaluates a higher density on boundaries and strong contours like corners.

### 4.4.2 Resolution

The resolution cannot be measured from a point cloud itself, but we can estimate the maximal possible resolution using the spatial bandwidth. This bandwidth was used for example in a method by Steeb (Steeb, 2005) and applied to point cloud by Graham (Graham, 2011).

$$f(x,y) = kf(kL_x, jL_y)sinc(2\pi B_x x - k\pi) \\ - sinc(2\pi B_y y - j\pi) \quad (5)$$

With:

$$B_x = \frac{1}{2L_x}, B_y = \frac{1}{2L_y}$$

$B_x$ and $B_y$ are the spatial bandwidth that maybe interpreted as line pairs per millimeter, the optical measurement of resolution during recording. However, this formulation must be handled with care and evaluated within local boundaries because the distribution of points from SfM and MVS is not uniform.

### 4.5 Error and robustness

#### 4.5.1 Warping

Warping, a geometrical distortion from rotation of single elements, of the 3D reconstruction is important when analyzing deviation in construction. The error of the reconstruction adds to the building errors of the construction site. If we want to identify warping in comparison to the ground truth, we can use a projective transformation matrix. The needed points can be found using corner detection on both datasets. While there are different algorithms for corner detection, we achieved good results with a *Harris Corner Detector (Harris and Stephens, 1988)*. We consider the non-euclidean factors as warping.

$$H_w = \begin{bmatrix} 0 & 0 \\ v^T & \vartheta \end{bmatrix} \quad (6)$$

#### 4.5.2 Spread of points

Each point cloud that is recorded spreads around the real value. We determine the natural spread by considering the largest $i$ patches. The patch is then reduced and isolated to isolate the patch from any other geometric entity. We fit a plane to this patch with PCA and measure the distance to this plane. This distribution $\delta_{ps}$ can be evaluated to define the minimal and maximal spread $\delta_{ps}$ of the point cloud as FWHM (full width half maximum). Figure 4 shows an example for a photogrammetric reconstruction with Photoscan.

#### 4.5.3 Overall correctness

It is possible to measure the overall distance of a point cloud to the ground truth. Before we can use the ground truth as reference model we had to align our reconstructed model (see paragraph 4.1) in a non-convex optimization. As an error estimate we took the overall normalized minimal distance during the fitting process (7 degrees of freedom). We follow the generalized approach (Segal et al., 2009). The minimal distances $s_{min}(p,q)$ and the distance

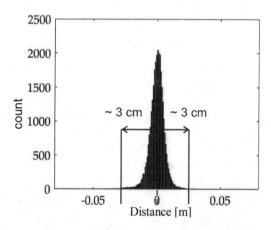

Figure 4. Distribution of distances of the selected points to the estimated plane. The width is a guiding indicator for the precision of the point cloud.

Table 2. Listing of all metrics that were tend to in this paper.

| Metric | Description |
|---|---|
| $\Delta H$ | Camera alignment error |
| $V_H^2$ | Variance of the camera alignment |
| $\rho_w$ | Weighted density of the point cloud |
| $V_{dm}^2$ | Variance of the intensity map difference |
| $\delta_{ps}$ | Distance distribution of the point spread |
| $s_{ps}$ | Width of the point spread [min, max] |
| $\delta_{pq}$ | Distance distribution of the overall model to ground truth |
| $f(x,y)$ | Recoverable Signal e.g. contour |
| $H_w$ | Warping of the model |
| $s_c, A_c$ | Coverage measure |

distribution of $d(p,q)$, $\delta_{pq}$ define the overall fit of the model. The distribution can be regarded as a Weibull distribution. Sometimes specific areas can contribute unintentionally large to this metric. If further investigation is needed, the overlay with a color coding of the distances has proven to be an intuitive tool.

## 5 CONCLUSION

In this paper, we summarize a list of quality measures that can be applied to different point clouds (see Table 2). We concluded that there are three types of metrics for 3D reconstructed point clouds from images: process metrics (Paragraph 4.1), self-contained information (paragraph 4.2) and relative precision (paragraph 4.3). While the first two types can be used to classify the point cloud of the photogrammetric process, the latter needs a ground truth and is therefore only suited to evaluate the toolchain of the user. This is a serious limitation and this third category cannot be used to pose requirements on the point cloud consuming analysis. In practice knowing the properties of the reconstruction pipeline, will help to produce point clouds of similar quality. Going even further, policies for the reconstruction could serve as requirements. This can be regarded as a minimum reconstruction system requirement.

With this concept of point cloud and toolchain evaluation we can estimate the properties of a newly generated point cloud a priori. This information does not provide any benefit on its own and must be further related to the automation processes.

In progress of object detection used for Scan-vs-BIM on construction sites, the detection of elements is done by local density on the number of the points in the vicinity of the building element. The chosen thresholds and recognition criteria will directly impose a certain density of points for the correct as-built recognition. Since most of the time the cloud has a sufficient density for the chosen thresholds (Bosché vincinity ±5 mm, Rebolji projected coverage of 0.5) an analysis with our proposed metrics needs to be made to establish a relation between our criteria and the performance of the object detection algorithm in question. By comparing the needed input quality for a successful detection with the output of the chosen reconstruction pipeline, we can investigate the metrics on their influence and define minimal requirements for the Scan-vs-BIM recognition. Separating these two steps is important for a generalized selection of processing parameters and the abstraction of the recognition process from the pipeline benchmark.

## 6 FUTURE WORK

We consider the definition of the criteria a first step in the benchmarking of algorithms and pipelines. The next step would be to create a selection of captures either with a mixture of synthetic data or/and precise measured data as ground truth.

With this benchmark, studies can be made that emphasize on the pipeline parameters, the needed processing size of the images and the selection of views during the capture. Further, defining the minimal requirements of proposed algorithms (see section 2) for Scan-vs-BIM and combining these insights with the input parameter analysis. This should lead to policies and best practices for the 3D reconstruction of construction sites for as-built recognition.

Other use cases like visualization, documentation, process tracking, quality management and digitalization of building stock to BIM should also be considered and opens up further requirement definitions.

Another research topic would be the use of these metrics for object recognition instead of the proposed algorithms (Bosché, Kim, Rebolji). Instead of solely using a local density it could be beneficial to relate to the weighted density, the coverage, and the resolution criteria $f(x,y)$.

## REFERENCES

Acute3D (2018) *Context Capture—Superior precision* [Online]. Available at https://www.acute3d.com/why-choose-contextcapture/ (Accessed 14 May 2018).

Angel Alfredo Martell (2017) *Benchmarking structure from motion algorithms with video footage taken from a drone against laser-scanner generated 3D models*, Master Thesis, University of Würzburg and Lulea University of Technology [Online]. Available at https://robotik.informatik.uni-wuerzburg.de/telematics/download/MSc_Angel_Martell.pdf (Accessed 22 February 2018).

Besl, P.J. and McKay, N.D. (1992) 'A method for registration of 3-D shapes', *IEEE Transactions on Pattern Analysis and Machine Intelligence*, vol. 14, no. 2, pp. 239–256.

Bosché, F. (2010) 'Automated recognition of 3D CAD model objects in laser scans and calculation of as-built dimensions for dimensional compliance control in construction', *Advanced Engineering Informatics*, vol. 24, no. 1, pp. 107–118.

Bosché, F.N., Haas, C. and Murray, P. (2008) *Performance of automated project progress tracking with 3D Data fusion*.

Braun, A., Tuttas, S., Stilla, U. and Borrmann, A. (2016) 'Classification of detection states in construction progress monitoring', *11th European Conference on Product and Process Modelling*.

Demetrescu, C., Halldórsson, M.M., Goldberg, A.V., Hed, S., Kaplan, H., Tarjan, R.E. and Werneck, R.F., eds. (2011) *Maximum Flows by Incremental Breadth-First Search: Algorithms—ESA 2011*, Springer Berlin Heidelberg.

Dmitry Semyonov (2011) *Algorithms used in Photoscan* [Online]. Available at http://www.agisoft.com/forum/index.php?topic = 89.0.

Dyer, C.R. (2001) 'Volumetric Scene Reconstruction from Multiple Views', in Davis, L.S. (ed) *Foundations of Image Understanding*, Boston, MA, Springer US, pp. 469–489.

Eickeler, F. and Jahr, K. (2017) 'Prediciton and Simulation of Crane Based Camera Configurations for Construction Site Monitoring', *Proc. of the 29th Forum Bauinformatik*. Dresden, Germany.

Fischer, M.A. and Aalami, F. (1996) 'Scheduling with computer-interpretable construction method models', *Journal of Construction Engineering and Management*, vol. 122, no. 4, pp. 337–347.

Furukawa, Y. and Ponce, J. (2007–2007) 'Accurate, Dense, and Robust Multi-View Stereopsis', *2007 IEEE Conference on Computer Vision and Pattern Recognition*. Minneapolis, MN, USA, 17.06.2007 - 22.06.2007, IEEE, pp. 1–8.

Golparvar-Fard, M., Bohn, J., Teizer, J., Savarese, S. and Peña-Mora, F. (2011) 'Evaluation of image-based modeling and laser scanning accuracy for emerging automated performance monitoring techniques', *Automation in Construction*, vol. 20, no. 8, pp. 1143–1155.

Graham, L. (2011) 'Aerial and Mobile Point Cloud Integration—An Industrial Perspective', *Photogrammetric Week*, no. 53.

Haala, N., Cramer, M. and Rothermel, M. (2013) 'Quality of 3D point clouds from highly overlapping UAV imagery', *ISPRS—International Archives of the Photogrammetry, Remote Sensing and Spatial Information Sciences*, XL-1/W2, pp. 183–188.

Harris, C. and Stephens, M., eds. (1988) *A combined corner and edge detector*, Citeseer.

Helge, L. and Nuechter, A. (2018) *Benchmarking Structure From Motion Algorithms of Urban Environments with Applications to Reconnaissance in Search and Rescue Scenarios* [Online], Chair of Computer Science VII—Robotics and Telematics, University Würzburg. Available at https://robotik.informatik.uni-wuerzburg.de/telematics/download/SSRR2018/ (Accessed 17 May 2018).

Holz, D., A.E. Ichim, F. Tombari, R.B. Rusu and S. Behnke (2015) 'Registration with the Point Cloud Library: A Modular Framework for Aligning in 3-D', *IEEE Robotics & Automation Magazine*, vol. 22, no. 4, pp. 110–124.

Huang, J. and You, S. (2013) 'Detecting Objects in Scene Point Cloud: A Combinational Approach', *2013 International Conference on 3D Vision*. Seattle, WA, USA, 29.06.2013 - 01.07.2013, IEEE, pp. 175–182.

Masuda, T., Sakaue, K. and Yokoya, N., eds. (1996) *Registration and integration of multiple range images for 3-D model construction*, IEEE.

Rebolj, D., Pučko, Z., Babič, N.Č., Bizjak, M. and Mongus, D. (2017) 'Point cloud quality requirements for Scan-vs-BIM based automated construction progress monitoring', *Automation in Construction*, vol. 84, pp. 323–334.

Rothermel, M., Wenzel, K., Fritsch, D. and Haala, N., eds. (2012) *SURE: Photogrammetric surface reconstruction from imagery*.

S.M. Seitz, B. Curless, J. Diebel, D. Scharstein and R. Szeliski, eds. (2006) *A Comparison and Evaluation of Multi-View Stereo Reconstruction Algorithms* (2006 IEEE Computer Society Conference on Computer Vision and Pattern Recognition (CVPR'06)).

Schönberger, J.L., Zheng, E., Frahm, J.-M. and Pollefeys, M. (2016) 'Pixelwise View Selection for Unstructured Multi-View Stereo', in Leibe, B., Matas, J., Sebe, N. and Welling, M. (eds) *Computer vision—ECCV 2016: 14th European conference, Amsterdam, The Netherlands, October 11–14, 2016: proceedings*, Cham, Springer, pp. 501–518.

Segal, A., Haehnel, D. and Thrun, S., eds. (2009) *Generalized-icp*.

Shannon, C.E. (1948) 'A Mathematical Theory of Communication', *Bell System Technical Journal*, vol. 27, no. 3, pp. 379–423.

Steeb, W.-H. (2005) *Mathematical tools in signal processing with C++ & Java simulations*, New Jersey, World Scientific.

Strecha, Tuytelaars and van Gool, eds. (2003) *Dense matching of multiple wide-baseline views*.

Toth, T., Rajtukova, V. and Zivcak, J. (2013) 'Comparison of optical and laser 3D scanners', *2013 IEEE 14th International Symposium on Computational Intelligence and Informatics (CINTI): 19–21 Nov. 2013, Budapest*. Budapest, Hungary, 11/19/2013 - 11/21/2013. Piscataway, NJ, IEEE, pp. 79–82.

Tuttas, S., Braun, A., Borrmann, A. and Stilla, U. (2017) 'Acquisition and Consecutive Registration of Photogrammetric Point Clouds for Construction Progress Monitoring Using a 4D BIM', *PFG—Journal of Photogrammetry, Remote Sensing and Geoinformation Science*, vol. 85, no. 1, pp. 3–15.

# HBIM acceptance among carpenters working with heritage buildings

S.A. Namork & C. Nordahl-Rolfsen
*Department of Civil Engineering and Energy Technology, OsloMet—Oslo Metropolitan University, Oslo, Norway*

ABSTRACT: Several researchers have suggested the use of BIM in projects concerning historic buildings and sites. Aspects like facility management, visualisation of historic development, monitoring changes over time and collecting qualitative assets (historic photographs, etc.) are highlighted. As more and more of the software in the AEC-industry are compatible with BIM-formats, a Historic Building Information Model, HBIM, can also be used to assess performance, environmental aspects and cost. This research aims to survey the acceptance of HBIM technology amongst craftsmen working with heritage buildings. Through interviews and testing of HBIM on-site during the rebuilding phase of a project concerning moving a historic building, the carpenter's perception is assessed. The data from the interviews were analysed based on the Technology Acceptance Model. The results reveal that the carpenters find the technology to be useful and easy to use. However, this depends on careful preparation by the site engineer.

## 1 INTRODUCTION

The Scandinavian region can be considered an international leader in BIM implementation. Statsbygg, a governmentally owned firm, is responsible for construction, management and developing state owned properties. Since 2007 Statsbygg have used BIM for the whole life-cycle of their buildings and has specified the use of BIM in all their property developments since 2010 (buildingSMART, 2012, p. 15). Due to Statsbygg being one of the largest construction developers in Norway, all large architecture, engineering and construction (AEC) firms operating in Norway have hands on experience in working with BIM (Merschbrock & Nordahl-Rolfsen, 2016).

### 1.1 Data collecting

The development in technologies such as laser scanning and photogrammetry combined with better software to handle the massive output data from these, makes BIM more attractive as a working platform for documentation compared to traditional systems. The concept of using BIM as a tool in projects containing heritage buildings and sites is called Historic Building Information Modeling (HBIM). As stated in the Historic England guide for developing an HBIM, it is a prerequisite to obtain accurate metric data. Laser scanning, photogrammetry (ground based, or drone mounted), lidar, close range scanning and mobile mapping are all survey techniques that provide a point cloud, which is supported by most BIM software. These 3D digital survey techniques are fast and non-contact methods. Combining the different techniques facilitates a total documentation of geometry with a higher level of detail compared to more traditional survey methods (Historic England, 2017).

### 1.2 Historic building information model

Most research on BIM focuses on the planning and construction of new buildings. Recently there has been more research focusing on using BIM technology on heritage building projects (Ma, Hsu, & Lin, 2017). This requires the production of an as-built BIM, or a digital twin, as a base for the HBIM (Murphy, McGovern, & Pavia, 2009). Assuming a good underlay from photogrammetry or laser scanning, an HBIM can serve as a base for data collection to make a more easily understandable and total presentation of the historic building and its history. From the fact that HBIM also can highlight the development and changes made to the building in a visual way, it can result in a new understanding of the building. One of the main advantages of HBIM is to collect both quantitative assets (intelligent objects, performance data) and qualitative assets (historic photographs, oral stories, music etc.) (Stephen Fai, Graham, Duckworth, Wood, & Attar, 2011). As more and more of the software used in the AEC-industry are compatible with BIM-formats, an HBIM can also be used for simulations to assess performance in terms of statics, environmental aspects and cost (Historic England, 2017, p. 3).

## 1.3 Documentation hierarchy and authenticity

The arguments for preserving buildings is to preserve something that has a value for future generations. It is essential to maintain a high level of authenticity. The building can tell about material used, tools used, joinery, construction techniques, style and marks of use. Godal divides the authenticity term in material authenticity, structural authenticity, historic authenticity and authenticity of a symbol (Godal, 2004). All of these are important for the total authenticity of the heritage building. The hierarchy in the documentation is essential. The establishment of an HBIM can never outdo the building itself but can work as a valuable supplement and an information hub.

## 1.4 BIM: On-site

Several papers discuss the use of BIM in the engineering phase However, there is limited research on the use of BIM on-site, and to which extent it will be accepted by the craftsmen (Bråthen & Moum, 2016; Merschbrock & Nordahl-Rolfsen, 2016). It is essential for the success of on-site BIM implementation that there exists a perceived advantage of use over the existing solution it replaces (Rogers, 2003).

## 1.5 The case and research question

In the case studied in this research both the implementation of HBIM in heritage building projects and the use of HBIM on-site are of interest. The aim is not to prepare and assess the full concept of HBIM, but the features of implementing historic photos in the HBIM is essential for the case studied. In this research we focus on the craftsmen's perception and acceptance of new technology. More specifically we focus on how a group of carpenters perceive the use of HBIM on-site in the project of relocating a heritage building. To assess their point of view a series of semi-structured interviews was considered appropriate. The data collected from the interviews were analysed based on the Technology Acceptance Model (TAM) developed by Davis (Davis, 1989). The research question this article aims to answer is:

> How will carpenters working with heritage buildings perceive the use of HBIM on site in the project of relocating a historic building?

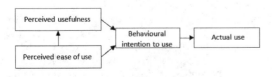

Figure 1. Technology acceptance model (Davis, 1989).

## 2 THEORY

*"BIM is an absolutely wonderful tool, and it has great potential to streamline costs and processes, to help different disciplines communicate effectively and to ensure little confusion on a job site. But to get to that promised land of benefits, you have to pass through the wilderness of adoption, which always seems to hinge on organizational change, not technology. This is the inconvenient truth."* (Dobelis, 2015).

As stated in the quote above by Dobelis the implementation of BIM depends more on the organizational change, than the technology. For an organization to be willing to change, the perceived benefits must outdo the cost of changing the organization. For successful BIM implementation the perceived benefits must be assessed. There is a wide range of different information system (IS) theory models which aim to predict and explain technology adoption.

This research is based on the Technology Acceptance Model (TAM), which posits to predict individuals' intention to use new information technology. The TAM-theory is based on the Theory of Reasoned Action (TRA) developed by Fishbein and Ajzen (Fishbein & Ajzen, 1975). The TRA aims to explain a persons' decisions and acceptance in a general manner. The TAM is an IS theory which suggests that the perceived ease of use and perceived usefulness of a new technology will influence the system acceptance, behavioral intention of use, and actual system use (Davis, 1989). According to Davis the usefulness can be defined to the "...extent they believe it will help them perform their job better." (Davis, 1989, p. 320). Even though the user finds the new system to be useful they may "[…] believe that the system is too hard to use and that the performance benefits of usage are outweighed by the effort of using the application" (Davis, 1989, p. 320), hence the assessment of ease of use. By analyzing data on individuals' stated perception of ease of use and usefulness, the likelihood of continued use can be elucidated.

## 3 CASE

The case concerned is the moving of three historic houses from Finnmark in northern Norway to at the Norwegian Museum of Cultural History in Oslo, Norway. First one need to know the history then we will tell more detailed about the case.

The Norwegian Museum of Cultural History in Oslo, Norway is the World's first open-air museum opened in 1894. The museum's collection of historic building have a great variety spanning from a Stave church from the 13th century to a modern

Table 1. Usefulness and ease of use for information systems.

| Usefulness | Ease of use |
|---|---|
| • Easy access register for information<br>• Full on-site access to detailed information<br>• Better visual understanding of the building<br>• Information flow to project participants<br>• More detailed information<br>• Time savings | • Software is intuitive to use<br>• Interaction with the system is clear and understandable<br>• Easy to find the information you want<br>• Challenges |

flat from 1999. The aim of the museum is to collect and visualize the built heritage from all over Norway. Until recently there have been no buildings representing the northern part of Norway.

### 3.1 The post war rebuilding of Finnmark

In 1944 most of the buildings in Finnmark and Nord-Troms were burned to the ground by the invading German forces, and the population was forcibly evacuated to the South. After the liberation the population started to return to their homes despite the governments' plans of only returning the men who could rebuild the societies. Everything was lacking: labour, carpenters, materials, transportation, accommodation and money. Many people lived in tents or sheds built of whatever they could find, and the need for houses was urgent. An architectural competition was announced for making a material-efficient and standardized house type suitable for prefabrication. A uniform type of house was made called "reconstruction house". This house was based on earlier typical features where the rooms are based around one chimney in the middle. The house was built with one and a half floor and with gable roof (Christensen, 1995).

### 3.2 Reconstruction house from Finnmark

The case studied in this research is an ongoing project at the Norwegian Museum of Cultural History in Oslo, Norway. The aim of the project is based on a desire to show building history and culture from all of Norway at the open-air museum. Through fieldwork done the autumn of 2015 the decision to move a house and shed from Olderfjord and a cowshed from Indre Billefjord to the museum was taken. Another field trip to the buildings were taken in 2016 to assess the state of the buildings and to make documentation for the rebuilding phase at the museum (Sandvik & Revold, 2017; Wennberg & Revold, 2016). During the autumn of 2016 the buildings was disassembled by a contractor and transported to Oslo in containers.

The norm for heritage preservation is to preserve buildings on site as can be understood from the title of an article by Arne Berg in 1951: "If one turns crazy and starts moving houses" (Berg, 1951, p. 3). This illuminates the conflict between heritage protection and open-air museums as the museums depend on moving buildings. A part of the museums assessment on which buildings to move, this conflict has been taken into consideration: "It's an exciting and challenging task to re-erect buildings with such extensive damages as these houses have. At the same time, it is a correct choice to not move houses that could have had a good life where they originally were built. This opportunities the chosen houses do not have where they stand now" (Wennberg & Revold, 2016, p. 41).

## 4 METHOD

A case study was considered appropriate since it allows for exploring "sticky practice-based problems where the experience of the actors is important, and the context of the action is critical" (Bonoma cited in Benbasat, Goldstein, & Mead, 1987, p. 370). According to Benbasat, Goldstein and Mead a case study have three main advantages: First it allows for answering "how" and "why"-questions. Second it allows for studying information systems in its natural setting. Thirdly it allows for research in areas in which few have studied before (Benbasat et al., 1987). The case was chosen since the project was at the re-erecting phase, and the HBIM could be tested on-site. Since the buildings were already taken down the documentation made before the disassembly needed assessment. The documentation consisted of several thousands of photos, sketches and drawings with measurements. Full access to the documentation was granted by the department responsible for the project and was found to be extensive enough for this research.

The purpose of making this BIM was to give the carpenters easy and understandable access to the BIM-model, with plans, sections, elevations and photos. An easy link from the model to photos was important for the presentation.

As no BIM-making software assessed were sufficiently facilitated for handling both quantitative and qualitative assets, the presentation to the carpenters was made in a different software. For this purpose, the BIM was made in Autodesk Revit and

Table 2. Interviews conducted.

| Interviewees | Role in project |
|---|---|
| Interview object (IO#1) | Head carpenter |
| Interview object (IO#2) | 2nd carpenter |
| Interview object (IO#3) | 3rd carpenter |

exported in Industry Foundation Classes(IFC) format to ensure the other software were able to read/adopt the model correctly (ISO 16739, 2013). The Dalux© software where the IFC-file and photos were linked together was chosen due to its compatibility with different operative systems on tablet computers to insure mobile and easy access on-site. The hardware used in the testing were Apple I-pads.

To ensure good follow-up the testing was performed one to one. The testing phase was done in three phases:

- Phase 1: Introduction
- Phase 2: Testing
- Phase 3: Interview

In the first phase the on-site Dalux© software was introduced to the professionals. This to visualize what the model contained and how to navigate between different information sources (e.g.3D-model, drawings and photos). The second phase was a series of made up cases the carpenters had to solve. The cases were made to simulate situations where the carpenters otherwise would have used paper drawings to find the information they needed (e.g. measurements, drawings and photos of details). The third and last phase of the testing was the interviews. They were conducted in the on-site construction office directly after the testing. All interviews were voice recorded and later transcribed. The reason for this was to keep the interviews anonymous as their perceptions as users are what is interesting to the research.

## 5 ANALYSIS

The analysis of the data collected follows the structure of the TAM-theory presented in chapter 2. First the results from the interviews are discussed in relation to the carpenters' *perceived usefulness*. Second the *perceived ease of use* of the BIM technology is discussed. Third the aspect of the carpenters' *intention to use* is discussed.

Although all the interview objects stated that they had heard about BIM before they were introduced to this research, they also acknowledged that they did not have a full overview of what it was: "I have heard the term [BIM], but without having experience or practice using it at all. [I have] partly heard about it through architect friends in social media, and partly through media to a certain extent." (Interview object# 1, IO#1). Another expressed: "Not enough, because I haven't understood what it is before it was introduced in this project." (IO#2).

### 5.1 Perceived usefulness

#### 5.1.1 Easy access register for information

In general, the carpenters stated that they perceived the BIM technology to be useful. All the interviewed carpenters had positive expectations prior to the on-site testing of the technology. Throughout the re-erecting phase the binder with drawings, sketches and photos were, in addition to the building elements themselves, the main source of information. As one of the carpenters stated after the testing of HBIM: "As it has been [shown] here I think it would have been useful as a register, that it simply is easier to locate information there than in our filing systems on computers and in the binder." (IO #1).

#### 5.1.2 Full on-site access to detailed information

A selection of the "[...] thousands of photos [...]" (IO#2) taken prior to the disassembly were chosen out as relevant. These were printed in small scale and filed in the binder for on-site use. To see the details in these photos the carpenters were required to find the right small-scale photo in the binder, note the ID and look up the photo on a computer in the construction office. On-site access prevents this type of time consuming trips back and forth between the construction site and office as it avoids a *full on-site access to detailed information*. It also presents the possibility to compare the photos to the construction on-the-go as the BIM-system provides full access to all photos in the project. One of the carpenters explains it in this way: "Well, but it's a bit like that because one always makes a selection of measurements and stuff and based on experience there is always something one misses that one hasn't thought that one needs, and [with BIM] then one would get it automatically." (IO#1). With these benefits in mind and the comment: "we look a lot at photos for details" (IO#1), the BIM-technology leads to time savings and easy access to highly detailed sources.

#### 5.1.3 Better visual understanding of the building

A key to succeed in projects concerning historic buildings is to have a comprehensive understanding of the building. Not only the surfaces, but the whole building needs to be examined. A great advantage of BIM is the possibility to change between visual filters. One of the carpenters finds it "[interesting] to be able to walk into a room, seeing the different walls, either in a photo or in the model, with the different layers that are in the wall." (IO#2).

### 5.1.4 Information flow to project participants

All projects have a list of stakeholders and the information flow between these are crucial. In the case studied in this research there has been one firm responsible for project management, one contractor responsible for disassembly of the buildings, one contractor responsible for foundation works and one contractor responsible for cladding, roofing and floors. All these firms are in addition to the carpenters at the museum who did all the work concerning documentation prior to disassembly and rebuilding of the timber frames and roof. Earlier research suggests that using BIM on-site "[...] provide far more effective communication than other types of synchronous communication" (Svalestuen, Knotten, Laedre, Drevland, & Lohne, 2017).

The carpenters from the museum have been involved in the project since the beginning and have a good overview of the buildings and the project. One of the carpenters reflected on this topic before he stated the following: "Yes, well in this project we have been involved in the disassembly of the building. So, we have seen the process and the building thoroughly. But let's say that we did not participate in disassembly of the building, and we had a complete system with BIM and photos, then it is clear that it would have been useful." (IO#3).

### 5.1.5 Level of detailing in the documentation

One of the disadvantages with traditional documentation methods is that a selection of building elements and details are not documented. As the professionals depend on the documentation, and especially the photos, one challenge is inevitable: the lack of photo documentation of many of the building elements and details. "The first thing you look for is not found, because it's not documented with photos." (IO#1). Considering that the building was disassembled before the research started, an on-site documentation with the use of digital data capture methods (i.e. laser scanning, photogrammetry etc.) was not possible. However, it seems that the carpenters were positive to this type of registration methods as the quote suggests: "I have throughout the whole process thought that it [BIM] would be a very helpful tool if one had joined the project and registered the building before dismantling. As it is now, the information has been retrieved from our registrations or what is already registered, and that has meant that I have thought it might have a more limited utility than it would otherwise have had." (IO#1).

Another carpenter shares his thoughts on the method used for data collection in the documentation phase of the project. "We have, in this project, done everything the traditional way with a bunch of sketches, drawings, elevations and thousands of photos. I think that if you, [the researcher], had joined the project from the very beginning and made a scanning of the building I think we could have saved time. Got it all gathered. Now there have been many who have made drawings and sketches. There is always room for sources of error." (IO#2). This visualize the advantage of HBIM as an interlaced registration system.

## 5.2 Perceived ease of use

### 5.2.1 Software is intuitively to use

After testing the carpenters' perception of BIM-software we concluded that it was *intuitive to use*. One of the carpenters, despite seeing himself as an inexperienced computer user, finds the software to be convenient and user-friendly: "I experience that it was quite intuitive to use. Without me being a skilled user, it went quite well. At least with some guidance." (IO#1). That the carpenter felt that he needed guidance can be seen as an early usage stage statement and must be seen in context with the timeframe for the testing.

### 5.2.2 Interaction with the system clear and understandable

The carpenters found the interaction with the system clear and understandable. The navigation between the different documentations (model, photos i.e.) were perceived easy and logic, despite the system were completely new for the professionals. "I think it was logic. Okay to navigate. If one were to use it fully, I assume one gets some more training. But I have never used a program like this or worked with it. But I think it was logic and okay to navigate." (IO#2).

## 5.3 Easy to find the information you want

The system of structuring the information of building elements in the IT-system follows the same system as the traditional registration done prior to moving the houses. Whether it is the structure or the functions in the software that is perceived easy to use is not assessed, but the carpenters find the system to be easy to navigate: "I would say it is fairly simple [to learn the system]" (OI#3). The workers also find "[...] that it is very easy to take measurements, assuming that we perceive that it is the right measurements we get." (IO#1). This underlines the importance of the integrity of the BIM. If the workers find the system to contain errors, the way back to the paper blueprints is short.

### 5.3.1 *Challenges*

Even though the carpenters in general are positive to the BIM-technology, the interviews have also illuminated possible challenges. As for instance one of the carpenters points out that the measurement tool in the software does not "attach" to elements and therefore the measurements become inaccurate. "The first thing [challenge] I think of is the measurements. I have used AutoCAD a bit and from there I am used to be able to "fix" it [the measurement cursor]. When measuring from somewhere, you can "fix" the cursor precisely in a corner, for example. I miss that a bit. Now it depends on who measures, and how careful they are." (IO#2). Another of the professionals points out another potential challenge with the Norwegian climate in wintertime: "There may be challenges with temperatures. That it stops working in low temperatures. But on the other hand, paper does not like rain either." (IO#3).

### 5.4 *Behavioral intention to use*

As the TAM-theory suggests, the 'behavioral intention to use' indicates how the information system is perceived by the interviewers. Given that this research has been limited to a brief introduction of BIM-technology to the workers, the interview objects, were asked about their thoughts on using it in future projects. Taken into consideration that the project of moving a house is rare, the carpenters were positive to using BIM in future projects. "If one can work with it as it has been introduced here, I think it can be an extremely good tool if one can use it through the whole process." (IO#1).

Another carpenter answered this when he was asked about the intention of using BIM in future projects: "Yes. Or just here we work, it's rare we have such types of projects with so modern houses and building techniques. But, certainly it had been very interesting to use it right from the start." (IO#2). This statement proves that they see a usefulness in this specific project concerning the moving of a relatively modern house compared to the rest of the buildings at the museum.

The testing of the BIM-technology was only conducted in the house from Olderfjord. The project at the museum also consists of moving a cow shed and a shed. One of the professionals finds the technology interesting and sees it as a potentially valuable tool for the upcoming part of the project: "Yes, as I'm saying. I would like to have it [BIM] on the upcoming work on the cow shed. […] If possible to have a screen here [on-site]. Yes, possibly larger than that one [iPad], instead of scrolling in printed photos. To be able to click directly to for example wall J and click on it and get all of the photos." (IO#3).

## 6 DISCUSSION

The concept of BIM was originally developed for designing new buildings, but in recent years it has been adopted for cultural heritage preservation and management. One of the reasons for this interest is the more accessible surveying techniques for remote recording of complex structures. (Dore and Murphy, 2012). Earlier research on HBIM suggests that the use of BIM in projects concerning historic buildings and places can be a valuable tool for creating, conserving, documentation and managing complete engineering drawings and information (Megahed, 2015).

This research aims to give a contribution to the rapidly increasing interest for HBIM-technology through a different lens. As opposed to earlier research, this research aims to illuminate how the HBIM-technology is perceived and used by the professionals working on-site with historic buildings. Based on the TAM-theory the findings shown in Table 3 from interviews indicates that there is a relatively strong interest for further use of the technology in other projects.

This research, as well as other research, points out several obvious advantages of using BIM on site. (Bråthen & Moum, 2016; Davies & Harty, 2013; Merschbrock & Nordahl-Rolfsen, 2016). The results show that the professionals find the HBIM-technology to be useful and user friendly.

Table 3. Key findings.

| TAM element | Findings in the "reconstruction house" case |
|---|---|
| Perceived usefulness | • Easy access register<br>• Time savings<br>• Access to highly detailed underlay<br>• Comprehensive understanding of the building<br>• Project information flow<br>• Benefits of digital registration<br>• Interlaced registration data |
| Perceived ease of use | • Software is intuitively to use<br>• Logic to navigate<br>• Important with high integrity of the HBIM<br>• Imprecise measurements<br>• Hardware vulnerable to low temperatures |
| Behavioral intention to use | • Strong interest of future use in similar project<br>• Possibly limited utility in more common project at the museum<br>• Desirable with further use in this project |

The HBIM allows for collecting all on-site necessary information in an easy access register. This reduces the unproductive time related to searching for and collecting detailed information from the site office. The digital platform also allows for easy handling of larger quantum of data. In the case studied this relates especially to the possibility to have full time access to high resolution photos, and the HBIM model on site at all times.

Despite the positive feedback from the professionals they also illuminated several challenges. One was that the BIM was based on the carpenters' own registration made prior to disassembly, and not a digital data collection (i.e. photogrammetry, laser scanning). Imprecise measurement tools in the software were also a weakness pointed out. As being one of the BIM's advantages of taking precise measurements anywhere, this proves that the software used on site still has potential for development. In line with what Bråthen and Moum point out in their research: "[…] there seems to be no appropriate tools to make use of BIM in rough environments in our part of the world." (Bråthen & Moum, 2016), the interviewed carpenters in this research also find it potentially problematic with low temperatures. This is also one of the findings in the research of Merschbrock and Nordahl-Rolfsen regarding the use of BIM on site for concrete reinforce workers (Merschbrock & Nordahl-Rolfsen, 2016).

The findings indicate that there is still a lot of potential in developing the HBIM software. As Megahed point out there are a number of barriers between academic research on HBIM and practical applications that prevent this development (Megahed, 2015). One of the main challenges found in this research during the preparation of the HBIM, was that the BIM software was not compatible with adding qualitative assets, and in particular photographs. In general, this may limit the potentials of implementing HBIM in future projects.

## 7 CONCLUSION

In the recent years there has been an increasing interest in using BIM in the heritage sector. As this type of technology is optimized for industrialized buildings, using it for cultural heritage anticipate challenges. This research focuses on how the carpenters working with heritage buildings perceive this HBIM-technology as a substitute to printed drawings sketches and printed photos on site. Through the theoretical lens of the TAM-theory, this paper aims to give an answer to the question: *How will carpenters working with heritage buildings perceive the use of HBIM on site in a project of relocating project of an historic building?*

As visualized in Table 3, the carpenters found the HBIM-technology to be useful and user friendly. However, this perception depends on the degree to which the engineers manage to prepare a satisfactory model with high level of credibility amongst the carpenters. Whether the cost of engineering hours to prepare the HBIM can answer to its benefits are still to be assessed, and it needs further research, preferably projects with full HBIM implementation. This could also illuminate how remote non-contact digital recording hardware and software would work towards making a HBIM suitable for on-site work. It is important to point out that the case considered in this research is just one of many potential usages of HBIM. One of the greatest challenges for implementing the use of HBIM in cultural heritage is to overcome the practical challenges of BIM software being optimized for industrialized building systems (S. Fai & Sydor, 2013). Full on-site implementation of HBIM depends on useful and user friendly HBIM software. The full potential of HBIM, in all stages, can only be shown by further research on HBIM in all phases of preservation of cultural heritage buildings and sites.

## REFERENCES

Benbasat, I., Goldstein, D.K., & Mead, M. (1987). The case research strategy in studies of information systems. *Management Information Systems Quarterly, 11*(3), 369–386. doi:10.2307/248684.

Berg, A. (1951). Om gale skal vera og hus må flyttast. *Museumsnytt*(1), 3–14.

Bråthen, K., & Moum, A. (2016). Bridging the gap: bringing BIM to construction workers. *Engineering, Construction and Architectural Management, 23*(6), 751–764. doi:doi:10.1108/ECAM-01-2016-0008.

buildingSMART. (2012). *National Building Information Modelling Initiative*. Retrieved from.

Christensen, A.L. (1995). *Den norske byggeskikken: Hus og bolig på landsbygda fra middelalderen og til vår egen tid*. Oslo: Pax Forlag A/S.

Davies, R., & Harty, C. (2013). Implementing 'Site BIM': a case study of ICT innovation on a large hospital project. *Automation in Construction, 30*, 15–24. doi:10.1016/j.autcon.2012.11.024.

Davis, F.D. (1989). Perceived usefulness, perceived ease of use, and user acceptance of information technology. *Management Information Systems Quarterly, 13*(3), 319–340. doi:10.2307/249008.

Dobelis, M. (2015). *BIM Education in Riga Technical University*. Paper presented at the The 13th International Conference on Engineering and Computer Graphics BALTGRAF–13.

Fai, S., & Sydor, M. (2013). *Building Information Modelling and the documentation of architectural heritage: Between the 'typical' and the 'specific'*. Paper presented at the 1st International Congress on Digital Heritage, DigitalHeritage 2013, October 28, 2013 - November 1, 2013, Marseille, France.

Fai, S., Graham, K., Duckworth, T., Wood, N., & Attar, R. (2011). *Building information modelling and heritage documentation*. Paper presented at the Proceedings of the 23rd International Symposium, International Scientific Committee for Documentation of Cultural Heritage (CIPA), Prague, Czech Republic.

Fishbein, M., & Ajzen, I. (1975). *Belief, attitude, intention and behaviour: An introduction to theory and research.*

Godal, J.B. (2004). Restaurering og autentisitet. *Fortidsminneforeningens årbok, 158*, 27–34.

Historic England. (2017). BIM for Heritage: Developing a Historic Building Information Model. In: Historic England.

ISO 16739, International Organization for Standardization (2013). Industry Foundation Classes (IFC) for Data Sharing in the Construction and Facility Management Industries. In. Geneva: International Organization for Standardization.

Ma, Y.-P., Hsu, C.C., & Lin, M.-C. (2017). *Combine BIM-based and mobile technologies to design on-site support system for the communication and management of architectural heritage conservation works.* Paper presented at the 2017 International Conference on Applied System Innovation (ICASI), 13–17 May 2017, Piscataway, NJ, USA.

Megahed, N.A. (2015). Towards a theoretical framework for HBIM approach in historic preservation and management. *ArchNet-IJAR: International Journal of Architectural Research, 9*(3), 130.

Merschbrock, C., & Nordahl-Rolfsen, C. (2016). BIM Technology Acceptance Among Reinforcement Workers—The Case of Oslo Airport's Terminal 2. *Electronic Journal of Information Technology in Construction, 21*, 1–12.

Murphy, M., McGovern, E., & Pavia, S. (2009). Historic building information modelling (HBIM). *Structural Survey, 27*(4), 311–327. doi:10.1108/02630800910985108

Rogers, E.M. (2003). *Diffusion of innovations* (5th ed. ed.). New York: Free Press.

Sandvik, B., & Revold, E. (2017). Gjennreisningshus fra Finnmark. *Museumsbulletinen, 2*(85), 4–11.

Svalestuen, F., Knotten, V., Laedre, O., Drevland, F., & Lohne, J. (2017). Using Building Information Model (BIM) Devices to Improve Infomration Flow and Collaboration on Construction Sites. *Journal of Information Technology in Construction, 22*, 204–219.

Wennberg, J., & Revold, E. (2016). Dokumentasjon av bolighus og uthus. *Museumsbulletinen, 1*(81), 38–41.

# Towards the application of the BIMgrid framework for aged bridge behavior identification

M. Polter & R.J. Scherer
*Institute for Construction Informatics, Technische Universität Dresden, Dresden, Germany*

ABSTRACT: To assess the current condition of a bridge and make predictions about its expected lifetime, many manual on-site work is still required today. For reasons of cost, characteristic values are not measured continuously but only on a random basis and are often replaced by assumptions. In this paper, we introduce the cyberBridge platform, which is currently being developed to create an online bridge assessment system with automatic data analysis and engineer-friendly visualization of bridge damage. The goal is to use the platform to model the condition of a bridge in near real-time, and thus to be able to make more accurate forecasts at low cost over the expected lifetime of a bridge.

## 1 INTRODUCTION

In times of increasing traffic and decreasing budgets of local authorities, bridges often have to be operated beyond their intended lifetime, which creates a high demand for assessment of bridge health, deterioration and increased traffic load behavior. Today, this is done primarily through visual inspections and with sonic impact equipment each 6–12 months, dynamic vibration analysis, continuous deflection and stress/strain measurements. The problem with these types of measurements is, that the specific related loads are unknown, which makes it impossible to detect early stage, still locally limited deterioration. The current methods do have some substantial problems that require improvement. First, current systems are not online identification and forecasting systems, so that engineers have to be present at the bridge what increases costs for bridge assessment. Currently used databases are parameter- but not system-oriented and only have alpha-numerical interfaces with some business graphics. A system- and element-oriented BIM database and a context-based 3D BIM view coupled with semantic information and a related 3D Graphical User Interface are yet missing. Furthermore, existing approaches are unable to reliably predict both the overall response and the behavior within a local region of the system. The analysis results with regard to material behavior and state of deterioration are uncertain; they rely heavily on assumptions about the deterioration process, the material behavior and the loads. ICT (Information and Communications Technology) systems are inflexible. New sensors or the relocation of sensors demand considerable adaption work on the ICT system. Life cycle prediction is an uncertainty task because currently only a few scenarios are investigated and strongly simplified probabilistic methods are applied. In addition, the degree of automation of today's approaches is insufficient. In this paper, we introduce the concept of the cyberBridge platform, a cyber-physical bridge assessment system that will at low cost allow continuous online monitoring and system identification on the level of crack propagation and hence improve prognosis of bridge deterioration. The platform is based on the BIMgrid framework (Polter & Scherer 2017) and aims at the composition of existing software into an integrated bridge assessment solution.

This paper is structured as follows: Section 2 introduces the cyberBridge platform. Therefore, section 2.1 gives a brief overview about the BIMgrid framework, upon which the platform was designed. In section 2.2 we first depict the intended workflow of the system and then describe the architecture. Section 3 evaluates how experiences from other BIMgrid-based software platforms can be transferred to the cyberBridge platform. Finally, section 4 concludes the paper.

## 2 THE CYBERBRIDGE PLATFORM

The cyberBridge platform is designed to provide long-term, automated and cost-effective bridge monitoring, with a focus on ease of operation by the engineer and meaningful presentation of collected and calculated data.

The following demands are made on the platform:

- Sensor data must be continuously collected, filtered, evaluated and stored
- The data and damage model of a bridge must be constantly updated, as automated as possible, based on the measured values
- For system identifications with a large number of unknown parameters, parameter values must be sensibly varied and corresponding input models generated
- The platform must provide graphical user interfaces for various user roles, e.g. for administrators, civil engineers and technicians
- A large number of complex calculations must be cost-effective, fast and automated
- The resulting large amounts of data must be stored inexpensively, automatically annotated with metadata (e.g., timestamps), and retrieved by appropriate mechanisms
- The platform components need to provide web service interfaces to communicate with each other online as they are spatially distributed (sensors at the bridge, platform on the server, GUI accessible by various office users, or accessible via mobile devices).

The next section gives a brief overview of the BIMgrid framework on which the platform was designed. Section 2.2 shows the modular structure of the cyberBridge system and the planned bridge assessment workflow. The focus in this paper is on the software architecture. Data management in the multi-model, the use of collected and computed data for bridge assessment, and advanced system identification approach details will be published separately.

### 2.1 The BIMgrid framework

The BIMgrid framework is a reference concept and blueprint for the implementation of integrated software platforms that are tailored to a company's established work processes and technical equipment. At the core of the concept is the combination of different technologies, such as calculation solvers, databases or hardware infrastructure, to improve workflows that include different software, such as: by automation or distribution to suitable hardware. Therefore, the software products to be integrated are encapsulated in modules and their functionality is integrated by adapters, which may either be existing libraries (such as JDBC for Java access to databases) or implemented by software developers. After development, the adapters and corresponding off-the-shelf software components can be reused to develop new composite applications, resulting in a significant reduction in software development costs. In addition to the development of new applications, BIMgrid also allows the extension of existing applications (especially web platforms) with new functionalities through the integration of new software and hardware. For a more detailed overview of the framework and its application, see (Polter & Scherer 2017). The next section describes the conception of the cyberBridge platform from the framework.

### 2.2 Application for bridge assessment

The designed platform is designed to continuously monitor bridge behavior under traffic load and to predict the propagation of bridge damage, especially of cracks. Figure 1 shows the workflow schematically. The starting point is a model of the bridge, which is present in IFC. The initial modeling of the bridge is done in a tool of choice and is not covered by the platform. The bridge model is uploaded and integrated into a multi-model. The concept of the multi-model is described in detail in Fuchs (2014) and is not explained here. The bridge model is then translated into a probabilistic analysis model based on a rule-based approach (not covered here). The sensors attached to the bridge continuously collect data about traffic loads, axle loads, bridge deformation etc. and send them to the platform where they are stored, annotated and attached to the multi-model. System identifications are triggered automatically and result in initial damage models. Based on a best fit method, the damage models are compared to real measured values (e.g., crack widths) and the model that best approximates the damage is selected. The continuously measured sensor values provide both input for the refinement of the damage model and for the generation of parameter values for further system identifications. The damage model can now be used for the deterioration forecast (for example, crack propagation), where the constant update allows for precision prediction that cannot be achieved with previous methods.

The application will contain the following top-level components (Fig. 2):

- Sensors
- Solver for system identification

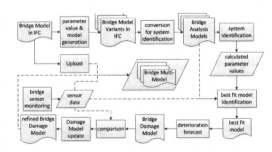

Figure 1. Workflow of the cyberBridge platform.

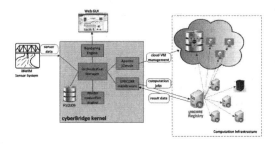

Figure 2. Top-level architecture of the cyberBridge platform.

Figure 3. iBWIM sensor system mounted on a bridge.

- Cloud Adapter
- Resource Management Middleware
- Orchestration Manager
- Data Storage
- Graphical User Interface.

Adapters from other projects already exist for most of the above-mentioned components, which can be reused without much adaptation off the shelf for integrating appropriate technologies into the platform. In the following, the individual components are examined in more detail.

### 2.2.1 Sensors

The iBWIM sensor system[1] is a third-party sensor system mounted on the target bridge and integrated into the platform via a web service interface. Communication takes place via the Web Application Messaging Protocol (WAMP). For this purpose, the platform registers as a client on the WAMP server and continuously receives new sensor data such as speed, axle load and weight of the vehicle according to the publish-subscribe principle. The data is used to create traffic flow models and input parameters such. As loads on a particular bridge section, to derive for the structural analysis of the bridge.

### 2.2.2 Solver

The Atena[2] solver for structural analysis has been used in previous projects and can therefore be reused without major adjustments. It can be completely controlled by Windows command line commands, which is a prerequisite for integration with BIMgrid applications. It is the core of the component that performs the system identifications. The solver kernel must be installed on each computation node, both in the private grid and on the cloud VMs.

### 2.2.3 Cloud access

In order to make cloud resources available to the platform, three functional groups must be provided:

- Management functions for cloud resources
- Transfer of computation jobs and input as well as result data
- Upload/download data to/from the cloud-based data store as well as filtered data queries

Group 1 includes methods for creating virtual machines (VMs) from images, starting and stopping VMs, status queries, and all cloud-based database control functions. For the implementation of the corresponding cloud component the library Apache jClouds[3] was used, which provides a vendor agnostic abstraction layer for the implementation of corresponding functions.

Functions of group 2 are not executed via jClouds, but instead pre-implemented methods of UNICORE Grid middleware[4] are used. This will be discussed in more detail in the next section.

While cloud-based calculations are performed on virtual machines that need to be configured to meet their needs, a ready-made cloud solution is used to manage the payload. In previous projects, Amazon Web Services was chosen as the cloud provider, which is why Amazon Simple Cloud Storage Service (Amazon S3) should be used for data management. Appropriate Group 3 database management and data transfer capabilities are also implemented with Apache jClouds.

### 2.2.4 Resource management

The UNICORE grid middleware originally designed for the management of private grids but can also be used to distribute jobs to virtual cloud images. This allows the abstraction of the physical

---

[1]http://www.petschacher.at/en/bwim/.
[2]https://www.cervenka.cz/products/atena/.
[3]https://jclouds.apache.org/.
[4]https://www.unicore.eu/.

location of the resources used and thus a unified view of the infrastructure connected to the platform (Fig. 4). Calculation nodes are provided with a cost function and metadata that Orchestration Manager can use to distribute jobs. UNICORE considers only the free resources of computers when distributing jobs. If a machine is busy, it will not be used for further calculations by the middleware. This allows the integration of locally-used workstations into the entire platform infrastructure, without employees being aware of or being restricted by PCs.

### 2.2.5 Graphical user interface

The graphical user interface (GUI) of the application must provide views for different roles. In addition to the engineering view for evaluating the collected data and calculations, management functions and control options for the connected infrastructure (sensors, calculation hardware) must be provided. To achieve maximum flexibility when using the application, the graphical user interface web browser is based. Communication between the distributed hosted application components is via the Hyper Text Transfer Protocol, where between the core of the platform and the GUI, lightweight data formats, such as e.g. JavaScript Object Notation (JSON), to be exchanged. Thus, the application can also be operated locally by mobile devices with low power and is not limited to workstations. Communication between distributed backup components uses the heavier SOAP protocol to exchange more complex data structures.

### 2.2.6 Data storage

Data accumulated in the system and its processing are subdivided into the areas a) control and administration data and b) user data. A) includes both data needed for system failure recovery and user and project metadata used for access management and user data mapping. These data are stored on server side on the local network in a Hyper SQL Database (HSQLDB) to ensure access speed and data security. User data is all data related to the assessment functions of the platform. This includes all model, damage, parameter and metadata stored in the multi-models. As these data require a large amount of memory and are generated continuously by the system, they are managed in a cloud-based database (see subsection Cloud Access). This has the advantage that, as long as the corresponding calculations also take place in the cloud, they do not burden the local network. It also prevents local storage space bottlenecks and associated migration and downtime requirements.

### 2.2.7 Orchestration manager

The Orchestration Manager is responsible for controlling all control and data flows between the components of the platform. He also distributes jobs to the calculation nodes of the connected infrastructure (Fig. 4). In doing so, cost functions and metadata assigned to the nodes are taken into account. For example, metadata includes information about whether it is a virtual cloud or physical network node, such as privacy. Sensitive data can thus be restricted to distribution in the corporate network. Cost functions include information about monetary costs of using a node and data such as measured latencies and other Quality of Service (QoS) parameters. Local machines, for example, have a monetary cost of 0, while cloud resources incur usage charges under the terms of the cloud vendor. Using the cost functions, calculation jobs can now be distributed in a more targeted and effective manner, taking into account QoS constraints in the infrastructure. The Orchestration Manager, which has already been developed, will be enhanced with functions for a more user-friendly definition of scheduling policies within the framework of the cyberBridge project.

Resource-intensive calculations and large amounts of data require efficient resource management to keep the system's operating costs as low as possible. Here, however, a balance between performance (near real-time monitoring), availability and costs has to be found. The cloud paradigm is one way to quickly make large computing resources available when needed. Similarly, cloud-based data storage enables high reliability and location-independent availability of data. However, with continuous operation with permanently high data throughput, a purely cloud-based system would cost a lot in the long run and would not be economical. Therefore, the platform operates on a hybrid composite hardware infrastructure consisting of public cloud and private grid resources.

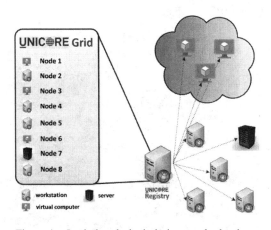

Figure 4. Logical and physical view on the hardware infrastructure.

The private grid includes computers from a local corporate network and a server that hosts the core of the platform. Tasks that consume only limited resources can thus be run at a company's local hardware at no additional cost. This includes, for example, the generation of model variations or data preparation and filtering operations. If the local resources are utilized, complex system identifications can be outsourced to powerful cloud resources. The system can be configured to automatically run calculations on the appropriate infrastructure. Corresponding policies take into account both the calculation software and user-defined categorization features of calculation jobs. More details will be published in future.

This section provided an overview of the architecture and intended use of the cyberBridge platform for continuous online bridge assessment. The next section estimates the expected performance of the system, based on the experiences made with a similar system.

## 3 EXPERIENCES FROM OTHER PROJECTS

As the cyberBridge project is still in the conception phase, no practical evaluation results are available yet. However, the BIMgrid framework has already been used in the SE-Lab project for the development of a similar platform for the parallel execution of a large amount of structural analysis (Polter & Scherer 2015). The following additional features of the SE-Lab platform are also requirements for the cyberBridge platform:

– Effective parallel execution of resource intensive calculations on a heterogeneous infrastructure with high demands on data management
– Automation of a sequence of calculation steps that previously had to be started manually
– Providing web-based graphical user interfaces
– Distributed hosting of application components.

Much of the experience from this project can thus be transferred to the cyberBridge platform. A detailed overview of the test parameters and results can be found in Polter et. al (2016). In summary, the following statements can be made: By using the grid infrastructure, it was possible to parallelize calculations and thus considerably shorten the calculation times. With 8 knots in the grid, an acceleration by a factor of 6 could be achieved. Workflows, consisting of preprocessing, simulation and aggregation of the results of many simulations, could be automated and thus faster and cheaper. In this case, unnecessary data transfer was avoided by scheduling interdependent operations on the same compute node, using the data locality principle. Most features of the platform can also be used on mobile devices, with the variable display size indicators still needing to be expanded to accommodate themselves. However, there are limits to the presentation of 3D models, which is due to the size of the files to be transferred, which are often too large for mobile network access and therefore associated with too long load times. In addition, unlike their desktop counterparts, the mobile versions of the web browser typically have limited functionality, such as the often lacking support for plug-ins or libraries for 3D rendering of objects. In some cases, completely alternative GUIs for mobile devices need to be developed. In summary, this project demonstrates the applicability of the BIMgrid framework for building composite applications based on existing applications to accelerate and automate engineers' workflows.

## 4 CONCLUSIONS

The bridge monitoring is still largely offline and therefore not continuously, but at longer intervals. The data management is mostly alphanumeric and without 3D representation. Important information is therefore not directly visible, but must be derived by the engineer from the data. Furthermore, the estimation of the material behavior of bridges is still based on many assumptions.

In this paper, we introduced the cyberBridge platform for online bridge monitoring. It was designed based on the BIMgrid framework and integrates third-party functional components into a composite application. The transparent integration of heterogeneous hardware, consisting of local grid and public cloud resources, enables the implementation of complex system identifications in near real time and the continuous prediction of bridge deterioration. The automation of calculation and evaluation processes reduces labor costs and human error sources. Collected and calculated data are managed in multi-models and can be visualized on the bridge model in 3D. Web browser-based graphical user interfaces allow the use of platform features even on low-power mobile devices.

Although there is no evaluation data available for the project yet, experience gained with another BIMgrid-based platform suggests the cyberBridge platform's performance and cost-saving potential over previous approaches to bridge assessment. The target groups of the platform will be bridge owners, monitoring and assessment companies as well as consultants. We expect the use of the platform to reduce the cost of bridge monitoring, increase the accuracy of damage predictions, and ultimately increase the actual life span of bridges.

ACKNOWLEDGEMENTS

This research is funded as an EUROSTARS project through the Federal Ministry of Education and Research in Germany and the European Union under the project (01QE1712C) 'BIM based Cyber-physical System for Bridge Assessment'.

REFERENCES

Fuchs, S. 2014. Erschließung domänenübergreifender Informationsräume mit Multimodellen. PhD Thesis.

Polter, M. & Scherer, R.J. 2015: An integrated web platform for grid-based advanced structural design and analysis. In: Proceedings of the 32nd CIB W78 conference 2015, October 27–29 2015, Eindhoven, The Netherlands.

Polter, M., Ismail, A., Scherer, R.J., Grille, T., Hamdan, A., Luu, N.T. 2016. Virtuelles Ingenieurlabor—SOA-basiertes Informationssystem zur Integration von Grid-/Cloud Rechenleistung und BIM Informationsmanagement (in german), *Project Report, Technische Informationsbibliothek (TIB). Available at: https://www.tib.eu/de/suchen/id/TIBKAT%3A872575926/Schlussberichtzum-Vorhaben-Virtuelles-Ingenieurlabor/.* Accessed: December 2017.

Polter, M. & Scherer, R.J. 2017. Towards an adaptive framework for customized civil engineering platforms. In: Proceedings of the 24th International Workshop on Intelligent Computing in Engineering (eg-ice 2017), July 10–12 2017, Nottingham, United Kingdom.

# Collaborative platform based on standard services for the semi-automated generation of the 3D city model on the cloud

I. Prieto, J.L. Izkara & A. Mediavilla
*Sustainable Construction, Tecnalia, Spain*

J. Arambarri
*Fundación Virtualware Labs, Spain*

A. Arroyo
*Estudios GIS, Spain*

ABSTRACT: The urban 3D model is increasingly recognized as the adequate support to integrate, harmonize and store the information of a city and make it accessible to all stakeholders (citizens, city managers, companies or researchers). These models facilitate the cooperation of experts in different areas, contributing their knowledge in the generation of a single model. Urban 3D models will allow transforming urban management processes. One example of such processes is the Special Interior Reform Plans (PERI). These urban management processes are developed by the administration, but require the collaboration of different agents. In addition, they represent a clear example of the need to advance in the integration of heterogeneous data at different scales (building and city). The solution proposed in this article presents a platform based on web services for the collaborative generation of urban 3D models. The platform is composed of an information system based on standard data models (IFC and CityGML) and a web services infrastructure that manages the information and relationships stored in the information system at different levels. Besides, a prototype of 3 applications based on the service infrastructure and developed to support the urban management process is presented. The platform is used and tested in the collaborative generation process of the 3D model of the historic district of the city of Vitoria-Gasteiz (Spain) during the definition, editing and monitoring of the PERI of the historic center of the city.

## 1 INTRODUCTION

New technologies offer increasingly advanced solutions for generating and collecting data. This situation generates a volume of data that grows significantly, containing information from different domains, nature and scale. In addition, technological advances are causing a relocation and distribution of information that was previously administered primary locally. The concept of the cloud is becoming more common and collaborative work emerges in the management of both personal and professional information (Grey et al. 2017). Collaborative platforms in the cloud value the concept of software as a service as opposed to the on-premise software.

The current context is ideal to transfer the paradigm of collaborative management to the field of urban management, significantly optimizing times and costs, accessibility to data, capability to take complex decisions, and include citizen participation within the decision cycle, leading to true digital municipal management. In order to respond to these demands, administrations need comprehensive city information systems that guarantee the compatibility of the different applications developed based on them with existing ones and streamline the management and maintenance of the information.

The definition of urban 3D models, the creation and collaborative editing of these models has several advantages in urban management processes; facilitating the work on the unique model always updated and the execution of works in parallel. The 3D component facilitates the representation of complete information easily understood by any user. These models facilitate the cooperation of experts in different areas, contributing their knowledge in the generation of a single model.

The urban management process with the greatest practical impact in Spain are the Special Interior Reform Plans (in Spanish, PERI – Plan Especial de Reforma Interior). The purpose of the PERI is the urbanistic improvement of the urban space: decongestion, creation of endowments, sanitation or circulation problems, among others. The PERI allows defining an integrated internal refurbishment operation, which attempts to cover all areas and problems of the urban environment. The administration is the one that develops a PERI however, the process of defining the PERI requires communication with citizens to gather their needs and problems in a first stage and then to collect feedback through improvements or suggestions after the communication of an initial proposal by the administration. Additionally, the interventions associated with this plan incorporate public subsidies that must be carefully controlled and justified, for which accurate information is required from both owners and buildings. To obtain such information it is also necessary to carry out technical and visual inspections. The process of generating the plan is long and tedious and requires the participation of different agents (administration, citizens and professionals).

The management process outlined above is a clear example of the need to advance in the development of urban information systems based on open models and standards that integrate the different scales (building – BIM and urban environment – GIS). A representation system that allows the integration of heterogeneous information, covering multiple domains and evolving over time. It also requires a great semantic capacity, which allows a reasoning on the model and evaluates the multi-domain interactions.

The rest of the article is structured as follows. First, a review of the state of the art of work related to collaborative generation of multiscale urban 3D models is presented. Section 3 describes the platform based on web services developed for the collaborative generation and edition of urban 3D models. Section 4 shows the process of developing a PERI, focused on information sharing activities, for the case of the historic center of the city of Vitoria-Gasteiz (Spain). For this process, the different components of the platform described in section 3 have been used. Finally, the main conclusions obtained from the work described in this article are presented.

## 2 RELATED WORK

The introduction of new technologies in the field of the city generates an associated vocabulary closely related to information and data (e.g. IoT, BIM, Big Data, SmartCity). By introducing these technologies, a growing volume of data is generated, which is also heterogeneous in nature. The urban 3D model is increasingly recognized as the adequate support to integrate, harmonize and store the information of a city and make it accessible to all stakeholders (citizens, city managers, companies or researchers) (Biljecki et al. 2015). Urban 3D models will allow transforming the way in which the city faces social, economic and environmental challenges.

Current technological advances are causing a relocation and distribution of information that was previously managed locally. A clear example are Google's collaborative office tools (for example, Docs, Forms or Calendar), tools for managing images (for example, Flickr or Panoramio), projects (for example, Podio or Evernote) or files (for example, Dropbox or Drive), among others.

This collaborative work has a set of common characteristics, such as: Collaboration, multiple users access/edit the information, with the appropriate permission levels; Access anytime and anywhere, from multiple platforms, desktop computers, mobile devices or Tablets (Dyer et al. 2017). Online collaborative platforms put the value of the software service concept as oppose to the product. And they allow software companies to offer their developments under new modalities (payment for use, access to services, annual subscription or depending on the volume of data).

The collaborative creation and edition of 3D urban model has several advantages, including more immediate results as work is done in parallel, or the possibility for experts in different areas to cooperate and contribute with their knowledge in same model. However, the creation and edition of 3D urban models is still an open issue (Budhathoki & Haythornthwaite 2012).

Open Street Map (OSM) project is an excellent example of collaborative creation and maintenance in relation to georeferenced 2D geographic models (Budhathoki & Haythornthwaite 2012). OSM has its own tools and data model, with integrated ad hoc support for multiuser editing and version control.

In the case of 3D cities, several proposals have been launched. One of them is a methodology for the collaborative and interactive development of 3D building models that includes walls, roofs, doors and windows as structural elements (Abbasi & Malek 2015).

Other authors have chosen to improve OSM data to make them compatible with three-dimensional visualization. For example, the OSM-3D platform has been developed precisely to offer an interactive 3D view of the data online (Goetz & Zipf 2012) (Goetz 2012). However, it must be borne in mind that the three-dimensionality of OSM is based on extrusion, so that it does not serve to represent complex geometries.

Another solution is 3D Repo, a pioneering open source work environment for version control that allows the coordinated management of large-scale 3D engineering data over the Internet, although at the moment it is not compatible with CityGML (Scully et al. 2015).

The integration of information at different scales is one of the main aspects to consider in the management of urban information. Currently there are systems and tools that address urban management from the geospatial perspective (GIS) and others from the perspective of the building (BIM). The integration of both domains is a current challenge for which solutions are proposed at different levels: data level, process level and application level (Liu et al. 2017). CityGML is currently the most widely used standard for the representation and exchange of urban information. This data model is especially relevant when an integration between GIS and BIM is required thanks to semantic modeling and 3D representation of geospatial information (Gröger & Plümer 2012). The most developed integration proposals are based on the use of standard and open data models (mainly CityGML and IFC) and integration at the process level through semantic technologies (Deng et al. 2016) or based on web services (Lapierre & Cote 2007).

## 3 THE PLATFORM

The collaborative platform has been developed within the framework of the U3DCLOUD project, whose objective is the development of a collaborative platform based on the cloud for the creation and edition of semiautomatic urban 3D models, supporting decision making involving multiple agents and scope of decision. It is a domain of great relevance and that synthesizes the multi-agent and multi-scale complexity that is intended to be undertaken.

The platform is based on data models (CityGML and IFC) and Open Geospatial Consortium (OGC) standards. The digital model ranges from the scale of the city to the details of the building, with a transparent transition for the user. The platform consists of a modular architecture divided into three main blocks (see Figure 1):

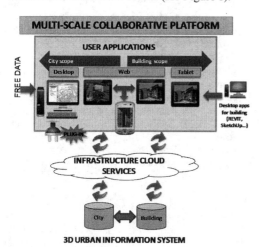

Figure 1. Architecture of the platform.

### 3.1 *3D Urban information system (3D-UIS)*

The 3D Urban Information System (3D-UIS) represents the centralized information model, based on the Common Data Environment philosophy, which provides a single point of access to urban information, that is properly structured and linked, and is generated from multiple external sources and connected to different tools. It allows multiscale representation and is based on standard data models (CityGML and IFC).

The 3D Urban Information System (3D-UIS) is split in 3 main components:

1. *City storage*, which is a PostgreSQL relational database that has been augmented with PostGIS extension (adding support for geographic referenced 3D objects). The data model storage is based on the 3D City Database[1], that allows representing and storing information modelled in CityGML.
2. *Building repository,* which is a cloud file storage. This repository includes all the IFC files, which contain most of the building information, as well as other file formats obtained from the IFC, which are used for visualization purposes (e.g., COLLADA, XML).
3. *City-Building link database*, which is a relational database that stores the relation between the different models (through the URI of the models) included in the City and Building repositories. The link database represents relationships between models, allowing linking buildings or individual objects such as windows or walls. The different users participating in a project can interact and collaborate uploading resources like pictures and sending messages or other issues. This information is also included and linked in this database. In the Figure 2 the structure of City-Building link database is represented.

### 3.2 *Infrastructure cloud services*

The cloud services provide access to the 3D-UIS through open standards (REST and OGC-based). This layer is divided into low-level services and high-level ones.

Among low-level services, highlight the services of uploading, downloading, accessing and consulting information contained in the model. The access to *City database* is performed using standard OGC WFS requests; *Building repository* is managed using upload/download web services and *City-Building link database* is managed using REST services.

---

[1]https://www.3dcitydb.org/3dcitydb/.

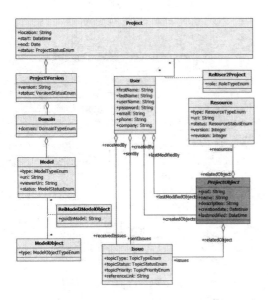

Figure 2. Structure of the city-building link database.

Among the high-level services are the services of 3D model generation, linkage and interoperability of data models. These services use previously explained low-level services:

1. *3D model generation.* This service creates a 3D city model in CityGML with geometric and basic semantic information. This service receives as input the footprint of buildings and LIDAR data (digital terrain model and digital surface model). After calculating objects height using LIDAR data the geometric generation is done, creating buildings, roads, vegetation, points of interest, etc.
2. *Semantic completion.* This service allows completing the semantic information of city objects in a massive way from a SHP file. A XML file that defines the relationships that are between the SHP file parameters and parameters in the CityGML database is also necessary as an input. The service processes the inputs and completes the CityGML with the semantic information within the SHP file.
3. *Geometry-based data processing.* This service provides the functionality to automatically complete semantic information that can be obtained directly from the geometric information. Some of the main features that are available with this geometric processing are the identification of adjoining walls of buildings, calculate the orientation of the facade and calculation of area and volume.
4. *IFC interoperability.* This service provides the functionality that allows visualizing and IFC file in a Tablet application. In order to do that, first the geometry of the IFC is exported to COLLADA format using the IfcOpenSheel library[2].

As only geometry is exported in the COLLADA file, a XML file is generated including all the semantic information contained in the IFC file. The semantic information is related to the geometry (defined in the COLLADA file) using the same IDs. Using this service, it is possible to visualize the IFC model in a Tablet application and show/edit the semantic information of the model. Both models IFC and XML are synchronized in the 3D-UIS through the link data base.

5) *Matching between IFC and CityGML.* This service provides the functionality to perform a link between an IFC and a specific building within CityGML. Knowing the building location in CityGML, the desired IFC model can be correctly geolocated (in location, rotation and elevation) with respect to it. So, first the IFCSite entity of the IFC is adapted in order to correctly geolocate the IFC model. Then the matching (relating the two buildings using their respective IDs) is stored in the *City-Building link database*.

### 3.3 User applications

Based on the defined infrastructure (model and services), different applications can be developed to generate and use the urban information. The developed prototype contains 3 applications:

- *City scope application:* A set of ArcGIS plugins that allow the generation of 3D city models in CityGML, based on the high-level services, which allow the automated 3D model creation from available sources (cadaster and LIDAR), and massive semantization
- *U3DCloud-Web application:* A 3D web application that allows city and building visualization with different levels of detail and include functionalities for layer based visualization and uploading and downloading of resources associated to elements in the model (e.g. 3D detailed models, images or claims)
- *Building scope application:* An application that allows 3D building model edition and visualization. The 3D model of the building can be generated with an external tool based on IFC standard, which will be uploaded into the 3D-UIS. A Unity-based application has been developed to visualize, interact and update the model based on the service infrastructure to keep the integrity of the model.

## 4 THE PROCESS

The collaborative platform has been tested for the generation of the 3D model of the historic district

---
[2]http://ifcopenshell.org/ifcconvert.html.

of Vitoria – Gasteiz (Spain). Based on the platform described above, different agents participate in the process of defining, editing and monitoring a Special Interior Reform Plans (PERI). The municipality is responsible for coordinating the process, but citizens and technicians from different domains can also share information and participate in the process.

The process based on the developed platform is divided into 4 stages. Each of these stages is divided into several activities in which different agents collaborate. To carry out the activities of each participant, different tools are used in different devices, which share the information model, and a single service infrastructure, which allows maintaining the integrity of the model and interoperability between the components of the platform.

The stages in which the process is divided are the following:

1. Generation of the urban 3D model
2. Including detail information of the buildings
3. Diagnosis of the current situation of the study area
4. Citizen participation for the definition of the Plan

The Figure 3 shows the detailed process including the stages, agents and the tools. The different tools and services or interfaces on which the flow of information is based are represented by different colors.

### 4.1 Generation of the urban 3D model

The district manager requests a 3D model of the historic district of Vitoria-Gasteiz. Next, the Geographic Information System (GIS) technician collects and processes information available in ArcGIS, to generate the model in LOD2 through plugins (*City scope application*) and services developed in the project, starting with the geometric generation and later completing the semantics of the CityGML model. Once the model is generated, it is published on the Web. Through the *U3DCloud-Web application* of the Vitoria-Gasteiz project it is possible to visualize the generated model, as well as navigate and make different queries about the information contained in the model. Figure 4 shows the plugin developed in ArcGIS for model generation.

As a result, it is obtained a CityGML model of the historic center of Vitoria—Gasteiz with buildings in LOD0, LOD1 and LOD2 (using the *3D model generation* service). The model has been extended with the required thematic data (using the *Semantic completion* service). The geometrical data of the model have also been processed to obtain new semantic parameters, such as area, main orientation or adjoining walls, among others (using the *Geometry-based data processing* service).

### 4.2 Including detailed information of buildings

The manager requests a detailed model (IFC) for a specific building. The building technician generates an IFC model using a BIM authoring tool (in this case REVIT) to respond to the manager request. The generated model is uploaded through the *U3DCloud-Web application* to the platform. The IFC model need to be validated by the manager and, once validated the resource appears visible in the web application. Figure 5 shows the building modeled in REVIT and the same building visualized through the *U3DCloud-Web application* of the project.

Once the model is uploaded and validated, it is linked to corresponding building in the CityGML (using the *Matching between IFC and CityGML* service) and the IFC is converted into other data formats (COLLADA and XML) for its visualization in the *Building scope application* (using the

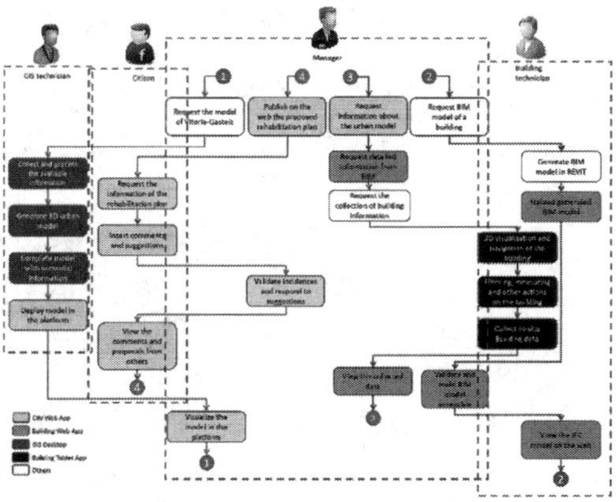

Figure 3. Process of the special interior reform plans.

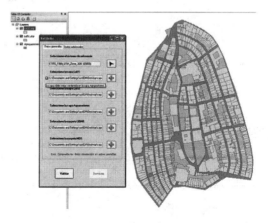

Figure 4. Plugin developed in ArcGIS for model generation.

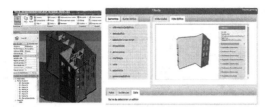

Figure 5. Generation of a IFC model in REVIT/ Visualization of the IFC model in the *U3Dcloud-web application*.

Figure 7. Viewing images of the selected building.

### 4.4 *Citizen participation for the definition of the plan*

Once all the relevant information has been collected, the manager defines and publishes on the web a PERI proposal to be shared with the citizens. Citizens consult the available information on the buildings of the study area through the *U3DCloud-Web application*. In addition to the information associated with each property, the plan will collect information on resources associated with the elements of the environment (buildings or others). Citizens can include their comments, incidents or photographic resources to complete the PERI, send feedback or propose modifications. The resources introduced by citizens must be reviewed and validated by the manager as a step prior to the publication. Finally, new resources uploaded by citizens are uploaded and associated to the corresponding project and building in the *City-Building link database*. Figure 7 shows an example of presentation of photographic resources associated with an element of the environment on the reference model for the definition of the PERI.

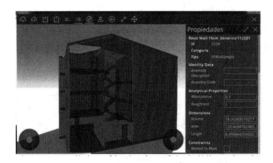

Figure 6. Compilation of in-situ data of the building using the tablet.

*IFC interoperability* service). In this stage new files are stored in the *Building repository* and the matching between the IFC and the CityGML is performed in the *City-Building link database*.

### 4.3 *Diagnosis of the current situation of the study area*

The manager analyzes the information of the urban model from the *U3DCloud-Web application* (both district-wide and building by building). Subsequently, the manager asks the building technician to complete the information that is not included in the building detail model. The technician moves to the building location and uses the *Building scope application* to visualize the 3D model on the Tablet, analyze it, take notes and measurements and modify or complete the semantic data of the model. The modifications are updated on the platform. A new XML file with the updates is generated and synchronized in the 3D-UIS (using the *IFC interoperability* service). The data collected by the technician during the inspection can be displayed by the manager from the *U3DCloud-Web application*. Figure 6 shows the appearance of the *Building scope application* running on a Tablet.

## 5 CONCLUSIONS

The described work presents a collaborative platform based on standard services for the generation and edition of multiscale urban 3D models.

The definition of a centralized information model based on standard data models (CityGML and IFC) allows to have a single model with structured and linked information that enables the collaborative generation and management of the model.

An important aspect of the model is that it can be completed over time. The simplified representation of the buildings in CityGML has been completed with the detailed representation of the IFC model of a building. It is also possible to extend or modify the semantic data of the buildings in the CityGML and IFC models, and the model can be completed with new resources associated with the elements of the model.

The use of standard data models facilitates the interoperability of the developments with existing tools.

The development of a cloud platform based on services allows the collaborative creation of the model by different users from any place and device.

Taking advantage of collaborative features of the platform it is possible to reduce time and costs in design, analysis and management of cities. Besides this facilitates the openness, transparency and having a "collaborative city for all" (managers, technicians or citizens).

ACKNOWLEDGMENTS

The work of this paper has been done as part of the project U3DCloud "Plataforma de servicios estandares en la nube para la gestion colaborativa del modelo digital 3D de la ciudad durante su ciclo de vida" with reference TSI-100400-2013-47.

REFERENCES

Abbasi, S. & Malek, M.R., 2015. Design and Modeling of a 3D Volunteered Geographic Information with an Interoperable Description for Fundamental Components of a Building. *Journal of Geomatics Science and Technology*, 4(4), pp. 15–28.

Biljecki, F. et al., 2015. Applications of 3D City Models: State of the Art Review. *ISPRS International Journal of Geo-Information*, 4(4), pp. 2842–2889. Available at: http://www.mdpi.com/2220-9964/4/4/2842.

Budhathoki, N.R. & Haythornthwaite, C., 2012. Motivation for Open Collaboration: Crowd and Community Models and the Case of OpenStreetMap. *American Behavioral Scientist*, 57(5), pp. 548–575. Available at: http://abs.sagepub.com/cgi/doi/10.1177/0002764212469364.

Deng, Y., Cheng, J.C.P. & Anumba, C., 2016. Mapping between BIM and 3D GIS in different levels of detail using schema mediation and instance comparison. *Automation in Construction*, 67, pp. 1–21. Available at: https://doi.org/10.1016%2Fj.autcon.2016.03.006.

Dyer, M., Gleeson, D. & Grey, T., 2017. Framework for Collaborative Urbanism. In *Citizen Empowerment and Innovation in the Data-Rich City*. Springer, pp. 19–30.

Goetz, M., 2012. Towards generating highly detailed 3D CityGML models from OpenStreetMap. *International Journal of Geographical Information Science*, pp.1–21.

Goetz, M. & Zipf, A., 2012. OpenStreetMap in 3D-Detailed Insights on the Current Situation in Germany. In *City, Proceedings of the AGILE 2012 International Conference on Geographic Information Science, Avignon, April*. pp. 24–27.

Grey, T., Dyer, M. & Gleeson, D., 2017. Using Big and Small Urban Data for Collaborative Urbanism. In *Citizen Empowerment and Innovation in the Data-Rich City*. Springer, pp. 31–54.

Gröger, G. & Plümer, L., 2012. CityGML—Interoperable semantic 3D city models. *ISPRS. Journal of Photogrammetry and Remote Sensing*, 71(0), pp. 12–33. Available at: http://www.sciencedirect.com/science/article/pii/S0924271612000779.

Lapierre, A. & Cote, P., 2007. Using Open Web Services for urban data management: A testbed resulting from an OGC initiative for offering standard CAD/GIS/BIM services. In *Urban and Regional Data Management. Annual Symposium of the Urban Data Management Society*. pp. 381–393.

Liu, X. et al., 2017. A State-of-the-Art Review on the Integration of Building Information Modeling (BIM) and Geographic Information System (GIS). *ISPRS International Journal of Geo-Information*, 6(2), p. 53. Available at: https://doi.org/10.3390%2Fijgi6020053.

Scully, T. et al., 2015. 3drepo.io: building the next generation Web3D repository with AngularJS and X3DOM. In *Proceedings of the 20th International Conference on 3D Web Technology*. pp. 235–243.

# Visualizing earthwork and information on a linear infrastructure project using BIM 4D

L. Schneider Jakobsen
*University College of Northern, Denmark*

J. Lodewijks
*Züblin-Strabag BIM 5D, Germany*

P. Nørkjær Gade
*University College of Northern, Denmark*

E. Kjems
*Aalborg University, Denmark*

ABSTRACT: Countries worldwide are facing the need for infrastructure improvements on a large scale. Projects concerning road, railway, tunnel and pipelines include a large amount of earthworks which is often delayed or has considerable cost overruns. This study presents a method to assist earthwork planning, by extracting separate 3D objects of cutting and filling, attaching schedule information and visualize these in commercial BIM 4D software. The 4D method was applied and verified in the re-scheduling of a linear infrastructure project which is one of the German governments selected pilot projects before implementing BIM requirements in 2020. The study indicated that the 4D visualization enable information flow between projects participants. The paper gives suggestions to the level of information detailing standardized open data format should be able to contain in regards of facilitating the use of BIM 4D information visualizing in future linear infrastructure projects.

## 1 INTRODUCTION

### 1.1 Background

Infrastructure has a large effect on the national economies, to keep up with population growth a global investment of 49,1 trillion USD until 2030 is necessary (Woetzel et al. 2016). Unfortunately, the construction of infrastructure is often delayed or has considerable cost overruns. Linear infrastructure projects such as roads, railways, tunnels, and pipeline constructions comprise a large amount of earthwork. Castro (2005) concluded that earthwork accounts for 20% of the overall cost of road construction projects. The earthwork requires expensive equipment, the scheduling of the equipment activities have large uncertainties due to variable site conditions, soil characteristics and meteorological factors. All these factors risk to result in cost overruns (Castro 2005). Considering the effects and constraints of the variables on the schedule, requires a detailed collection of information, especially considering that the necessary data are currently scattered across multiple documents and information sources (Kang et al. 2013). Project decisions based on scattered data cause assumption and experience-based decision-making which is prone to errors. According to Mazza (2009), data must be organized processed and suitably presented-to bear proper meaning for the users.

Building information modeling (BIM) is an approach to organizing, containing and visualizing data in the construction industry that potentially can reduce the time and cost of infrastructure projects. McKinney & Fischer (1998) have developed a BIM method to combine 3D models with schedule data in a graphic visualization of the construction process called 4D cad (3D CAD+time). The 4D method enriches traditional schedule representations by adding information concerning location and geometry which has shown to be effective for optimizing the planning and scheduling process (Ghaffarianhoseini et al. 2017). However, the application and research of 4D methodologies have been predominantly found in non-linear construction projects (Bradley et al. 2016), To the best of our knowledge, there are no previous studies describing how 4D visualization can be used to visualize earthwork data, applied to 3D cut and fill objects by using commercial software. The aim of this paper is to investigate the possible use of BIM

4D to visualize earthworks information in a linear infrastructure project.

## 1.2 Earthwork scheduling

Earthwork activities in linear infrastructure projects involve significant costs associated with the movement of soil and have a large influence on the later proceeding construction process (Askew et al. 2002). Furthermore, the earthwork is most uncertain and unpredictable in linear construction projects (Creedy et al. 2010). Earthwork can be divided into three main categories cutting, transport from cut to fill section and earth filling (Shah Dawood, & Castro 2008). Earthwork is done with expensive machinery, delay of the machinery can have a large impact on the construction cost (Mahamid 2013). According to Mohammadian et al. (2007), The costs of the machinery accounts for 80% of the total cost of earthwork in linear construction projects. Furthermore, the use of machinery has a large impact on the total $CO_2$ emissions caused by the project (Muresan et al. 2015).

The scheduling of earthwork is generally based on geometrical data such as volume, area, and location. Furthermore, earthwork scheduling includes many non-geometrical variables such as earth type, construction phasing, equipment availability, haulage distance, site constraints, and geotechnical conditions (Shah et al. 2008). The non-geometrical variables make earthwork scheduling unique for every project and it requires individual scheduling skills to include and consider the impact of the many variables on the construction schedule (Molenaar 2005). The earthwork is traditionally scheduled based on segmented 100–200-meter workstations (Burdett et al. 2015).

According to Jrade & Markiz (2012), the main decisions taken during earthwork scheduling are:

- Transport distance of the earth
- Machine use
- Construction Phasing
- Deposit area
- Crew assigned
- Earth type classification
- Site constraints

According to Shah & Dawood (2011), The decision data is currently scattered in the following two-dimensional representations methods:

- Geotechnical reports
- Worksheet – workforce instruction or method statement,
- Gantt chart schedules
- Linear schedules (location-based)
- Two-dimensional drawings

A method is needed to collect and visualize the schedule data in linear infrastructure projects.

## 2 BUILDING INFORMATION MODELING (BIM)

Four-dimensional computer-aided design (4Dcad) has shown its potential for improving the communication of schedule data in construction projects (Ghaffarianhoseini et al. 2017). The method of linking 3D building models with schedule data was described in the 1990's with early case studies (Retik et al. 1990). McKinney & Fischer (1998) describe the advantage of removing the mentally imagined abstractions of schedule data by visualizing 3D models linked with schedules in a case study. Building construction projects progress vertically and by component-based representations using components such as walls, slabs, and beams. Non-geometrical information can be added in a meaning full way which then can be linked with schedule data in a 4D model (Eastman et al. 2013). The open data format IFC has been developed for the construction industry. The Aim has been to standardize and organize the data and enhance interoperability between software platforms (Laakso & Kiviniemi 2012) Kang, Anderson, & Clayton, (2007) describe the advantage of visualizing the schedule progression and the underlying information which was previously scattered. Moon et al. (2014) have shown in studies how the 4D method can be further used in workspace modeling, spatial conflict detection and coordination in construction projects.

### 2.1 BIM 4D in linear infrastructure projects

4D in linear infrastructure projects differs greatly from 4D in building projects (Kim et al. 2011). Linear infrastructure projects progress with nonrepetitive horizontal works over a large geographical area (Zanen et al. 2013). A large amount of variables and changes in site conditions causes earthwork to be less organized compared to a building construction project (Mawlana et al. 2015). The segmentation of earthwork in 3D objects is not defined by real-world "objects" but rather by planned volumes and quantities. Differentiation between 4D visualization in building projects and infrastructure projects is required. Liapi (2003) has applied 4D-methods to road construction projects focusing on the visualization and communication, the results indicating the need for segmentation of geometrical objects. Dawood & Castro (2009) Shah et al. (2008) Castro & Dawood (2005) Chen et al. (2011) has during studies described methods of creating algorithms and automated simulations of earthwork schedules in visual profiles. The benefit of representing the earthwork schedule visually is described to reduce unnecessary machine movement. The industry hesitates to adopt automated scheduling due to a large number of variables and dependencies in earthwork scheduling (Arayici et al. 2012), Leite et al. (2016). During a case study Platt (2007) experienced the large advantage of using commercial software to visualize the earthwork schedule because it allowed the project team to take manual decisions based on the schedule visualization. The study clarified a need for segmenting the earthwork in the 3D model into linear 100-meter grid system the project team used as a planning reference, preferably also the ability to apply non-geometrical data to the objects. Kang et al. (2013) solved the segmentation problem with the use of own software and demonstrated the possibility for segment earthwork in 3D objects.

## 3 THE CONCEPTUAL APPROACH

The company Autodesk has developed an upgrade for the software Civil 3D with possibilities for extracting cut and fill as separate 3D objects shown in Figure 1. Attaching non-geometrical data to the 3D objects enable its use as a container for information. The 3D objects can be linked with schedule data in an external software which makes it possible to apply the 4D method to earthwork in linear infrastructure projects. There is no standard for organization of data in the 3D objects of earthwork in linear infrastructure projects. According to Porkka & Kähkönen (2007) the use of the data organization standard IFC in BIM 4D has possible benefits experienced in the construction industry. The data organization in this study is based on specific company standards.

## 4 TEST STUDY ON A 7 KILOMETER HIGHWAY PROJECT

A 7-kilometer highway project in Northern Germany was used as a test project to the 4D method in linear earthwork projects. The project contained a large amount of earthwork due to the widening of lanes and new bridge constructions. One of the main constraints was traffic lanes remaining open during the construction period which required a large amount of coordination and communication. The project was planned and executed with traditional scheduling methods and the new 4D visualization enhancements were tested by the same project team to validate the advantage of applying 4D visualizations.

### 4.1 The composition of a 3D model

The first step in the 4D method is to create a 3D model of the earthwork. The survey data is used to create the 3D model. The model was created with the software Civil 3D, composing a triangulated terrain network surface. Several layers of data are modeled to represent the existing surface and accordingly the newly built surface. The triangulated network layers were extracted and identified as cut or filling and represented in separate 3D objects. Since linear infrastructure projects align horizontally across a large geographical area, it is of great importance to segment the model. The earthwork is usually divided into sections of 100 or 200 meter (Burdett et al. 2015). The 3D model was divided into 100-meter sections as separate cutting and filling 3D objects. The earthwork is defined with color coding in the Civil 3D model shown in Figure 2. It is possible for the foreman to visually see the earthwork volume as 3D objects. Geometrical data extracted from the objects can be used in the early scheduling process. The following geometrical data is available:

- Orientation
- Size
- Volume
- Shape
- Height

### 4.2 Information objects

The schedule data was scattered in several document types and it was then necessary to create

Figure 1. 3D object of cutting and filling.

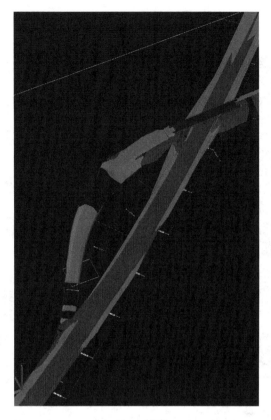

Figure 2. 3D model of cut and fill earthwork.

workflows to ensure the correct data was collected and was attachable to the 3D object of cutting and filling in DWG file format.

The followings information was argued important by the project team and manually added to the 3D objects:

- Earth type classification
- 100-meter alignment stationing
- Construction Phase
- Crew
- Deposit area
- Machine use
- Transport
- Work type (Cut, transport or fill)

The procedure for adding the collected data to the 3D object does function as a collaborative task between 3D modeler and work preparator. In Figure 3 the non-geometrical data is added to the 100-meter fill section in the congested work area where the road works intersect with the bridge work. It is possible to review decisions taken during the scheduling process and visually communicate this to project participants, so they can review the decisions underlying the estimated time of

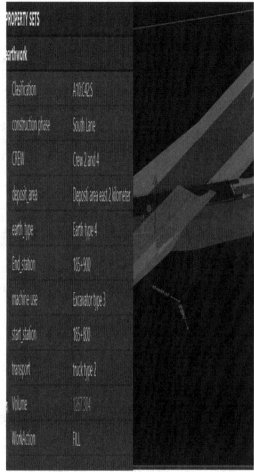

Figure 3. 3D fill earthwork object with information.

the work by selecting the 100-meter section. The Earth type classification and description of the data is made based on the project team company standards. The classification in other companies might be different and misunderstanding is a risk. (Heikkila 2013).

## 5 IMPLEMENTATION IN THE TEST PROJECT

To validate the 4D method, it was tested in a 7-kilometer German Private public partnering project. The 4D visualization was prepared in the software Navisworks based on new workflows. The project team was using the 4D method in an information flow between departments during rescheduling of the project. Figure 4 shows how decisions taken in a linear scheduling tool are collected and visualized as information in the 4D object, because the

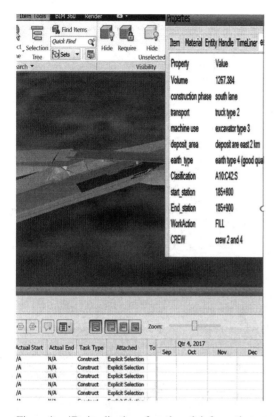

Figure 4. 4D visualization of earthwork information.

1. Segmentation of cut and fill earthwork as separate 3D information objects in linear infrastructure projects is shown possible in Figure 3 with the use of commercial software.
2. The information objects visualized and linked with the earthwork schedule shown in Figure 4 did function as a visual container in an information flow between projects participants. The project team argued that the use of such method could be generally beneficial to overall project collaboration although the large amount of manual work in the rescheduling workflows should be reduced.
3. The study has by literature review, interviews and 4D implementation in a project shown in Figure 4 suggested which information and detail level of these future standardized open data formats for earthwork as minimum is to transport between software used to 4D in linear infrastructure projects

4D object relate to the time schedule the information is placed in the correct location of the project, the earth type is visible in the 4D object which makes it possible to ensure the optimal machine and deposit area is chosen. The scheduling department was able to describe the machine use under a bridge for the cost estimation department which creates a cost estimation based on the actual scheduled machine use.

## 6 RESULTS

The aim of this paper was to investigate the possible use of BIM 4D to better visualize earthworks information in a linear infrastructure project. A 7-kilometer German Private public partnering project was used as a test study. A 3D model of the project was used and the 3D objects of cutting and filling were extracted and added information from work preparation. The project team was part of the process and contributed with their experience regarding advantages and challenges of the implementation in future projects, the results of the study were:

## 7 DISCUSSION

Result 1 shows that the software possibilities today enable the previous constraints described by Kang et al. (2013), Platt (2007) regarding segmenting cut and fill as separate 3D objects and attaching information. The result 2 shows that Castro & Dawood. (2005) suggestions of enhanced collaboration caused by visualizing earth work information are experienced by the project team in the test study. Workflows during rescheduling and linking are realized labor intensive, Automation could be a solution. Advantages to creating the segmented 3D model attaching information and link this with schedule within one commercial 4D software was suggested by the project team, this must be further investigated. The result 3 give a suggestion to the infrastructure industry and software companies regarding which level of information detail 4D earth objects future open data format should be able to consist. This study is done with the perspective of one company, future studies are to be done regarding finding a standardized organization of data in the earthwork objects which could improve exchange between companies. Especially a common agreed earth classification for the 4D earthwork objects is suggested.

## 8 CONCLUSION

There is currently a large research gap between The BIM enhancements of the building industry and the infrastructure industry. The project team in the test study has suggested a possible large enhancement of information sharing with the use of the BIM 4D method described in this paper. It is important that the segmentation of the 3D earthwork objects takes place uniformly in open format as the IFC standard. The Segmentation must be part of the early design phase and therefore more test which apply the method from the start of the projects is needed. The schedule parameters are to be linked to the earthwork segments, which is possible in IFC format and thus the type of information can be exchanged between all involved. In this way, the process will not be one-way, with conversions and risk of data loss to follow, but may be one of a single model that can be accessed by all the project participants.

REFERENCES

Arayici, et al. (2012). Building information modelling (Bim) implementation and remote construction projects: Issues, challenges, and critiques. *Electronic Journal of Information Technology in Construction*.

Askew, et al. (2002). Planning linear construction projects: Automated method for the generation of earthwork activities. *Automation in Construction*, 11(6), pp. 643–653.

Bradley et al. (2016). BIM for infrastructure: An overall review and constructor perspective. *Automation in Construction*, 71, pp. 139–152.

Burdett, et al. (2015). Block models for improved earthwork allocation planning in linear infrastructure construction. *Engineering Optimization*, 47(3), pp. 347–369.

Castro & Dawood (2005). Roadsim: Simulation modelling and visualisation in road construction. *Construction Research Congress 2005: Broadening Perspectives – Proceedings of the Congress*, pp. 1037–1045. Available at: https://www.engineeringvillage.com/share/document.url?mid=cpx_18a992f107b36de7e2M57072061377553&database=c px.

Castro (2005). "Integrated simulation models applied to road construction management", PhD Thesis, University of Teesside, Middlesbrough, UK.

Chen et al (2011). 3D Animation Applications in Highway Design. *Advanced Materials Research*, 341–342, pp. 878–882. Available at: http://www.scientific.net/AMR.341-342.878.

Creedy, et al (2010). Evaluation of Risk Factors Leading to Cost Overrun in Delivery of Highway Construction Projects. *Journal of Construction Engineering and Management-Asce*.

Dawood, & Castro (2009). Automating road construction planning with a specificdomain simulation system. *Electronic Journal of Information Technology in Construction*, 14, pp. 556–573.

Eastman, C. et al., 2013. *BIM Handbook*.

Ghaffarianhoseini, et al. (2017). Building Information Modelling (BIM) uptake: Clear benefits, understanding its implementation, risks and challenges. *Renewable and Sustainable Energy Reviews*, 75, pp. 1046–1053.

Heikkila, (2013). Development of BIM and Automation based total process for Infra construction – The state of the art in Finland. *Bim*, pp. 1216–1222.

Jrade & Markiz (2012). A Decision-Support Model Utilizing a Linear Cost Optimization Approach for Heavy Equipment Selection. In *International Journal Of Structonics & Mechatronics*. pp. 100–109. Available at: http://ascelibrary.org/doi/abs/10.1061/97 80784412329.011.

Kang, Anderson, & Clayton, (2007). Empirical Study on the Merit of Web-Based 4D Visualization in Collaborative Construction Planning and Scheduling. *Journal of Construction Engineering and Management*, 133(6), pp. 447–461. Available at: http://ascelibrary.org/doi/10.1061/%28 A SCE%290733- 9364%282007%2913 3%3 A6%28447%29.

Kang, et al. (2013). Development of a 4D object-based system for visualizing the risk information of construction projects. *Automation in Construction*, 31, pp. 186–203.

Kang, et al. (2013). Development of Improved 4D CAD System for Horizontal Works in Civil Engineering Project. *Journal of Computing in Civil Engineering*, 27(3), pp. 212–230.

Kim, et al. (2011). Applicability of 4D CAD in Civil Engineering Construction: Case Study of a Cable-Stayed Bridge Project. *Journal of Computing in Civil Engineering*, 25(1), pp. 98–107. Available at: http://ascelibrary.org/doi/10.1061/%28 A SCE%29CP.1943–5487.0000074.

Laakso & Kiviniemi (2012). The IFC standard – A review of history, development, and standardization. *Electronic Journal of Information Technology in Construction*, 17, pp. 134–161.

Leite, et al. (2016). Visualization, Information Modeling, and Simulation: Grand Challenges in the Construction Industry. *Journal of Computing in Civil Engineering*, 30(6), p. 4016035. Available at: http://ascelibrary.org/doi/10.1061/(ASC E)CP.1943–5487.0000604.

Liapi, (2003). 4D visualization of highway construction projects. In Proceedings of the International Conference on Information Visualisation.

Mahamid, (2013). Effects of project's physical characteristics on cost deviation in road construction. *Journal of King Saud University – Engineering Sciences*, 25(1), pp. 81–88. Available at: http://dx.doi.org/10.1016/j.jksues.2012.0 4.001.

Mawlana, et al., (2015). Integrating 4D modeling and discrete event simulation for phasing evaluation of elevated urban highway reconstruction projects. *Automation in Construction*, 60, pp. 25–38.

Mazza, (2009). *Introduction to Information Visualization*, Available at: http://www.springerlink.com/index/10.1 007/978–1–84800–219–7.

McKinney, & Fischer (1998). Generating, evaluating and visualizing construction schedules with CAD tools. *Automation in Construction*, 7(6), pp. 433–447. Available at: http://linkinghub.elsevier.com/retrieve/pii/S092658059 8000533.

Mohammadian, et al. (2007). Estimation of Costs of Cars and Light Truck Use per Vehicle-Kilometer in Canada. *Transportation Research Board 86th Annual Meeting Transportation Research Board*, 0, p. 22. Available at: http://pubsindex.trb.org/orderform.html%5Cnhttp://ovidsp.ovid.com/ovidweb.cgi?T=JS&PAGE=reference&D=tspt&NE WS=N&AN=01044334.

Molenaar, (2005). Programmatic Cost Risk Analysis for Highway Megaprojects. *Journal of Construction Engineering and Management*, 131(3), pp. 343–353.

Available at: http://ascelibrary.org/doi/10.1061/%28 ASCE%290733- 9364%282005%29131%3 A3%28343%29.

Moon, Dawood, & Kang (2014). Development of workspace conflict visualization system using 4D object of work schedule. *Advanced Engineering Informatics*, 28(1), pp. 50–65.

Muresan, et al. (2015). Key factors controlling the real exhaust emissions from earthwork machines. *Transportation Research Part D: Transport and Environment*, 41, pp. 271–287.

Platt (2007). 4D CAD for highway construction projects. Computer Inte-grated Construction Research Program Department of Architectural Engineering The Pennsylvania State University 104 Engineering Unit A University Park, PA 16802 Retik Warszawski & Banai (1990). The use of computer graphics as a scheduling tool. *Building and Environment*, 25(2), pp. 133–142.

Porkka, & Kähkönen (2007). Software Development Approaches and Challenges of 4D Product Models. *VTT Technical Research Centre of Finland*, pp. 85–90.

Available at: https://pdfs.semanticscholar.org/3401/39 a554f498b834344f4e47b1a677f8994c6c.pdf.

Russell, et al. (2009). Visualizing high-rise building construction strategies using linear scheduling and 4D CAD. *Automation in Construction*, 18(2), pp. 219–236.

Shah, & Dawood, (2011). An innovative approach for generation of a time location plan in road construction projects. *Construction Management and Economics*, 29(5), pp. 435–448.

Shah, Dawood, & Castro, (2008). Automatic generation of progress profiles for earthwork operations using 4D visualisation model. *Electronic Journal of Information Technology in Construction*, 13, pp. 491–506.

Woetzel, et al. (2016). Bridging global infrastructure gaps. *McKinsey Global Institute*, (June), p. 60.

Zanen, et al. (2013). Using 4D CAD to visualize the impacts of highway construction on the public. *Automation in Construction*, 32, pp. 136–144.

*Interoperability and standardization of data structures and platforms*

# Multi-LOD model for describing uncertainty and checking requirements in different design stages

J. Abualdenien & A. Borrmann
*Chair of Computational Modeling and Simulation, Technical University of Munich, Germany*

ABSTRACT: The design of a building is a collaborative process between multiple disciplines. Using Building Information Modeling (BIM), a model evolves throughout multiple refinement stages to satisfy various design and engineering requirements. Such refinement of geometric and semantic information is described as Levels of Development (LOD). So far, there is no method to explicitly define an LOD's requirements nor any specification of its uncertainty. Furthermore, despite the insufficient information available in early design stages, a BIM model appears precise and certain. This can lead to false assumptions and model evaluations, for example, in the case of energy efficiency calculations or structural analysis. Hence, this paper presents a multi-LOD meta-model to explicitly describe an LOD's requirements taking into consideration the information uncertainty. This makes it possible to check the consistency of the geometric, semantic, and topologic coherence across the different LODs. The model is implemented as a webserver and user-interface providing a means for managing and checking exchange requirements between disciplines.

## 1 INTRODUCTION

The *architecture, engineering,* and *construction* (AEC) industry is a collaborative environment which requires an iterative and cooperative exchange of models information (Chiu 2002). For example, developing the structural design demands the architectural design information as an input. In this kind of collaboration, the information quality, such as compliance with regulations and analysis requirements, is essential for exchanging, coordinating and integrating the partial designs at the different stages. The design of a building evolves throughout multiple stages, each characterized by a set of consecutive and calibrated actions, to satisfy the different design and engineering requirements.

*Building Information Modeling* (BIM) is a promising approach that supports managing and exchanging semantically rich 3D-models between the project disciplines (Eastman et al. 2011). Recently, BIM has been widely adopted in the AEC industry (Young et al. 2009), it improves the process' efficiency and quality by promoting the early exchange of 3D building models. Through the different phases of a construction project, the building model is gradually refined from a rough conceptual design to highly detailed individual components. The sequential refinement of geometric and semantic information is described as *Level of Development* (LOD) (Hooper 2015). LOD is a concept that describes the different stages of the project life-cycle by providing definitions and illustrations of BIM elements at the different stages of their development (BIMForum 2017).

In the early design stages, BIM model information is not yet accurate as it is subject to multiple changes in the subsequent design stages (Knotten et al. 2015). Presently, model-based planning techniques are incapable of managing multiple levels of development including a description of their geometric and semantic information uncertainty. Neither is there a formal definition of a building component's level of development nor is there an explicit description of the fuzziness of information. On the contrary, a BIM model appears precise and certain which can lead to false assumptions and model evaluations, as in case of energy efficiency calculations or structural analysis, which affect the design decisions taken throughout the design stages.

The research project MultiSIM aims to develop methods for evaluating building design variants in early design stages. The variants may have different LODs as well as incomplete and uncertain information. The main approach focuses on providing:

- Consistent management of multiple LODs
- Describing the information uncertainty
- Consistent management of design variants
- Supporting model analysis at the early design stages.

As part of this research group, we propose the development of a multi-LOD meta-model, which explicitly describes the LOD requirements of each individual building component type taking into consideration the possible uncertainties.

The multi-LOD meta-model introduces two layers, *data-model level* and *instance level*, which offers high flexibility in defining per-project LOD requirements and facilitates formally checking their validity, such as defining and checking required information to support the *Embodied Energy* calculations at different design stages. This paper discusses the advantages in representing the uncertainties at early design stages and highlights the benefits of systematically managing and checking exchange requirements between disciplines. In order to ensure the model's flexibility and applicability, its realization is based on the existing *Industry Foundation Classes* (IFC). IFC is an ISO standard, which is integrated into a variety of software products (Liebich et al. 2013).

The paper is organized as follows: Section 2 discusses the background and related work of our research. Section 3 provides an overview of the multi-LOD requirements and describes the design concepts, and Section 4 presents the meta-model design. In order to evaluate the multi-LOD model, Section 5 illustrates how it can be used to define and check the requirements of the *Embodied Energy* calculations, and a prototype implementation is discussed in Section 6 in terms of usability and possible integration in the design process. Finally, Section 7 summarizes our progress hitherto and presents an outlook for future work.

## 2 BACKGROUND & RELATED WORK

### 2.1 Level of Development (LOD)

The concept of LOD is employed to manage the model evolvement through the different stages of the building life-cycle. It organizes the iterative nature of the design process which enhances the quality of the decision taken (Hooper 2015). An LOD describes the BIM elements on a particular stage providing definitions and illustrations (BIMForum 2017) which represents their information quality, i.e. certainty and completeness. The LOD scale increases iteratively from a coarse level of development to a finer one. Consequently, the associated characteristics' quality of the exchanged model elements is increased.

The *American Institute of Architects* (AIA) introduced a definition of the term LOD that comprises six levels, starting from LOD 100 reaching LOD 500. Additionally, it adds more flexibility by defining intermediate stages, like LOD 350 which requires the representation of the interfaces between the different building system (BIMForum 2017). Several guidelines have been proposed in an attempt to define the available information at each LOD. Most popularly, *Level of Development Specification* follows the AIA definitions, and *Level of Definition* in the UK (BSI 2017) consists of seven levels and introduces two components: *Levels of model detail* (LOD) representing the graphical content of the models, and *Levels of model information* (LOI) representing the semantic information. Recent approaches propagate the terms *Level of Information* and *Level of Geometry* to clearly distinguish semantic from geometric detailing grades (Hausknecht, Liebich 2017).

In this paper, the abbreviation *LOD* stands for the *Level of Development*, which represents the composition of both *Level of Geometry* (a.k.a. Level of Detail) and *Level of Information* (semantics).

### 2.2 Refinement of LODs

Multiple efforts have been conducted for describing the LODs refinement through the project lifecycle. The main idea is the attempt to represent and formalize the model maturity. Either by explicitly defining relationships or by controlling the amount of added details within an LOD, which makes it possible to check the model's consistency. (Biljecki et al. 2016) argue that five LODs are not enough to capture the building model's development, as the information ambiguity is high. Thus, they restrict the LODs refinement by allowing less specification and modelling freedom using a set of 16 stages. Similarly, (van Berlo, Bomhof 2014) looked into producing a more suitably refined set of LODs for the Dutch's AEC industry, they developed seven LODs after performing multiple geometric tests and analyzing the industrial practices.

From another perspective, (Borrmann et al. 2014) presents a methodology for creating and storing multi-scale geometric models for shield tunnels by explicitly defining the dependencies between the individual levels of detail. For this purpose, a multi-scale product model is developed including a geometric-semantic description of five levels; where the levels 1–3 describe the outer shell in terms of boundary representation of the tunnel volume, boundary surface as well as openings, and the fourth level includes the modeling of the tunnel's interior structure. It is shown how the LOD concept can be integrated into the IFC data model. In order to model the relationship between the different levels and maintain their aggregation, a new relationship class *IsRefinedBy*, a subclass of *Aggregates*, is introduced. The proposed multi-scale model makes use of the parametric modeling techniques to preserve the consistency among the different levels of detail by interpreting and processing the procedural geometry representations. Consequently, the change of a geometric object is propagated by updating all the dependent representations.

In this paper, we adhere the BIMForum's definition, starting from LOD 100 reaching to LOD 500, while making use of its flexibility by introducing intermediate LODs, including LOD 120 and LOD 250, to capture the refinement relationships of the semantic-geometric information.

## 2.3 Interoperability

The design and construction of a building is a collaborative process between multiple disciplines, each expert, such as architect and structural engineer, uses different authoring tool and requires custom specifications to support a particular type of simulations and analysis. With increasing the projects specialization and heterogeneity, the building industry requires a high level of interoperability.

The US national institute of standards and technology confirmed the high annual costs, around $16 billion, resulting from the lack of interoperability between the AEC industry software systems (GCR 2004). Over the last decade, numerous methods of exchanging data in the domain of AEC have been investigated. The aim is to define a common interface for lossless geometric as well as semantic data exchange. Therefore, buildingSMART is promoting the development of the industry standard, *Industry Foundation Classes* (IFC) which was published as ISO standard in 2003 (Liebich 2013). IFC is a free vendor-neutral standard and includes a large set of building information representations, including a variety of different geometry representations and a large set of semantic objects modeled in a strictly object-oriented manner. To allow for dynamic (schema-invariant) extensions and adaptation to local or national requirements, the IFC data model provides the *PropertySet* (PSet) mechanism, which relies on dynamically definable name-value pairs.

Besides exchanging data using IFC, dealing with different kinds of building information, e.g. property sets and definitions, requires a standardized terminology. Thus, the *buildingSmart Data Dictionary* (bsDD) (buildingSMART 2016) was developed as a central repository that stores multilingual definitions of the IFC entities and common schema extensions, for instance, an *IfcWall* entity description and *Pset_WallCommon*. Additionally, bsDD integrates multiple classification systems, including *OMNICLASS* (OmniClass 2012) and *UNICLASS* (Chapman 2013), which are widely adopted for structuring the building information. Each object in the dictionary is identified by a *Globally Unique ID* (GUID) which makes it computer-readable and independent of the object name and language (Bjorkhaug, Bell 2007).

As the IFC data model is too large for authoring tools to handle (Bazjanac 2008), buildingSMART developed the *Model View Definition* (MVD) mechanism as a standard approach for IFC implementation, which reduce the size of models through filtering. An MVD represents a subset of the IFC schema that specifies the requirements and specifications of the exchanged data between the authoring tools (Hietanen, Final 2006). In order to ensure the exchanged data completeness, the required information for each discipline scenario needs to be documented and defined as computer-executable rules (Yang, Eastman 2007). Hence, MVD and its open standard mvdXML (Chipman et al. 2012) can be used to structure the exchange requirements with specific IFC types, entities, attributes (Karlshøj et al. 2012).

So far, the IFC model supports neither the notion of LOD, nor a description of its uncertainty. However, as it is a very widespread and well-established format, we will show how an external meta-model can be used to enrich IFC data by these aspects.

## 3 MULTI-LOD META-MODEL

### 3.1 Requirements analysis

The efforts and costs required to make changes in a building model in the early stages are relatively lower than in the subsequent stages (Kolltveit, Grønhaug 2004). However, the lack of adequate information impedes taking informed decisions. Hence, it is crucial to maintain the individual component's LOD requirements. Especially in the process of designing a building, the components are associated with diverse levels of development within the same phase, such as load-bearing components can be described with a higher LOD than the interior fittings in the early design stages.

To our knowledge, there is no approach for formally defining and maintaining multiple levels of development throughout the design stages. Neither is there a formal definition of a building component's level of development nor is there an explicit description of the fuzziness of its geometric and semantic information. Therefore, the multi-LOD meta-model is proposed in order to:

– Define component types' LOD requirements
– Model information uncertainty
– Represent a building model on multiple stages
– Describe the relationships between LODs
– Check the consistency between LODs.

To manage the requirements of the individual building component types for a specific LOD, a component type is associated with multiple LOD definitions. An LOD definition consists of two separate groups: one for defining the geometric representation and alphanumerical attributes, and another for specifying the semantic alphanumerical attributes. This separation helps to achieve and to maintain the semantic-geometric coherence of the overall model (Stadler, Kolbe 2007; Clementini 2010). Finally, the building model is presented by creating multiple instances from the defined component types.

### 3.2 *Separation of geometry and semantics*

The multi-LOD meta-model aims to maintain a clear separation between the building components' semantic and geometric requirements. In terms of geometry representation of a building component, it is refined along with increasing the level of development. For example, as demonstrated in Figure 1 at LOD 100, an *external wall* is presented as a *centerline*, since in the next LODs additional information is available, such as a *thickness* and *material*, it is possible to render the wall *solid* model in its 3D shape and dimensions. This kind of hierarchical development of a *centerline* towards a *solid* model defines the dependencies between the geometric representations on the different levels of development. Accordingly, the relationships between the semantic requirements are determined, which supports checking the consistency between the multiple LODs.

### 3.3 *Alphanumerical attributes and fuzziness*

With incrementing the LOD, additional attributes become available, for example, the *construction type* and *material* information can be determined starting from LOD 200. In some cases, it is uncertain whether a specific attribute is available or can be estimated from a specific LOD. Thus, the multi-LOD model provides the ability to specify whether an attribute is mandatory or optional as well as offering a level of precision in specifying the attribute's assigned value in case of uncertainty. The level of precision in assigning the attribute's value is related to its type; it might be achieved by specifying an abstract value, such as a classification, or a fuzziness range. With that said it is possible to model and analyze the known uncertainties of the building model at the early design stages where uncertainty is at its highest.

Figure 2 provides geometric and semantic attributes of an *External Wall* component type for LODs 120 to 300. The *surface* dimensions exist starting from LOD 120 with a permissible fuzziness range of ±10 cm, while no fuzziness is

| Attributes | LOD 120 existing | LOD 120 fuzziness | LOD 200 existing | LOD 200 fuzziness | LOD 300 existing | LOD 300 fuzziness |
|---|---|---|---|---|---|---|
| Dimensions | ✓ | ±10 cm | ✓ | - | ✓ | - |
| Wall thickness | | | ✓ | ±10 cm | ✓ | ±5 cm |
| Opening position | | | ✓ | ±10 cm | ✓ | ±5 cm |
| Opening percentage | | | ✓ | ±5 cm | ✓ | - |
| Material | | | ✓ | material group (wood, concrete,...) | ✓ | material |

Figure 2. Example of assigning geometric-semantic attributes and fuzziness of an *external wall*.

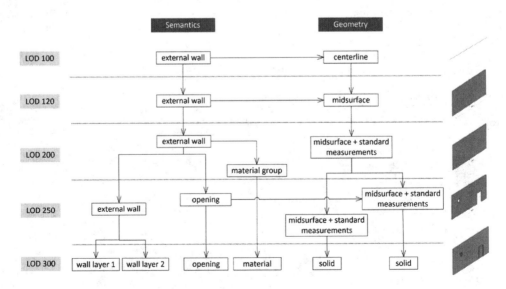

Figure 1. Separation of geometry and semantics on different LODs of an *external wall*.

permitted afterward. Additionally, the information describing *wall thickness* and *opening position* are available starting from LOD 200 with ±10 cm of fuzziness and then reduced to ±5 cm on LOD 300. Considering a different type of fuzziness, the information about *material* can be available from LOD 200, where at this level; it is defined by specifying the *material group*, such as *Ceramic*, whereas afterward on LOD 300 the exact material value, like *Brick*, should be assigned.

## 4 META-MODEL DESIGN

The multi-LOD meta-model design provides means for defining a project-specific data-model, incorporating formal LOD definitions for individual component types. It introduces two layers: *data-model level* defines the component types as well as their geometric and semantic requirements for each LOD. The *instance level* represents the building model by instantiating multiple instances of the component types defined on the data-model level.

The meta-model design complies with the object-oriented modelling principles, which offers high flexibility and extensibility. It allows for a dynamic definition of any component types as well as their attributes for the different LODs. This provides the flexibility required when dealing with different construction types, different domains, and different analysis tools. At the same time, the meta-model provides a consistent way to query information about LOD definitions on both the *data-model level* as well as the *instance level*. Thereby, as illustrated in Figure 3, a component type definition is represented as a separate class, where it is linked to an IFC type, *IfcWall* as an example, and associated with a list of LOD definitions. The component types are mapped to instances of the IFC data model, on the one hand, to make use of the geometry representations defined there and on the other hand to experiment with real-world data produced by IFC-capable BIM authoring tools.

An LOD definition is produced out of two objects, geometric and semantic requirements. Both requirements are explicitly described in the form of properties. The details of each property are determined in addition to the permissible fuzziness and geometry representation. The properties are managed by means of grouping, the *PropertySet* class. A *PropertySet* includes multiple *PropertyDefinition* instances defining property details but excluding its fuzziness. At the same time, the fuzziness type and maximum percentage as well as whether the property is mandatory are specified when assigning a *PropertyDefinition* to an LOD property. This brings multiple advantages, including decoupling the property definition from the LOD requirements, and flexibility of using the same property definition in multiple LODs along with different fuzziness.

In some cases, multiple components fall under the same category, such as *Heating, Ventilation, and Air Conditioning* (HVAC) systems, and share several properties. Hence, the *ComponentType* class supports defining sub-types of a specific component type through inheritance. This means

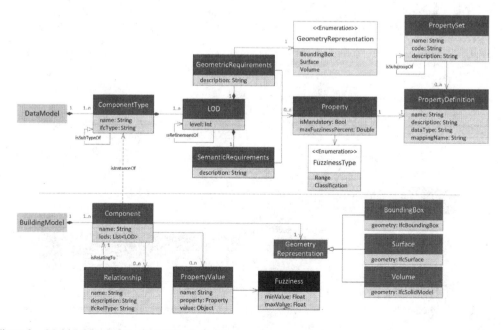

Figure 3. Multi-LOD meta-model (UML diagram).

a sub-type inherits the parent component type's requirements in addition to specifying additional specific requirements.

After defining the component types' requirements, the building model is represented by multiple instances of the available types. An instance is assigned to a geometry representation, which complies with IFC, such as *IfcSurface*, and its properties are filled with values. In terms of fuzziness, its range is automatically transferred from the maximum fuzziness percentage defined at the component type level, for example, 4% and an attribute value of 250 cm is translated into a range of ±10 cm. Moreover, at the instance level, it is possible to increase the limitation of the range values, such as to be between −5 cm and +7 cm. Finally, the connections between the individual components within the same LOD, including aggregation and association, are presented through the *Relationship* class. With that said, the meta-model allows checking if the instance of a given type on a particular LOD complies with the requirements defined in terms of semantic fuzziness and geometric representation.

## 5 USE CASE: EMBODIED ENERGY ANALYSIS

*Life Cycle Assessment* (LCA) is one of the most established and well-developed methods for assessing the potential environmental impacts and resource consumption throughout a product's life-cycle (Ness et al. 2007). As one of its applications, LCA is used to calculate the *Embodied Energy* which is represented as the sum of non-renewable energy consumption during the life cycle (Merkblatt 2010). Performing the LCA calculation involves multiple geometric and semantic information of the building model, including the building location, dimensions, number of storeys and window-to-wall ratio. Additionally, custom energy-related attributes are required for each component and need to be transferred when exchanging the model, such as the *Thermal transmittance* (U-value).

In order to include this information in the model, it has to be provided in a correct way in the BIM authoring tool. Here, the multi-LOD meta-model comes into play where it allows defining the data-model of the individual component types. For instance, Table 1 lists the component types and their required attributes for the LCA calculation on LODs 120, 200 and 300.

LOD 120 is limited to the building model's generic information. Whereas, at LOD 200 and 300, information about the windows and walls including the *thickness* and *material* become available. Therefore, a *U-value* of each component type can be provided. Having these attributes specified at the data-model level guarantees their association at the instance level and provides a way to check the model's validity. Based on the data listed in Table 1, Figure 4 illustrates a data-model level of an external wall at LOD 200. The geometric requirements include the *height, width,* and *thickness*, where the thickness attribute permits a fuzziness range of 10%. For the semantic requirements, the *U-value* and the material group are required. As the *window-to-wall* ratio is expected to be precise at LOD 300, it is not required and allows a fuzziness range of 5%.

Table 1. Required components and attributes for LCA calculation in different LODs.

| LOD | Available components | Required attributes |
|-----|----------------------|---------------------|
| 120 | building | dimensions, location |
| 200 | floor, roof, wall | thickness, material, U-value |
| 300 | windows | material, U-value |

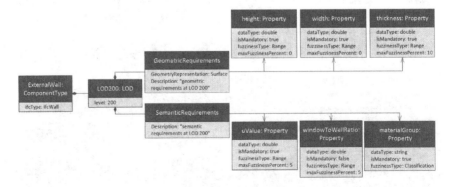

Figure 4. Data-model level example of an *external wall* at LOD 200 supporting embodied energy calculation (UML diagram).

## 6 IMPLEMENTATION

To evaluate the proposed multi-LOD model for practical use, the data-model level is implemented as a webserver and *User Interface* (UI). The UI

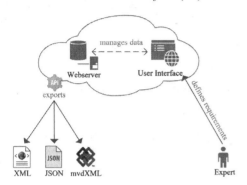

Figure 5. Data-model level system design.

Figure 6. Component details screen of *WallType* (UI prototype).

provides a user-friendly way for defining disciplines, levels of development, property sets and component types. Figure 5 provides an overview of the system design. The main idea is that every discipline is capable of defining its own property sets, and then assigning particular properties to a specific component type's LOD. The property sets management screen is demonstrated in Figure 7. A property set can have sub-sets in order to minimize the properties redundancy. Additionally, a property is assignable to multiple disciplines. Finally, the properties are associated to an LOD at the component types' screen. Figure 6 shows the *WallType* component details screen. The *General* tab is for defining the component name, IfcType and description. Whereas the second tab *Requirements* facilitates associating every LOD with properties including a specification of their fuzziness. The properties are grouped based on their *Property Set* name, following the naming scheme *Pset_\**, for instance, *Pset_ThermalWall*. For improving the usability and increase the data integrity, the bsDD's *Application Programming Interface* (API) is employed. It assists the process by listing the commonly known IFC elements, properties, and classifications to the user. Consequently, this mapping to the bsDD's *GUID* provides additional context and meaning to each value, which improves interoperability between different disciplines and assists the model analysis.

The multi-LOD webserver stores the component types' requirements into a relational database and exports them as XML and JSON formats using *REpresentational State Transfer* (REST) API as shown in Figure 8. To facilitate using these requirements as exchange requirements and validate their existence, the webserver exports them into the common formats supported by the BIM authoring tools, such as *PropertySets* file provided

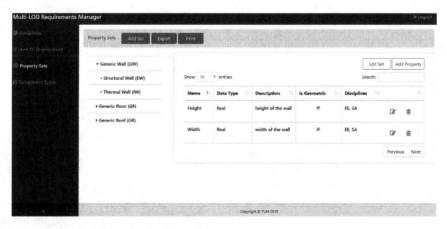

Figure 7. Property sets management screen (UI prototype).

```
<ComponentTypes>
▼<ComponentType id="1" name="WallType" ifcType="IfcWall">
  ▼<LOD id="2" componentId="1" level="200">
    ▼<PropertySets>
      ▼<PropertySet id="1" name="Pset_GenericWall">
          <Property id="1" name="maxHeight" description="Maximum height"
          dataType="Real" isMandatory="true" fuzzinessType="1"
          fuzzinessTypeName="Range" maxFuzzinessPercent="5.0"
          propertyDefinitionId="1" isGeneric="true"/>
          <Property id="2" name="maxWidth" description="Maximum Width"
          dataType="Real" isMandatory="true" fuzzinessType="1"
          fuzzinessTypeName="Range" maxFuzzinessPercent="5.0"
          propertyDefinitionId="2" isGeneric="true"/>
          <Property id="3" name="maxLength" description="Maximum Length"
          dataType="Real" isMandatory="true" fuzzinessType="1"
          fuzzinessTypeName="Range" maxFuzzinessPercent="5.0"
          propertyDefinitionId="3" isGeneric="true"/>
          <Property id="5" name="openingPercentage" description="Windows
          Openings percentage" dataType="Real" isMandatory="true"
          fuzzinessType="1" fuzzinessTypeName="Range"
          maxFuzzinessPercent="10.0" propertyDefinitionId="5"
          isGeneric="false"/>
      </PropertySet>
      ▼<PropertySet id="4" name="Pset_StructuralWall">
          <Property id="6" name="materialGroup" description="The main material
          classification" dataType="Text" isMandatory="true" fuzzinessType="2"
          fuzzinessTypeName="Classification" maxFuzzinessPercent="0.0"
          propertyDefinitionId="6" isGeneric="false"/>
      </PropertySet>
      ▼<PropertySet id="2" name="Pset_ThermalWall">
          <Property id="4" name="thickness" description="Wall thickness"
          dataType="Real" isMandatory="true" fuzzinessType="1"
          fuzzinessTypeName="Range" maxFuzzinessPercent="10.0"
          propertyDefinitionId="4" isGeneric="true"/>
          <Property id="7" name="thermalTransmittance" description="Thermal
          Transmittance (u-value)" dataType="Real" isMandatory="true"
          fuzzinessType="1" fuzzinessTypeName="Range" maxFuzzinessPercent="5.0"
          propertyDefinitionId="7" isGeneric="false"/>
      </PropertySet>
    </PropertySets>
  </LOD>
  <Parent/>
</ComponentType>
</ComponentTypes>
```

Figure 8. Example of a *WallType* LOD200 XML format.

by Autodesk Revit, and translates the requirements into mvdXML rules. Hereby, it is possible to use the requirements for external services, such as a Revit plugin, for automatically generating and ensuring the exchanged building models attributes completeness.

## 7 CONCLUSION AND FUTURE WORK

The multi-LOD meta-model offers a high-level interface that provides a consistent way for defining and querying LODs in term of their semantic and geometric requirements. As the LOD requirements take into account the permissible fuzziness, the known uncertainties are explicitly modelled, which delivers great advantages in assessing and verifying the model consistency in the early design stages. The meta-model introduces two layers, *data-model level* and *instance level*. This offers a high degree of flexibility in defining per-project LOD requirements and facilitates formal checking of their validity, such as requiring specific information for *Embodied Energy* calculations, *Building Performance* simulations, or *Structural* analysis.

As part of evaluating the model, the *data-model level* is implemented in a webserver and a corresponding user-interface. The system provides a means for managing the exchange requirements between the project disciplines for every LOD. In this way, the requirements' consistency, correctness, and completeness are maintained. Additionally, the system exports the exchange requirements into JSON, XML, and an automatically generated mvdXML rules to encourage their integration in the modeling process. As a next step, further research is necessary to develop a methodology for describing the detailed refinement relationships of the building elements and checking their consistency.

## ACKNOWLEDGEMENTS

We gratefully acknowledge the support of German Research Foundation (DFG) for funding the project under grant FOR 2363.

## REFERENCES

Bazjanac, Vladimir (2008): IFC BIM-based methodology for semi-automated building energy performance simulation. Ernest Orlando Lawrence Berkeley National Laboratory, Berkeley, CA (US).

Biljecki, Filip; Ledoux, Hugo; Stoter, Jantien (2016): An improved LOD specification for 3D building models. In *Computers, Environment and Urban Systems* 59, pp. 25–37.

BIMForum (2017): 2017 Level of Development Specification Guide. Available online at http://bimforum.org/lod/, checked on 1/11/2018.

Bjorkhaug, L.; Bell, H. (2007): Ifd: IFD in a Nutshell. In *SINTEF Building and Infrastructure*.

Borrmann, André; Flurl, Matthias; Jubierre, Javier Ramos; Mundani, Ralf-Peter; Rank, Ernst (2014): Synchronous collaborative tunnel design based on consistency-preserving multi-scale models. In *Advanced Engineering Informatics* 28 (4), pp. 499–517.

BSI (2017): PAS 1192–2:2013 Specification for information management for the capital/delivery phase of construction projects using building information modelling. Available online at http://shop.bsigroup.com/Navigate-by/PAS/PAS-1192–22013/, checked on 1/11/2018.

buildingSMART (2016): Data Dictionary. Available online at https://www.buildingsmart.org/standards/standards-tools-services/data-dictionary/, checked on Jan 2018.

Chapman, I. (2013): An introduction to Uniclass2. Available online at https://www.thenbs.com/knowledge/an-introduction-to-uniclass-2, checked on 3/22/2018.

Chipman, T.; Liebich, T.; Weise, M. (2012): mvdXML: Specification of a standardized format to define and exchange Model View Definitions with Exchange Requirements and Validation Rules: BuildingSmart.

Chiu, Mao-Lin (2002): An organizational view of design communication in design collaboration. In *Design studies* 23 (2), pp. 187–210.

Clementini, Eliseo (2010): Ontological impedance in 3D semantic data modeling. In Eliseo Clementini (Ed.): Ontological impedance in 3D semantic data modeling. 5th 3d Geoinfo Conference (Ed. by TH

Kolbe, G. König & C. Nagel), ISPRS Conference (Vol. 38, p. 4)., vol. 38, p. 4.

Eastman, Charles M.; Eastman, Chuck; Teicholz, Paul; Sacks, Rafael (2011): BIM handbook: A guide to building information modeling for owners, managers, designers, engineers and contractors: John Wiley & Sons.

GCR, NIST (2004): Cost analysis of inadequate interoperability in the US capital facilities industry. In *National Institute of Standards and Technology (NIST)*.

Hausknecht, Kerstin; Liebich, Thomas (2017): BIM-Kompendium: Building Information Modeling als neue Planungsmethode: Fraunhofer IRB Verlag.

Hietanen, Jiri; Final, Status (2006): IFC model view definition format. In *International Alliance for Interoperability*, pp. 1–29.

Hooper, Martin (2015): Automated model progression scheduling using level of development. In *Construction Innovation* 15 (4), pp. 428–448.

Karlshøj, Jan; See, R.; Davis, D. (2012): An Integrated Process for Delivering IFC Based Data Exchange. Available online at https://bips.dk/files/integrated_idm-mvd_processformats_14_0.pdf., checked on 4/15/2018.

Knotten, Vegard; Svalestuen, Fredrik; Hansen, Geir K.; Lædre, Ola (2015): Design management in the building process-a review of current literature. In *Procedia Economics and Finance* 21, pp. 120–127.

Kolltveit, Bjørn Johs; Grønhaug, Kjell (2004): The importance of the early phase: the case of construction and building projects. In *International Journal of Project Management* 22 (7), pp. 545–551.

Liebich, T.; Adachi, Y.; Forester, J.; Hyvarinen, J.; Richter, S.; Chipman, T. et al. (2013): Industry Foundation Classes IFC4 official release. In *Online: http://www.buildingsmart-tech.org/ifc/IFC4/final/html*.

Liebich, Thomas (2013): IFC4—The new buildingSMART standard. In Thomas Liebich (Ed.): IC Meeting. IFC4—The new buildingSMART standard. bSI Publications Helsinki, Finland.

Merkblatt, S.I.A. (2010): 2032 (2010): Graue Energie von Gebäuden. In *Zürich, Schweizerischer Ingenieur-und Architektenverein*.

Ness, Barry; Urbel-Piirsalu, Evelin; Anderberg, Stefan; Olsson, Lennart (2007): Categorising tools for sustainability assessment. In *Ecological Economics* 60 (3), pp. 498–508. DOI: 10.1016/j.ecolecon.2006.07.023.

OmniClass, A. (2012): A strategy for Classifying the Built Environment: USA. Available online at http://www.omniclass.org/, checked on 1/11/2018.

Stadler, Alexandra; Kolbe, Thomas H. (Eds.) (2007): Spatio-semantic coherence in the integration of 3D city models. Proceedings of the 5th International Symposium on Spatial Data Quality, Enschede.

van Berlo, LAHM; Bomhof, F. (2014): Creating the Dutch national BIM levels of development. In Van Berlo, LAHM and Bomhof, F (Ed.): Creating the Dutch national BIM levels of development, pp. 129–136.

Yang, Donghoon; Eastman, Charles M. (2007): A Rule-Based Subset Generation Method for Product Data Models. In *Computer-Aided Civil and Infrastructure Engineering* 22 (2), pp. 133–148.

Young, N.W.; Jones, Stephen A.; Bernstein, Harvey M.; Gudgel, John (2009): The business value of BIM-getting building information modeling to the bottom line. In *Bedford, MA: McGraw-Hill Construction* 51.

# NovaDM: Towards a formal, unified Renovation Domain Model for the generation of holistic renovation scenarios

A. Kamari, C. Schultz & P.H. Kirkegaard
*Department of Engineering, Aarhus University, Aarhus, Denmark*

ABSTRACT: A central, fundamental challenge within the renovation field is handling enormous complexity, both at the level of an individual building project (consisting thousands of building components that are amenable to renovation, i.e. effectively managing the combinatorial explosion of renovation decisions) and at the AECO-community level of knowledge about what renovation options are available today, and how each such renovation alternative impacts criteria (i.e. energy efficiency, indoor comfort, spatial quality etc.). The aim of this paper is to present an innovative approach for addressing these challenges by developing a Renovation Domain Model: a formal (logic-based) domain-specific language for expressing and capturing key concepts for renovation alternatives, which is derived from empirical information. The domain model provides a structured, direct, and formal way of expressing an extensive set of renovation alternatives as a renovation "knowledge base", to manage the complexity of decision problems regarding the selection and use of various renovation alternatives.

## 1 INTRODUCTION

The topic of renovation of existing buildings is receiving ever-increasing attention in many European countries (Jensen & Maslesa 2015) as well as facing a variety of challenges (Galiotto et al. 2015). There is a drive towards addressing climatic aspects, energy efficiency, environmental impacts, life-cycle cost, indoor climate, spatial quality etc. as a response to the urgent need for significantly more sustainable societies, and the challenges of rapidly increasing urbanization (BPIE 2011).

In essence, we might describe the renovation design task as follows: given a built environment, we can repair, replace, remove, modify and add building facilities and architectural elements. We may be tempted to expect that existing building information models such as Industry Foundation Classes—IFC (buildingSMART 2008), or other initiatives such as the Danish Sigma Enterprise database (Molio 2016) or the buildSMART data dictionary (buildingSMART 2017) give a comprehensive overview of renovation alternatives.

Unfortunately, real renovation options are highly context—and project-specific, ruling out the direct and sole use of monolithic, universal product databases. For example, *windows* have properties such as glazing, material, dimensions, and a opening mechanism type. While these properties can be gleaned from an existing domain model of "windows", this is not sufficient for solving the renovation task: certain materials may not be available or not appropriate; the costs of products can vary greatly between countries and regional economic circumstances; certain product features may not be applicable due to legal reasons, e.g. safety requirements in large public buildings may forbid certain window opening mechanisms. Thus, while BIMs and product databases are a crucial resource that must be utilized, we require a broader framework within which we can easily specify real, project-specific renovation options to address the needs of the project at hand.

There is now a large body of research and knowledge on renovation as both a *design task* (specifying and choosing amongst an enormous number of renovation options) and as a construction process. There is an urgent need to bring these different lines of research together in a unified formal framework so that decision support software tools for renovation can be developed based on a standard, uniform "perspective" on what renovation is, that can also leverage and exploit existing product databases and building information models.

In this paper we present our first steps towards a unifying renovation domain model, called NovaDM. Our key contribution is raising, and making in-roads in addressing, the following fundamental research questions:

1. Does *renovation* as a concept have a unifying abstract structure that we can formally express? (i.e. the NovaDM meta-model)

2. Can we develop a formal language for modelling reusable libraries of renovation concepts that is also easily customizable? (i.e. the NovaDM subject-feature knowledge base)
3. Can we make our renovation language compatible with other data models to facilitate reuse, e.g. IFC, buildSMART data dictionary, etc.?
4. Given the highly combinatorial character of renovation design tasks, does our language enable project managers and civil engineers to compactly express the set of concrete, distinct, mutually exclusive renovation choices? (i.e. NovaDM action trees)
5. How can we operationalize such a modelling language, e.g. can we query an instance of the model by searching for actions that have certain properties, or checking if all actions are semantically consistent with domain knowledge? E.g. *aluminum* is not a valid *glazing* value, a *hinged-opening mechanism* does not apply to floor slabs, "indoor comfort" is a value criterion and not an architectural element.

We have developed our model based on many years of research in the topic of renovation, and through many use cases of renovation projects (Sections 2, 3). Figure 1 illustrates how we envision NovaDM to be employed in practice.[1]

In this paper our specific contributions are:
– presenting an initial version of NovaDM (Sections 4.1–4.3)

– an extensible prototype NovaDM parser tool and prototype querying engine implemented in Answer Set Programming (Brewka et al. 2011), (Section 4.4).

## 2 RESEARCH CONTEXT AND RELATED WORK IN RENOVATION

SBi (2014) states that renovation initiatives can often be more cost-effective than new building projects. The extent of the potential for renovation can be described and made up in several ways. Existing buildings can benefit from adopting a more broad approach to sustainability, which seeks to decrease operation and maintenance costs; reduce environmental impacts; and can increase the building's adaptability, durability, and resilience towards future challenges. Ultimately, buildings may be less costly to operate, may grow in value, last longer, and contribute to a preferable, healthier, more convenient environment to the occupants (Kamari et al. 2017). When all of these interventions are summated, they can move the renovation case towards the goal of overall sustainability, which demands more holistic renovation approaches.

Recent research in this field seeks methods for the development of more holistic renovation scenarios that live up to a broader set of sustainability objectives and criteria. These methods can potentially be exploited in two major steps: current condition assessment and future upgrade strategies (Juan et al. 2010). Most of the methods focus on the first step of the improvement process (Nielsen et al. 2016), understanding or exploration of a renovation project (i.e. energy usage), however, without addressing the challenging task of generation of renovation scenarios. This latter task concerns the proposal of future upgrade renovation solutions via the development of effective renovation scenarios. A central, fundamental challenge within this is handling enormous complexity, both at the level of an individual building project (consisting of thousands of building components that are amenable to renovation, i.e. effectively managing the *combinatorial explosion* of renovation decisions) and at the AECO[2]-community level of knowledge about what renovation options are available today, and how each such renovation alternative impacts objectives and criteria (i.e. energy efficiency, indoor comfort, spatial quality etc.).

An effective method (Ferreira et al. 2013) for generating holistic renovation scenarios should initially be able to deal with the complexity of decision-making due to the existing large number of renovation

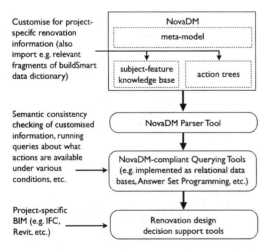

Figure 1. Our envisioned role of NovaDM in renovation projects.

---

[1]Our prototype NovaDM C++ parser and comprehensive set of renovation action trees may be downloaded from: http://think-spatial.org/NovaDMParser.zip.

[2]Architecture, Engineering, Construction, & Operation.

alternatives and various sustainability objectives and criteria, which need to be embraced (Kamari et al. 2018c). In this framework, the aim of this paper is to present an innovative approach for addressing these challenges by developing a Renovation Domain Model: a formal (logic-based) domain-specific language for expressing and capturing key concepts for renovation alternatives, which is derived from empirical information. This paper builds upon our previous work related to the RE-VALUE[3] research project on development of *a hybrid Decision Support System (DSS) for the generation of holistic renovation scenarios* (Kamari et al. 2018b). The domain model provides a structured, direct, and formal way of expressing an extensive set of renovation alternatives as a renovation "knowledge base", with the aim of managing the rapidly increasing complexity of decision problems regarding the selection and use of various renovation alternatives.

## 3 METHODOLOGY FOR DEVELOPING THE RENOVATION DOMAIN MODEL

We use the term *domain model* to address a collection of concepts and relationships between concepts, similar to the notion of a *schema*; i.e. we are aiming to develop a "schema" or domain model of renovation concepts, which we refer to as NovaDM. Concepts include cultural, spatial, environmental, societal notions that have either a direct or indirect impact on building renovation. We have developed NovaDM following the IDEF5 methodology for knowledge engineering and ontology development (Benjamin et al. 1994). IDEF5 is organized into five stages of domain model development. The following subsections elaborate on how we have undertaken these stages.

### 3.1 *Organization and scoping*

Although NovaDM is applicable for renovation of different types of existing buildings, in our present study we focus on large scale residential buildings (residential complexes) and social housing. The NovaDM development approach is driven by collecting, exploring, classifying, and evaluating physical and technical renovation components, as well as functional characteristics. As part of a previous extensive research program, we employed a seven stage research methodology (Kamari et al. 2017). To address functional characteristics, we employed a four stage research methodology (Kamari et al., 2018c). The results are now formalized in our NovaDM.

---

[3] http://www.revalue.dk

### 3.2 *Data collection*

The terms *sustainable renovation* and *holistic renovation scenarios* in this paper are adopted from (Kamari et al., 2018a). These terms serve to underline a holistic approach where various objectives linking to the sustainability in its full sense are targeted and achieved in a balanced way. This is attained by selecting and mixing various types of renovation approaches through the development of holistic renovation scenarios.

**Sustainability objectives and criteria.** There are a wide array of advantages that can be obtained as an outcome of a holistic and sustainable retrofitting to higher energy performance standards. Many are tangible and possible to quantify, while others are less so and may be difficult to allocate a monetary value. These renovation goals must be identified and targeted early in the design process while renovation scenarios are generated. Regarding the full scope of this discussion, Kamari et al. (2017) addressed a new "Holistic sustainability decision-making support framework for building renovation" by applying Checkland's Soft Systems Methodologies – SSM (Checkland, 2000) beside Keeney's Value Focused Thinking – VFT (Keeney 1992). As such, sustainability was defined and represented in its full sense from three categories including *Functionality*, *Accountability* and *Feasibility* (18 sustainable value oriented criteria, and 118 sub-criteria, have been identified) for holistic/ deep building renovation purpose (Table 1).

**Renovation approaches.** There are a broad range of renovation approaches that can be applied for the renovation of existing buildings including insulation approaches, replacement of existing windows, integration or replacement of existing equipment, heating/cooling system, building envelope implementation of roof and partially of facades to avoid thermal bridges, total building envelope implementation, volumetric additions, partial replacement of

Table 1. List of sustainability objectives and criteria for building renovation from (Kamari et al. 2017).

| Functionality | Accountability | Feasibility |
|---|---|---|
| – Indoor comfort | – Aesthetic | – Investment cost |
| – Energy efficiency | – Integrity | – Operation & maintenance cost |
| – Material & waste | – Identity | – Financial structures |
| – Water efficiency | – Security & safety | – Flexibility & management |
| – Pollution | – Sociality | – Innovation |
| – Quality of services | – Spatial quality | – Stakeholders engagement & education |

existing windows, partial building envelope implementation, integration of PV and solar collectors on the roof/facades etc. (Boeri et al. 2014). Following the previous work by the authors (Kamari et al., 2018c), for the development of a database of renovation approaches, the relevant data have been collected from literature (Boeri et al. 2014; Baker 2009; Burton 2012), evaluation of the 10 European renovation research projects (Kamari et al., 2018a), as well as investigation of the SIGMA database by Molio (2016). Moreover, a renovated building project (the Section 3 of Skovgårdsparken located in 8220 Brabrand, Denmark) has been studied. It was a Residential building (including nine blocks), Modernistic in terms of typology, built during 1968/72. It has been renovated by Brabrand Housing Association. The case has been selected due to a comprehensive renovation scenario (i.e. insulation of walls, renovation of foundation, installation of PV etc.) that has been applied for the renovation purpose. The results in total led us to compile a list of 26 renovation approaches (Table 2) including 139 renovation alternatives, and about 3831 renovation actions.

Figure 2 demonstrates the terminology that has been used in this paper to distinguish between a renovation scenario, a renovation *approach*, a renovation *alternative*, and a renovation *action*.

Table 2. A–Z renovation approaches from (Kamari et al. 2018c).

| | |
|---|---|
| Insulation approaches | A |
| Envelope (exterior finishes) | B |
| Window (replacement) | C |
| Doors (replacement) | D |
| Airtightness and damp proofing approaches | E |
| Waste facilities | F |
| Building security approaches | G |
| Building site | H |
| Structural system | I |
| HVAC system | J |
| Renewable energy sources | K |
| Energy storage | L |
| Electrical system | M |
| Plumbing system | N |
| Controls | O |
| Flooring | P |
| Interior finishes—Ceiling | Q |
| Interior finishes—Walls | R |
| Increasing solar gain | S |
| Avoiding overheating | T |
| Re-designing of external and internal spaces | U |
| Common areas (interior) | V |
| Individual building elements | W |
| Sanitary appliances | X |
| Fixed furniture [essential] | Y |
| Movable furniture [optional] | Z |

The term renovation scenario used in this study means a selection and combination of some different renovation *approaches* (i.e. insulation of the external walls, replacement of windows etc.) consisting of a specific *alternative*, and *action* that together build renovation scenarios and subsequently is applied in a renovation project. As an example, Figure 3 represents a renovation approach (in this example: Insulation approaches), and its related *alternatives*, and a list of actions for one of its alternatives.

3.3 *Data analysis*

We have classified our collected renovation approaches according to the evaluation of the 10

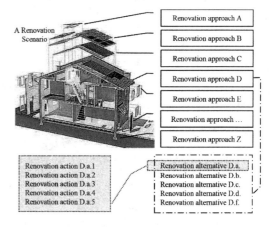

Figure 2. An example of a renovation scenario including various renovation approaches, renovation alternatives, and renovation actions.

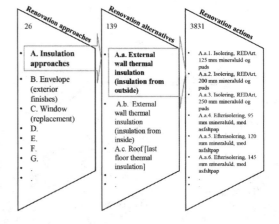

Figure 3. An example of a renovation approach including several alternatives and actions.

European renovation research projects (Kamari et al., 2018a), as well as investigation of the SIGMA database by Molio (2016). Likewise, we categorized the renovation objectives and criteria within a consensus-based process, based on extensive analysis of various sustainability assessment methods and methodologies, i.e. LEED, BREEAM, DGNB-DK etc.

### 3.4 Initial domain model development

The initial domain model has been built iteratively from all prominent concepts that appear while generating renovation scenarios. That is, we incorporated the selection of various types of renovation approaches, renovation alternatives, renovation actions, building elements, sustainability criteria and sub-criteria. Moreover, we iteratively integrated concepts from a variety of characteristics and parameters regarding the renovation approaches and their interdependencies with sustainability criteria and sub-criteria. As semantic distinctions emerged, more general concept classes were introduced. The concepts have been cross-checked representing various actions as *trees* and within different layers, that we present in detail in Section 4, i.e. a meta-model layer, subject-feature knowledge base, and renovation actions (encoded as NovaDM trees).

### 3.5 Refinement and validation

The initial validation stage consisted of an unstructured expert review by partners in the RE-VALUE project from the Aarhus University. We are currently undertaking a second validation stage following the process described in (Perez et al. 1995) of reviewing the domain model performance in case-specific scenarios and through competency questions. This includes the creation of instances inside the domain model classes and checking if the model adequately captures the inserted scenario information, and then running queries to check response validity.

## 4 NOVADM

In this section we present the NovaDM renovation domain model, through which we aim to formally organize renovation information, at many layers of abstraction, about: *what* can be changed (the *subject*), in *what ways* (the *features*) what combinations of decisions can be made (the *alternatives*), and for each combination, what decisions are mutually exclusive (the *actions*).[4]

---

[4]Our prototype NovaDM C++ parser and comprehensive set of renovation action trees may be downloaded from: http://think-spatial.org/NovaDMParser.zip.

We emphasize that the scope and focus of this paper is on the theory and operational foundations of NovaDM. We are not addressing high-level usability issues here, e.g. in a practical setting we do not intend architects to write queries directly in low-level languages such as SQL or ASP, but rather, an intuitive interface would be developed to stream-line this process.

### 4.1 Meta-model layer

The meta-model layer consists of concepts that describe the abstract structure of renovation actions *in general*. Figure 4 illustrates a UML diagram of general renovation concepts, and Figure 5 illustrates a UML diagram of the general structure of a specific, concrete renovation plan (which we refer to as a renovation *scenario*).

Firstly, a renovation *action* has exactly one subject (e.g. windows), zero or more features of the subject (e.g. double glazing), and zero of more criteria that will be impacted (e.g. thermal comfort). Actions may involve repairing, replacing or introducing new elements, thus we may also (optionally) assign an

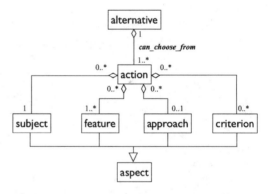

Figure 4. UML class diagram of meta-model renovation concepts.

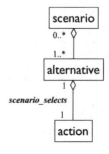

Figure 5. UML class diagram showing meta-model relationship between *scenarios*, *alternatives* and *actions*.

*approach* category (such as "repair") to actions. Collectively, the components of an action (*subject, features, approaches,* and *criteria*) are renovation *aspects*.

Groups of actions are typically mutually exclusive, e.g. we can only select one of "single" or "double" glazing for a given window. We introduce the concept of a renovation *alternative* to organize such groups of mutually exclusive actions, e.g. the set of all ways in which we can renovate windows, for a particular project, would be organized in the project as the window renovation *alternative,* in contrast to (for example) the door renovation *alternative*. In Figure 5 we refer to the relation that every alternative is associated with a mutually exclusive set of actions as the "**can_choose_from**" relation.

A renovation *scenario* is a specific detailed plan that could be carried out, i.e. for every *alternative* we have committed to either zero *actions,* or exactly one *action* (in the case that we select zero actions then the alternative is ignored, e.g. perhaps we decide not to renovate windows in any way). We refer to the relationship between alternatives and selected actions for a scenario as "**scenario_selects**". Thus, we can formally express the renovation design task as exploring the large number of possible renovation *scenarios*.

### 4.2 Subject-feature knowledge base

The particular set of subjects and their associated features and actions depend on the context (country, economic circumstances, laws etc.) and the specific characteristics of the project at hand. The *subject-feature knowledge-base* (KB) consists of the particular concepts that refine and specialize the meta-model, i.e. subclasses of *subject, feature,* etc. Each project manager can, in principle, use their own customised subject-feature KB to suit their own project. However, our intention is that "libraries" of useful and common KBs will be readily available, e.g. the "Denmark Residential" KB that has all *subject-feature* concepts relevant to the Danish context of residential renovation. Users can then select the KB fragments that are relevant for their project.

Moreover, our intension is that existing detailed and comprehensive schemas and knowledge bases such as the buildSMART data dictionary can be **directly employed** in our domain model. For example, every object in the buildSMART data dictionary can be defined as a subject, and every property of that object can be used to define features.

### 4.3 Action trees

Action trees are collections of project-specific, mutually exclusive renovation options that can be selected. We have developed a formal tree-language for compactly specifying sets of actions. A set of actions is defined by traversing the tree, starting from the root node. Due to space limitations we will only briefly describe our tree language here. Each node assigns zero or more attributes to an action. Tree nodes can be either *xor*-nodes "–" (meaning that the current action must be built by traversing **exactly one** child), or *and*-nodes "+" (meaning that the current action must be built by traversing **all** children).

**Example.** Figure 6 illustrates an action tree for describing the set of mutually exclusive actions for renovating windows. To read the actions from the tree, we start at the tree root (the node at the top with no "parent"). This node assigns the subject of the action to be *windows*. The node is an *xor*-node so we traverse the only child. The next child assigns a feature of *plastic material* to the action. It is an *and*-node, and so we must traverse every child. First we traverse the left branch and arrive at an *xor*-node, and thus must select exactly one child to traverse. We again choose the left branch and assign a feature of *double glazing* to the current action. There are no further children and so we *backtrack* to the last *and*-node, and now traverse the right branch, which eventually leads us to assign the feature *fixed mechanism* to the action. We have no more children to traverse, and no more *and*-nodes to backtrack to, and thus we have finished constructing an action: *subject* = window; *features* = plastic material, double glazing and fixed mechanism. There are six distinct, mutually exclusive actions specified in this tree (i.e. two choices for glazing and three choices for the mechanism).

### 4.4 NovaDM parser and prototype querying tool

We have implemented a NovaDM parser (in C++) that:

1. reads a text file description of *(a)* action trees and *(b)* subject-feature knowledge bases;

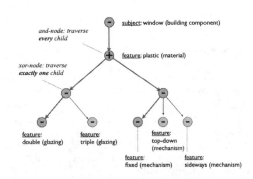

Figure 6. Example of an action tree describing six mutually exclusive actions. The red edges correspond to one particular action with: subject = window; features = plastic material, double glazing, fixed mechanism.

2. generates all actions (grouped by *alternatives*);
3. exports the sets of actions in a variety of formats. Currently we support export to CSV and the logic programming language Answer Set Programming (similar to Prolog) to facilitate the querying of actions.

Moreover, as both a proof of concept of our NovaDM and as itself a formal, reusable renovation library, we have encoded a comprehensive set of alternatives and actions from (Kamari 2018) in our NovaDM language, as presented in Section 3.2. This corresponds to 127 alternatives, and a total of 278,131 actions with 127 subjects, 324 approaches, and 286 features. The power of our language is that the text file expressing this renovation information is only approximately 200 lines long.[5]

The parser can be readily extended to export renovation information into various formats, e.g. into a relation database format to facilitate querying via SQL. Currently we have implemented an option to generate either a CVS file or an ASP file.

**CVS.** Every CSV row specifies the set of renovation actions that share an aspect type and value. E.g. the Table 3 represents an extract of a generated CVS file showing that:

- actions *1,2,3,4* for alternative *R0* have the window building component as the *subject*;
- only action *1* for alternative *R0* has plastic material as a feature.

**ASP.** As a demonstration of querying with NovaDM we also export to Answer Set Programming (ASP). ASP is a logic programming language that has its foundations in first-order logic (Brewka et al. 2011). Similar to Prolog, ASP has a knowledge base of facts and rules, and has a built-in reasoning engine that we use to query and make inferences with renovation actions e.g. the following ASP code expresses that (a) the *window* building component is a valid action *subject*, and (b) *plastic*, *steel* and *wood* materials are valid action *features*.

Table 3. An extract of a generated CVS file.

| Alternative | Actions | Aspect | Type | Value |
|---|---|---|---|---|
| R0 | 1;2;3;4 | subject | building_component | window |
| R0 | 1 | feature | material | plastic |
| ... | ... | ... | ... | ... |

---

[5]Our parser traverses the action tree file and generates a corresponding 23 MB CSV file within approximately 1 second on a standard laptop computer (MacBook Pro, Intel i7 CPU). This fast runtime demonstrates the practicality of our approach in tackling large, real-world renovation projects.

```
novaDm(property(subject), type(building_component), value(window)).
novaDm(property(feature), type(material), value(plastic; steel; wood)).
...
```

We encode the relationship between renovation actions and their properties as ASP facts, e.g. the following fact states that actions *1,2,3,4* of alternative *r1* have the *window* building component as a *subject*:

```
novaAction(alternative("r1"), action(1;2;3;4),
        property(subject), type(building_component), value(window)).
```

We then use ASP to execute queries, e.g. the following ASP rule checks for semantic inconsistencies: "*Find actions where the property type has not been defined in the subject-feature knowledge base*".

```
inconsistent_property_type(R,A,P,T) :-
    novaAction(alternative(R), action(A), property(P), type(T), _),
    not novaDm(property(P), type(T), _).
```

In this example ASP identifies an inconsistency and generates a new fact expressing the inconsistent information that: *action 3* of alternative *r1* has glazing incorrectly specified as a *subject* (whereas *glazing* is only specified as a *feature* in the subject-feature knowledge base):

```
inconsistent_property_type("r1",3,subject,glazing)
```

Below we give a range of further query examples:
**Query**: "*Which alternatives involve windows?*"

```
window_alternative(R) :-
    novaAction(alternative(R), _, property(subject),
                type(building_component), value(window)).
```

**Query**: "*Which actions use triple glazing?*"

```
triple_glazing_action(R,A) :-
    novaAction(alternative(R), action(A),
                property(feature), type(glazing), value(triple)).
```

**Query**: "*Which building components impact the indoor comfort criterion via some action?*"

```
indoor_comfort_component(C) :-
    novaAction(alternative(R), action(A),
            property(criterion), value(indoor_comfort)),
    novaAction(alternative(R), action(A),
            property(subject), type(building_component), value(C)).
```

Finally, we may also express a BIM instance in ASP and then execute further queries relating the renovation alternatives to a specific building model. As an example, we parse IFC into ASP facts (Schultz, Bhatt 2013). The following ASP fact states that object with identifier *"6$2noaszh"* is a type *IfcWindow* (these facts are automatically generated after parsing an IFC file in our separate tool *InSpace3D*):

```
ifcObject(id('6$2noaszh)uddm'), type(ifcWindow)).
```

We add facts that relate IFC object types with NovaDM object types, as specified in the custom subject-feature knowledge base, e.g. the following fact states that "IfcWindow" and NovaDM's "window" type are the equivalent:

```
same_types(ifc(ifcWindow), novaDm(window)).
```

We then execute a query: "*What replacement actions can be applied to the current IFC building model? i.e. given the objects in the current IFC building model, find replacement actions that operate on those object types*".

```
viable_repair_actions(R,A) :-
  ifcObject(_, type(T)),
  same_type(ifc(T), novaDm(NovaT)),
  novaAction(alternative(R), action(A),
             property(subject), type(building_component),
             value(NovaT)),
  novaAction(alternative(R), action(A),
             property(approach), value(replacement)).
```

The result is that ASP will find all actions that are assigned with the *replacement approach*, and that have a *subject* value (e.g. window) that matches at least one IFC object type in the given model. In this example there is: at least one IFC window object in the model; *actions 1,4* of *alternative r1* are *replacement approaches*, and have a *window subject*:

```
viable_repair_actions(r1,1)
viable_repair_actions(r1,4)
```

## 5 CONCLUSION

This paper presents an overview of our initial version of the NovaDM, designed to cope with the complexity of various concepts related to the generation of holistic renovation scenarios. A comprehensive list of renovation approaches, alternatives, and actions was collected and integrated from a large variety of sources. The NovaDM is a first step towards formally structuring this information in a general, extensible manner, with the aim of (a) enabling the research community to more easily unify, share and collaborate on renovation topics, and (b) enabling practitioners to more easily utilize this collective renovation knowledge developed within both the research community and industry. Future research concerns the integration of the NovaDM parser to a general optimization framework for exploring the extremely large search space of renovation scenarios. We are also investigating the development of intuitive interfaces for working with NovaDM (specifying actions, querying, etc.).

## ACKNOWLEDGMENT

The authors of the paper would like to show their gratitude to the Danish Innovation Foundation for financial support through the RE-VALUE research project.

## REFERENCES

Baker N. 2009. *The Handbook of Sustainable Refurbishment: Nondomestic Buildings.* London: Earthscan.

Benjamin, P.C., Menzel, C.P., Mayer, R.J., Fillion, F., Futrell, M.T., deWitte, P.S. & Lingineni, M. 1994. IDEF5 method report. *Knowledge Based Systems, Inc.* Retrieved from https://www.scss.tcd.ie/Andrew.Butterfield/Teaching/CS4098/IDEF/Idef5.pdf.

Boeri, A., Antonin, E., Gaspari, J. & Longo, D. 2014. *Energy Design Strategies for Retrofitting: Methodology, Technologies, Renovation Options and Applications.* Southampton: WIT Press.

Brewka, G., Eiter, T., & Truszczynski, M., 2011. Answer set programming at a glance. Commun. ACM 54(12), 92–103.

BPIE [Buildings Performance Institute Europe] 2011. Europe's buildings under the microscope. Retrieved from http://bpie.eu/wp-content/uploads/2015/10/HR_EU_B_under_microscope_study.pdf.

BuildingSMART 2008. Industry Foundation Classes. Retrieved from http://www.buildingsmart-tech.org/specifications/ifc-overview.

BuildingSMART 2017. buildingSMART Data Dictionary. Retrieved from http://bsdd.buildingsmart.org/#

Burton, S. 2012. *The Handbook of Sustainable Refurbishment: Housing.* Abingdon: Earthscan.

Ferreira, J., Pinheiro, M.D. & Brito, J.D. 2013. Refurbishment decision support tools review—Energy and lifecycle as key aspects to sustainable refurbishment projects. *Energy Policy*, 62: 1453–1460. doi:https://doi.org/10.1016/j.enpol.2013.06.082.

Galiotto, N., Heiselberg, P. & Knudstrup, M. 2015. Integrated Renovation Process: Overcoming Barriers to Sustainable Renovation. *Journal of Architectural Engineering*, 22. doi:10.1061/(ASCE)AE.1943–5568.0000180.

Jensen, P.A. & Maslesa, E. 2015. Value based building renovation—A tool for decision making and evaluation. *Building and Environment*, 92: 1–9.

Juan, Y., Gaob, P. & Wangc, J. 2010. A hybrid decision support system for sustainable office building renovation and energy performance improvement. *Energy and Buildings*, 42(3): 290–297. doi:10.1016/j.enbuild.2009.09.006.

Checkland, P. 2000. Soft Systems Methodology: A Thirty Year Retrospective. *Systems Research and Behavioral Science*, 17: 11–58. doi:10.1002/1099-1743(200011)17:1+<::AID-SRES374 > 3.0.CO;2-O.

Kamari, A., Corrao, R. & Kirkegaard, P.H. 2017. Sustainability focused Decision-making in Building Renovation. *International Journal of Sustainable Built Environment*. 6(2): 330–350. doi:10.1016/j.ijsbe.2017.05.001.

Kamari, A. 2018. A multi-methodology and sustainability-supporting framework for implementation and assessment of a holistic building renovation: Implementation and assessment of a holistic sustainable building renovation [PhD thesis]. Aarhus: Dept. of Eng., Aarhus University. doi: 10.7146/aul.257.181.

Kamari, A., Corrao, R., Petersen, S. & Kirkegaard, P.H. 2018a. Tectonic Sustainable Building Design for the development of renovation scenarios—Analysis of ten European renovation research projects. In SER4SE 2018 (seismic and Energy Renovation for Sustainable Cities) conference, Catania, Italy (ISBN: 978-88-96386-56-9), 645–656.

Kamari, A., Jensen, S., Christensen, M.L., Petersen, S. & Kirkegaard, P.H. 2018b. A Hybrid Decision Support System (DSS) for Generation of Holistic Renovation Scenarios—Case of Energy Consumption, Investment Cost, and Thermal Indoor Comfort. *Sustainability*, 10(4): 1255. doi: 10.3390/su10041255.

Kamari, A., Corrao, R., Petersen, S. & Kirkegaard, P.H. 2018c. Towards the development of a Decision Support System (DSS) for building renovation: Dependency Structure Matrix (DSM) for sustainability renovation criteria and alternative renovation solutions. In SER4SE 2018 (seismic and Energy Renovation for Sustainable Cities) conference, Catania, Italy (ISBN: 978-88-96386-56-9), 564–576.

Keeney, R.L. 1992. *Value-Focused Thinking*. Cambridge: Harvard University Press.

Molio 2016. Molio Price data. Retrieved from https://molio.dk/molio-prisdata/prisdata-footer/brug-molio-prisdata/.

Nielsen, A.N., Jensen, R.L., Larsen, T.S. & Nissen, S.B. 2016. Early stage decision support for sustainable building renovation: A review. *Building and Environment*, 103: 165–181. doi:10.1016/j.buildenv.2016.04.009.

Perez, A.G., Juristo, N. & Pazos, J. 1995. Evaluation and assessment of knowledge sharing technology. *Towards very large knowledge bases*, 289–296.

Schultz, C & Bhatt, M. 2013. InSpace3D: A Middleware for Built Environment Data Access and Analytics. In: Int Conf on Computational Science (ICCS 2013), pp. 80–89.

# Software library for path planning in complex construction environments

K. Kazakov, S. Morozov, V. Semenov & V. Zolotov
*Ivannikov Institute for System Programming RAS, Moscow, Russia*

ABSTRACT: Recently spatial-temporal (4D) modeling is becoming increasingly important in the construction industry due to better planning of project works. One of the advantages not reachable through the usage of traditional methods is the detection of spatial-temporal conflicts in construction schedules. In the paper mathematical methods and software tools for the detection of a special type of conflicts—path conflicts are discussed. A software library for path planning in complex construction environments with non-trivial topology and dynamic behavior is presented. The main attention is paid to its general organization, functionality and object-oriented design. Computational experiments confirm the efficiency of the library and its advanced capabilities for spatial-temporal validation of construction schedules against path conflicts.

*Keywords*: 4D modeling, motion planning, collision detection, software engineering

## 1 INTRODUCTION

In recent decades, spatial-temporal modeling is becoming increasingly important in the construction industry due to better planning of project works, better communication among project stakeholders, and better coordination [1]. One of the key advantages not reachable through the usage of traditional planning methods is the detection of spatial-temporal conflicts in construction schedules. Ultimately, it enables to anticipate potential problems at earlier planning phases and to reduce risks at final construction phases often undergone to delays and reworks. To achieve this goal, building information modeling should be applied in a wider context of simulation of construction processes and conflict resolution.

As opposed to the problems of collision detection and workspace congestion, path conflicts have not been given much attention by researchers. Such conflicts are usually caused by the inability to deliver construction elements, materials or equipments to their eventual positions along collision-free paths. Modern 3D geometric kernels, popular game engines and spatial reasoning tools do not provide necessary functionality.

To identify path conflicts the motion planning theory and software tools should be applied [2]. Unfortunately, motion planning problems are PSPACE-hard. Even being formulated in local statements, these problems can cause serious computational difficulties. Popular software libraries such as Motion Planning Kit (MPK), OpenRave, Open Motion Planning Library (OMPL) are basically intended for such local formulations [3, 4, 5]. Mathematical arsenal of the libraries is mainly based on sampling and searching techniques such as RRT (Rapidly Exploring Trees) and PRM (Probabilistic Roadmaps). These demonstrate high efficiency in disparate applications such as humanoid robotics, automotive manufacturing, computational geography, computer graphics, computational biology, but fail in complex construction environments with non-trivial topology and dynamic behavior.

Such construction environments consist of a large number of geometric objects with individually prescribed behaviors. Although at each time step, the environments are usually subject to minor changes, the search of global paths remains a serious problem that is insoluble for the mentioned software libraries. The libraries do not provide global planning tools and prevent the implementation of new promising methods suitable for these purposes. Such methods should integrate both the metric and topological paradigms to represent and explore complex construction environments [6, 7]. While metric representations produce accurate models of the environment, they consume a huge data volume and are computationally hard for analyzing. On the other hand, topological maps can be processed in a more efficient way, but they are typically difficult to extract from metric representations and to update if the representations have been changed.

In the paper a software library for path planning in complex construction environments is presented. In Section 2 we give a brief overview of the library with the emphasis on its general organization and underlying principles. Sections 3, 4, 5 provide explanatory descriptions of the key subsystems intended for collision detection, local and global path planning. The consideration is supplemented by UML diagrams illustrating functionality, design and implementation features. The results of performed computational experiments are presented in Section 6. The benefits of the library are shortly summarized in Conclusions.

## 2 ORGANIZATION AND PRINCIPLES

The developed software library is intended for automatic routing in static and dynamic environments for an object moving based on a priori knowledge of the environment. The library not only provides ready-to-use tools for solving various motion planning problems, but also assumes advanced capabilities to simplify and unify the implementation of new models, methods and applications. By these reasons the library has been implemented as an object-oriented framework. Because of the C++ language the library allows the development of high-performance applications by using procedural, object-oriented, and generic programming paradigms. The selected abstractions with high-level semantics play role of hotspots for further extensions of the object-oriented library and its functional evolution.

The design of the library was preceded by the conceptualization and systematization studies covered various mathematical problems, computational methods and algorithms of the motion planning theory. Two fundamental formulations and computational strategies were proposed to distinguish.

Global path planning assumes a priori knowledge about the modeled environment and admits multiple path planning requests. Typically a topological map of free zones is preliminary extracted from the geometry representation of the environment and then it is used for fast search of perspective routes using graph theory algorithms. An accurate validation of the selected routes and, if necessary, their correction are accomplished by local path planning methods.

Local path planning implies the resolving of single requests for a particular solid or mechanical system with account of its specific geometric representation and kinematic constraints. The mentioned sampling methods have proven to be effective for such purposes even if the system induces a multi-dimensional configuration space.

Once a topological map is obtained for the entire environment, multiple requests can be resolved for different objects using graph theory algorithms. However, the validation and, if necessary, the correction of the routes is performed individually for each object or system in its own configuration space.

Thus, to satisfy the requirement of effective path planning in complex construction environments, the software library must support alternative metric and topologic representations of the explored environment and implement combined computational strategy leveraging both global and local motion planning methods. The advantage of such approach is that the environment exploration can be performed in a fast and efficient way by global planning at topological level and local planning at accurate geometric level. In a certain sense, this can be regarded as a reduction of the original computationally hard problem to a sequence of relatively simple computations. Our previous studies have shown the applicability of the approach to static construction environments [7]. The presented library makes it possible to apply the approach to complex dynamic environments too.

The library is organized as a system of classes which can be subdivided into WorkSpace, ConfigurationSpace, DiscreteSpace packages. Being properly combined, the classes form CollisionDetection, LocalPlanning and GlobalPlanning subsystems.

The WorkSpace package includes interfaces and implementations to access the explored dynamic environment. Obstacles and movable objects are considered as integral parts of the environment. The ConfigurationSpace package consists of classes which enable to specify allowable configurations for the movable object, to generate and validate sample configurations as well as to execute local path requests. All the problems are resolved in the object configuration space. The DiscreteSpace package is intended to represent different sorts of topological graphs, roadmaps, fast growing trees and to execute global path requests. Generic implementations make it possible to reuse the package classes for the problems formulated in both three-dimensional workspaces and object configuration spaces.

The CollisionDetection subsystem enables to determine clashes between a movable object and the environment obstacles. Because of the unified interfaces, this can be done both internally and by tools provided by third-party systems (e.g. CAD/CAM/CAE, 4D BIM), into which the library is embedded. The LocalPlanning and GlobalPlanning subsystems provide tools for solving corresponding path planning problems. The LocalPlanning subsystem implements popular families of sampling techniques. The GlobalPlanning subsystem implements the general computational strategy mentioned above.

Because of basic operations such as nearest neighbor search, exact intersection, penetration depth determination, connectivity analysis are expensive and intensively called, the library deploys and keeps advanced spatial-temporal indexes and auxiliary data structures for the entire dynamic environment.

The event tree plays a role of primary temporal index which is computed once and then be updated only if the events occur. The event tree is a binary search AVL tree ordered by the event timestamps. It allows fast lookup, efficient retrieval of events in a given time interval as well as quick updates when adding, removing or changing events. Nested octrees and occupancy tree are secondary spatial indexes which are recomputed on a given focus time. These trees are well suited for spatial indexing of dynamic hierarchically organized scenes because of non-expensive incremental updates when some events happened [8].

In addition, bounding volume hierarchies (BVH) of axis-aligned bounding boxes (AABB) and oriented bounding boxes (OBB) are computed and cached for the individual objects. These caches are important for local planners which subject movable objects to multiple collision checks. Topological graphs, roadmaps, trees are implemented as relation networks which allow performing qualitative analysis and computing assertions about the connectivity of free zones and their attainability. Thereby, being indexed, cached and qualified the 4D modeling data allows effective processing.

## 3 COLLISION DETECTION SYSTEM

The CollisionDetection subsystem implements a well-known two-phase computational scheme combining both cheap negative tests for localization of potential collisions (wide phase) and precise intersections selectively applied for localized pairs of objects (narrow phase) [9]. The efficiency is significantly increased if the spatial indexes and BVH caches are utilized.

The ISpaceIndexStructure interface defines methods to build, to incrementally update and to apply a spatial index for the nearest neighbor

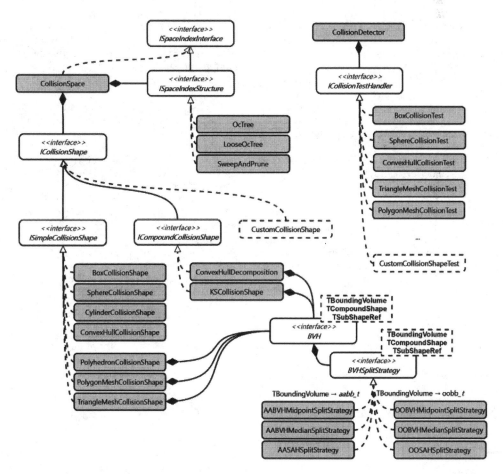

Figure 1. Class diagram of the collision detection subsystem.

search and collision localization. The methods are implemented in inherited classes with account of underlying data structures and particular algorithms. The developer is given the opportunity to use alternative spatial indexes based on regular octrees, octrees with boundary relaxation and sorted lists of bounding volumes (see classes OcTree, LooseOcTree, and SeepAndPrune in Figure 1). If necessary, the developer can provide own implementations of the spatial indexes.

The index behavior is delegated to the CollisionSpace class that provides access to the modeled environment as a composition of shaped objects of the type ICollisionShape and its numerous specializations. To establish the intersection of a given object with the environment obstacles a special method is provided. The method identifies neighboring obstacles using a preset spatial index and then performs precise check using a callback function. This allows the calling code to abort the operation when the first intersection is detected and to increase the total efficiency of requests.

The CollisionDetector class implements the narrow phase using so-called collision handlers. Each handler is responsible for analyzing a pair of geometric objects of corresponding types. Being pre-registered in a hash table the collision handlers can be quickly selected and activated when analyzing heterogeneous objects. For this purpose the hash key is defined as a pair of the geometry types. The CollisionDetector class is implemented in such a way that developers can support own geometric types and registry handlers in the subsystem.

The handlers share the ICollisionTestHandler interface that defines methods to perform intersection checks between a given pair of the objects of the abstract type ICollisionShape. For simple analytic geometries, the handler implementations are rather straightforward. For polygonal objects containing a large number of faces, BVH hierarchies are utilized to perform checks effectively. The hierarchies enable to localize potentially intersected pairs of faces and to minimize the total number of precise and expensive intersection checks. BVH structures are implemented in a generic way with the use of the class template TBoundingVolumeHierarchy. The processing of convex polygonal objects can be performed even more effectively if the extended polytopes or well-known Gilbert-Johnson-Curti algorithm are applied.

## 4 LOCAL PLANNING SYSTEM

The abstract class IProbabilisticPlanner is intended for the unified implementation of local planners and their families as inherited classes providing proper path finding methods. Their input parameters are the initial and target object configurations. Output parameters are the computed path and a success status of the obtained result. Each path is an ordered collection of conflict-free configurations with the interpolant predetermining intermediate non-conflicting states between adjacent configurations. Since the representation of paths depends on the configuration space of a particular object, the motion of which they specify, the class template TPath <TStateSpace> is used for this purpose.

The IProbabilisticPlanner class defines methods to execute both single and multiple requests. The explored configuration space and already deployed sampling structures (e.g. graphs, trees, and roadmaps) are accessed through corresponding associations and are maintained throughout the entire modeling sessions. Noteworthy that a general computational scheme with preprocessing of input data, invocating of a sampling algorithm, post-processing and optimizing of found paths is implemented directly in the abstract class, while particular algorithms are refined in the concrete inherited classes. It makes possible to implement new algorithms by modifying available ones with account of the coverage density, the relative number of successful propagation attempts, the dynamic sampling area, and the total cost estimations [10].

The library provides an advanced family of local planners. Classical algorithms of Rapidly Exploring Random Trees, Expansive-Spaces Trees and Iterative Diffuse Path Planner are implemented by the classes RRT, EST, and IPP respectively. Improved versions of the algorithms based on online optimization of trees and bi-directional propagation are implemented by the classes TRRT, RRTStar, RRTConnect, SBL, BTRRT Finally, the PRM and LazyPRM classes provide probabilistic roadmap algorithms effective for multiple requests. References on the algorithm descriptions and implementation details can be found in [2, 10].

Conflict-free configurations are represented using auxiliary templates StateTree <TNodeData>, StateForest <TNodeData>, and StateRoadmap <TVertexData, TEdgeData>. Because of random nature of sampling algorithms, the paths found are usually unsuitable for practical purposes and need to be improved to satisfy the requirements of minimum length, smoothness, remoteness from obstacles [10]. For this purpose the library provides the path optimizers PathShortening, PathSmoothing and PathRetractor with the common interface IPathOptimizer. A particular path optimizer can be preset through the IProbabilisticPlanner interface and then be started automatically when post-processing of conflict-free paths found by the main sampling algorithms.

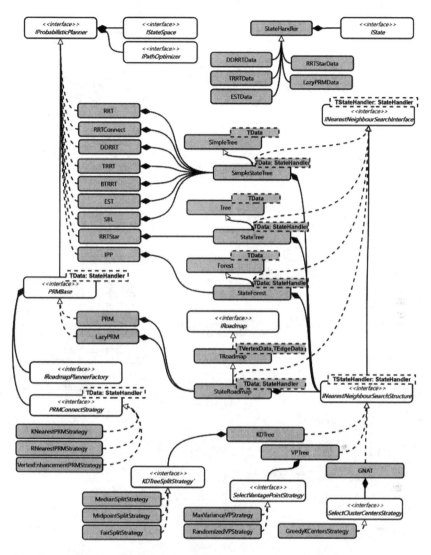

Figure 2. Class diagram of the local path planning subsystem.

The nearest neighbor search is one of the most critical operations called by sampling algorithms. To increase the operation efficiency and to improve total performance, spatial indexes such as kD-tree, Geometric Near-neighbor Access Tree, Vantage Point Tree should be applied [11]. These structures are similar to those used in collision detection, but are more effective for multi-dimensional configuration spaces. The library provides proper implementations as concrete classes KDTree, GNAT, and VPTree (see Figure 2). The indexes are updated automatically whenever a sampling data is changed. The interface INearestNeighbourSearchInterface assumes methods for search of the nearest point, k nearest points and the points lying inside a sphere with a given radius. All the methods imply the use of a metric function provided by the object configuration space. Such organization allows developers to separate the interface and its implementation, taking into account the features of particular sampling algorithms and spatial distributions of the object configurations.

## 5 GLOBAL PLANNING SYSTEM

The global planning subsystem employs the computational strategy which consists in reducing the original path planning problem to a relatively simple search for routes in a topological graph.

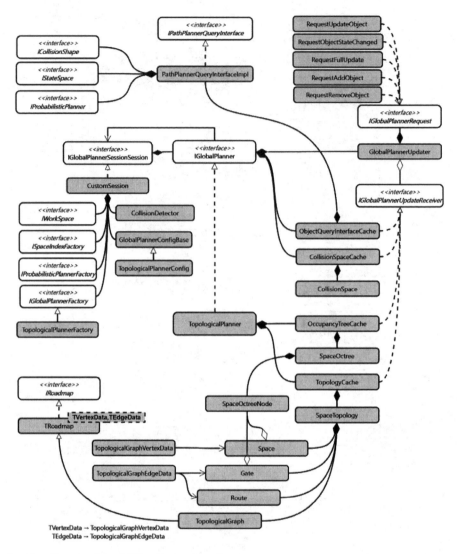

Figure 3. Class diagram of the global path planning subsystem.

The graph can be formed by using spatial decomposition methods, Voronoi diagrams, or manually drawn plans. The strategy implies the deployment of the graph directly in a three-dimensional space of the explored environment and then its use to make decisions about most promising routes while resolving single and multiple requests. The selected routes are validated and, if necessary, corrected by assigned local planners with account of the individual object geometry and the imposed kinematic constraints.

The abovementioned strategy was successfully validated for static construction environments [7]. The implementation presented allows complex dynamic environments due to effective incremental updates of all data structures (e.g. metric, topological, indexing, sampling) as the events occur in the explored dynamic environment. Let us consider the subsystem in more details.

Basically, the subsystem is represented by the abstract classes IGlobalPlanner, IGlobalPlannerSession, IGlobalPlannerRequest, IPathPlannerUpdateReceiver and their concrete specializations. The computational strategy is implemented in the IGlobalPlanner class with account of environmental changes and updates that need to be applied to the data to keep them consistent throughout the modeling session. For this purpose a message notification and processing mechanism is supported by the subsystem. The messages are represented by the class IGlobalPlannerRequest and are connected with the object creation, deletion, or modification events. The messages are queued and dispatched by the subsystem for further processing by receiver handlers. A response to received messages is processed by the handlers driven by the class IPathPlannerUpdateReceiver. To activate the mechanism, particular classes of messages and handlers must be preliminary registered in the subsystem.

To embed the global planning subsystem in a target application, the developer provides own specialization of the class IGlobalPlannerSession with the implemented methods to access to the explored environment and to get preset collision detectors, local and topological planners. These components in turn are configured from the classes representing various data structures and implementing different algorithms. Such organization enables the developer to configure a global motion planner from available components and to adjust it in the most efficient way. The developer does not need to worry about the availability and consistency of the data because of the provided mechanisms already built in the subsystem. The changes occurring in the dynamic environment are controlled and the data are incrementally updated in the process of modeling the environment and executing single and multiple path planning requests.

## 6 COMPUTATIONAL EXPERIMENTS

To prove the efficiency and flexibility of the developed software library, a series of computational experiments has been carried out. The experiments related to the spatio-temporal validation of construction schedules and the identification of path conflicts. The explored dynamic environments were originated from building projects with the schedules prescribing at each time step the installation of the next construction element. It was required to confirm that there is at least one collision-free path for every installed element and thereby the schedule is trustworthy and feasible in practice.

Local probabilistic planners were not applicable to this computational problem because of the impossibility to complete computations in an acceptable time. Global planners being configured properly demonstrated satisfactory efficiency, which made it possible to conclude on path conflicts in a reasonable time. In order to extract topological graphs, a special multi-step method was implemented and applied [7]. Based on the environment geometry, an occupancy octree is built with the octants enriched by the distance to the nearest obstacle. Then spatial zones are identified and qualified as spaces and gates using the distance distribution over adjacent octants. Finally, being connected the spaces and gates form a bipartite topological graph which is applied for resolving single and multiple requests.

The validation of the schedules was carried out in two ways. The first one was to check the schedule events using single requests. It took significant CPU time because of the need to prepare a topological graph anew at each time step of the schedule. In the case of dynamic environments, this led to high overhead costs. An important advantage of the library is that it allows effective incremental updates of the internal data and avoidance of the overheads. When the environment changes, its topological graph can be effectively updated and new multiple requests can be executed without redundant computations.

The path validation by local planners took negligible CPU time in comparison with the construction and updating of the topological graph. This can be explained by effective use of the spatial indexes that significantly accelerated the work of collision detectors and local planners, while the computational costs of the construction and updates of the topological graph remained fixed. The computational experiments carried out on a typical computer configuration Core 2 Duo E8600 processor (2.13 GHz), 2GB of RAM (800 MHz). Middle-size building models of a hospital, a steel construction, a bridge, and a warehouse were used (see Figure 4).

The total numbers of construction elements and triangles in their polygonal boundary representations are given in Table 1. The next row

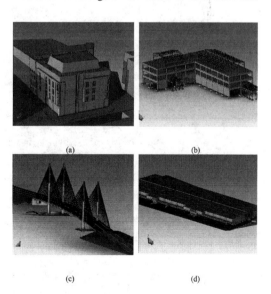

Figure 4. Building models: hospital (a), steel construction (b), bridge (c), warehouse (d).

Table 1. Total CPU (milliseconds) spent on the schedule validation.

| Building model | Hospital | Steel construction | Bridge | Warehouse |
|---|---|---|---|---|
| Total elements | 1319 | 595 | 811 | 1452 |
| Total triangles | 62268 | 14416 | 61524 | 259226 |
| Deployment | 10608 | 320118 | 21505 | 287243 |
| Incremental validation | 59873 | 1693584 | 149584 | 4603657 |
| Complete validation | 27168523 | 63111483 | 16341413 | 129034623 |
| Gain factor | 453 | 37 | 109 | 28 |

shows CPU milliseconds required to deploy all data structures, including topological graphs, necessary to perform single and multiple requests. It can be seen that in all the cases it took reasonable time not exceeding a few minutes. Total CPU milliseconds required to validate schedules against path conflicts are presented in the next two rows. For the considered projects it took less or slightly more than an hour, which is quite acceptable for practical work. Incremental validation of entire schedules by multiple path planning requests was much more efficient than complete validation of separate events by single requests as evidenced by the last row in the table with a factor expressing the performance gain. However, both complete and incremental methods are supported by the library because of the particular situations when a large number of events occur simultaneously and it is easier to recalculate all the data again.

## 7 CONCLUSIONS

Thus, the software library for path planning in complex construction environments has been presented. The library provides advanced tools for automatic routing in static and dynamic environments for an object moving based on a priori knowledge of the environment. The performed experiments have confirmed the efficiency of the library for executing single and multiple requests in global dynamic environments and the applicability to the problems of validation of construction schedules against path conflicts. Being implemented as an object-oriented framework the library allows further extensions and implementations of new models, methods and applications of the motion planning theory.

## REFERENCES

[1] Dawood N., Sikka S. Measuring the effectiveness of 4d planning as a valuable communication tool, ITcon, Vol. 13, 2008, pp. 620–636.
[2] LaValle S.M. Planning Algorithms. Cambridge University Press, 2006.
[3] Gipson I., Gupta K., Greenspan M. MPK: An open extensible motion planning kernel. J. Robot. Syst. 2001. Vol. 18, № 8. pp. 433–443.
[4] Diankov R. Automated Construction of Robotic Manipulation Programs Architecture. 2010. Vol. Ph.D. pp. 1–263.
[5] Sucan I., Moll M., Kavraki L.E. The Open Motion Planning Library IEEE Robot. Autom. Mag. 2012. Vol. 19, № 4. pp. 72–82.
[6] Poncela A., Perez E.J., Bandera A., Sandoval F. Efficient integration of metric and topological maps for directed exploration of unknown environments, October 2002, Robotics and Autonomous Systems 41(1):21–39.
[7] Semenov V., Kazakov K., Zolotov V. Global path planning in 4D environments using topological mapping.//eWork and eBusiness in Architecture, Engineering and Construction, eds. G. Gudnason and R. Scherer, CRC Press, Taylor & Francis Group,London, UK, 2012, pp. 263–269.
[8] Morozov S., Semenov V., Tarlapan O., Zolotov V. Indexing of Hierarchically Organized Spatial-Temporal Data Using Dynamic Regular Octrees. In: Petrenko A., Voronkov A. (eds) Perspectives of System Informatics. PSI 2017. Lecture Notes in Computer Science, vol. 10742, pp. 276–290.
[9] Jimenez P., Thomas F., Torras C. 3D collision detection: a survey. Computers and Graphics, 2001, vol. 25, pp. 269–285.
[10] Kazakov K., Semenov V. Object-oriented framework for motion planning in complex dynamic environments. Trudy ISP RAN/Proc. ISP RAS, vol. 29, issue 5, 2017. pp. 185–238. DOI: 10.15514/ISPRAS-2016-1(2)-33.
[11] Samet H. Foundations of Multidimentional and Metric Data Structures. Morgan Kaufmann, 2006.

# Delivering COBie with ProNIC—compliance and implementation

P. Mêda & J. Moreira
*Instituto da Construção—CONSTRUCT—GEQUALTEC, Porto, Portugal*

H. Sousa
*Faculdade de Engenharia da Universidade do Porto—CONSTRUCT—GEQUALTEC, Porto, Portugal*

ABSTRACT: Continuous improvement on information consistency and interoperability are main concerns for some public owners in Portugal. The present research benefits from a large implementation on more than 100 projects where standard procedures, bill of quantities and specifications were used by a large number of agents during design and construction. Facility management is a main concern and rework during handover constitutes a waste of effort. Through the use of COBie it is possible to foster interoperability, extending the integration of processes from design to use. The scope of the research was to observe the ability of ProNIC to deliver COBie drops. The results were beyond the expectations, meaning that ProNIC can provide COBie drops during design and with improvements deliver a more satisfactory drop at the end of construction/during handover, fostering the demands of the industry in terms of efficiency and contributing to the competitiveness.

## 1 INTRODUCTION

The origin of ProNIC (Portuguese acronym for Construction Information Standardization Protocol) comes from the Portuguese Government awareness that there was lack of standardization on construction projects at the bill of quantities level, as well as flaws at the specifications level supporting the design stage documents (Sousa, H.; Moreira, J.; Mêda 2008).

In addition to the problems more related with the design stage and design teams, public owners were struggling on the definition of common procedures for the design and construction stages as well as the collection of information to develop project indicators and perform economic analysis (Sousa, H.; Moreira, J.; Mêda 2012).

This state of the art led the Portuguese Government to the promotion of the ProNIC project.

During 2008, the publication and implementation of the transposition for national domain of the EU Directive on Public Procurement highlighted these needs, namely through the mandatory use of e-procurement (Ministério das Obras Públicas 2007) (Soares L. 2017).

ProNIC standard contents and procedures were implemented on a cloud based tool, gathering all stakeholders on a collaborative environment for project development linked with e-procurement platform.

The application of ProNIC on a large public project geared for the refurbishment of buildings, led to improvements, extension of functionalities and new integration challenges, namely the link between construction project stages, as design, procurement and construction, as well as the achievement of high levels of information consistency and streamlined information flows between the mentioned stages and across all involved agents (Mêda, P.; Sousa 2014).

Information exchange protocols with other tools, as public procurement platforms, ERP—Enterprise Resource Planning software of the public owner and proprietary tools used by designers, contractors and supervision teams, constituted the following investment in terms of improvements, reducing the information lost and rework (Sousa and Mêda 2012).

ProNIC nowadays counts with more than 1100 million euros in public works being managed and with more than 1500 users with different roles, from designers, to contractors, owners, supervisors and managers (Sousa, H.; Moreira, J.; Mêda 2016).

Continuous improvement and alignment with construction industry trends, namely in terms of construction process stages integration, procedures and information standardization, agents engagement, information consistency and integrated delivery, led the developers to work further on interoperability and information exchange protocols.

The above mentioned public owner main concerns regarding facility management, maintenance costs and sustainability analysis/standardization of construction technologies and products characteristics set new scope for future improvements (Vieira, A., Marques Cardoso 2014).

Among them and related with the scope of the research being presented, was questioned the ability of ProNIC to provide COBie drops at the design stage and at the end of the construction/ during handover stages, setting through it the link with a facility management tool (Beltrão 2015).

This established the goals for the present research. Evaluate the readiness of ProNIC to deliver COBie drops on both moments as well as identify changes/improvements/requirements (East, E., Nisbet, N., Liebich 2013) (British Standard Institution (BSI) 2014) (Limited 2012).

To develop this analysis a three step methodology was drawn to provide distinct outcomes, related with the project definition:

– early design COBie worksheets;
– detailed design worksheets;
– and Operations and Maintenance worksheets.

In first, a survey on ProNIC database was made to identify a group of relevant construction works used on projects, both in terms of cost and number of times used/instances. These construction works supported the detail level analysis (Detailed Design and Operations and Maintenance).

Following, a comparison between ProNIC project setup menus and different COBie spreadsheets, namely "Facility", "Floor" and "Contacts", was made to evaluate top level compliance.

The last step consisted on the analysis of elements produced and COBie requirements towards the drop at the end of construction/handover stage.

Through this approach it was possible to draw different conclusions related with the compliance/ability of ProNIC to provide COBie drops in different stages, as well as the changes/requirements in terms of future developments.

## 2 BACKGROUND

### 2.1 Implementation context and needs

ProNIC, as mentioned, was implemented on a public program for the refurbishment of buildings. This process occurred at the beginning of the third stage of the program, and coincided with the publication of a new public procurement legal framework (Ministério das Obras Públicas 2007). With this diploma several procedures were stabilized and higher requirements for public owners were set. One of the most relevant aspects was the clear definition of construction process stages that other developed study proved that are quite similar and adaptable to those set by the RIBA Plan of Work 2013 (Ministério das Obras Públicas 2008) (Architects 2013) (Mêda, P., Sousa, H., Moreira 2015). Given this, several aspects of the law were embedded in ProNIC as the construction stages, types of procurement procedures, structures and organization of design disciplines, as well as agent's roles.

From the side of owner's, higher requirements in terms of contract control during construction led to the implementation of improvements during the construction stage, namely reports, methodologies for quantities follow up and information exchange procedures for invoice processing.

This owner is also the manager of the facilities during use. For this purpose, a facility maintenance tool was adopted in parallel and at the end of construction it was placed the need of transposing all the information from one system to other.

Rework was made in few situations as an appeal solution and it was identified the need to establish a way to perform a more agile and automatic exchange.

### 2.2 Implementing construction trends

The industry is being pushed to sustainability, efficiency and competitiveness (Comission 2012). Digitalization is a common aspect in these strategic guidelines, where collaboration, cloud environments, integration of processes and BIM—Building Information Modeling assume high relevance (Mckinsey Global Institute 2017) (Forum 2016).

In the context of this program, the improved and developed tools led to very good results validated by the Portuguese Court of Auditors. Several targets in terms of construction trends were achieved, but the lack of interoperability was highlighted as a situation to improve. Direct integration with other tools such as 3D modeling software and COBie were found to be the next steps towards the expected industry achievements.

## 3 METHODOLOGY

The backbone of ProNIC is the information structure, namely the breakdown structure that enables the development and delivery of standardized bill of quantities for specific construction projects. The information structure follows the main rules set on both versions of ISO 12006-2, and for the development of bill of quantities it is used a breakdown of construction works (International Standards Organization 2015) (Mêda, P.; Hjelseth, E.; Sousa 2016). They consist of standard text and customized fields to be set on the context of a project/

attributes, composing the high detail of the structure and in addition with the measurement units constitute the bill of quantities items.

The projects follow up during the construction stage, allowed the control of the quantities as well as the items and global costs, once the winning bid is placed on the project, replacing the design budget estimate, in order to deliver the monthly measurement reports. Quantity and cost indicators were drawn from this for very different kinds of analysis, namely life cycle costs.

From the more than 10.000 items that constitute the breakdown structure, the 100 projects used more than 3500, more or less 1/3 of the database. From these and with the purpose of performing a compliance check with some COBie spreadsheets it was set a selection criteria, based on the number of times/instances the work was used, the unitary price and global price/relevance on the database, as well as different types of works/different chapters/systems to promote a complete insight.

This resulted on a group of 10 items (that represent more than 8,5% of the global cost of the projects managed) presented on Figure 1.

This group of items was analyzed in two different perspectives/construction process stages and observing different COBie spreadsheets.

The first was geared for the design stage and COBie drops at the end of design, where due to the most common type of public procedure used in Portugal, the construction works need to be specified without mentioning trademarks. At this level, several examples were explored.

The second was at the end of construction/during handover, where all the information is defined. For this situation a specific situation, on the context of a project, was selected to develop the analysis.

Other part of the study worked the high level aspects of the projects. There are COBie spreadsheets and data that must be placed/defined at the beginning of the project. In ProNIC this occurs when the project is created. To evaluate the compliance at this level, several checks were made between ProNIC initial project definition menus and the "Facility", "Floor", "Spaces/Zones" and "Contacts" COBie spreadsheets.

Following the natural sequence of construction projects, these are the first to be presented and explored on the next point.

## 4 ANALYSIS/RESULTS

### 4.1 *Organization*

As previously mentioned, the analysis consisted of several steps addressing to different aspects between COBie and ProNIC. In order to streamline the organization, the construction process sequence will be followed.

In first, spreadsheets related with the project or "Early Design drop" will be explored, followed by the compliance check of "Components" and "Types" sheets for "Detailed Design drop" and the results from the analysis of "Construction/Handover drop".

### 4.2 *Early design drop*

#### 4.2.1 *Facility*

All projects in ProNIC must be linked to a facility. Facilities are defined by the owners from internal registers or assets classification. To start a new project in ProNIC is necessary to access to the "Create Work" menu where several fields must be completed. Figure 2 presents the COBie "Facility" sheet, as the example set on BS 1192–4:2014, and the correspondence between the fields of ProNIC menu.

As it is possible to observe, most of the data to be placed on the "Facility" sheet must be introduced in ProNIC in order to create a project. Elements such as Name, ProjectName, SiteDescription find direct correspondence. The Category corresponds to the

| ProNICconstructionWorkDescription | instances | totalSumPrice |
|---|---|---|
| Metal frame exterior window | 3877 | 22.524.413,37 € |
| Interior false ceilings | 1213 | 19.399.355,60 € |
| Standard salient luminaires | 2339 | 9.966.572,29 € |
| Reinforced concrete slabs | 348 | 8.617.250,22 € |
| Leaning roof coating | 370 | 8.104.233,37 € |
| HVAC ducts with rectangular section | 722 | 7.012.023,02 € |
| Roofing waterproofing with prefabricated membranes | 541 | 6.093.233,13 € |
| Electric cables on conveyors | 5385 | 5.825.889,46 € |
| Interior wood doors | 1637 | 5.182.785,10 € |
| Water Supply - PP Piping | 3457 | 2.731.149,42 € |
| | 8,7% | 95.456.904,98 € |

Figure 1. Top 10 items in terms of cost, instances, and different types of construction works/chapters, selected for analysis development.

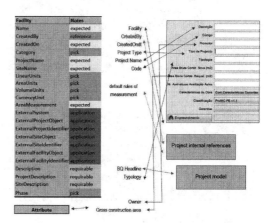

Figure 2. Compliance with "Facility" sheet.

Project Type that is, selecting the type of construction according with EUROSTAT classification or other (Mêda 2014). There are several information elements that are related with the users; as it will be further mentioned at "Contacts" sheet; and are just available in back office, as CreatedBy or CreatedOn. Similar situation occurs with the different Units.

The project is created and it will pass from stage to stage, therefore the Phase field will be automatically placed, following ProNIC stages.

SiteDescription is found to be the only without correspondence. ProNIC has, at this level, the identification of the Owner and other characteristics related with the facility, such as gross construction area. The first will be placed on the "Contacts" sheet, once it is associated to an agent role. All additional elements regarding the facility can be placed on the "Attributes" spreadsheet. Figure 3 presents the above mentioned example for a specific project.

### 4.2.2 *Floor*

During the study there was a discussion regarding the use of "Floor" (region) or "Space" sheets for the type on information that it will be explored. This came due to COBie philosophy applied on buildings or infrastructures. The identified situation in ProNIC is that most projects consist of service use facilities composed by several buildings.

To improve the management of the projects, it was set a division defined as "Construction Units", corresponding to buildings; example: Library, or to regions/defined areas, as parking lot or garden. This division is also found useful to separate the works related with yard and preliminary activities. These divisions are defined on the "Project model". This functionality allows the establishment of the

| Attribute | Example | Notes |
|---|---|---|
| Name | Gross construction area | expected |
| CreatedBy | pmeda@fe.up.pt | reference |
| CreatedOn | 2018-03-29T12:51:37 | expected |
| Category | | pick |
| SheetName | Facility | pick |
| RowName | n/a | reference |
| Value | | expected |
| Unit | squaremeters | pick |
| ExtSystem | | application |
| ExtObject | | application |
| ExtIdentifier | | application |
| Description | Area of construction for KPI | requirable |
| AllowedValues | n/a | requirable |

Figure 3. Example of facility gross construction area placed on "Attribute" sheet.

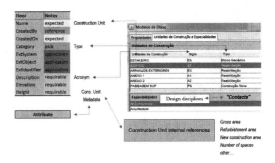

Figure 4. Compliance with "Floor" sheet.

"Construction Units", their characterization, design disciplines to be set for each, as well as the definition of the project documents headlines.

On the basis of the discussion was the decision of placing COBie interpretation for infrastructures on buildings.

At the end, it was decided to go ahead with the use of "Floor" sheet along with a group of additional parameters to be placed on "Attributes" sheet; Figure 4.

As it is possible to observe, most of the information fields are common. Elevation and Height are applicable if the Construction Unit is a building. For each one, and for the purpose of delivering indicators, the owners can define several parameters/additional information/characteristics. Data as Use (type of), GrossArea, NetArea are, among those, the more requested and are endorsed to "Attributes".

### 4.2.3 *Space and zones*

These two sheets will be explored together due to the origin of the information contained in them. "Spaces" and "Zones", for the purpose of the bill of quantities come often on the detailed measurement sheets and sometimes on the construction work description. This means that this information is placed on the project on a detailed level than the construction work when, in fact, it should be placed on an upper level.

Without entering in this detail, that it will be explored in Discussion/Conclusions, in terms of data fields the requirements are quite similar and simple to achieve without significant effort.

### 4.2.4 *Contact*

The establishment of a collaborative environment requires that all agents must have an access provided by a username and password. Additional information regarding the person and its company is also required. On the context of a specific project, each user must have a role related with the tasks/processes that will perform/follow during the process (owner, coordination, designer, supervision, contractor, other). All agents with the role of design will have an additional configuration on the access depending of the design discipline(s) that are authorized to work on. To attend this singularity it was found interesting and suitable the use of "Department" field. As it is possible to observe from Figure 5, all contact fields have correspondence with the user definition menu.

## 4.3 Detailed design drop

### 4.3.1 Systems

The breakdown structure top level corresponds to chapters. In many situations these are equivalent to systems, as plumbing, HVAC or structure. This interpretation of built object, as a sum of systems, is more natural to mechanical and electrical engineers than for civil engineers and architects. Notwithstanding, it is very useful for the purpose of developing analysis related with construction costs, maintenance, among others. In addition, ProNIC has the ability of transforming the main breakdown structure on a mirror structure organized in accordance with a specific owner needs/requirements. The item Category can work very well with the default structure or with others to be set for specific projects.

From the analysis it was found to be a complete correlation at this level as shown in Figure 6.

The field for system description is just visible in ProNIC back office (#), but it can be directly placed on COBie without any changes. This field can also work combined with Category if the default structure is not being used.

### 4.3.2 Component

Following the above mentioned, the construction works correspond to components (elements or part of) that are parameterized on the context of a project by a design discipline, meaning that compliance is assured for the CreatedBy and CreatedOn fields. Notwithstanding, a tap will always be a component from the water supply system, independently if it is specified by the Architect or by the MEP Engineer.

For the purpose of this analysis it was selected the construction work with the higher number of instances and, at the same time, one that presents significant economic relevance; metal frame exterior window. This work was used more than 3800 times and represents more than 22,5 million euros. Figure 7 presents the results.

Most of the information complies with the requirements. The field marked with grey are those that cannot be set until the moment of the construction. Yet, during construction, some of that information is not "worked" by ProNIC, see in the discussion.

### 4.3.3 Type

The same example was used, at this stage, to analyze the "Type" spreadsheet. The results are placed on Figure 8.

This evidences four types of situations marked with colors and (*). In white are the fulfilled fields based on construction works data. In grey are the fields that can only be set during construction, and in black are the ones missing on construction work description. The ones marked with (*) are related with information that is being worked on, meaning that it will be in short provided by ProNIC.

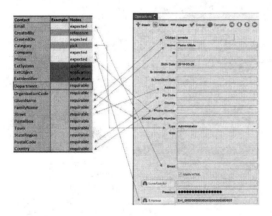

Figure 5. Compliance with "Contact" sheet.

Figure 6. Compliance with "System" sheet.

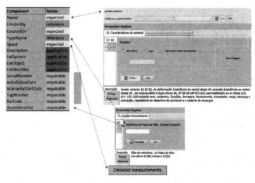

Figure 7. Compliance with "Component" sheet.

| Type | Example | Notes | |
|---|---|---|---|
| Name | Exterior window | expected | |
| CreatedBy | pmeda@fe.up.pt | reference | |
| CreatedOn | 2018-03-29T17:20:11 | expected | |
| Category | Metallic elements | pick | |
| Description | Metallic frame exterior windows | expected | |
| AssetType | | pick | (*) |
| Manufacturer | | reference | |
| ModelNumber | | expected | |
| WarrantyGuarantorParts | | reference | |
| WarrantyDurationParts | 5 | expected | |
| WarrantyGuarantorLabor | | reference | |
| WarrantyDurationLabor | 5 | expected | |
| WarrantyDurationUnit | Year | pick | |
| ExtSystem | | application | |
| ExtObject | | application | |
| ExtIdentifier | | application | |
| ReplacementCost | | requirable | (*) |
| ExpectedLife | | requirable | (*) |
| DurationUnit | Year | pick | |
| WarrantyDescription | | requirable | |
| NominalLength | $2 - 4,00 | expected | |
| NominalWidth | n/a | expected | |
| NominalHeight | $3 - 2,10 | expected | |
| ModelReference | | requirable | |
| Shape | | requirable | |
| Size | 4,00m x 2,10m | requirable | |
| Color | | requirable | |
| Finish | $2 - anodized | requirable | |
| Grade | | requirable | |
| Material | $1 - steel | requirable | |
| Constituents | | requirable | |
| Features | safety device | requirable | |
| AccessibilityPerformance | | requirable | |
| CodePerformance | | requirable | |
| SustainabilityPerformance | | requirable | |

Figure 8. "Type" sheet and fulfilled fields.

For this specific construction work eleven characteristics are to be placed on "Attribute" sheet. These are the deformation, pressure, water tightness and impact energy classes, sound isolation, heat transfer, solar factor, air tightness, operation strength, mechanical durability and twist resistance. All comply with the requirements of the mentioned sheet.

4.3.4 *Other examples*

The same analysis developed in point 4.3.3 was applied to the items presented on Figure 1.

They all present the same behavior, in terms of compliance and information from Name to WarrantyDescription, as well as Sustainability Performance.

Significant differences were observed on the 14 items from NominalLeght to CodePerformance.

The construction work that evidenced high compliance was Electric cables on conveyors, with just a single information missing; shape. In fact, for this specific work all have the same shape, which can lead to a default definition for the item. The same situation occurs for many other construction works. In addition, two characteristics were identified to be placed on "Attribute" sheet.

The item that evidenced the worst compliance was Roofing waterproofing with prefabricated membranes. It was only possible to set 4 fields, meaning that 10 were left in blank. Although some might still that way, the item needs to be changed in order to place more requirements on its specification. Two characteristics were identified to be placed on "Attribute" sheet. The standard salient luminaires is the item that requires more characteristics on "Attribute" sheet; 20. In terms of compliance analysis, 6 fields were left in blank, mainly those related with dimensions.

From the analyzed items it was possible to identify the information that is always or most of the times present, and the information that lacks the most on the construction works. CodePerformance was the most absent with 9 missing's on 10 items. This is followed by Shape and Color where 7 miss out of 10. On the other hand, Material and Features were always identified, followed by Constituents and ModelReference and Size.

From this it is possible to state that most of the construction works can provide reasonable information to comply with the COBie "Type" spreadsheet, but an extensive review during the implementation shall be made in order to best adapt some items.

4.4 *Construction/Handover drop*

As previously mentioned, in terms of procedure (type of procurement), all projects followed the traditional procedure, meaning that all bill of quantities were developed without mentioning any trademarks or respecting the rules applicable by the public procurement code. During construction, ProNIC deals with changes in terms of quantities and additional works or work suppressions, performing mainly the construction work economic balance and work development. There is no data input besides the quantities related with the work process, for the development of the monthly measurement reports.

Given the present situation and for the purpose of providing drops at the end of construction/during handover few information can be added to the one presented for the design drop.

Notwithstanding, and in close link with the above mentioned report there is space for some evolution without the need of extensive developments as it will be discussed.

5 DISCUSSION

From the analysis it worth's to highlight that there is a total compliance between ProNIC and COBie on the "Contact", "Facility", "Floor" and

"System" sheets, working in addition with the "Attribute" sheet.

Regarding the "Space" and "Zone", the absence of visualization contributes to fewer systematization of this type of information, contributing to difficulties towards compliance. In brief, the development of the bill of quantities follows different assumptions regarding this information. Notwithstanding and in practical terms, these fields exist already/are implemented and it is possible to enhance the ability of establishing improved compliance if new tables, where this information can be settled/organized, are foreseen. The possibility of predefine the spaces and zones or organize them at the end of design and with link to the "Construction Units" is very easy to achieve. This can also benefit from the studies to establish links between ProNIC and 3D modeling tools.

The "Component" sheet has several items where direct correspondence is missing. This absence/mismatches are directly related with the type of information and can be grouped in two different situations. One, is the information related with its placement on the facility and handover date, InstalationDate, WarrantyStartDate. These information elements can be easily set through the use of the monthly measurement report and the date of the end of works/provisional handover date, as previously mentioned.

Other information SerialNumber, TagNumber, BarCode is more related with reception in the yard and commissioning, involving contractor and supervision. The value of this information for other purposes must be evaluated before implementing supporting functionalities in ProNIC.

In what relates to the analysis of the "Type" sheet, fewer compliance was achieved. This has, again, direct relationship with the information that is provided during design and during construction. The ones marked in black imply changes on the construction works specification during design in order to best fit to the requirements. These are found to be easy to implement, considering the effort of evaluating all the items of the breakdown structure. In what concerns those marked in grey the same two situations described for "Component" occur, leading to the same principles and based on the same implementation solutions.

It worth's to highlight that it was recently implemented in ProNIC an additional classification to mark all works with warranty terms according with the law (European Parliament 2014). Related with this, there are ongoing developments to improve the ability to support the new public procurement requirements in terms of Life Cycle Analysis (LCCA). This will have positive impact on the typification of assets, expected life definition and replacement costs.

## 6 CONCLUSIONS

The results evidence that the compliance between ProNIC and COBie is very high at all levels, namely for the production of drops prior to design and at the end of design.

It is possible to say that the results went positively beyond the initially expected.

Considering the results it can be stated that ProNIC is suitable to deliver COBie drops, mainly for the Early design and for Detailed design.

In what concerns providing drops at the end of construction/handover, several data is missing. These mismatches can be characterized in two groups. One where, with low changes on ProNIC and on the user's functionalities/actions it is possible to implement and provide additional data. The other is more related with the construction stage development and validation processes between contractor and supervision, meaning that for the purpose of providing/placing this information in ProNIC specific functionalities must be developed.

In brief, the process of implementing COBie drops in ProNIC can be easily achieved, with minor changes regarding the adaptation of the construction works to the requirements of the "Type" sheet.

One additional aspect that was not addressed, but that was observed during the analysis and is relevant for the conclusions, is the compliance with the "Document" sheet. The results evidenced complete compliance.

In relation with this, additional studies must be made in order to identify the link between Product Data Sheets, ProNIC and COBie.

Future study to evaluate compliance between COBie and the facility management system used by the public owner should also be developed.

This study evidences that with low effort it is possible to contribute further to the goals of integrating processes/interoperability, promote efficiency and achieve high levels of information consistency.

## REFERENCES

Architects, RIBA – Royal Institute of British. 2013. *Guide to Using the RIBA Plan of Work 2013*. RIBA Publi. ed. RIBA Publishing. London: RIBA Publishing.

Beltrão, Alexandre. 2015. 'Professional Activity Report: A Standardization Approach for the Design Enhancement'. Porto University – Faculty of Engineering. https://catalogo.up.pt/F/?func=find-b&request=A+standardization+approach+for+the+design+enhancement&find_code=WRD&x=36&y=14&filter_code_1=WLN&filter_request_1=&filter_code_2=WCN&filter_request_2=&filter_code_3=WYR&filter_request_3=&filter_code_5=WFMT&filt.

British Standard Institution (BSI). 2014. British Standards Institution (BSI) *Collaborative Production of Information Part 4 : Fulfilling Employer' s Information Exchange Requirements Using COBie – Code of Practice*. UK. http://shop.bsigroup.com/forms/BS-1192-4/.

Comission, European. 2012. *Strategy for the Sustainable Competitiveness of the Construction Sector and Its Enterprises*. Brussels.

East, E., Nisbet, N., Liebich, T. 2013. 'Facility Management Handover Model View'. *Journal of Computing in Civil Engineering* 27(1): 7.

European Parliament. 2014. Official Journal of the European Union *Directive 2014/24/UE of the European Parliament and of the Council*.

Forum, World Economic. 2016. *Shaping the Future of Construction - A Breakthrough in Mindset and Technology*. Geneva. https://www.weforum.org/reports/shaping-the-future-of-construction-inspiring-innovators-redefine-the-industry.

International Standards Organization. 2015. ISO Standards *ISO 12006-2: 2015 – Building Construction – Organization of Information about Construction Works – Part 2 : Framework for Classification*. https://www.iso.org/standard/61753.html.

Limited, Bryden Wood. 2012. *COBie Data Drops – Structure, Uses & Examples*. London.

Mckinsey Global Institute. 2017. *Reinventing Construction: A Route to Higher Productivity*.

Mêda, P., Sousa, H., Moreira, J. 2015. 'Processo Construtivo – Fases, Atribuições, Objetivos Gerais e Requisitos de Informação'. In *Simpósio Gequaltec 2015 – Tarefas e Funções Futuras Da Engenharia Civil: Sustentabilidade e Energia*, Porto: Engineerng, Gequaltec – Porto University Faculty of Engineering.

Mêda, P., Hjelseth, E., Sousa, H. 2016. 'Construction Information Framework-the Role of Classification Systems'. In *ECPPM2016 – 11th European Conference on Product and Process Modelling*, ed. To Define. Limassol: Define, To, Define, To.

Mêda, P., Sousa, H. 2014. 'ProNIC Contributions for Building Refurbishment – Procedures and Technology'. In *REHABEND 2014*, Santander.

Mêda, Pedro. 2014. 'Integrated Construction Organization – Contributions to the Portuguese Framework'. Porto University Faculty of Engineering.

Mêda, P., Sousa, H., Moreira, J. 2012. 'Avaliação de Custos de Construção Com Base Na Normalização de Trabalhos de Construção – Contributos Para a Sustentabilidade'. In *CINCOS' 12 – National Congress of Innovation on Sustainable Construction*, Aveiro: Cluster Habitat.

Mêda, P., Sousa, H., Moreira, J. 2016. 'ProNIC on the Schools Refurbishment Program – Contributions for the Construction Process Improvement'. In *REHABEND 2016 – Euro-American Congress on Construction Pathology, Rehabilitation Technology and Heritage Management*, Burgos.

Ministério das Obras Públicas, Transportes e Comunicações. 2007. *Decreto-Lei n.º 18/2008*. Portugal: Diário da República, 1.ª série – N.º 20–29 de Janeiro de 2008.

Ministério das Obras Públicas, Transportes e Comunicações. 2008. Diário da República n.º 145 *Portaria n.º 701-H/2008*. https://dre.pt/pesquisa/-/search/575341/details/maximized.

Soares L., Carvalho A. 2017. 'E-Procurement and Innovation in the Portuguese Municipalities: When Change Is Mandatory'. In *State, Institutions and Democracy*, eds. Norman Schofield and Gonzalo Caballero. Springer International Publishing, 338.

Sousa, H., and P. Mêda. 2012. 'Collaborative Construction Based on Work Breakdown Structures'. In *EWork and EBusiness in Architecture, Engineering and Construction – Proceedings of the European Conference on Product and Process Modelling 2012, ECPPM 2012*, ed. Raimar Gudnason, Gudni, Scherer. London, 839–45.

Sousa, H., Moreira, J., Mêda, P. 2008. 'ProNIC© and the Evolution of the Construction Information Classification Systems'. In *GESCON 2008 the Best Papers*, ed. João (Porto University Faculty of Engineering) Abrantes, Vitor; Sousa, Hipólito; Martins Poças. Porto: GEQUALTEC, 180.

Vieira, A., Marques Cardoso, A.J. 2014. 'Maintenance Conceptual Models and Their Relevance in the Development of Maintenance Auditing Tools for School Buildings' Assets – an Overview'. In *Proceedings of Maintenance Performance Measurement and Management Conference*, ed. Pombalina. Coimbra, 10.

# Automatic development of Building Automation Control Network (BACN) using IFC4-based BIM models

R. Sanz, S. Álvarez & C. Valmaseda
*Fundación CARTIF, Energy Division, Boecillo (Valladolid), Spain*

D.V. Rovas
*Institute for Environmental Design and Engineering, University College London, London, UK*

ABSTRACT: Nowadays Building Information Modelling (BIM) tools are widely used in the process of design and construction works for new building and renewals. Architects and engineers usually design the architectural characteristics without paying attention to other facilities such as Building Automation Control Networks (BACN) which are normally defined, developed and deployed after the building is finished by a third party company. Then the model of the building does not match with the final status as it is built. Taking advantage of the fact that IFC standard tries to cover the problematic of interchanging information among building model design software (interoperability), the standard will be used to automatically design and deployment of BACN (facing problematics related to geometry and location of devices) and its possibility to act on control facilities.

## 1 BUILDING AUTOMATION CONTROL NETWORK AND BIM MODELLING

Building Information Modelling often encapsulate geometric and building services information, but rarely is information related to the Building Automation and Control Network (BACN) captured within these models. This is partly because design and commissioning of the BACN happen at the later design stages and there is little-perceived value in adding this information to the model. A second aspect is that often it is not possible to capture all relevant BACN parameters and configurations within existing open data models, e.g. IFC4 (BuildingSMART 2018) – there are other data models and ontologies, e.g. SAREF (ETSI 2015) or BRICK (BRICK Schema 2016) as well as the data models linked to specific automation protocols, that better capture all pertinent parameters and configurations of the building automation domain. Still, in line with lifecycle interoperability considerations, there is value and partial model support, in keeping the IFC-based building model updated with BACN information.

In this paper a methodology is presented whereby a BACN network can be configured semi-automatically and, once this process is complete, the IFC model of the building is updated to include information about the control network. Thus, this information becomes available and queryable for facility management or other operational purposes. Core to our approach is a product library of commercially available devices that includes sensor, controllers and actuators to be integrated into an automation network. We chose to work with devices that support the LonWorks protocol—but many of the ideas would be applicable, with necessary adaptations, to other standard building communication protocols.

The devices database stores various types of sensors, controllers (focusing in this paper on controllers that conform to Space Comfort Controller Profile), actuators, from different vendors and associated functional features, as well as, purchasing costs. In LonWorks functionalities are derived from the different Functional Blocks which a device is composed of and associated Network Variables. Once device selection is complete, these Blocks need to be connected between them for information exchange to be possible; this is translated to the ways the Network Variables of each Functional Block in devices are linked to others in other Functional Blocks. Once the automation network is completely developed, it is necessary to update the BIM model with the geometric and non-geometric characteristics of the devices included in the network. In the first case, of geometrical characteristics, devices have been modelled as rectangular prisms using a simplified geometric representation; in this case, characteristics related to the bounding box, location, dimensions and orientation are needed. A more detailed elaboration of the automation components' geometry is also possible, but this has no impact to our study.

Replicating expert knowledge, the geometrical placement of each device in the model is done following a set of rules: sensor, controller and actuator can be placed in 1) a wall in the same space as the HVAC system is placed in the model, or 2) in the ceiling in which the HVAC system is placed, or 3) sensors are placed in the same space as the HVAC system and the remaining devices in alternative locations such as a mechanical room or an electric cabinet. In all cases the BIM model is interrogated to identify the walls or ceilings where the devices can be potentially placed. Other non-geometrical characteristics regarding vendor, device type (temperature sensor, controller, actuator) unique identification number, electrical needs, price, and so forth, are also included in the model of the automation network when the devices are selected. As a proof of concept, a tool implementing the proposed methodology has been developed. The user through a simple interface can setup or change configurations of the control network, see historical data, and find the inter-connected devices. Tests in building case studies demonstrate the relevance of both strands of the proposed work: (i) in the automated creation of the Building Automation Control Network—which significantly improves the commissioning process of the BACN network; and, (ii) in the automated updating of the BIM model, addressing the issue of not having an up-to-date model updated after the control network has been developed. By way of example, a case study is presented in this paper, whereby the automation network has been developed for temperature regulation of each space in a building, assuming an that building is serviced by individual split-type air conditioning systems. This can be expanded to other types of control, to include lighting and indoor air-quality aspects.

## 2 IFC4 AND LONWORKS

### 2.1 *Building controls domain in IFC4*

The Industry Foundation Classes data schema have been in development for more than 20 years and this ambitious project under the oversight of BuildingSmart has expended and improved over the various iterations of the schema. Now an ISO standard (ISO 16739) IFC is widely accepted as the *de facto* openBIM standard. In view of the above, the use of IFC is particularly appealing due to its open nature. Moreover, as stated in the standard the IFC development aspires to capture aspects along the building life cycle to include beyond design, operation and maintenance. Building services and facility management are explicitly within the scope of the definition, and particularly relevant for our purposes.

During this evolution of IFC, the schema has expanded to include definitions of new elements (entities), systems (domains). The latest version of IFC, IFC4 Add2 includes timely definitions related to the representation of building control systems. In the latest version new entities have been added or expanded in the Buildings Control Domain (*IfcBuildingControlDomain* schema with forms part of the Domain Layer of the IFC model) to capture information on pertinent elements (sensors, controllers and actuators) using the IfcSensor, IfcController and IfcActuator entities. The concepts of building automation, control, instrumentation and alarm are defined to support ideas including types and occurrences of: actuator, alarm, controller, sensor, flow instrument, unitary control element. Components that physically perform the control action such as valves and dampers are subtypes of distribution flow elements located within the *IfcHvacDomain* and *IfcElectricalDomain* schemas.

### 2.2 *LonWorks control networks*

LonWorks networks are based on the use of the LonTalk protocol which is an open standard developed for LonMark mainly used for Building Automation. LonWorks is based on a hierarchical level architecture, see Figure 1. (Echelon Corporation 1997a, 1997b, 1999)

The Object Server object contains a collection of Network objects, each of which represents a defined network. Each Network object contains

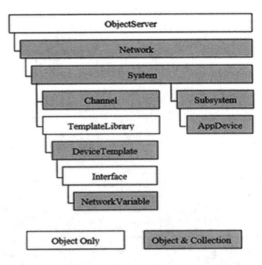

Figure 1. Hierarchy of the LonWorks.

a System object representing the network system, and each System object contains a set of Subsystem objects that represent logical or physical partitions of that particular network. Each Subsystem object contains a collection of AppDevice objects which represent the application devices defined in that subsystem. Thus, LonTalk protocol provides a common applications framework that ensures interoperability using powerful concepts called network variables and Standard Network Variable Types (SNVTs). Functional device models have been developed by the Lonmark Interoperability Association to assure plug and play compatibility.

Communication between nodes on a network takes place using the network variables that are defined in each node. The product developer defines the network variables when the application program is created as part of the Application layer of the protocol. Network variables are shared by multiple nodes. Some nodes may send a network variable while others may receive. By only allowing links between inputs and outputs of the same type, network variables enforce an object-oriented approach to product development. This greatly simplifies the process of developing and managing distributed systems. In addition, Network Variables are part of Functional Blocks which are used as the device interface including also configuration properties that are used to perform a determined task.

## 3 MAPPING BETWEEN IFC AND LONWORKS DATA MODEL

### 3.1 IFC and LonWorks matching

Connection among control devices in IFC4 is based on the idea that the output of a Sensor is connected to an input of a Controller, which can have outputs connected to other controller or connected to an actuator as it is shown in Figure 2. IFC4 also defines several types of sensors, depending the physical characteristics of the measured signal, controllers, depending on whether they are programmable, proportional, etc and actuator such as electric, hydraulic, pneumatic, etc.

Although there are several physical connection topologies for control networks using the LonWorks protocol (ring, bus, star, mixed) the fundamental idea of connecting between devices (for logical connection and interchange of information in the network) follows the sensor-controller-actuator paradigm even though the link does not necessarily have to be one-to-one.

As shown, both data models, IFC and LonWorks are pretty similar, allowing the user to make the assumption of using LonWorks Network Variables as Ifc Outputs and Ifc Inputs and LonWorks Devices through Functional Blocks as IFC Devices.

## 4 BUILDING AUTOMATION CONTROL NETWORK GENERATION AND IFC UPDATING

### 4.1 Automated generation of the building automation control network

As explained in the introduction, a software tool/application has been developed to use on this work, also other previous developed tools such as an exporter from LonWorks to SQL database, and several SQL databases which support the storage of historical (dynamic database) data and architecture of LonWorks Control Network (static database). The static database is an exact copy of the structure of the LonWorks network including all devices and connections between them. This database is deployed to have a possibility to access this configuration from any other software module, and it is performed once the Building Automation Control Network is totally functional and deployed. A software module reads information from a xml file (exported using LONMAKER software) and automatically creates the overall structure of the database, relationships between tables, etc. On these tables are also stored all the connection between Functional Blocks and Network Variables of the control network Devices.

In case of automatic development of the Building Automation Control Network the application follows the sequence that is shown in the following Figure 4.

Starting from a building model (IFC4 file), the tool uses information contained in the model to identify spaces inside the building. Once the

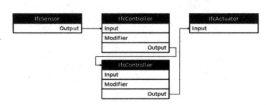

Figure 2. Control linking paradigm in ifc.

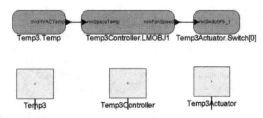

Figure 3. Basic LonWorks control connection.

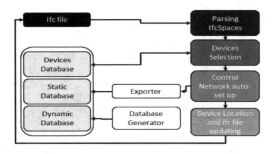

Figure 4. Flowchart on generation of BACN.

division into spaces is done, next step is Device Selection, this is carried out using a Device Database (a product catalogue) in which many different devices from several vendors, applications, and types are stored. This catalogue is also used to store price data for each device, as well as, the number and type of functional blocks. The selection is then performed using different criteria to cover the needs of the heating system (in this approach based on individual split-type air conditioning system for each space). At this point the control network set-up is automatically done, linking Functional Blocks of devices through Network Variables (SNVT's). ). An example of this linking can be seen in Figure 3. In this Figure an example is shown of the Temperature sensor (Temp3), temperature controller (Temp3Controller) and actuator for the split system (Temp3 Actuator) which in this case is an ON/OFF switch. Once selection of devices and the links between them have been established the LonWorks Building Automation Control Network definition can be exported to a formatted xml file. This can in turn be imported to the commissioning and integration tool LonMaker for deployment of the configurations to the physical devices. In the first stage, the information about unique identification number (NeuronID in Lonworks) is fulfilled with a "dummy" number due to this number is not known until the devices are purchased, then the Lonworks control network needs to be updated (by hand) with this information to be entirely operative.

### 4.1.1 Device location and ifc file updating

In IFC4 the placement of objects is achieved using IfcObjectPlacement which is an abstract super type of placement subtypes that define the location of an item, or an entire shape representation, and provide its orientation. Any IfcObject has an IfcLocalPlacement that can be related to other/others IfcLocalPlacement and a RelativePlacement which configures the local coordinates system using IfcAxisToPlacement3D. IfcAxis2Placement3D provides location and orientations

to place items in a three-dimensional space. The attribute Axis defines the Z direction, RefDirection the X direction. The Y direction is derived.

Using a coordinate transformation matrix, each position of any of the IfcObject is calculated using "Bounding Box" properties. As result coordinates for ($X_{min}$, $Y_{min}$, $Z_{min}$) and ($X_{max}$, $Y_{max}$, $Z_{max}$) of each component (IfcSpatialStructure, IfcWall, IfcObject, etc.) can be obtained. Depending on the results of the mapping between the local and global coordinate systems four different cases are possible regarding numerical value of $X_{min}$ and $Y_{min}$. It is supposed to have the origin reference the same as the global origin reference.

To select the correct placement for the devices, the relative location and orientation between spaces and walls needs to be checked due to several possibilities existing depending upon the orientation of space and walls. Knowing if a wall is part of a space the easiest method is comparing if any of the lines R1, R2, R3 and R4 are common between wall and space. Implicit to this assumption is that we have a good quality model, so model checking, and proper guidelines are an essential pre-requisite.

The schematic above shows how in IFC4 spatial containment is generated to include the devices of the control network in spatial contexts. According to IFC, structural elements such as walls may not be directly related to spaces. This relationship can be inferred using geometric properties and coordinates in space. To identify whether a wall surrounds or is part of a space, in addition to using the criterion explained above in which some of the lines part of their "Bounding Box" are common it is necessary to know if the wall is part of the space and not part of

Figure 5. IFC4 placement for an object.

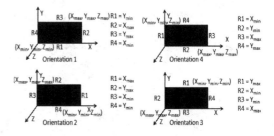

Figure 6. Orientation depending on coordinates.

the prolongation of it. Wall and space are projected on the vertical and horizontal axes and then the tool decides if the wall is surrounding the space by comparing the coordinates of both.

For the example of the previous figure when projecting in X axes; Space1 $X_{min}$ < Wall 1 $X_{min}$ and Space 1 $X_{max}$ >= Wall 1 $X_{max}$ and Wall 1 $X_{min}$, then Wall 1 surrounds space 1. Wall 2 $X_{min}$ and Wall 2 $X_{max}$ < (Space 1 $X_{min}$ and Space 1 $X_{max}$). In this case wall 2 does not surround Space 1. In this case wall 2 does not surround Space 1. As result of the projection and using the common lines criteria, the application makes a list including all possible walls where the devices of the automation control network can be placed. It is necessary to note here that this work is developed for square-based layout buildings on which spaces are defined as rectangular areas, in case the layout is other, the tool needs to be updated to include new capabilities. Now is necessary to know if the wall has openings (doors or windows) and, in this case, delete them from the list of possible locations. When the selection of the location is finished (right now the tool selects the first wall of the non-opening containing walls) the model of the building is updated including the information of the selected devices such as name, type of sensor, controller, actuator, graphical characteristics, etc. (in this work, both, sensors, controllers, actuators have been modelled as rectangular polygons, due to the graphical characteristics not being considered as critical at this stage).

Tests have been done using the next building model (IFC4 file). This model has several spaces, considered as individual zones, shown in Figure 10 below.

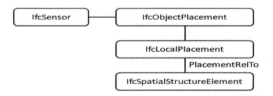

Figure 7. Relation between ifc control device and ifc spatial structures.

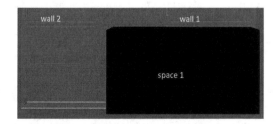

Figure 8. Example of spaces surrounded by walls.

Figure 9. Building model used for tests.

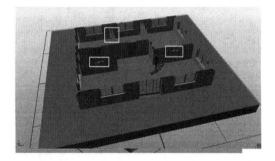

Figure 10. IFC4 placement for sensor-controller-actuator.

Following application of the tool developed and described in this paper, an enhanced IFC4 file is generated whereby the control network characteristics are included in the model and devices are located in a proper position (white squares are surrounding devices).

It should be admitted that the spatial placement of these devices might have little correspondence with their placement in reality. The procedure just described encapsulates expert knowledge regarding the placement of the sensors at various locations. If more precision on the placement in the model is required, then using and IFC editor (capable of handling placement of IfcSensor objects) can be used. But the model thus generated has value in relation to queries like number and types of sensors in a specific space, sensors connected to a specific building service (e.g. split-type air conditioning), controller linked to a conditioning of a space, and so on. All this information is particularly relevant in the context of hard facility management and therefore there is great value to the information even if the placement is approximate. Obviously, as mentioned before, manual and automated ways to establish a correct sensor placement can be established, but this is beyond the scope of this work.

## 5 IFC4 FOR BUILDING CONTROL

When the Building Automation Control Network has been defined, included in the IFC4 file and physically deployed we can use the application to monitor and control the building facilities—the flowchart in Figure 11 illustrates how this is accomplished.

First the spaces contained in the building are obtained from the model and, for each space, sensors contained are identified (in this case temperature). The remaining devices of the automation and control network and their connectivity (in terms of communication flow) are queried from the "Static Database" which maintains an accurate representation of the control network, information about devices, network variables, Functional Block and how all these are inter-connected (called "Targets" in LonWorks terms).

When this information has been retrieved the user can access historical data for any of the variables in the network (values are stored in the dynamic database which stores time-series data), change reference values, switch on/off devices, etc. To access these data and store them, the application uses a web—services based interface, which is deployed in a device of the control network actuating as server and network interface, then the client retrieves data and store them in the dynamic database which is composed by three tables which include information on name of signal, value, and date time log between others. Thus, the user can control the facilities of the building directly using a simple interface (application) and the BIM model of the building through its IFC4 file.

By selecting one of the devices (sensor, controller, actuator) we can access to other window which allows the user to change any reference on the network variables.

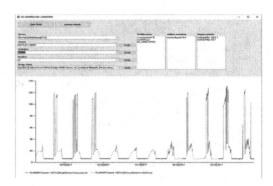

Figure 12. Historical data of a variable.

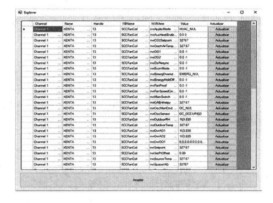

Figure 13. Variables changing window.

These read and write values of variables on the Network and their instant values are accessed using the capability of the network of communicating with other systems using Web Services (Client/Server based).

## 6 FUTURE WORK

Once the proof of concept of the methodology, the first step in the development of the tool is clearly the addition of some capabilities that increase the utility such as automatic parsing of devices (modeled in IFC4), extracting geometrical and structural characteristics, in order to have a library of components more extended including the exact graphical representation, avoiding rectangular prism and having as result an exact model of the building and facilities.

Integrate the information of the Building Automation Control Network in a 3D viewer with adding to it the capabilities of control, historical data representation, alarming, etc. making more attractive to customers and systems integrators

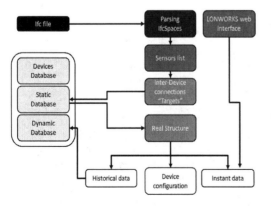

Figure 11. Flowchart for direct control from Ifc4 files.

to implement this solution in their products as an interface to the BEMS. This needs to be completed by extending the work to other Building Automation Control Network protocols (BACNET, etc) to increase the interoperability and standardization of the tool.

Expand the use cases to other more complex automatic systems for heating and cooling and other facilities as well as make the tool smarter in the process of device selection and design of the network, adding other criteria i.e. length of cabling, number and type of Functional Blocks (in case there are devices with several Functional Blocks for different operations), etc.

Improve the placement of devices procedure, adding possibilities such as technical rooms, electrical cabinets, etc. and different morphologies for building spaces (Manhattan-World-type structures).

## 7 CONCLUSIONS

The integration of Building Automation Control Networks in the BIM through IFC4 is possible in an automated way firstly by developing the network in a BACN-specific data model and secondly by including devices and their geometric location in the IFC4 file.

The possibility to control from the BIM model once IFC4 format includes elements relating to the network is quite clear.

Relations between devices can be found automatically, thus knowing from a user-friendly interface, regardless of where and how the control network has been developed.

Control of the facilities without knowing the network topology of the automation network, which offers the possibility to error detection of devices and alarming.

As long as all the information of the Building Automation Control Network can be accessed, devices can be set. It is also possible the representation of historical and trends, analysis of the behaviour of the building, etc.

Benefit from both the architectural part and the part of engineering for the control, since that is not two separate developments in which both have that design and integrate the network in their corresponding platform, but it saves much time on these tasks since that occurs automatically and completely transparent.

Building model (IFC4 based) is updated after selecting devices including the Building Automation Control Network information.

This work is based on the use of LonWorks control networks but it is expandable to other protocols and architectures such as BACNET, MODBUS, etc.

## ACKNOWLEDGEMENTS

Part of the work presented in this paper is based on a research developed in an internship at the Institute of Environmental Design and Engineering, University College London (UCL), London. This internship has received funding from the Agencia de Innovación, Financiación e Internacionalización Empresarial de Castilla y León.

## REFERENCES

BRICK Schema. 2016. Brick: A Uniform Metadata Schema for Buildings. Available online: https://brick-schema.org/ (accessed: May 2018).
BuildingSMART. 2018. IFC4 Release. Available online: http://www.buildingsmart-tech.org/specifications/ifc-releases/ifc4-release (accessed: January 2018).
Echelon Corporation, Copyright © 1997–2011. Lonworks Network XML Programmer's Guide.
Echelon Corporation, Copyright © 1997–2009. i.LON ® SmartServer 2.0 Programming Tool User's Guide.
Echelon Corporation, Copyright © 1999. Introduction to the Lonworks System.
ETSI. 2015. SAREF: the Smart Appliances REFerence ontology. Available online: http://ontology.tno.nl/saref/ (accessed: May 2018).
Karavan A., Neugebauer M, Kabitzsch K. Mergin Building Automation Network Design and IFC2x Construction Projects. ITC Digital Library 2003.
Karavan A. Neugebauer M, Kabitzsch K, Integration of Building Automation Network Design and 3D Construction Tools by IFC Standard. Proc. FeT2005 6th IFAC International Conference on Fieldbus Systems and their applications, Puebla, Mexico, November 2005, pp 247–254.

# Design-to-design exchange of bridge models using IFC: A case study with Revit and Allplan

M. Trzeciak & A. Borrmann
*Chair of Computational Modeling and Simulation, Technical University of Munich, Germany*

ABSTRACT: This paper presents a case study of exchanging bridge models between two commercial modeling applications using the vendor-neutral format Industry Foundation Classes (IFC). Particular emphasis is put on investigating the modifiability of the received geometry, a major aspect of design-to-design exchange scenario. The case study is based on two BIM models of a bridge at two different design stages provided by professional engineering consultancies. The paper describes the current IFC configuration options in the analyzed BIM design tools and the related mapping mechanisms. Next, the executed case study is presented, including the preparation of the BIM models for exchange, configuration of the IFC interfaces and mappings, and recognized geometric modifiability cases of imported building elements. Due to lacking support, the IFC 4 Design Transfer View cannot be used, yet. However, in the Revit-to-Allplan design-to-design exchange scenario, the IFC $2 \times 3$ Coordination View 2.0 serves as a suitable fallback solution. On the other hand, the exchange in the opposite direction (Allplan-to-Revit) does not seem viable for now. In both cases, the coordination scenario using the coordination views IFC $2 \times 3$ CV 2.0 is realistic, resulting however in limited modifiability of the received geometry.

## 1 INTRODUCTION

### 1.1 Transition period towards BIM

The AEC industry is on its way towards adopting Building Information Modeling (BIM) thanks to the involved increase in efficiency and quality in project execution (Eastman et al., 2011). The traditional drawing-based approaches are being gradually replaced by modern procedures based on information models.

To avoid economic damage caused by a too rapid implementation of BIM at a national scale, a transition period is usually planned so that the necessary standards, guidelines, and contract templates can be established (Borrmann et al., 2016). The prominent example is the United Kingdom, which made the use of BIM mandatory starting from 2016 (Cabinet Office, 2011). Among other countries, where government-driven initiatives have been planned, Germany is in the process of the step-wise introduction of BIM methods in public infrastructure projects in response to the governmental BIM road map named "BIM-Stufenplan" set up by the Ministry of Transport (BMVI, 2015). According to the roadmap, BIM in planning and construction is supposed to be the standard method for all federal infrastructure projects by 2020. For data handover to the public client, only standardized, vendor-neutral data formats are supposed to be applied.

### 1.2 BIM4INFRA 2020

BIM4INFRA ("BIM4INFRA2020" n.d.) is a project aimed at developing and supporting the implementation of Building Information Modeling in infrastructure projects in Germany.

As part of the BIM roadmap, a number infrastructural pilot projects are being executed across Germany. They are supposed to gradually introduce the necessary practical experience to the diverse road agencies, engineering consultancies, construction companies, and other parties. In one of the pilot projects (Highway A99, Bridge 27-1), the conceptual design for a bridge was developed and the corresponding BIM model was created using Revit (Autodesk). In the next project phase, the detailed design was prepared on the basis of the conceptual design using Allplan (a Nemetschek Company). This a typical design-to-design exchange scenario which requires the modifiability of the exchanged geometry. To transport the model information between the two design applications, the vendor-neutral data format Industry Foundation Classes (IFC) was used.

In the context of the project, a detailed analysis of the implementation of the IFC standard in both Revit and Allplan was performed, focusing on the modifiability of geometry after the exchange of professional BIM models. To this end, models were exchanged in both directions. Additionally, IFC configuration options of these BIM design applications and the related mapping mechanisms have been analyzed in detail. The insights gained are documented in this paper.

### 1.3 *Structure of the paper*

This paper is organized as follows. Section 2 provides background including the latest advances in IFC in the context of infrastructure projects, geometry, exchange scenarios, and Model View Definitions. Section 3 discusses the current state of the IFC configuration options of the analyzed BIM design tools and the related mapping mechanisms. Section 4 presents the executed case study, including the preparation of the BIM models, configuration of the IFC export and import functions and mappings, recognized geometric modifiability cases and the results. Finally, Section 5 discusses the presented results.

## 2 BACKGROUND

### 2.1 *Industry foundation classes*

Interoperability is the ability to exchange data between software applications, which eliminates the need to manually copy data already created by one application. In the construction industry, exchanging data is of paramount importance due to its high level of fragmentation. Data can be exchanged using Application Programming Interfaces, proprietary formats of software providers, or vendor-neutral formats (Eastman et al., 2011). The latter ones play the integral role in the construction industry because they allow to preserve fair competition on the software market and to prevent the market from vendor lock-in (Borrmann et al., 2018).

Maintained by buildingSMART, Industry Foundation Classes is the most prominent vendor-neutral, international standard for exchanging BIM data in the AEC industry. It is widely supported by BIM authoring and downstream tools. In a broader sense, IFC is a Product Data Model (often called "schema") which provides general definitions of objects to address all building information throughout the whole building lifecycle. In this paper, the reader will come across a term "IFC file" which is a population of the IFC schema. An IFC file follows the patterns and constraints stipulated by the schema and contains the actual instances of the IFC classes. Generally, such models are called populated data models, however, if their content is construction-specific, they are named Building Information Models.

The existing versions of IFC do not yet fully support the modeling of infrastructure assets, such as roads, bridges and tunnels. The respective extension is currently being developed in the frame of the buildingSMART Infrastructure Room (bSI, 2018a) and will results in the full support of these facilities, enabling the data exchange and open access in the context of planning, realization and maintenance of infrastructure facilities. However, a number of BIM use cases (e.g. visualization, coordination, quantity take-off) can already be well realized by means of the existing (and implemented) IFC versions $2 \times 3$ and 4.

### 2.2 *Geometry*

In Building Information Modeling, two different approaches can be distinguished in modeling volumetric bodies (so called Solid Modeling). The first one—Explicit Modeling, known as Boundary Representation, describes a body in terms of its bounding surfaces. The basic principle is that the bounding surfaces, called Faces, are described by Edges, and they, in turn, are depicted by Vertices. The whole system of relationships between them is denoted as the topology of the modeled body. An alternative approach to Solid Modeling—Implicit Modeling—is based on a sequence of construction steps to describe a resulting volume. The approach is also known as Procedural Method. The available construction operations include Constructive Solid Geometry (CSG) operations as well as sweeps and extrusions. CSG employs the predefined geometric primitives (such as cubes, cylinders and pyramids) and combines them using the Boolean operators (such as union, intersection, and difference). The output is a more complex body, which can be used in further construction steps. Another example is an extrusion which is defined by a planar profile extruded along a desired vector (Borrmann and Berkhan, 2018).

When it comes to data exchange, these two approaches result in fundamental differences. Boundary Representation is straightforward to process and visualize in the receiving software application. It is thus well suited for use cases such as visualization, coordination, and quantity take-off. Implicit models, in turn, require the receiving applications to precisely reproduce all the modeling steps, which may become challenging for more complex geometry. At the same time, implicit models much better support the modifiability of the geometry on the receiving side as all construction steps and their parameters are available. For example, a column can be represented by a circular profile which is extruded for a given distance (height). However, this approach makes the applications' export and import interfaces far more complex to implement (Borrmann and Berkhan, 2018). Both approaches, explicit and implicit modeling, are supported by the IFC data exchange format.

## 2.3 Exchange scenarios

In the construction industry, it has been recognized that interoperability between authoring tools should support the use cases defined by practice-based workflows resulting in precisely specified exchange scenarios. They are particularly helpful as the exporting application knows what is required (and also what is not required) and the receiving application knows what to expect (Eastman et al., 2011). Accordingly, only a sub-part of the full IFC schema has to be implemented.

One of the most demanding exchange scenarios is "Design-to-Design". It requires to transport the geometry in a way that allows its modifiability in the receiving application. In this respect, explicit representations which are based on representing the objects' surface by means of triangles for example, are of limited use. Instead, the use of implicit geometry is required, including sweeps and Boolean operations.

On the contrary, the exchange scenarios "Reference" or "Coordination" are fully supported by explicit geometry, as the geometry of the model exchanged is not subject to modifications. This means that the geometric representation of entities should be explicit so that the receiver can analyze and extract the necessary information. The scenario can be applied for clash detection between two domain specific BIM models, or for performing quantity take-off.

## 2.4 Model view definitions

In order to meet the requirements of different exchange scenarios, buildingSMART has developed the Model View Definitions (bSI, 2018b), which define the subset of the IFC schema required to be implemented by software vendors to support a given exchange scenario in terms of geometry and semantics. This significantly reduces the effort required for the software vendors and the users better understand what information they should exchange.

The IFC 4 Design Transfer View (DTV) refers to the "Design-to-Design" scenario, and IFC 4 Reference View (RV) is designed to meet the expectations of the "Coordination" exchange scenario. The previous version—IFC 2 × 3 – introduced the Coordination View 2.0 (CV 2.0), Structural Analysis View and Facility Management Handover View. For the time being, the IFC 2 × 3 Coordination View 2.0 remains the most widespread version. It is a default and certified version which is generally supported by other BIM authoring applications, including Allplan and Revit. Such a BIM model is not supposed to be re-editable by the receiving application. It includes the definition for spatial structure, building, and building service elements with shape representations, including both, parametric (implicit) shapes for a limited range of standard building elements, and the ability to also include non-parametric (explicit) shape for all other elements (bSI, 2018b).

## 3 ANALYSIS OF IFC INTERFACES

Besides the concepts of exchanging building information models using IFC, there remains a matter of the actual implementation of the IFC interfaces in the target software applications and their proper configuration so that the receiver can reuse the imported file in case of the DTV for example. This section discusses the configuration of the IFC interfaces in the two BIM design applications: Revit and Allplan. Figure 1 shows an example of an exchanged bored pile between Revit and Allplan using the IFC 2 × 3 Coordination View 2.0 and assigning different IFC classes to the same building element. On the left side, it can be seen that the imported pile is modifiable by a dialog box (1) when its geometric representation in the IFC file is an extrusion (IfcExtrudedAreaSolid object) – behavior similar to the native elements in BIM design applications; (2) shows that the pile cannot be modified even though it has the same geometric representation as in case (1); (3) presents an imported pile which—in principle—can be modified using control points placed on all its surfaces (imported from the IFC file as IfcFacetedBrep object). However, this is an impractical option because the diameter cannot be easily changed if needed. In addition, (4) shows a heavily tessellated pile (IfcFacetedBrep object) which can be modified by a large number of control points, however not in a meaningful manner. This example shows that the configuration of the IFC export interfaces has an influence on the imported elements in the target BIM authoring tools.

### 3.1 IFC interfaces

BIM design applications support data exchange via IFC. This means that they provide import and export modules, which translate the internal

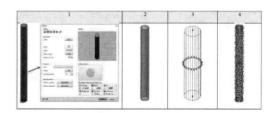

Figure 1. Different cases of geometric modifiability based on an imported bored pile.

proprietary models (geometry and semantics) into the IFC-compliant model and vice versa. An example of an IFC export configuration interface can be seen in Figure 2. The configuration interfaces offer to export project models using different MVDs. In addition, they allow the user to choose certain project-specific options such as export only specific types of elements, support of free-form geometry, level of tessellation and others.

### 3.2 Mappings

When a precise configuration of the IFC export is necessary, the options provided by the GUI are not sufficient for most real-world cases. The missing aspect relates to mappings of building elements and properties. Because of differences in data models of individual software products and the IFC schema, the users must specify how the building elements and properties must be translated into IFC-compliant objects. For the needs of the users, software vendors have developed template and mapping mechanisms to automate this process.

#### 3.2.1 Building elements

The receiver of an IFC file usually expects that all objects are of the correct IFC type—for example, a pile is of type IfcPile. An issue is the mapping of internal data types onto external IFC entities, as sometimes a 1:1 mapping does not exist. For example, IFC types might not exist in a BIM authoring tool or vice versa. In this case, the BIM authoring applications should provide the users with means for the configuration of how the proprietary classes of their data models are mapped to the respective IFC types. In case of building elements, a fallback solution is the use of IfcBuildingElementProxy—a generic type—which however can cause problems in the receiving applications as the object type is unknown and it is hard to find the correct internal type automatically.

In the analyzed BIM design applications, there are two levels of mappings related to how building elements are mapped to IFC classes. The first one—general, which maps internal types onto IFC classes (using a respective mapping file to which the user might have an access), and the other—specific, which maps only selected internal types or individual building components to desired IFC classes (usually using a software-specific attribute assigned to selected building elements). The settings in the specific level overwrite the ones in the general level. An example can be seen in Figure 3, where an element of internal type "cylinder" is mapped onto the IFC type "IfcPile" by means of assigning an "IFC Object type" attribute with the proper value.

#### 3.2.2 Properties

It is often required by clients to provide respective properties assigned to building elements (such as compressive strength of concrete or exposure class). Besides the static attributes defined in the IFC schema (e.g. Globally Unique Identifier), user-defined properties can be created and added to an IFC project model. Properties must be grouped into Property Sets and then assigned to building components. Because of differences in data models of individual software products and the IFC schema, the Property Sets created in specific BIM authoring applications must be translated so that they comply with the rules stipulated by the IFC schema. Accordingly, the software vendors have developed mapping mechanisms to automate this process. An example of such mapping mechanism for properties can be seen in Figure 4. The mapped properties correspond to the ones shown in Figure 3.

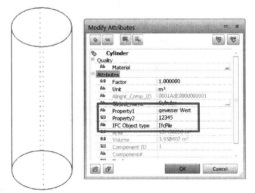

Figure 3. Allplan—example of a mapping of an internal Allplan type onto an IFC type.

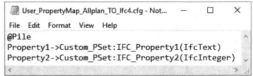

Figure 4. Property Set Configuration File for the custom properties in Allplan.

Figure 2. Allplan: Example of an IFC interface.

## 3.3 Geometric modifiability cases

This section discusses geometric modifiability cases, i.e. it specifies ways of how building elements imported from an IFC file can be geometrically modified in the respective BIM design applications (what is their "behavior" when the user tries to geometrically modify the imported elements).

In the design-to-design exchange scenario, the desired manner is that all imported building elements can be geometrically modified in a similar way to how they are modified in the original BIM design application from which they have been exported. In a bridge model, for example, a bored pile should be imported as a cylinder whose length and diameter are directly modifiable. For wing walls, in turn, it should be possible to change the thickness and the position of the corner points. It should be possible to add openings/voids to these elements.

## 4 CASE STUDY: EXCHANGE OF BRIDGE MODELS

### 4.1 Model exchange

In this case study, two different BIM modes are exchanged. The first one, created as a conceptual design in Revit, is exported to an IFC file and then imported into Allplan. The other one, created as a detailed design in Allplan is exported to an IFC file and then imported into Revit as presented in Figure 5.

### 4.2 Software versions

Since the actual implementation of the presented concepts in the BIM design applications is driving the outcome of this paper, their versions used in

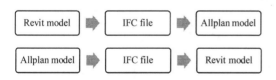

Figure 5. Two directions of the transfer of the BIM models in the executed case study.

Table 1. Software applications and their versions used in this paper.

| Software | Version |
| --- | --- |
| Allplan (A Nemetschek Company) | 2018-0-2 |
| Solibri Model Viewer | 9.8.17 |
| Revit (Autodesk) | 2018.2 |

this paper are important. Table 1 presents the BIM design applications and their versions used in this paper. It must be understood, however, that after finishing this paper, some improvements have been released which could possibly change the outcome presented here.

### 4.3 Model preparation

Because the analysis done in this paper is element-wise, the two project models need to be first preprocessed so that only selected building elements can be exported to an IFC file and then imported into the other BIM design application. Accordingly, such types of building elements as: (1) beams, (2) abutments, and (3) piles are distinguished and the further analysis is based on the individual building elements of these types (for each type of a building element separately) and not on the whole project models as shown in Figure 6 and Figure 7.

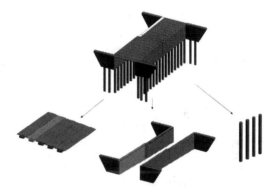

Figure 6. The Revit model is first preprocessed by distinguishing separete types of building elements.

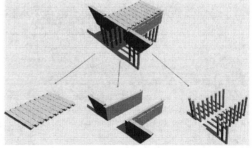

Figure 7. The Allplan model is preprocessed by distinguishing separete types of building elements.

Figure 8. Revit: Different IFC classes assigned to individual instances.

## 4.4 IFC export configuration

### 4.4.1 Mapping of building elements

The preprocessed models consisting of the same building elements (as described in section 4.3) are assigned with different IFC classes by means of the mapping mechanisms described in section 3.2.1 and exemplarily shown in Figure 8.

### 4.4.2 Settings of IFC export interface

The building elements prepared as described in the previous section are exported many times using different options provided by the IFC export interfaces of the BIM design applications. It must be understood, however, that these options (besides the MVD) are exclusive to BIM authoring applications and differ among them.

In the Revit-IFC-Allplan transfer, the following options are considered: (1) Revit models are exported to an IFC file using the IFC $2 \times 3$ Coordination View 2.0; (2) Mixed "Solid Model" representation (turned on and off); (3) Low level of tessellation. In the Allplan-IFC-Revit transfer, the models are exported with the following options provided by Allplan's IFC interface: (1) IFC $2 \times 3$ Coordination View 2.0; (2) IFC $2 \times 3$ data (a newer version of the Allplan IFC interface); (3) IFC 4.

## 4.5 IFC import configuration

In Revit, preliminary tests have been done in order to check if the mapping file which maps IFC classes to Revit categories while importing has any influence on the modifiability of imported building elements. Since the outcome of these tests is negative, the default Revit settings apply.

Because none of the IFC import options in Allplan refer to geometric representation of entities and the mappings between IFC classes and Allplan internal types are not explicitly accessible to the users, the default settings apply.

## 4.6 Intermediate checks

Before loading the IFC files into the BIM design application, all the exported building elements in these files are inspected using the Solibri Model Viewer in order to check their types of geometric representation.

## 4.7 Modifiability check in importing applications

The IFC files are loaded into the BIM design applications and the building elements are checked against their modifiability of geometry. The different modifiability cases can be seen in earlier presented Figure 1 and in Figure 9, Figure 10, Figure 11, Figure 12.

Figure 9. Revit: Imported wall and beam—unmodifiable.

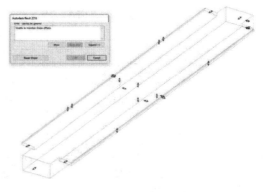

Figure 10. Revit: An imported beam—unmodifiable due to the inability to maintain its shape while any attempt of geometric modification.

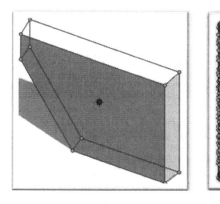

Figure 11. Allplan: Imported wing wall and a tessellated column modifiable by control points—the tessellated column represents the undesired case.

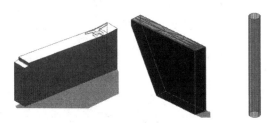

Figure 12. Imported unmodifiable elements (dump objects)—undesired cases.

### 4.8 Result Presentation

The outcome of the exchanges of building elements between Revit and Allplan using IFC can be seen in Table 2 and Table 3. More details regarding the geometric modeling of the building elements (reference to the second column in the tables) can be found in the Technical Report (Trzeciak and Borrmann, 2018).

## 5 DISCUSSION

For the analyzed models, the desired design-to-design exchange using IFC4 Design Transfer View which allows geometric modifications of a model after its hand-over to another design application could not be tested. A number of reasons contribute to this. In the Revit-to Allplan exchange, the problem is that Allplan (version 2018–0-2) does not support the import of IFC4. In the Allplan-to-Revit exchange, Allplan does not offer to specify a model view definition during the IFC 4 export and the geometry of the entities is translated into explicit forms.

However, as a workaround, in the Revit-to-Allplan design-to-design exchange, the IFC 2 × 3 Coordination View 2.0 can be used, although CV 2.0 is not intended to support the design-to-design scenario. This ex change has shown that all the elements imported into Allplan could be made

Table 2, The outcome of the transfer of a BIM model from Revit to Allplan using IFC.

| Revit | | | IFC File | | Solibri | Allplan | | |
|---|---|---|---|---|---|---|---|---|
| Building element | Geometry internal type | IfcExportAs | Geometry class | Geometry type | Component type | Modifiability case | Geometry remarks |
| Pile | Extrusion | IfcBuilding-ElementProxy | IfcExtruded-AreaSolid | Extrusion | Smart Symbol | Unmodifiable | – |
| | | IfcSlab | | | Slab | By dialog box | – |
| | | IfcPile | | | Smart Symbol | Unmodifiable | – |
| | | IfcColumn | | | Column | By dialog box | – |
| | | IfcBeam | IfcFaceted-Brep | Brep | User-def. arch. Elem. | By control points | Triangulated |
| Abutment | "Free-form" element | IfcBuilding-ElementProxy | IfcFaceted-Brep | Brep | Smart Symbol | Unmodifiable | Some faces triangulates |
| | | IfcWall | | | User-def. arch. Elem. | By control points | |
| | | IfcColumn | | | | | |
| Beam | Combination of extrusion and sweep | IfcBuilding-ElementProxy | IfcFaceted-Brep | Brep | Smart Symbol | Unmodifiable | A face triangulated |
| | | IfcBeam | | | User-def. arch. Elem. | By control points | |
| | | IfcSlab | | | Smart Symbol | Unmodifiable | |
| | | IfcColumn | | | User-def. arch. Elem. | By control points | |
| | | IfcWall | | | | | |

Table 3. The outcome of the transfer of a BIM model from Allplan to Revit using IFC.

| Allplan | | | IFC File | Solibri | Revit | | |
|---|---|---|---|---|---|---|---|
| Building element | Geometry internal type | IFC Object type | Geometry class | Geometry type | Component Type | Modifiability case | Geometry Remarks |
| Pile | Cylinder | IfcBuildingElementProxy | IfcFacetedBrep | Brep | Generic Models | By control points | – |
| | | IfcSlab | | | Slab | | – |
| | | IfcPile | | | Structural Foundation | | – |
| | | IfcColumn | | | Not found | – | – |
| | | IfcBeam | | | Structural Framing | By control points | Triangulated |
| Abutment | General 3D object | IfcBuildingElementProxy | IfcFacetedBrep | Brep | Generic Models | Unmodifiable | Triangulated |
| | | IfcWall | | | Walls | | |
| | | IfcColumn | | | Not found | – | – |
| Beam | General 3D object | IfcBuildingElementProxy | IfcFacetedBrep | Brep | Generic Models | Unmodifiable | Triangulated |
| | | IfcBeam | | | Structural Framing | | |
| | | IfcSlab | | | Floors/Generic Models | | |
| | | IfcColumn | | | Not found | – | – |
| | | IfcWall | | | Walls | Unmodifiable | Triangulated |

modifiable (via control points) by exporting them with a proper component type even though their geometric representation is Brep. This means that Allplan converts the imported Brep elements into internal objects which are geometrically modifiable by control points, which makes it possible to change the height, however not the diameter of the pile. The problem arises when the imported elements are (heavily) triangulated, which makes the geometric modifiability way harder.

The exchange in the opposite direction (Allplan-to-Revit) is possible using both IFC 4 or IFC 2 × 3 as Allplan supports exporting its model to both IFC 4 and 2 × 3 versions and Revit supports their import. However, the Model View Definition for the export using IFC 4 in Allplan is unknown. This exchange has shown that geometric modifications of imported elements in Revit seem to depend solely on the type of geometric representation of entities (and not on their component types as it is the case for imported elements in Allplan). The piles modeled in Allplan as cylinders have proved to be the only objects which could be modifiable by control points in Revit even though their geometric representations were Breps. The other elements could not be modified because either Revit did not provide this possibility (in case of tessellated elements) nor they were able to maintain their shape while any attempt of geometric modification (in case of non-tessellated beams).

The "coordination" exchange scenario (it was not the main subject of this case study though) using IFC 2 × 3 CV 2.0 seems to be fully viable between the analyzed software applications, as they both provide means for properly referencing building elements and accessing their properties.

BIM authoring tools are in constant development and the current state of their IFC interfaces presented in this paper can change in the near future. After finishing this paper, Allplan released a new version of the IFC4 import module, which could very likely change the outcomes presented here.

6 SUMMARY

This paper presents a case study of exchanging BIM models between Allplan and Revit, in particular emphasizing the limitations of the modifiability of geometry. The case study is based on two BIM models of a bridge at two different design stages provided by professional engineering consultancies.

For the time being, the IFC 4 Design Transfer View cannot be realized between these software application. Instead, in the Revit-to-Allplan design-to-design exchange, the IFC 2 × 3 Coordination View 2.0 can serve as a fallback solution for now. The exchange in the opposite direction

(Allplan-to-Revit) does not seem viable for now. In both cases, the coordination exchange scenario using the mentioned IFC 2 × 3 CV 2.0 is realistic.

Since the BIM authoring applications are constantly upgraded, the results of the executed exchanges in this paper might very likely change in the near future. Another important aspect is the availability of the Bridge extension of IFC which is expected to be released in early 2019. It will provide additional entities and geometry representation tailored to the needs of bridge model exchanges.

## REFERENCES

"BIM4INFRA2020." n.d. Accessed February 23, 2018. http://bim4infra.de/.

BMVI. 2015. "Digitales Planen Und Bauen – Stufenplan Zur Einführung von Building Information Modeling (BIM)." 2015. http://www.bmvi.de/SharedDocs/DE/Artikel/DG/digitales-bauen.html.

Borrmann, A., J. Beetz, Ch. Koch, and T. Liebich. 2018. "Industry Foundation Classes – A Standardized Data Model for the Vendor-Neutral Exchange of Digital Building Models." In *Building Information Modeling, Technological Foundations and Industry Practice*, 81–228. Springer.

Borrmann, A., and V. Berkhan. 2018. "Principles of Geometric Modeling." In *Building Information Modeling, Technological Foundations and Industry Practice*, 27–41. Springer.

Borrmann, A., M. Hochmuth, M. König, T. Liebich, and D. Singer. 2016. "Germany's Governmental BIM Initiative – Assessing the Performance of the BIM Pilot Projects." In *16th International Conference on Computing in Civil and Building Engineering*. Osaka.

bSI. 2018a. "Infrastructure Room." 2018. www.buildingsmart.org/standards/rooms-and-groups/infrastructure-room/.

———. 2018b. "Model View Definition Summary." 2018. http://www.buildingsmart-tech.org/specifications/ifc-view-definition.

Cabinet Office. 2011. "Government Construction Strategy." 2011.

Eastman, Ch., P. Teicholz, R. Sacks, and K. Liston. 2011. *BIM Handbook: A Guide to Building Information Modeling for Owners, Managers, Designers, Engineers, and Contractors*. 2nd Editio. New Jersey: John Wiley & Sons, Inc.

Trzeciak, M., and A. Borrmann. 2018. "Technical Report: Model Exchange between Revit and Allplan Using IFC: A Case Study for a Bridge Model." Munich. https://www.cms.bgu.tum.de/publications/reports/2018_ModelExchangeBetweenRevitAndAllplanUsingIFC_BIM4INFRA.pdf.

*IoT, sensor and industrialized production*

# Modeling construction equipment in 4D simulation

R. Amrollahibuki
*Department of Building, Civil and Environmental Engineering, Concordia University, Montreal, Quebec, Canada*

A. Hammad
*Concordia Institute for Information Systems Engineering, Concordia University, Montreal, Quebec, Canada*

ABSTRACT: Construction equipment constitutes an important portion of the resources used on construction sites. Every year, many construction workers lose their lives at work due to being struck by equipment or caught between objects. In the construction planning phase, equipment movements and workers' performance can be simulated in a 4D model with a high Level of Detail (LOD) to analyze the safety, productivity, and constructability of the construction project. This type of high 4D-LOD modeling requires a very detailed schedule, which can capture micro-tasks (e.g. the swing movement of the boom of a crane). Although many research works discussed the possibility of improving the safety of construction sites using sensors and analyzing the equipment workspaces, these works cannot be linked with available 4D Building Information Model (BIM) tools, which cannot easily represent equipment movement. The objective of this paper is to compare the available commercial and research platforms in terms of visualizing, animating, and simulating equipment movements, and to identify the potential improvements.

*Keywords*: 4D, equipment, BIM, LOD, construction

## 1 INTRODUCTION

Construction equipment constitutes an important portion of the resources used on construction sites. Every year, many construction workers lose their lives at work due to being struck by equipment or caught between objects. 4D simulation is developed by linking a 3D Building Information Model (BIM) and the corresponding activities in the schedule (Getuli et al., 2016). In the construction planning phase, equipment movements and workers' performance can be simulated in a 4D model with a high Level of Detail (LOD) to analyze the safety, productivity, and constructability of construction projects. This type of high 4D-LOD modeling requires a very detailed schedule, which can capture micro-tasks (e.g. the swing movement of the boom of a crane). Although many research works discussed the possibility of improving the safety of construction sites using sensors and analyzing the equipment workspaces, these works cannot be linked with available 4D BIM tools, which cannot easily represent equipment movement.

Five LODs are defined for 3D BIM models (BIMForum, 2017; Biljecki et al., 2016): LOD 100, LOD 200, LOD 300, LOD 350, and LOD 400. LODs for schedules are classified into 5 levels (Stephenson, 2007). However, there is no specific definition for 4D-LOD. Guevremont and Hammad (2018) proposed a new guideline for defining 4D-LODs at different phases of a project and considering the needs of different stakeholders. The highest 4D-LOD can be used for representing equipment tasks. Linking the detailed 3D BIM model with the micro-tasks executed by equipment (e.g. cranes) can lead to better simulation of construction activities (Akhavian & Behzadan, 2015).

4D simulation can be used to analyze safety measures in construction sites (Hammad et al., 2012) and to automatically check safety rules (Zhang et al., 2011; Zhang et al., 2012). Furthermore, 4D simulation can be used to efficiently identify and manage conflicts (Akinci & Fischer, 2000).

4D simulation of articulated equipment can generate workspaces, which can be used to analyze safety and improve productivity. Akinci et al. (2002) categorized spaces into three main groups: macro-level spaces, micro-level spaces, and paths. The spaces for equipment, crew, protected, and hazard areas were included in micro-level spaces. Akinci and Fischer (2000) suggested a prototype system, 4D WorkPlanner, which automatically generates micro-level spaces. Some researchers focused on the automated workspace generation for crane (Tantisevi & Akinci, 2006). Vahdatikhaki and Hammad (2015) proposed an approach for producing dynamic equipment workspaces for increasing earthwork safety. In this approach, both the present state and position of the equipment and the speed of each object are considered. Choi et al. (2014) developed a 4D workspace planning framework for clash-detection considering the workspace of each activity. Shang and Shen (2016) developed a spatial-temporal workspace conflict matrix to detect conflicts on construction sites for site safety assessment considering hazard frequency and severity indices. They developed a method that visualized workspaces of equipment and workers, categorized into static and dynamic workspaces.

The objective of this paper is to compare the available commercial and research platforms in terms of visualizing, animating, and simulating equipment movements, and to identify the potential improvements. The structure of the paper is as follows: First, the requirements for modeling construction equipment are explained. Then, the available commercial and research platforms are compared with respect to their functions in modeling equipment. Finally, the conclusions are presented with suggestions for potential improvements of equipment modeling within BIM tools.

## 2 REQUIREMENTS FOR CONSTRUCTION EQUIPMENT MODELING

In order to have effective modeling of construction equipment, the BIM tools should satisfy the following requirements:

1. BIM compatibility: Compatibility between construction equipment modeling and BIM results in better visualizing the progress of the construction activities by simulating equipment movements and the changes in workspaces at the micro-schedule level.
2. Equipment library: BIM tools usually have a library of common objects, ranging from materials and furniture to equipment. However, most of the tools do not have a library of construction equipment. The vehicle objects, which are available in most of these libraries, lack the inverse kinematics feature. Ideally, the equipment library should include not only the set of equipment commonly used in construction projects (e.g. excavators, cranes, compactors, etc.) but also the specific types of the equipment based on different brands, sizes, etc. Some crane simulation tools (e.g. 3D Lift Plan, 2018) have a detailed library of cranes including the crane models and the load charts. Furthermore, the 3D model of the equipment should be as detailed as possible. It should be noted that equipment is not considered in the current version of IFC, except as an abstract type of IfcResource (2018).
3. Ability to model multiple Degrees of Freedom (DOFs): The DOFs of articulated equipment with moving parts define the pose of the equipment. As a result, to properly control the equipment movements, its DOFs should be easily controllable in the simulation tool. As shown in Figure 1, a typical boom lift (cherry picker) could be controlled through eight DOFs. Two or more DOFs might be combined at any point in time. However, the movements and rotations of some parts are constrained. For example, the boom lift's bucket should be kept horizontal.
4. Ability to model at high 4D-LOD: With the purpose of full analysis of construction equipment activities, 4D-LOD for equipment simulation should be able to link micro-tasks of the movement of the equipment components with the detailed schedule of these tasks (e.g. the swinging of the boom of a crane). Therefore, tasks related to equipment should be broken down to the highest LOD to be able to efficiently manage the performance. Modeling and visualizing equipment operation can be inaccurate and unrealistic if the detailed LOD lacks all the DOFs (Akhavian & Behzadan, 2015).
5. Path and motion planning: Path and motion planning of the construction equipment is an important part of 4D construction simulation. This planning has two parts: planning the relocation path of equipment (e.g. relocation of an excavator between two locations on the site), and the planning of the movements of the parts of equipment (e.g. the boom swing, trolley and hook movements of a tower crane). Although the former planning may be available in some simulation tools, the later planning of the movements of equipment parts is still done manually

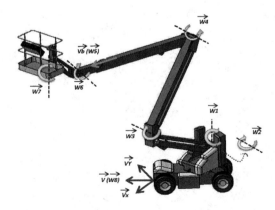

Figure 1. DOFs of a cherry picker (Model of the cherry picker is retrieved from (Graydon, 2016)).

in these tools, which is time-consuming. Also, it is important to be able to simultaneously plan the movements of several parts of the equipment and to support the motion planning of complex repetitive tasks (e.g. for cranes and excavators).

6. Collision avoidance: Congested site conditions often result in poor safety and productivity performance (Zhang et al., 2015). Congestion causes objects to move slowly, which leads to productivity reduction. On the other hand, the probability of accidents, either between pieces of equipment or between workers and equipment, increases when pieces of equipment do not have enough spaces to do their tasks. If equipment conflicts can be identified in advance, productivity and safety would be enhanced through rescheduling. The same approach used for time-dependent clash detection in 4D simulation can be utilized for equipment conflict detection. The conflict detection and collision avoidance can be based on the actual 3D model of the equipment or based on its dynamic workspace, which can be generated automatically (Vahdatikhaki & Hammad, 2015).

7. Other requirements: It is important to set a specific speed for each equipment movement because this affects the cycle time and productivity. Moreover, equipment activities may take minutes or even seconds. Project managers can create an efficient plan if they model equipment on construction sites based on the accurate activity time. The physical behavior of objects, such as wind effects, gravity, and terrain following, have direct effects on equipment performance. Additionally, the ability to generate dynamic workspaces for each component leads to improving safety.

## 3 COMPARISON OF TOOLS

In this section, several modeling and simulation tools are compared based on equipment-related features, such as their capabilities to simultaneously move equipment components, visualize dynamic workspaces, detect equipment and workspace conflicts, animate complex repetitive tasks, etc. These features are compared for the following tools: (1) 3D animation software: 3DS Max and Lumion; (2) 4D BIM software: Synchro, Navisworks, and Fuzor; (3) Game engine: Unity; and (4) Equipment training simulators. In some cases, more than one tool should be used to fulfill the desired purpose. Table 1 shows the comparison of these tools.

### 3.1 3D Animation software

#### 3.1.1 3DS max

Since 3DS Max (Autodesk, 2018) is commonly used for producing animation in the game industry, it has a large number of options and a library of modifiers, facilitating the creation of objects with different types of motions. 3DS Max can produce objects with kinematics and equipment animation with workspace linked to it. In order to animate the detailed movements of equipment in 3DS Max, kinematics should be added to it. Figure 2 shows the steps for generating an animation of articulated construction equipment with a dynamic workspace.

#### 3.1.2 Lumion

Lumion (2018) is a powerful 3D visualization engine for architectural design. However, it can be used to visualize construction processes along with all the details of the construction site. Lumion supports path planning of equipment. However, it does not provide inverse kinematics for equipment parts. It can import keyframe equipment animations from any 3D modeling software, such as 3DS Max, Revit, AutoCAD, SketchUp, and ArchiCAD. Figure 3 shows a model of a boom lift and its workspace, which was animated in 3DS Max and then imported into Lumion via FBX file format.

### 3.2 4D BIM software

#### 3.2.1 Synchro

Synchro Professional (Bentley, 2018) is 4D BIM software for construction project management. It provides workspace generation, animation of equipment path planning, conflicts detection, and clash reports. This platform enables straight-

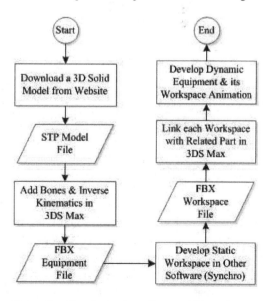

Figure 2. Flowchart of creating movable equipment and its workspace in 3DS Max.

Table 1. Comparison between purposed platforms.

| | 3D Animation | | 4D BIM | | | Game engine | Equipment training simulators |
|---|---|---|---|---|---|---|---|
| Requirements | 3DS Max | Lumion | Synchro | Navisworks | Fuzor | Unity | CM Labs |
| 1 BIM—Supported | × | × | ✓ | ✓ | ✓ | × | × |
| 2 Equipment Library | × | × | × | × | ✓ | × | ✓ |
| 3 Determining DoF | ✓ | × | × | × | ✓ | ✓ | ✓ |
| 4 Simultaneously Moving Parts | ✓ | Possible by imported animation | ✓ | ✓ | ✓ | ✓ | ✓ |
| 5 Path Planning | ✓ | ✓ | ✓ | ✓ | ✓ | ✓ | ✓ |
| Simultaneously Moving Parts and Equipment | ✓ | Possible by imported animation | × | × | ✓ | ✓ | ✓ |
| Complex Repetitive Tasks | Limited | Limited | × | × | ✓ | Possible by programming | ✓ |
| 6 Collision Avoidance | × | × | ✓ | ✓ | ✓ | Possible by pro-gramming | ✓ |
| 7 Setting Speed of Movements | ✓ | Possible in path planning | Possible in path planning | Possible in path planning | ✓ | Possible by programming and non-programming | ✓ |
| Time Scale (Days vs. Minutes) | × | Activities can be set in sequence but time frame is not available | Minute/days | Minute/days | Days | Possible by pro-gramming | ✓ |
| Physics (Weight, Wind, etc.) | × | × | × | × | Ability to turn on and off gravity/collision | ✓ | ✓ |
| Terrain Following | Limited | Limited | Only solid workspace | × | – | ✓ | ✓ |
| Dynamic Work Spaces | ✓ | Possible by imported animation | Only solid workspace | × | Import FBX & 3DS | ✓ | – |

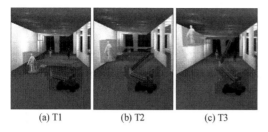

(a) T1　　　(b) T2　　　(c) T3

Figure 3. Several frames of animated boom lift and its workspace in Lumion.

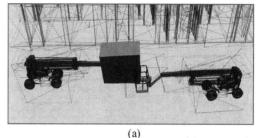

(a)

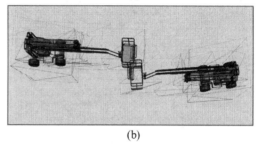

(b)

Figure 5. (a) Conflicts between equipment and workspace, (b) Clashes between two Equipment.

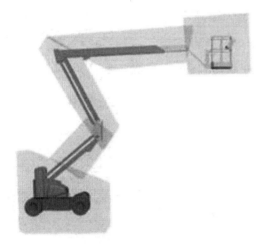

Figure 4. Workspaces created in Synchro.

forward creation of objects' workspaces statically as shown in Figure 4. Synchro can detect several types of soft and hard clashes: (1) conflicts between equipment workspaces and other types of workspaces; (2) conflicts between equipment and workspaces.

3.2.2 *Navisworks*

Navisworks Manage (Autodesk, 2018) is 4D BIM software widely utilized in the construction industry. It supports the animation of path planning (relocation), equipment/workspace collision avoidance, and conflict reports. This tool can be used to identify, inspect, and report the following construction clashes in a 3D model: (1) clashes between a couple of equipment, (2) conflicts between equipment and workspaces. Figure 5 shows two pieces of equipment moving toward each other. Their movements caused conflicts between both equipment/equipment and workspace/equipment. The 3D model of a boom lift is downloaded from a website (Graydon, 2016) in STP format, before importing it into Synchro software to add workspaces. After both the equipment and the workspaces around

them are prepared, the model is saved in FBX format and imported into Navisworks software. After creating the animation, a hard clash test is applied between two boom lifts and their workspaces.

3.2.3 *Fuzor*

Fuzor (Kalloc Studios, 2018) is the new generation of Virtual Design Construction (VDC) applications in the construction sector. It enables the user to walk around the project in virtual reality while observing the BIM data and information of every component in real-time. In the software environment, construction site logistics can be designed with simulated workers and articulated equipment, which leads to better visualization of the entire project. Fuzor VDC library is classified into three main categories: (1) Vehicles; (2) Foliage; (3) and Entourage. Vehicles are sub-classified into the standard library, construction vehicles and equipment, and temporary construction equipment. The vehicle category has about hundred construction equipment, providing a user-friendly interface for interactively modeling keyframed articulated equipment. Table 2 shows some important equipment available in construction equipment library.

As shown in Figure 6, a tower crane is animated while grabbing a beam from the location it should be loaded and releasing it at the actual location in the BIM model. This process is linked to the construction schedule and has an adjustable speed of simultaneous movements. As shown in Figure 6(a), the crane hook goes down to grab the beam and

Table 2. Equipment library in Fuzor.

| Cranes | Trucks |
|---|---|
| Tower Crane | Box Truck |
| Luffer Crane | Dump Truck |
| Mobile Crane | Drill Truck |
| Crawler Crane | Flatbed Truck |
| Derrick Crane | Liffting Truck |
| ECO Crane | Pickup Truck |
| Mini Crane | Semi-Dump Trailer |
| Rough Terrain Crane | Tipper Truck |
| Roof Crane | Trailer Truck |

| Excavators | Others |
|---|---|
| Backhoe | Road Roller |
| Compact Excavator | Bulldozer |
| Demolition Excavator | Transit Mixer |
| Electric Rope Shovel | Pile Driver |
| Loader | Hydrofraise |

| Lifts |
|---|
| Fork Lift |
| Telescopic Boom Lift |
| Scissor Man Lift |

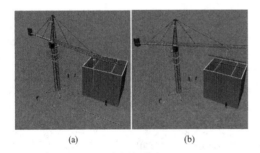

Figure 6. Two frames of crane animation generated in Fuzor.

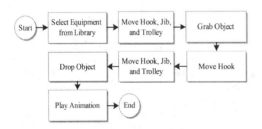

Figure 7. Flowchart of modeling movements of a tower crane.

hoists it before the jib slews toward the location the beam should be unloaded. Then, after moving the trolley to the final location, the hook is lowered down to place the beam, as shown in Figure 6(b).

Figure 7 shows the flowchart of modeling the activities of the tower crane, with attention to its DOFs and the range of equipment movements. Additionally, Fuzor can apply clash management.

### 3.3 Game engine

#### 3.3.1 Unity 3D

Unity3D (2018) can be used to produce both 2D and 3D games. Scripting is a crucial ingredient in all types of games, even the simplest game engines. Unity3D supports two types of scripting languages: C# and JavaScript. The program allows users to see how both the construction equipment as a whole and its components are moving and accomplishing the complex repetitive tasks on the construction site. Articulated construction equipment and processes can be animated by the game engine using scripts, although it is not compatible with BIM. Figure 8 shows the functions of earth digging and dumping in a truck of an excavator.

Furthermore, this program is identified as one of the most well-known virtual reality tools. Users can interface with computer-simulated surroundings through Virtual Reality (VR) technology. Built-in physics engines of Unity control the physical behavior of objects, such as gravity, collisions, and wind effects. Additionally, Unity3D has the terrain following feature, which ensures that the equipment moves on the terrain based on its characteristics (e.g. elevation and slope).

(a) Earth digging

(b) Dumbing in a truck

Figure 8. Loading and dumping motions of an excavator animated in Unity3D (Langari, 2015).

### 3.4 Other applications

#### 3.4.1 Lift planning applications

3D Lift Plan (2018) has a library of cranes containing over nine hundred cranes and load charts. It finds the most appropriate crane based on the following characteristics: dimensions and weight of the object to be lifted, and the location and size of obstacles.

#### 3.4.2 Equipment training simulators

There are different types of equipment training simulators, such as Vortex simulators (CM Labs, 2018). These simulators model complex construction equipment performance considering all conditions available on the real site, such as terrain characteristics, clash occurrences, and laws of physics. They prepare trainers to operate heavy equipment based on the available conditions of worksites, make better decisions in tough conditions, and operate vehicles more accurately and safely.

## 4 CONCLUSION AND FUTURE WORK

This paper compared the strengths and limitations of several tools in terms of visualizing, animating, and simulating equipment characteristics. In addition, the paper identified the main requirements related to high 4D-LOD and equipment DOF for improving safety, productivity, and constructability on construction sites. There are some tools that enable the visualization of detailed equipment activities although they are not BIM-based. Several 4D BIM tools provide basic functions for visualizing equipment movements. However, these applications have the following limitations: (1) time-consuming process for modeling articulated objects; (2) limited capabilities to visualize equipment movements; (3) lack of ability to schedule equipment activities at micro-level; and (4) difficulties for generating dynamic object workspaces. Therefore, further research is needed to enable the simulation and scheduling of equipment activities at the micro-level.

## REFERENCES

3D Lift Plan. 2018. Retrieved from http://www.3dliftplan.com/ default.aspx.
Akhavian, R., & Behzadan, A. 2015. Construction equipment activity recognition for simulation input-modeling using mobile sensors and machine learning classifiers. *Advanced Engineering Informatics*, 29: 867–877.
Akinci, B., & Fischer, M. 2000. 4D Workplanner—A prototype system for automated generation of construction spaces and analysis of time-space conflicts. *Computing in Civil and Building Engineering ICCCBE*, 740–747.
Akinci, B., Fischer, M., Kunz, J., & Levitt, R. 2002. Representing work spaces generically in construction method models. *Journal of Construction Engineering and Management*, 128(4): 296–305.
Autodesk 3DS Max. 2018. Retrieved from https://www.autodesk.ca/en/products/3ds-max/features.
Autodesk Navisworks. 2018. Retrieved from https://www.auto desk.com/products/navisworks/features.
Biljecki, F., Ledoux, H., & Stoter, J. 2016. An improved LOD specification for 3D building models. *Computers, Environment and Urban Systems*, 59: 25–37.
BIMForum. 2017. Level of development specification.
Choi, B., Lee, H., Park, M., Cho, Y., & Hyunsoo, K. 2014. Framework for work-space planning using four-dimensional BIM in construction projects. *Journal of Construction Engineering and Management*, 140(9): 04014041.
CM Labs. 2018. Retrieved from https://www.cm-labs.com/.
Getuli, V., Peretoli, G., Sorbi, T., Kindinis, A., & Capone, P. 2016. Adding construction workspaces modeling and planning to a 4D BIM-based Simulation Model. *Italian society of Science, Technology Engineering of Architecture ISTEA*, 259–268.
Graydon, M. 2016. Cherry picker boom lift genie. (Grabcad Community) Retrieved from https://grabcad.com/library/cherry-picker-boom-lift-genie.
Guevremont, M., & Hammad, A. 2018. Defining levels of development for 4D simulation of major capital construction Projects. *Proceedings of the CIB W78 Conference*.
Hammad, A., Setayeshgar, S., Zhang, C., & Asen, Y. 2012. Automatic generation of dynamic virtual fences as part of BIM-based prevention program for construction safety. *Proceedings of the 2012 Winter Simulation Conference*, 686–695.
IfcResource. 2018. Retrieved from http://www.buildingsmart-tech.org/ifc/IFC2 $\times$ 3/TC1/html/ifckernel/lexical/ifcresource.
Kalloc Studios. 2018. Retrieved from https://www.kalloctech.com/.
Langari, M. 2015. Enhanced path planning method for improving safety and productivity of excavation operations, Master thesis, Concordia University. Montreal, Canada.
Lumion. 2018. Retrieved from https://lumion.com/.
Shang, Z., & Shen, Z. 2016. A framework for a site safety assessment model using statistical 4D BIM-based spatial-temporal collision detection. *Construction Research Congress*, 2187–2196.
Stephenson, L. 2007. Scheduling management: classifications vs. levels. *AACE International Transactions*.
Synchro. 2018. Retrieved from https://www.synchroltd.com/ products-2/synchro-pro/.
Tantisevi, K., & Akinci, B. 2006. Automated generation of workspace requirements of mobile crane operations to support conflict detection. *Automation in Construction*, 16: 262–276.
Unity 3D. 2018. Retrieved from https://unity3d.com/.
Vahdatikhaki, F., & Hammad, A. 2015. Dynamic equipment workspace generation for improving earthwork safety using real-time location system. *Advanced Engineering Informatics*, 29: 459–471.

Zhang, S., Lee, J., Venugopal, M., Teizer, J., & Eastman, C. 2011. Integrating BIM and Safety: An automated rule-based checking system for safety planning and simulation. *Proceedings of CIB W099*, 99.

Zhang, S., Teizer, J., Lee, J., Eastman, C., & Venugopal, M. 2012. Building Information Modeling (BIM) and safety: automatic safety checking of construction models and schedules. *Automation in Construction*, 29: 183–195.

Zhang, S., Teizer, J., Pradhananga, N., & Eastman, C. 2015. Workforce location tracking to model, visualize and analyze workspace requirements in building information models for construction safety planning. *Automation in Construction*, 60: 74–86.

# IoT 2.0 for BIM, Bluetooth technology applied to BIM facility management oriented

S. Giangiacomi
*Politecnico di Milano, Milan, Italy*

R. Seferi
*Politecnico delle Marche, Ancona, Italy*

ABSTRACT: This research focuses on a method that will simplify the management of buildings. This will be demonstrated through detailed analysis of state of the art BIM applied to facility management in the Italian context and identification of the critical points of the most commonly used facility management software. Smart solutions to create intelligent buildings are proposed through Bluetooth technology which allow us to communicate the data of BIM model to insiders involved in maintenance and viceversa. An online platform, working with machine learning algorithms, could be the central point of communication of data and could make the ticketing process easier and faster, one of the core and complex tasks of facility management systems. This research will also show that wireless or Bluetooth beacons could have a key role in this process and if applied on a large scale could create smart cities with secure communication of large amounts of data.

## 1 INTRODUCTION

### 1.1 Research aim

The goal of this research is to give a virtual identity to any building to simplify its future management through IoT systems. The research focuses on Bluetooth Smart BLE 4.0-protocol applied to the BIM. The method used also provides tools that allow the application of the same techniques to any system regardless of the telecommunications protocol used. The main approach is to create a link between the physical real world of the constructed building and the virtual one represented by the BIM Database of the building itself.

### 1.2 Physical-Bim

The wireless technology proposed, through a web interface from a smartphone, allows quick and effective access to the BIM database, profiles the user, and allows access only to the data the user needs. It is possible to update the data from the smartphone and send it to the database. A different approach similar to blockchain, that use strings can be used We call this PhyString.

This system of data manipulation, that we will call from now on PhysicalBim, will ensure the synchronization between the physical maintenance and the update of the representative computer model of our building. As smartphones and internet are now part of our daily life, PhysicalBim should become part of buildings.

## 2 BACKGROUND

### 2.1 Facility management, Italian panorama

In today's interconnected and online world there is currently a saturation of smart devices. At the same time the BIM method is being consolidated in the building sector. Analyzing the Italian framework we find that the new law, UNI 11337, is still under development. Part 4 of the law defines LOD (level of detail) looking at the English PASS and shows the difference between graphic LOD and Level of Information. This difference is very important for BIM orientation to Facility Management because if a BIM Model has a high level of detail in geometry, it risks being too heavy to run in a facility software. Italian UNI 11337 has an alphabetic scale to define LOD and identify the LOD G as the "updated object" during the lifecycle.

We also look in Italy at the first attempts to adjust the price list of the Municipality of Milan with the #TagBim to facilitate the flow of information in the BIM model, Cianciulli (2018). More and more attention is dedicated to BIM for Facility Management and many Italian public administrations as well as large private companies request it. An example from Italy is the Municipality of

Turin, which has asked for the digitalization of BIM in 40 existing public buildings to manage the information database with traditional tools, Osello & Ugliotti (2017). Another case study is the Bank of Italy that has requested BIM to manage its building properties. The process has just started and the next step for the Bank of Italy will be to define the standards, Corti (2018).

### 2.2 Software CAFM

Software CAFM (Computer Aided Facility Management) is designed to manage the building during the lifecycle, helping us creating the maintenance plans and estimating costs. After a rigorous examination of FM software availability, we found that some of the most used software send and receive data directly to the BIM model in various format files.

If a BIM model is done with Revit, for instance, Archibus could read the BIM model in an rvt extension and take from it all the data and geometry the Facility Manager needs for simulations and maintenance, (Archibus website). The same happens with Allfa and the BIM Allplan, (Allplan website).

In both cases, starting from an as-built BIM model, simplifying the geometry and adding the useful parameters for the maintenance, we can map the parameters according to the required standard and synchronize the BIM model with the FM software. This kind of link is "two-way" which makes it possible to add data and parameters to the BIM model and update the FM software, or vice-versa. This process assumed a strong control before the synchronization. The Facility Manager must know what is the most recent data set as well as the BIM model or FM software to avoid information overwriting.

If we want to use the open format IFC to manage the BIM data, we can also use openMAINT, an open Facility Software. It can read the IFC standard and could be a useful tool for facility managers, (openMAINT website). Once the BIM model is updated we would need to once again export the IFC, making it impossible to put in the BIM native extension via the data settings written through the software openMAINT.

All the FM software considered have some critical points to take in consideration concerning the collaboration with BIM models. First there is the need to control which data is updated and in which software. The risk to lose the updated data due to incorrect overwriting is very high. Secondly, if we are working with the native extension, the choice of the FM software is very limited while if we are working with the IFC it is difficult to have two-way sharing of data.

Last, but not least, none of the FM software offers to the property an easy interface to query and look at the data management. The software mentioned above are based on hierarchies and functions are not easy understandable. The building owner will need personnel with a deep knowledge of both software and management skills.

This is one of the most critical points because the owners cannot have the knowledge or just an overview of their data by themselves.

### 2.3 BIM & facility management, state of art

Several researches have been made in the field of Facility Management linked to the BIM model. The main themes are the algorithm of machine learning to optimize mep design and sensors used to obtain data directly from the rooms, such as temperature or humidity, as shown in the case study of Kerry Plaza (Hu et al. 2018). In this case, a web platform is used to store the data captured and to update the BIM model and data inside the FM software. Another research highlights the heterogeneous source of data from the documents of the as-built to the live data from the sensors installed into the building. The consequence is that the storage of data varies in difficulty for the facility manager in having a complete and linked database (Pourzolfaghar et al. 2017). The research also proposes a process model to have a consistent database. According to the researcher, an open storage could be built using the standard ISO for construction industry. After the construction of the BIM model that ensures the link of the data of the project, this data could be integrated with the live data inside the open storage, resulting in a complete database to manage.

Another field analyzed in the international panorama is the requirements of data for FM (Hosseini et al. 2018). A critical approach to COBie is explained, stating that the requirement for FM is strictly connected to the user and the task of the management. The research was based on a focus group that answered several questions about the data need for their specific task.

As a result, the researcher proposed a matrix which clearly identified the quality and quantity of data needs depending on the users. The matrix is intended as support for the existing FM process that could help people identify only the useful information and could give the consistency and accuracy of the FM database.

### 2.4 Improvement of FM system

The system proposed in this research work would like to improve the problem of the interface, making it easy and user friendly for the properties. This work has the aim to simplify the ticketing process, discover of the issue, advise the right people to solve the problem, and update the BIM object history, all while keeping track of each intervention made in real time.

The research study is also oriented to find a better solution to a QR code reader. This means placing a sensor (beacon) in the building to provide data.

We also want to use an open format for the data exchange. For this reason, we have choosen the IFC format to export the BIM model made in an authoring software (in our case study Autodesk Revit). Our aim is to have a process model working with all of the BIM models.

## 3 RESEARCH ACTIVITIES

### 3.1 *Method*

After analyzing the main topic, the research method will use these sub questions to look forward and answer the main research question.

What is the best technology to optimize information exchanges between the physical and the digital world? How to make BIM data queries more efficient? How do users interact with BIM data intuitively? How to update the BIM database automatically?

We proceeded to study and apply the bluetooth technology in static system Bnet and then try to capture the behavior of a dynamical system DBNet in order to meet the real life requests and multiusers scenario. A series of practical tests will about the signal strength and the microcontroller response will take place in order to identify the best network solution and understand how we can implement it. The bluetooth beacons system applied to BIM works through a web interface, which was developed to test the actual functionality via smartphone.

### 3.2 *Why bluetooth?*

Bluetooth is a wireless technology for short distance data exchange based on the use of short wavelength UHF radio waves. Considering the Open System Interconnect (OSI) standard, Bluetooth does not exactly match the standard. Bluetooth devices operate at 2.4 GHz, ISM license-free band (the operating band is divided into spaced 1 MHz channels). The modulation scheme is Gaussian Frequency Shift Keying (GFSK).

All Bluetooth devices can operate in two modes: Master or Slave. Each Bluetooth device has a unique address and a Bluetooth clock. The baseband part of Bluetooth specifications describes an algorithm that can calculate the frequency hopping sequence. After each packet is transmitted, both devices reconnect their radio to a different frequency, effectively "hopping" from radio channel to radio channel (FHSS with frequency Hopping Spread Spectrum).

In this way, Bluetooth devices use the whole ISM band so that if a transmission is compromised by interference on one channel the retransmission will always be on another channel (with more probability that is available) Bray & Struman (2002).

The Bluetooth® v4.0 Bluetooth specification standard allows two wireless communication systems

– BR Basic Rate
– BLE

The BLE system allows the transmission of small data packets with reduced power consumption compared to BR. It introduces new technology to the Bluetooth Core Specification, enabling new Bluetooth devices that can operate for months or even years on tiny, coin cell batteries. These are forecasted to ship in nearly 4 billion devices in 2018 according to the Bluetooth website.

### 3.3 *Bluetooth, useful features*

A useful feature of the BLE system is the functionality of advertising/broadcasting.

Broadcasting enables short burst wireless connections and uses multiple network topologies, including a broadcast topology for one to many (1:m) device communications. The Bluetooth LE broadcast topology supports localized information sharing and is well suited for beacon solutions, such point of interest (POI) information and item finding and wayfinding services. It also supports a mesh topology for establishing many to many (m:m) device communications.

The mesh capability is optimized for creating large scale device networks and is ideally suited for building automation, sensor network, and asset tracking solutions. According to Bluetooth website, only Bluetooth mesh networking brings the proven, global interoperability and mature, trusted ecosystem associated with Bluetooth technology to the creation of industrial grade device networks.

### 3.4 *Why beacons?*

The beacon is a small device (the size of a sim card) which uses wireless technologies to associate physical objects with the digital world. Microcontroller with wireless enabled will run Arduino code and can do everything like an Arduino, as the Rfduino website shows. Rfduino presents the additional library to the Arduino IDE. Signed as a beacon or sometimes Ibeacon, we therefore consider the terms Beacon and Rfduino as equivalent except in the cases where it will be specified. The Bluetooth interface is present on 99% of the devices of the world, therefore, additional costs are not required to prepare receiving devices.

The physical device, the beacon, is already on the market, it needs only customized program for the end customer. The beacon needs electricity to work, it can be connected to a 5 v power supply or it can have its own battery. It is low energy consumption, a small battery could work for about a year without recharging. The beacon can be thought of as the successor of the QR-code, presenting many advantages with respect to the latter.

- Proximity radius: the beacon emits a signal with a beam from a few centimeters up to a maximum of about 30 meters in urban environment, radius that can be programmed according to of the client's needs. The QR-code, on the other hand, has a radius limited to the distance of the smartphone's camera, from 10 to 20 centimeters.
- Automatic information: the beacon transmits actively a signal. It is the information that the user searches for and is received instantly depending on the radius of proximity. The QR-code, on the contrary, it works passively, it must be seen and approached by the user looking for the information.
- Dynamic information: the beacon allows to change the information to which it is associated, while the QR-code, once generated, will always remain tied to the same url. If you want to change the information associated with a QR-code is necessary to generate one other and replace it in the place where it was placed.
- Selected frequency of information: the beacons makes it accessible and sends different messages depending on the time value set inside the code. The QR-code provides the same single information to anyone who photographs the image.

### 3.5 BNet system and DBNet system

The static system, that we call BNet system, provides the same information every time you enter the system (when we are near the first beacon).

Input => X1 => X2 => X3 ... => Exit

Our system will therefore be able to translate each letter to a Pop-up response for the individual user.

Lm = abc......;

Using the event-driven paradigm for control, communication and optimization. Each letter is an event and represents a past beacon Cassandras C.G & Lafortune (2008).

We have created a relationship between each Beacon/Rfduino and any event. We have studied a language for this system, a language that allows us at any time to create patterns and to classify user behaviors without knowing them. Through this language the system understands in which group the single user belongs to. This way we create connections between similar users and behaviors. Trying to calculate each time for cause-effect the next move inside the building or simply the next most useful information to that specific user or user group. Artificial intelligence is about algorithms enabled by constraints exposed by representations that model targeted thinking, perception, and action, as defined by Professor Patrick Winston, MIT. We define from now The DBNet system, as a dynamic system.

The information provided by the DBNet System will change based on previous paths, behavioral properties or groups with different users. A combination of MAC address and UUID, date or time, allows the system to collect data.

Based on this data, the system will change the response. By optimizing the process using Machine Learning algorithms, each time the system will become more useful for the individual user or user group. With this choice the system can calculate each route and will know not only that the user is near a specific point, but also the time spent near it. By improving each time the positioning (e.g. notifications) inside a given environment, the notices are updated, allowing users, e.g. to seek help, to send "WALL" messages to others, or to give feedbacks. This algorithm translates the process of movement and proximity inside the "BNet" in a string that is a combination of events.

We can think the physical building as a set A of different physical elements. The BIM model as a set B with virtual elements. The main goal here is to create a bijective mapping between A and B. That is essentially what a BIM model should do. The extra layer of complexity here is another set C formed from elements of DBNet that is equipotent with B.

So if A is equipotent with B and B is equipotent with C then A is equipotent with C. Now that we create a bijective mapping from the physical word to our system, we can benefit from different techniques of a Software Design approach to manage our physical building. This is because in certain ways our physical building is now also a software or more precisely a web application.

### 3.6 Users and selection of data sets

The web site will be the users or insiders main UI. The system recognizes the user after the first login and will show them only the data they need. Users identified for the first prototype are:

- Administrator/Facility manager: will have access to all the dataset.
- Maintainer: will read the data selected by type of system.
- Security: will read data connected to the security of the system itself.
- Common user: will be able to read the user handbooks and to report a fault.

– Groups. A temporary user that needs access to the database.

All these users identified have different access to the data connected to the BIM model. The systems itself is able to query the data through a selective hierarchy showing them depending on the user's needs.

The hierarchy to have the selection is:

– Spaces (each beacon is referred to a space that could be a room, multiple rooms, or a floor).
– Users (as defined above)

By placing the beacon inside a room and approaching with the smartphone it will be possible to have a first access by the user who will create his profile.

Subsequently the System can recognize the user and link him with his relevant data, while the space is already defined by the coordinates of the beacon itself.

## 3.7  WEB interface

Web Interface The interaction between any user and software or in our case the web application represent the web interface or shortly the user interface UI. The Browser base application is the combination of JavaScript HTML and CSS all are execute via web browser. Our choice is based on the cross-platform model that the Web applications are known for. Our web application should respond to user's behavior and different screen sizes. This kind of approach is called responsive design. Any single Bluetooth beacon will broadcast the most useful page to the user.

The web pages in HTML are created on the fly and the only way to access them will be by the proximity to the Bluetooth beacon. The access will be guaranteed everywhere and anytime only for the admin. Imposing a physical layer of security between the web application and its users will improve the security of cyber physical systems and some of vulnerability inherent from using the Internet. The use of Bluetooth or any other type of broadcasting technology will limit some of vulnerability inherent from the wireless networking and will not compromised any sort of gateway, router or modem inside the physical building. The use of Bluetooth beacons inside the building will emulate the smart sensors. The web application will be the main operating system of the building itself. Both use of sensors and a unique updatable O.S will guaranteed

## 3.8  Tests completed

We have designed a prototype to test the method, concerning a small BIM model of a modular apartment simplifying geometry and distribution layout.

We have used the authoring software Autodesk Revit to place the relevant objects.

We have created the shared parameter named "PhBIM_ObjHistory", an instance parameter assigned to all the Revit categories, with the exception of views, schedules and sheets.

Then we have assigned a value to this parameter, "T0", because without an initial value the exportation of our shared parameter from Revit to IFC will not work. The value "T0" means "Time zero", the starting point for the maintenance operations.

We have also investigated what kind of IFC export settings work best for our purpose. We have found that we can have better performance and lighter IFC file if we don't export the schedules, avoiding multiple places in the IFC schema for the same parameter.

In our parameter "PhBIM_ObjHistory" the algorithm mentioned above is able to write the history of the object, when the failure has happened, who is responsible for the reparation, what kind of problem it is, and when the intervention is made.

## 3.9  Data pick list, starting from COBie and typology matrix for data and information

According to the critical approach to COBie (Hosseini et al. 2018) mentioned above, our research started from identifying the useful data for the final user of our system, considering the quality and quantity of data needs. We supposed to work for a company that is both the owner and facility manager of the building, only for the purpose of understanding what data we need. Our aims are to use BIM model FM oriented to improve the ticketing process, to keep track of changes and to give to the owner a clear, easy and smart view of his building. Then, starting from COBie we have made our spreadsheets containing only information that are actually needed. This process and our results could be supported by interviews and focus group team in future researches, or implemented by the client itself according to the requirement.

We have filled out three datasets. The first dataset is a spreadsheet that contains each item and its own relevant features. This list is useful when designing the BIM model, having the right dataset to insert in each BIM object is essential, because these data will appear on the web interface. The second is a picklist for each object of the intervention planned, associated with the user Id or insider Id. This kind of data will be selected from the system itself and shown as simple task. The third consist in relevant data with respect of the maintenance process, such as the frequency of the intervention. These data will be the input value inside the BIM model, exported as IFC, linked on web storage and selected according the spatial hierarchy and users, as explained in the next paragraphs.

## 4 CONCLUSIONS

The system proposed, that we call "PhyBIM" (Physical BIM), is the connection between the physical world and the digital representation.

It works with the IFC in an easy and understandable way, with the ability to create a pattern of users and of all the BIM and intervention data managed. It is also very efficient for owners to have control of their asset and update data in real time. It is a useful tool that could significantly simplify the FM process, specifically the ticketing, storage, and analysis of data. The IoT solution proposed with the beacons system ensures low energy cost, dynamic information, and selects information depending on the user and the need.

"The event-driven paradigm offers an alternative, complementary look at control, communication and optimization. The key idea is that a clock should not be assumed to dictate actions simply because a time step is taken; rather, an action should be triggered by an "event" specified as a well-defined condition on the system state or as a random state transition. Note that such an event could be defined to be the occurrence of a "clock tick", so that this framework may in fact incorporate time-driven methods as well" Cassandras (2014).

## 5 NEXT DEVELOPMENT

We have just made a hypothesis of items, attributes, and interventions needed to test the system, however the dataset has to be implemented. This research has several subsystems that need to be studied in depth.

First, the standardization of data requirements needs to be completed. In near future, we would like to create spreadsheets starting from COBie but selecting only the data requested from the committee and taking into consideration the typology matrix of information (Hosseini et al. 2018). To do this we will need to validate these data spreadsheets with interviews, focus groups, and practical tests on a real case study.

Secondly, the initial BIM test conducted was using only a few rooms to validate the method. To improve the system itself, we need to test it on an entire existing building.

It will be also possible in future development to improve the selective hierarchy and to have more customized data by adding systems and objects as noted below:

- System (kind of system analyzed, such as piping, electrical).
- Object: It will be possible having the inventory and datasets available in real time, just with the proximity to the physical object.

## REFERENCES

Bijedic, N., Gimbert, J., Miret, J.M. & Valls, M. 2007. Elements of discrete mathematical structures in computer science, Mostar: Univerzitetska knjiga.

Bray, J. & Sturman, C.F. (eds) 2002. *Bluetooth 1.1: Connect Without Cables*, Prentice Hall.

Cassandras, C.G. & Lafortune S. (eds) 2008. *Introduction to Discrete Event Systems*, Springer.

Cassandras, C.G. 2014 (Received 31 December 2013; accepted 17 January 2014). The event-driven paradigm for control, communication and optimization. In *Journal of Control and Decision*. Boston.

Chomsky, N. 2012. Linguistic and philosophy study of language, MIT *http://videolectures.net/noam_chomsky.*

Cianciulli, G. 2018. BIM 5D e prezzari BIM: il Comune di Milano è il primo in Italia ad utilizzare #TAGBIM, *https://www.ingenio-web.it/18703-bim-5d-e-prezzari-bim-il-comune-di-milano-e-il-primo-in-italia-ad-utilizzare-tagbim.*

Corti, G. 2018. Banca d'Italia accetta la sfida del BIM: Come un'importante istituzione affronta il passaggio al Building Information Modeling, *http://www.monitorimmobiliare.it/banca-d-italia-accetta-la-sfida-del-bim_20184261525.*

Cufi, C., Davidson, R., Townsend K, & Wang, C. (eds) 2014. *Getting Started with Bluetooth Low Energy: tools and techniques for low-power networking*, USA: O'Reilly.

Hosseini, M.R., Roelvink, R., Papadonikolaki E., Edwards, D. J. & Parn, E. Integrating BIM into facility management: typology matrix of information handover requirements. In *International Journal of Building Pathology and Adaptation* 36 (1): 2–14.

Hu, Z.Z., Tian, P.L., Li, S.W. & Zhang, J.P. BIM-based integrated delivery technologies for intelligent MEP management in the operation and maintenance phase. In *Elsevier, Advances in Engineering Software* 115:1–16.

Pourzolfaghar, Z. & McDonnell, P. & Helfertand, M. 2017. Barriers to benefit from integration of building information with live data from IOT devices during the facility management phase. In: *CITA BIM Gathering 2017 23–24 Nov 2017*, Dublin.

Osello, A. (eds) 2015. *Building information modelling geographic information system augmented reality per il facility management*. Palermo: D. Flaccovio.

Osello, A. & Ugliotti, F.M. (eds) 2017. *BIM Building Information Modelling: verso il catasto del futuro, conoscere, digitalizzare, condividere*. Roma: Gangemi.

Official Allplan website, *https://www.allplan.com/references/facility-management/.*

Official Archibus website, *https://archibus.com/products/extensions-framework/smart-client-extension-for-autocad-revit/.*

Official Bluetooth website, *http://www.Bluetooth.com*

RFDuino website, *http://www.rfduino.com.*

UNI 11337–4 2017, *Building and civil engineering works—Digital management of the informative process—Part 4: Evolution and development of information within models, documents and objects.*

Winston, P. 2010. MIT intro 6.034 AI Instructor, *http://ocw.mit.edu/courses/electricalengineering-and-computer-science/6-034-artificial-intelligence-fall-2010/.*

# A specialized information schema for production planning and control of road construction

Eran Haronian & Rafael Sacks
*Virtual Construction Lab, Faculty of Civil and Environmental Engineering, Technion—Israel Institute of Technology, Israel*

ABSTRACT: Building Information Modelling (BIM) and the IFC information schema are well developed to suit building construction projects. Not only do they represent the building as a designed product, they can also be used to support Lean production planning and control. Road construction, a sub-type of civil infrastructure construction, is fundamentally different to building construction, in terms of its products, the types of work and operations, and the resources used. One of the key differences from the point of view of production flow, is that roads are composed of geometrically continuous courses rather than discrete 'products', making work packaging difficult. We propose a product schema which models *road sections* with distinct *road course segments* that are dynamically defined aggregations of *roadels* (fine-grained vertical triangular prism objects). The schema represents the continuous nature of road construction. Its discrete entities also enable computations of as-made work using the raw data obtained from sensors and surveys, thus enabling systematic analysis not only of machine productivity and utilization rates, but also of Lean production flow metrics. Finally, we demonstrate how the proposed schema can be implemented with existing IFC entities, but conclude that extending the schema with new entities is preferable for both semantic and practical reasons.

## 1 INTRODUCTION

Many Building Information Modelling (BIM) tools have native schema that represent buildings and building objects in ways that support not only design, analysis and detailing, but also project management (Sacks et al. 2018). Task-specific BIM tools are also available for the detailed production planning and control functions that are part and parcel of Lean Construction tools such as the Last Planner System® (Ballard 2000). Like most BIM tools, the IFC information schema was also developed originally for building construction projects, with hierarchical and well defined products (elements and spaces), and its schema contains the entities and relationships needed for exchange of production planning and control information (ISO 2013).

Road construction is fundamentally different to building construction and is characterised by high volume and continuous products. For production planning and control of road construction, the information schema must enable dynamic subdivision of the project into sub-products for definition of work packages and for aggregation into as-built sub-product representations that represent the work done, as measured by field surveys.

Existing BIM tools for design and construction of roads do not provide the level of detail and the relationships needed in their native schema for effective production control of earthworks. Likewise, the IfcAlignment (the first extension of the IFC schema for infrastructure, adopted in 2015) enables location of objects with relation to road, rail or bridge alignment curves (Amann et al. 2015), but it does not address the representation of road courses or production segments that is needed for production planning and control.

The hypothesis of this work is, therefore, that the unique production processes and operations that characterize road construction projects, as well as the unique products they create, call for a thorough reconsideration of how those products should be modelled, and what data representations are needed to support production planning and control.

In this paper, we propose a product schema for lean construction production planning and control of the earthworks and surfacing operations of road construction projects. An analysis of the principles of production flow provides the foundation for definition of the use cases considered, and the use cases in turn lead to development of an information model. We also discuss considerations for binding the schema in terms of the IFC standard.

## 2 BACKGROUND

### 2.1 Production control for road construction

Ballard and Howell (1998) asked "what kind of production is construction?" They proposed that "construction is essentially a design process, but one in which the facilities designed are rooted-in-place, and thus require site assembly". Bertelsen (2003) extended the definition by discussing complexity, stating that construction is an "assembly-like process, which is complicated, parallel and dynamic, and thus more complex and dynamic than project management often envisages". While these definitions are accurate regarding building construction projects, their suitability, especially the reference to assembly, is questionable regarding road construction projects.

Building construction projects can be conceived of as discrete products that can be divided into sub-products. In most cases, there is correlation between the location breakdown structures (LBS) and the product breakdown structure. For instance, a residential building's LBS will include a hierarchy of floors, apartments and rooms—all of which constitute discrete components of the final product. These sub-products contain discrete elements that in many cases are manufactured in off-site factories and only assembled on site. Accordingly, various production strategies can be implemented; for the production of exterior plaster, for example, it might be appropriate to work in batch flow, while for production of customized kitchens it might be appropriate to work in a job shop, and so on. Understanding the uniqueness of each product and process leads to choosing the most appropriate strategies.

In contrast to building construction projects, in road construction, products are not discrete, and accordingly the LBS must be continuous (Kenley and Seppanen 2010). When analyzing the production process of roads, characteristics of continuous flow and process manufacturing are found:

- Material processing is a core component of production: excavating, grinding, compacting, paving, etc.
- Real time quantity calculation systems are required for controlling consumed resources and for monitoring production status.
- Finished products can be disassembled only through demolition (unlike assembled components in a building).

Although there are other sub-processes as well, the dominant process in road production is continuous flow and process manufacturing. Table 1 summarizes some of the key differences between building and road production systems.

According to Schmenner (1993), production strategies located on the main diagonal of the product-process matrix shown in Figure 1 represent an alignment between the product and the process. Deviation from the main diagonal reflects inferior strategies that increase costs and wastes. A downward deviation leads to "out-of-pocket costs", where unnecessary resources are invested in the process. For example, automating production for a "one of a kind" product would be wasteful due to the extensive overhead required. An upward

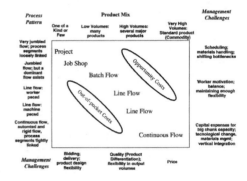

Figure 1. Product-process matrix (Schmenner 1993).

Table 1. Comparison of building and road construction.

|  | Road construction | Building construction |
|---|---|---|
| Product | One of a kind.<br>Continuous.<br>No customization. | One of a kind.<br>Discrete.<br>High customization. |
| Process | Materials produced on site (by digging, crushing and mixing)<br>Much of the work is "material processing".<br>Few task types, few subcontractors.<br>Location breakdown structure is governed by external constraints.<br>Technological and safety aspects prevent execution of several tasks simultaneously in the same working area. | Primarily components manufactured in a factory, transferred to site and installed.<br>Much of the work is "assembly".<br>Many task types, many subcontractors.<br>The location breakdown structure is defined by the sub-products (floors, apartments, etc.).<br>Several tasks can be performed simultaneously in the same working area. |
| Resource | Heavy machinery is the main resource. | Professional workers are the main resource. |

deviation from the diagonal leads to "opportunity costs", where the proper resources are not invested in equipment or in automation.

Regarding the *products* in road construction, on one hand, each project constitutes a unique "one of a kind" product (in terms of design, external conditions and constraints). On the other hand, this unique product consists of many standard continuous sub-products, so perhaps the definition "very high volumes, standard products" is appropriate too. Regarding the *production process* in road construction, there is a well-defined, inflexible sequence of operations, and each operation requires dedicated equipment involving considerable investments. From this aspect, "line flow: paced by machines" or "continuous flow, tightly linked operations" may be an appropriate definition for the process. However, in practice, field management and control relies mostly on the superintendent's professional experience and leadership (like a craftsman in a workshop).

Reliance on the superintendent's ability is apparently well-suited to the "one of a kind" aspect of the product. Similarly, capital investment in heavy machinery and automation systems is apparently suitable for "very high volumes, standard products". The result of this combination is a complex and dialectical production system that requires special awareness for the wastes that may arise in the process. The use of heavy machinery and automation systems in a dynamic production system may lead to wastes related to "out-of-pocket costs". These systems require large investment and may not work continuously as desired. On the other hand, relying on the superintendent's abilities for production control of "high volumes, standard products" may lead to wastes related to "opportunity costs" when heavy equipment cannot be applied effectively.

For this field to significantly benefit from new technological innovations (e.g. information technologies, remote sensing and automation), we argue that the essence of the products, the way they are represented, and how they are divided into sub-products must be reconsidered. The fundamental difference between the products and processes in road construction compared to building construction, substantially impacts the representing information schema. In this context, our focus is on aspects of production management and their impact on the information schema.

## 2.2 Monitoring and automation technologies

We are witness recently to significant industry adoption of monitoring and automation technologies for road construction. Despite this, productivity in road construction has not improved significantly over the period 2002–2016 (Sveikauskas et al. 2018). Local and operational use of sophisticated monitoring technologies for construction equipment may lead to a locally efficient optimum in operations, but in order to achieve a global optimum, a different approach is needed. According to Shingo "*Process analysis examines the flow of material or product; operation analysis examines the work performed on products by worker and machine* (Shingo and Dillon 1989)." We propose that one root cause for the stagnation in productivity lies in the focus on operations alone, and the failure to consider, monitor and analyse the production process.

Beyond specific operational aspects, the new technologies provide an enormous amount of raw data, yet questions arise regarding what to do with this data, and how to analyse it, especially in the context of process flow. The extensive geographic distribution and the continuous nature of the products and operations in road construction projects require a systematic analysis of the raw data obtained from the sensors in the project. First, the raw data must be linked to the information schema of the project. Only then can the process be analysed systematically. However, the current IFC information schema does not represent the product and process information needed for production control of road construction.

## 2.3 Information schema for infrastructure

Development of information schema for infrastructure construction is not a new topic (Rebolj et al. 2008). LandXML, for example, has been used for civil engineering design and for analyses of survey measurement data for almost two decades. The LandXML schema provides an object-oriented multi-domain data interoperability instrument, including representation of solid model geometry (landxml.org). Extensions to the IFC data schema have also been proposed to represent infrastructure projects, such as roads, bridges and tunnels (Liebich et al. 2013). IFC Alignment, introduced in 2015, is the first extension of IFC for infrastructure and it is designed to deal with particular aspects of linear projects by enabling the association of physical elements with continuous vertical and horizontal alignments (Amann et al. 2015), as shown in Figure 2.

Infrastructure facilities include a verity of element types, comprising unique products that require spatial product modelling approaches. The work of (Vilgertshofer et al. 2016), which proposes extension of the IFC schema for shield tunnels, is an example of the type of research required in this area. Following their work, in this paper we propose an extension for production management in road construction projects.

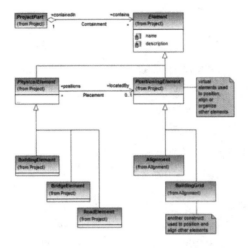

Figure 2. Alignment information schema (buildingSMART 2015).

## 2.4 *Summary*

Building Information Modelling (BIM) technologies (Sacks et al. 2018), and the IFC information schema that supports them, were developed for building construction projects. Their schemas represent buildings with hierarchical compositions of discrete, distinct and well-defined product entities. This representation does not reflect the continuous nature of the courses (layers) of roads, whose geometric extents need to be defined and redefined dynamically to adapt to the ways in which the work itself is structured and restructured as the process emerges.

Unfortunately, the IFC Alignment extension does not add this functionality. As shown in Figure 2 the primary spatial hierarchy does not extend beyond discrete, distinct and well-defined product entities.

The more precisely and appropriately the product is defined to suit the process, the more efficient production management can be. Sacks et al. (2010) described Lean Construction (LC) and BIM as two concepts whose simultaneous implementation can improve construction projects further than they can when adopted separately. They identified 56 positive interactions between LC and BIM practices, including aspects of visualization, automation, stability, reliability and more.

Yet, as described above, implementation of these principles is not trivial for continuous products. This work deals with the dynamic subdivision of continuous elements into finite sub-elements from a production perspective. The subdivision supports dynamic aggregation of the sub-elements to represent changing work-packages and ad hoc as-built sections. This is relevant not only for Lean production planning and control, but also for quality management (assignment of laboratory tests to specific areas), road maintenance (rehabilitation of certain areas on the road), and possibly for other similar projects.

## 3 USE CASES

A set of four use cases, based on Lord Kelvin's well-known Plan-Do-Check-Act process, defines the extents of the Lean production planning and control processes for road construction. The purpose of these use cases is to describe the main components of production planning and control in road construction. As can be seen in the UML use case diagram (Figure 3), these components include concepts from the Last Planner System® (Ballard 2000) such as weekly work planning, a make ready process, look-ahead planning and the percent plan complete (PPC) metric.

The first use case is "Plan". The basic requirement for planning is an accurate definition of work packages. In continuous and layered elements like roads, the main element (the road or the road courses) must be divided into layered sub-components that can be used to define work packages. The different layers of a road require specialized heavy equipment and are carried out at different accuracy levels according to the layer's function and material. For each layer, several operations are needed (such as dispersing, compacting, finishing, etc.), and each operation requires a corresponding set of heavy equipment. The work package sizes are determined by the characteristics of the operations and the equipment, and vary significantly in composition and size. In addition, physical constraints such as infrastructure lines along the route of the road must also be considered when setting work packages. Therefore, the flexibility to divide the main elements into segments that will change over time throughout the project is a key requirement.

The second use case is "Do", in which work is executed. The information flows in this use case include delivery of the work package information to the equipment operators and monitoring the work in real-time. This includes collection of raw data obtained from GPS or grade control systems installed on heavy equipment, as well as raw data collected by other sensors, laser scanners or aerial imagery. The ability to produce value from this data lies in the ability to represent it in relation to the planned work, which requires appropriate entities and relationships in the information schema.

Table 2 illustrates the resolution of data obtained from sensors in road construction projects. This data was received from the Trimble GCS900 Grade Control System. The table presents the location of a road roller (x, y, z), and additional data regarding the compacting operation at a given timestamp. This data must be linked to the project's informa-

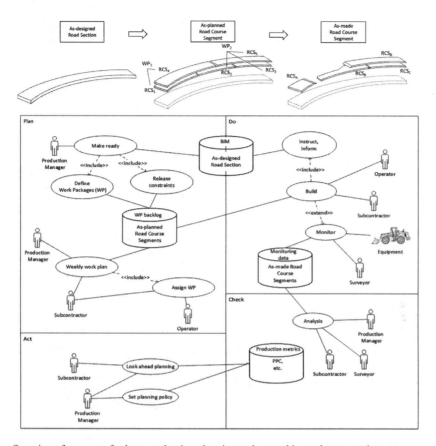

Figure 3. Overview of use cases for lean production planning and control in road construction.

Table 2. Resolutions of the raw data obtained in a road construction project.

| Time | CellN_m | CellE_m | Elevation_m | Pass count |
|---|---|---|---|---|
| 2016/Apr/07 09:53:24.290 | 750605.25 | 210163.35 | 6.41 | 3 |
| 2016/Apr/07 09:53:29.290 | 750605.25 | 210163.69 | 6.426 | 4 |
| 2016/Apr/07 09:54:30.326 | 750592.33 | 210182.05 | 6.017 | 2 |

tion schema in order to enable systematic analysis of the work done.

The third step is to "Check". Once the work has been performed and monitored, managers need to analyze the status of the project. The most basic analysis is to compare the planned work packages against the "as-built" work packages. Distinct object instances are needed for each of these, one reflecting the planned work and the other reflecting work performed, so that the production process can be assessed using quantitative measures calculated.

The final stage of the PDCA cycle, "Act", constitutes the required managerial policy decisions. At this stage, the nature of the operations, the sizes of the work packages and the levels of equipment resources may be changed.

These four use cases lay the foundation for definition of a supportive information schema, with an emphasis on the ability to dynamically divide a continuous road element into differential entities that can be aggregated into work packages.

4 PROPOSED INFORMATION SCHEMA

Based on the production process analysis described in the use cases, we suggest additional levels of product hierarchy (Figure 4) that will enable dynamic distribution of continuous road elements into sub-products and their aggregation into work packages, while maintaining the continuous nature of the parent product element. The conceptual data schema, defined using UML class diagrams, is shown in Figure 5.

Table 3 lists the classes of the data schema. The first two classes define the road as it is designed. The *Road Section* is the primary physical designed object. The road section in this schema is the equivalent of the RoadElement object in the Ifc Alignment extension, snown in Figure 2, which is a subtype of *PhysicalElement*. Roads are built in layers, called courses. The *Road Course* class allows definition of the layers—their material, thickness, degree of compaction and other operational parameters. The road sections have geometry and defined extents—the road courses only define the thickness of each layer within a road section.

The next two classes enable the specific functionality for Lean planning and control that was defined in the use cases. *Road Course Segments* define the physical extents of the segments used for work planning and monitoring, and they are designated either 'as-planned' or 'as-made'. In planning, their size—extents and depth—are set to suit the operations that are needed to build the road (e.g. spreading, compacting, grading, surfacing, each of which are done by different machines with differet optimal section extents). They do not necessarily have uniform depth, but they do not cross the boundaries of road courses.

The geometry of a road course segment is defined as a solid aggregation of road elements, or *Roadels*. Roadels are discrete parts of a road segment, just as a picture element—a pixel—is a discrete part of a digital image). A roadel is a vertical prism with a

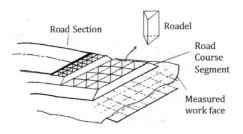

Figure 4. A schematic illustration of the data schema.

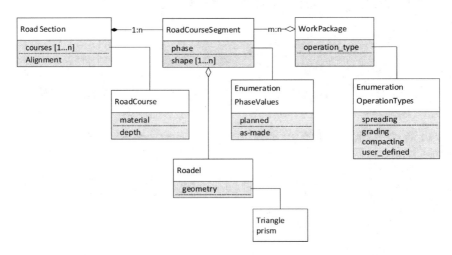

Figure 5. UML class diagram depicting the conceptual data schema.

Table 3. Definition of classes for road production planning and control.

| Class | Definition |
| --- | --- |
| Road Section | A road section represents the product as designed. |
| Road Course | A designed layer of a road section, e.g. 'base course'. A set of road courses define a road section's structure in the same way that a wall is defned by its composite layers. |
| Road Course Segment | A segment of a road section that lies within a single course. |
| Roadel | A triangular prismatic primitive element of a road course segment. |
| Work Package | An aggregation of road course segments on which a single particular operation type (e.g. compaction) is to be executed as a distinct package. |
| Task | An aggregation of work packages assigned to be executed by a particular crew in a particular time frame. |

triangular cross-section. The top surfaces of a set of roadels aggregated together in a road course segment define the upper surface of a road course segment as a Triangular Irregular Network (TIN). The Boolean union of the solid volumes of the roadel prisms define the volume of a road course segment, which can have a curved surface and irregular depth. This geometry is ideally suited to modelling the solid volume shapes that are typically assigned to work packages and measured in as-made surveys of a road's construction, as the courses are progressively layered one on top of the other. A roadel may belong to only one road course segment. However, because 'as-planned' road course segments and 'as-made' road course segments may overlap, roadels of different road course segments may overlap.

Road course segments are collected in *Work Packages*. Work packages may include work of one or more operation types to be performed on one or more road course segments. Work Packages in turn are aggregated in *Tasks*, which are assigned to work crews and other resources. Tasks are the vehicle for planning and control. When work is performed and then measured, the resulting geometry is represented as 'as-made' road course segments composed of roadels. The surfaces of the roadels are derived directly from the triangles of the measured surfaces, provided as triangular irregular networks (TIN), and their depths are computed from the current surface level to that of the previous measured surface at any given point. In this way, production metrics for work packages can be computed using Boolean operations on the volumes of road course segments (RCS) associated with work packages, such as a daily volume performance index shown in Equation 1(a). This index can exceed 100% when work performed exceeds planned. Lean metrics, similar to the PPC index used in the Last Planner System to reflect planning reliability, can also be calculated by comparing the proportion of the volume planned in any given work package that was in fact executed as-planned, as shown in Equation 1(b).

$$Performance\ index = \frac{\bigcup_{i=1}^{n} RCS_i^{as-made}}{\bigcup_{j=1}^{m} RCS_j^{planned}} \quad (1a)$$

$$PPC(WP_k) = \frac{\bigcup_{i=1}^{n} RCS_i^{as-made} \cap \bigcup_{j=1}^{m} RCS_j^{planned}}{\bigcup_{j=1}^{m} RCS_j^{planned}} \quad (1b)$$

For all $RCS_j^{planned} \in WP_k$

Equation 1. Examples for gross performance and lean production metrics.

Moreover, discrete information entities will enable calculation of complex metrics related to Little's Law (cycle times, work in progress and throughputs), as all the required information is available in a structured manner.

## 5 CONSIDERATIONS FOR IFC BINDING

After presenting the conceptual model to capture the information schema requirements for road production, we consider the feasibility of implementing this information schema using existing IFC schema entities. Appendix 1 provides links to example instance files for this implementation.

A Road Element represents the basic product, conforming precisely to the definition of *IfcCivilElement*, a subtype of *IfcElement*. This is analogous to a slab, wall or column in a building.

A Road Course is used for as-designed description of a road's layered structure and material and can be implemented using *IfcMaterialLayerSet* in the same way that layers in a concrete slab are defined. Note that this does not define the physical geometry of the road, only the thicknesses of its layers.

Road Course Segments, with different boundaries, compose a Road Element similarly to the way in which layers with different geometry representations can compose a precast concrete sandwich wall. Road Course Segments can therefore be implemented using *IfcBuildingElementPart*. Thus, each segment is represented by a discrete instance of a *IfcBuildingElementPart* with its own geometry and properties.

Roadels represent the "finite elements" that form a Road Course Segment. The aggregation of their geometry creates the geometry of a Road Course Segment. Thus, roadels can be implemented using *ifcProductDefinitionShape*, as they identify and form the product's shape.

As can be understood from the above explanation, since the IFC schema is well developed, concepts that were not conceived of in advance can be implemented using existing information entities. In practical terms, this is done through the mechanism of a formal Model View Definition.

However, the risk of abusing the IFC schema arises when doing so. Semantics also have great importance, and entity, relationship and property names should reflect their subjects as precisely as possible. For example, use of an *IfcBuildingElementPart* to represent a road course segment is inappropriate, and a new specific entity would be preferable. Similarly, the use of *IfcProductDefinitionShape* to represent a roadel is a workaround. A definition which allowed the roadels to be aggregated as parts of TIN surface network geometry, as well as triangular prisms, would be more suitable for volumetric computation. In other words, the current IFC Alignment and IFC Roads efforts should consider the specific needs of production and control of infrastructure, with a view to including appropriate entities in future versions of the IFC standard.

Table 4. Possible IFC bindings for the proposed road production and control schema.

| Entity | Possible IFC binding |
| --- | --- |
| Road Section | IfcCivilElementType |
| Road Course Segment | ifcBuildingElementPart |
| Road Course | Ifc Material Layer Set |
| Roadel | IfcProductDefinitionShape IfcShapeRepresentation IfcSweptAreaSolid + IfcArbitraryClosedProfileDef |

## 6 CONCLUSION

The information schema presented maintains the continuous nature of road construction, using discrete elements that can be aggregated dynamically into sub-products and work packages. The information classes defined in the schema create a product hierarchy that is continuous, differential and supports aggregation. The concepts of a 'roadel' and of a 'road course segment' are its primary new contributions.

Roadels constitute the lowest hierarchical level in the structure of the road and are formed both during production planning and production monitoring. A road course segment is located at the intermediate level in the hierarchical structure of the road. A Boolean union of these entity instances constitutes the primary product, as-planned and as-made.

These new information concepts are specifically designed to support dynamic Lean production planning and control of earthworks operations, including compilation of as-made work packages from the raw data obtained from machine sensors, drones and traditional surveys. As such, the proposed information model lays the foundation for technological development of tools that support Lean production planning and control for road construction and enable integration and exploitation of the information available from the array of IoT sensors and survey technologies that have become available.

## REFERENCES

Amann, J., Singer, D., and Borrmann, A. (2015). "Extension of the Upcoming IFC Alignment Standard with Cross Sections for Road Design." *2nd International Conference on Civil and Building Engineering Informatics*, Tokyo, Japan.

Ballard, G. (2000). "The Last Planner System of Production Control." PhD Dissertation, The University of Birmingham, Birmingham, U.K.

Ballard, G., and Howell, G. (1998). "What Kind of Production is Construction?" *6th Conference of the International Group for Lean Construction*, Guaruja, Brazil.

Bertelsen, S. (2003). "Construction as a Complex System." *11th Annual Conference of the International Group for Lean Construction*, Blacksburg, Virginia, 11–23.

buildingSMART. (2015). "Ifc Alignment."

ISO. (2013). "ISO 16739:2013 Industry Foundation Classes (IFC) for data sharing in the construction and facility management industries." International Standards Organization.

Kenley, R., and Seppanen, O. (2010). *Location-Based Management for Construction: Planning, Scheduling and Control*. Spon Press, London, UK.

Liebich, T., Adachi, Y., Forester, J., Hyvarinen, J., Richter, S., Chipman, T., Weise, M., and Wix, J. (2013). *Industry Foundation Classes: Version 4*. buildingSMART International.

Rebolj, D., Tibaut, A., Čuš-Babič, N., Magdič, A., and Podbreznik, P. (2008). "Development and Application of a Road Product Model." *Automation in Construction*, 17(6), 719–728.

Sacks, R., Eastman, C.M., Lee, G., and Teicholz, P. (2018). *BIM Handbook: A Guide to Building Information Modeling for Owners, Designers, Engineers, Contractors and Facility Managers*. John Wiley and Sons, Hoboken, NJ.

Sacks, R., Koskela, L., Dave, B.A., and Owen, R. (2010). "Interaction of Lean and Building Information Modeling in Construction." *Journal of Construction Engineering and Management*, 136(9), 968–980.

Schmenner, R.W. (1993). *Production/Operations Management: From the Inside Out*. Maxwell Macmillan International, New York.

Shingo, S., and Dillon, A.P. (1989). *A Study of the Toyota Production System: From an Industrial Engineering Viewpoint*. Productivity Press, Portland, Oregon, USA.

Sveikauskas, L., Rowe, S., and Mildenberger, J.D. (2018). *Measuring Productivity Growth in Construction : Monthly Labor Review*. Bureau of Labor Statistics, U.S. Department of Labor.

Vilgertshofer, S., Jubierre, J.R., and Borrmann, A. (2016). "IfcTunnel—a Proposal for a Multi-scale Extension of the IFC Data Model for Shield Tunnels under Consideration of Downward Compatibility Aspects." *Proceedings of the 11th European Conference on Product and Process Modelling*, CRC Press, Limassol, Cyprus, 175.

## APPENDIX 1

IFC file examples will be made available on https://bit.ly/2n8hqwl.

# Formwork detection in UAV pictures of construction sites

Katrin Jahr, Alexander Braun & André Borrmann
*Chair of Computational Modeling and Simulation, Technical University of Munich, Germany*

ABSTRACT: The monitoring of the construction progress is an essential task on construction sites, which nowadays is conducted mostly by hand. Recent image processing techniques provide a promising approach for reducing manual labor on site. While modern machine learning algorithms such as convolutional neural networks have proven to be of sublime value in other application fields, they are widely neglected by the CAE industry so far. In this paper, we propose a strategy to set up a machine learning routine to detect construction elements on UAV photographs of construction sites. In an accompanying case study using 750 photographs containing nearly 10.000 formwork elements, we reached accuracies of 90% when classifying single object images and 30% when locating formwork on multi-object images.

## 1 INTRODUCTION

The digitization of the construction industry offers various new possibilities for the planning, monitoring, and design process of buildings. In recent years, many research projects are focusing on using methods of computer-aided engineering, such as building information modeling or structural simulations, to facilitate and enhance the planning process. However, as of now, not many of the advantages of using digital support are used after the planning of a construction has been finished. While monitoring of the construction progress by comparing planned conditions to the actual situation is a labor-intensive task, it is still mostly conducted by the workforce on site with little technical support.

During the last decades, image processing techniques have been increasingly adopted by the construction industry, greatly improving and facilitating the process of construction monitoring. These methods gained new potential due to the more affordable and precise acquisition devices like unmanned aerial vehicles (UAVs) or laser scanners. Using the resulting 3D point clouds and information retrieved from the building information model, the possibility to track the progress of construction sites arises (Golparvar-fard, Pena-Mora, and Savarese 2009; Braun et al. 2015).

A detailed geometric as-planned vs. as-built comparison allows to track the current progress of a construction site, assess the quality of the construction work, and to check for construction defects such as cracks.

To generate high-quality point clouds, a significant number of consecutive photographs covering the monitored area is needed, requiring extensive image capturing and processing. However, most monitoring tasks do not entail the need for detailed 3D information. These include the monitoring of the quantity and positions of site equipment, of externally stored construction material, and major construction phases.

The image analysis and object detection on aerial photographs, which can be taken with relatively low effort, offers an alternative to expensively generating 3D point clouds. The scientific field of computer vision provides different solutions to process and, to a certain extent, understand images.

In this contribution, we use two state-of-the-art techniques of image processing to analyze aerial photography of construction sites. On the example of formwork elements, we demonstrate an artificial intelligence approach to recognize and locate construction elements on site. In the first part of the paper, we give an overview of the state of the art in image analysis as used on construction sites today, followed by a further description of the used methodology. We conclude the paper with a proof of concept and a summary of our results.

## 2 STATE OF THE ART

Computer Vision is a heavily researched topic, that got even more attention through recent advances in autonomous driving and machine learning related topics. Image analysis on construction sites, on the other hand, is a rather new topic. Since one of the key aspects of machine learning is the collection of large datasets, current approaches focus on data gathering. In the scope of automated progress monitoring, Han et al. published an approach for Amazon Turk based labeling (Han and Golparvar-Fard 2017). (Kropp, Koch, and König 2018) tried to detect indoor construction elements based on similarities, focusing on radiators.

For effective and efficient image analysis and object recognition, machine learning algorithms have been increasingly used during the last decades. In 2012, the convolutional neural network (CNN) "AlexNet" (Krizhevsky, Sutskever, and Hinton 2017) achieved a top-5 error of 15.3% in the prestigious ImageNet Large Scale Visual Recognition Challenge (Russakovsky et al. 2015). These results were surprisingly accurate at the time, proving the advantages of using CNN. On this account, the software industry shifted towards using CNN for all machine learning based image processing tasks (LeCun, Bengio, and Hinton 2015).

There are different tasks to be solved by image processing algorithms. Well known problems include classification, where single-object images are analyzed, object detection, where several objects in one image may be classified and localized within the image, and image segmentation, where each pixel of an image is classified (Buduma 2017). In this paper, we focus on image classification and object detection.

CNNs are structured in locally interconnected layers with shared weights. Each layer comprises multiple calculation units (called neurons). The neurons of the first layer (input layer) represent the pixels of the analyzed image, the last layer (output layer) comprises the predictable object classes.

In between input and output layer, any number of hidden layers can be arranged. While AlexNet contained 8 hidden layers, GoogLeNet (Szegedy et al. 2015), and Microsoft ResNet (He et al. 2016) use more than 100 hidden layers. The layers are usually convolution layers (sharpening features), pooling layers (discarding unnecessary information), or fully connected layers (enabling classification) (Buduma 2017; Albelwi and Mahmood 2017).

To adapt to different problems, such as recognizing formwork elements on images, CNNs must be trained. During training, the connections between certain neurons are increased, while the connections between other neurons are reduced–the weights connection consecutive layers are weighted. The training is usually carried out using supervised backpropagation, meaning that the network is fed with example input-output pairs (Buduma 2017). The correct solution for each input is called ground truth. To train a CNN towards reliable predictions, a significant amount of training data is required, which has to be prepared in a preprocessing step. ImageNet provides around 1.000 images per class, for example (Russakovsky et al. 2015). To accelerate the training processes, weights of previously trained CNNs can be used. To adapt pretrained CNNs, the fully connected layers are replaced with layers representing the new data and trained with the new data.

## 3 METHODOLOGY

In the context of the introduced research topics, the paper focusses on the image-based detection of temporary construction elements such as formwork. The detection of recurring, similar objects can be solved by machine-learning approaches. Several tools support the image analysis regarding automated detection of pretrained image sets.

### 3.1 Image classification using CNN

During image classification, which is also known as image recognition, images that contain but exactly one object are classified. Each class, that the CNN can detect, is represented by one output neuron. The activity of the neurons is read as the probability that the image contains an object of the corresponding class. Image classification algorithms will fail on images containing multiple objects. As images of construction sites contain more than one object, image classification algorithms can only be applied after preprocessing of the data. However, they can be very useful to confirm certain questions, e.g. if a wall with a known position is missing, currently shuttered or finished.

### 3.2 Object detection using CNN

The evident solution to analyze multi-object images is using a sliding window on the image and run an image classification on each window, which is computationally very expensive. Different proposals have been made to reduce the computational effort, e.g. region-proposal networks (e.g. R-CNN, (Girshick et al. 2014), (Girshick 2015), (Ren et al. 2017)), which intelligently detect regions of interest within an image and analyze those further, and single shot detectors (e.g. DetectNet (Tao, Barker, and Sarathy 2016) and YOLO (Redmon et al. 2016), (Redmon and Farhadi 2017), (Redmon and Farhadi 2018)), which overlay the image with a grid and analyze each cell.

### 3.3 Evaluation of CNN

To measure the performance of an image classifying CNN, the top-1 error and top-5 error are used. The top-1 error represents the fraction of images,

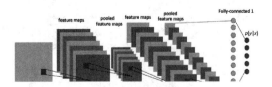

Figure 1. Structure of a sample CNN containing convolutional, pooling and fully connected layers.

for which the correct class has been predicted with the highest probability. The top-5 error is the fraction of images, for which the correct class is within the 5 classes that have been predicted with the highest probability, accordingly.

To measure the performance of an object detecting CNN, precision $p$, recall $r$ and mean average precision $mAP$ can be used. They are calculated using the number of true positives $TP$, false positives $FP$ and false negatives $FN$:

$$p = \frac{TP}{TP+FP} \qquad r = \frac{TP}{TP+FN}$$

In object detection tasks, a prediction is counted as true positive, if it has an intersection over union $IoU$ of a distinct value, usually over 0.5, meaning that more than 50% of the predicted bounding box should overlap the ground truth bounding box (see Figure 2):

$$IoU = \frac{area\ of\ overlap}{area\ of\ union}$$

For object detection, the mAP is the average of the possible precision at different recall values across all classes. To calculate the AP for each class, (Russakovsky et al. 2015) propose to consider 11 recall values according to the proposal by ImageNet:

$$AP = \frac{1}{11} \sum_{r \in \{0.0,\dots,1.0\}} p_i(r)$$

With $p_i$ = maximum precision for any recall value exceeding $r$.

### 3.4 Labeling

Labeling defines the approach of marking all regions of interest in a set of pictures and defining the type of the marked region. A subset of labeled pictures is depicted in Figure 5 a). The labels are marked with green bounding boxes.

As the labeling work takes a lot of time, a novel approach for automated labeling has been

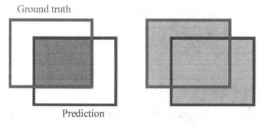

Figure 2. Area of overlap and area of union for predicted and labeled bounding boxes.

Figure 3. Reprojected bounding box of a column on a picture gathered during acquisition.

introduced by (Braun et al., 2018). In the frame of the research project ProgressTrack focusing on automated progress monitoring with photogrammetric point clouds, an algorithm has been developed to validate detection results of the as-built vs. as-planned comparison. As depicted in Figure 3, the projected 2D geometry of construction elements can be transformed from the building information model's coordinate system into the 2D coordinate system of each picture, the element is included in. This is possible, as the pictures were aligned and oriented during the photogrammetric process and thus making it possible to know the exact position in relation to the Building Information Model.

The process of labeling can benefit from this work because by this method, labels for all building element can be marked in all pictures, that were taken and aligned accordingly. Future research will focus on this method to extract labels for all construction elements and train a CNN accordingly without the time consuming, manual labeling work to be done.

## 4 CASE STUDY

In the following sections, we present an image analysis routine including data preparation as well as the training of convolutional neural networks to be able to recognize formwork elements. We focus on two different image analysis tasks: image classification and object detection.

### 4.1 Data preparation

As an initial dataset, 9.956 formwork elements were labeled manually on pictures of three construction sites that were collected during different case studies in the recent years. The images contain formwork elements from two different, German manufacturers and vary in size (30 cm up to 2,70 m length) as well as color (red, yellow, black, grey). They were taken at varying weather conditions on partly cloudy, as well as sunny days. The image acquisition was achieved with aerial photography by different UAVs, but also from the ground with regular digital cameras, resulting in image sizes from 4000 × 3000 px up to 6000 × 4000 px. The manual labeling process for this data set took around 130 h to complete.

The gathered data is processed as plain text files for each picture and processed for the various neural networks according to their respective requirements.

### 4.2 *Image analysis*

For image analysis, we used the Nvidia Deep Learning GPU Training System DIGITS (Yeager 2015), which provides a graphical web interface to the widespread machine learning frameworks TensorFlow, Caffe, and Torch (NVIDIA 2018). It enables data-management, network design and visualization of the training process.

#### 4.2.1 *Image classification*

We used a standard GoogLeNet CNN implemented in Caffe for the image classification task. The training is performed using the Adam Solver (Kingma and Ba 2014). We retrieved a classification dataset of formwork elements from the labeled data (Section 4.1) by automatically trimming the images around the bounding boxes of the labeled formwork elements (see a subset in Figure 4 b)). The automation was achieved by a self-written tool that takes all labeled data and images as input and crops them automatically. The tool is made avail-

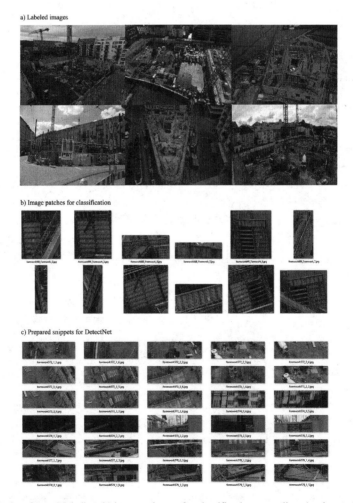

Figure 4. Sample data from a) labeling, b) image snippets for classification, as well as c) snippets for DetectNet.

Table 1. Classes and number of images per class used for training of an image classification CNN.

| Class | Origin | Number of images |
|---|---|---|
| Barrel | Caltech 256 | 47 |
| Bulldozer | Caltech 256 | 110 |
| Car | Caltech 256 | 123 |
| Chair | Caltech 256 | 62 |
| Formwork | Own dataset | 1410 |
| Screwdriver | Caltech 256 | 102 |
| Wheelbarrow | Caltech 256 | 91 |
| Wrench | Caltech 256 | 39 |

able on GitHub as an OpenSource solution[1]. To assure relatively even image sizes with sufficient detailing, we removed all images with resulting dimensions under 200 × 200 pixels.

To train the algorithm not only on formwork elements but on several classes, we added seven classes (see Table 1) that are related to construction sites from the Caltech 256 dataset (Griffin, Holub, and Perona 2007). The Caltech 256 provides single object images of 256 classes that need no further preprocessing for image classification.

As GoogLeNet requests input images of 256 × 256 pixels, all images are resized to that dimensions by DIGITS. For image classification, DIGITS automatically splits the data into training and validation data.

The CNN converged quickly towards high accuracies (top-1-error) around 85% (Figure 4) and stagnated at 90% after 100 epochs, which is a satisfying result. To achieve even higher accuracies throughout all classes, the number of images per class could be evened out by adding additional images to the underrepresented classes of the training data in future work.

### 4.3 Object detection

As next step, an object detection algorithm is introduced, to exactly detect certain elements in images and also precisely find the position of these elements. For this purpose, the dataset depicted in Figure 4 c) is used. To detect several formworks within an image of a construction site, we used a CNN with DetectNet architecture, implemented in Caffe. To reduce training time, we used the weights of the "BVLC GoogleNet model"[2], which has been pretrained on ImageNet data.

---

[1]https://github.com/tumcms/Labelbox2DetectNet
[2]Released for unrestricted use at https://github.com/NVIDIA/DIGITS/tree/master/examples/object-detection

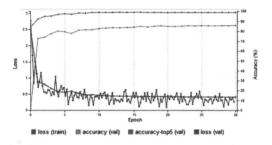

Figure 5. Loss and accuracy of the GoogLeNet after 30 epochs of training for classifying images of formwork elements and objects typically found on construction sites.

Table 2. Number of images and number of formwork elements contained in that images for training and validation of the object detection.

| Purpose | Nr. of images | Nr. of formwork elements |
|---|---|---|
| Training | 646 | 8429 |
| Validation | 99 | 1487 |

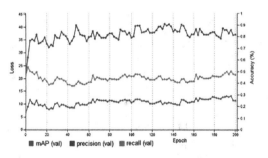

Figure 6. Precision, recall and mAP of the DetectNet after two rounds of 200 epochs of training for detecting formwork on images of construction sites.

The training again is performed using the Adam Solver.

We split the labeled images into 85% of training data and 15% of validation data. The images were recorded at a high resolution between 4000 × 3000 and 6000 × 4000 pixels. To minimize the necessary computational effort, we split the images into smaller patches with a size of 1248 × 384 pixels.

We trained the CNN twice with 200 epochs each. While the precision went up to 90%, the recall stagnated around 50% and the mAP at under 30% (Figure 5). From the results, we can conclude that the network detects only a few of the present formwork elements reliably. The high precision indicates low numbers of false positives, while the lower recall and mAP indicate a high number of false negatives. All in all, while the CNN does not tend to make many mistakes, it fails to recognize

Figure 7. Detected bounding box for formwork elements on a photography of a construction site.

many instances of formwork. Further steps to improve the object detection algorithm entail more extensive preprocessing of the data, longer training periods and adjustments of both the network architecture and the solving algorithms.

In Figure 6, the resulting bounding box for one example image is depicted. For this image, a very good result was retrieved.

## 5 SUMMARY

The presented research focusses on image analysis of construction site images. To make automated assumptions on the construction elements depicted on an image, machine learning tools need to be trained. First, the current state of the art for machine learning approaches is introduced and examined for their suitability of application in the domain of construction.

Then, these approaches are tested on construction site elements. For the training, 750 images of construction sites were labeled, resulting in nearly 10.000 labeled formwork elements. The images were used as input to various classification and detection algorithms, resulting in very high success rates for the classification of single object images and mediocre success rates for object detection on multi-object images. However, as object detection is a highly demanding task concerning a large community of researchers, the results give a promising starting point for future improvements.

## ACKNOWLEDGMENTS

This work is supported by the Bavarian Research Foundation under grant 1156-15.

We thank the Leibniz Supercomputing Centre (LRZ) of the Bavarian Academy of Sciences and Humanities (BAdW) for the support and provisioning of high-performance computing infrastructure essential to this publication.

## REFERENCES

Albelwi, Saleh, and Ausif Mahmood. 2017. "A Framework for Designing the Architectures of Deep Convolutional Neural Networks." *Entropy* 19 (6): 242. doi:10.3390/e19060242.

Braun, Alexander, Sebastian Tuttas, André Borrmann, and Uwe Stilla. 2015. "A Concept for Automated Construction Progress Monitoring Using BIM-Based Geometric Constraints and Photogrammetric Point Clouds." *ITcon* 20: 68–79.

Buduma, Nikhil. 2017. *Fundamentals of Deep Learning: Designing Next-Generation Machine Intelligence Algorithms*. Vol. 44. doi:10.1007/s13218–012–0198-z.

Girshick, Ross. 2015. "Fast R-CNN." In *2015 IEEE International Conference on Computer Vision (ICCV)*, 1440–48. IEEE. doi:10.1109/ICCV.2015.169.

Girshick, Ross, Jeff Donahue, Trevor Darrell, and Jitendra Malik. 2014. "Rich Feature Hierarchies for Accurate Object Detection and Semantic Segmentation." In *2014 IEEE Conference on Computer Vision and Pattern Recognition*, 580–87. IEEE. doi:10.1109/CVPR.2014.81.

Golparvar-fard, Mani, F Pena-Mora, and S Savarese. 2009. "D4 AR—a 4 Dimensional Augmented Reality Model for Automation Construction Progress Monitoring Data Collection, Processing and Communication." *Journal of Information Technology in Construction* 14 (June): 129–53.

Griffin, G., A. Holub, and P. Perona. 2007. "Caltech-256 Object Category Dataset." http://www.vision.caltech.edu/Image_Datasets/Caltech256/.

Han, Kevin K., and Mani Golparvar-Fard. 2017. "Potential of Big Visual Data and Building Information Modeling for Construction Performance Analytics: An Exploratory Study." *Automation in Construction* 73 (January): 184–98. doi:10.1016/j.autcon.2016.11.004.

He, Kaiming, Xiangyu Zhang, Shaoqing Ren, and Jian Sun. 2016. "Deep Residual Learning for Image Recognition." In *2016 IEEE Conference on Computer Vision and Pattern Recognition (CVPR)*, 770–78. IEEE. doi:10.1109/CVPR.2016.90.

Kingma, Diederik P., and Jimmy Ba. 2014. "Adam: A Method for Stochastic Optimization," December.

Krizhevsky, Alex, Ilya Sutskever, and Geoffrey E. Hinton. 2017. "ImageNet Classification with Deep Convolutional Neural Networks." *Communications of the ACM* 60 (6): 84–90. doi:10.1145/3065386.

Kropp, Christopher, Christian Koch, and Markus König. 2018. "Interior Construction State Recognition with 4D BIM Registered Image Sequences." *Automation in Construction* 86 (February): 11–32. doi:10.1016/j.autcon.2017.10.027.

LeCun, Yann, Yoshua Bengio, and Geoffrey Hinton. 2015. "Deep Learning." *Nature* 521 (7553): 436–44. doi:10.1038/nature14539.

NVIDIA. 2018. "Nvidia Digits—Deep Learning Digits Documentation," no. May.

Redmon, Joseph, Santosh Divvala, Ross Girshick, and Ali Farhadi. 2016. "You Only Look Once: Unified, Real-Time Object Detection." In *2016 IEEE Conference on Computer Vision and Pattern Recognition (CVPR)*, 779–88. IEEE. doi:10.1109/CVPR.2016.91.

Redmon, Joseph, and Ali Farhadi. 2017. "YOLO9000: Better, Faster, Stronger." In *2017 IEEE Conference on*

*Computer Vision and Pattern Recognition (CVPR)*, 6517–25. IEEE. doi:10.1109/CVPR.2017.690.

———. 2018. "YOLOv3: An Incremental Improvement," April.

Ren, Shaoqing, Kaiming He, Ross Girshick, and Jian Sun. 2017. "Faster R-CNN: Towards Real-Time Object Detection with Region Proposal Networks." *IEEE Transactions on Pattern Analysis and Machine Intelligence* 39 (6): 1137–49. doi:10.1109/TPAMI.2016.2577031.

Russakovsky, Olga, Jia Deng, Hao Su, Jonathan Krause, Sanjeev Satheesh, Sean Ma, Zhiheng Huang, et al. 2015. "ImageNet Large Scale Visual Recognition Challenge." *International Journal of Computer Vision* 115 (3): 211–52. doi:10.1007/s11263-015-0816-y.

Szegedy, Christian, Wei Liu, Yangqing Jia, Pierre Sermanet, Scott Reed, Dragomir Anguelov, Dumitru Erhan, Vincent Vanhoucke, and Andrew Rabinovich. 2015. "Going Deeper with Convolutions." In *2015 IEEE Conference on Computer Vision and Pattern Recognition (CVPR)*, 1–9. IEEE. doi:10.1109/CVPR.2015.7298594.

Tao, Andrew, Jon Barker, and Sriya Sarathy. 2016. "DetectNet: Deep Neural Network for Object Detection in DIGITS." https://devblogs.nvidia.com/detectnet-deep-neural-network-object-detection-digits/.

Yeager, Luke. 2015. "DIGITS : The Deep Learning GPU Training System." *ICML AutoML Workshop*.

# Fostering prefabrication in construction projects—case MEP in Finland

R.H. Lavikka, K. Chauhan, A. Peltokorpi & O. Seppänen
*Department of Civil Engineering, Aalto University School of Engineering, Finland*

ABSTRACT: Prefabrication increases construction project performance. However, the construction sector has not widely adopted it in Finland. This study focuses on understanding the barriers, enablers, value addition, and value capture of different construction project stakeholders when implementing mechanical, electrical, and plumbing (MEP) prefabrication. Based on a literature study and semi-structured interviews, the paper proposes that the implementation of MEP prefabrication should be seen as a challenge of implementing a systemic innovation. The implementation of a systemic innovation necessitates the buy-in of main project stakeholders, namely the owner, designers, general contractor, MEP sub-contractors, and MEP fabricator. This study provides evidence of the value added activities of each stakeholder to implement MEP prefabrication. The study implies that MEP prefabrication can be a 'win-win' situation for all stakeholders of a project network, which at its best, can be described as a project ecosystem. The research proposes ways to boost MEP prefabrication in construction projects.

## 1 INTRODUCTION

Already thirty years ago, Tatum et al. (1987) showed that pre-fabrication, preassembly, modularisation, and off-site fabrication (PPMOF) increases construction project performance. More recent studies have also demonstrated that PPMOF can provide many other benefits, such as reduced time to construct, improved product quality, reduction of waste, and improved worker safety (Babič et al., 2010; Li et al., 2014). As a result, the use of PPMOF had already doubled in 2000 (Haas et al., 2000) and it has been increasing ever since worldwide (Steinhardt and Manley, 2016). PPMOF has been successfully adopted by many countries, such as Sweden, the UK, Australia, the United States, Malaysia, and China (Li et al., 2017). Also, lately, new technology-based companies have started to enter the construction sector with their innovative PPMOF solutions. For example, Bone Structure, InterModal Structures, Nautilus Group, Project Frog, and ConXtech to name a few from the United States.

However, in our domestic market Finland, PPMOF plays a minor role in commercial construction. Our motivation to study the enablers, hindrances, value capture and value addition in implementing PPMOF in the Finnish commercial construction projects stems from this practical observation.

Despite the growing interest towards PPMOF, the construction sector is lagging behind in the use of PPMOF and automation compared to the manufacturing industry (c.f. Vähä et al., 2013). Previous studies have found several challenges in the implementation of PPMOF, such as uncertainty stemming from the multiple project environment, short-term buyer-supplier relationships, lack of trust between contractors and suppliers, the reluctance of suppliers to adopt new standards, and the lack of design-production interfaces (c.f. Bekdik et al., 2016). However, there are success stories of companies moving from traditional construction to PPMOF. For example, Bekdik et al. (2016) report of a mechanical, electrical, and plumbing (MEP) building system company in the US construction sector. The company has industrialised part products and provides installation services of prefabricated parts.

In contrast, MEP prefabrication is not a common practice in the Finnish construction sector, although point solutions exist. However, two factors drive the growing interest towards MEP prefabrication; Firstly, general contractors have realised that MEP prefabrication can bring the benefits of less hassle on site, improved project efficiency, and reduced project schedule. Secondly, the technology behind Building Information Modelling (BIM) is developed enough to support the creation of 'prefabrication-level' BIM models. However, the market for prefabricated MEP is still relatively small and unknown for the majority of construction stakeholders. The interesting question is how to implement MEP prefabrication into the fragmented and risk-averse project-based construction business. Inspired by the successful US case, we decided to focus on understanding how to increase MEP prefabrication in Finland. We will study the barriers, enablers, value addition, and value capture of different project stakeholders in implementing MEP prefabrication. A recently conducted study on prefab practice in the electrical construction industry also suggests that future research should focus on investigating how to increase the level and magnitude of prefabrication in construction (Hanna et al., 2017). The study conducted by Hanna et al. (2017) recommends that electrical contractors could use prefabrication as a way to enhance their projects' performance.

## 2 LITERATURE ON PPMOF: DRIVERS AND INHIBITORS FOR IMPLEMENTATION

Prefabrication is the practice of manufacturing and assembling the components of a structure in a factory, and transporting complete assemblies to the construction site (Lidelöw et al., 2015). The implementation of PPMOF has been slow in the construction sector. Several reasons exist for the slow adoption, for example, the fragmented structure and risk-averse culture of the sector (c.f. Bekdik et al., 2016) as already discussed. Goodier and Gibb (2007) have found that one inhibitor to implementing PPMOF is the belief that PPMOF is more expensive than traditional construction. Also, complex interfaces between systems and inability to freeze the design early on are reported as inhibitors for implementing PPMOF (Pan et al., 2008).

The reasons to start implementing PPMOF, however, are also numerous, such as improved project quality and productivity, reduced time and costs, and reduced health and safety risks (Pan et al., 2008). For example, a study on a large hospital construction project revealed that the general contractor could maintain a safe and efficient site throughout the construction of MEP systems because most systems were prefabricated and only installed onsite (Khanzode et al., 2008). For example, all the plumbing and low-pressure ductwork were prefabricated. The same study showed that the MEP work productivity improved between 5% and 25% because of MEP prefabrication. These benefits were partly gained through efficient work coordination using BIM (3D model). (Khanzode et al., 2008)

Many studies have focused on the drivers and inhibitors of prefabrication (e.g., Gibb and Isack, 2003; Goodier and Gibb, 2007). These studies show that PPMOF needs to be considered very early in the design process; otherwise, the benefits of PPMOF are not achieved (Pan et al., 2012). Some researchers advocate the integration of design and construction processes and logistics as a way to help in the implementation of PPMOF (e.g. Pan et al., 2008).

Goodier and Gibb (2007) studied the opinions of clients, designers, contractors, and off-site suppliers on off-site technologies in the UK construction industry through a questionnaire. They found out that the main inhibitors to increased PPMOF were the lack of transparent information about the actual costs of PPMOF and the lack of multi-skilled labour to work in the off-site factories. The authors suggested that the construction project stakeholders need to be brought together through improved communication, more experience, and education.

Hedgren and Stehn (2014) studied the impact of clients' decision-making on their adoption of PPMOF. They suggest that it is important for the clients to embrace uncertainty and equivocality as means to overcome organisational and cognitive barriers that are hindering the adoption of PPMOF. In practice, this means that the decision-making processes are built on dialogue and relationships between the stakeholders, which enable the creation of multiple meanings and interpretations to interact with decision-making.

The most recent study, which we were able to find, on the prefabrication practices, studied the impact of prefabrication on project performance as perceived by electrical contractors (Hanna et al., 2017). The study found that prefabrication enables contractors to reduce labour wages and expedite the construction process by performing more activities in parallel. The study also showed that prefabrication could be widely used in many electrical activities, and prefabrication often results in cost and schedule savings. Based on their study, Hanna et al. (2017) suggest that companies could implement more PPMOF practices by developing training programs for personnel and by establishing protocols for working with suppliers.

## 3 LITERATURE ON IMPLEMENTING SYSTEMIC INNOVATIONS

According to Hedgren and Stehn (2014), PPMOF often necessitates changes in the process and the product dimensions of a construction project. The process dimension is affected by off-site production and new forms of organisation, such as the more intensive integration of the value chain which includes the onsite practices, procurement, and logistics. PPMOF can also induce new technical and business practices. (Hedgren and Stehn, 2014) Hence, the implementation of PPMOF resembles the implementation of a systemic innovation in construction. A systemic innovation in a construction project is an innovation that impacts the inter-company processes (Taylor and Levitt, 2004). The locus of innovation is in the linkages between subsystems, whereas the entities affected by the systemic innovation are the multiple companies. In other words, a systemic innovation is an innovation that usually is not contained within the control of an implementer. Instead, it necessitates that other stakeholders within the 'influence domain' of the innovation also take action to adjust to the needed changes (Taylor and Levitt, 2004; Harty, 2005; Alin et al., 2013).

A properly implemented systemic innovation enables the project network to share knowledge. A systemic innovation also improves the operational performance in the long term. The changes required by the systemic innovation may create switching or start-up costs for some project stakeholders and reduce or even eliminate the role of some stakeholders. (Taylor and Levitt, 2004) The implementation of systemic innovations has been found to be difficult because these innovations span organisational boundaries and are often misaligned with the structures of the project network (Hartmann et al., 2009).

As an example of the implementation of a systemic innovation, we found a study conducted by Alin et al. (2013). They studied the implementation of Building Information Modeling (BIM) into a construction project network and found evidence for the idea that a task sequence alignment between the project companies leads to a knowledge-base alignment and a subsequent work allocation alignment. The task sequence alignment refers to the process that the project network applies to change the work task sequence. Knowledge-base alignment takes place by integrating knowledge and keeping knowledge current, whereas work allocation is aligned by changing work tasks, creating new tasks, and changing the roles of specialists. (Alin et al., 2013)

In sum, the implementation of a systemic innovation necessitates the coordination of inter-organisational work tasks and knowledge flows, often through mutual adjustment. Taylor and Levitt (2004) report that the magnitude of this coordination is a function of organisational variety, the degree of interdependence, boundary strength, and organisational span. An example of the organisational span is the interface between the plumber and mechanical contractor. As a result, Taylor and Levitt (2004) suggest four focus areas for project managers to implement a systemic innovation; Firstly, to reduce the organisational variety of specialist contractors. Secondly, to monitor the degree of interdependence of work tasks to know where the potential problems lie. Thirdly, to reduce boundary strength through an environment that creates inter-organisational trust. Moreover, fourthly, to decrease the span of the systemic innovation by using integrator firms, such as an MEP contractor that can integrate the work of multiple specialist firms, in this case, the mechanical, electrical, and plumbing companies. (Taylor and Levitt, 2004)

Table 1. Interview data.

| Organization | Interview | Expertise | Date |
| --- | --- | --- | --- |
| Bathroom module fabricator | Project manager | Installation of kitchen/ bathroom modules | 18.12.2017 |
| General contractor | Production unit manager | Installation of MEP systems | 20.12.2017 |
| MEP union | Branch manager | MEP systems installation, MEP contracts | 3.1.2018 |
| MEP system provider | Head of projects | Procurement and installation of prefabricated MEP systems | 5.1.2018 |
| Precast concrete producer | Design manager | Precast concrete design | 8.1.2018 |
| MEP design consultant | Technology director | Design of MEP solutions | 9.1.2018 |
| Construction union | Negotiations manager | MEP installation contracts | 12.1.2018 |
| MEP fabricator | Production manager, Chief engineering officer | Design and production of MEP elements | 23.1.2018 |
| Client (business) | Senior vice president | Procurement of construction process professionals | 24.1.2018 |
| Client (government) | Specialist | Indoor air, energy consumption | 30.1.2018 |
| Client (university) | Director of construction | Procurement of design and construction work | 31.1.2018 |
| Client (government) | Director of construction | Procurement of design and construction work | 9.2.2018 |
| General contractor | Project manager, MEP manager | Management of construction processes, MEP procurement | 13.2.2018 |
| Bathroom module fabricator | Operations director | Installation of bathroom modules | 14.2.2018 |
| MEP contractor | Mechanic | MEP installation | 20.2.2018 |
| Client (university) | Premise manager | Procurement of modules | 21.2.2108 |
| University | Professor | Prefabrication of mechanical engineering | 8.3.2018 |
| Modular MEP mounting systems | Key account manager, project manager | MEP mounting systems | 15.3.2018 |
| Architectural firm | Architect | MEP prefabrication design process | 23.3.2018 |

## 4 METHOD

Our hypothesis based on the literature study was that the implementation of MEP prefabrication necessitates the willingness of all organisational project stakeholders to change their current process practices. This hypothesis was derived from the understanding that the implementation of MEP prefabrication resembles the implementation of a systemic innovation in the construction sector.

To identify the motives and perspectives of all construction project stakeholders towards implementing MEP prefabrication, we wanted to conduct a multi-perspective analysis of the barriers, enablers, value addition, and value capture of different stakeholders. In practice, we needed to find interviewees representing construction project stakeholders. First, we approached a few fabricators and general contractors that we knew were interested in the topic. After that, we applied snowball sampling (Biernacki and Waldorf, 1981) with the help of the already interviewed informants.

We conducted semi-structured interviews from December 2017 to March 2018; altogether, we had twenty-two interviewees from eighteen organisations. Table 1 presents our interview data; the organisations we interviewed, the role of the interviewee, her/his expertise, and the date of the interview. We selected industry interviewees that were experienced in the use of PPMOF, either in the residential or commercial construction projects. Each interview was voice-recorded by the permission of the interviewees, and each interview took about one hour.

We followed the interview data analysis recommendations of Miles and Huberman (1994). First, after each interview, a transcription service provider transcribed the interview verbatim. Second, we read the transcription to get a preliminary understanding of the data. Then, we encoded the transcription and analysed the chosen quotes using a qualitative data analysis software, Atlas. ti. We used six codes in the analysis: 1) Enablers of MEP prefabrication, 2) Hindrances of MEP prefabrication, 3) Benefits of MEP prefabrication, 4) Planning process, 5) Technical solutions of MEP prefabrication, and 6) Other interesting. Two researchers conducted the data analysis, and the analysis findings were compared to receive a consensus on the meaning of data.

## 5 FINDINGS

Table 2 presents the findings of the multi-perspective analysis regarding barriers, enablers, value addition, and value capture in implementing MEP prefabrication. Bathroom modules with pipework, electrical cables and ductwork for building services provide one example of MEP prefabrication. Other examples of MEP prefabrication include prefabricated pipeline manifolds and corridor elements that include ductwork, pipework and electrical cables in MEP racks. These prefabricated multi-service 'modules' are usually insulated, pressure tested and mounted in the ceiling or under the floor.

The findings show that several barriers hinder the implementation of MEP prefabrication in Finland. However, the interviewees also provided many suggestions for enabling the implementation of MEP prefabrication. Next, we will discuss the four main barriers to implementing MEP prefabrication. After that, we provide the four main enablers for implementing MEP prefabrication. Finally, we discuss the value addition and value capture of each stakeholder through the implementation of MEP prefabrication into commercial house-building projects.

### 5.1 *Main barriers to implementing MEP prefabrication*

The first barrier is that current business models and contract boundaries between different trades do not support collaboration, which is needed for implementing MEP prefabrication. For example, one example of MEP prefabrication is the 'common hanger system' where all MEP hangers are prefabricated and installed in one MEP rack. A 'common prefabricated hanger system' would mean that only one MEP sub-contractor would install a hanger system. However, the implementation of this 'common hanger system' is against the current business model where each sub-contractor, performing a part of the MEP work, counts its hangers into their contracts because the business model is partly based on selling building parts. An MEP sub-contractor explains his company's viewpoint in the following way, "If everything is prefabricated, it means less work to our plumbing employees." Thus, one problem lies in the 'switching costs' of companies; how to agree on a common hanger system between several companies that all have the same kind of business model that is being challenged by the prefabrication practices.

The second barrier is that designers, i.e., architects and MEP designers, are still used to designing one-of-a-kind products, whereas prefabrication necessitates repetition in design solutions. At its best, for example, bathrooms and kitchens in commercial construction could be standardised to some extent. A general contractor provides one solution to the challenge, "The design process should be the other way around. First, we need to have the design modules which we can use to build the facility, the same way as Legos work." This challenge relates to the way

Table 2. The views of different stakeholders.

| Stakeholder | Barriers | Enablers | Value addition | Value capture |
|---|---|---|---|---|
| Client | Lack of prefab procurement knowledge; Lack of knowledge about the timing of fixing client requirements | Relational contracts | Know-how in facility management; Change agent | Reduced schedule and cost; Improved quality |
| Designers | Rigid contracts; Industry's resistance to change; Rigid division of responsibilities for MEP design and installation; | Design collaboration with MEP sub-contractor; Relational contracts; Changes in sub-contractor responsibilities; Changes in business models/trade union requirements | Installation-level BIM model; | Designing only 'one time'; More design work 'for construction' |
| MEP sub-contractor | Tight schedule; Risk-averse culture; Lack of resources; Bad designs; Unions' agreements for prefabrication payments; Lack of repeatability | Relational contracts; Installation-level BIM; Workshops for prefabrication | Installation knowhow | Improved quality; Reduced labour costs; Reduced throughput time; Project efficiency; Improved work safety |
| General contractor | Lack of MEP prefabrication procurement knowledge | Showcases of good practices for prefab | Change agent | Site productivity improvement; Fewer logistics; |
| Fabricator | The market is missing; Design revisions; Detailed MEP design made too late | Client requirements/ freezed design early on | Less material waste on site; Better quality; Reduced schedule | Market development |

designers are educated in schools and what they learn during their work years, i.e., what kind of experience and attitude towards learning new approaches they possess.

The third barrier to MEP prefabrication is that the quality of MEP design is not yet supporting MEP prefabrication or the installation work of MEP prefabricated building parts. An MEP designer suggests, "We need to involve the MEP contractor or fabricator early in the design process to produce designs for prefabrication." This solution could be quite easily implemented in the design process, for example, as consultation hours either from the MEP contractor or the fabricator.

The fourth barrier is that clients are required to change their procurement practices because the traditional practice of separately procuring each building part does not support the procurement of, for example, a prefabricated wall system that includes subcomponent systems of MEP. The comparison between procuring building parts separately versus procuring a prefabricated wall system with MEP is not straightforward because the benefits of prefabrication are spread to the entire construction supply chain, and the benefits are realised, for example, as lower logistics and material waste costs. An MEP fabricator explains it in the following way, "The client should see MEP as one package to be procured and not as separate building parts". In fact, half of our interviewees mentioned that prefabrication reduces the costs of the whole value chain through quicker on-site installation (shorter turnaround time) and fewer logistics; thus the procurement should consider the whole construction supply chain.

5.2 *Main enablers for implementing MEP prefabrication*

The four main enablers of MEP implementation follow almost straight from the barriers. The first enabler is relational contracts, such as alliance models that support collaboration between the trades. A designer sheds light on the benefits of alliance models "In an alliance, we can think about the benefit of the project, instead of only my direct costs."

The second enabler for MEP prefabrication is the standardisation of design solutions, "technically it would be easy to agree on design standards on pipes and ducts and interfaces between the trades", explains a general contractor.

The third enabler for MEP prefabrication would be the installation-level BIM, "We could design more detailed level BIM through cross-trade design collaboration between the designer, the MEP contractor, and the fabricator", reveals an MEP designer.

The fourth enabler relates to the procurement of MEP prefabrication. A fabricator suggests, "The general contractor and the client should change their procurement practices to support MEP prefabrication. The workers' union agreements do not support MEP prefabrication, but some workers are under a different union."

### 5.3 Value capture and value addition of each stakeholder

Table 2 reveals that each stakeholder can capture value through MEP prefabrication, which means that there is ground for working together in implementing MEP prefabrication practices. For example, the client receives a better-quality facility at reduced time and schedule, whereas the MEP sub-contractor benefits from improved quality, reduced labour costs, reduced throughput time, and improved worker safety to name a few. MEP prefabrication allows the MEP designers to design only once and not having to revise their designs later on as prefabrication necessitates designs to be 'freezed' early on. Prefabrication could potentially also increase design work for MEP designers as they need to provide more detailed designs. General contractor, on the other hand, benefits from site productivity improvements and lower logistic costs. The fabricator will most likely benefit from the market development.

Each stakeholder also adds value to the organisational project ecosystem through implementing MEP prefabrication. For example, the client can provide facility management knowhow to MEP designers and work as a 'change agent' in implementing best MEP prefabrication practices. Also, the general contractor can act as a change agent. At its best, the fabricator can provide better quality products at reduced price and schedule, whereas the MEP designer can provide an installation-level BIM model. The MEP sub-contractor can provide its knowledge of installations to the designs. When all the project ecosystem stakeholders work towards the common goal of MEP prefabrication, they can reduce material waste and inter-organisational on-site work, which often includes challenges. At its best, MEP prefabrication can help in the production of a high-quality facility for the owner at a fast pace and a fixed cost.

## 6 DISCUSSION

The findings of our study hint that the implementation of MEP prefabrication resembles the implementation of a systemic innovation that necessitates changes in the design and construction practices of the project network that is delivering the project. To effectively implement MEP prefabrication, it is important that all the project stakeholders commit to a design and construction process that supports MEP prefabrication. This process resembles a manufacturing process in the sense that design requirements need to be freezed quite early on. At its best, all stakeholders add value to the design and construction process but also capture value from it. The systemic innovation approach supports understanding the sustainable implementation of MEP prefabrication in the 'MEP prefabrication' ecosystem.

Based on our results, the clients seem to play a key role in enabling MEP prefabrication as already previous studies have found (Gibb and Isack, 2003). First, the clients need to be aware of the right timing for making decisions, especially when to freeze design requirements so that detailed design can start and expenses due to late design changes are avoided. Second, it seems that many general contractors and clients are still using price as the main procurement criteria instead of value, which often precludes PPMOF. This finding was discovered already over ten years ago (Blismas et al., 2006), which means that not much change has taken place. The reduced costs of MEP prefabrication do not usually come in the form of a cheaper MEP prefabricated product, but the costs of the whole supply chain are reduced through the quicker onsite installation, reduced logistics costs, less material waste, and fewer worker injuries.

Our study confirms the previous finding by Song et al. (2005) and Goodier and Gibb (2007) that the construction stakeholders lack knowledge on the benefits of prefabrication. Our study reveals that both construction industry stakeholders and clients are unaware of the financial impacts of applying MEP prefabrication, which diminishes their willingness towards implementing prefabrication. The stakeholders also lack knowledge about the design and construction phase practices that would support MEP prefabrication. Also, industry stakeholders still tend to postpone decisions on the application of MEP prefabrication until the very end of the design phase. The reason is that the customers do not understand that the decisions concerning prefabrication have to be made early in the design to receive the benefits of reduced schedule and price. Previous findings show that the design needs to be 'freezed' early on (Gibb and Isack, 2003; Pan et al., 2012).

Our results support the earlier findings by Li et al. (2017) that the construction sector has not focused on standardising the product and its components, but it has rather focused on optimising the current design and construction processes and organisation. Based on our results, we argue that the sector and its customers should systematically start using standardised components and products, which need to be considered early in the design and procurement processes to capture the value potential for each stakeholder.

We contribute to the discussion on implementing MEP prefabrication in construction by showing how the resources and activities of each stakeholder can be combined towards achieving MEP prefabrication more widely into the Finnish construction sector. The findings suggest that the design process needs to be changed to provide the potential value capture for each stakeholder. For example, designers, contractors and fabricators should collaborate in the early design phases to receive a design that supports prefabrication, installation, and onsite work. This finding relates to the previous finding by Arif et al. (2012) that PPMOF necessitates an understanding of the manufacturing process. In fact, already the design phase requires effective interaction with manufacturing to ensure the production of designs that can be manufactured. A survey in 2009 revealed that the use of BIM for direct fabrication was limited, even though 25% of the respondents told to utilise BIM for direct fabrication (Becerik-Gerber and Rice, 2010).

Our empirical case of MEP in Finland is limited, and further empirical data collection in other countries would allow us to conduct a comparative case study of MEP prefabrication between countries. Cross-country comparison is needed to understand the implementation of PPMOF more widely in the world.

## 7 CONCLUSIONS

This paper has focused on understanding the barriers, enablers, value addition, and value capture of different construction project stakeholders. The paper presents the key findings of twenty-one interviewees. Based on the literature study and the analysis of the findings, the paper proposes that the implementation of MEP prefabrication should be seen as a challenge to implementing a systemic innovation in the construction sector. The implementation of a systemic innovation necessitates the buy-in of main project network stakeholders, namely the owner, the designers, the general contractor, the MEP sub-contractors, and the fabricator.

This study provides evidence for the value-added activities of each stakeholder to help in the implementation process of MEP prefabrication. The study also reveals that each stakeholder can capture added value from MEP prefabrication. Thus, in practice, MEP prefabrication seems a 'win-win' situation to all the stakeholders of the project network, which at its best can be described as a project ecosystem for MEP prefabrication (c.f. Walrave et al., 2018).

Two practical implications can be suggested based on the study. First, the study suggests that the client and the general contractor should start nurturing a culture that embraces MEP prefabrication. For example, best project practices for MEP prefabrication should be documented and shared with a broader audience in the industry magazines and conferences. The client and the general contractor can also help in the creation of a 'trusting' atmosphere at the beginning of a construction project through collaborative practices that increase communication between the stakeholders. The early involvement of the general contractor and MEP designers in assisting roles could also be one solution to increase shared understanding about MEP prefabrication practices. At best, they can provide input to decision making processes concerning MEP prefabrication and operations and logistics management. In the US, design-assist subcontracting is used because it allows the subcontractors to assist designers during the early design phase (Khanzode et al., 2008).

The second practical implication of this study is that the clients should change their current procurement practices of 'separate building parts' to 'MEP as one package' which better supports MEP prefabrication. The financial benefits of MEP prefabrication should also be quantified and reported to industry practitioners. For example, there is a tool named IMMPREST which attempts to help decision makers in comparing and costing solutions (Blismas et al., 2003). In other words, the project decision makers need transparent information about the costs of MEP prefabrication compared with traditional construction methods. These activities will, hopefully, help to minimise the resistance to change in the industry.

## REFERENCES

Alin, P., Maunula, A., Taylor, J.E. and Smeds, R. (2013), "Aligning misaligned systemic innovations: Probing inter-firm effects development in project networks", *Project Management Journal*, Vol. 44 No. 1, pp. 77–93.

Arif, M., Goulding, J. and Rahimian, P.F. (2012), "Promoting off-site construction: Future challenges and opportunities", *Journal of Architectural Engineering*, Vol. 18 No. 2, pp. 75–78.

Babič, N.C., Podbreznik, P. and Rebolj, D. (2010), "Integrating resource production and construction using BIM", *Automation in Construction*, Vol. 19 No. 5, pp. 539–543.

Becerik-Gerber, B. and Rice, S. (2010), "The perceived value of building information modeling in the U.S. building industry", *Journal of Information Technology in Construction (ITcon)*, Vol. 15, pp. 185–201.

Bekdik, B., Hall, D. and Aslesen, S. (2016), "Off-site prefabrication: What does it require from the trade contractor?", *International Group for Lean Construction*, No. 43, pp. 43–52.

Biernacki, P. and Waldorf, D. (1981), "Snowball sampling: Problems and techniques of chain referral sampling", *Sociological Methods and Research*, Vol. 10 No. 2, pp. 141–163.

Blismas, N., Pasquire, C. and Gibb, A. (2006), "Benefit evaluation for off-site production in construction", *Construction Management and Economics*, Vol. 24 No. 2, pp. 121–130.

Blismas, N.G., Pasquire, C.L. and Gibb, A.G.F. (2003), *IMMPREST*, Loughborough.

Gibb, A.G.F. and Isack, F. (2003), "Re-engineering through pre-assembly: Client expectations and drivers", *Building Research and Information*, Vol. 31 No. 2, pp. 146–160.

Goodier, C. and Gibb, A. (2007), "Future opportunities for offsite in the UK", *Construction Management and Economics*, Vol. 25 No. 6, pp. 585–595.

Haas, C.T., O'Connor, J.T., Tucker, R.T., Eickmann, J.A. and Fagerlund, W.R. (2000), *Prefabrication and preassembly trends and effects on the construction workforce*, Rep. No. 14, Center for Construction Industry Studies, Austin, Texas.

Hanna, A.S., Mikhail, G. and Iskandar, K.A. (2017), "State of prefab practice in the electrical construction industry: Qualitative assessment", *Journal of Construction Engineering and Management*, Vol. 143 No. 2.

Hartmann, T., Fischer, M. and Haymaker, J. (2009), "Implementing information systems with project teams using ethnographic-action research", *Advanced Engineering Informatics*, Vol. 23 No. 1, pp. 57–67.

Harty, C. (2005), "Innovation in construction: A sociology of technology approach", *Building Research & Information*, Vol. 33 No. 6, pp. 512–522.

Hedgren, E. and Stehn, L. (2014), "The impact of clients' decision-making on their adoption of industrialized building", *Construction Management and Economics*, Vol. 32 No. 1–2, pp. 126–145.

Khanzode, A., Fischer, M.A. and Reed, D.A. (2008), "Benefits and lessons learned of implementing building virtual design and construction (VDC) technologies for coordination of mechanical, electrical, and plumbing (MEP) systems on a large healthcare project", *Journal of Information Technology in Construction*, Vol. 13, pp. 324–342.

Li, X., Li, Z. and Wu, G. (2017), "Modular and offsite construction of piping: Current barriers and route", *Applied Sciences*, Vol. 7 No. 6, p. 547.

Li, Z., Shen, G. and Xue, X. (2014), "Critical review of the research on the management of prefabricated construction", *Habitat International*, Vol. 43, pp. 240–249.

Lidelöw, H., Stehn, L., Lessing, J. and Engström, D. (2015), *Industriellt Husbyggande*, Författarna och studentlitteratur AB, Lund, Lund, Sverige.

Miles, M.B. and Huberman, A.M. (1994), *An Expanded Sourcebook – Qualitative Data Analysis*, Sage Publications, California, USA.

Pan, W., Gibb, A.G.F. and Dainty, A.R.J. (2008), "Leading UK housebuilders' utilization of offsite construction methods", *Building Research and Information*, Vol. 36 No. 1, pp. 56–67.

Pan, W., Gibb, G.F. and Dainty, A.R.J. (2012), "Strategies for integrating the use of off-site production technologies in house building", *Journal of Construction Engineering and Management*, Vol. 138 No. 11, pp. 1331–1340.

Song, J., Fagerlund, W.R., Haas, C.T., Tatum, C.B. and Vanegas, J.A. (2005), "Considering prework on industrial projects", *Journal of Construction Engineering and Management*, Vol. 131 No. 6, pp. 723–733.

Steinhardt, D.A. and Manley, K. (2016), "Adoption of prefabricated housing-the role of country context", *Sustainable Cities and Society*, Vol. 22, pp. 126–135.

Tatum, C.B., Vanegas, J.A. and Williams, J.M. (1987), *Constructability improvement using prefabrication, preassembly, and modularization*, Bureau of Engineering Research, University of Texas at Austin, Austin, Texas.

Taylor, J.E. and Levitt, R.E. (2004), "Understanding and managing systemic innovation in project-based industries", in Slevin, D., Cleland, D. and Pinto, J. (Eds.),*Innovations: Project Management Research 2004*, Newtown Square: Project Management Institute, pp. 83–99.

Vähä, P., Heikkilä, T., Kilpeläinen, P., Järviluoma, M. and Heikkilä, R. (2013), *Survey on Automation of the Building Construction and Building Products Industry*, Espoo, Finland.

Walrave, B., Talmar, M., Podoynitsyna, K.S., Romme, A.G.L. and Verbong, G.P.J. (2018), "A multi-level perspective on innovation ecosystems for path-breaking innovation", *Technological Forecasting and Social Change*, Vol. in press.

# RenoBIM: Collaboration platform based on open BIM workflows for energy renovation of buildings using timber prefabricated products

A. Mediavilla, X. Arenaza & V. Sánchez
*TECNALIA, Derio, Spain*

Y. Sebesi
*Dietrich's, Strasbourg, France*

P. Philipps
*Dietrich's, Munich, Germany*

ABSTRACT: AEC industry is by far one of the most fragmented industries, with still important data and process silos. This paper presents an innovative platform developed in H2020 funded BERTIM project to overcome this barrier in the field of building energy renovation using timber prefabricated products. It supports the overall process from BIM creation using laser scanning techniques, assessment of the renovation project feasibility and selection of the most cost-effective alternative to be then produced by CNC machines. It follows the decision-making methodology developed in the project, involving several stakeholders (service provider, designer, manufacturer and potential client of the renovation process), as support their collaboration, with relevant time savings and better knowledge sharing. The platform implements Open BIM workflows through the interoperability of tools using IFC files, linking custom developed decision support tools, existing energy simulation engines (Energy Plus) and CAD/CAM tools, with secondary objectives like running cloud energy simulations by non-expert users. Simulation models are automatically created in two approaches: procedurally created models from user inputs (for early decisions), as well as an advanced IFC to Energy Plus conversion in detailed phases. Additionally, the user can quickly define the façade splitting configuration to be added to the existing IFC and conveyed to the CAD/CAM tool for manufacturing. 3D model visualization is supported using Web3D technologies. The final platform is easily adaptable to different manufacturers' processes and products.

## 1 INTRODUCTION

The energy renovation of buildings consists on many stages: current building data gathering, technical or legal feasibility assessment of the project, energy savings and return on investment estimation and implementation of the solution. When the solution consists on prefabricated products, the existing geometry must be accurately captured, to avoid potential problems when installing the prefabricated panels.

BERTIM project aims to systematize the process by developing a methodology supported by a cloud-based platform (RenoBIM), addressed to manufacturers of prefabricated timber products for renovation of residential buildings. In any case, the concept of the tool can be easily extended to other building uses and product types.

Figure 1 shows the two main aspects covered by RenoBIM, namely the Decision Support Tool (DST) and Design Configurator, and its interaction with external tools in the overall BERTIM process.

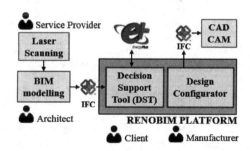

Figure 1. RenoBIM in the overall BERTIM workflow.

The rationale of the followed approach is that a huge set of tools already exist for various uses (such as design, simulation, fabrication and costs) and rather than creating new ones, the goal is to integrate existing and reliable tools by developing smooth links via Open BIM standards. IFC has been chosen as the most widespread data exchange format.

The DST aims to be a marketing tool to show to potential customers the benefits of renovating their buildings using the prefabricated timber panels produced by the company.

The Design Configurator is targeted to a technical person, which quickly generates a first façade splitting layout, further exported to CAD/CAM tools. In any case, the platform allows the use of each module by different user roles.

## 2 RENOBIM WORKFLOW

Building energy renovation is a complex process, involving different decisions at different stages by different actors, especially complex when based on prefabrication.

RenoBIM enables a sequential decision making, flexibly adapting its functionalities to the level of detail of the available data in each stage. Initially, the technical feasibility of the project is verified after evaluating if the imposed constraints are met. This can avoid going into a detailed cost-energy analysis when the project is not technically feasible.

Then, the cost-effectiveness is evaluated by calculating energy and cost indicators, comparing the current situation of the building and one or more renovation scenarios based on different product configurations for the façade renovation. RenoBIM is flexible enough to adapt to different situations. Thus, it implements an IFC-based energy simulation and cost calculations, but also enabling the definition of a virtual building geometry, based on which the simulations are run. The motivation for this double approach is allowing very early estimations, before actually engaging to the real project. In this stage, for existing residential buildings, no digital data is usually available. However, without the IFC file the full range of functionalities of RenoBIM will not be available. In both cases (IFC and procedural) an embedded 3D viewer is supported using WebGL technologies and Energy Plus™ engine is used for automatically launching cloud energy simulations on the cloud.

Finally, an easy-to-use configurator is implemented using a web 2D drawing functionality. This supports technical users in designing and optimizing the conceptual façade splitting layout, which is converted to a collection of panels in IFC and merged back into the original IFC of the building. The merged file is then imported by the CAD-CAM tool from Dietrich's, also a partner in BERTIM project and leading software vendor in the timber industry.

A private section for manufacturers allows managing their specific restrictions and the catalogue of products used for both the DST and the configurator.

The platform also enables the collaboration of actors in the process. All projects created by a user in the company are accessible by the rest. Projects created by external users are automatically notified to the manufacturer. A shared document repository and a messaging system are also implemented for collaboration and time-saving purposes.

## 3 FEASIBILITY PHASE

The feasibility phase has two steps: a technical feasibility test and a conceptual energy and cost analysis.

### 3.1 Check technical and legal feasibility

The first step in the workflow, done in the DST, is to assess the technical and legal feasibility of the project. Two types of restrictions are implemented in the platform: generic restrictions of the BERTIM process and specific ones from the manufacturer.

Generic restrictions include legislation (possibility to extend the building, to modify the façade or use timber for instance), type of structure or the status of the building, among others. They are applied regardless of the targeted manufacturing company.

On the other hand, specific restrictions vary from one company to another and can be customized by its users in the private area of the DST. They refer to restrictions imposed by the company's products or processes (e.g. related to the building geometry or the surrounding context).

A restriction is defined by a name, a value type (real, integer, yes/no or choice of options) and target value and a condition (higher, lower or equals, etc.).

The user of the DST is requested to enter the values for both types of restrictions in the building under analysis and the feasibility check is run, showing a report with the fulfilled and not fulfilled ones. When there is no restriction not fulfilled the user can proceed with the next step: the energy analysis.

Additionally, a restriction can be left unchecked when the value is unknown or simply don't want to consider it, as shown in Figure 2. This allows skipping part of the feasibility check and go directly to the early energy estimation, especially important when the manufacturer wants to show the value of the prefabricated products.

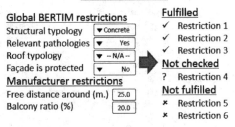

Figure 2. Feasibility check, with sample restrictions.

## 3.2 Conceptual energy and cost analysis

Once the project is considered feasible, an energy savings estimation is performed. In a more advanced stage, the required data is taken from the IFC model, but since in preliminary stages no BIM models are usually available, the DST offers a quick and easy way to define a building geometry.

The user selects the shape type that better fits the building from a list of options (see Figure 3) and conure the values of some lengths and heights of that template, as well as average glazing ratios per side and orientation towards the real north. The district context can also be characterized by defining shading surfaces per façade, specifying height and distance (0 if adjacent).

If the building latitude and longitude are provided (which can be easily picked from many mapping services), RenoBIM DST automatically displays the virtual model using Web3D technologies based on Cesium. The underlying representation format used is Google's KML. Vertical surfaces are used to represent shadowing buildings.

For energy simulation, the selected engine is Energy Plus, one of the most popular and reliable engines, released as Open Source. Two main innovations are present in the DST:

- Use Energy Plus for both approaches (simplified and IFC-based, explained later).
- Automatic running of parallel cloud simulations for each alternative, invoked through REST web services.

The first point provides a very flexible and robust simulation approach, using the same engine for all phases, adapting it to different levels of detail of the inputs (real geometry vs procedurally created). Since the same engine is used, the results are comparable.

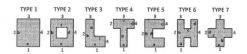

Figure 3. Simplified geometry definition.

Figure 4. Web visualization of the virtual geometry.

The second point allows using powerful dynamic engines by non-experts. The DST is the middleware between the end user and the simulation engine and a web service based architecture enables standard calls to the engine. Recent EU projects in the topic of collaboration platforms for cloud energy simulation have been taken as reference, e.g. HOLISTEEC (Romero, 2016, Pruvost et al., 2016).

The simulation process consists on three steps:

- Create the Energy Plus input file (IDF) for the current situation and each renovation scenario in 3 steps: (i) geometry, (ii) materials and (iii) usage patterns.
- Automatically run the models.
- Process the results and display to the user.

The automation of the first step is crucial, as well as the most complex. In relation to constructions, a collection of generic templates is created for main categories (roof, walls, slabs and windows), which, in the case of walls and slabs are split in external, internal and in contact with the ground. The user selects the combination for his building, and in the background physical properties (thermal transmittance or specific heat) are mapped from the templates to the 'Construction' object in the IDF file.

For each renovation solution, the timber product from the manufacturer is modelled as an additional 'Material' layer on top of the existing 'Construction' object, with its own physical properties.

The schedules and usage patterns are taken from default existing IDF templates for residential buildings (not modifiable by the end user).

Finally, for the creation of the IDF geometry the following logic has been implemented (U.S. Department of Energy, 2018):

- One 'Zone' object per building storey.
- One 'BuildingSurface:Detailed' of type Wall for each external wall, with embedded 'FenestrationSurface:Detailed' objects, if they have windows inside. The area depends on the glazing ratio expressed by the user.
- One 'ShadingSite:Detailed' for each external shadow. If it is attached to the building, no windows are created in the wall.
- One 'BuildingSurface:Detailed' of type Floor per storey, being all thermal zone separators, except the lowermost one, to which ground conditions are assigned.
- One 'BuildingSurface:Detailed' of type Roof on top of the last storey.

Since a single thermal zone is defined for the whole storey (no internal partitions), an area usage ratio is defined to account for the fraction of unheated spaces, e.g. staircase or common external areas.

Finally, an approximate cost is then calculated considering the total area covered by timber panels and the unitary cost of each, plus some constant factors. Some correction factors are added when the panel has installations or its put in place involves cutting the balcony. Then, from the yearly energy reduction estimated, we can deduce the money saved and with some energy and financial scenarios calculate the return on investment.

## 4 DETAILED DESIGN: OPEN BIM COLLABORATION WITH IFC

The feasibility phase can give the end user quick estimations with few data, but it is in the detailed design phase where, using the IFC model, the full capabilities of RenoBIM are accessible: a more accurate energy and cost analysis, a base layout for the design configurator tool and the template for designing the panels' fabrication details in the CAD/CAM tool. The requirements imposed to the IFC creation must account for all these uses of the model.

### 4.1 *Building model creation*

As the BERTIM timber modules will be prefabricated, the singularity of the 3D acquisition process of the building is the high accuracy required. This specific point represents a big difference with other 3D acquisition methods for conventional onsite renovations, with many adaptations to the real building.

In prefabrication, the clear goal is to install the modules and carry out inside finishes, especially around windows, with no onsite alterations to the prefabricated modules or cutting existing concrete walls. For that goal, the two sources of tolerances, from surveying (TS) and fabrication (TF), must be aggregated. Construction rules, such as operating clearances (OC), will be required to manage them in all situations.

Many devices and technologies can be used for the survey of the existing building (UK BIM Task Group, 2013), but considering accuracy goal and completeness, total stations or photogrammetry with drones have been abandoned and 3D laser scan has been selected. These devices are expensive and the process to survey buildings followed by 3D modelling with BIM software using point clouds require specific skills: this demonstration on a real building to be renovated with BERTIM modules has been achieved through collaboration with a specialist provider for such services.

Considering our accuracy goal, 56 of the 70 merged scans to form a single point cloud have been obtained by positioning the device inside the building in front of all opened windows to avoid shadows in outside scans from ground. This geo-located point cloud and a series of 360° panoramic photos constitute the primary deliverables.

Then, producing the 3D model from point clouds in BIM authoring tools is certainly the most important phase of the whole building-acquisition operation, both in terms of the time required (five to ten longer than the data capture time) and the skills involved. This 3D volumetric model produced by a CAD tool then serves as the basis for other secondary deliverables, such as 2D drawings (storeys, outside walls, cross sections etc.), lists and quantities (surface area of facades, number of windows in each category, etc.).

The resulting model and its usefulness fully depends on the modelling method used by the CAD operator. First, geolocation must be carefully handled. The way site coordinates are captured and expressed in Revit may not match with how survey points and project base points are defined in BIM and how they are exported to IFC. For example, using rectangular coordinates (UTM-like) and referring to the project origin leads to very big coordinate values which cause problems in IFC. The correct solution was to capture Latitude/Longitude of a reference point and express the rest of the project values as relative to this 0,0,0 point in cartesian coordinates.

The first step in the Revit workflow consists in preparing a structure with several floors, to obtain elevation sections for work in the point cloud and to define a structure for classifying the 3D objects created subsequently.

To facilitate subsequent steps in modelling prefabricated BERTIM modules, it was decided that the outside walls of the existing building must form for each façade, a vertical plane parallel to the plane occupied by the future prefabricated modules. The verticality of the future module plane is an absolute constraint to enable ease of assembly at the corners. This problem was also addressed in the similar TES Energy Façade project (Larsen et al., 2011).

To position this single vertical plane, it is necessary to define the horizontal direction of each outside

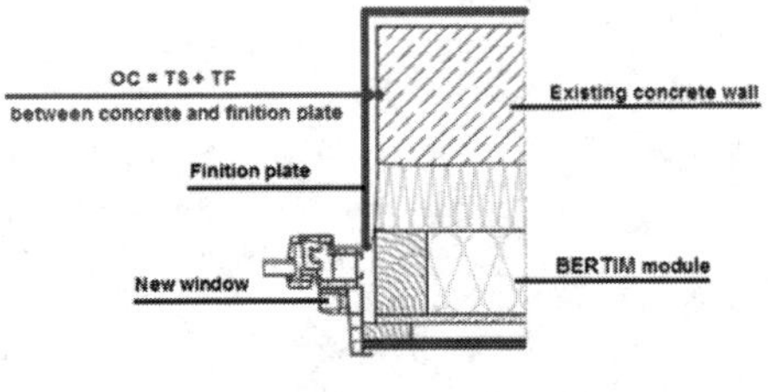

Figure 5. Example of Operating Clearance (OC).

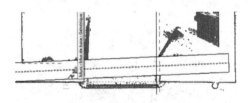

Figure 6. 3D wall creation on a line added in point cloud.

wall: the bottom of the first floor has been used as baseline to avoid additional shop fronts in ground floor. In the model this section corresponds to the top, rather than the bottom, of the first horizontal BERTIM modules to be fitted to the walls. Each wall is positioned in terms of its outer face, by defining its direction on the section in the point cloud by the two points observed as jutting the furthest out from the building (bumps on the outside wall). Measured in the point cloud, several thicknesses of walls have been reduced to a few average standards.

For windows modelling, the final strategy has been to model the widest, highest rectangular opening that would fit in the opening in the point cloud. This approach considers any points that may jut out into the opening. Moreover these 'salient' points are not only defined on the external plan of the outside wall, but over the whole visible thickness of the reveals, up to the existing window. It is thus possible to make allowance for any faulty squaring of the rendered reveals. To achieve this goal the operator creates working sections to restrict the field of visibility of the point cloud as shown in Figure 7 (left).

At the end, in the second vertical section of Figure 7 (right), it becomes possible to plot, passing through these salient points four reference lines, to which the opening may now be fitted. As the salient sills will be hidden by the BERTIM modules, only the measurement of how far they jut out is of any use, in order precisely to define the position of the rear plane of the modules.

To prepare the implantation of the modules on façades with a total station, topographic targets have been fixed on all facades. They have been also modelled as volumes to be correctly exported to IFC.

The level of detail of the internal structure of the building is not so critical, since it was not the object of BERTIM to consider internal renovation works. However, for energy simulation purposes a basic spatial zoning and divisions must be defined in the selected BIM tool, but with a significantly lower geometrical accuracy. This is analysed next.

Additionally, the surrounding buildings are modelled as conceptual mass blocks to consider shadowing effects.

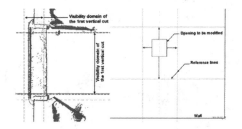

Figure 7. Visibility domain of 2 vertical work cuts (left) and adjustment of a 3D opening on reference lines (right).

### 4.2 IFC to IDF conversion and simulation

Compared to the procedural way of creation simulation models described earlier, the automatic conversion of IFC to IDF has two main advantages: much more accurate inputs (and consequently output results) and much less manual inputs from the user.

Energy Plus, like other dynamic simulation engines rely on the concept of Building Energy Model (BEM), a mathematical abstraction of the building geometry and topology, by modelling it as a graph, where the nodes are thermal zones, whereas connectors between nodes represent different heat transfer paths (i.e. opaque or glazed surfaces which enclose the zone), with a resistance depending of thermal properties. These analytical surfaces are represented as conceptual 2-dimensional polygons.

Thus, the main challenge is to automatically create this topological zone-surface model from the physical BIM geometry (solid 3D building elements and spaces). To be able to automate the BIM-2-BEM transformations for any kind of model is a highly demanding and still unsolved issue and for RenoBIM a minimum quality conditions are imposed to the input IFC so that the conversion is possible.

First, many BIM modelling tools (such as Autodesk Revit, used in BERTIM demo cases) are already able to generate the space boundary or analytical energy model from a given BIM model (BuildingSMART, 2013). This model is then exported to IFC, although not yet readable by Energy Plus, so specific algorithms have been developed for such purpose (hereinafter described).

Secondly, a modelling guideline has been created to ensure the most proper IFC files on input, based on existing guidelines (e.g. See & Welle, 2010). However, these requirements are quite generic and in BERTIM they have been particularized for the case of Energy Plus. Some of the requirements provided are the following ones:

- Proper orientation of the building with respect to the real North of the world.
- Proper classification of the building element functions (internal, external, in contact with ground, etc.), especially for walls and slabs.
- All internal locations of the building must belong to a given space (excluding the interior of walls/slabs). No empty areas should be left unmapped.
- No internal location of the building can belong simultaneously to two spaces, i.e. spaces should not overlap.
- External shadowing buildings are modelled as mass blocks (exported as IfcBuildingElementProxy).

In relation to the thermal zoning model two approaches are supported: the user can define the zones and group the spaces to the zone where they belong (mapped with a IfcRelDefinesByGroup relationship in IFC) or only define the spaces, in which case there is a one-to-one mapping between an IFC space and an IDF Zone. Since we deal with residential buildings, a recommendation is two consider each individual dwelling as a single zone with no further partitions, since the purpose is a decision support for renovation and not a highly detailed simulation model.

Once the IFC is imported, the IFC-2-IDF workflow is triggered following the sequence depicted in Figure 8, as follows:

- Detect adjacent spaces: IFC defines for a Space Boundary (SB) the space to which it belongs (S1) and the related building element GUID. However, we need to know the adjacent space (S2), detected after triggering the procedure described in Figure 9, using the fact that SBs come in pairs when they are internal (interzone).
- Although single SBs correspond to external surfaces, there could be a case where they correspond to internal ones, if the twin SB has not been properly created. RenoBIM can detect it by geometric analysis.

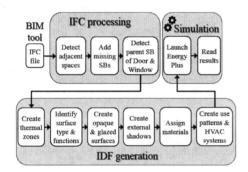

Figure 8. Complete BIM-2-BEM workflow.

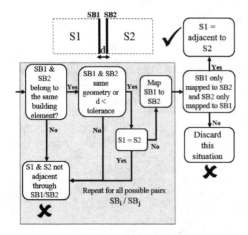

Figure 9. Detection of space adjacencies.

- Then a parent-child relationship is created between the SBs corresponding to doors and windows and the wall the belong to.
- With the previous information Zone objects and Surfaces (walls, floors, roofs and doors/windows) can already be created in IDF.
- For each of these surfaces the function must be defined in IDF (external, adiabatic, interzone or in contact with ground). Al algorithm (reflected in Figure 10) has been developed to derive this information from: (i) the SBs conditions (internal or external), (ii) the type and conditions of the adjacent space (if any) and (iii) the IFC class of the physical element mapped to the SB.
- Finally, the assignment of constructions and materials is done, according to this classification and the values entered by the user, the same way as for the simplified method.
- The assignment of usage pattern and HVAC templates is also the same as before.

The next figure shows the procedure to identify BEM functions of surfaces (ground, outdoors, adiabatic, surface) and BEM surface types (roof, floor, wall, ceiling, door, window) from BIM space boundaries and building element classes. Similar approaches have been carried out in other researches (Jeong, W., Son, J. 2016).

Regarding HVAC installations, three approaches are possible:

- Obtain just a demand estimation.
- If the dwellings have individual heating, apply a factor that accounts for efficiency losses considering the status or age of the building.

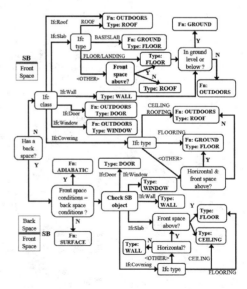

Figure 10. Algorithm for identifying surface type and function.

- Additionally, for typical central production systems (e.g. central boiler and individual radiators in each zone) Energy Plus templates has been created. Depending on the number of zones (dwellings) the algorithms automatically replicates the radiator instances and maps them to a given zone and applies some user-defined configured values (flow rates, pipe lengths, design temperatures, boiler volumes...). Default values are assumed when not provided by the user.

The final IDF file for simulation is a merged model containing the IDF fragment for geometry, the IDF fragment for HVAC and a default IDF fragment for scheduling and usage patterns (residential template).

For the simulation of energy savings potential, each timber module is characterized as a layer in Energy Plus with their physical properties (thermal conductivity, thickness, specific heat and density), using the 'Material' object. Then, to the original Construction objects in the existing buildings an additional material layer is added (which corresponds to a given timber module).

### 4.3 Design configurator module

Once the cost-energy analysis is done and the renovation is considered worth undertaking, RenoBIM offers a module for quickly configuring different façade splitting options, implementing a drawing functionality based on Web 2D technologies. It optimizes the pre-design process of the prefabricated panels, by saving a lot of time and quickly selecting the best configuration, which can be done in a collaborative way between the architect and the manufacturer.

The final result can be visualized within the web environment and exported to a new IFC file which

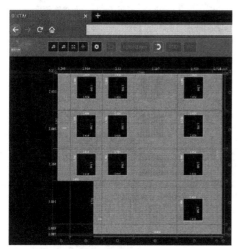

Figure 11. RenoBIM design configurator module.

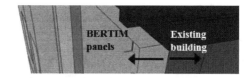

Figure 12. Output IFC file from the configurator.

contains the original building plus the extra layer of building elements corresponding to the prefabricated timber modules. This enhanced IFC is imported in Dietrich's CAD/CAM tool, where all the required details are added (transoms, layers, internal framing...) and finally launch fabrication orders to a CNC machine.

For each panel, additional information is added in custom IFC Property Sets, including physical properties or panel type (with or without internal installations) or indications whether it encloses a balcony or not, which can affect cost analysis.

The complete procedure is as follows:

- The user provides the IFC file of the building and a DWG file of a façade view.
- A section or working area is defined. All panels inside a section share horizontal and/or vertical bounds, which are configured using a grid tool.
- Openings can be created inside the section, which usually enclose windows in the real building.
- The dimensions are checked against the product database configured by the manufacturer. Thus, grid sizes exceeding maximum or minimum values are notified. A warning is also provided for non-optimal sizes.

The IFC export process creates additional IfcBuildingStorey entities, since the panels may not respect the actual storeys of the building. Then each panel is exported as an IfcWallStandardCase, with their IfcOpeningElements, when needed.

For the development of the logic three programming libraries have been used: fabric.js, paper.js and CADLib4.0. No plugins are required in the client web browser.

### 4.4 Integration with CAD/CAM software

By exploiting the high potential of the IFC standard, Tecnalia and Dietrich's work in defining the best solutions to be universally exploitable by all timber manufacturing CAD tools. In parallel, the work has been focused on obtaining the largest capacities with Dietrich's existing productivity tools during the final detailed design phase of the timber panels imported from RenoBIM.

For the best exploitation of CAD, new stories are created to contain the BERTIM panels and differentiate them from existing building stories. Simple rules have been defined to manage these new stories:

- Same storey for all panels/walls with a similar bottom level on all façades (as architecture rule).
- Two horizontal walls, one over the other belong to two different storeys.
- Horizontal panels mixed with vertical panels belong to the storey of their bottom line and two horizontal walls, one over the other belongs to two different storeys.

It has been defined at an early stage that these BERTIM modules should be IfcWall instances (instead of IfcPanel or IfcCovering) to facilitate next stages and the best productivity for detailed design in Dietrich's. The final choice has been to create them as IfcWallStandardCase, because IfcMaterialLayer-Set described further is not available for IfcWall.

Many solutions to create geometries in IFC files and Dietrich's first suggested to describe geometry with SweptSolid (CSG: extrusion and then clipping if needed). Finally, Brep (boundary representation based on faces) has been chosen.

IfcMaterialLayerSet is used to define the construction type of the wall (BERTIM panel). Currently, a single layer is defined, but additional layers can easily be added if needed in manufacturing CAD with their material properties. The name of IfcMaterialLayerSet is an important property: it comes directly from RenoBIM products database, which can be accessed from the private area for manufacturers.

Windows and doors are defined with IfcOpeningElement. This class only describes the hole in the wall. As the information for an opening to be a window or a door is required, it is defined using the IfcRelFillsElement to assign an IfcWindow or an IfcDoor to the opening.

Even if Dietrich's is the target tool in the project, integration tests with other manufacturing CAD tools for wood production are ongoing, such as Cadwork, always based on IFC file exchanges.

## 5 CONCLUSIONS AND FUTURE WORK

This paper shows an example of integration of actors and tools in complex processes like building energy renovation with prefabrication, using Open BIM as the main driver, but at the same time flexible by supporting alternative simplified approaches.

As such, the objective is to enable collaboration and integration of various domains (restriction checking, energy simulation, cost analysis, web based configuration and export to fabrication tools) rather than focusing on a specific domain (e.g. BIM-2-BEM transformations). Thus, some assumptions are done for the imported models, documented in guidelines and recommendations. The modules developed are intuitive enough to be used by non-experts in early decision making, but benefiting from the reliability of professional engines like Energy Plus, run in the cloud.

Even if the algorithms developed are demonstrated on specific use-cases and building types, imposed by BERTIM project, they offer a great potential for further extensions and enhancements, for example by extending the concept to non-residential buildings, consider renovation of HVAC systems, allow user-configurable building use pattern scenarios, etc.

The web configurator of the panels design could also be oriented to minimize user interaction and increase automation, based on a previous audit of the IFC file (e.g. suggest a pre-design), although rather complex when the façade geometry is very irregular.

## ACKNOWLEDGEMENTS

BERTIM project has received funding from the European Union's Horizon 2020 research and innovation program under grant agreement No 636984.

## REFERENCES

buildingSMART International Ltd. 2013. Industry Foundation Classes Release 4 (IFC4). http://www.buildingsmart-tech.org/ifc/IFC4/final/html/schema/ifcproductextension/lexical/ifcrelspaceboundary.htm.

Jeong, W., Son, J. 2016. An algorithm to translate building topology in building information modeling into object-oriented physical modeling-based building energy modeling. *Energies, 9(1),50.*

Larsen, K.E., Lattke, F., Ott, S., Winter, S. 2011. Surveying and digital workflow in energy performance retrofit projects using prefabricated elements. *Automation in Construction 20(8), pp. 999–1011.*

Romero, A.; Izkara, J.L.; Mediavilla, A.; Prieto, I; Pérez, J. 2016. Multiscale building modelling and energy simulation support tools. *eWork and eBusiness in Architecture, Engineering and Construction: ECPPM 2016. ISBN: 978-1-138-03280-4:* 316–322.

Pruvost, H; Scherer, R.J.; Linhard, K.; Dangl, G.; Robert, S.; Mazza, D.; Mediavilla, A.; Van Maercke, D., Michaelis, E.; Kira, G.; Häkkinen, T.; Delponte, E.; Ferrando, C., 2016. A collaborative platform integrating multi-physical and neighbourhood-aware building performance analysis driven by the optimized HOLISTEEC building design methodology. *Proceedings of the ECPPM 2016 – eBusiness and eWork in Architecture, Engineering and Construction. ISBN: 978-1-138-03280-4.*

See, R. & Welle, B. 2010. Information Delivery Manual (IDM) for BIM Based Energy Analysis as part of the Concept De-sign BIM.

UK BIM Task Group. 2013. Client-Guide-to-3D-Scanning-and-Data-Capture.

U.S. Department of Energy. 2018. Energy Plus Input Output Reference: https://energyplus.net/sites/all/modules/custom/nrel_custom/pdfs/pdfs_v8.9.0/InputOutputReference.pdf.

*Modelling of design, construction, operation,
maintenance management processes*

# BIM model methods for suppliers in the building process

A. Barbero, M. Del Giudice & F. Manzone
*Politecnico di Torino, Turin, Italy*

ABSTRACT: Building Information Modelling (BIM) process is always considered a collaborative method to exchange information between all the actors of building Industry. As this method is based on 3D parametric models, different strategies are available to develop graphical databases and manage information during the operational step. In these terms, collaboration between the involved actors of the building process is one of the main focuses of this research that analyses features that influence data management of a BIM model, testing its flexibility in term of data exchange between actors and tools.

## 1 INTRODUCTION

Nowadays, society is influenced by the wave of transformation encouraged by Information and Communication Technologies (ICTs) based on sharing of data. In these terms, AEC Industry is going to innovate itself adopting forefront methodologies that optimize data management related to new and existing buildings. Connected to this aspect, all the actors involved in the building process play a key role in each field based on specific needs. For this reason, it is important to identify, at the beginning of the process, goals that could be reached to satisfy each users' needs through the employment of proper tools.

Currently, BIM is a methodology that allows professionals reaching their objectives, optimizing their work according to worksharing. As in many other industrial sectors, a major difficulty that Architectural, Engineering and Construction (AEC) companies are currently facing with ICT is lack of interoperability between software applications to manage and progress in their business (Grillo et al. 2010). As BIM process is always considered a collaborative method to exchange information between all the actors of building Industry, it is based on a 3D parametric model as a common platform where different users fill in and extract data.

Many researchers investigated its potentialities related to multidimensional and multi-disciplinary models to use data stored in BIM graphical Data-Base for different application such as Time, Costs and Facility Maintenance (Ding et al. 2014), useful for the whole life cycle building. This goal requires, in each different topic, the employment of proper applications based on closed standard exchange formats, with lot of issues in interoperability workflow. To overcome this issue, International Alliance of Interoperability (IAI) provided Industry Foundation Classes (IFC) standard to overcome lack of data with the aim to solve collaboration difficulties due to proprietary language (Renaud, et al. 2008).

Actually, several model collaboration systems are available on the market to enrich, exploit and reuse data stored into multi-disciplinary models during the operational step (Shafiq et al. 2013). However, these platforms do not specify how the model have to be managed, offering only a shared space between the various actors, according to the project requirements (Grilo et al. 2010), not mentioning the collaboration workflow.

Therefore, it is necessary to establish the project goals, its use and the file sharing to optimize data communication developing a flexible BIM model ready for the operational step that is one of the main topics to optimize the exchange process between different actors involved. For this reason, this paper focuses on finding best collaboration methods to generate a consolidated BIM model useful during the building operational step related to the Facility Management (FM) field. For this research an existing building such as a football arena was selected as a case study. The stadium is located at the border of Turin and Venaria municipality on a total area of 355000 $m^2$. Its refurbishment was completed in 2011 with the idea to create an innovative example based on ICTs to optimize the efficiency of the structure and bring it closer to supporters.

## 2 METHODOLOGY

### 2.1 *Application framework*

Nowadays, evaluating optimal BIM process depends on the choice of uses and objectives of the model. A hypothetical workflow that summarizes the whole process is presented in Figure 1.

This process starts manly by defining information requirements that have to be adopted by the Project Information Model (PIM) and subsequently by the Asset Information Model (AIM) (PAS, 2013), until the end of the building's lifecycle.

Each actor (e.g. owner, constructor, supplier, end user) have to satisfy lot of needs through specific tools able to reach different goals such as model creation, FM organization, visualization and simulation.

The creation of a 3D parametric model is based on availability of data that can be inserted, displayed three-dimensionally and modified easier than using traditional approaches. For this reason, developing a BIM model has been set in this research in two different ways that can be associated to two different steps of the building process. The first one is related to the creation of the BIM model itself, while the second one concerns the operational phase of the building related to maintenance activities. Certainly, an important objective of a BIM process is the collaboration between different actors. So it is essential to establish the right rules and the project workflow to create a multidisciplinary BIM model able to interact each other. As, currently BIM models can be worked on separately and combined in a consolidated model, an interoperable model collection procures for information to be accessible and retractable continuously. In this way an operational update of drawings, time and cost estimation can be made from the information in the 3D building geometry (Thomassen, 2011). One of the major difficulties to apply efficiently BIM process is management of a large quantity of design files where information is often stored and represented in different ways that can be not aligned each other, generating more revisions that slow down the building progress. Basing on what said above, this study investigated two different ways to manage a general consolidated BIM model, oriented to the operational step of the case study: i) integrated model; ii) federated model, focusing on data and work-sharing between maintenance's suppliers. This is due to the fact that choose one of these two strategies implies a different use of data, based in the first case on a single multi-discipline model where information is aggregated together, while in the second case on a shared area where several mono-discipline models are coordinated and linked together. (Eastman et al. 2008). In both cases, information exchange is coordinated via 3D parametric model differently, where each actor has the ability to involve in the design by their continuous analysis from adaptive geometries and information.

## 2.2 Modelling approach

For this case study, particular attention was paid on the analysis of main factors that influence the modelling approach, employing Autodesk Revit as BIM software. This one takes into account the archival document research and the survey step that align documents to reality starting from a real structure.

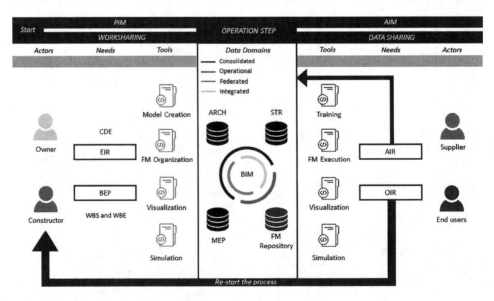

Figure 1. BIM methodology workflow.

Both integrated and federated model present different characteristics about modeling steps useful for model development, considering availability of data among the involved actors. In this term, while the first strategy is based on a unique database managed with worksets tool (Fig. 2), the second one focuses on the use of multiple models that can be overlapped in a single coordination model where different disciplines are joined (Fig. 3).

Then, part of the stadium was modelled focusing on lighting systems following both the strategies in order to develop the consolidated model through visualization and simulation tools, testing programming and planning maintenance activities for the new lighting plant of the stadium. So, the mentioned two strategies were selected to investigate collaborative agreements and evaluate interexchange between stakeholders.

Table 1 summarizes and explains identified features such as model creation, viewing, reporting and system administration, allowing the evaluation of these two strategies' flexibility. The two strategies have been compared assigning a score for each feature (1 = coarse, 2 = medium, 3 = high), generating a synthetic matrix, as a BIM method evaluation based on the owner goals. Each factor was evaluated considering the effort on geometry generation (based on work sharing) and its usability

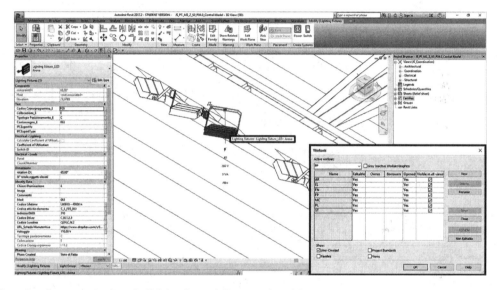

Figure 2. Axonometric view of a lighting fixture in Integrated model.

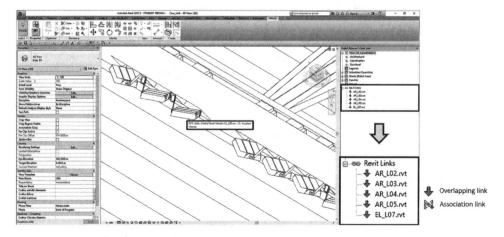

Figure 3. Axonometric view of a lighting fixture in Federated model.

293

Table 1. List of project features with Yes/No assigned values.

| n. | Features | Integrated model | Value | Federated model | Value |
|---|---|---|---|---|---|
| 1 | Objects | YES | 1 | YES | 1 |
| 2 | Object interaction | YES | 2 | NO | 1 |
| 3 | Embedded schedules | YES | 3 | NO | 2 |
| 4 | Reference level | YES | 1 | YES | 1 |
| 5 | Object editing | YES | 2 | YES | 1 |
| 6 | Plant systems | YES | 3 | YES | 2 |
| 7 | Host | YES | 3 | YES | 1 |
| 8 | Schedule | YES | 1 | YES | 1 |
| 9 | Reliability | YES | 1 | YES | 1 |
| 10 | Clash (Revit) | YES | 2 | NO | 1 |
| 11 | Clash (NWD) | YES | 1 | YES | 3 |
| 12 | Small file size | NO | 1 | YES | 3 |
| 13 | Inter-exchange response | YES | 1 | YES | 3 |
| 14 | FM synchronization | YES | 2 | YES | 1 |
| 15 | Database upgrade | YES | 3 | YES | 2 |
| 16 | File number (>1) | NO | 2 | YES | 1 |
| 17 | Shared model number (>1) | NO | 2 | YES | 1 |
| 18 | Use of model (by supplier) | NO | 1 | YES | 3 |
| 19 | Facility with VR | YES | 2 | NO | 1 |
| 20 | nD application | YES | 1 | YES | 1 |

during the operational step (data sharing). Thus, an evaluation rank was adopted to create a hierarchy based on each level of importance, in function of the modelling techniques.

At first, each feature was evaluated considering its usage in both integrated and federated models assigning a Yes/No value, taking into account the real application into the modelling environment.

Considering only the total amount of Yes value compared with No one, the choice should be oriented on integrated model but this choice may be wrong in relation to the model's aim. So, this evaluation was implemented adding a level of importance related to this value.

Moreover, taking into account the two different steps of the whole building process (operation and consolidated one), these features have been implemented assigning a percentage value able to enhance their characteristics (Table 2).

As an example, features n. 18 underlines how the use of federated model allows the contemporary use of the model without synchronization issues, considering the Common Data Environment (CDE) workflow. However, this difference is mitigated by the assigned score related to both creation and operation steps, according to its usage in the project. These assessments are based on the proper Level of Geometry (LOG) and Level of Information (LOI), related to each 3D object in the workflow definition, according to Work Breakdown Structure (WBS) and Work Breakdown Element (WBE). Assigned scores considered the goodness of each modelling way, enhancing their characteristics and highlighting their weaknesses.

Table 2. List of assigned parameters value related to creation and operation steps.

| n. | Features | Value creation % | Value operation % |
|---|---|---|---|
| 1 | Objects | 0,25 | 0 |
| 2 | Object interaction | 0,5 | 0,5 |
| 3 | Embedded schedules | 0,25 | 0,5 |
| 4 | Reference level | 0 | 0,25 |
| 5 | Object editing | 0,5 | 0,25 |
| 6 | Plant systems | 0,25 | 0,75 |
| 7 | Host | 0,5 | 0,25 |
| 8 | Schedule | 0,25 | 0,5 |
| 9 | Reliability | 0,25 | 0,75 |
| 10 | Clash (Revit) | 0,5 | 0,25 |
| 11 | Clash (NWD) | 0,25 | 0,5 |
| 12 | Small File size | 0,5 | 1 |
| 13 | Inter-exchange response | 0,5 | 1 |
| 14 | FM synchronization | 0,25 | 1 |
| 15 | Database upgrade | 0,5 | 1 |
| 16 | File number (>1) | 0,75 | 0,75 |
| 17 | Shared model number (>1) | 0,25 | 0,25 |
| 18 | Use of model (by supplier) | 0,75 | 1 |
| 19 | Facility with VR | 0,25 | 0,5 |
| 20 | nD application | 0,25 | 0,5 |

The achievement of model flexibility purposes can change depending on the selected tools and the project's goals.

## 3 RESULTS

### 3.1 Comparative analysis

Following the two modelling strategies, two types of BIM models have been developed, underlining their strengths related to work-sharing and data sharing to improve and extract graphical 3D model for different uses (e.g. training, FM execution, visualization, simulation).

The comparative matrix (Table 1) can be considered the first result as it provides different outputs in relation to the importance of each feature that every time can change their value relating to the project goals (Table 2).

The two spider charts (Figs. 4–5) highlight the main differences between these two model strategies, underlining features that reach the highest score. Comparing the two diagrams, some features of the operation strategy (e.g. 18, 15, 13, 12) reach the highest evaluation score, while in the creation one the highest score is 2,25. This analysis, highlights that creation step is influenced mainly by the choice of a proper BIM workflow. In these terms, starting from the importance of feature 18, as one of the main factors of this research, the optimal solution provided by these tests is federated approach.

While integrated model allows an easier information update, federated one is more effective with

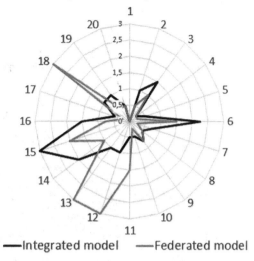

Figure 5. Spider chart of the operation step.

suppliers who can work in different way using different models with small size.

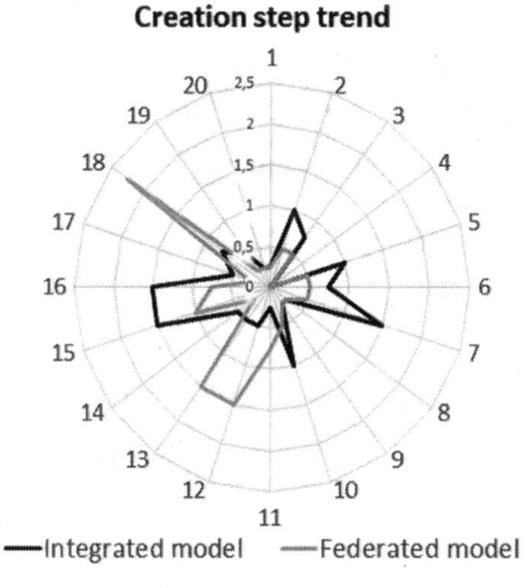

Developing a federated model for operation step implies the adoption of the same inter-exchange model strategy also in the creation step, taking into account some possible issues visible in Table 1. The total amount of the assigned score is quite similar in both strategies, demonstrating that analytical area is the same, while graphical databases reached different specific peaks. In addition, federated approach allows the employment of specific discipline models that belong to target tools, providing a replicable methodological approach through open BIM languages.

Clearly, this aspect is strictly related to the wild field of interoperability that is an actual research challenge.

### 3.2 OpenBIM overview

As illustrated in introduction paragraph, IFC standard exchange format should offer a possible solution to overcome interoperability issues related to the employment of different tools that generate possible data loss during exchange data flow. Unfortunately, this language is not yet completed and ready to describe FM estate. As an example, stadium lighting features were exported from a discipline model of the federated workflow and automatically were grouped in the correct class, basing on a custom IFC activity.

However, as visible in Figure 6, some geometrical components are missed, maintaining the characteristic of BIM categories.

Figure 4. Spider chart of the creation step.

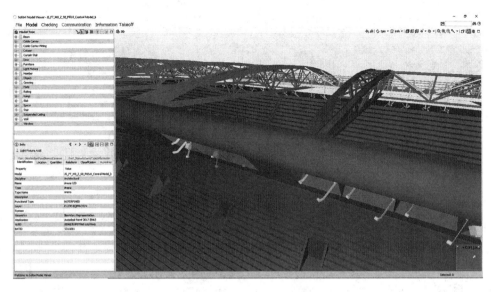

Figure 6. 3D view of IFC model.

## 4 CONCLUSION

Currently, work-sharing and data management are one of the most important challenges of the AEC industry that is going to innovate itself using BIM as a ICT method. So, this study demonstrates that several ways can be followed to achieve the objectives of a certain project using BIM methodology that can be customized relating to specific needs, allowing building Industry to be renovated, enhancing the value of sharing information, optimizing time and costs resources.

Moreover, Open BIM language will allow the achievement of federated model requirements inside the work-sharing area without restrictions related to proprietary software. For this reason, this represents one of the major real industrial challenge.

Finally, a next step of this study will consider the connection of the BIM dataset of the entire building with a Computer Aided Facilities Management (CAFM) software, introducing the value of interoperability and DB bidirectionality to updates information.

## REFERENCES

Ding, L., Zhou, Y., & Akinci, B. 2014. Building Information Modeling (BIM) application framework: The process of expanding from 3D to computable nD. *Automation in Construction, 82–93.*

Eastman, C., Teicholz, P., Sacks, R., & Liston, K. 2008. BIM Handbook. *A guide to Building Information Modeling for Owners, Managers, Designers, Engineers, and Contractors.* New Jersey: Jonn Wiley & Sons, Hoboken.

Grillo, A., & Jardim-Goncalves, R. 2010. Value proposition on interoperability of BIM and collaborative. *Automation in Construction, 522–530.*

PAS. 2013. 1192–2:2013 Specification for information management for the capital/delivery phase of construction projects using building information modelling.

Renaud, V., Christophe, N., & Christophe, C. 2008. IFC and building lifecycle management. *Automation in Construction, 70–78.*

Shafiq, M.T., Jane, M., & Lockley, S.R. 2013. A study of BIM collaboration requirements and available features in existing model collaboration systems. *Journal of Information Technology in Construction, 148–161.*

Thomassen, M. 2011. BIM & Collaboration in the AEC Industry. *Master degree thesis.*

# Implementation framework for BIM-based risk management

I. Björnsson, M. Molnár & A. Ekholm
*Division of Structural Engineering, Lund University, Sweden*

ABSTRACT: There is currently an imbalance with regards to the way that risks are treated in construction projects. Risks associated with structural failures as well as construction safety risks are often in focus while building performance risks (e.g. related to moisture or energy performance) receive less attention. These types of risks often disproportionately affect the end users/owners and can in many cases be avoided given adequate access to relevant knowledge by the process actors. The modern construction process, which relies heavily on digitization and automation, provides an opportunity for improving the awareness and management of building performance risks. The current paper presents a conceptual framework for the implementation of BIM-based risk management in the modern construction process. A knowledge delivery system is envisioned which will make risk relevant information available to the process actors thereby improving risk awareness and enabling risk informed decision making. In contrast with earlier implementation attempts, risk information will be made available to construction objects structured according to existing, commercially available, building classification system(s). An implemented system should provide automated support to identification of potential risks associated with different construction solutions and enable the process actors to make informed decisions concerning the treatment of these risks in construction projects.

## 1 INTRODUCTION

### 1.1 Background

The management of risks in the construction process entails asking the right questions concerning potential problems with technical solutions and determining appropriate measures to avoid or reduce their impacts. When it comes to building performance risks (moisture safety, indoor environment, energy performance, etc) the impacts often disproportionately affect the end users/owners. Access to relevant knowledge concerning potential building performance problems is a first major step towards facilitating risk informed decision making in the construction process. This knowledge, although known by some, may not be known by the process actors, leading to a recurrence of errors which could have otherwise been avoided (Schneider 1997).

The modern construction process relies heavily on digitization and in some instances automation. This paradigm shift concerns a range of activities including the planning, design and production processes. As it stands, there are a number of BIM tools and approaches available which can be utilized throughout these processes. The digitized environment enables an improved access to relevant knowledge for informed decision making (Sellaka et al. 2017); providing an opportunity to improve the management of risks and reduce potential problems related to building performance caused by poor or un-informed decisions. A highly relevant issue relates to the process actors' role in exploiting this environment to achieve production results that fulfill design criteria with due consideration of the potential risks and uncertainties involved.

A number of approaches exist for managing risks in the construction process. In recent decades, focus has been on structural safety or construction/operational risk management (Stewart & Melcher 1997, Kirchsteiger 2002, Zou et al. 2017). In fact, the modern design process is guided by codes which have safety formats that are calibrated with the aim of consistently delivering safe structures (Nowak & Collins 2000). However, the treatment of risks related to poor building performance (e.g. moisture problems or poor energy performance) receives much less attention. These risks entail less severe or long term consequences which directly affect end users/owners and often manifest decades after the structure has been built. Thus there is less incentive for the construction process actors to regard these risks in the same as e.g. structural or construction site safety.

Building failures—generally defined as any unwanted deviation from design expectations—often result from unfavorable influences being subjectively unknown, inadequately treated or overlooked during the planning, design or production process (Schneider 1997, Breysse 2012). This highlights a poor knowledge/experience transfer within the architectural/engineering/contractor (AEC) community in construction projects. The knowledge may be documented in books, reports, articles and databases but may not be readily available, difficult to access or improperly communicated. The modern digitized construction process, however, offers an opportunity for providing access to this type of knowledge and as such facilitates essential decision support to process actors for managing building performance risks more effectively (Ding et al. 2016).

### 1.2 Aim

The overall aim of the research presented in this paper is to improve the management of building performance risks in the modern construction process. To achieve this, a process which integrates risk based design with modern BIM tools and approaches is envisioned. A conceptual framework for BIM based building performance risk management is presented in this paper. The focus is on making information concerning technical solutions with potential building performance risks available in BIM-software, aiming to provide decision support for improved risk management. A crucial part of this development concerns its implementation. The following success factors are considered relevant for the implementation of the framework in practice (Björnsson & Molnár 2018):

– Improved risk awareness and critical reflection by process actors.
– Improved access to existing risk relevant knowledge throughout all project phases.
– Improved transparency and traceability.
– Improved risk communication & knowledge transfer throughout all project phases.

A plan for implementation and further development is provided with these success factors in mind.

### 1.3 Method

Existing efforts to integrate risk management in AEC through BIM and BIM-related technologies include automatic rule checking as well as reactive and proactive IT-based systems to manage safety risks (Zou et al. 2017). In terms of application, a vast majority of the existing approaches have focused on construction personnel safety risks (Ding et al. 2016, Malekitabar et al. 2016). There are some exceptions, however, such as Pruvost & Scherer (2017) who have focused on risks in the building life cycle through quantified uncertainty modelling. There are, however, limitations of existing approaches to consider building performance risks (Björnsson & Molnár 2018).

The development of the conceptual framework for BIM-based risk management and a plan for its implementation is presented in this paper. The development is supported by a review of relevant sources (books, databases, etc.) which contain knowledge for improved risk management in the construction process. Specific attention is paid to sources which are relevant for the Nordic region. A knowledge delivery system is also envisioned as an application to BIM software. In contrast with earlier implementation attempts (Ding et al. 2016), risk information will be made available to construction objects structured according to existing, commercially available, building classification system(s). An implemented system should also provide automated support to identification of potential risks associated with different construction solutions and enable the process actors to make informed decisions concerning the treatment of these risks in construction projects.

## 2 CONCEPTUAL FRAMEWORK FOR BIM-BASED RISK MANAGEMENT

### 2.1 Overview

The conceptual framework, which was previously introduced in Björnsson & Molnár (2017), integrates the process of risk management with a BIM platform as illustrated in Figure 1. The treatment of building performance risks is accomplished through a cognitive loop which is divided into three cyclic stages:

1. Risk filter of design decision (e.g. choice of wall design).
2. Feedback to user/actor to improve risk awareness.
3. Risk mitigation strategies and possible alteration of design decisions.

Initially, a design decision within a construction project (e.g. the choice of wall design) results in a BIM model whose objects are automatically evaluated using a risk filter. The filtration process reviews attributes of the BIM objects, which are codified based on existing classification systems described in later sections, alongside a digitized database containing risk relevant information. In this way, the BIM objects are considered input to the risk filter while the identification of risk, which provides the basis for risk informed decision making, is the main output. The result of the risk filter is feedback to the user, information which can be documented and utilized for the treatment of

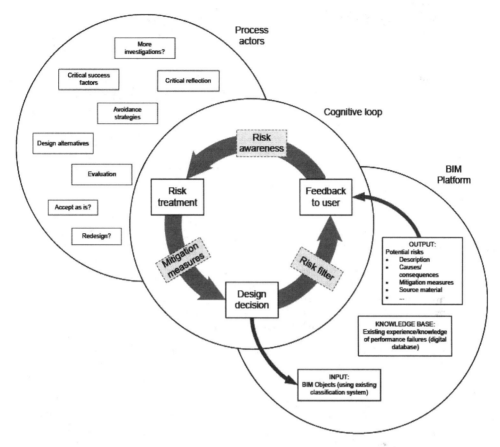

Figure 1. Illustration of conceptual framework showing cognitive loop connecting BIM platform with process actors for integrated risk informed decision making.

building performance risks in a proactive manner. The primary output is identified risk scenarios providing information concerning the following aspects:

Description of potential risk (e.g. damage type)

- Potential causes and factors influencing initiation
- Possible consequences of the identified risk
- Possibilities for risk avoidance and mitigation strategies
- Link to source material and reliability of information

To encourage risk informed decision making, the process actors shall provide input concerning what measures were arrived at to address the risks identified by the risk filter; i.e. mitigation measures. Such input may include neglecting the risk if it is considered insignificant or for whatever reason assumed insignificant. An important factor in any case is to improve risk awareness amongst the process actors and provide a system for documenting decisions concerning the management of building performance risks in the design phase of construction projects.

The first two stages of the cognitive loop in Figure 1 are integrated within the BIM platform while decisions concerning mitigation measures are at the discretion of the decision maker. The cyclic process may be repeated in cases of significant design changes, in which case an additional filtration of risk may be necessary. Currently, the development of the framework is at a conceptual level, while work is ongoing concerning its implementation. An implementation plan is provided in the next section.

## 3 IMPLEMENTATION OF FRAMEWORK

### 3.1 Overview

The conceptual framework from Figure 1 is intended to improve the treatment of risks related to poor building performance in construction projects. An integral requirement towards achieving this goal relates to its implementation in practice. An implementation plan is thus developed which considers:

– Description of risk filter including:
  o Classification system for BIM objects
  o Development and management of digitized knowledge base
– Application and evaluation of the approach in practice

These aspects are discussed further in the subsections which follow.

## 3.2 Risk filter

The risk filter should enable a review of construction objects within a BIM model to identify potential risks related to poor building performance. This identification is accomplished via an IT database which contains relevant existing experience and performance failures; i.e. a database containing risk relevant information. To facilitate the connection between the BIM objects and the database, a classification system can be utilized. Generally, there are two approaches to encoding BIM objects in a way to enable risk filtration based on an established digital knowledge base. Firstly, by using a bespoke or project based classification system, utilizing e.g. semantic web technologies (Ding et al. 2016), and secondly, utilizing existing building classification systems, which allows integration with other analysis applications, e.g. for cost calculation and technical specification.

### 3.2.1 Classification of BIM objects

In a building information model, BIM, the objects may represent different parts of a building and their relations. In a BIM application in order for the risk information knowledge base and the building model to be compatible, they should use the same object naming standard. This makes it easy to identify relevant building parts and relate these to similar parts in the knowledge base, from which information about risk characteristics for different technical solutions can be retrieved.

Building classification systems are used for naming building parts in many applications in the construction industry, e.g. technical specifications, drawings, and cost and quantity calculations. For this reason, the use of a building classification system for object naming is advantageous and a way to ensure semantic interoperability of construction information.

Most building classification systems are national but often share a view on building parts first developed in the Swedish SfB-system (The Swedish Building Centre 1999:14) and later formalized as the ISO 12006–2 standard (ISO 2015). The standard presents principles for building classification that, if used nationally, may support international information exchange as long as its principles for class definitions are followed. In the context of this research, the Swedish BSAB-system and the British Uniclass, both adhering to ISO 12006-2 are applied.

These standards classify building parts in two different ways, one using a view including function and main geometrical form, mainly referring to larger parts of the building, the other using a view including construction and used material, mainly referring to the smaller parts that are used in the construction of the larger parts. These parts are presented in two different tables: Construction Elements and Work Results respectively. In Uniclass the latter are referred to as Systems, defined as results of work by different trades. The subdivision is intended to support design, where in an early stage a Construction Element may be defined and represented in a drawing or a geometry model, and which in a later stage may be given a technical solution defined by relevant Work Result (BSAB) or Systems parts (Uniclass).

The knowledge base is structured according to the classification system such that for a given Construction Element (e.g. an external wall) the relevant technical solutions of Work Results (e.g. brick masonry façade) and Construction Products (e.g. clay brick) are listed with relevant construction risk information presented.

The usefulness of applying an established classification system depends on several factors, e.g.:

– A common understanding of the meaning of the concepts
– Relevance to the applications using the classification system
– A responsible publisher of the classification system
– Continued update and review of the classification system.

### 3.2.2 Development and management of knowledge base

A considerable part of today's knowledge concerning building technology failures is documented in various printed or digital sources. Making this knowledge accessible for processing by a BIM application requires creation of an adequate information structure and extraction of relevant information from the knowledge source.

The information structure is determined by the objectives of the risk management system and the requirement to be compatible with the chosen building classification system. The following information structure is proposed, see also Figure 2:

– Description of the failure, indicating the involved Construction Elements and Work Results

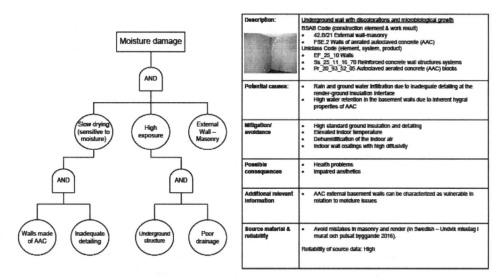

Figure 2. (Left) Fault tree for discoloration and microbiological growth on external basement walls and (Right) risk relevant information.

- Potential causes of the failure
- Measures to avoid or mitigate the effects of the failure
- Possible consequences of the failure
- Additional relevant information, such as reference to building and testing standards or frequency of the failure
- Information about the source material and its reliability.

It is anticipated that extraction of the relevant information from the various knowledge sources must be carried out manually, also in the case of digital sources. This is probably the most work intensive part of the implementation of the planned risk management system. Reproducing information items such as photographic material might further require obtaining the consent from authors and copyright owners.

A risk management system can be established as an independent organization under the auspices of the AEC sector of a specific country, see the Danish BYG-ERFA, or several countries within a region with similar building technology (BYG-ERFA). The experiences from the BYG-ERFA-system indicate, that:

- State funding during a start-up period might be a prerequisite in order to get in place a viable risk manage system.
- When the risk management system is established and accepted by the AEC sector, funding can be raised by subscriptions from the potential users.
- Updating and development may still be dependent on voluntary contributions from sector experts and researchers.
- Possible advantages with this option are broad outreach and consensus in the AEC community. Yet, a lengthy process to reach consensus might be a major drawback.

An alternative use of a risk management system is its adoption and management by a specific organization, such as a private company. This alternative might facilitate a fast implementation, but results in a limited impact for the AEC community as a whole.

3.3 *Application and evaluation of the approach*

To apply the approach in practice a beta version of the tool (software) for BIM-based risk management is planned based on the conceptual framework in Figure 1. The approach to structuring risk information could be implemented as a Semantic Web where construction objects (classified according to established construction classification systems like BSAB or Uniclass) are related to relevant risk information as well as other properties of interest. Statements (triples) in a semantic web may be expressed in the web ontology language OWL using a software tool such as Protegé (https://protege.stanford.edu/). For an overview of the concepts related to the Semantic Web, see e.g. https://en.wikipedia.org/wiki/Semantic_Web. A similar approach is presented by Ding et al. (2016).

The BIM-based risk management tool shall be tested by potential users (i.e. process actors) to provide feedback for further development and possible alterations. Ultimately the development and testing should consider how effectively the software improves process actors' treatment of building performance risks in construction projects. To accomplish this, a qualitative study may be carried out using questionnaires and interviews. This study should be grounded on the success factors mentioned in section 1.2.

Another important aspect for evaluation of the approach in practice concerns the reliability and traceability of the information presented in the knowledge base of the risk management system. References shall be provided to the sources for the risk relevant information. Furthermore, a ranking of the quality of the information sources is highly relevant. The ranking might take into account aspects such as:

– The type of verification of the information (e.g. newspaper article, incident report, peer reviewed scientific paper);
– The status of the issuing body, potential bias (e.g. private company, research organization, university);
– Consensus in the AEC and researchers' community.

When reliable, open source, information is unavailable, warnings may be issued concerning presumed risks associated with a technical solution. Such a case may be relatively frequent in practice since many litigations concerning construction industry failures, at least in Sweden, are solved by arbitration; making the underlying information unavailable for the public. On the other hand, if the risk management system is adopted for use within an organization, the aforementioned considerations concerning reliability and traceability of the risk information can be relieved.

4 ILLUSTRATIVE EXAMPLE

An illustrative example is provided which highlights the type of risk information that could be useful for improving the treatment of risks in the design of external basement walls. The specific risk concerns discoloration and microbiological growth on aerated autoclaved concrete (AAC) walls. These problems might occur due to the combined effects of rain and ground-water infiltration and high water retention due to inherent hygral properties of AAC. Figure 2a shows the fault tree for this damage case and Figure 2b provides the risk relevant information. A fault tree starts with a potential undesirable event (i.e. damage or failure) and works backwards to identify fault events or conditions that may contribute to the top event (Stewart & Melcher 1997). While a fault tree can be used in quantitative risk assessments to calculate risks they are also very useful in that they graphically provide a logical structure describing conditions necessary for potential initiation of a risk; information which is valuable towards identifying viable risk mitigating strategies. The fault tree also contains two nodes which concern the designation of the BIM object within an existing, commercially available, building classification system, BSAB (https://bsab.byggtjanst.se/). Uniclass codes are also provided in Figure 2b for information.

5 DISCUSSIONS

The paper presents a conceptual framework for the implementation of BIM-based risk management in the modern construction process. Such an approach should facilitate risk informed decision making in construction projects and improve risk awareness among process actors. There are, however, some challenges towards implementing such a system in practice. One major issue is the incentivisation of the AEC industry to adopt a more risk aware approach.

Adoption of a BIM-based risk management system is considered to have some significant advantages for the AEC industry as well as end users. To start, implementing the framework in practice has the potential to minimize costly building failures that result from poor building performance. These costs impact all stakeholders and there is an obvious benefit in mitigating these risks. In addition, environmental impacts from construction activities are potentially reduced since risk aware design and construction can optimize resource expenditure, improve building performance (e.g. energy use) and avoid unnecessary operational costs and actions as a result of building performance failures. From an AEC industry perspective, incentives to act in a more risk responsible way might come from considerations such as brand protection, good-will improvement, supporting Agenda 2030 objectives, and building regulations. Consumers' organizations push for protection of end-users' rights through extended warranties and consumers' increasing interest for a sustainable built environment might constitute further incentives.

Interest in environmental and climate change issues among individuals in the AEC community is a major potential facilitator of the adoption of a BIM-based risk system since it is believed to promote critical reflection and risk informed design. There is a common perception among AEC professionals that sustainability issues get systematically lower priority than costs, construction time or promotion of specific products. Thus there is an accumulated need for more knowledge and access to transparent and traceable information concerning the performance of building materials and technical solutions. Risk aversion in this sense might constitute a pull for adoption of the risk management system. Risk aversion might, however, act as a hindrance as well. Warnings issued by the risk filter might not be possible to mitigate unless major changes are made in e.g. choice of materials, detailing of the building envelope, construction method. Potential users of the risk management tool might consider that warnings left without measures can raise legal issues in the future. Being subjectively unaware might be considered a better alternative.

There is a continuous development taking place in the construction sector, with new production methods and products being introduced on the market. While introduction of innovative products and methods are necessary to improve building performance and productivity, new uncertainties and risks arise. Innovative, complex and unusual technical solutions have been identified as common factors to failures in the built environment (Wood 2012). This indicates that innovative products and solutions could be introduced stepwise into the knowledge base and through an open monitoring of their performance. Deficiencies and incidents should be reported and relevant information made available to the public. Unfortunately, this is not necessarily the case in practice. In this context, the proposed risk management system has the potential to 'fill the gap', e.g. through the creation of a forum for users' suggestions and feedback. The occurrence of failures might be reduced by intercepting early indications of malfunction. Such a dynamic and pro-active approach is essential to keep pace with continual developments within the AEC industry.

Although the proposed risk management system is conceived as a tool for the AEC community, it might be possible to involve the end users as well. Generally, end users have limited knowledge concerning building technology; a fact which hinders their ability to ask relevant questions during negotiations. Empowerment of the end users has the potential to increase risk awareness within the AEC community. Through own suggestions and incident reports, end users are potential contributors to the development and dynamic nature of the risk management system.

REFERENCES

Björnsson I. & Molnár M. 2018. Conceptual framework for improved management of risks and uncertainties associated with the performance of the building enclosure. *7th International Building Physics Conference, 23–26 Sept. 2018*, Syracuse, NY.

Breysse, D. 2012. Forensic engineering and collapse databases. *Forensic Engineering* 165 (FE2): 63–75.

BYG-ERFA. Accessed 2018–05–08 at https://byg-erfa.dk/ (in Danish).

ISO (2015). ISO 12006-2. Building construction – Organization of information about construction works – Part 2: Framework for classification. ISO, Geneva.

Ding, L.Y., Zhong, B.T., Wub, S. & Luo, H.B. 2016. Construction risk knowledge management in BIM using ontology and semantic web technology. *Safety Science* 87: 202–213.

Kirchsteiger C. 2002. International workshop on promotion of technical harmonisation on risk-based decision-making. *Safety Science*, 40(1–4), 1–15.

Malekitabar H., Ardeshir A., Sebt M.H. & Stouffs R. 2016. Construction safety risk drivers: A BIM approach. *Safety Science*, 82, 445–455.

Nowak A.S. & Collins K.R. 2000. *Reliability of structures*. McGraw-Hill, New York.

Protegé. Accessed 2018-05-14 at https://protege.stanford.edu/.

Pruvost H. and Scherer R.J. 2017. Analysis of risk in building life cycle coupling BIM-based energy simulation and semantic modeling. *Creative Construction Conf. 2017, Primosten, Croatia*.

Schneider, J. 1997. *Introduction to safety and reliability of structures*. Zurich: International Association for Bridge and Structural Engineering.

Sellaka, B., Ouhbia, B., Frikhb, B. & Palomares, I. 2017. Towards next generation energy planning decision-making: An expert based framework for intelligent decision support. *Renewable and Sustainable Energy Reviews* 80: 1544–1577.

Semantic Web. Accessed 2018-05-14 at https://en.wikipedia.org/wiki/Semantic_Web

Stewart M.G. & Melchers R.E. 1997. *Probabilistic risk assessment of engineering systems*. Chapman & Hall, London.

The Swedish Building Centre. 1999. *BSAB 96. The Swedish construction industry classification system*. AB Svensk Byggtjänst, Stockholm.

Uniclass. 2015. Accessed 2018-04-17 at https://toolkit.thenbs.com/articles/classification.

Wood J.G. 2012. Combating myths, oversimplifications and misunderstandings about the causes of failure. *Forensic Engineering 2012* (p. 871–80) ASCE, San Francisco.

Zou Y., Kiviniemi A. & Jones S.W. 2017. A review of risk management through BIM and BIM-related technologies. *Safety Science*, 97, 88–98.

# BIM-based model checking in a business process management environment

P.N. Gade
*University College Northern Denmark, Aalborg, Denmark*
*Aalborg University, Aalborg, Denmark*

R. Hansen
*University College Northern Denmark, Denmark*

K. Svidt
*Aalborg University, Denmark*

ABSTRACT: BIM-based model checking has the potential to improve the building design process concerning efficiency and consistency by allowing for automatic assessment of BIM-models. However, BMC is infrequently used the building design practice. A fundamental challenge in applying BIM-based model checking in practice is the chaotic and dynamic nature of the building design process, which is subject to many changes, like changing requirements or constraints. BIM-based model checking systems have been criticized for having poor flexibility, as well as not being able to adapt to these changes sufficiently. To improve the flexibility, we developed a proof-of-concept prototype based on requirements of flexibility from process-aware information systems theory. The prototype was used to assess a sustainability criterion on a BIM-model to explore the possibilities of developing BIM-based model checking systems with improved flexibility. Based on this demonstration we discuss the limitations and opportunities for future research and development of more flexible BIM-based model checking solutions.

## 1 INTRODUCTION

The design of a building can be considered a composition of design choices optimized for satisfying the goals, requirements, and constraints in the environment of the design, as well as the technology used (Ralph & Wand, 2007). The designers are challenged to optimize the buildings in a process that is considered highly inefficient, due to complex rules and troublesome methods of organization (Kuben Management, 2016). The finished buildings contain many flaws due to design errors (Lopez & Love, 2012) and are poorly optimized (Flager, Welle, & Bansal, 2009). Cost of design errors is estimated to be approximately 6,85% directly, or 7,38% indirectly, of the total contract sum of building projects (Lopez & Love, 2012).

### 1.1 BIM-based model checking

Hjelseth (2016) defines Building Information Modelling (BIM)-based Model Checking (BMC) as a grouping of concepts that focuses on using BIM as an information source and algorithms for processing the information based on rules. The rules can be derived from various rulebooks, such as building codes or sustainability assessment methods. A rule derived from building code, for example, could be defining the width of hallways related to fire exits. BMC has the potential to improve designers work to ensure compliance with regulative or performance-based rules through enabling rapid, consistent, and precise feedback on compliance and performance (Dimyadi & Amor, 2013; Eastman, Lee, Jeong, & Lee, 2009).

Currently, BMC has found limited use in the design practice (Dimyadi, Clifton, Spearpoint, & Amor, 2014; Dimyadi & Amor, 2013; Hjelseth & Nisbet, 2010; Cornelius Preidel & Borrmann, 2015). Large, ambitious national BMC projects and commercial BMC-systems have proven difficult to integrate into the design practices (Khemlani, 2017; Refvik, Skallerud, Slette, & Bjaaland, 2014; Solihin & Eastman, 2016). Studies indicate that it is related to the "black-boxing" of the BMC-system's processes (Beach, Rezgui, Li, & Kasim, 2015; Dimyadi & Amor, 2013; C. Preidel & Borrmann, 2017).

### 1.2 *A lack of flexibility challenges the practical use*

Many scholars have commented on the unpredictability and complexity of the construction projects, which are subject to many changes in scope during its progression (Bertelsen, 2003; Dubois & Gadde, 2002). These dynamics make it difficult to apply tools like BMC in the design practice because people improvise, make errors, and are subject to changing requirements. Bertelsen (2003) characterized the construction projects as: "a nonlinear and dynamic phenomenon, which often exists on the edge of chaos." The dynamics lead to changes that systems like BMC needs to accommodate. The changes are related to the evolution of the rules and the design.

One of the main challenges has been the lack of standardization of rule and design representation. An ongoing effort is made by the organization buildingSMART to develop open-standards for representing BIM-models to increase interoperability, lessening the need for managing information (Golabchi & Kamat, 2013). Moreover, various open-standards for rule formulations are available, but need to be manually updated (Dimyadi & Amor, 2013).

There exist attempts to automate the interpretations of rules using natural language processing techniques. Zhang and El-Gohari (2016) have attempted to improve the flexibility to changes of the rules by applying Natural Language Processing technology to automate a machines interpretation of rules. However, regulatory tests are not written for computer interpretation and have not been very successful (Dimyadi & Amor, 2013).

Dimyadi and Amor (2013) argued that there was a lack of research focusing on semi-automation in the domain of BMC accommodating both qualitative and quantitative aspects. Later studies attempting to close this gap has been conducted by Preidel and Borrmann (2017; 2015, 2016) by using visual languages to improve transparency of processes in BMC to allow for domain experts to easier adapt to the real-life changes. The emphasis of allowing domain experts to understand the automotive processes better has been suggested to ease the management of the information used in the BMC processes.

Preidel and Borrmann (2017) argued that the limitations and challenges of using visual languages are the complexity of the BMC process representation. Their findings indicated that the complex representation needs to accommodate better handling of the information involved such as errors and iterations. Handling such issues are related to changes happening in the construction industry that the BMC system needs to mimic in order to provide the users of BMC with reliable feedback. Therefore, BMC systems must be better to accommodate such changes and become more flexible.

### 1.3 Improving the flexibility of BMC

To improve the flexibility, we use theories related to Process-Aware Information Systems (PAIS) and Business Process Management (BPM). Both theories are interrelated, and emphasize improvement of productivity of business practices through a process-centered approach (Reichert & Weber, 2012b; Van der Aalst, 2013). These methods have been used to improve the structure of information related to handling processes to accommodate the challenges pointed out by Preidel and Borrmann (2017) that currently limits the use of visual language to improve the flexibility of BMC systems. The increased flexibility allows processes to easier and more rapidly be adapted to real-life changes. BPM systems have previously been applied to various practices that required information systems to respond effectively to changes (Van der Aalst & Basten, 2002).

We conducted a screening of the scientific literature to identify previous research in relation to the construction industry, PAIS, and BPM, finding only a few studies. Few studies were found related to application of PAIS/BPM into the domain of construction industry, for example, Bergman, Gessinger, and Bergman (2016) used a PAIS approach to improve deficiency management. This approach was achieved by improving flexibility to handle actions related to assigning personnel to address the deficiencies. However, we were unable to identify any efforts of PAIS related to either BIM or BMC.

### 1.4 The aim of the article

In this article, we aim to explore how it is possible to improve the flexibility of BMC. To achieve this flexibility, we use the PAIS/BPM theories to set technical requirements for a prototype that integrates BIM-model information in a BPM environment. The prototype is tested to investigate the applicability of flexibility of BMC. The test case is a criterion from the sustainability assessment method Deutsche Gesellschaft für Nachhaltiges Bauen (DGNB). The investigation is used to discuss the opportunities and limitations of improving BMC flexibility with PAIS and BPM.

## 2 THE NEED OF FLEXIBILITY IN THE CONSTRUCTION INDUSTRY

Reichert and Weber (2012b) argue that knowledge-intensive real-world environments, like building design, cannot be completely prespecified but require maneuvering room for the users, due to the drivers of flexibility. The need for flexibility stems from the relationship between the real world and the digital environment. In the real world, the designers need to get feedback on their design according to rules. When BMC needs to provide this feedback, it requires that both the rules and the design is represented in the digital environment. Asynchrony between the real world and the digital environment makes the feedback irrelevant. Additionally, the digital environment can cause errors on its own, and that system would require adjustment. Making the digital environment of BMC flexible to both changes will provide for more relevant feedback.

The digital environment of BMC is specified according to Eastman et al. (2009)'s defined classes of BMC functionalities: 1) translation of natural language rules into computer executable rules, 2) BIM-model verification, 3) BMC execution with a checking mechanism and 4) dissemination of the checking results. Each of these functionalities is subject to changes from the different organizational layers. At the company level, the steps must be aligned with the company's business goals and resources, for example, when a company has specified quality of documentation of the checking results. The BMC tools are operational at the project level, where it is subject to project-specific conditions and context-dependent rule interpretation. Figure 1 illustrates the relationship between the real-world environment of building design and the digital environment of BMC.

The digital and real-world environment are affected by drivers of change. Van der Aalst & Basten (2002) argue that the drivers are caused by either ad-hoc or evolutionary change. Ad-hoc change is defined as the changes happening on a project basis, which is a result of errors, uncommon events or special demands from the customer. Evolutionary change is of more structural nature, related to changes in company policies, regulations, or a change in market demands (Van der Aalst & Basten, 2002). Reichert and Weber (2012b) separate drivers of changes to be either external or internal. The real-world environment is affected by changes in legislation, technology, and the context of business. When technology improves, new possibilities emerge for a new application of, for example, faster computers. Companies change due to the forces of the markets and changing needs of the customers. Customers' requirements shift along with, e.g., trends that can affect the choice of materials. Finally, the real-world environment can be affected by the organizational learning, for example, when learning creates opportunities to optimize the processes.

The digital environment is affected by internal drivers of change, which include design errors, technical problems, and poor model quality. Design errors can cause problems due to missing information in, e.g., either the BIM-model or in the checking mechanism. Technical problems could be performance degradation due to increasing amount of data. Poor model quality could either be related to the BIM-model or the process model. The process model is the steps of the digital process in the digital environment and can relate to the quality of processes, like amount redundant processes. Poor model quality can also relate to poorly created BIM-models due to inconsistencies in classification or collision of building objects.

The inability of the digital environment of BMC to adapt to external and internal changes can cause inconsistencies and flaws, making the feedback from BMC irrelevant to the design practice. We argue that addressing these dynamics of the construction industry is of importance to make BMC more practically applicable. This would thereby allow for more automation of the design practice and reduce inefficiencies and flawed building designs.

## 3 MAKING BMC MORE FLEXIBLE WITH A PROCESS AWARE INFORMATION SYSTEMS APPROACH

We suggest making BMC more flexible by applying a Process-Aware Information Systems (PAIS) approach to BMC. PAIS is used to increase flexibility in the development of information systems in complex and dynamic domains and is also used in workflow information systems, case handling tools, and service orchestration engines (Reichert & Weber, 2012b).

Key characteristics of PAIS are the separation of process logic (formalized in process models) and application code (formalized in data models). Also, by the separation of build-time and run-time components. Build-time components are used when the processes are developed and maintained by process specialists. Run-time components are used by the users like designers (not necessary process specialists) to execute the process models to obtain feedback on their building design. Reichert and Weber (2012b) argue that by splitting monolithic

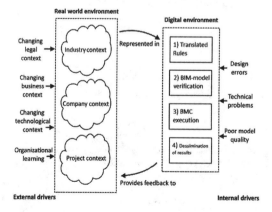

Figure 1. External and internal drivers are affecting the real-world and digital environment, adapted from (Reichert & Weber, 2012b).

applications into smaller components, as well as a separation between build-time and run-time components can make information systems more flexible than traditional ones.

The application code is the statements of code formulated in applications specialized in executing certain functions. The processing logic is typically formalized in process modeling language such as Business Process Model Notation 2 (BPMN2) (Object Management Group, 2011). Formulating business processes with language like BPMN2 allows for more expressiveness of the processes that are used as reliable schemes of executions. It is through the process models that the logic is represented through control and data flow dependencies. These dependencies specify the dynamic behavior of the process, for example, visualizing the process of validating the information of a BIM model. This is done to ensure that the business processes are executed in a specified order and allow for easier development, maintenance, and monitoring of business processes (Reichert & Weber, 2012b).

## 4 REQUIREMENTS OF A FLEXIBLE BMC PROTOTYPE

We developed a prototype to explore and address the previously mentioned flexibility needs. This led to a three-part architecture that consisted of a BIM-model server platform, a Business Process Management platform, and a Custom REST service. The development of this architecture aimed to allow to conduct more flexible BMC. This architecture was created to accommodate the flexibility needs of BMC based on the PAIS theory (Dumas, Aalst, & Hofstede, 2005; Reichert & Weber, 2012a; Weber, Reichert, & Rinderle-Ma, 2008).

We focused on: 1) separation between logic and application code, 2) separation between build-time and run-time components 3) accommodation of looseness, adoption, and evolution. The logic and application code must separate in different data objects representing the logic of rules and represent the application code. The build-time and run-time components must be separated to allow specialists to manage information related to the data-objects representing specialized information not needed by the run-time users. The accommodation of looseness, adoption, and evolution is presented below:

### 4.1 *Looseness*

If changing processes become comprehensive and complex they require much maintenance to be kept relevant and still will they fail to match the uniqueness required in e.g., specific design projects. Moreover, can complex processes defer technical issues. A solution to this is the approach of looseness. Loosely specified processes that are made as simple and general as possible require more input by the run-time user but are both easier to maintain and applicable to more situations.

### 4.2 *Adaption*

A process of assessing the sustainability of a building, for example, must be able to continue even if not all required information is present in the BIM-model. In this case, it cannot provide comprehensive feedback on the performance of the design, but it must still give a partial insight of the BIM-models sustainability performance and score. If deviations are not permitted to the run-time users, it can lead to rejection of the system. Therefore, a PAIS must allow for ad-hoc changes for the run-time users and be able to handle exceptions in order for the run-time user to progress with executing the checking.

### 4.3 *Evolution*

A system will never reach perfection, but instead act as a continuous optimization and is subject to changes happening in the real-world environment such as changes in laws, materials and such. Changes of processes (due to the evolutionary changes) need to be supported for refactoring (the process of restructuring existing code for optimization) without the run-time users noticing and for users to apply older versions of processes. Especially older versions of such processes are necessary in the construction processes because building project typically span over a long period of time and often only needs to be assessed to a set of rules governing at the point of initiation. Such changes entail the need for the ability to make versioning of the processes and rules and the ability to manage to refactor the information used in the system. Moreover, the build-time users must be supported with feedback through monitoring, analyzing, and mining of process performance information, which can be used to improve the processes.

## 5 THE ARCHITECTURE OF THE BMC PROTOTYPE

We selected and developed different sub-systems components to accommodate the previously set requirements to create the architecture of the new BMC prototype. To address the main functionalities of flexibility, i.e., a separation between logic and application code, a separation between build-time and run-time components, and accommodation of looseness, adoption, and evolution, we applied a BPM system Bizagi Studio 11 (2017). The choice of BPM system was based upon a functional screening according to the requirements earlier specified. The screening included various software presented in a comparative study by Koncevics and Penicina (2017).

Bizagi Studio contains the functionality of separation of application and logic. Also, it allows a separation between run and build-time users. Moreover, it accommodates the requirements of looseness, adaption, and evolution through various functionalities.

Bizagi accommodates looseness by allowing BMC processes to be loosely formulated and require extensive but constrained user input. For example, by creating as simple and general as possible processes. This allows the run-time users to adjust the process to their specific contextual demands, thus requiring more run-time input.

Bizagi also accommodates adaption by making exceptions in the processes to the issues that are possible to foresee. Moreover, it allows evolution by handling versioning of the processes evolved over time and enabling refactoring of existing processes. It also supports organizational learning through extensive mining, analysis and monitoring elements of the processes.

Bizagi represents the business logic in both rule objects and the BPMN2 notation form, which is widely used by academics and professionals to express business processes visually. We created a process activity (expressed visually as rectangle that describes a task) for a web service to get the BIM information) named Service, and one process to assess the information named Assess. The information required in the processes was specified in the data model in Bizagi. The data in Bizagi Studio was connected with a custom-made web service to the BIM-model server platform. The data model from the BIM-model server was mapped to the data model in Bizagi Studio. The data were assessed according to logic translated from DGNB and specified in Bizagi Studio. Bizagi Studio then presented the result of the automated assessment of room heights of the building.

The BIMserver software parses IFC models and stores the data revisions in a non-relational database. The server can be accessed directly by the BIMserver software and the underlying web service implementation via the HTTP protocol. The data exposed by the BIMserver web service implementation are formatted as either JavaScript Object Notation (JSON) or Simple Object Access Protocol (SOAP) XML data structures.

In Bizagi Studio it is possible to model business processes and create a data-model that specifies the information containers needed for the automation. It is possible to integrate information from external sources through either REpresentational State Transfer (REST) or SOAP web services. We developed a RESTful web service that acts as middleware between our hosted BIMServer.org server and Bizagi Studio.

This web service makes it possible for Bizagi Studio to request data contained in the BIMserver-hosted IFC-models. We chose to implement a piece of middleware because the structure of the data exposed from BIMservers own web service implementation was unfit for direct integration with Bizagi Studio. Therefore, the custom web service acts as a data scrubbing middleware, which solves the structural mapping issues between Bizagi Studio and BIMerver data structures. The data scrubbing web service was implemented using Microsoft ASP.NET Web API 2.0, written in C#. The actual scrubbing of the JSON formatted BIM data consumed from BIMserver was done manually by querying BIMservers' API and using Newtonsoft.Json (de)serialization library. The main work of the web service is the querying for BIMserver data, deserializing the data and reconstructing it in an easily readable format, which is then serialized again, and exposed as JSON to Bizagi Studio.

The different tiers in the architecture can be geographically separated and are only communicating via message passing (HTTP requests). Building a middleware web service has the benefit of providing multiple numbers of clients with a shared interface, which exposes well-formed data to client applications. In our case, we have developed a system that can make BIM-model data

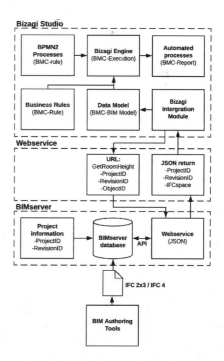

Figure 2. A high-level overview of the systems functionalities.

available to a BPM system. The custom web service, however, is not limited to any client application, since it communicates over a standardized protocol, HTTP. The architecture is visualized in Figure 2.

## 6 THE CASE OF SUSTAINABILITY ASSESSMENT

We used the prototype to test its functionality of automatically assess a DGNB criterion. In Denmark, the preferred method for assessing sustainability is named "DGNB system Denmark manual for Office buildings 2016" (DK-GBC, 2016). The method is originally German but was adapted to a Danish context. It is a second-generation sustainability assessment method, which means that it contains a broader range of sustainability criteria than first generation assessment methods. The method is very comprehensive, containing 40 criteria with 140 sub-criteria covering aspects of social, economic, environmental, and process aspects of the design and construction of an office building. The criteria are weighted differently, and they vary in comprehensiveness and complexity. The designers struggled in handling the comprehensiveness and complexity of the sustainability assessment methods in general and this was a key reason why this was chosen as a case to test the prototype. The case tested a sub-criterion of the criterion ECO2.1 Flexibility and Adaptability named ECO2.1–2 Room Height. The sub-criterion is used to assess the possible flexibility of the rooms by measuring the height of "primary" rooms. However, this limited case is considered a test and example of the use of the prototype. The score of the criteria is given as checklist points, TLP (Danish: TjekListe-Points). The description of the sub-rule is made both in prose and in a table, specifying the possible score according to the room heights of the building design. The sub-criterion specifies the following (important outtakes):

- The average height of the room is used
- Room height is defined as the distance between "top floor" and "bottom ceiling."
- Room height is measured in meters
- The ECO2.1-2 score is named TLP and is calculated (average Room height => 3) = 10 TLP

We used Autodesk Revit (2017) to create a simple test model showed in Figure 3, containing walls, floors, ceiling, a roof, and three rooms. The ceilings restricted the height of the rooms and had various heights. One room had the height of 2600 mm; one had 3100 mm and one 3000 mm. The test BIM-model was exported in IFC version 2 × 3 – Coordination View 2.

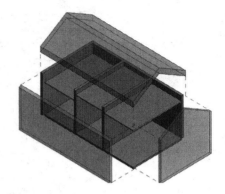

Figure 3. An isometric, displaced, and transparent view of the test model created in Autodesk Revit 17.

## 7 TESTING RESULTS

We developed a prototype accommodating for the technical requirements regarding flexibility. This prototype was functional and was able to use the information from a BIM-model located at the BIMserver with the business logic represented in the BPM environment of Bizagi Studio. This connection allowed for a BMC check of the room heights in Bizagi Studio. The connection between the two servers succeeded with a custom-made web-service, wherein the data model requirements were specified (i.e., project information and room heights) and needed to be assessed according to the ECO2.1–2 criterion. The information was represented in Bizagi Studio and was automatically assessed based on the formalized logic. The logic was specified in a BPMN2 process diagram and the rules of ECO2.1–2 were specified in the "Actions & Validation" functionality, where control flow statements (If, Then, Else) could be represented. The test of using the information from the BIM-model resulted in the correct TLP score of 0. The logic successfully assessed the room heights and calculated the average room heights of three room's which was lower than 3 meters. The results are shown in a screenshot from Bizagi Studio in Figure 4.

The processes formalized in Bizagi Studio was loosely formulated. This was done with a focus to simplify the process and rely on run-time user input. This meant that the method of assessment of the room height was measured based on the room height information in the IFC model, and were not conducted as a post calculation.

Also, we included exceptions to control the quality of information used both through BIMserver and Bizagi studio to ensure that the right information was present to conduct the assessment. For example, in Bizagi, the information was validated before the assessment commenced.

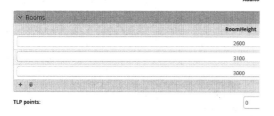

Figure 4. A screenshot from Bizagi Studio's run-time web server showing calculated room heights and TLP score.

The process management functionalities of Bizagi Studio allowed for managing versions of the automated BMC-processes. In the BIMserver it was possible to manage the uploaded BIM-models through versioning, which can help manage the BIM-model evolution throughout the progress of the projects.

The prototype allows for versioning of the processes enabling the handling of deferred evolution due to the external drivers such as rule updating (e.g., new versions of DGNB). Updating the processes can be done real-time through refactoring in Bizagi Studio to continuously improve the quality of the process without making problems for the run-time users.

The versioning is especially important in building project cases, where the time determines, which rules are applicable, for example, if a project started in 2014, it is often the rules that apply to that year and not the most recent.

Bizagi Studio allowed for data mining for various process-related parameters, such as instances run, time to complete, errors, exceptions, and the possibility to include and integrate resource overview and management of the people executing the processes. These insights can be presented real-time for build-time users to enable integration of optimizations. The integration allows companies using such as setup to facilitate organizational learning by obtaining critical data about how the processes are executed and allowed for actions to be made for improvements.

## 8 CONCLUSION AND FUTURE WORK

In this paper, we explored the possibility to improve practical use of BMC by improving flexibility with the use of BPM software. We used PAIS theories to identify the functional requirements for BPM software in a BMC context. The requirements were used to create a prototype with BIMserver to host the BIM-model information and REST web service to enable a transfer of BIM data into the BPM software. With this prototype, we were able to assess a test BIM-model according to sustainability assessment method sub-criterion ECO2.1–2 from DGNB. Allowing to conduct BMC in a BPM environment provide several features of flexibility. Such environment could potentially improve the operationalization of BMC allowing to adapt better to the dynamics of the design projects, shifting business goals of the designer's companies and changing legislation of the industry. However, during our development and test of the prototype we discovered the following limitations of our prototype:

### 8.1 Poor integration of the BIM-model information

The integration of a data model from BIMserver requires that the webservice is required to be continuously adapted according to what is going to be checked in the BIM-model. Each adaption of the web service requires a specialist web service programmer, which complicates the maintainability and flexibility of the system. Instead, we suggest that future investigation on the use of BPM should emphasis better integration of BIM-model information enabling both the run-time and build-user on retrieving information from the BIM-model.

### 8.2 Lack of information processing transparency for the run-time user

The run-time user needs to understand how the information is processed to address potential adaptions. Instead, the run-time user is restricted to modifying certain parameters. The limitations of adapting the run-time processes could potentially hamper the practical use of such as system due to the large uncertainty in each project.

However, we suggest that future research should include further evaluation of the practical usability of flexible approaches to BMC in realistic settings. This would further explore the potential to utilize BMC in BPM environments.

## ACKNOWLEDGMENTS

We are thankful for the feedback of professor Barbara Weber from the Technical University of Denmark.

## REFERENCES

Autodesk. (2017). Revit Family | BIM Software | Autodesk. Retrieved April 30, 2017, from https://www.autodesk.com/products/revit-family/overview

Beach, T.H., Rezgui, Y.R., Li, H., & Kasim, T. (2015). A rule-based semantic approach for automated regulatory compliance in the construction sector. *Expert*

Systems with Applications, 42(12), 5219–5231. https://doi.org/10.1016/j.eswa.2015.02.029.

Bergmann, R., Gessinger, S., & Bergmann, R. (2016). Flexible Process-Aware Information Systems Deficiency Management in Construction Flexible Process-Aware Information Systems Deficiency Management in Construction, (January 2015), 330–338.

Bertelsen, S. (2003). Complexity – Construction in a New Perspective. *International Group of Lean Construction*, 12.

Bizagi. (2017). Bizagi - Digital Transformation &amp; Business Process Management BPM. Retrieved August 16, 2017, from https://www.bizagi.com/.

Dimyadi, J., Clifton, C., Spearpoint, M., & Amor, R. (2014). Regulatory Knowledge Encoding Guidelines for Automated Compliance Audit of Building Engineering Design. *Computing in Civil and Building Engineering (2014)*, 536–543. https://doi.org/10.1061/9780784413616.067.

Dimyadi, J., & Amor, R. (2013). Automated Building Code Compliance Checking – Where is it at? *Proceedings of CIB WBC 2013*, 172–185. https://doi.org/10.13140/2.1.4920.4161.

DK-GBC. (2016). DGNB system Denmark manual for kontorbygninger 2016. Retrieved from http://www.dgnb-system.de/en/.

Dubois, A., & Gadde, L.-E. (2002). The construction industry as a loosely coupled system: implications for productivity and innovation. *Construction Management and Economics*. https://doi.org/10.1080/01446190210163543.

Dumas, M., Aalst, W.M. van der, & Hofstede, A.H. ter. (2005). *Process Aware Information Systems: Bridging People and Software Through Process Technology*. Hoboken: Wiley-Interscience. Retrieved from http://sfx.aub.aau.dk/sfxaub?+Aalst,+Wil+M.+van+Der=&+Hofstede,+Arthur+H.+Ter=&+Ter+Hofstede,+Arthur+H=&+van+Der+Aalst,+Wil+M.=&ctx_enc=info:ofi/enc:UTF-8&ctx_tim=2017–08–14T13:10:35IST&ctx_ver=Z39.88–2004&req.language=dan&rfr_id=info:sid/primo.exlibrisgr.

Eastman, C., Lee, J., Jeong, Y., & Lee, J. (2009). Automatic rule-based checking of building designs. *Automation in Construction*, 18(8), 1011–1033. https://doi.org/10.1016/j.autcon.2009.07.002.

Flager, F., Welle, B., & Bansal, P. (2009). Multidisciplinary process integration and design optimization of a classroom building. *Journal of Information …*, 14(August), 595–612. Retrieved from http://www.researchgate.net/publication/259221519_Multidisciplinary_Process_Integration_and_Design_Optimization_of_a_Classroom_Building/file/9c96052a815e0d5e67.pdf.

Golabchi, A., & Kamat, V. (2013). Evaluation of Industry Foundation Classes for Practical Building Information Modeling Interoperability. *International Association for Automation and Robotics in Construction*, 17–26. Retrieved from http://www.iaarc.org/publications/fulltext/isarc2013 Paper109.pdf.

Hjelseth, E. (2016). Classification of BIM-based Model checking concepts. *Journal of Information Technology in Construction*, 21(July), 354–370.

Hjelseth, E., & Nisbet, N. (2010). Exploring Semantic Based Model. In *Proceedings of the CIB W78 2010: 27th*.

Khemlani, L. (2017). Automated Code Compliance Updates.

Koncevics, R., & Penicina, L. (2017). Comparative Analysis of Business Process Modelling Tools for Automated Compliance Management, 21(May), 22–27. https://doi.org/10.1515/acss-2017–0003.

Kuben Management. (2016). *Værdiskabelse og effektivitet i dansk byggeri*.

Lopez, R., & Love, P.E.D. (2012). Design Error Costs in Construction Projects. *Journal of Construction Engineering and Management*, 138(5), 585–593. https://doi.org/10.1061/(ASCE)CO.1943–7862.0000454.

Object Management Group. (2011). BPMN 2.0. Retrieved September 27, 2017, from http://www.omg.org/spec/BPMN/2.0/.

Preidel, C., & Borrmann, A. (2015). Integrating Relational Algebra into a Visual Code Checking Language for Information Retrieval from Building Information Models. *Icccbe*, (June). https://doi.org/10.13140/RG.2.1.4618.5201.

Preidel, C., & Borrmann, A. (2016). TOWARDS CODE COMPLIANCE CHECKING ON THE BASIS OF A VISUAL PROGRAMMING LANGUAGE, 21(July), 402–421.

Preidel, C., & Borrmann, A. (2017). Refinement of the visual code checking language for an automated checking of building information models regarding applicable regulations. In *Congress on Computing in Civil Engineering, Proceedings*.

Preidel, C., Daum, S., & Borrmann, A. (2017). Data retrieval from building information models based on visual programming. *Visualization in Engineering*, 5(1), 18. https://doi.org/10.1186/s40327-017-0055-0.

Ralph, P., & Wand, Y. (2007). A Proposal for a Formal Definition of the Design Concept. *Design Requirements Engineering: A Ten-Year …*, 1–8. Retrieved from http://link.springer.com/chapter/10.1007/978-3-540-92966-6_6.

Refvik, R., Skallerud, M., Slette, P.A., & Bjaaland, A. (2014). *ByggNett – Status survey of solutions and issues relevant to the development of ByggNett* (Vol. 1). https://doi.org/10.1017/CBO9781107415324.004.

Reichert, M., & Weber, B. (2012a). Enabing Flexibility in Process-Aware Information Systems.

Reichert, M., & Weber, B. (2012b). *Enabling Flexibility in Process-aware Information Systems Challenges, Paradigms, Technologies*. Springer. Retrieved from http://www.springer.com/computer/database+management+%26+information+retrieval/book/978-3-642-30408-8.

Solihin, W., & Eastman, C. (2016). A knowledge representation approach in BIM rule requirement analysis using the conceptual graph, 21(March), 370–402.

Van der Aalst, W. (2013). Business Process Management: A Comprehensive Survey. *ISRN Software Engineering*, 2013, 1–37. https://doi.org/10.1155/2013/507984.

Van der Aalst, W., & Basten, T. (2002). Inheritance of workflows: An approach to tackling problems related to change. *Theoretical Computer Science*, 270(1–2), 125–203. https://doi.org/10.1016/S0304–3975(00)00321–2.

Weber, B., Reichert, M., & Rinderle-Ma, S. (2008). Change patterns and change support features Enhancing flexibility in process aware information systems. *Data & Knowledge Engineering*, 66(3), 438–466. https://doi.org/http://dx.doi.org/10.1016/j.datak.2008.05.001.

Zhang, J., & El-Gohary, N.M. (2016). Semantic NLP-Based Information Extraction from Construction Regulatory Documents for Automated Compliance Checking, 30(1), 1–5. https://doi.org/10.1061/(ASCE)CP.1943–5487.0000346.

# Building Information Modelling (BIM) value realisation framework for asset owners

M. Munir & A. Kiviniemi
*School of Architecture, University of Liverpool, Liverpool, UK*

S. Jones
*School of Engineering, University of Liverpool, Liverpool, UK*

ABSTRACT: The paper is presenting a value realisation framework for asset owners based on an exploratory study. The study is descriptive in nature and adopting a qualitative approach towards data collection. The paper adopts the viewpoint of BIM business value measurement considering that; (i) if the process is better as a result of BIM-based processes, then it is different in some relevant way; (ii) if it is different in some relevant way as a result of certain BIM properties or characteristics, then the change is observable; (iii) if the change is observable because of certain direct BIM benefits, then it is countable; (iv) if it is countable using defined measurement metrics, then it is measurable; (v) if it is measurable using established measurement techniques, an organisation can value each unit and therefore, realise the benefits of BIM. The specific contribution of paper is to improve asset owners' understanding of BIM-business value measurement techniques and approaches.

## 1 INTRODUCTION

Building Information Modelling (BIM) in Asset Management (AM) is an area that has not been given much attention by researchers. There has been more focus on BIM business value realisation in pre-construction and construction stages rather than the post construction stage, which has a longer life of the asset (Love et al., 2014). Although, BIM is claimed to provide an efficient tool to asset managers in improving building performance and management of operations, there are very few case studies on the real use of BIM in the operations and use phase (Codinhoto & Kiviniemi, 2014).

BIM investments like other Information Technology (IT) based business initiatives are continuously questioned on the level of impact they have on organisational business value. Many clients worry that the value that BIM delivers may not be as high as expected. Similarly, like other IT-based investments, BIM suffers from the *'productivity paradox'* (Brynjolfsson 1993, Willcocks & Lester 1996). Some asset owners find themselves adopting BIM but cannot find sufficient economic justification. The investment in BIM in the Architectural, Engineering and Construction (AEC) industry is increasing and there are still doubts that the benefits may not be as high as expected. The difficulty in the realisation of BIM benefits can be related to weaknesses in measurement techniques and business value realisation practices of the AEC industry. These factors make it difficult to evaluate the benefits of IT-based tools or methodologies such as BIM (Vass & Karrbom Gustavsson, 2014).

As a result, asset managers constantly have to justify IT-based investments such as BIM because of the huge capital outlays and are compelled to appraise value at the strategic and operational levels (Irani, 2010). Many owner-operator organisations tend to approach the lifecycle management of BIM in an *ad-hoc* or unstructured manner. Investment evaluation should be conducted by asset owners in the same way projects are managed (Irani, 2010). One of the predominant issues is that managers tend to measure those activities that are easily identifiable, thereby, creating a tendency of ignoring and undervaluing those that are not (Dawes, 2010). There should be a parallel activity, where investment decisions are reviewed in relation to cost, risks and benefits. The conduct of this activity will help asset owners evaluate the success of BIM and the business value it delivers.

Benefits realisation management is a significant business process for asset owners to derive value from BIM. Defining requirements, measuring, analysing and monitoring the entire process is important for asset owners to be able to identify BIM business value (Lin et al. 2007, Love et al. 2014). A number of studies have attempted to measure the benefits of BIM (Giel et al. 2010, Kreider et al. 2010, Barlish & Sullivan 2012, McGraw-Hill 2012, Love et al. 2013, Love et al. 2014, Walasek & Barszcz 2017), but more research is needed to clarify the difficulty in measurement of BIM benefits in the operations and use phase.

## 2 METHODOLOGY

### 2.1 Research question

This study presents a framework on how an asset owner can realise BIM business value in the operations and use phase of built assets. It also seeks to demonstrate how owner-operator organisations can link intangible to tangible value for easy measurement. The study will address the following research questions:

- What are the techniques and strategies of measuring the business value of BIM in AM processes?
- How can intangible value be linked to tangible value?

### 2.2 Research methods

This study adopts exploratory and descriptive methods of research. The study is divided into two phases. The first phase is the literature review, where the study explores existing research on techniques of measuring business value for BIM and other IT-based initiatives. The reviewed literature was used to identify elements of the framework for measuring BIM business value. The second phase comprises of development of the BIM business value realisation framework (Figure 1) and intangible business value linkage map (Figure 2).

### 2.3 Data collection

The study adopts a qualitative approach towards data collection. The review of literature was conducted to identify relevant existing studies on the business value of BIM, IT value realisation frameworks and techniques for measuring business value of IT-based methodologies. The elements of the proposed business value realisation framework were also identified through this exercise.

In developing the framework, the main factors were drawn from three main theoretical foundations, those are: the AM-FM business processes; the value realisation concepts of Gliderman (2000) (tangible); and the theoretical concepts of Carayannis (2004) and Nogeste & Walker (2005) (intangible). Each of the aspects contribute to the framework with specific types of information.

This review of secondary data sources led to the development of a BIM business value realisation framework (Figure 1). The framework provides a procedural model for approaching BIM business value realisation for asset owners. The framework organised concepts such as outputs, result evaluation and business value realisation dimensions that the study explored directly during data collection.

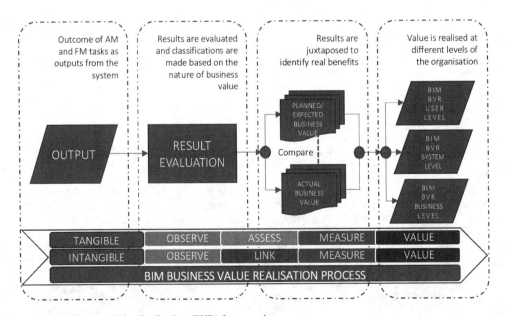

Figure 1. BIM Business Value Realisation (BVR) framework.

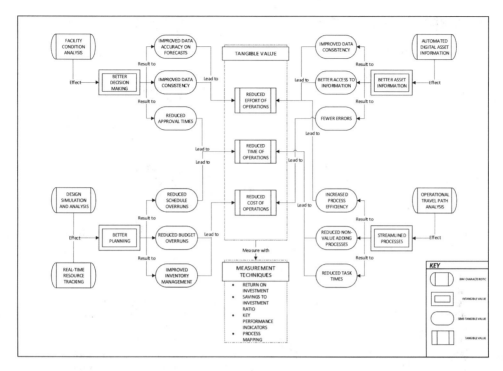

Figure 2. Concept map for linking intangible value to tangible value for BIM-based processes.

Finally, the concept map for linking intangible value to tangible for BIM-based processes (Figure 2) was drawn from the theoretical proposition of Bakis et al. (2006).

## 3 LITERATURE REVIEW

### 3.1 Building Information Modelling (BIM) in Asset Management (AM)

BIM is a technology-based process that enhances performance and information delivery in the lifecycle of a built asset. BIM as defined by Succar et al. (2007) is '*a set of interacting policies, processes and technologies producing a methodology to manage the essential building design and project data in digital format throughout the building's life-cycle*'. On the other hand, ISO 55000 (2014), defines AM as a '*coordinated activity of an organisation to realise value from assets*'. Certainly, AM facilitates a holistic methodology that manages an asset from inception to disposal. The implementation of BIM in AM is an opportunity for asset owners to attain value propositions in their organisations through effective management of business processes (Love et al., 2014). With handover standards such as COBie that deliver structured information of the facility to the client, asset managers have the ability to leverage asset data through BIM. It is however worthy to mention that these tools and techniques do not fully support the asset owner in realising all the benefits that BIM generates in AM processes (Love et al., 2014).

BIM implementation in the operations and use phase will require changes in business processes and development of new roles for asset owners to achieve desired benefits (Ayyaz et al., 2012). These business modifications and resultant benefits continuously change throughout a facility's lifecycle. Some of the challenges asset owners face are cultural and operational in nature. That is, managing the change process and putting in appropriate measures for smooth implementation of BIM in AM. Without addressing these issues an organisation may not be able to track the business value BIM generates. Similarly, Codinhoto & Kiviniemi (2014) suggest that asset owners need to be aware of their organisational inefficiencies in the first place before they can address technology related challenges associated with BIM implementation.

### 3.2 Value realisation management

IT business value is the sustainable benefit realised by an organisation through IT-based systems, either by collective or individual systems, assessed from an organisational perspective (Cronk & Fitzgerald, 1999). Consequently, value realisation is the process of evaluating these benefits. Value realisation management is defined as *'the process that realises the benefits that are achieved and manages the unexpected ones'* (Farbey et al., 1999). It is however the strategy adopted by an organisation to determine how benefits are realised, at what level and when. The main purpose of value realisation is not to forecast benefits but to make them come true (Ward et al., 1996).

Many studies on value realisation management have tried to address the issue of realising the benefits of IT-based investments, but the problem is a dynamic one. Andersen, et al., (2000) present a procedural benefits measurement framework for IT-related investments in the AEC industry. Also, in a bid to address the complex issues of value realisation, Sapountzis et al. (2007) review four value realisation management approaches that may support organisations in the optimisation of benefits from investment programmes. Ashurst & Doherty (2003) suggest best practice for value realisation through the development of a conceptual framework. Similarly, Ward et al. (1996) adopt a procedural approach and explains the steps within the value management process. On the other hand, Leyton (1995) propose a model which approached value realisation from the perspective of business change. Sapountzis et al. (2007), then propose a value realisation framework integrated with business and investment processes.

The main challenge of realising BIM business value is that asset owners do not plan to realise benefits in the first place. Value realisation has to be done deliberately and consciously. Some asset owners harbour the *'silver bullet thinking'* on BIM investments (Thorp, 1998). That is, if they invest in BIM, the benefits will come automatically. Lin & Pervan (2003), suggests the need for asset owners to change their strategy of value realisation management from a passive approach to a more proactive one. Love et al. (2014) further assert that BIM alone cannot deliver business outcomes and that the process of its implementation has to be proactively managed to ensure that the organisation realises the business value it expects. Irrespective of the primary strategic objective for adopting BIM, an organisation will have to understand its capability and maturity before it can realise any value from the whole process.

3.3 *BIM business value*

In identifying BIM business value, the main issue of contention for asset owners is how to identify the benefits and the methods with which to measure them. A number of studies have attempted to identify the benefits if BIM in the operations and use phase. Ding et al. (2009) find that BIM enabled facilities management yields a 98% reduction in time used to update asset databases. Similarly, Codinhoto & Kiviniemi (2014) identify thirteen (13) metrics of BIM benefits in the operations and use phase.

In a bid to evaluate the business value of IT-based investments, Willcocks & Lester (1996) propose a balance score card (BSC) approach to examine the contributions of the IT-based investment from the financial, internal business, innovation and learning, and customer perspectives. However, this method does not provide for an aggregate system for these factors, as the decision of worthiness of an investment still remains with the asset manager. On the other hand, Construct IT (1998) identify different IT business value and classified each benefit against three factors; (i) efficiency assigned with a financial value, (ii) effectiveness defined with a subjective value, and (iii) performance qualified with qualitative accounts, however, with no quantification. The benefits were also classified according to the business processes they support. Also, Gartner (2003) present a five-pillar benefit realisation framework; strategic alignment, business process impact, architecture, direct payback, and risk. The framework determines the overall business value expected to be created by an IT-enabled business initiative. It uses a standard set of concepts for quantitative and qualitative value methods. The Gartner framework provides an aggregate score card for IT-based investments. Melville et al. (2004) propose an IT business value model that uses a resource-based view (RBV) to focus on the impact of IT-based investments such as BIM on organisational resources and business processes. Similarly, Love et al. (2014) present a framework that asset owners can utilise to realise value from investing in BIM. The framework adopts governance, change management, performance measurement, and stakeholder management as factors that enable the strategic alignment of the asset owners' business strategy. However, the framework does not provide a scoresheet for aggregating the key factors. Furthermore, Sanchez et al. (2016) present an eight-step BIM value realisation framework that introduces a methodology together with a benefits, metrics and enablers dictionary to aid measurement. However, the framework focuses mostly on tangible and semi-tangible benefits.

Deriving business value from an IT-based system can be difficult and depends largely on many different complex factors that cannot be controlled or isolated for formal experiments (Bakis et al., 2006). This is because the workings of an organisation are a collection of various integrated systems to perform tasks and deliver outputs. It is not possible to isolate BIM from other closely interconnected processes, in order to assess it, or one independently of the other. Another reason why it is difficult to objectively prove business benefits of a BIM-based investment is because an IT system only has the potential to create value, and not direct value in its own right (Mooney et al. 1995, Farbey et al. 1999, Remenyi 2000, Love et al. 2014). Another argument is that an IT-based investment may not yield the desired benefits simply because it is not well implemented (Brynjolfsson & Hitt, 1998).

effectiveness (Ağra, 2011). When applied to BIM, an asset owner can determine whether the proposed savings in a BIM business case justifies the cost of the total investment. SIR involves the following processes; Determining the cost of the project; Determining the useful life of the asset; Determining the savings associated with the project; and Calculating the ratio. SIR is calculated as:

$$SIR = \frac{PV(\text{Internal Project Cost Savings}) + PV(\text{Programme Cost Savings})}{PV(\text{Initial Investment})}$$

### 3.4 BIM value measurement techniques

There are many value measurement techniques that can be used by asset owners to identify the value that BIM affects. Some techniques are generic and some may be unique to certain organisations. The study will focus on four techniques; Return on Investment (ROI), Savings to Investment Ratio (SIR), Key Performance Indicators (KPI) and Process Mapping. It is worth mentioning that an organisation may utilise multiple methods to measure BIM benefits. Also, Love et al. (2013), argues that if financial techniques such as ROI are only used to justify investment in BIM, then the entire process is limited to financial management.

#### 3.4.1 Return on Investment (ROI)

ROI is defined as the ratio of resources gained or lost in a process or investment, as against the total amount of resources provided (White, 2007). When applied to BIM, it is suggested that positive ROI means business value to clients, contractors, consultants and other stakeholders and measurement to be calculated as a ratio of benefit to cost (Giel et al., 2010). ROI is calculated as:

$$ROI = \frac{\text{Gain from Investment} - \text{Cost of Investment}}{\text{Cost of Investment}} \times 100$$

One of the challenges of using this method to measure business value, is the lack of industry-wide accepted benchmark for measuring BIM ROI (Giel et al., 2010). Another problem is the inapplicability to generalise or compare ROI data because it is hardly ever possible to find two organisations using the same business processes and accounting policies.

#### 3.4.2 Savings to Investment Ratio (SIR)

SIR represents the ratio of savings in relation to investment. The important factors when using SIR are investment cost and functional asset effectiveness (Ağra, 2011). When applied to BIM,

One of the challenges of adopting this method to measure business value is the meticulous nature of identifying savings in relation to the total investment and comparing it with alternative options. This task requires a great deal of skill to execute but can be very beneficial if done correctly.

#### 3.4.3 Key Performance Indicators (KPI)

KPIs are a set of data measures used to evaluate the performance of a system, task or operation (Cox et al., 2003). Evaluations using KPIs usually compare actual performance against estimated with reference to efficiency, effectiveness and quality of outcome and workmanship. KPIs can be used to assess both tangible and intangible value through quantitative and qualitative performance indicators.

One of the challenges of applying this method for measuring business value is the amount of resources required to develop performance benchmarks in owner-operator organisations. The benchmarking data for comparison has to be available for the asset manager to be able to evaluate whether targets have been met. Another weakness of using KPIs is that the models fail to identify actual parameters that represent change in performance (Cox et al., 2003).

#### 3.4.4 Process mapping

Process mapping is a technique that involves identifying, documenting, analysing and developing an improved process (Anjard, 1996). It can be used to identify where improvements can be made or to compare where improvements have been made. Process mapping is a useful tool for identifying business process problems such as errors, repetitive processes, delays and inefficiencies. A process map is a visual aid that helps show how inputs, outputs and tasks are interlinked (Anjard, 1996).

One of the challenges of implementing this method for measuring value is achieving the required level of detail and accuracy in process mapping. In some cases, mapped processes are not representative of the actual task. Another issue is the sourcing of skilled labour to draft process maps (Anjard, 1996). Similarly, patience to draft a process maps poses another drawback because the task of producing process maps organisation-wide can be overwhelming.

## 4 BIM BUSINESS VALUE REALISATION (BVR) FRAMEWORK

This section presents a framework for measuring business value of BIM-related outcomes for asset and facility managers (Figure 1). It suggests the process in a sequence and describes various aspects to be taken care of while observing outcomes, evaluating results, comparing planned and actual benefits, and finally, realising business value at the user, system or business dimension. The model is based on the methodology of BIM business value measurement considering that; (i) if the process is better as a result of BIM-based processes, then it is different in some relevant way; (ii) if it is different in some relevant way as a result of certain BIM properties or characteristics, then the change is observable; (iii) if the change is observable because of certain direct BIM benefits, then it is countable; (iv) if it is countable using defined measurement metrics, then it is measurable; (v) if it is measurable using established measurement techniques, an organisation can value each unit and, therefore, realise the benefits of BIM (Glideman, 2000). Furthermore, Cronk & Fitzgerald (1999), propose three dimensions to IT business value, they are: user, system and business levels.

### 4.1 *Measuring tangible value*

The framework proposes a four-step process in measuring tangible value (Glideman, 2000).

- Observe: The change to the system is to be observed closely and properly documented. This step should assess the value realisation plan produced at the beginning of the investment lifecycle and review the targeted benchmarks. It is important that benefit identification is closely connected to the value realisation plan and business case so that the benefits that BIM could deliver are aligned with the organisational business strategy.
- Assess: The next step is to assess the nature of the change, whether it is positive or negative. This can only be determined if the outcomes are properly classified to be measured against organisational performance benchmarks.
- Measure: At this stage, a drawdown list of expected benefits from the value realisation plan and business case should be compared with the outcomes that may materialise from the system. After identification, a suitable measurement technique is selected. Value is then measured using the identified technique.
- Value: The benefits or dis-benefits are realised at the end of the process. The nature of the BIM business value and its dimension is determined at the user, system or business level. The organisation will have to document the value realised for learning and continuous improvement.

### 4.2 *Measuring intangible value*

This is a procedural technique of measuring intangible value of BIM business value in the operations and use stage. A four-step process of measuring intangible benefits are observe, link, measure and value was adopted for this model (Carayannis, 2004, Nogeste & Walker 2005, Bakis et al. 2006).

- Observe: The change to the system is to be observed closely and documented properly. Identifying intangible outputs is challenging and complicated, hence, managers will have to cast a wide net over many outcomes from the system. The value realisation plan and business cases should be reviewed so that performance benchmarks are identified.
- Link: The next step will involve linking the observed phenomenon to measurable organisational metrics. This process is further explained in Section 4.2.1 (Figure 2). Intangible outcomes are to be linked to tangible outcomes for measurement. This process is exploratory as organisations have to develop appropriate pairing techniques.
- Measure: The next step is to compare the new observed phenomenon with already existing organisational standards or results. This is the stage where benefits planned for are monitored for realisation. The appropriate measurement technique is adopted and value is identified.
- Value: The benefits or dis-benefits are realised at the end of the process. The nature of the BIM business value and its dimension is determined at the user, system or business level. Finally, the process is properly documented.

#### 4.2.1 *Linking intangible value to tangible value*

One of the difficult tasks of measuring intangible value is identifying a metric with which to measure it. Also, measuring intangibles is very difficult because it is not always possible to quantify those values in absolute terms without any degree of subjectivity. Whether intangibles are assigned with value or not, these benefits still remain significant to achieving organisational objectives (Remenyi, 2000). Bakis et al. (2006) evaluate the inherent problems of quantifying business value and the difficulties associated with intangible value and demonstrated a business value linkage diagram for the benefits of IT-based investments.

In order to measure or quantify intangible business value of BIM, intangible value will have to be linked with tangible outcomes. Business value linkage is a technique which assists asset managers can utilise in identifying value and understanding the process through which value is created.

This study proposes a concept map used to link intangible to tangible value through a concept map. The BIM capability of the system is observed in order to identify the processes it affects and the value it delivers. Subsequently, the intangible benefits are identified. The possible semi-tangible benefits derived as a result of the intangible benefits are also acknowledged. The linking of semi-tangible benefits to the tangible benefits is done for ease of measurement. Finally, intangible value linked to tangible derived from the BIM-based process may be evaluated using any value measurement technique. The concept map proposes a simple four step process but, in some cases, it can be shorter or longer. Whilst using this process, asset owners need to establish benchmarks so as to improve linkage and measurement metrics over time.

## 5 CONCLUSION

The measurement of the BIM business value has been the subject of considerable debate within the normative literature. The difficulties in measuring benefits and costs are often the cause of uncertainty about expected benefits, particularly in AM. Thus, how then can an asset owner obtain business 'value' from investing in BIM?

In addressing this issue, a framework for measuring business value from an investment in BIM is proposed. Furthermore, the study proposes the use of concept maps in identifying intangible value and linking it to tangible based on BIM capability that an organisation may attain from its implementation. The proposed model is conceptual in nature but provides the underlying foundation for developing a strategy for asset owners to consider how BIM can create value in their organisations. Whilst using this process, asset owners need to establish benchmarks so as to improve linkage and measurement metrics over time.

Tangible and intangible value have significant impact on the attainment of organisational business objectives. However, the measurement of intangible value is not a straightforward one. No matter how unclear the measurement is, it is still of value if the organisation learns and understands its processes more than they did prior to the process. It is worth repeating that asset owners need to plan for these benefits to be able to properly realise them. Having a value realisation plan is significant for asset owners to be able to track the benefits that BIM brings.

## 6 FUTURE WORK

Research focusing on business value of BIM has not been forthcoming. Therefore, the proposed framework in this study provides impetus for future research in this area.

## REFERENCES

Ağra, Ö., 2011. Sizing and selection of heat exchanger at defined saving-investment ratio. *Applied Thermal Engineering*, 31(5), pp. 727–34.

Andersen, J., Baldwin, A., Betts, M., Carter, C., Hamilton, A., Stokes, E. and Thorpe, T., 2000. A framework for measuring IT innovation benefits. *Electronic Journal of Information Technology in Construction*, 5(1), pp. 57–72.

Anjard, R.P., 1996. Process mapping: one of three, new, special quality tools for management, quality and all other professionals. *Microelectronics Reliability*, 36(2), pp. 223–25.

Ashurst, C. & Doherty, N.F., 2003. Towards the formulation of 'a best practice' framework for benefits realisation in IT projects. *Electronic Journal of Information Systems Evaluation*, 6(2), pp. 1–10.

Ayyaz, M., Emmitt, S. & Ruikar, K., 2012. Towards understanding BPR needs for BIM implementation. *International Journal of 3-D Information Modeling archive*, 1(4), pp. 18–28.

Bakis, N., Kagioglou, M. & Aouad, G., 2006. Evaluating the business benefits of information systems. In *In Proceeding of 3rd International SCRI Symposium, Salford Centre for Research and Innovation, University of Salford, Salford, U.K.* Salford, 2006.

Barlish, K. & Sullivan, K., 2012. How to measure the benefits of BIM—a case study approach. *Automation in Construction*, 24, pp. 149–59.

Brynjolfsson, E., 1993. The Productivity Paradox of Information Technology: Review and Assessment. *Communications of the ACM*, 36(12), pp. 66–77.

Brynjolfsson, E. & Hitt, L.M., 1998. Beyond the productivity paradox: computers are the catalyst for bigger changes. *Communication of the ACM*, 41(8), pp. 49–55.

Carayannis, E., 2004. Measuring intangibles: Managing intangibles for tangible outcomes in research and innovation. *International Journal of Nuclear Knowledge Management*, 1(1), pp. 333–38.

Codinhoto, R. & Kiviniemi, A., 2014. BIM for FM: a case support for business life cycle. In *IFIP International Conference on Product Lifecycle Management. Advances in Information and Communication Technology. July 2014.* Yokohama, 2014.

Construct IT, 1998. Measuring the Benefits of IT Innovation. *Construct IT Centre of Excellence*.

Cox, R.F., Issa, R.R.A. & Ahrens, D., 2003. Management's perception of key performance indicators for construction. *Journal of Construction Engineering and Management*, 129(2), pp. 142–51.

Cronk, M.C. & Fitzgerald, E., 1999. Understanding 'IS Business value': derivation. *Logistics Information Management*, 12(1), pp. 44–49.

Dawes, S.S., 2010. Stewardship and Usefulness: Policy Principles for Information-Based Transparency. *Government Information Quarterly*, 27(4), pp. 377–83.

Ding, L., Drogemuller, R., Akhurst, P., Hough, R., Bull, S. and Linning, C., 2009. Towards Sustainable Facilities Management. In *Technology, Design and Process Innovation in the Built Environment*. London: Taylor and Francis. pp. 373–92.

Farbey, B., Land, F. & Targett, D., 1999. The moving staircase—problems of appraisal and evaluation in a turbulent environment. *Information Technology and People Journal*, 12(3), pp. 238–52.

Gartner, 2003. TVO methodology: valuing IT investments via the gartner business performance framework. [Accessed 09 November 2017].

Giel, B., Issa, R.R. & Olbina, S., 2010. Return on investment analysis of building information modeling in construction. In *Proceedings of the International Conference on Computing in Civil and Building Engineering*. Nottingham, United Kingdom, 2010. Proceedings of the International Conference on Computing in Civil and Building Engineering, Nottingham University Press, 30 June–2 July.

Glideman, C., 2000. Total economic impact workbook: performing a TEI study to evaluate technology initiatives.

Irani, Z., 2010. Investment evaluation within project management; an information systems perspective. *The Journal of the Operational Research Society*, 61(1), pp. 917–28.

ISO, 2014. Asset management—Overview, principles and terminology—ISO 55000. *International standard ISO 55000*, Available at: HYPERLINK "http://www.irantpm.ir/wp-content/uploads/2014/03/ISO-55000-2014.pdf" http://www.irantpm.ir/wp-content/uploads/2014/03/ISO-55000–2014.pdf [Accessed 08 February 2017].

Kreider, R., Messner, J. & Dubler, C., 2010. Determining the frequency and impact of applying BIM for different purposes on projects. In *Innovation in AEC Conference. The Pennsylvania State University, University Park, PA*. Pennsylvania, 2010.

Leyton, R., 1995. Investment appraisal: the key for IT? In B. Farbey, F.F. Land & D. Target, eds. *Hard Money, Soft Outcome*. Alfred Waller Ltd, in association with Unicom, Henley on Thames.

Lin, S., Gao, J., Koronios, A. & Chanana, V., 2007. Developing a data quality framework for asset management in engineering organisations. *International Journal of Information Quality*, 1(1), pp. 100–25.

Lin, C. & Pervan, G., 2003. The practice of IS/IT benefits management in large Australian organizations. *Journal of Information Management*, 41(1), pp. 13–24.

Love, P.E.D., Matthews, J., Simpson, I., Hill, A. and Olatunji, O.A., 2014. A benefits realization management building information modeling framework for asset owners. *Automation in Construction*, 37(1), pp. 1–10.

Love, P.E.D., Simpson, I., Hill, A. & Standing, C., 2013. From justification to evaluation: Building information modeling for asset owners. *Automation in Construction Volume*, 35(1), pp. 208–16.

McGraw-Hill, 2012. SmartMarket Report. pp. 1–72. Available at: HYPERLINK "https://www.icn-solutions.nl/pdf/bim_construction.pdf" https://www.icn-solutions.nl/pdf/bim_construction.pdf [Accessed 2017 November 08].

Melville, N., Kraemer, K. & Gurbaxani, V., 2004. Review: Information technology and organizational performance: an integrative model of IT business value. *MIS Quarterly*, 28(2), pp. 283–322.

Mooney, G.J., Gurbaxani, V. & Kraemer, K.L., 1995. A process oriented framework for assessing the business value of Information Technology. In *The Sixteenth International Conference on Information Systems, Amsterdam, ICIS 1995*. Amsterdam, 1995.

Nogeste, K. & Walker, D.H.T., 2005. Project outcomes and outputs: making the intangible tangible. *Emerald Publishing Limited*, 9(4), pp. 55–68.

Remenyi, D., 2000. The elusive nature of delivering benefits from IT investment. *The Electronic Journal of Information Systems Evaluation*, 3(1).

Sanchez, X.A., Mohamed, S. & Hampson, D.K., 2016. BIM Benefits Realisation Management. In X.A. Sanchez, D.K. Hampson & S. Vaux, eds. *Delivering value with BIM: A whole-of-life approach*. 1st ed. London: Routledge.

Sapountzis, S., Harris, K. & Kagioglou, M., 2007. Benefits realisation process for healthcare. In *4th International Research Symposium (SCRI), March 26–27, 2007*. Salford, 2007.

Succar, B., Sher, W. & Aranda-Mena, G., 2007. A proposed framework to investigate Building Information Modelling through knowledge elicitation and visual models. In *Proceedings of the Australasian Universities Building Education Association, 4–5 July*. Melbourne, Australia, 2007.

Thorp, J., 1998. *The Information Paradox—Realizing the Business Benefits of Information Technology*. Toronto, Canada: McGraw-Hill, Inc.

Vass, S. & Karrbom Gustavsson, T., 2014. The perceived business value of BIM. In Scherer, M.a., ed. *Proceedings at the 10th European Conference on Product and Process Modelling, ECPPM 2014, 17 September 2014 through 19 September 2014*. Vienna, 2014.

Walasek, D. & Barszcz, A., 2017. Analysis of the adoption rate of Building Information Modeling (BIM) and its Return on Investment (ROI). *Procedia Engineering*, 172, pp. 1227–34.

Ward, J., Taylor, P. & Bond, P., 1996. Evaluation and realization of IS/IT benefits: an empirical study of current practice. *European Journal of Information System*, 4(1), pp. 214–25.

White, L.N., 2007. An old tool with potential new uses: return on investment. *The Bottom Line*, 20(1), pp. 5–9.

Willcocks, L. & Lester, S., 1996. Beyond the IT productivity paradox. *European Management Journal*, 14(3), pp. 279–90.

# BIM solutions for construction lifecycle: A myth or a tangible future?

E. Papadonikolaki
*Bartlett School of Construction and Project Management, University College London, London, UK*

M. Leon
*Scott Sutherland School of Architecture and the Built Environment, Robert Gordon University, Aberdeen, UK*

A.M. Mahamadu
*Department of Architecture and the Built Environment, University of the West of England, Bristol, UK*

ABSTRACT: Building information Modelling (BIM) lies at the centre of construction industry's interest nowadays, with a revolutionary impact on the ways that professionals work, collaborate and conduct business. The application of BIM is not as straightforward as it sounds though, with numerous software solutions available, various implementation processes across the project lifecycle, which challenges the interoperability and how information flows throughout the various project stages. This paper performs and presents a systematic review of the BIM software landscape currently available for the construction industry across the various project phases and in alignment with the 2013 RIBA Plan of Works. A gap analysis is conducted among these BIM solutions to examine the different software application areas, software architecture, interoperability possibilities, accessibility and affordability, by applying descriptive statistics. Surprisingly, the BIM software ecosystem is fragmented across the different project stages and highly proprietary and further hindered by the large number of specialised and highly sophisticated solutions addressed to advanced computer users. To this end, the paper aims to inform the industry's stakeholders, policy makers and software vendors, while shedding light on the extent that sophisticated BIM solutions can be disseminated to the market.

*Keywords*: Building Information Modelling (BIM), lifecycle, digital twin, software, interoperability

## 1 INTRODUCTION

Building information Modelling (BIM) and information digitisation are ubiquitous within construction projects for overall project efficiency and effectiveness. BIM and digitisation have been touted as revolutionary forces in the Architecture, Engineering and Construction (AEC) industry. BIM and digitalisation initiatives are further supported in the United Kingdom (UK) and European markets due to various government mandates for implementation (GCCG, 2011). Such mandates and the general market demand for BIM require an inclusive solution to the many persistent challenges of the construction sector, i.e. low productivity, fragmented information flows, poor collaboration and inefficiencies in time and costs. The construction industry is increasingly implementing various BIM software to tackle these challenges and so far, the results have been more than promising.

At the same time, and in addition to being an inclusive solution to construction challenges, BIM has been presented as a technology and process that can radically digitise the construction lifecycle, from inception to operation of an asset (HMG, 2015). The BIM software ecosystem is abundant with both commercial and non-commercial solutions with functionalities that support various tasks across the lifecycle phases. We define BIM software ecosystem as the set of commercial and non-commercial (i.e. non-profit or freeware) software tools available for the generation, sharing and management of building information in the AEC. However, no sufficient effort has been placed in providing a holistic BIM solution encompassing all the life-cycle functions (Hallberg and Tarandi, 2011). Nwodo et al. (2017) identified challenges in BIM for life-cycle assessment at early stages, highlighting the disconnect between BIM for design and whole life-cycle.

Nevertheless, BIM has been placed at the forefront of digital transformation with the promise that it greatly improves construction life-cycle (Eadie et al., 2013, Rezgui et al., 2013). Apart from the recent BIM-related mandates in the UK and the various Publicly Available Specifications

(PAS) that specify BIM use, various professional associations such as the Royal Institute of British Architects (RIBA) have adjusted their processes to align with BIM with recommendations of functions needed to be delivered at the various lifecycle phases of facilities, i.e. the 2013 Royal Institute of British Architects (RIBA) Plan of works overlay (RIBA, 2013). Despite these propositions, there is lack of knowledge of how existing BIM solutions address the functional needs of the lifecycle of an asset. This study sought to bridge this gap by reviewing BIM software to establish the extent to which these functionalities are relevant at the various lifecycle phases of construction projects.

This paper is structured as follows. Following this introduction, firstly, it will present the research and industry background of this work and relevant research. Secondly, it will present the research methodology and subsequently, it will present the data, results and findings. In the ensuing section, the findings will be discussed with reference to relevant scientific literature and state-of-the-art in the industry. Afterwards, the paper will conclude by summarising the main points, outlining implications for practice and policy and setting the agenda for further research.

## 2 BACKGROUND AND RELATED RESEARCH

### 2.1 Digitisation in the built environment via BIM

The fourth industrial revolution, also known as industry 4.0, is changing the way manufacturing and construction industries are perceiving efficiency, productivity and data exchange. The UK BIM Level 2, which supports data interoperability within a project's lifecycle, and Level 3, which focuses on the smooth data transition from concept to operational stages, are another manifestation of the impact of this revolution within construction and infrastructure (HMG, 2015). The interconnected cyber-physical systems that allow the life-cycle representation, monitoring and re-calibration define the progression of the industry, and, as a result, represent the future of the everyday practice for the sector, towards the so called *Digital Twin*.

According to Whyte and Hartmann (2017) a number of national construction standards have being developed worldwide for managing delivery, operations, handover and data classification with the main target being to achieve smooth information transitions and fluidity. For example, BIM is not entirely new for construction as it has emerged through long-standing institutional processes and efforts for structuring and consistently representing initiatives and knowledge about building artefacts (Papadonikolaki, 2017), which was a predominant line of thought in the 1970s (Eastman, 1999).

### 2.2 Standardisation

The statement "*BIM is about sharing structured data*" is the motto of BuildingSMART (2018), the organisation which is the leader in defining, identifying and supporting the implementation of the construction industry standards related to the application of BIM. Standards are set to define best practice, usability, safety and promote greater efficiency, as decided by an extensive engagement with different group of experts, government bodies, businesses, trade associations, etc. (BSI, 2017). In the case of BIM, the standards ensure the ways construction professionals share, structure and define information and data.

Sharing information could be paralleled with passing the baton in a relay race, with the different construction professionals and stakeholders exchanging information among them (ISO/FDIS 29481-1:2010). BuildingSMART (2018) is responsible for maintaining the structure of data related to the Industry Foundation Classes (IFC), an object-based file format that is open, neutral and available for the OpenBIM initiative (ISO 16739:2013). OpenBIM promises work in a BIM environment not dictated by the software solutions used, but based on open file formats. BuildingSMART (2018) is also providing the template for the adaptation of the IFC format for the construction industry. As a result, compliance with IFC standards allows products and applications that can operate in any platform and device and that are compatible with other systems that are developed with the same standards. Furthermore, IFC format can facilitate both geometric and contextual/ non-geometric data.

### 2.3 Types of interoperability

Interoperability is divided into organisational, semantic, syntactic and technical, according to the European Telecommunication Standards Institute (ETSI), while there is a strict hierarchy, which means that in order to achieve the organisational one, all the others have to be in place (Veer and Wiles, 2008). Technical interoperability "*covers the technical issues of linking computer systems and services. It includes key aspects such as open interfaces, interconnection services, data integration and middleware, data presentation and exchange, accessibility and security services*" (Kubicek et al., 2011). Syntactic interoperability is focused on data formats, and it supports the use of well-defined syntax and messages encoding. Semantic interoperability concerns the precision of the exchanged information for it to be understood in a meaningful manner by other applications that do not share the same developers. Finally, according to Kubicek et al. (2011) organisational interoperability focuses on the common descriptions of inter-organisational processes and can be achieved through common enterprise architectures and securing technical, syntactic and semantic interoperability.

According to BuildingSMART (2018), interoperability, that is the systems' property to exchange information in a shared data schema, is key aspect of working with BIM. OpenBIM is based on open standards and workflows and it promotes and supports these aspects by ensuring data interoperability among project teams and collaborators irrespectively of types of software they use and by applying non-proprietary, neutral file formats. It also contributed to the requirement and development of the IFC format and the Construction Operations Building Information Exchange (COBie).

According to the National BIM Survey (NBS, 2018), 72% of UK construction professionals have adopted the IFC format, to achieve projects' coordination, in terms of models, documents and overall information. The reason for the extensive level of adoption is that conflicting information can risk a project's realisation, thus, causing potential disputes (thus, loss in cost and time) among different stakeholders.

COBie is another non-proprietary data format that includes non-geometric information, which can be easily published, usually with a spreadsheet format. The COBie output is typically focused on informing the client regularly regarding the project progression and the operation and it also ties with the project delivery and the asset operation and management data. Only 41% of the UK construction industry professionals are actually producing COBie data, and according to NBS (2018) the reason for this is the lack of clients' awareness and the fact that BIM Level 2 mandate concerns only public projects.

### 2.4 Research gap, research aim and question

Shafiq et al. (2013) reviewed available BIM collaboration systems in order to identify their ability to support intra-disciplinary collaboration as well as integrated practice. Their review concluded that while BIM solutions for construction industry collaboration exist, they offer inter-disciplinary functionality to different extents and capacities. Furthermore, no solution provided a comprehensive functionality for inter-disciplinary integration. Given this study was conducted before 2013 it is unclear whether the landscape has changed in terms of functionalities BIM solutions offer.

Some of the criteria used for assessing collaboration and integration functions of BIM solutions in the work by Shafiq et al. (2013) were: multiple user supported model content management, content creation, viewing and reporting, and system administration including data exchange protocols access control among others.

However, this study was focused on server related BIM solutions rather than all potential applications. Some other studies have reviewed BIM software capability and functionality for only specific disciplines including quantity surveying (Wu et al., 2014), risk management (Zou et al., 2017) and safety management (Martínez-Aires et al., 2018). Despite these developments, however, no study has comprehensively reviewed a wide range of BIM software in relation to the relevance of their functions for each lifecycle phase of a facility.

This paper complements previous work done by Papadonikolaki et al. (2014), which attempted to map out the relation between the acclaimed benefits and usability of BIM software for construction Project Management (PM) and the actual impact of BIM solutions on delivering these benefits. In a similar spirit, this study focuses on the whole lifecycle of the AEC to investigate the extent to which BIM solutions support a whole lifecycle consideration of digitisation in the built environment. The main research question can be thus formulated as follows:

*To what extent the existing ecosystem of BIM solutions addresses the promise of a whole lifecycle BIM?*

## 3 METHODOLOGY

### 3.1 Rationale and research setting

This paper presents a systematic review of the BIM software ecosystem currently available for the construction industry. Due to the practical nature of the research aim and question, the review will not focus on scientific literature, but on software instead. This study places interoperability in the epicentre of BIM work, given that it enables various actors to collaborate with BIM across project lifecycle. Thus, only BIM solutions which allow interoperability and the generation and exchange of open industry standards are reviewed. Additionally, the type of software, e.g. stand-alone, plug-in and their platform, e.g. mobile or desktop affect the collaborative potential of BIM tools and create 'hard' transition points. Nevertheless, the study will employ scientific methods to collect and analyse the data.

As the construction industry moves gradually from paper-based to data-driven, this review will increase the understanding of the degree to which the BIM software ecosystem can support the promise of a whole lifecycle and fully interoperable AEC. To operationalise the concept of lifecycle thinking in the AEC, the 2013 RIBA Plan of Works stages (RIBA, 2013) have been used as a guideline. The various solutions of the BIM software ecosystem will be analysed against their applicability to the 2013 RIBA Plan of Works, which are as follows:

The alignment between the different software and their applicability within the RIBA Plan of Work stages is based on identifying the core objectives of each stage and ensuring that the different types of software can provide solutions to these objectives, as presented in Table 1. The study features a gap analysis between the proclaimed BIM solutions and BIM application areas throughout the construction lifecycle and their availability in commercial solutions. To this end, the study will attempt to highlight any lack of applications focusing on specific stages of the construction projects' lifecycle with the ultimate aim to propose new areas for Research and Development (R&D), knowledge transfer and ad-hoc solutions for efficient management of construction projects.

Table 1. RIBA plan of work stages and objectives.

| RIBA plan of work stages | Core objectives |
| --- | --- |
| Stage 0: *Strategic Definition* | Business Case, Strategic Brief |
| Stage 1: *Preparation and Brief* | Project objectives, Quality Objectives, Project Outcomes, Sustainability Aspirations, Initial Project Brief, Feasibility studies, Site Information. |
| Stage 2: *Concept Design* | Concept Design, Cost Information, Project Strategies, Design Programme, Final Project Brief. |
| Stage 3: *Developed Design* | Developed Design, Cost Information, Project Strategies, Design programme. |
| Stage 4: *Technical Design* | Technical Design, Design Responsibility Matrix, Project Strategies, Design Programme. |
| Stage 5: *Construction* | Construction, Construction Programme, Design Queries |
| Stage 6: *Handover and Close Out* | Building Contracts |
| Stage 7: *In Use* | In Use, Schedule of Services, Post-occupancy and Project Performance evaluation |

### 3.2 Data and methods

The data on BIM solutions are collected from readily publicly available information from relevant databases, e.g. certified software by BuildingSMART (2018), and from the webpages of the software manufacturers. This database was selected because it contains a record of the BIM-tools that allow the import and export of IFC format, the only widely-used open data format. Before the analysis, this dataset was 'cleaned'. From the 205 BIM software in the database, 31 tools were discontinued, as no information could be found about them online or irrelevant as no IFC import/export functionality was supported. On the contrary, 2 new tools were added to the dataset. In total, 173 BIM applications took part in the analysis.

The data analysis was performed through coding of the data on BIM software using a priori sets of codes. These codes were set according to the BIM software application areas, relevant RIBA stages within they might be applied, their software architecture, interoperability and collaboration possibilities, business model, accessibility and affordability, by applying descriptive statistics. All authors were involved in the coding for internal validation of the analysis and performing two rounds of coding.

## 4 DATA PRESENTATION AND ANALYSIS

### 4.1 Demographics of IFC-compliant software

The first level of analysis of the BIM software ecosystem relates to the descriptive data about the tools as derived by BuildingSMART (2018). This data is presented in Figure 1, which consists of four parts. Figure 1(a) includes the BuildingSMART (2018) categorisation of the software into architectural, building performance energy analysis and simulation, building services, construction management, data servers, development tools, facility management, modelling tools, Geographic Information Systems (GIS), model viewer and structural BIM tools. Based on this analysis, the majority of BIM software is of structural use (25 tools), followed by architectural (22 tools).

Drawing upon the data from BuildingSMART (2018), Figure 1(b) presents the IFC functionality of the BIM tools. Most of the tools (n = 83) allow both import and export functionality, followed by tools that allow only import (n = 72) and only 18 tools allow only export of IFC files. Figure 1(c) illustrates whether the tools are proprietary or free. Out of the 173 tools, 148 are proprietary with commercial interests and requiring license, and only 25 tools are freeware, either completely open source of free versions of limited functionality of commercial

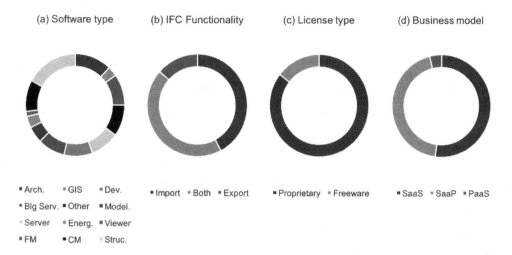

Figure 1. Demographics of IFC-compliant BIM software ecosystem per (a) software type, (b) functionality of IFC, (c) license type and (d) business model.

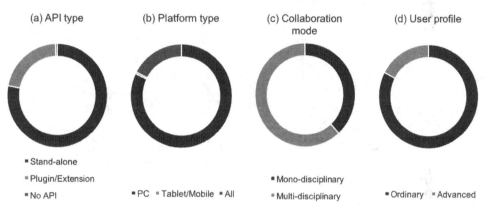

Figure 2. Usability profile of BIM-related software per (a) application programming interface (API), (b) platform type, (c) potential use collaboration mode and (d) expected application user profile.

software. Figure 1(d) presents the type of license of the BIM tools, that is whether they have a Software-as-a-Product (SaaP), Software-as-a-Service (SaaS) or Platform-as-a-Service (PaaS) business model, the majority of which have a SaaS model.

### 4.2 User-friendliness of BIM-related software

Apart from analysing the types of license, and the fairly descriptive characteristics of how BIM tools could be procured and accessed, the study also focused on the usability profile of the various BIM solutions. In particular, the Application Programming Interface (API) of the various BIM software is mostly stand-alone software (n = 135), followed by extensions and plug-in solutions and a handful of software did not have API, as shown in Figure 2(a). Figure 2 (b) shows that the majority of BIM solutions are based on personal computers (PC) (n = 142), followed by several applications that run on both PC and tablets and mobile (n = 30), whereas one application is only mobile.

As BIM has been widely acknowledged for its collaborative way of working, the data on the rest of Figure 2 focus on the users. In Figure 2(c), the software was analysed based on its capacity to allow multi-disciplinary work. From the sample, 110 BIM software applications were found to be able to support collaborative work among more than one disciplines, whereas the rest, was software primarily addressed to specialists. Following upon this point, the software was also analysed with regards to the computer or digital skills of the users that the BIM software was addressed to. Figure 2(d) shows that the BIM software was addressed primarily to ordinary computer users, whereas a part of the sample is addressed to advanced computer users (n = 27), who might need to be familiar with specialised technical knowledge to set-up data servers or engage to additional programming.

### 4.3 *Phasing and lifecycle fragmentation*

Figures 3 and 4 show how different types of software support the various RIBA stages in a bar chart and a heat map respectively. A staggering 96.5% of the software (n = 167) is actually targeting the Developed Design Stage 3 of RIBA Plan of Works, as illustrated in Figure 3, followed by a 71% (n = 123) of the software targeting Technical Design Stage 4 and 69.4% (n = 120) focusing on Concept Design Stage 2. Only 2.3% (n = 4) of the software is actually tackling the strategic definition Stage 0 of a project, which resonates with the fact that each project is tackled as a business case while this stage is mostly focused on the client and investor, information typically hosted in Data Servers. However, only 25% of these data servers concern the whole project lifecycle, which means that the information within Stage 0 is not followed throughout the project (see Figure 4). That translates into issues with time and cost, as the initial decisions are not followed through the project.

Furthermore, a mere 13.9% (n = 24) is supporting Preparation and Brief Development Stage 1, which also questions the focus and understand-

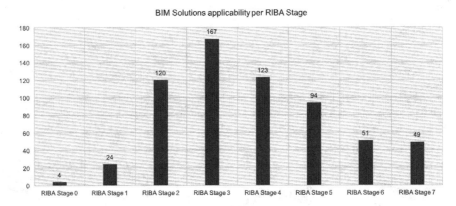

Figure 3. Number of BIM software applicable across the various RIBA stages.

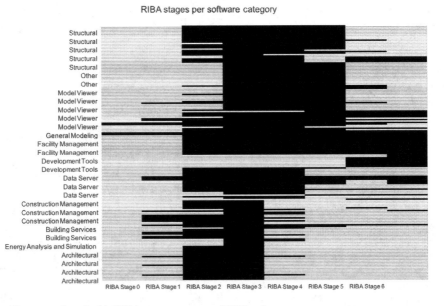

Figure 4. Heat map of applicable RIBA stages per types of BIM software.

ing of the project team on the objectives of the project. Following the technical design, a 54.3% of software (n = 94) is dealing with Construction, project completion and mobilisation Stage 5, an aspect of the industry that is typically the most time consuming, where typically issues with cost and time occur. A 29.5% of the software (n = 51) is tackling the Handover and Closeout Stage 6, and a 28.3% (n = 49) examines the "In-Use" and post-occupancy evaluation Stage 7.

## 5 DISCUSSION

### 5.1 *BIM as a digital platform across the lifecycle*

The analysis shows that the construction sector is experiencing a major digital transition underpinned by numerous solutions of BIM-related software available along with a number of supporting software and infrastructure, e.g. servers. Although the mapping of the BIM software across RIBA stages shows that there are available software solutions across the construction lifecycle, only 18 tools allow both import and export of IFC files (Figure 1), thus hindering interoperability among software. Also, only 4 of the total software concern the Stage 0, which means that the information is not followed throughout the project (Figure 3). While construction disputes and issues with cost and time typically occur within Stage 5 and 6, the fact that Stages 0 and 1 are barely supported does not help with projects' efficiency, as the initial decisions are not followed throughout the project. The support of BIM across the project lifecycle remains fragmented into specialised pieces of software (Figure 4) some of which are addressed to advanced users not allowing for multi-disciplinary work.

Small software developing firms enter the market, to provide specialised or more affordable solutions. Whereas this plethora of BIM tools and the top-down policy push in the UK cause a radical innovation in construction (Papadonikolaki, 2017) the industry is ready from disruption by new players entering the market and changing the BIM software ecosystem. Importantly, there is a major difference between radical and disruptive innovation, the former being at a micro level, whereas the latter at a macro level (Hopp *et al.*). To this end, construction is more susceptible to new technologies and business innovation from other industries now more than ever, as reorganisation and restructuring for the commercialisation of breakthrough ideas takes place at a higher rate. According to Figure 1d, the BIM ecosystem transits towards servitisation, where new business models reshape BIM software provision as Software-as-a-Service, rather than the traditional SaaP model.

### 5.2 *Bridging the policy and industry mismatch*

The contribution of the study is twofold. Firstly, the data could be utilised for decision-making on BIM software from construction industry stakeholders, such as contractors, consultants and clients. Secondly, after confronting the data with relevant public mandates and perceptions about BIM solutions from existing scientific and market research (NBS, 2018), especially from the UK where BIM use is mandated in public procurement, the findings would potentially mobilise software vendors and industry consortia towards expanding Research and Development (R&D) on BIM. Importantly, the study targets policy-makers and enriches the BIM debate by presenting a new evidence-based paradigm on the accessibility and affordability of BIM solutions, while shedding light on the extent that sophisticated BIM solutions can be disseminated to the market.

This research also identified a mismatch in relation to the adherence with the industry standards and the available BIM software solutions, especially within the UK. Not only there is a lack of clear link between BIM standards and their actual application, most importantly, the lack of clear regulations on BIM is highly contradicting the BIM mandate. As a result, the UK industry is falling behind in BIM implementation within the supply chain (NBS, 2018).

### 5.3 *Research limitations and future research*

A number of limitations are acknowledged within this research with first and foremost concerning the fact that not all types of projects and procurement methods can follow the RIBA stages, nor these Stages are applicable in a worldwide scale, thus the relevance of this study is limited to building projects. The analysis is also considering that the RIBA stages are aligned with Level of Details (LoDs); thus, the software alignment with the different stages is compliant not only to the core objectives of each stage but also to the expected LoDs. However, LoDs are quite often dependant of the different companies' policies and procurement methods. An additional and important barrier is the potential researchers' bias, as there has been no usability testing applied as such. As a result, future research will include triangulation of data from other sources, e.g. users and software manufacturers.

## 6 CONCLUSIONS

How can we ensure that we do not end up with a mythical "*chimera*" of BIM software? It seems that there is no easy answer to that. Data and systems' interoperability and usability, especially among interdisciplinary professional project teams, have been proclaimed for a long time but not fully achieved. This paper performed a BIM software review and revealed the gaps in provision of BIM software across project lifecycle. Whereas interoperability via IFC has been the strength of various BIM software, it is hardly achieved due to the high specialisation and sophistication of the solutions and the fragmentation of software packages across lifecycle stages.

To this end, the Digital Twin development is becoming more and more of an elusive goal for construction although it is feasible in manufacturing. At the same time, BIM strategies are often government-led like in the UK, thus revealing a disconnect not only between practice and policy, but also between AEC and software vendors. Finally, this underlines the need for the informed client, who is aware of the value of data, over and above the COBie requirements for realising the full potential of built assets through whole lifecycle BIM and digitalisation.

## REFERENCES

Bsi, 2017. *What is a standard – Benefits* [online]. British Standards Institution. Available from: https://www.bsigroup.com/en-GB/about-bsi/uk-national-standards-body/about-standards/what-is-a-standard-benefits/ [Accessed Date].

Buildingsmart, 2018. *All applications by category: IFC-Compatible Implementations Database* [online]. http://www.buildingsmart-tech.org/implementation/implementations [Accessed Date].

Eadie, R., Browne, M., Odeyinka, H., Mckeown, C. & Mcniff, S., 2013. BIM implementation throughout the UK construction project lifecycle: An analysis. *Automation in Construction*, 36, 145–151.

Eastman, C., 1999. *Building Product Models: Computer Environments, Supporting Design and Construction* Boca Raton, Florida, USA: CRC Press.

Gccg, 2011. *Government Construction Client Group: BIM Working Party Strategy Paper.*

Hallberg, D. & Tarandi, V., 2011. On the use of open bim and 4d visualisation in a predictive life cycle management system for construction works. *Journal of Information Technology in Construction (ITcon)*, 16, 445–466.

Hmg, 2015. *Digital Built Britain, Level 3 BIM Strategic Plan* [online]. HM Government. Available from: http://digital-built-britain.com/DigitalBuiltBritainLevel3BuildingInformationModellingStrategicPlan.pdf [Accessed Date].

Hopp, C., Antons, D., Kaminski, J. & Salge, T.O., Perspective: The Topic Landscape of Disruption Research A Call for Consolidation, Reconciliation, and Generalization. *Journal of Product Innovation Management.*

Kubicek, H., Cimander, R. & Scholl, H.J., 2011. Layers of interoperability. *Organizational Interoperability in E-Government.* Springer, 85–96.

Martínez-Aires, M.D., López-Alonso, M. & Martínez-Rojas, M., 2018. Building information modeling and safety management: A systematic review. *Safety science*, 101, 11–18.

Nbs, 2018. *The National BIM Report 2018* [online]. National Building Specification (NBS). Available from: https://www.thenbs.com/knowledge/the-national-bim-report-2018 [Accessed Date 2018].

Nwodo, M., Anumba, C. & Asadi, S., Year. BIM-Based Life Cycle Assessment and Costing of Buildings: Current Trends and Opportunitiesed.^eds. *Computing in Civil Engineering 2017*, Seattle, Washington ASCE, 51–59.

Papadonikolaki, E., 2017. Grasping brutal and incremental BIM innovation through Institutional Logics. *Proceedings of the 33rd Annual ARCOM Conference.* Association of Researchers in Construction Management, 54–63.

Papadonikolaki, E., Koutamanis, A. & Wamelink, J.W.F.H., 2014. The Utilisation of BIM as a Project Management Tool. *Proceedings of the 10th European Conference on Product and Process Modelling (ECPPM 2014).* Vienna, Austria.

Rezgui, Y., Beach, T. & Rana, O., 2013. A governance approach for BIM management across lifecycle and supply chains using mixed-modes of information delivery. *Journal of Civil Engineering and Management*, 19, 239–258.

Riba, 2013. *RIBA Plan of Work 2013 overview* [online]. Royal Institute of British Architects. Available from: https://www.ribaplanofwork.com/Download.aspx [Accessed Date].

Shafiq, M.T., Matthews, J. & Lockley, S., 2013. A study of BIM collaboration requirements and available features in existing model collaboration systems. *Journal of Information Technology in Construction (ITcon)*, 18, 148–161.

Veer, H. & Wiles, A., 2008. Achieving Technical Interoperability-the ETSI approach, European Telecommunications Standards Institute.

Whyte, J.K. & Hartmann, T., 2017. How digitizing building information transforms the built environment. Taylor & Francis.

Wu, S., Wood, G., Ginige, K. & Jong, S.W., 2014. A technical review of BIM based cost estimating in UK quantity surveying practice, standards and tools. *Journal of Information Technology in Construction (ITCon)*, 19, 534–562.

Zou, Y., Kiviniemi, A. & Jones, S.W., 2017. A review of risk management through BIM and BIM-related technologies. *Safety science*, 97, 88–98.

# Schema-based workflows and inter-scalar search interfaces for building design

P. Poinet, M. Tamke & M.R. Thomsen
*Centre for Information Technology and Architecture, Copenhagen, Denmark*

F. Scheurer
*Design-to-Production, Zürich, Switzerland*

A. Fisher
*Buro Happold, London, England*

ABSTRACT: Today, scattered design processes and manual interventions are almost inevitable within the AEC industry, especially for 3D modelling processes and data management during the post-tender phases of large-scale and geometrically complex architectural projects that involves the participation of many different trades, which have to communicate efficiently between themselves. This research paper presents a state of the art in managing modelling processes of geometrically complex architectural projects and proposes a theoretical framework aiming to simplify, improve and standardize those processes. This framework will be illustrated by practical experiments using schema-based workflows and inter-scalar search interfaces enabling the assembly, visualization and query of the produced data.

## 1 INTRODUCTION

### 1.1 *The non-scalable nature of Directed Acyclic Graphs (DAG)*

Within the realm of highly complex and large-scale architectural projects, quitting the tender phase does not necessarily implicate that all digital objects have been actually "frozen", meaning that the geometry is fully generated and accessible. Those will be further design-processed during the post-tender phase, once the materials used and available machines are perfectly known and the fabricator actually needs specific data (before and during construction). This data is generally very expensive, since it takes time to generate, organize and share amongst all trades.

The usually tight time-line framing such complex architectures has an important influence on how the team manages, handles and delivers the latter. Hard deadlines require high productivity preventing therefore the possibility to organize all computational processes within a neatly designed Directed Acyclic Graph (DAG) based workflow (the underlying logic on which are based many existing generative design tools, such as Grasshopper and Dynamo, for example). Indeed, trying to encapsulate the full complexity of the project's database within one single pipeline would most probably take more time than completing the project itself, especially when we know that the AEC industry is fragmented through many different software platforms used by the various stakeholders. Therefore, the different trades need to come up with their own custom methods, interoperability tools and quick "shortcuts" in the design process (such as exported sub-models and email exchange processes), producing redundant models and files in order to communicate with others and across different software environments. Consequently, computational design specialists and consultancy practices have been developing their own internal methods in order obtain the best control on the generated data (and corresponding meta-data) through all stages.

We will illustrate such practices by describing the workflows developed both at Front and Design-to-Production, both being well known practices specialized in rationalizing very complex geometries at late stages in the design process, specialists in rationalizing geometrically complex architectural projects for further delivery of precise fabrication data to construction companies.

## 2 STATE OF THE ART

### 2.1 *Front*

During the realization of the City of Dreams Casino Hotel in Macau (conceived by Zaha Hadid Architects), the consultancy practice Front developed a modelling strategy ("Building Information Generation") enabling parallel generation of information and attributes (through their own custom Rhino3d Plug-in *Elefront*) necessary for further fabrication. The whole modelling process consisted of a strategic alternation between the generation of objects in Grasshopper and their subsequent storage and classification within Rhino3d models (in which the geometry was "locked", or "dead") from which was generated further information through a next iteration of Grasshopper sessions (within which parametric linkage was kept), and so on. The computational design process was therefore scattered into several parts, allowing manual interventions, proof checking and potential corrections from the expert user, before generating a new set of data based on the previous one, and thus through all scales until the highest modelling resolution (Van Der Heijden, 2015).

2.2 *Design-to-Production*

The consultancy practice Design-to-Production organizes its objects through specific layer tables within Rhino3d, enabling both its storage and classification. Similar to a directory tree structure, a layer table is organized into different levels (or depths) that communicate from a root layer to all its successor sub-layers. Generally, both the root-layer (containing the component's label) and the leaf-layers (containing its geometries) are defined through the protocol generating the information. The intermediate layer levels are identified by the component's name and mainly serve to structure the information in order to make it more human-readable, defined by the expert user him/herself. Therefore, the layer tree is independent from the "semantics" of the component and just a UI-feature. Depending on the nature of the component (being a beam, an opening, a facade element, etc.), it will be stored within a specific submodel (or worksession) that refers to a larger master file from which data communication with other submodels becomes possible (Scheurer, 2012).

Both modelling processes developed at Front and Design-to-Production carefully curate and fragment the design process into separate parts in order to allow manual interventions, classifying the generated objects and passing important information between the different (sub-)models. However, this segregation of the design process and its current curation could be improved and better supported by the development of specific tools and interfaces, enabling the user to experience a more seamless workflow at late stages in the project.

While section 3 will focus on inter-scalar search interfaces, section 4 will describe different methods to build and queries custom schemas. Finally, section 5 will tackle the problem of interoperability and schema-based workflows.

3 INTER-SCALAR SEARCH INTERFACES

3.1 *Dataviz tools: Sunburst diagrams*

In order to get a better understanding of Design-to-Production's data structure, the layer table has been extracted and displayed through the form of a Sunburst Diagram (Fig. 1). Each level is represented by a circle, the latter being divided into smaller arcs corresponding to the sibling layers, the offset or each arc corresponding to the number of objects contained within each layer. Such diagram allows the user to obtain a better insight into the general structure and intricacy of the layer table, and quickly spotlight the specific layers that contain a high number of objects.

3.2 *LayerStalker: An inter-scalar search interface*

The functionality of the corresponding Sunburst Diagrams (acting so far as pure data visualization tools) has been extended by introducing interaction features through "LayerStalker" (Fig. 2), a developed imbedded interface within Rhino3d from which the user is able to call multiple layers based on generic strings or keywords (e.g. "detailed volume", "dowel", "drill", "axis", "connector", etc.). This is exemplified in Figure 2, where the value "drill" is called, displaying all objects whose respective layer names contain the exact same tag, and hiding all the others. The multi-scalar model on the right displays the selected layers and highlights the gathered centralized information within the Rhino3d viewport, on the left. Such example highlights the fact that implicit relationships exist between objects that are situated within unrelated leaves of the data structure. Indeed, different ways of organizing information exist, and one classification strategy might actually hide another. Therefore, objects need to be properly referenced by meaningful attributes, user strings or User Dictionaries, in order to be able to operate efficient queries later on during the design process. The interface described here has been developed within Rhino3d, relying specifically on the respective software's geometrical database. However, one might speculate to generalize such approach by querying objects from a more holistic database which could gather as well data coming from external software platforms. Therefore, section 4 describes how to assemble and query cross-platform custom

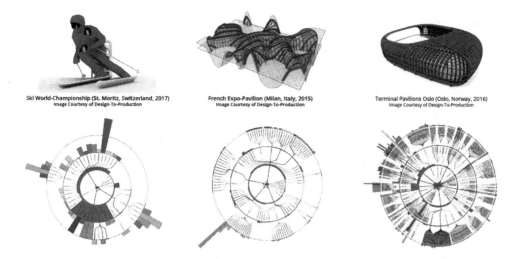

Figure 1. Different projects modelled by Design-to-Production with their respective Layer Table represented as a Sunburst Diagrams.

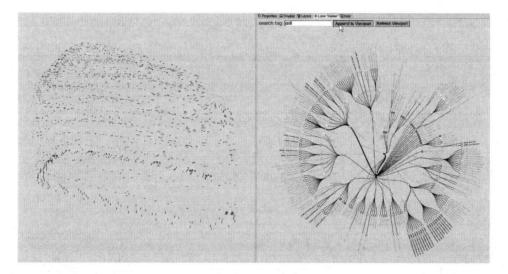

Figure 2. One of the terminal oslo pavilions modelled by Design-to-Production and its respective interactive Layer Table represented as a Sunburst Diagram. Here, the value "drill" is called, displaying all objects whose respective layer names contain the exact same tag.

schemas based on existing tools, workflows and technologies.

### 3.3 LayerExplorer: A zoomable search interface

Because the layer hierarchy might be too intricate and complex to be grasped and comprehended as a whole, I would also suggest that a zoomable interface would be required, letting the user navigate between the different levels of resolution that he or she defined upstream. "LayerExplorer" (Fig. 3) has been developed and built within the Processing environment platform, based on an existing sketch (http://www.generative-gestaltung.de/1/M_5_5_01_TOOL) and adapted for this specific scenario. It takes as input the existing Rhino3d's LayerTable data structure and translates it within the Processing's interactive environment.

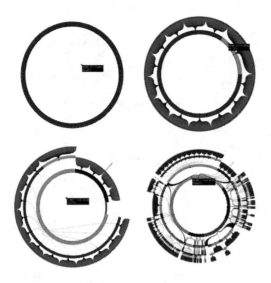

Figure 3. LayerExplorer, a prototype interface allowing the user to navigate between the different hierarchical levels from the Rhino3d's Layer Table.

## 4 BUILDING AND QUERYING CUSTOM SCHEMAS

### 4.1 *Initial experiment with speckle*

The next experiment (Fig. 4) has been built on top of Speckle, a main data communication protocol platform which offers at its core a database generic enough to host geometrical objects coming from various software platforms. The minimum amount of information necessary for the future reconstruction of those objects is stored along with customized user dictionaries containing parallel information defined by the user. This enables the construction of custom schemas embodying a neat hierarchy of objects and sub-objects, allowing external stakeholders to operate meaningful queries from the database at later stages. In order to exemplify this process, we take here as a case study a "MainBeam" object (Fig. 3, left), created by Design-to-Production. This "MainBeam" is defined by a group of sub-objects stored across many different layers and sub-layers from the software's layer table, the latter acting as UI feature embodying the object's schema itself. This nested layer hierarchy has been translated into a Rhino3d's UserDictionary (Fig. 3, middle) and further converted into a JSON format that has been sent to the Speckle's database, currently relying on MongoDB (Fig. 3, right), from which specific targeted queries can be operated upon, both structured and unstructured. In the first case, the user can enter a specific path corresponding to the object's custom schema, and retrieve its corresponding geometry (Fig. 5). In the second case, the user can enter a specific type or attribute to query all concerned objects. For example, Figure 6 represents the data transfer of a Design-to-Production's "MainBeam" that is composed from many different data-rich object types (placed on different layers and grouped together), from its original model environment (left) to a new one (right) within which can be retrieved filtered information. In this case, all objects of type "Curve" are called and re-generated along with their respective attributes (Layer, Group, Color, etc.).

### 4.2 *SchemaBuilder*

Based on the previous experiments, a more generic UI/UX interface, called "Schema Builder" (Fig. 7), has been developed, allowing to seamlessly build custom nested hierarchies of geometries, ready to be sent to the Speckle's database (along with their respective schemas and properties/attributes). It would prove to be extremely useful for the many consultancy practices that interact with data-rich models at late stages in the design process. Ideally, this interface should be able to process both software specific object schemas (e.g. a RhinoObject with its respective properties and attributes) as well as the user-defined customized schemas (e.g. the D2P's "MainBeam") that usually host a multitude of different software specific objects. The next sections describe the current UI/UX flowchart as well as its tree interface and speculates on different features that could improve the current version of SchemaBuilder.

#### 4.2.1 *SchemaBuilder's UI/UX flowchart*

First, the user selects an object located within the software environment. Then, after clicking on the "Attach Properties" button, the corresponding object's schema unfolds as a checklist from which it is possible to manually check/uncheck the properties/attributes that need (or not) to be kept and attached to the object. The user might encounter a situation where the attributes that he or she wishes to attach are not directly accessible from the software object's schema. Therefore, the latter is still editable after the first generation its inherent properties, so the user can manually add other custom ones. The interface could also embed some small pipelines helping the user to automatize such workflow, if attributes non-accessible from the software object's schema reveal to be important ones. Because specific sets of attributes might need to be re-applied onto a multitude of different objects, the user should be able to save different schema's

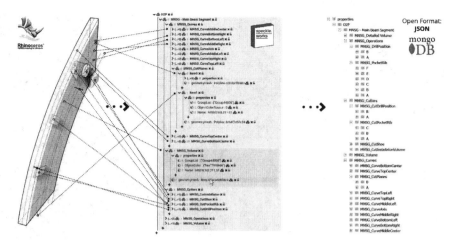

Figure 4. A "MainBeam" modelled by Design-to-Production containing a hierarchy of different objects and sub-objects translated successively into a UserDictionary and a JSON string.

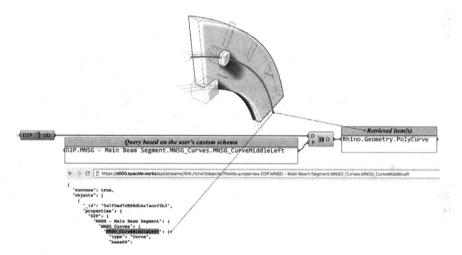

Figure 5. The "MainBeam" object modelled by Design-to-Production in Figure 4, from which user-defined structured queries can be operated upon. In this case, a specific path is given as an input in order to retrieve one object in particular: "MNSG-CurveMiddleLeft".

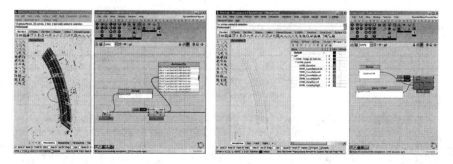

Figure 6. A "MainBeam" object modelled by Design-to-Production is sent to the Speckle's database along with its rich metadata (left). An unstructured query is operated (right), retrieving all objects of type Curve along with their respective layers and attributes.

333

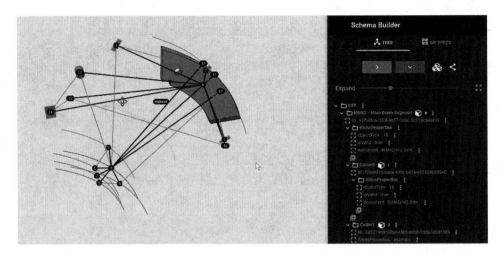

Figure 7. SchemaBuilder's current user-interface.

versions embedding all previously checked/unchecked properties/attributes in order to re-apply them later onto other objects. Therefore, the UI/UX should also contain a "Skeleton" section where could be saved the different produced skeleton versions and load the saved ones to SchemaBuilder. Finally, we must also consider the scenario where the user wants to embed its own custom schemas, classes and subclasses, containing at the leaves of the data structure geometrical, software specific objects. Therefore, the SchemaBuilder interface should also allow the user to load its own custom objects. The SpeckleAbstract class is allowing to do so (along with the SpeckleCore Converter), by serializing and deserializing standard .NET classes. Once the user has finished to build his object with SchemaBuilder, he or she could choose between two options: either sending it directly to the database or adding it to a queuing system, another part of the UI where nested objects would be fully saved. This would allow the user to save "Object A" on the side while building "Object B", before nesting "Object A" within "Object B". Finally, the SchemaBuilder's interface constantly serializes the customized object into a UserDictionary or a JSON string that can be interpreted by the Speckle's local converter and database, from which can be retrieved any stored custom object by external trades.

4.2.2 *SchemaBuilder's tree interface*
To navigate across the schema, the user can use the tree interface that keeps track of the parent-child relationships between the objects. UI-UX features have been prototyped, such as highlighting the object's geometry while hovering its location within the tree, folding/unfolding branches, and pairing the objects indices between the tree interface and the 3d environment through preview features when selection events are being triggered.

## 5 SHARING SCHEMAS THROUGH COMMON DESCRIPTIONS

We have described in the previous sections different prototypes, methods and workflows to visualize, cross-reference, build, explore and share schemas online. All those methods could enable more transparency in the design process and therefore improve interoperability between the different trades working together on a same architectural project. In collaboration with Design-to-Production and Buro Happold (an international engineering practice partly based in London), a last experiment has been developed in order to assess how far we can enhance communication processes between different companies through schema-based workflows, using as an example a specific timber assembly. A common schema has been shared between the two practices that are working towards different ends, Design-to-Production generating full geometrical description of each architectural component and Buro Happold focusing on obtaining precise structural analysis results. The retained common schema has been discussed by both trades, which agreed on the minimum information necessary to generate both of their respective data. This scenario enables a bi-directional workflow from which data can be transferred seamlessly between the respective working environments of the two companies (Fig. 8).

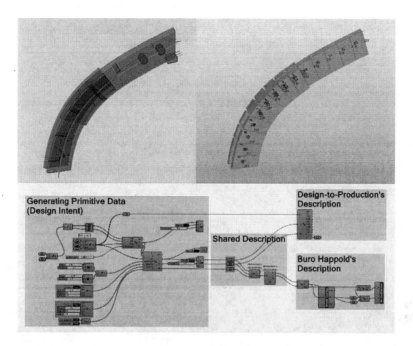

Figure 8. A Grasshopper definition demonstrating a digital design workflow between data generated by Design-to-Production within Rhino3d (right) and shared through a common schema via Speckle to Buro Happold, from which can be retrieved just the necessary information to run a structural analysis within Robot (right) using their BHoM (Bureau Habitat Object Model) interface.

## 6 CONCLUSION

This paper presented both a state of the art in managing modelling processes of large scale and complex architectural projects, as well as different methods aiming at improving contemporary workflows in the AEC industry. Even though the approach described here only focused on the hierarchical relationships between objects, we believe that parallel solutions could also be developed and applied in the future on the time scale and versioning aspect of the project (e.g. keeping track of data transfer between the trades throughout the whole timeline). All the experiments and workflows described in this paper do not constitute a single piece of software, but aim at presenting a holistic approach, or paradigm, within which it could be possible to both build, query, visualize and explore models in more meaningful ways.

## ACKNOWLEDGEMENT

This research is undertaken as part of Inno-Chain. This project has received funding from the European Union's Horizon 2020 research and innovation program under the Marie-Sklodowska-Curie grant agreement No. 642877. We would also like to acknowledge the support of Design-to-Production, Buro Happold and McNeel, and thank all of their involved consultants for their insightful comments and advices for the current and future development of this research.

## REFERENCES

Scheurer, F. 2012. From Thinking to Modeling to Building, *Digital Workflows in Architecture: Design-Assembly-Industry*, 110–131. 12.

Speckle (https://speckle.works/) is an extensible Design & AEC data communication protocol and platform, initiated and developed by Dimitrie Stefanescu, Marie Curie Fellow within the international Innochain Training Network.

Van Der Heijden, R., E. Levelle, and M. Reise. 2015. Parametric Building Information Generation for Design and Construction.

# 4D BIM model adaptation based on construction progress monitoring

K. Sigalov & M. König
*Chair of Computing in Engineering, Ruhr-Universität Bochum, Bochum, Germany*

ABSTRACT: The construction process is usually characterized by delays, deviations and rescheduling. An efficient control of the construction process requires a timely and continuous detection of the construction status and possible deviations. Subsequently, the construction progress should be analyzed quickly and in detail. As a result, responses to emerging issues and rescheduling can be performed timely. Currently, such adaptation is a mostly manual procedure, which is labor-intensive and time-consuming. This contribution examines the possibility of an intelligent adaptation of 4D models for construction scheduling by using pattern recognition methods. The primary focus is on the reducing of manual intervention and improving of progress monitoring.

## 1 INTRODUCTION

The detection of delays or of deviations from schedules are essential for project control. In order to achieve these objective, documentation of as-built status should be sufficiently frequent and accurate. Traditional manual performance monitoring, which involves collecting and analyzing paper-based progress reports, is a very labor-intensive, subjective and error-prone process. This procedure alone causes project delays and cost overruns (Skibniewski 2014, Omar & Nehdi 2016). Consequently, on-site inspections cannot be performed as frequently as required. To overcome the limitations of the conventional method, a lot of approaches for automated as-built data collection have been developed and presented in the last time (Kropp et al. 2018, Omar & Nehdi 2016).

Due to the advanced development of technologies based on Building Information Modelling (BIM) and widespread use of BIM models significant progress has been made in the area of construction scheduling as well. 4D BIM (i.e. 3D BIM model enriched with time allocation and construction processes of the schedule) enables an efficient communication, an early-stage analysis and a visual validation of schedules and increases the transparency of the planning. Adopting 4D BIM for planning, design and progress tracking help to solve a large number of problems even before construction starts. A continuous and frequent acquisition of the actual state of construction processes using (partially) automated progress monitoring in combination with 4D BIM provides an unprecedented opportunity for a fast and accurate comparison of the as-built data with the as-planned data.

Because of the complexity of constriction projects and the involvement of a large number of stakeholders, deviations from the actual schedule are inevitable. It is all the more important to identify any delays and their causes and to take timely counter-measures or to profit from potential accelerations. This implies fast, easy and accurate update of the underlying 4D model. Contrary to unforeseeable events, such as unfavorable weather or technical failures, delays caused by the uncertainties associated with the planning process are avoidable for the most part. For this purpose, the obtained information about the actual construction states can be used to carry out detailed deviation analyzes. Such evaluations could lead to improvement of 4D models in the future.

The aim of this contribution is to examine how the 4D BIM model can be intelligently adapted by means of pattern recognition methods based on the results obtained during progress monitoring. Some possible planning factors causing discrepancies between actual and as-planned performances are investigated. It is important to note that there is no intention to make either the optimal adaptation of the as-planned 4D model to a new as-built 4D model, nor to generate a new optimized construction schedule based on actual data. The focus is rather on finding out how the as-planned model can be adapted appropriately in order to avoid a repeat of the planning errors and to provide more efficient progress monitoring in the future. Moreover, special consideration is given to the simplicity and accessibility of the adaptation, by automating the procedures and eliminating the manual steps. Investigations and evaluations are carried out in a case study with a number of processes related to interior finishing.

## 2 RELATED WORK

In order to prevent adverse events from reoccurring it is crucial to understand the reasons for lateness in construction schedules. Hsu et al. (Hsu et al. 2017) investigated in depth the roots of construction delays and how various causes are connected logically and in time. A systematic examination of a chain of events has revealed that the roots of most delays in the construction are related to design and planning. Schedule deviations due to unfavorable weather, inaccurate deliveries, lateness in transportation and productivity fluctuation have been identified to be of less importance. The authors suggest improving this situation by using a more sophisticated scheduling method.

The application of approaches based on 4D BIM helps to overcome some of the drawbacks of the conventional scheduling methods. Nevertheless, process planning is often carried out separately from the underlying BIM model and the linking and sequencing of construction processes is realized manually later on. Apart from the fact that manual linkage is error-prone and tedious, the initial decoupling of the planning and design models is a potential source of shortcomings in the schedule. Due to that the automatic generation of construction schedules has gained an enormous significance in recent years (Hartmann et al. 2012, Kim et al. 2013, Melzner & Hanff 2016, Mikulakova et al. 2010, Nepal et al. 2012, Tauscher et al. 2014, Wang et al. 2014). Progress in this field has led further to integration of 4D BIM into established software products (e.g. Synchro Software, NAVISWORKS).

With the increasing availability of 4D BIM models these could be advantageously integrated into construction progress monitoring applications. Among the most commonly-used methods that aim to achieve a higher degree of automation in progress tracking are laser scanning (Bosché et al. 2015, Turkan et al. 2012, Tuttas et al. 2017), photogrammetry (Han & Golparvar-Fard 2014, Skibniewski 2014), videogrammetry (Gong et al. 2011, Ibrahim et al. 2009, Kropp et al. 2018, Park & Brilakis 2012) and geospatial technologies like radio frequency identification (RFID) (Li et al. 2012, Razavi & Haas 2011), ultra-wide band (UWB) (Cheng et al. 2013, Shahi et al. 2013) and global positioning systems (GPS) (El-Omari & Moselhi 2011, Pradhananga & Teizer 2013). An extensive overview of existing methods can be found in (Kropp et al. 2018, Omar & Nehdi 2016). Each of these technologies has its advantages and limitations. However, visual sensing (images and videos) has significant advantages over other alternative methods (Han & Golparvar-Fard 2017). Technological progress, increasing image quality and portability but also accessibility of smart devices, such as smart phones or tablets and camera-equipped vehicles (UGV and UAV) allows for a faster and more cost-effective progress tracking on a daily basis.

One of the key challenges associated with visual progress recognition is alignment of collected data with respect to as-planned BIM model (Han & Golparvar-Fard 2017). In recent years, several research results have been published dealing with this problem (Hamledari et al. 2017, Ibrahim et al. 2009, Skibniewski 2014, Kropp et al. 2018). In the approach presented in Kropp et al. 2018 the registration of the image data to the building model is an essential part of the proposed framework and thus no subsequent alignment is necessary. Relevant details of the method are described in the next section, as it is the basis for underlying progress monitoring.

Registration of acquired as-built data with the underlying 4D BIM model can be used for a thorough schedule deviation analysis. According to actual studies most research in this area focuses on documentation and visualization of progress and deviations (Chou & Yang 2017, Han & Golparvar-Fard 2017). Only a few approaches address an effective update of the initial 4D BIM model (Hamledari et al. 2017, Hyojoo et al. 2017, Turkan et al. 2012). The schedule update, however, is limited to the automation of setting up a schedule. Further options to adapt the 4D model and the possibility to improve subsequent construction performance monitoring based on undertaken adjustments are not investigated.

## 3 OVERALL APPROACH

For determining the deviations of schedules it is necessary to compare the actual data to the original as-planned data embedded in 4D BIM model. In this research, the actual state of construction processes are determined using the approach introduced in (Kropp et al. 2018). In this method a variety of algorithms from the fields of computer vision, image processing and machine learning have been applied to facilitate the automation of progress monitoring. The acquisition of as-built status is made by means of video frames captured on construction sites (Figs. 1, 2). During the registration the pose of each image of a sequence is estimated according to the coordinate system of the underlying 3D model (Fig. 1). After the mapping of the sensed data to the BIM objects, the identification of the progress is determined in the recognition block. At the beginning several preparation steps, including the reduction of the search space within the video sequences, are performed. The relevant tasks of the expected state of the 4D BIM model are projected onto the image space. The resulting image regions of interest are then taken

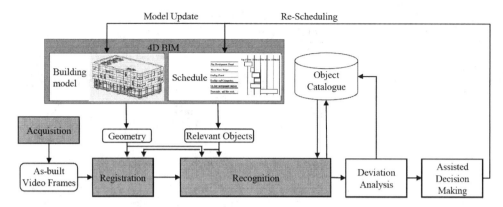

Figure 1. Concept overview (cf. Kropp et al. 2018).

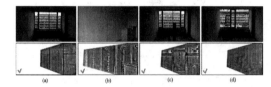

Figure 2. Classification results for drywall works from Kropp et al. 2018: (a) recognized installed drywall state, (b–c) recognized plastered drywalls state, (d) recognized painted drywall state.

as input for the determination of the activity state. This method enables a high degree of automation and is well suited for indoor construction progress monitoring, as not only the presence or absence of construction objects but also activity states can be identified (Fig. 2).

A necessary prerequisite for the described approach is the availability of a 4D BIM model. In particular, construction schedules should be very detailed and contain information about all possible states of the objects. Furthermore, to ensure an efficient deviation analysis and fast responses it should provide easy adaptability to changing requirements. Therefore, schedules should be generated as automated as possible based on BIM models. The used software framework is based on the modelling concept described by Tauscher et al., Mikulakova et al. and Hartmann et al (Hartmann et al. 2012, Mikulakova et al. 2010, Tauscher et al. 2007). Within the framework construction processes are defined including relevant constraints to be fulfilled. Process constraints represent prerequisites and results. They are described by physical or virtual elements in certain states. A prerequisite describes the as-is state that has to be fulfilled before a process can be started. During the process the as-is state is transformed into the target state to produce the results. Process durations are calculated automatically based on the specified production rates. The constraints are directly linked with corresponding elements of the building model. It allows for a flexible definition of arbitrary detailed states, including intermediate states. The schedule is generated automatically by putting the processes into a correct order so that all constraints are considered. The resulting 4D model is than used as an input for the acquisition of as-built data.

After a successful progress monitoring a list containing all identified objects in the BIM model and their actual states is generated. This data can be passed back to the scheduling framework without any difficulties, as IDs of the objects and the state designations are known. Next, a comparison of as-built and as-planned states is carried out automatically. The deviations are reported to the user by color-coding. The distinction is made between the processes that should have already been completed at that time (*completion*) and processes with insufficient percentage of completion (*under-construction*). Construction processes that are behind schedule are highlighted on the Gantt chart bars and corresponding BIM elements of the 3D model. Furthermore, all subsequent processes and objects affected by delays are also highlighted. This condition is called *pre-construction* and represent a potential delay if no measures are taken in order to solve this issue. The insights gained make it possible to better understand the actual or possible impact of the deviation and form the basis for a subsequent adaptation of the 4D model.

## 4 DEVIATION ANALYSIS AND 4D BIM MODEL ADAPTATION

A detailed analysis of the status quo facilitates the detection of planning inaccuracies or errors. Unnecessary waiting times in the schedule can be caused by an insufficient level of detail (LOD) of a 3D model, which needs to be sufficiently improved

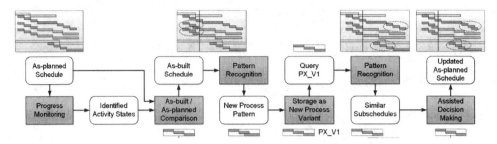

Figure 3. Assisted schedule update using pattern recognition method.

in this case. For example, if several rooms in a model have a shared wall-object, subsequent activities can start only after completion of the certain activity for the entire wall, even though the following work could be carried out room by room. In such a case, it would be necessary to create several wall-objects or to divide the wall into separate sections and to assign an activity to each of them. Increasing the LOD of the model will also have a positive impact on the video-based progress monitoring, as it will provide more details.

Another reason for discrepancies are the assumed production rates that do not necessarily correspond to the actual workload. As the duration of activities is calculated automatically, the adjustment of the production rates results in an automatic rescheduling. In general terms, automatic generation enables to examine various time lags and analyze their effects on the overall delay. It is useful to integrate the findings obtained during this analysis into project discussions and to look for possible solutions, such as provision of additional resources, rescheduling or consideration of construction alternatives. Various scenarios can be therefore easily run through before the decision for the best solution is made.

A further significant source of deviations from the initial schedule is the fact that for various reasons the planned order of activities is not always strictly complied with. A detailed scheduling requires a broader range of experience and expertise and a lot of information about the on-site situation. Therefore, inaccurate or imprecise planning is not uncommon. Apart from that, a certain variability is always given. However, deviations in the work order have a negative impact on the progress recognition, as the underlying method expects the correct sequential order of activities (or rather states) as an input. For example, drywall installation process is usually finished after completing three successive states *build stud partitions*, *install insulation* and *fix plasterboards*. Generally these steps are executed in succession, while each activity is performed only after total completion of the preceding one. Certainly, the parallel execution of construction work is also conceivable. In such a case stud partitions, insulation and plasterboards would be visible at the same time. If the underlying 4D model provides a purely sequential processing, it will lead to incorrect results by the visual recognition of the degree of completion.

A one-time discrepancy does not pose a problem and can be fixed manually. For frequent identical modifications based on as-built status it would be desirable to save the differing sequence of works as a process variant. In this way various alternatives can be considered in the further course of the actual or future projects. At this point an approach for automatic recognition of process patterns in BIM-based construction schedules presented in (Sigalov & König 2017) can be integrated appropriately. After adapting the schedule to as-built state, the search for new process patterns is performed (Fig. 3). In order to detect recurring process structures, construction schedules have to be compared with each other. It is not feasible to compare whole schedules, as process patterns are rather small substructures, containing only a few activities. Thus, modeled schedules must be broken down into a set of subschedules first. In this context, a subschedule is a part of a schedule consisting of closely related activities. This means a set of connected activities meeting certain criteria. It could be activities that belong, for example, to one and the same building element, working section or trade. Such decomposition can be realized manually or automatically. In the proposed approach both alternatives have been realized (Sigalov & König 2017 for details).

The actual pattern recognition includes the determination of similar subschedules by applying graph-based methods. An indexing technique based on features (small characteristic fragments of the graphs, in this context) is used to solve this problem more efficiently. Generally, one of the subschedules from the data base is selected as a query and it similarity to all other subschedules is estimated using the defined similarity metrics. Based on this, irrelevant subschedules are filtered out and only those classified as similar remain.

In this application case, subschedules differing from the planned condition build a query set. If the as-planned schedule was adjusted in the same manner several times, this new subschedule occurs repeatedly in the as-is schedule and will be recognized as a new process pattern (Fig. 3). The planner can decide whether it appears reasonable to store the identified process pattern as a new process variant to the initial subschedule (Fig. 3 PX-V2 and PX-V1 for Process X (PX) respectively). In the positive case, it may be necessary to examine the planned schedule and to consider whether the new defined process variant should be applied instead of the initial one. This decision-making process can be favorably supported by the pattern recognition once again. This time the search is conducted for subschedules similar to the initial process variant (PX-V1), taking it as a query. After a renewed pattern recondition the planer gets a direct overview where in the schedule the initial process variant is used and can make an informed choice regarding the possible options in each individual case. Due to the automatic schedule generation any further adjustments are required, and the updated as-planned schedule is produced immediately.

In addition to the adaptations given above, frequent registered visual characteristics, such as simultaneous visibility of three states of the drywall from the example, should be also systematically saved for a proper state recognition. Since the underlying method needs a test set image data for each material as an input, it is required to extend the object catalogue with inclusion of the new visual information (Fig. 1). If several alternatives are concerned, it has a positive impact on the progress monitoring. After updating the 4D as-planned BIM and the object catalog, the state of progress can be recognized successfully. Such transformation of possible deviations from 4D as-planned model into process variants will bring more flexibility in the planning process. Furthermore, it will help to make visual state recognition for interior finishing more robust and efficient.

## 5 CASE STUDY

For the evaluation of the framework for construction progress monitoring presented in Kropp. et al. 2018 a lot of visual data was recorded during on-site inspections at the renovation of the IC building on the campus of the Ruhr-Universität Bochum. The activity state recognition was carried out, inter alia, on the differentiation of stages of a drywall installation (Fig. 2). The collected data was used as a basis for deviation analysis and model update within the scope of this contribution. A detailed schedule containing various construction works for drywall installation, finishing and electrical installation was generated for a given 3D BIM model within the presented scheduling framework (Figs. 4, 6). Exemplary sensed data was mapped to the wall objects from the model providing in this way their identified states at a certain date.

First, the automatic comparison of as-built and as-planned condition and the identification of discrepancies was implemented. Any recognized delays are represented on the 3D model and Gantt chart using color-coding (Fig. 4). The color scheme is based on a research presented in Chang et al. (Chang et al. 2009). It was determined after a systematic procedure and recommended as an

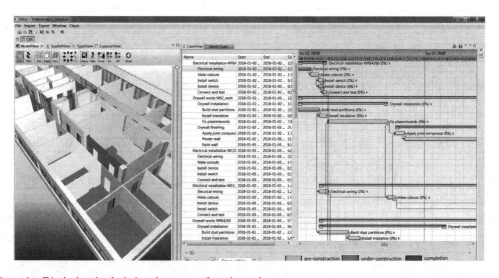

Figure 4. Displaying the deviations by means of a color-code.

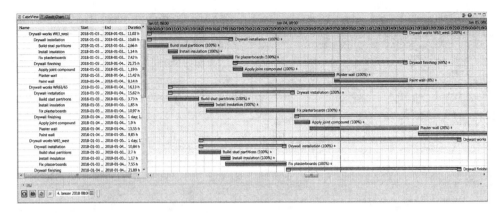

Figure 5.  As-built schedule with manually modified subschedules.

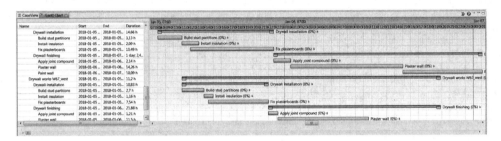

Figure 6.  Updated as-planned schedule with the new process variant applied.

ideal color scheme for construction states of 4D models. Activities behind the schedule are marked in red (*completion*) and orange (*under-construction*) color. The colored marking for all concerned subsequent activities (*pre-construction*) makes the impacts of the delay more obvious. Thanks to automatic generation of the schedule new dates of completion are calculated consistently based on the updated activity durations.

In order to test the practical applicability of pattern-based approach a scenario similar to the one presented in the last section was implemented. Three successive activities for the preparation of a drywall (*build stud partitions → install insulation → fix plasterboards*) occur several times in the designed schedule. The analysis of the video footage showed, that it was sometimes started with the installation of the plasterboards before the insulation was completely installed. However, the simultaneous presence of these two states has not yet been implemented in the progress monitoring approach. For testing purposes, the sequence *build stud partitions → install insulation → fix plasterboards* was modified several times manually in such a manner that the last activity (fix plasterboards) started after 50% of the insulation is installed (Fig. 5). After a successful recognition of the modified sequence as process pattern, it was saved as a new process variant. In the next step, pattern recognition was applied to find all occurrences of the initial sequence in the as-planned schedule and to replace it with the identified process pattern. A part of the resulting schedule that was adapted in this way is shown in the Figure 6.

## 6 CONCLUSIONS AND OUTLOOK

The latest achievements in the field of progress monitoring allows for a high degree of automation for this laborious process. Thus, the acquisition of the actual state and inspection updates can be performed at frequent intervals. It requires a detailed and easily adaptable 4D model, which can be provided by BIM-based tools for the automatic generation of schedules. The immediate localization of the visual data with respect to BIM enables the automatic comparison of as-built and as-planned states and the identification of discrepancies. This information can be used to carry out further analyzes and to appropriately adapt the underlying 4D BIM model. If certain as-is process sequences are identified frequently, this information should be integrated into the planning context. In many situations it might be necessary to save the identified process sequences as a construction alternative and to adjust the schedule accordingly. This paper contributes a promising method, how pattern recognition in construction schedules can be effectively applied to support project managers in this task.

First research results indicate that this approach is a useful measure for the increase of efficiency of the planning process. For further improvements in progress monitoring methods for integration of frequent registered deviating visual characteristics are to be developed next. A renewed inspection in order to reflect the modifications and extension made is certainly needed. A further possible area for future research is a development of an enhanced visual communication with the scheduler during the deviation analysis. Because a clear and comprehensive presentation of possible features is certainly is of great significance for a continuous improvement process.

## ACKNOWLEDGEMENTS

The authors gratefully acknowledge the financial support by the German Research Foundation (DFG) for this work under the grants KO 4311/4-1 and KO 3473/8-1.

## REFERENCES

Bosché, F., Ahmed, M., Turkan, Y., Haas, C.T., & Haas, R. 2015. 'The value of integrating Scan-to-BIM and Scan-vs-BIM techniques for construction monitoring using laser scanning and BIM: The case of cylindrical MEP components', *Automation in Construction*, 49: 201–13.

Chang, H.S., Kang, S.C., & Chen, P.H. 2009. 'Systematic procedure of determining an ideal color scheme on 4D models', *Advanced Engineering Informatics*.

Cheng, T., Teizer, J., Migliaccio, G.C., & Gatti, U.C. 2013. 'Automated task-level activity analysis through fusion of real time location sensors and worker's thoracic posture data', *Automation in Construction*, 29: 24–39.

Chou, H.Y., & Yang, J. Bin. 2017. 'Preliminary Evaluation of BIM-based Approaches for Schedule Delay Analysis'. *IOP Conference Series: Materials Science and Engineering*.

El-Omari, S., & Moselhi, O. 2011. 'Integrating automated data acquisition technologies for progress reporting of construction projects', *Automation in Construction*, 20/6: 699–705.

Gong, J., Caldas, C.H., & Gordon, C. 2011. 'Learning and classifying actions of construction workers and equipment using Bag-of-Video-Feature-Words and Bayesian network models', *Advanced Engineering Informatics*, 25/4: 771–82.

Hamledari, H., McCabe, B., Davari, S., & Shahi, A. 2017. 'Automated Schedule and Progress Updating of IFC-Based 4D BIMs', *Journal of Computing in Civil Engineering*.

Han, K., & Golparvar-Fard, M. 2014. 'Automated Monitoring of Operation-level Construction Progress Using 4D BIM and Daily Site Photologs'.

Han, K.K., & Golparvar-Fard, M. 2017. 'Potential of big visual data and building information modeling for construction performance analytics: An exploratory study', *Automation in Construction*.

Hartmann, V., Beucke, K.E., Shapir, K., & König, M. 2012. 'Model-based Scheduling for Construction Planning'. *14th International Conference on Computing in Civil and Building Engineering*. Moscow.

Hsu, P.Y., Aurisicchio, M., & Angeloudis, P. 2017. 'Investigating Schedule Deviation in Construction Projects through Root Cause Analysis'. *Procedia Computer Science*.

Hyojoo, S., Changwan, K., & Yong, K.C. 2017. 'Automated Schedule Updates Using As-Built Data and a 4D Building Information Model', *Journal of Management in Engineering*, 33/4: 4017012.

Ibrahim, Y.M., Lukins, T.C., Zhang, X., Trucco, E., & Kaka, A.P. 2009. 'Towards automated progress assessment of workpackage components in construction projects using computer vision', *Advanced Engineering Informatics*, 23/1: 93–103.

Kim, H., Anderson, K., Lee, S., & Hildreth, J. 2013. 'Generating construction schedules through automatic data extraction using open BIM (building information modeling) technology', *Automation in Construction*, 35: 285–95.

Kropp, C., Koch, C., & König, M. 2018. 'Interior construction state recognition with 4D BIM registered image sequences', *Automation in Construction*, 86: 11–32.

Li, N., Calis, G., & Becerik-Gerber, B. 2012. 'Measuring and monitoring occupancy with an RFID based system for demand-driven HVAC operations', *Automation in Construction*, 24: 89–99.

Melzner, J., & Hanff, J. 2016. 'Automatic Generation of 4D-Schedules for reliable Construction Management'.

Mikulakova, E., König, M., Tauscher, E., & Beucke, K. 2010. 'Knowledge-based schedule generation and evaluation', *Advanced Engineering Informatics*, 24/4: 389–403.

Nepal, M.P., Staub-French, S., Pottinger, R., & Webster, A. 2012. 'Querying a building information model for construction-specific spatial information', *Advanced Engineering Informatics*, 26/4: 904–23.

Omar, T., & Nehdi, M.L. 2016. 'Data acquisition technologies for construction progress tracking'. *Automation in Construction*.

Park, M.W., & Brilakis, I. 2012. 'Construction worker detection in video frames for initializing vision trackers', *Automation in Construction*.

Pradhananga, N., & Teizer, J. 2013. 'Automatic spatio-temporal analysis of construction site equipment operations using GPS data', *Automation in Construction*, 29: 107–22.

Razavi, S.N., & Haas, C.T. 2011. 'Using reference RFID tags for calibrating the estimated locations of construction materials', *Automation in Construction*, 20/6: 677–85.

Shahi, A., West, J.S., & Haas, C.T. 2013. 'Onsite 3D marking for construction activity tracking', *Automation in Construction*, 30: 136–43.

Sigalov, K., & König, M. 2017. 'Recognition of process patterns for BIM-based construction schedules', *Advanced Engineering Informatics*, 33: 456–72.

Skibniewski, J.M. 2014. 'Construction Project Monitoring with Site Photographs and 4D Project Models', *Organization, technology and management in construction: An international journal*, 6/3.

Tauscher, E., Mikulakova, E., König, M., & Beucke, K. 2007. 'Generating Construction Schedules with Case-Based Reasoning Support', *Computing in Civil Engineering (2007)*, 2007: 119–26.

Tauscher, E., Smarsly, K., König, M., & Beucke, K. 2014. 'Automated Generation of Construction Sequences using Building Information Models'. *Computing in Civil and Building Engineering*, pp. 745–52. Orlando.

Turkan, Y., Bosche, F., Haas, C.T., & Haas, R. 2012. 'Automated progress tracking using 4D schedule and 3D sensing technologies'. *Automation in Construction*.

Tuttas, S., Braun, A., Borrmann, A., & Stilla, U. 2017. 'Acquisition and Consecutive Registration of Photogrammetric Point Clouds for Construction Progress Monitoring Using a 4D BIM', *Photogrammetrie, Fernerkundung, Geoinformation*.

Wang, W.C., Weng, S.W., Wang, S.H., & Chen, C.Y. 2014. 'Integrating building information models with construction process simulations for project scheduling support', *Automation in Construction*, 37: 68–80.

*Ontology, semantic web and linked data*

# A novel workflow to combine BIM and linked data for existing buildings

M. Bonduel, M. Vergauwen & R. Klein
*Department of Civil Engineering, Technology Cluster Construction, KU Leuven, Ghent, Belgium*

M.H. Rasmussen
*Department of Civil Engineering, Technical University of Denmark, Kgs. Lyngby, Denmark*

P. Pauwels
*Department of Architecture and Urban Planning, Ghent University, Ghent, Belgium*

ABSTRACT: Combining conventional Building Information Modeling (BIM) tools and Linked Data technologies improves the options to connect building models to external datasets. Existing workflows in this regard expect a conventional BIM model—including object's geometry—as a starting point. This paper presents a novel, alternative workflow oriented towards existing buildings, including an initial implementation. Modeling the building topology using the BOT ontology is done first, allowing a Linked Data modeler to enrich this initial graph from the start of a project without being dependent on a BIM with (detailed) geometry. Later, a conventional BIM—including objects' geometry—of the existing building can be made, starting from the shared building topology. At the end of this more flexible workflow, both the initial RDF graph and the BIM-based RDF graph are directly connected to each other, combining both datasets.

## 1 INTRODUCTION

### 1.1 BIM, linked data and linked building data

The acronym BIM or Building Information Model(ing) has multiple definitions depending on the context in which it is used. In its broadest sense, it is a concept where a database of a building is created, by defining the building elements and their semantics such as the classification of elements, element properties and relations towards other objects. When applying this broad vision, geometry becomes an optional property and is no strict requirement to define something as BIM. In this paper, the term 'geometric BIM' will be used to denote a building database where (3D) geometry and semantics are combined. Conventional BIM authoring tools such as Revit, Tekla, ArchiCAD, etc. are all examples of geometric BIM software. Such tools typically use a proprietary data schema to structure building information. In order to improve data exchange between different conventional BIM tools, buildingSMART International published several ISO standards, such as IFC (Industry Foundation Classes), IDM (Information Delivery Manual) and IFD (International Framework for Dictionaries). The IFC standard defines an open data schema and neutral format for BIM models, respectively based on EXPRESS and SPFF (Step Physical File Format).

Linked Data has its origins in the Semantic Web domain and is based on several W3C standards such as RDF (Resource Description Framework), SPARQL (SPARQL Protocol and RDF Query Language) and vocabulary/ontology languages such as RDFS (RDF Schema) and OWL (Web Ontology Language). Linked Data in essence consists of RDF triples with a subject node, predicate (relation) and object node, forming a directed graph. The subject and object node can be defined by a URI (Uniform Resource Identifier) or a so-called blank node. The object node can also be a literal value, while the predicate is always a URI. These RDF triples form a data layer (Abox) containing the actual data instances, and a terminology layer (Tbox) based on the applied ontologies. Reasoning engines can be used to infer implicit knowledge based on the used ontologies in a standardized manner.

The principles of Linked Data are used in a broad range of knowledge domains to create and use data via the web that is both human and machine readable. Because of its basic principles, it is extremely well suited to connect elements of different datasets. Linked Building Data (LBD) is the application of Linked Data in the building domain. Initial research in the LBD domain was mainly oriented towards conventional BIM, and more specifically the conversion of the neutral IFC schema into an ifcOWL Linked Data ontology (Pauwels & Terkaj 2016). New initiatives emerged to make the ifcOWL ontology better suited for usage in Linked Data applications, by adhering best practices from the semantic web. Within the W3C LBD Community Group, the Building Topology Ontology or BOT[1] was created, to serve as a central building ontology where other modular ontologies can relate to (Rasmussen, Pauwels, et al., unpubl.). The BOT ontology defines classes for spatial zones (bot:Site, bot:Building, bot:Storey and bot:Space) and building elements (bot:Element), topological relations between instances of these classes and some constraints on the usage of these classes and properties. Following BOT, both the development of an ontology for building products (PRODUCT[2]) and for building-related properties (PROPS[3]) has been initiated within the same W3C group. Using these new LBD ontologies, it becomes easier to create and use Linked Data Abox graphs of buildings as query writing is simplified significantly (Bonduel et al., in press). An LBD Abox graph can be defined as BIM if the earlier mentioned broad definition is used, even if a graph does not contain any geometry of the building elements.

### 1.2 *Existing workflows for combining linked data and BIM*

In literature, three implicit workflows exist for combining Linked Building Data (LBD) and conventional BIM data, either available in proprietary or neutral, standardized data formats. These workflows all start with the creation of a conventional BIM model, including often detailed geometry, which is afterwards converted into a Linked Data Abox graph. Following the workflow in the BPMN process map of Figure 1a, the proprietary BIM data is converted to Linked Data with a dedicated converter plugin for the BIM authoring tool. A limited demo implementation of such a workflow is demonstrated in (Rasmussen, Hviid, et al. unpubl.) for Autodesk Revit and the BOT ontology.

The BPMN (Business Process Model and Notation) standardized process map in Figure 1b depicts two other workflows, where the proprietary BIM data is first exported to the IFC format, which is then converted to an ifcOWL- or BOT/PROPS/PRODUCT-based Abox graph, using respectively the IFC-to-RDF[4] or IFCtoLBD[5] converter (Bonduel et al. 2018). In the first work-

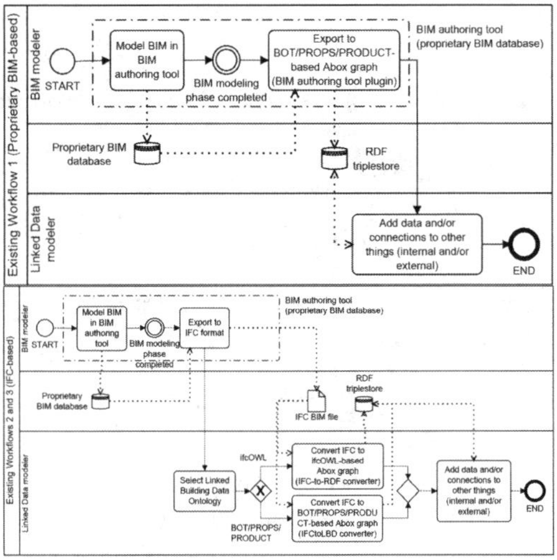

Figure 1. Existing workflows to combine BIM and Linked Data (above: workflow 1, below: workflow 2 and 3).

flow (Fig. 1a), a new, software specific plugin has to be made and maintained for every BIM authoring tool. The intermediate step via IFC used in workflow two and three (Fig. 1b) means an extra conversion step (proprietary BIM to IFC, and IFC to RDF Abox graph) and thus an increased chance for conversion errors. The first approach however, remains closer to the original data source and is therefore preferred over the other approaches via IFC. From a practical point of view, the two workflows using IFC can be used directly in practice as all necessary tools and mappings already exist. In general, challenges to map the internal data scheme of the proprietary BIM software to the Linked Data counterpart defined in the used ontologies, might occur in all three workflows.

### 1.3 *Data management for existing buildings*

Current conventional BIM modeling tools are typically oriented towards newly built projects, and are less suited for modeling existing buildings, with their irregular geometries and specific elements (Volk et al. 2014). Modeling an existing construction in conventional BIM tools, often demands a survey campaign to assemble accurate geometric information reflecting the as-is state of the building, before the BIM modeling can start. In everyday practice, the above can have a serious impact on the price and time to complete a conventional BIM of an existing building. Therefore, multiple researchers investigate the automation of the BIM modeling process based on survey data, e.g. scan-to-BIM focusing on the automatic conversion of point clouds to BIM geometry (Volk et al. 2014).

---

[1]https://w3id.org/bot.
[2]https://github.com/w3c-lbd-cg/product.
[3]https://github.com/w3c-lbd-cg/props.
[4]https://github.com/pipauwel/IFCtoRDF.
[5]https://github.com/jyrkioraskari/IFCtoLBD.

The three earlier described workflows to combine BIM and Linked Data are all dependent on the availability of this geometric BIM, while it could be useful to start describing the building topology, individual building elements and their properties without having to care for the geometry of the building objects. Such a workflow would allow Linked Data modelers to reconstruct the building's topology and to connect it with general information, historical, geographic, public authorities', sensor and material Linked Data, as well as existing files such as 2D plans and photographs. In this way, a non-geometric BIM is created. When at a certain moment, the need arises to create a geometrical BIM, it should be possible to connect the elements of the earlier created RDF graph to the corresponding instances of the BIM-derived RDF graph. Additionally, it makes sense to reuse the modeled building topology from the initial graph, while creating the geometric BIM.

The subject of this study, is to propose a novel and flexible workflow for combining Linked Data and BIM, and is supported by the implementation of (1) a web application to create a Linked Building Data graph and (2) a plugin for a conventional BIM authoring tool to use the building topology of the initial graph as a template while modeling the geometric BIM. The same plugin also assists the user to connect each geometric BIM object to its counterpart in the initial RDF graph.

## 2 A NOVEL WORKFLOW FOR COMBINING LINKED DATA AND BIM

### 2.1 Proposal for a new workflow

The new workflow is documented in a BPMN process map (Fig. 2) and clearly visualizes the fact that the Linked Data modeler and the (geometric) BIM modeler can work in parallel, as long as the building topology is synchronized between both environments. As the complex and elaborated ifcOWL ontology is not very practical to create a graph of the building topology (Terkaj & Pauwels 2017), the earlier mentioned BOT ontology was selected for this workflow, supported by the emerging PROPS and PRODUCT ontologies.

The Linked Data modeler starts with defining the building topology using BOT (site, building, storeys, spaces and elements). Note that not all `bot:Element` instances have to be defined in this graph, but only the ones that will be used during the Linked Data enrichment phase. Afterwards, links to external Linked Data instances (datasets related to materials, sensor data, geographic information, etc.) as well as properties with literal values can be made (e.g. relevant dimensions of a wall, name of a storey, etc.), using PROPS or any other suitable ontology. It is also possible to add classifications for the `bot:Element` instances using the PRODUCT ontology.

The BIM modeler uses first the dedicated BIM plugin to establish a connection between the BIM file and the initial graph, followed by the selection of the correct `bot:Site` and `bot:Building` from the same graph. Both the database name, URI of the selected site and building instance are stored in the BIM project. Afterwards, the building topology of the initial graph is used by the same BIM plugin to automatically create storeys and placeholder spaces in the geometric BIM tool. They automatically get a URI parameter assigned upon creation, which value relates to the corresponding `bot:Storey` or `bot:Space` instance. In a last phase, the BIM modeler has the option to use the BIM plugin to match modeled objects to their Linked Data counterpart and to add the URI of the corresponding instance to the BIM element as a property. The plugin assists the user in finding the right instance, by looking at its relation to the

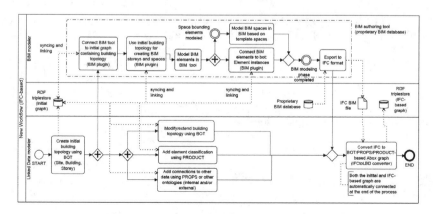

Figure 2. Novel workflow for combining linked data and BIM.

other building elements (storeys and spaces) and the kind of object (e.g. a wall, floor, etc.). When a geometric BIM modeling phase is finished, the proprietary BIM is exported to IFC.

The Linked Data modeler uses the earlier mentioned IFCtoLBD converter—similar as in the third mentioned existing workflow—to translate the IFC building model to a BOT/PROPS/PRODUCT-based Abox RDF graph (Bonduel et al. 2018). If the BIM modeler selected the option to export the earlier mentioned URI properties to IFC, it is trivial to connect the instances of the initial graph to the ones of the IFC-based graph. A control by the Linked Data modeler is recommended, to check if the IFC-based RDF instances are linked correctly to their counterparts in the initial graph, if they are modeled there.

## 2.2 Comparing existing and new workflows

With this new workflow, the geometric and the semantic modeling phases are decoupled allowing the Linked Data modeler to define the building topology without having to model the geometry of each object. This topology graph can be used as a backbone for linking other building-related data by project stakeholders such as the owner, the facility manager, the local government, the architect, etc. The geometric BIM modeling can happen parallel with the modeling of the topology and it can reuse the building topology in Linked Data form as a template.

When the geometry modeling in the BIM tool is finalized, the conventional BIM model is converted to IFC, before being converted to an Abox RDF graph. The step via IFC is added, as it allows us to reuse the existing (batch) IFCtoLBD converter, instead of programming a new, specific LBD exporter for the used conventional BIM authoring tool.

# 3 PART 1: STAND-ALONE LINKED BUILDING DATA APPLICATION

## 3.1 The LD-R framework

For the implementation of a web application to create LBD independent of a conventional BIM tool, the open source Linked Data Reactor (LD-R) framework[6] was used. Other custom web applications using different programming frameworks can be developed as well and still manipulate the same Linked Data. LD-R offers application developers a generic, but flexible plug-and-play Linked Data web application,

as the GUI can be configured to the user's preferences. The framework is based on adaptive, reusable components for viewing, editing and browsing Linked Data (Khalili et al. 2016). Such an approach, allows component developers to create new or modify existing components that can be reused in other applications. The LD-R framework assists the application developer by managing the connection and data flow between the application and the SPARQL endpoint, while the developer can focus on the UI and thus the end-user of the application.

The framework consists of two main parts: a reactor part where Linked Data can be viewed and modified in a human-friendly way, and a browser part where the end-user uses facets. The application uses SPARQL queries behind the scene to communicate with the SPARQL endpoint of the triplestore.

## 3.2 Implementation details

The newly proposed workflow demands a Linked Building Data application that is unrelated to conventional BIM software. To make such an application with LD-R, its configuration files had to be adapted for Abox graphs structured according the BOT, PROPS and PRODUCT ontologies. Besides the above discussed settings, it is also possible to add URI prefixes and to modify the autocomplete function that is used when new properties are added to a resource. These modifications assist the user of the application to create a more consistent Linked Data graph at a faster rate.

For the implementation of the LBD web application, LD-R version 1.2.3 was used in combination with a Stardog RDF triplestore (v. 5.2.1). The source code of the LD-R application, including the configuration files and the modified prefix and autocomplete functions, is available on Github[7].

## 3.3 Modeling LBD graphs using LD-R

The Linked Data modeler first creates the building topology by using BOT. It is advised to start the modeling from the bot:Site instance towards the individual elements, as the reactor part of LD-R only visualizes the outgoing triples of a resource. Based on the domain and range of the used predicate (e.g. bot:hasBuilding), the built-in Stardog reasoning engine infers the class of the subject and object of the triple (e.g. in the case of bot:hasBuilding, the subject is a bot:Zone and the object is a bot:Building). This not only speeds up the general modeling process, it also helps the LD-R UI to know how a certain resource

---

[6]https://github.com/ali1k/ld-r.

[7]https://github.com/mathib/ld-r.

should be visualized in the application, based on its inferred class. As the reactor part only visualizes one subject and its properties per webpage, the overview of the modeled graph can be lost. At this point, the faceted browser part of LD-R can be used in parallel to understand the created graph and to find earlier modeled elements. Buildings can contain thousands of objects, which can again aggregate multiple subobjects. It is not the goal of the Linked Data (or BIM) modeler to model every object of the building: it only makes sense to model the objects that are relevant for the project.

When the building topology is documented, the Linked Data modeler can add PRODUCT classes to the bot:Element instances and add extra properties to resources using PROPS or any other ontology. It is also possible to enrich the graph with links to external datasets available as Linked Data related to e.g. general knowledge, building materials, sensors, images, geographic information, etc. In order to help the BIM modeler, a props:name_simple or rdfs:label property for both the bot:Storey and bot:Space instances has to be added as a minimum. The Linked Data modeler is strongly recommended to use meaningful URIs and/or labeling properties when modeling building objects (e.g. props:name_simple or rdfs:label for storeys and spaces), not only for the Linked Data modeler but also to make it easier to connect the bot:Element instances of the initial graph to the corresponding building elements modeled in the conventional BIM tool.

### 3.4 *Effectiveness of LD-R for modeling LBD graphs*

The LD-R framework was successfully applied to setup a LBD application. As its modular architecture allows to reuse existing UI components, a serious amount of time was saved to create the UI. The reactor part can be highly customized, but this flexibility increases the complexity of the corresponding configuration file. As an alternative to the generic Linked Data UI of LD-R, it might be worthwhile to invest in a dedicated UI for creating and viewing LBD graphs.

The modeling of the building topology has to happen in one direction because of the current reactor UI implementation. An advantage is that the modeler is forced to connect at least one bot:Zone or bot:Element instance to the bot:Zone instance that contains it or is adjacent to it. As a result, there will be less instances that are not connected to the main building topology branch of the graph. The current version of BOT allows reasoning engines to correctly infer bot:containsZone and bot:contains Element relations. If a bot:adjacentZone or bot:adjacentElement relation is modeled, the respective relation cannot propagate via inferencing to other bot:Zone instances that contain the zone with the adjacency relation. The modeler has to be aware of this and explicitly model the correct relation. Depending on the situation, it is either a bot:adjacentZone or bot:containsZone, or either a bot: adjacentElement or bot:containsElement relation. An example of such a situation is given in Listing 1.

The current LD-R version was mainly designed for viewing and modifying RDF datasets, instead of creating new datasets from scratch. As a result, the UI components in the reactor part of LD-R are not optimized for the batch creation or modification of elements (e.g. create five storeys of a building at once). Additionally, some buttons of the out-of-the-box reactor UI take too much space or are not placed at an accessible location in the window, which can cause the modeler to lose the overview. When a new triple is made, the user cannot select an option to indicate that the object resource is new or already exists in the graph. Such an option could make the LD-R application able to assure the Linked Data modeler that the resource is respectively unique (there is no other resource coincidentally having the same URI) or already exists in the graph. As a result, a more consistent graph is created.

Listing 1. RDF triples containing bot:containsElement and bot:adjacentElement relations that can either be inferred or have to be modelled explicitly.

```
###Sample triples
inst:storey1 bot:hasSpace inst:spaceA .
inst:spaceA bot:containsElement inst:desk1 .
inst:spaceA bot:adjacentElement inst:wall1 .

###The following triple will be inferred and
#should not be modelled explicitly
inst:storey1 bot:containsElement inst:desk1 .

###The following triples CANNOT be inferred.
#One triple has to be modelled explicitly as it
#can be either a bot:containsElement
#(internal wall)
inst:storey1 bot:containsElement inst:wall1 .
#or a bot:adjacentElement (external wall)
inst:storey1 bot:adjacentElement inst:wall1 .
```

## 4 PART II: CONNECTING LINKED BUILDING DATA AND BIM

### 4.1 *Implementation details*

For the execution of the second part, the Autodesk Revit (v. 2018.3) BIM authoring tool was selected. A Revit plugin, named StardogRevit-synchronizer, was made using Dynamo visual programming (v. 1.3) and built-in IronPython. In the future, the plugin might be extended to exchange data with other SPARQL endpoints. The plugin consists of four numbered Dynamo graphs. Their source code and an overview of the used Dynamo packages is listed on a dedicated Github repository[8].

### 4.2 Creation of BIM template based on initial LBD graph

The second part of the new workflow, executed by the BIM modeler, can start from the moment an initial building topology is available in the RDF triplestore. In a first step, the Dynamo script 'PART1 – site and building' will prompt the user to first select the correct Stardog database, followed by the correct site-building combination. The name of the selected Stardog database and the URIs of the selected site and building are stored in the Revit Project Information respectively as a shared text property ('stardog_db') and URL parameters ('site_URI' and 'building_URI'). Such a workflow is allowed, as each Revit project is recommended to contain only one (part of a) building, neglecting any linked Revit models.

From the moment the initial connection between the Revit project and the RDF triplestore is made, the second script 'PART2 – storeys' is launched. The script creates Revit Levels based on the `bot:Storey` instances connected to the earlier selected `bot:Building` instance of the shared Stardog database. If these storeys have a props:name_simple and/or a props:elevation_simple property, the values of these properties are added to the automatically created Revit Levels, together with the URI of the corresponding `bot:Storey` in a Revit shared property named 'storey_URI'. If no storey name or elevation was defined in the initial LBD graph, a default name and elevation is created by the plugin as Revit Levels need this information as a minimum to exist.

In the last phase of the BIM template creation, a series of dummy Revit Spaces is made by the third script ('PART3 – spaces'). These new Revit Spaces are based on the `bot:Space` instances in the graph related to the earlier connected `bot:Storey` instances. The Revit Level of these Spaces is the same as the `bot:Storey` of the corresponding `bot:Space` instance. Similar as the Levels, the Revit Spaces get the values of the props:name_simple and/or props:number_simple property of the corresponding `bot:Space`, if they exist. The URI of the `bot:Space` is stored as a 'space_URI' shared parameter of the Revit Space.

### 4.3 Connecting BIM objects to Linked Data entities of initial LBD graph

When the BIM template is completed according to the building topology in the connected Stardog database, the BIM modeler can start with placing BIM objects such as walls, floors, doors, etc. If all space bounding elements are modeled, the Revit Space dummies should be moved to the right location to become actual 3D spaces. The modeler can already start to connect Revit building elements to their counterpart in the connected Stardog graph, but it is advised to wait until the Revit Spaces are modeled as this will help the matching algorithm in finding the corresponding element.

The plugin (script 'PART4 – elements') first queries the selected Revit element for (1) its Revit Category, (2) its Level, (3) its surrounding Levels and (4) the Revit Spaces it intersects or touches.

The Revit Category is converted to a corresponding class of the PRODUCT ontology, according to a mapping file available in the Dynamo Revit plugin. As the mapping file only contains the most accurate corresponding PRODUCT class for each Category, the potential superclasses of this class are not included directly. A SPARQL query on the PRODUCT ontology completes this information. Revit Levels however, are defined as horizontal planes and each Revit element is modeled on one Level, with an optional vertical offset. This forces the BIM modeler to allocate only one Level to an element that spans multiple floors (e.g. a stair). The plugin also scans the Revit project for other Levels that are potentially related to the object. More specifically, it searches every Level that is the highest Level below the lowest bounding box point, the highest Level below the highest bounding box point, and all Levels in between. Each Revit Level contains a 'storey_URI' property that was filled in during the creation of the BIM template in the earlier phase. In a final step, the plugin finds every Revit Space that intersects or touches the geometry of the selected element. Each Space has a 'space_URI' parameter of the corresponding `bot:Space` instance.

In parallel, the plugin queries the graph shared by the Linked Data modeler for `bot:Element` instances, their optional PRODUCT class(es) and their topological relation (either `bot:containsElement` or `bot:adjacentElement`) to `bot:Storey` and `bot:Space` instances.

The matching algorithm performs five tests for each `bot:Element` instance of the shared graph. The first test, checks for each `bot:Element` if is already connected to a Revit element, while in the second test the PRODUCT classes of the Revit element are compared with the classes of each `bot:Element`. If no direct match is found, the PRODUCT superclasses of the Revit element are compared—from most precise to most generic—with the classes of each `bot:Element`. In the third test, the 'storey_URI' parameter of the element's Revit Level is compared with the URI of the `bot:Storey` of each `bot:Element` instance. The fourth test is similar, but compares the 'storey_URI' parameter value of the other related Revit Levels with the `bot:Storey`

---

[8]https://github.com/mathib/StardogRevit-synchronizer.

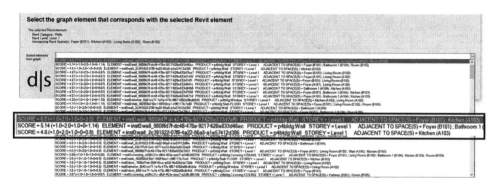

Figure 3. Results of matching algorithm assisting the BIM modeller while connecting Revit elements to bot:Element instances (screenshot of UI of Dynamo Revit plugin).

URI of each bot:Element. The final test compares the 'space_URI' of the Revit spaces related to the element with the URI of the bot:Space instance(s) related to each bot:Element. For each positive test, the weight of the test—which is user configurable—is added to the total score of the analyzed bot:Element. The BIM modeler gets a list of all instances of bot:Element in the graph, ranked according to their total score over the five tests, together with human-readable labels of the bot:Element and its topological relations (Fig. 3). The BIM modeler, assisted by the suggestions of the matching algorithm, then selects the corresponding bot:Element. The URI of the selected element is automatically stored in the Revit element as a shared property named 'element_URI'.

### 4.4 Converting conventional BIM data to a LBD graph

If a BIM modeling phase is finished, the model is exported from the conventional BIM authoring tool to the neutral IFC format. The option to include custom properties of the conventional BIM modeling tool has to be selected to include the earlier mentioned properties: 'stardog_db', 'site_URI', 'building_URI', 'storey_URI', 'space_URI' and 'element_URI'.

The resulting IFC file of the building model, is handed over to the Linked Data modeler who uses the IFCtoLBD converter to translate the IFC model as a whole into a BOT/PROPS/PRODUCT-based Abox graph. It is essential to select the PROPS module, to also include the above mentioned properties into the created graph. The objects of the connection triples contain by default URIs as string literals and a trivial SPARQL query is necessary to translate these literals into real URIs. The combination of the 'site_URI' and 'building_URI' property appears twice as properties of both the bot:Site and the bot:Building. This comes because Revit defines no site and building object in a Revit project, as IFC and BOT do. The 'building_URI' property of the bot:Site and the 'site_URI' property of the bot:Building have to be removed by a simple SPARQL DELETE query. At the end of this process, both the initial RDF graph and the IFC-based RDF graph are connected and detailed data exchange between the Linked Data modeler and the BIM modeler becomes possible.

### 4.5 Effectiveness of the synchronization plugin

The modeling process in the conventional BIM authoring tool can start quickly by using the 'template' derived from the shared building topology in Linked Data form. The usage of the Dynamo plugin demands some additional steps in the modeling process, but this results in a better synchronization of the building topology between both the RDF triplestore and the Revit environment.

In order for the element matching algorithm to perform well, it is necessary to have the same building topology in both the Revit environment and the initial RDF graph. As there is no geometry in the initial graph of the Linked Data Modeler, it can be difficult to know the one corresponding bot:Element of a Revit element if there is no clear difference in the topology of multiple bot:Element instances. The BIM modeler stays in charge of the matching process and is assisted by the plugin's suggestions of best matching bot:Element instances of the initial graph. Human readable labels or meaningful URIs can help in finding the right bot:Element.

The internal processes of the plugin are designed in the most software independent way, but it is inevitable that certain Revit specific issues had to be resolved. Comparable, but slightly different issues might appear when a similar plugin for other conventional BIM authoring tools is made.

## 5 CONCLUSION AND FUTURE WORK

A novel workflow to combine Linked Building Data and a conventional BIM environment for existing buildings, is proposed and compared to currently used methods. A two-part implementation demonstrates how this workflow might be turned into practice. For the first part, the flexible and open-source LD-R framework was used while developing a web application to create Linked Building Data graphs based on the BOT, PROPS and PRODUCT ontologies. The second part involved an implementation for Autodesk Revit as the conventional BIM authoring tool to model geometrical BIM objects. A plugin for Revit was described to transfer the Linked Data building topology of the initial RDF graph into the Revit environment as a kind of template. Based on this shared topology, the plugin can further assist the BIM modeler while connecting each Revit building element to its corresponding element in the initial RDF graph. In a successive research phase, the presented workflow will be tested thoroughly with real projects related to existing buildings.

Notwithstanding the clear advantages of the new workflow, its flexibility also increases the complexity of the overall workflow as an extra plugin for the conventional BIM environment is necessary. In the future however, it might not be unrealistic that major CAD and BIM software will have a similar functionality built-in by default, similar as they now have built-in support for IFC import and export. A reliable synchronization between both the proprietary BIM database and the RDF triplestore will be crucial for an effective workflow.

Future work includes either optimization of the LD-R web application or the creation of a new, dedicated LBD application for creating building-related graphs. The LD-R framework was judged to be useful when quickly developing a practical application for Linked Data modeling. A UI that is more oriented to the creation of new graphs from scratch can be developed as the LD-R framework (built on react.js) is oriented towards interchangeable UI components.

The matching of building elements between both environments can be challenging as the initial RDF graph contains no geometry, location or orientation of the modeled object. In the case of existing buildings, there might be plan documents available to locate bot:Element instances in the building. Methods such as semantic annotation of plan documents can be investigated to assist the BIM modeler later on in the workflow, during the matching process.

Besides optimization of the implementation, other workflows might be worthwhile to investigate. Instead of trying to connect a BOT/PROPS/PRODUCT-based Abox graph with elements of a conventional BIM tool, one might explore the possibility of disconnecting 3D geometry and model semantics by using a 3D CAD environment in combination with LBD-based semantics. Our proposed method can be seen as a flexible, transitional workflow where the use of already existing BIM authoring software and connected tools can be continued in combination with the benefits of a Linked Data enriched workflow.

The proposed workflow has resulted in an increased flexibility for the Linked Data modeler as he/she is no longer dependent on the availability of a complete geometric BIM model, to create and/or enrich the Linked Building Data graph of the particular building. The BIM modeler can synchronize its new BIM project seamlessly with the building topology that was modeled earlier using the BOT ontology. At the end of the workflow, both the initial RDF graph and the IFC-based RDF graph are linked together, to allow fine-grained data exchange between stakeholders.

ACKNOWLEDGEMENT

This research is funded by the Research Foundation Flanders (FWO) in the form of a personal Strategic Basic research grant.

REFERENCES

Bonduel, M. et al., (In press). The IFC to Linked Building Data Converter - Current Status. In *6th Linked Data in Architecture and Construction Workshop (LDAC), CEUR Workshop Proceedings.* London, UK.

Khalili, A., Loizou, A. & van Harmelen, F., 2016. Adaptive Linked Data-driven Web Components: Building Flexible and Reusable Semantic Web Interfaces. In H. Sack et al., eds. *The Semantic Web. Latest Advances and New Domains. ESWC 2016. Lecture Notes in Computer Science.* Springer, Cham, pp. 677–692.

Pauwels, P. & Terkaj, W., 2016. EXPRESS to OWL for construction industry: Towards a recommendable and usable ifcOWL ontology. *Automation in Construction*, 63, pp.100–133. Available at: http://dx.doi.org/10.1016/j.autcon.2015.12.003.

Rasmussen, M.H., Pauwels, P., et al., (Unpublished). Recent changes in the Building Topology Ontology. In *5th Linked Data in Architecture and Construction Workshop (LDAC 2017).* Dijon, France.

Rasmussen, M.H., Hviid, C.A. & Karlshøj, J., (Unpublished). Web-based topology queries on a BIM model. In *5th Linked Data in Architecture and Construction Workshop (LDAC 2017).* Dijon, France.

Terkaj, W. & Pauwels, P., 2017. A Method to generate a Modular ifcOWL Ontology. In *8th International Workshop on Formal Ontologies meet Industry (FOMI).* Bolzano, Italy.

Volk, R., Stengel, J. & Schultmann, F., 2014. Building Information Modeling (BIM) for existing buildings - Literature review and future needs. *Automation in Construction*, 38, pp.109–127. Available at: http://dx.doi.org/10.1016/j.autcon.2013.10.023.

# Linking sensory data to BIM by extending IFC—case study of fire evacuation

R. Eftekharirad & M. Nik-Bakht
*Department of Building, Civil, and Environmental Engineering, Concordia University, Montreal, Quebec, Canada*

A. Hammad
*Concordia Institute for Information Systems Engineering, Concordia University, Montreal, Quebec, Canada*

ABSTRACT: Building Information Modeling (BIM) is becoming a repository of building related data. To represent the building's real-time information, a dynamic BIM for recording and storing timely and accurate sensory data of building's components, spaces and occupants is required. Such a dynamic BIM can improve the building emergency management and enhance effective survival services in emergency conditions. This study aims to link sensory data to BIM, using the Industry Foundation Classes (IFC) standard, to capture, record and update the state of building elements, spaces and occupants. The objectives of this paper are: (1) extending IfcSensor entity to include occupant's sensors; (2) defining the relationships between sensors, occupants, time series, and spaces; and (3) creating dynamic BIM for tracking occupants and environmental states. A case study to highlight the feasibility of the proposed model is presented.

*Keywords*: BIM, IFC, sensory data, occupant, fire, emergency management

## 1 INTRODUCTION

Building Information Modeling (BIM) is becoming a rich source of building related data. Current models in BIM are developed based on static information. However, to represent the building's real-time information, a dynamic BIM will be required, capable of recording and storing the "state" of building's components, spaces and occupants. Such a dynamic/statful BIM will contain timely and accurate sensory data, which can be extremely helpful for supporting rapid responses to improve emergency conditions. In these conditions, detecting the real-time location of occupants, and the state of building elements and spaces can improve the building emergency management and enhance effective survival services. Integrating these two aspects can support evacuation decisions. In addition, visualizing the real-time locations of occupants and mapping them to the building condition can assist first responders to make effective decisions for evacuation planning.

In order to pursue such goals, BIM can be enriched with sensory data related to occupants' information—including ID, location, and time—as well as the environmental hazard conditions. Industry Foundation Classes (IFC), as an open BIM standard, includes IfcSpace and IfcBuildingElement entities, which can be employed to represent physical/spatial components of a building (such as walls, doors, stairs, rooms and exits) as well as their conceptual attributes (such as volume, use, accessibility, etc.). The current IFC 4 version is also offering basic definitions of sensors, time series, and occupant entities under IfcSensor, IfcTimeSeries, and IfcOccupant, respectively. However, these entities are missing details to provide the desired level of dynamism.

Our current study aims to link sensory data to BIM, using the IFC standard, to capture, record and update the state of building elements, spaces, and occupants. The specific objectives of this paper are: (1) extending IfcSensor entity to include occupant's sensors; (2) defining the relationships between sensors, occupants, time series, and building components; and (3) creating dynamic BIM for tracking occupants and environmental states.

In this paper, an approach for adding the real-time sensory data to BIM is presented. First, the available information in IFC 4 is checked. Then, the required attributes and the relations are introduced to IFC to support the dynamic BIM for occupant tracking. IfcTimeSeries is used to link the sensors to the sensory data and update the state of the BIM model. We present a case study to highlight the implementation and feasibility of the proposed method. In this regard, the collected sensory data for occupants are fused into the IFC model. The sensory data are visualized in the BIM tool and state updates of the BIM are presented over various time-frames.

The implementation part of the paper provides a proof-of-concept prototype for making the dynamic/stateful BIM by linking sensory data. A case study is used to validate the proposed method.

## 2 REVIEW OF RELATED WORKS

Smart buildings use sensors to measure the parameters that indicate the conditions related to the buildings, such as building assets, occupants, and incidents. In emergency conditions, the timely critical information measured by sensors can decrease the fatalities and damages. For instance, when a fire occurs in the building, having access to real-time information for occupants' location, fire propagation and dangerous zones can increase the awareness of occupants and first responders. Consequently, the building can be evacuated effectively and efficiently.

Recently, there has been an increase in the use of sensing technologies and the Internet of Things (IoT) in the construction and building management industry. Thus, the construction industry deals with large volumes of sensory data (also known as "Big Data") to support decision making (Davila Delgado et al., 2018; Bilal et al., 2016). However, several challenges limit an effective use of data (including compiling, organizing and analyzing) in the construction and building management domain.

As explained in Section 1, sensors as hardware specifications can be mapped by existing standards, but there is a limitation to map the sensory data. The limitations are: (1) BIM is mainly used for design and construction phases (Becerik-Gerber et al., 2011). (2) BIM cannot store and manage the large data sets. (3) Updating models based on sensory datasets is not possible by BIM (Davila Delgado et al., 2018). Therefore, current research efforts have been focusing on linking sensory data to BIM (Chen et al., 2014; Davila Delgado et al., 2015; Davila Delgado et al., 2016; Dávila et al., 2018) to have a dynamic BIM (Abrishami et al., 2014; Agdas & Srinivasan, 2014; Cahill et al., 2012; Park & Cai, 2017; Volkov & Batov, 2015).

In the context of an emergency condition, the changes of the occupants' locations must be captured, analyzed and visualized in real-time or near real-time. However, linking occupants' sensory data (for tracking movements) to BIM via IFC has not been fullyexplored.

### 2.1 Linking sensory data to BIM

Several studies have tried to visualize the changes sensed in a facility, through BIM and by the aid of colors, adding charts, or even animations. For instance, Chen et al. (2014) offered an approach to connect the sensory data to a BIM model. They graphed changes of temperature sensed by sensors locations on a bridge deck. Also, Davila Delgado et al. (2018) presented a bridge monitoring system including fiber optic based strain sensors to monitor the structural performance. A dynamic BIM environment for structural monitoring was suggested that enabled displaying the dynamic strain response of main girders in a bridge under trains load. In this work, the collected data (including the strains and stresses) were visualized by using colors and charts in several time steps. Moreover, they partially developed the IFC for defending the fiber optic sensors for structural health monitoring.

Similarly, Attar et al. (2011) proposed a model to combine BIM with sensors for the captured data of occupant-centric performance in a building. They mapped the sensed values of temperature by color coding, and those of light sensors through charts. Also, they provided a user-friendly interface to communicate with different stakeholders. However, the output of the mentioned research study is not a fully semantic BIM.

Since BIM is originally a static information source of building data, using it to react to emergencies in a timely and effective manner is a challenge. Also, BIM software tools are initially developed for the design and construction phases and lack capabilities to record and use sensory data. As an example, most of the software tools and IFC editors commercially available fail to accurately process/interpret data stored as IfcTimeSeries. In response to this gap, several studies have addressed some of the limitations by developing dynamic BIM (Ajayi et al., 2014; Al-Shalabi & Turkan, 2015; Francisco et al., 2018; Srinivasan, et al., 2014). To implement a dynamic BIM, the real-time information, such as data received from various types of sensors should be infused in the current static BIM. For this purpose, two main approaches have been suggested: (1) creating a stand-alone application; and (2) developing a plug-in or add-in to the BIM software (Davila Delgado et al., 2018).

In the first approach, a dynamic BIM is developed as a stand-alone application that does not require another software to run in parallel. This option can provide more capabilities to BIM, although it needs considerable programming experience and it is time-consuming (Davila Delgado et al., 2018).

In the second approach, the capabilities of the current BIM are expanded by using APIs (Application Programming Interfaces) to add additional features to the software. For instance, a dynamic BIM was developed by using a Revit's .NET API.

Although the two approaches can be used for visualizing the information of changes in sensory data, such as temperature, strain, light, and $CO_2$, they have some limitations to demonstrate the changes in occupants' locations and the fire. The methods are not fully semantic BIM and they do not facilitate data exchange because the generating models are not compliant with IFC.

## 2.2 *Extending IFC*

Providing Interoperability between different platforms for exchange of information is one of the main high level goals of BIM. IFC standard currently contributes to achieving such goal. IFC is an open standard developed by BuildingSMART to share physical and functional features of buildings among stakeholders and various platforms (Liebich et al., 2013). IFC has a hierarchical and modular framework containing different entities with the different attributes. Also, there are several relationships between the entities to model data related to different phases of building life cycle (Zhiliang et al., 2011).

IFC specifications can be extended to describe new use cases (Akinci & Boukamp, 2003; Eftekharirad et al., 2018; Motamedi et al., 2016; Zhiliang et al., 2011). For example, there are extensions for infrastructures assets, such as IFC Bridge, IFC Road, and IFC Tunnel. However, there are few research projects that extended the IFC (Eftekharirad et al., 2018; Fairgrieve & Falke, 2011; Motamedi et al., 2016; Weise et al., 2009). Also, extending IFC for emergency management to represent the live occupants' sensory data (e.g., occupant's location) has not been yet explored.

With regards to real-time monitoring in the emergency condition, IFC has some main entities, such as sensor, occupant and time series. However, new entities and relationships should be identified to consider the challenges of real-time monitoring. Those challenges are: (1) size of data, (2) accuracy of sensory data, (3) data storing, and (4) data interoperability with existing formats (Gerrish et al., 2015).

Three mechanisms can supposedly extend IFC: (1) defining new entities; (2) using proxy elements; and (3) using property sets. Extending the IFC through the first method can be considered the best option, because the new entities can be used in the same way as the existing ones; however it normally takes a long time for the new entities to be approved by Model Support Group of the y Building SMART Alliance BSA (Weise et al., 2009). The other two alternative mechanisms are more practical because IFC can be extended without changing its schema. However, they require additional implementation agreements about the definition of property sets and proxies when they are used to exchange data with other software (Motamedi et al., 2016; Weise et al., 2009; Zhiliang et al., 2011).

## 3 PROPOSED APPROACH

To have a real-time monitoring for the fire emergency management, comprehensive data related to fire, including location, smoke, temperature, toxic gases, and occupants should be integrated in BIM. Occupants' general attributes (such as ID, age, gender, etc.); location, disability condition (such as physical disability, vision or hearing impairment, and health condition) should be connected to BIM. For this purpose, the data are gathered from two main sources: sensors and external database. Table 1 shows the classification of occupant and fire information based on their source.

Figure1 represents two main methods to visualize the real-time locations of occupants in BIM. The first alternative is creating a link between BIM software tools and sensory data stored in an external database. The link can be created using APIs, plug-ins or add-ins normally offered by software tools. For instance, the occupants' information can be linked to BIM using tools such as visual programming plug-in Dynamo for Autodesk Revit. The output of this method is a real-time model that visualize the sensory data related to fire, building spaces, and occupants. However, the model is not fully semantic. Also, using this method can cause challenges of exchanging the data and interoperability because it is not standardized.

The second alternative is integrating the sensory da-ta, such as the real-time location of occupants, with IFC, as IFC objects. In IFC, basic occupants' infor-mation, such as ID, name, and type of actor are defined in IfcPerson and IfcOccupant classes. However, other detailed occupants' information should be added to IFC. In addition, IfcTimeSeries is an entity for a time-stamped data entries that allows a natural association of data collected over intervals of time (Liebich et al., 2013). This entity does not currently have a relationship with IfcSensor and IfcOccupant.

Therefore, defining the new relationships between IfcSensor, IfcOccupant, and fcTimeSeries is needed to facilitate the development of real-time occupant tracking in BIM. Figure 2

Table 1. Classification of the occupant and fire information based on their source.

| Data source | Occupant info. | Fire info. |
| --- | --- | --- |
| Sensory data | ID<br>Location<br>Time | Smoke<br>Temperature<br>Toxic gases<br>Location<br>Time |
| External database | Age<br>Gender<br>Health condition<br>Type of disability | |

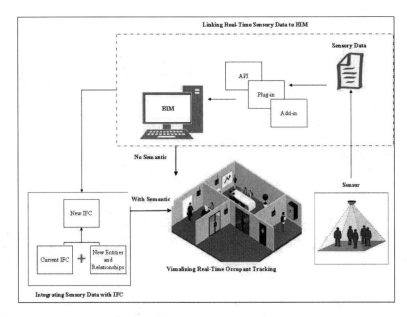

Figure 1. Two methods for real-time tracking of occupants' locations in BIM.

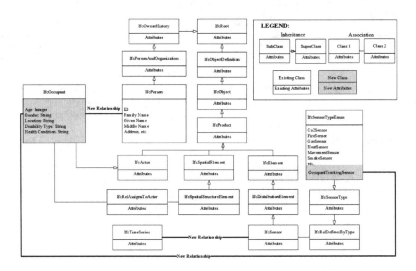

Figure 2. Modified IFC model for sensor, occupant, and time series (Eftekharirad. et al., 2018).

shows the relationships between existing and extended entities. The output of this method is fully semantic because sensory data sets are integrated with IFC. However, programming and APIs are used to integrate sensory data with IFC.

## 4 IMPLEMENTATION AND CASE STUDY

In order to illustrate the proposed method, a shared office for graduate students on the 9th floor of the EV (Engineering & Visual Arts) building at Concordia University was chosen. To collect the real-time sensory data of the occupants' locations, a Quuppa Intelligent Locating System was installed at the office. Quuppa system uses Bluetooth Low Energy (BLE) technology to gather the real-time tracking data, such as ID, date/time and X-Y-Z coordinates with accuracy of 50 centimeters.

A fire alarm went on in EV building on April 24th, 2018 at 3:10 PM. The locations of occupants in the office were tracked using Quuppa system and the data were stored as time series. The mentioned data were linked to the 3D model of the office (created in Autodesk Revit Architecture 2018). For the linking and visualizing the occupants' locations, the following three main steps were taken: (1) adjusting the office model to Quuppa system; (2) defining the time, occupants' locations and occupants' attributes including ID and mobility limitation; and (3) visualizing the dynamic model.

Firstly, the coordinate system of the 3D model was changed to match the local coordinate system used by Quuppa. The information of three different occupants was then added to the model, as shown in Figure 3.

Secondly, the real-time locations of occupants were visualized in BIM by developing a Python scripts in Dynamo. Dynamo, as a visual programming plugin, helped to link the sensory data of Quuppa system to Autodesk Revit. In Dynamo, the following steps were followed: (1) the Excel file containing occupants' data was selected. This file has several parameters including ID, date/time and X-Y-Z coordinates, and health condition of occupants. (2) A certain color and shape were assigned to each occupant. (3) A relationship between location point (X-Y coordinates) and each occupant was defined. (4) ID, time, and disability conditions of occupants were defined using the Python script in Dynamo. (5) A filtering system was generated in Revit, using Dynamo, to filter occupants based on their attributes (such as age, gender, and most importantly, disabilities and physical conditions). The filtering system can be particularly helpful to the first responders to facilitate finding the location of disabled occupants. (6) The real-time tracking of the occupants was automatically visualized during the fire in both 2D and 3D views as shown in Figures 4 and 5. These figures show the locations of three occupants that were present in the office.

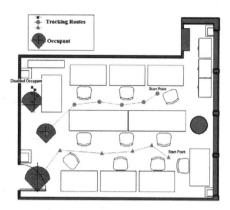

Figure 4. 2D view of tracking the occupants, evacuating the office during the fire alert.

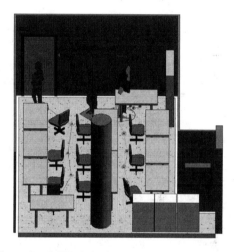

Figure 5. A 3D view of tracking the three occupants' locations at the 6th second of the evacuation.

The 3D model can be used to find the real-time condition of occupants. For instance, firefighters can find the occupants who are injured and/or laying down on the ground under the fire products such as heat, smoke and toxic gases.

5 CONCLUSION AND FUTURE WORK

This paper elaborated on the motivations and needs for modeling additional occupants' attributes, to have comprehensive real-time awareness in the case of disasters (such as a fire) in a building.

An extended IFC model formerly suggested by the authors, was updated here to identify the occupants' attributes, and relationships between sensors, occupants, occupancy, time series, and building components. A proof-of-concept prototype was described to make the dynamic

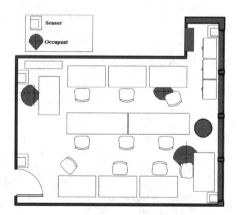

Figure 3. Revit model of the office and three occupants' locations before the fire alert.

model for a fire emergency by linking the real-time information to a BIM. The dynamic BIM, which can link the occupants' sensory data to BIM and visualize real-time occupants' tracking under a fire, was verified through a case study.

The proposed method improves the emergency management by visualizing real-time information related to the occupants. Furthermore, this research can improve the emergency first responders' perception by providing 3D visualization of the dynamic conditions of the building under fire and occupants' condition. Specially, the capability of creating filters on such a data-rich model, e.g., to identify the occupants with health condition or disabilities, can add a significant value to the process of safe evacuation. Introducing smarter filters to detect the occupants' mobility limitations by processing their motion tracks is an immediate extension of this work reported here.

Future work will also include the implementation and testing of the proposed method, including the interaction between the occupants and the simulated fire propagation during the building evacuation, as well as developing a platform for the next generation of intelligent building emergency management that can seamlessly integrate with BIM and IoT.

REFERENCES

Abrishami, S., Goulding, J., Rahimian, F.P., & Ganah, A. 2014. Integration of BIM and generative design to exploit AEC con-ceptual design innovation. *Information Technology in Construction*, 19: 350–359.

Agdas, D., & Srinivasan, R.S. 2014. Building energy simulation and parallel computing: Opportunities and challenges. *Simulation Conference (WSC)*, 3167–3175.

Ajayi, S.O., Oyedele, L.O., Akinade, O.O., Bilal, M., Owolabi, H.O., & Alaka, H.A. 2014. Ineffectiveness of construction waste management strategies: Knowledge gap analysis. *Smart, Sustainable and Healthy Cities*, 261.

Akinci, B., & Boukamp, F. 2003. Representation and integration of as-built information to IFC based product and process models for automated assessment of as-built conditions. *Nist Special Publication*, 543–550.

Al-Shalabi, F., & Turkan, Y. 2015. A novel framework for BIM enabled facility energy management: A concept paper. *Construction Specialty Conf. (ICSC 15)*.

Attar, R., Hailemariam, E., Breslav, S., Khan, A., & Kurtenbach, G. 2011. Sensor-enabled cubicles for occupant-centric capture of building performance data. *ASHRAE Transactions*, 117(2).

Becerik-Gerber, B., Jazizadeh, F., Li, N., & Calis, G. 2011. Application areas and data requirements for BIM-enabled facilities management. *Journal of Construction Engineering and Management*, 138(3): 431–442.

Bilal, M., Oyedele, L.O., Qadir, J., Munir, K., Ajayi, S.O., Akinade, O.O.,Pasha, M. 2016. Big data in the construction industry: A review of present status, opportunities, and future trends. *Advanced Engineering Informatics*. 30(3): 500–521.

Cahill, B., Menzel, K., & Flynn, D. 2012. BIM as a center piece for optimized building operation. *EWork and EBusiness in Architecture, Engineering and Construction ECPPM 2012*, 549–555.

Chen, J., Bulbul, T., Taylor, J.E., & Olgun, G. 2014. A case study of embedding real-time infrastructure sensor data to BIM. *Construction Research Congress 2014: Construction in a Global Network*, 269–278.

Davila Delgado, J.M., Butler, L.J., Gibbons, N., Brilakis, I., Elshafie, M.Z., & Middleton, C. 2016. Management of structural monitoring data of bridges using BIM. *Proceedings of the Institution of Civil Engineers-Bridge Engineering*, 170(3): 204–218.

Davila Delgado, J.M., Butler, L.J., Brilakis, I., Elshafie, M.Z., & Middleton, C.R. 2018. Structural performance monitoring using a dynamic data-driven BIM environment. *Journal of Computing in Civil Engineering*, 32(3), 04018009. Doi: 10.1061/ (ASCE) CP.1943-5487.0000749.

Davila Delgado. J.M., Brilakis, I., & Middleton, C. 2015. Open data model standards for structural performance monitoring of infrastructure assets. *Proceedings of the 32nd CIB W78 Conference.*

Eftekharirad, R., Nik-Bakht, M., & Hammad, A. 2018. Extending IFC for fire emergency real-time management using sensors and occupant information. *Proceedings of the International Symposium on Automation and Robotics in Construction.*

Fairgrieve, S., & Falke, S. 2011. Sensor web standards and the internet of things. Proceedings of the 2nd International Conference on Computing for Geospatial Research & Applications, 73.

Francisco, A., Truong, H., Khosrowpour, A., Taylor, J.E., & Mohammadi, N. 2018. Occupant perceptions of building information model-based energy visualizations in eco feedback systems. *Applied Energy*, 221: 220–228.

Gerrish, T., Ruikar, K., Cook, M.J., Johnson, M., & Phillip, M. 2015. Attributing in-use building performance data to an as-built building information model for lifecycle building performance management. *Proceedings of the 32nd CIB W78 Conference.*

Liebich, T., Adachi, Y., Forester, J., Hyvarinen, J., Richter, S., Chipman, T., Wix, J. 2013. Industry foundation classes release 4 (IFC 4). *Model support group.*

Motamedi, A., Soltani, M.M., Setayeshgar, S., & Hammad, A. 2016. Extending IFC to incorporate information of RFID tags attached to building elements. *Advanced Engineering Informatics*, 30(1): 39–53.

Park, J., & Cai, H. 2017. WBS-based dynamic multi-dimensional BIM database for total construction as-built documentation. *Automation in Construction*. 77: 15–23.

Srinivasan, R., Thakur, S., Parmar, M., & Ahmed, I. 2014. Toward a 3D heat transfer analysis in dynamic-BIM workbench. *iiSBE Net Zero Built Environment*, 356.

Volkov, A.A., & Batov, E.I. 2015. Dynamic extension of building information model for "Smart" buildings. Doi: https://doi.org/10.1016/j.proeng.2015.07.157.

Weise, M., Liebich, T., & Wix, J. 2009. Integrating use case definitions for IFC developments. *eWork and eBusiness in Architecture and Construction*. London: Taylor & Francis Group, 637–645.

Ma, Z., Wei, Z., Song, W., & Lou, Z. 2011. Application and extension of the IFC standard in construction cost estimating for tendering in China. *Automation in Construction*, 20(2): 196–204.

# Semantic BIM reasoner for the verification of IFC Models

M. Fahad
*Experis IT, Valbonne, France*

N. Bus & B. Fies
*Centre Scientifique et Technique du Bâtiment, Valbonne, France*

ABSTRACT: Recent years have witnessed the development of various techniques and tools for the building code-compliance of IFC models. Indeed these are great efforts, but, still there is a gap for the fully automatic building code-compliance. This paper presents our research and development of *Semantic BIM Reasoner (SBIM-Reasoner)* which employs semantic technologies to meet the requirements of semantic verification of an IFC model. *SBIM-Reasoner* employs several preprocessors (IFC to RDF converter, Geometry Extractor) to build the semantic repository from the input IFC model. Once all the triples are generated from the initial data (.ifc file), Stardog is used to build a knowledge graph for the semantic verification. All types of inference and reasoning mechanisms for the semantic verification are applied over this knowledge graph to meet the requirements of verification. Knowledge graph over triplets enables freedom of extending RDF based Semantic IFC model, creation of newer vocabulary and formation of newer rules, concatenation of triplets to build rules with condition and constraints over IFC data, dynamic reasoning over the triplets based on the initial data of IFC model, etc. Finally, we tested our prototype by using several online IFC models. We conclude that semantic technologies provide more rich mechanisms and answer vast types of queries for the verification of IFC models. It provides powerful features based on SPARQL libraries and serves best for the automated code compliance and verification of IFC models.

## 1 INTRODUCTION

*Building Information Modeling[1] (BIM)* is to understand a building through the usage of a digital model which draws on a range of data assembled collaboratively before, during and after construction [1]. BIM with its interoperability properties is intended to facilitate exchanges and handovers between different stakeholders. Whereas the visualization and geometric representation are intrinsic to the digital building model, the fields of quality requirements, evaluation and regulatory contextualization (destination, named areas, threshold values, certified data, evidence of compliance, etc.) need higher level of maturity [2]. *Industry Foundation Classes* (IFC), based on a neutral format, is a complete and fully stable open and international standard for exchanging BIM data [3]. *Code Compliance* checking of BIM is necessary in order to provide stake-holders a high quality *IFC* model that ensures accurate, consistent and reliable results in the entire life-cycle of BIM. Verification of IFC models for the code compliance checking is one of the hot challenges of the present decade. Different approaches and tools are already contributed for the automated code compliance checking [4].

Our enterprise, *CSTB*, through its research aims at automating *French Building Code Compliance* as much as possible, or at least improves the control of regulations from a digital model design phase. Its goal is to provide automatic requirements verification to warn the non-conformities with the associated 3D visualization, or to provide access to the technical documentation for a given digital model based on its sophisticated contextual information. This paper presents our several contributions towards this research. First, we analyze literature review regarding verification of IFC models and conclude that there are vast research works in this field, but, still there are many open challenges to address. Therefore, we present a need of an approach that can easily be extended, configured and deployed for the dynamic and changing environment having broad spectrum of functionalities for the verification of IFC models.

The main contribution of this paper is about the development of *Semantic BIM Reasoner (SBIM-Reasoner)* which employs semantic technologies to meet the requirements of semantic verification

---

[1] Open BIM, http://www.buildingsmart.org/openbim/.

of an IFC model. *SBIM-Reasoner* employs several preprocessors (elaborated in next sections) to build the semantic repository based on RDF [25] from the input IFC model. Once all the triples are generated from the initial data (.ifc file), it uses Stardog[2] to build a knowledge graph for the semantic verification. All types of inference and reasoning mechanisms for the semantic verification are applied over this knowledge graph to meet the requirements of verification, and in addition to discover additional information that is not explicitly stated in the initial data of the IFC model. Our semantic preprocessor uses both forward chaining and backward chaining mechanisms (where appropriate) to build the semantic repository. SPARQL[3] rules (statements and materialization) are applied to enrich the underlying semantic repository with several newer and high level concepts as per demand of regulation texts and verification rules by the end-users. Finally, SPARQL queries are performed over the semantic repository for the verification and code compliance of an IFC model. Once, *SBIM-Reasoner* finds non-compliant objects in the IFC model, it presents them to the end-user. Later in this paper, we also present our analytical results on several online IFC models[4] and also on four IFC models developed at our enterprise. We discuss our experimental finding on different analysis parameters, such as number of triplets in the RDF (turtle file) equivalent to IFC model, number of triplets in the semantic model (filtered turtle file) in the Stardog, estimated time taken by the conversion and geometric pre-processor, etc. On the basis of analysis from these parameters, we show encouraging results by several tests on the knowledge graph from the initial version of *SBIM-Reasoner*.

The rest of paper is organized as follows. Section 2 discusses related work. Section 3 presents our *SBIM-Reasoner*, its architecture, subcomponents, and as a semantic service to end-users. This section also presents statistical analysis of our implemented prototype, empirical results based on various IFC models and highlights various important points. Section 4 concludes the paper and presents future directions.

## 2 RELATED WORK

Over the last few years, many methods and techniques have been proposed for the verification of IFC models. There are three ways for the conformance checking of IFC models as discussed by Pauwels and Zhang [5]. The subsections elaborate each of them.

### 2.1 Hard coded rule checking

First, we have the *hard coded rule checking* mechanism for the verification of IFC models, which is similar to the approach adopted by Solibri Model Checker [6]. This tool loads a BIM model, considers rules stored natively in the application and performs rule checking against the BIM for the architectural design validations. This approach is fast as rules are integrated inside the application, but there is no flexibility or customization possible as rules are not available outside the actual application.

The traditional approach of compliance checking is with the *IfcDoc* tool [7] developed by buildingSMART International for generating MvdXML rules through a graphical interface. It is based on the MvdXML specification [22] to improve the consistent and computer-interpretable definition of Model View Definitions as true subsets of the IFC Specification with enhanced definition of concepts. This tool is widely used as AEC specific platform in the construction industry. MvdXML Checker [27] is a great contribution for the automatic verification of IFC models and to detect the non-conformities with the associated 3D visualization, or to provide access to the technical documentation for a given digital model based on its sophisticated contextual information. At our enterprise, we proposed several extensions and implemented those into a new research prototype. But after these extensions, still we analyze that this traditional approach of verification by the use of MvdXML is very limited and has narrow scope for the verification of IFC models. There are many drawbacks of MvdXML for extracting building views such as: lack of logical formalisms, solely consideration of IFC schema and MVD-based view constructors are not very flexible and dynamic [23]. In addition, major limitations exists such as restricted scope of applying conditions and constraints on several branches of an IFC model, poor geometric analysis of an IFC model, lack of mathematical calculations, support of only static verification of a model, etc. On the other hand, when we practice semantic technologies such as SPARQL, we think their suitability due to wide range of functions, intermediate calculations, and support of dynamic creation of verification rules at ease.

### 2.2 Query based rule checking

The second approach is *query based rule* checking of an IFC model. In this approach, BIM is interrogated by rules, which are formalized directly into

---
[2]Stardog triplestore: https://www.stardog.com/.
[3]SPARQL http://www.w3.org/TR/rdf-SPARQL-query/.
[4]IFC test Data https://github.com/opensourceBIM/TestFiles/tree/master/TestData/data.

SPARQL queries. As an example, Bouzidi et al. proposed this approach to ease regulation compliance checking in the construction industry [8]. They reformulated the regulatory requirements written in the natural language via SBVR, and then, SPARQL queries perform the conformance checking of IFC models.

### 2.3 Semantic rule checking approach

The third is a *semantic rule checking* approach with dedicated rule languages, such as SWRL [24], Jess [9] or N3 Logic [10]. There are few projects in AEC industries that use this approach for the formal rule-checking, job hazard analysis and regulation compliance checking. Wicaksono et al. [11] built an intelligent energy management system for the building domain by using RDF representation of a construction model. Then, they formulated SWRL rules to infer anomalies over the ontological model. Later, they also developed SPARQL interface to query the results of rules. Pauwels et al. built acoustic regulation compliance checking for BIM models based on N3 Logic rules [12]. They use N3logic rules with an ontology to reason whether a construction model is compliant or not with the European acoustic regulations. Another project that was built on the ontological framework for the rule-based inspection of eeBIM-systems was developed by Kadolsky et al. [13]. They used rules to query an IfcOWL ontology that captured a building.

Besides these projects that build an ontology for the IFC, recent years revealed some contributions based on *Semantic Web Technologies*. SWOP-PMO project is one of recent contributions that uses formal methodology based on the Semantic Web standards and technologies [14]. It uses OWL/RDF to represent the knowledge, and SPARQL queries and *Rule Interchange Format* (RIF) to represent the rules. The RDF/OWL representation is not derived from the written knowledge but has to be remodeled in accordance with the rules of OWL/RDF. There are some other works for the semantic enrichment of ontologies in the construction and building domain. Emani et al. proposed a framework for generating an OWL Description Logic (DL) expression of a given concept from its natural language definition automatically [15]. Their framework also takes into account an IFC ontology and the resultant DL expression is built by using the existing IFC entities. Fahad et al. have contributed a framework for mapping certification rules over BIM to enable the compliance checking of repository through the digital building model [16]. They aimed to align several specialized indexations of building components at both sides, by extending IfcOWL ontology with bSDD vocabulary (i.e., synonyms and description) as enriched IfcOWL ontology to deal with the same abstract concepts or physical objects. Fahad et al. also investigated semantic web approach by using SWRL and traditional approach by the use of IfcDoc tool and analyzed that the semantic web technique represent more global scope with larger visibility of querying for the validation of IFC models [17]. Ontologies play a vital role for the rule based semantic checking, therefore in the next subsection, we mention some of the important ontologies in the IFC domain.

#### 2.3.1 Ontologies in the IFC domain

To achieve the benefits of ontologies, there are many efforts to build an ontology for the IFC construction industry. One of the outcomes can be seen as an IFC-based Construction Industry Ontology and Semantic web services framework [18]. With simple reasoning built over the ontology, their information retrieval system could query the IFC model into XML format directly. The BuildingSMART Linked Data Working Group has developed IfcOWL ontology to allow extensions towards other structured data sets that are made available using semantic web technologies [19]. There are many versions of IfcOWL ontology since the work has been started. We have been working on an ontology IFC4_ADD1.owl that came on 25 Sept. 2015. We have enriched this ontology with English-French and IFC vocabulary (synonyms, descriptions, etc.) from bSDD semantic data dictionary in our research project where we map regulatory text and certification rules over BIM [16]. In addition, we assigned concepts of IfcOWL ontology with Global Unique Identifier (GUID) to serve as a unique language-dependent serial number from the bSDD. There are some other ontologies as well such as the ontology defining the core concepts of a building named Building Topology Ontology (BOT) [20] and the ontology for CAD Data and Geometric Constraints named OntoBREP [21].

## 3 SEMANTIC BIM REASONER

This section presents our research and development towards building *Semantic BIM Reasoner*. The following subsections elaborate its various aspects.

### 3.1 SBIM-Reasoner Architecture

*Semantic BIM Reasoner* (*SBIM-Reasoner*) deploys many preprocessors for building semantic repository for the verification of IFC models (see Figure 1). Primarily it has three pre-processors, i.e., IFC to RDF Converter, Geometry Extractor, and Semantic Preprocessor which has further sub-components named IfcOWL Ontology sub-graph, SPARQL Rules, SPARQL Queries and TripleStore.

### 3.1.1 IFC-to-RDF Converter+Filter

It is necessary for the semantic reasoner to convert IFC into RDF for building the semantic repository. *IFC-to-RDF* is a set of reusable Java component that allows parsing IFC files and converts them into RDF graphs. Our system deploys modified version of IFC-to-RDF conversion plug-in provided by Pauwels & Oraskari [26]. After conversion, underlying RDF acts as a foundation stone to execute all the verification rules. Therefore, we did filtration to get only relevant triplets from the IFC model. Generally there are two ways to get filtered model. First to get full RDF equivalent of IFC model and then apply *SPARQL Construct query* to get small graph of only wanted IFC classes. In this approach we found overhead of creating full graph and then extracting a sub-graph. Second, which we adopted is to integrate *IF-Then-Else* statements inside the code of Pauwels to filter unwanted elements like IFC classes {*Person, Address, MaterialList, SwitchingDeviceType, ColourRgb*, etc.}. By this filtration, we have noted that we got filtered RDF model which is 10 times smaller in size as compared to full RDF (equivalent IFC) model. Table 1 shows the comparison between RDF files, i.e., RDF equivalent IFC and Filtered RDF.

### 3.1.2 Geometry preprocessor

There are two geometry render engine plugins available with the *BIM Server* named *IFCOPENSHELL* and *IFC Engine DLL*. These are helpful to extract geometry data about the IFC objects. The outputs of this preprocessor are the RDF triplets which are formed from the extracted geometry data of relevant IFC objects. Table 2 shows the statistics of IFC objects present in the IFC models.

### 3.1.1 IfcOWL ontology sub-graph

As the standard IfcOWL ontology has a very large set of IFC elements, therefore, we deal with the sub-graph to achieve better processing and querying performance.

### 3.1.2 SPARQL Rules—Statements

We have created a large set of SPARQL rules, i.e., statements. In fact, these statements are shortcuts over the long chain of triplets to enable simplicity. For instance, we created *In_Storey* shortcut over the RDF triplets via relatedObjects and relatingObjects between IFC elements (see Figure 2). Likewise *Boundary* statement is created as a shortcut over the RDF triplets via relatingSpace and relatedBuildingElement. These statements promote readability, understandability and enable simplicity when creating SPARQL rules and queries. Otherwise the chain of triplets make things complex and ambiguous.

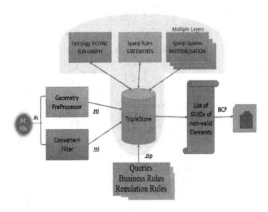

Figure 1. Top level architecture of *SBIM-reasoner*.

Table 1. Comparison between RDF files (Equivalent IFC vs. Filtered RDF).

| | | RDF size (Mb) | | No of triplets | |
| --- | --- | --- | --- | --- | --- |
| IFC model | IFC size Mb | Equivalent IFC | RDF – Filtered ttl | RDF Equivalent IFC | RDF – Filtered ttl |
| HITOS | 62.5 | 404.6 | 16.6 | 5,054,202 | 205,631 |
| AC9R1-Haus | 4.4 | 39.1 | 0.856 | 490,961 | 10,982 |
| BIM_EM | 1.9 | 21.3 | 0.149 | 275,927 | 2,138 |
| Candidat-23_04 | 28.9 | 56.6 | 11.4 | 694,709 | 132,115 |
| Chanteloup | 17.1 | 63.2 | 12.3 | 807,250 | 142,668 |
| LcD | 32.4 | 130.7 | 11.4 | 1627,750 | 132,845 |
| Bat_CSTB | 14.9 | 79.5 | 1.0 | 1047,216 | 13,973 |
| Maquette Test Checker | 11.2 | 122.6 | 7.7 | 1,572,792 | 100,230 |
| HAixFlowCtrl | 13.1 | 155.6 | 1.7 | 1,970,366 | 21,549 |
| Liberty_V3 | 12.2 | 137.2 | 6.6 | 1,760,207 | 86,545 |

Table 2. Statistics of IFC objects in IFC models.

| IFC model | Door | Wall | Wall (sd) | Window | Space | Stair | Slab | Covering | Open-El | Flow Seg | Flow Trml | Flow Ctrl | Railing |
|---|---|---|---|---|---|---|---|---|---|---|---|---|---|
| HITOS | 198 | 4 | 842 | 226 | 243 | 14 | 363 | 0 | 519 | 0 | 67 | 0 | 6 |
| AC9R1-Haus | 25 | 0 | 60 | 35 | 26 | 4 | 35 | 0 | 67 | 0 | 0 | 0 | 16 |
| BIM_EM | 9 | 0 | 33 | 48 | 17 | 0 | 16 | 0 | 58 | 0 | 0 | 0 | 0 |
| Candidat | 204 | 0 | 392 | 434 | 17 | 0 | 43 | 0 | 649 | 0 | 0 | 0 | 0 |
| Chanteloup | 153 | 0 | 328 | 77 | 166 | 3 | 46 | 0 | 228 | 0 | 0 | 0 | 21 |
| LcD | 256 | 0 | 634 | 74 | 360 | 5 | 6 | 0 | 424 | 0 | 0 | 0 | 63 |
| Bat_CSTB | 65 | 0 | 208 | 102 | 59 | 0 | 17 | 0 | 176 | 0 | 0 | 0 | 5 |
| HAixFlowCtrl | 0 | 0 | 0 | 0 | 0 | 0 | 0 | 0 | 0 | 0 | 0 | 1647 | 0 |
| Liberty_V3 | 85 | 0 | 599 | 136 | 68 | 5 | 12 | 47 | 321 | 2 | 15 | 5 | 25 |

*Wall(Sd) = StandardCaseWall, Open-El = OpeningElement, FlowSeg = FlowSegment, FlowTrml = FlowTerminal, FlowCtrl = FlowController.

```
[] a rule:SPARQLRule ;
      rule:content """
            PREFIX rdf: <http://www.w3.org/1999/02/22-rdf-syntax-ns#>
            PREFIX ifcowl: <http://www.buildingsmart-tech.org/ifcOWL/IFC2X3_TC1#>
            PREFIX rule: <tag:stardog:api:rule:>
            PREFIX list: <https://w3id.org/list#>
            PREFIX express: <https://w3id.org/express#>

            IF {
                  ?agr ifcowl:relatedObjects_IfcRelDecomposes ?element1 .
                  ?agr ifcowl:relatingObject_IfcRelDecomposes ?element2 .
            } THEN {
                  ?element1 ifcowl:inStorey ?element2  .
            }
"""  .
```

In_Storey Equivalent Shortcut in Stardog

```
[] a rule:SPARQLRule ;
      rule:content """
            PREFIX ifcowl: <http://www.buildingsmart-tech.org/ifcOWL/IFC2X3_TC1#>

            IF {
                  ?rel ifcowl:relatingSpace_IfcRelSpaceBoundary  ?space .
                  ?rel ifcowl:relatedBuildingElement_IfcRelSpaceBoundary  ?element .
            } THEN {
                  ?space ifcowl:boundary ?element  .
            }
""" .
```

Boundary Equivalent Shortcut in Stardog

Figure 2. Examples of statements over triplets.

```
CONSTRUCT {?s1 a ifcowl:circulationHorizontale}
{
  SELECT ?s1 ( COUNT(?s2) AS ?cnt )
  WHERE {
     ?s1 a ifcowl:IfcSpace ; ifcowl:boundary ?element ; ifcowl:inStorey ?storey .
     ?element rdf:type ?type .
     ?s2 a ifcowl:IfcSpace ; ifcowl:boundary ?element ; ifcowl:inStorey ?storey .
     VALUES ?type { ifcowl:IfcDoor ifcowl:IfcOpeningElement } .
     FILTER ( ?s1 != ?s2 )
         }
  GROUP BY ?s1
HAVING (?cnt > 2)
}
```

```
CONSTRUCT {?e a ifcowl:degagement}
WHERE {
            ?e a ?type .
   VALUES ?type { ifcowl:circulationVerticale ifcowl:circulationHorizontale
ifcowl:IfcDoor }
}
```

Figure 3. Examples of forward chaining SPARQL queries.

### 3.1.3 *SPARQL queries—materialization*

During the analysis of rules specification, we come across various types of vocabulary (introduced by regulatory texts) during building code compliance application. This vocabulary is composed of high level concepts present in business rules and regulation texts which are familiar by the stakeholders of BIM. There are two methods to build such vocabulary of newer high-level concepts, i.e.; via forwarding chaining and/or backward chaining. Based on the SPARQL rules, we have built SPARQL queries to introduce high level concepts based on the primary IFC vocabulary by using both forward and backward chaining where applicable. Figure 3 shows high level concepts *'circulationHorizontale'* and *'Degagement'* in our case study of building French code compliance via forward chaining. In these examples we have used our defined *inStorey* and *boundry* concepts along with the pre-defined IFC owl terminologies such as *IfcOpeningElement*, *IfcSpace*.

Backward chaining consists of ontology statements that align IFC concepts with regulatory concepts, whereas forward chaining consists of insert statements that create supplementary triplets. Forward chaining is a good at an implementation stage to save memory and CPU resources. From the machine point of view, backward chaining is processed each time a semantic query is submitted whereas forward chaining is executed each time the data changes. At this stage, this choice is a compromise between effective queries (forward chaining is more appropriate for complex and numerous queries) and model update frequency (backward chaining is more appropriate when data changes frequently). We can even say that it is a compromise between the amount of triplet (considering triplet generated by forward chaining statements) and the ontology complexity. Therefore, we have mixed both approaches Backward and Forward chaining to provide the optimal setting that minimizes response time and maximizes ontology consistence. With the help of these techniques, we have simplified several IFC patterns such as classifications, predefined types, properties, geometry, topology, etc.

### 3.1.4 TripleStore—stardog

Although IFC is an open standard; its complex nature makes information retrieval difficult from an IFC model as the size of IFC model grows. Therefore, we have used Stardog as a triplestore to build our semantic model. Querying semantic model is faster and gives a good run-time. When the application starts, an end-user provides an IFC model and the set of SPARQL queries which are the verification rules for checking code compliance of desired IFC model. As a result, our system converts IFC file into filtered RDF model. It loads that converted-filtered IFC equivalent RDF into stardog. After triplets concerning geometry are added to capture geometrical information in the triplestore. Then the semantic model is enriched with IfcOWL basic vocabulary, i.e., sub-graph of IFC ontology. Then, it adds SPARQL rules into the triplestore. Finally, it executes our project specific forward chaining SPARQL queries which creates high level vocabulary and builds further RDF graphs over the existing triplets. With reference to above examples, our semantic preprocessor is illustrated in Figure 4 based on the basic IFC vocabulary, shortcuts and constructs.

### 3.1.5 End-User Queries

Stardog provides fast access to triplets to fetch data to validate IFC models. All the end-user verification queries are executed on the top of final stardog triple store which is built successively by our reasoner. For example if an end-user needs to check whether a Room in the IFC model is accessible by a wheelchair, then it can be done easily by using above explained construct_degagement where the value of IfcDoor must be equal or greater than 90 cm. An end-user may apply SPARQL ASK and DESCRIBE queries to retrieve relevant information regarding the verification rules. Instead of using IfcDoc tool where there is no intermediate state and no explanation for the reason of non-compliance, we use SPARQL DESCRIBE Queries. The SPARQL DESCRIBE query does not actually return resources matched by the graph pattern of the query, but an RDF graph that "describes" those resources. It is up to the SPARQL query service to choose what triples are included to describe a resource. Therefore, SPARQL queries serve best by concatenating desired triplets for building verification rules to check the code compliance.

## 3.2 SBIM-Reasoner as a semantic service

We have developed *SBIM-Reasoner* as a semantic service inside a KROQI platform[5]. As it is developed especially for the *French building code compliance*, therefore our web interface is also in French targeting French community. When the applica-

Figure 5. Result page containing status of verification rules.

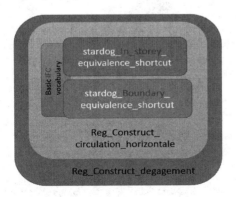

Figure 4. Example of construction of semantic preprocessor.

Figure 6. List of non-compliant IFC objects.

[5] https://svc-bim-semchecker.dev.coplus.fr/CheckerService-/v2/ui?file_id= test_20180302_NR_LibertyLoft_OK.

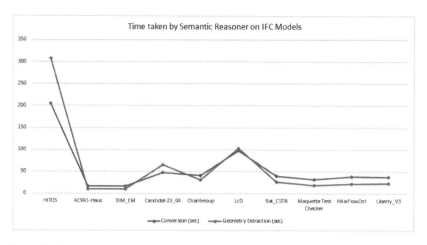

Figure 7. Time taken by preprocessors (Converter and geometry extractor).

tion starts, an end-user has to configure IFC input model by clicking under the synchronization button on the very first tab 'Maquette'. Then an end-user selects the set of rules to be verified on this input model by selecting/browsing their set of rules on the second tab 'Protocoles'. Then our web service starts by calling semantic reasoner which computes the set of rules and redirects to the result page 'Résultats'. Each of the rule is highlighted as green or red color depending on its status of compliance (see Figure 5).

When *SBIM-Reasoner* has detected non-compliant objects (in case of red status), an end-user can further analyze them by clicking on the corresponding row. It fetches and displays the list of IFC non-compliant objects containing Name, GUID and Type of each IFC object (see Figure 6). One can also export PDF and BCF files to analyze their results in detail.

### 3.3 Testing IFC models

We have used several IFC models from the online repository (see url above) to test our *Semantic BIM Reasoner*. These IFC models vary in size, number of IFC objects, free spaces, etc. We have also used four IFC models developed at our CSTB enterprise named Bat_CSTB (14.9 MB), HAixFlowCtrl (13.1 MB), Maquette Test Checker (11.2 MB) and Liberty (12.2 Mb). Above in Tables 1 and 2 we have shown statistics about these IFC models. We have also measured time taken by the *Converter Preprocessor* that does conversion and filtration to produce the initial RDF semantic model on a 'quad core i5 CPU at 2,5Ghz' machine. In addition, we have also measure time taken by the *Geometry Preprocessor* that extracts geometry data of our desired IFC elements (ref. Table 2) from the IFC models. Figure 7 shows the estimated time taken by the conversion and geometry pre-processors. When we see the time graph, we observe that *SBIM-Reasoner* took less than a minute by both the preprocessors to achieve their objectives when the size of IFC model is under 15 MB. But when the size of IFC model is 62 MB (in case of huge IFC model Hitos) then it took almost 3.4 minutes to convert and filter, and more than 5 minutes to extract geometry data. Here, we also mention that although preprocessors took time to build semantic model at the first time, but querying for the verification of IFC models are processed fast and a good run-time is achieved. On the other hand on traditional IFC model, it takes much time to verify each of the individual rule. In addition, we are not able to execute all types of rules as per our desire due to narrow scope of IFC tools available online. This is only with the semantic model that we are flexible enough to fetch any triplets and build rules according to our will for the verification of IFC models. On the basis of this analysis, we conclude that semantic model serves best for the verification of IFC models. We also revealed encouraging results via several tests from the initial version of *SBIM-Reasoner*.

## 4 CONCLUSIONS

There are many techniques for the automatic verification of IFC models, but, still there are many open challenges. In this paper, we have presented a semantic approach that can easily be extended, configured and deployed for the dynamic and changing environment having broad spectrum of functionalities for the verification of IFC models. We presented *SBIM-Reasoner* that builds semantic model by converting IFC model into RDF and also extracting geometry data as a set of triplets. Then, it loads this semantic RDF data into Stardog triplestore, and enriches the primary semantic model with our defined SPARQL Rules (shortcuts overs long chain of triplets) and Queries (regulation and business rules). End-user queries are formalized as SPARQL queries which are executed on the top of this final triplestore. *SBIM-Reasoner* responds their status of compliance along with the BCF representation of non-compliant IFC objects. We demonstrated several test models on *SBIM-Reasoner* and presented its efficiency and efficacy with empirical results. We conclude that the semantic model based on the semantic web technology is a good compromise between development efforts and opportunities. The graphical representation of RDF allows rules to be more intuitive and more efficient to reason and execute. Concatenation of triplets allows flexibility of making wide range of verification rules with condition and constraints at ease. SPARQL has a global scope with larger visibility of querying with the built-in functions and support of intermediate calculations for the validation of IFC models.

REFERENCES

[1] Rebekka, V., Stengel, J. & Schultmann, F. 2014. "Building Information Modeling (BIM) for existing buildings—Literature review and future needs." Automation in construction, vol. 38, pp. 109–127.

[2] Eastman, C., Teicholz, P., Sacks, R. & Liston, K. 2008. "BIM Handbook: A Guide to Building Information Modeling for Owners, Managers, Designers, Engineers, and Contractors", Hoboken, New Jersey, Wiley.

[3] Thein, V. 2011. Industry Foundation Classes (IFC), BIM Interoperability Through a Vendor-Independent File Format, A Bentley White Paper, September 11.

[4] Ismail, A.S., Ali, K.N. & Iahad, N.A. 2017. "A Review on BIM-based automated code compliance checking system," 2017 International Conference on Research and Innovation in Information Systems (ICRIIS), Langkawi, pp. 1–6.

[5] Pauwels, P. & Zhang, S. 2015. "Semantic Rule-Checking for regulation compliance checking: An overview of strategies and approaches". Proc. of the 32nd CIB W78 Conference, Netherlands.

[6] Khemlani, L. 2009. "Solibri model checker", AECbytes Product Review 31st March 2009.

[7] IfcDoc Tool, available at: http://www.buildingsmart-tech.org/specifications/ specification-tools /ifcdoc-tool/ifcdoc- help-page-section/IfcDoc.pdf, 2012.

[8] Bouzidi, K.R., Fies, B., Faron-Zucker, C., Zarli, A. & Thanh, N.Le. 2012. "Semantic Web Approach to Ease Regulation Compliance Checking in Construction Industry". *Future Internet*. 4 (3). pp. 830–851.

[9] Friedman-Hill, E. 2003. "Jess in Action: Rule Based Systems in Java". Manning Publications. ISBN 1-930110-89-8.

[10] Berners-Lee, T.I.M., Connolly, D.A.N., Kagal, L., Scharf, Y. & Hendler, J.I.M. 2008. "N3 Logic: A logical framework for the World Wide Web", Theory and Practice of Logic Programming. 8 (3), doi:10.1017/S1471068407003213.

[11] Wicaksono, H., Dobreva, P., Häfner, P. & Rogalski, S. 2013. "Ontology development towards expressive and reasoning-enabled building information model for an intelligent energy management system". Proc. of the 5th KEOD, pp. 38–47. SciTePress,

[12] Pauwels, P., Deursen, D.Van, Verstraeten, R., Roo, J.De, Meyer, R.De, De-Walle, R.Van, & Van-Campenhout, J. 2011. "A semantic rule checking environment for building performance checking". Automation in Construction. 20 (5). pp. 506–518.

[13] Kadolsky, M., Baumgärtel, K. & Scherer, R.J. 2014. An ontology framework for rule-based inspection of eeBIM-systems. Procedia Engineering. vol. 85, pp. 293–301.

[14] Josefiak, F., Bohms, H., Bonsma, P. & Bourdeau, M. "Semantic product modelling with SWOP's PMO, eWork and eBusiness in AEC", pp. 95–104

[15] Emani, C.K., Ferreira Da Silva, C., Fiès, B., Ghodous, P. & Bourdeau, M. 2015. "Automated Semantic Enrichment of Ontologies in the Construction Domain". Proc. of the 32nd CIB W78 Conference, Netherlands.

[16] Fahad, M., Bus, N. & Andrieux, F. 2016 "Towards Mapping Certification Rules over BIM", Proc. of the 33rd CIB W78 Conference, Brisbane, Australia, 2016.

[17] Fahad, M., Bus, N. & Andrieux, F. 2016, "SWRL Towards Mapping Certification Rules over BIM", Proc. of the 33rd CIB W78 Conference, Brisbane, Australia.

[18] Zhang, L. & Issa, R.R. 2011. "Development of IFC-based Construction Industry Ontology for Information Retrieval from IFC Models", International Workshop on Computing in Civil Engineering, Netherlands, Vol. 68.

[19] Terkaj, W. & Pauwels, P. 2014. "IfcOWL ontology file for IFC4", Available at: http://linkedbuildingdata.net/resources/IFC4_ADD1.owl.

[20] Rasmussen, M.H., Schneider, G.F. & Pauwels, P. Building Topology Ontology (BOT), https://github.com/w3c-lbd-cg/bot.

[21] Perzylo, A.C., Somani, N., Rickert, M. & Knoll, A. 2015. An ontology for CAD data and geometric con-straints as a link between product models and se-mantic robot task descriptions. In proceedings of IROS'15: 4197–4203E.

[22] Chipman, T., Liebich, T. & Weise, M. 2012. "mvdXML specification of a standardized format to define and exchange MVD with exchange requirements and validation Rules", version 1.0.

[23] Mendes de Farias, T., Roxin, A. & Nicolle, C. 2016. "A Semantic Web Approach for defining Building Views", buildingSMART Summit Jeju, Korea.

[24] Horrocks, I., Patel-Schneider, P.F., Boley, H., Tabet, S., Grosof B. & Dean, M. 2004. "SWRL: A Semantic Web Rule Language Combining OWL and RuleML".

[25] Brickley, D., Guha, R.V. & McBride B. 2004. "RDF vocabulary description language 1.0: RDF Schema". W3C Recommendation 2004.

[26] Pauwels, P. & Oraskari, J., "IFC-to-RDF Converter" https://github.com/IDLabResearch/IFC-to-RDF-converter.

[27] MvdXMLChecker, available at: https://github.com/opensourceBIM/mvdXMLChecker.

# Modular concatenation of reference damage patterns

A. Hamdan & R.J. Scherer
*Technische Universität Dresden, Dresden, Germany*

ABSTRACT: Although, catalogues and collections exist which contain knowledge about damage and common patterns that could enable the application of software assessment methods, recorded damage data must still be evaluated by experts. For this reason, an approach is developed and discussed in this article that allows the semi-automatic conclusion of damage causations by the input of inspection data and vice versa. Therefore, an ontology is created by using the Web Ontology Language (OWL) which is structured in the three knowledge domains for+ damage-, structure-, and causational elements, so that unidentified information could be retrieved by using semantic reasoning. Furthermore, the damage is structured in reference damage patterns which allow computer-aided modelling of damage based on pre-defined conditions or parameters. Additionally, retrieved data can be mapped to a generic damage model, which can be linked with external data sources such as measurement datasets as well as BIM models.

## 1 INTRODUCTION

Through innovations in the fields of laser scanning, photogrammetry and drone technology, new ways for discovering damage to structures are developed which supplement existing inspection methods. By using these technologies inhomogeneities in the structures surface are detected as defects which then could be processed into a graphical representation or specific models, e.g. finite-element models via photogrammetric 3D reconstruction for further evaluation (Mongelli, et al., 2017). However, an automated evaluation of the recorded damage is only limited to numerical analysis for assessment of the structural behavior. Consequently, an interpretation with regard to the type of damage or its causation is not supported and must be performed manually by experts, despite comprehensive collections of damage knowledge exist, such as (Bundesministerium für Verkehr und digitale Infrastruktur, 2017) or (Javor, 1991). Based on these catalogues that contain not only relevant information and rules, but also common damage patterns, an automated retrieval of damage objects or the conclusion of its causes would be possible. To enable a computer-aided classification and reasoning of detected structural damage, templates could be utilized for describing damage patterns based on the reference model approach.

Therefore, the purpose of this research is the development of a method for utilizing existing knowledge about damage and its causations to identify predefined reference damage patterns. In order to accomplish this, the patterns are stored in an ontology together with information about the structure as well as the causations and influence factors. By using rule engines, such as Drools (Red Hat Inc., 2018), the damage patterns in the ontology could be retrieved.

In this paper the concept of reference damage patterns is presented as well as their application and retrieval in an ontology written by using the Web Ontology Language OWL (W3C, 2012).

## 2 REFERENCE DAMAGE PATTERN

The individual defects that appear during the life cycle of civil structures, e.g. cracks or spalling, normally merge with other damages to form damage patterns. These damage patterns often consist of a complex geometry, so that for modeling damage in structural analysis, simplification methods need to be used, such as the smeared crack concept. The utilization of more accurate approaches, such as the discrete crack concept requires much more processing power (Espinosa, et al., 1998)and effort in modeling which is the reason why these methods in general are only used for damage of simple geometry. A possible approach for reducing the effort of modeling damage patterns would be the utilization of reference modeling techniques. Reference models are information models, which are developed with the purpose of being reused for different but similar purposes (Becker, et al., 2007). By using reference models of damage patterns, named Reference Damage Patterns (RDP), it would be possible to predesign the geometry as well as other relevant attributes in a parameterized model. Initialized versions of this model can then be used to represent the damage patterns in an accurate way without modeling each individual damage repetitively.

## 2.1 Basic structure of reference damage patterns

The concept developed in this research uses a modular structure in which an RDP *rdp* can contain multiple other elements. In this regard, these elements can either be subordinated RDP that describe the damage in a more detailed level or Elementary Damages *ed* that contrary to RDPs can no further subdivided into smaller elements and therefore define the basic damage elements in the system. Additionally, Connection Elements *ce* are necessary for linking the contained data objects with each other.

$$rdp := (\{rdp_1,...,rdp_n\},\{ed_1,...,ed_m\},\{ce_1,...,ce_m\})\,|\,$$
$$(n+m) \geq 2, m \geq 1$$
(1)

All three element types are defined as Data Elements *de* and must therefore possess a unique identificatory *id* to enable the possibility of referencing them by software applications or other external objects.

$$rdp, ed, ce \in de := \{id\}$$
(2)

$$\forall id = id(de): id_i \neq id_j \,|\, i \neq j$$
(3)

An Elementary Damage consists of multiple datasets that should define a damage in its smallest scale inside the macro level of the components structure, so that no further subdivisions are necessary. It defines through multiple geometrical parameters *geo* a damage representation. Additional properties, such as mechanical parameters or meta-data can be added by defining an arbitrary number of attributes *a*. Since Connection Elements also possess geometrical parameters as well as attributes, both element types are assigned to a super type called Geometrical Element *ge*.

$$ed, ce \in ge := (\{geo_1,...,geo_n\},\{a_1,...,a_m\})\,|\,$$
$$n \geq 1, m \geq 0$$
(4)

An example for an Elementary Damage would be a single linear crack inside a structure which possesses geometrical data in the form of one start—and end-point as well as a parameter for the width. By concatenating these elements with each other, an RDP can be created.

For combining Elementary Damages as well as RDPs with each other, additional Connection Elements are used. An RDP must consist of at least one Connection Element which works as a link between 2...n objects that can be RDP or Elementary Damages elements. Therefore, each Connection Element defines the linked objects through their identificators. Furthermore, a geometric representation of the Connection Element must be set. This could be for example a single point for connecting two adjacent cracks with each other. Additional attributes, similar to those in Elementary Damages, can also be defined in Connection Elements.

$$ce := (\{rdp_1(id),...,rdp_n(id)\},\{ed_1(id),...,ed_m(id)\})$$
$$(n+m) \geq 2$$
(5)

The aforementioned data objects form the basis for the concept of RDPs (see Fig. 1) which enables the modular concatenation of an arbitrary amount of predefined Elementary Damages through corresponding Connection Elements.

The resulting RDP can then be linked to other Elementary Damages or RDPs according to the same principle.

## 2.2 Elementary damage classification

In general, RDPs are only container elements that store existing damage data or link them by using Connection Elements. Consequently, the only elements that provide concrete damage information are Elementary Damages. In order to support the damage classification which is essential in the context of the assessment, it would make sense to perform categorizations at the level of Elementary Damages. Therefore, multiple subclasses are designed which classify damage based on the grade of material degradation. Contrary to other categorization systems such as in (DIN EN 1504–9:2008-11, 2008) or (Bertolini, et al., 2013) these damage types are not based on the external influences, e.g. mechanical or chemical effects. Instead, the damage is classified into four types based on physical symptoms in the structure of the assessed component (see Table 1).

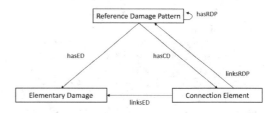

Figure 1. Basic concept of defining a reference damage pattern.

Table 1. Subclasses of elementary damage, including further exemplary subdivisions.

| Elementary damage subclass | Subtypes |
|---|---|
| Degradation damage | Crack, Spalling, ... |
| Deterioration of material parameters | Corrosion, Rot, ... |
| Component deformation | Shrinking, Swelling, ... |
| Biological infestation | Root, Animals, ... |

This classification is more suitable for modeling damage, when its cause is still not identified which happens regularly in the early phase of the damage assessment. Since the goal of this research is also to conclude the cause of damage by a reasoning system, it must be assumed that only the physical behavior and visual features of the damage are observed and therefore transformed into a digital representation. Similar to the classification scheme of (Baradan, et al., 2010) these classes can be referred to external influences if sufficient information about the damage is available. The four listed classes of Elementary Damage can be subdivided further into more differentiated sub-classes.

Degradation Damage includes all damage that causes a defect in a component, where material has been ablated. Examples for subclasses of this damage type could be Crack or Spalling. This type of damage occurs often in concrete structures, but other materials such as masonry or steel can be affected as well.

By using Deterioration of Material Parameters for categorizing damage, mostly chemical effects such as corrosion or rot in wooden materials are covered. The main characteristic of this class is that the damage does not ablate the material or at least not to any significant extend, thus reduce the strength. This kind of damage can be modelled at best through a reduction of mechanical parameters in a predefined area, similar to the smeared crack concept.

Component Deformations include damage that leads to an alteration of the geometrical parameters of the affected component, without ablating the material. This could be for example the hogging or sagging of concrete but also shrinking or swelling is covered. In general only plastic deformations are assessed, when analyzing structural damage. Since deformations often lead to degradations such as cracks, it can also be counted as a mechanical effect. Nevertheless, deformations always result from other external effects, mostly through changes in the environment such as temperature or through higher traffic loads.

Lastly, Biological Infestation covers damage which leads to vegetation in the component, ver-

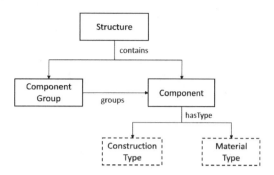

Figure 2. Structure of the components ontology.

min infestation or similar. Often these damage is combined with Degradation Damage, since the structure is damaged through roots or petrification from organisms.

A scheme that subdivides the four damage types into further subclasses is shown in Fig. 2. Nevertheless, the scheme is not final and can be extended, depending on the level of detail which the user needs. Additionally, these subclasses can be also divided into more types, e.g. cracks can be categorized into types such as hairline, diagonal or horizontal cracks.

### 2.3 Serialization of reference damage patterns

Since the primarily focus of the RDP ontology is not the creation of detailed damage models but the structuring and modular concatenation of predefined defects it makes sense to transform the retrieved Elementary Damages into an appropriate data model. In order to utilize the aforementioned concept for computer-aided modeling and evaluation of damage a machine-readable format must be used for serialization. Open data exchange formats such as the Extensible Markup Language (XML), the JavaScript Object Notation (JSON) or even the standardized BIM format, the Industry Foundation Classes (IFC) are suitable for serializing information which can be processed by software applications. Nevertheless, for supporting semantic reasoning and storing knowledge data, as aimed by this research, a format for defining ontologies is preferable. Therefore, several languages for describing ontologies exist, e.g. the Knowledge Interchange Format (KIF), Resource Description Framework (RDF) or the Web Ontology Language (OWL). The structure of RDPs in this research is based on ontologies formalized in OWL (W3C, 2012) which according to (Kalibatiene, et al., 2011) claims to be the most popular, spreading and applicable ontology language.

Models for digitalizing damage in civil structures have already been developed in previous studies, although the aspect of semantics as well as modularizing damage was not in focus. In this context (Hammad, et al., 2006) developed a software prototype that links all information about the life cycle stages of a bridge to a 4D model. A specific domain for storing damage information, e.g. videos or images is implemented and can be accessed through corresponding instruments in the graphical user interface.

In addition, as part of the SeeBridge project, an IFC-extension has been developed which supports the modelling and visualization of bridge defects and damages in a software prototype. Therefore, the relevant information is determined by data from previously recorded point clouds (Sacks, et al., 2018).

Furthermore, a generic damage model for the digitalization of structural damage has been created by (Hamdan, et al., 2018), which supports the recording of inspected damage elements in a data format serialized in XML. The damage model consists of three layers in a hierarchical data structure on which damage can be modelled. The top-level consists of representative elements for the building components which can be linked with an BIM model and are used for describing the overall damage state and grade. Damage areas are assigned to these elements at the second level which define the damage through reduced mechanical parameters, similar to the Deteriorated Material Parameters in the RDP system. Inside these areas, additional elements can be defined that describe individual damage in a more detailed level. Additional documents, images and other external files can be linked with these elements, allowing for using the damage model also as a database for damage data.

In this research, the RDPs in the OWL ontology are linked with elements from the generic damage model by (Hamdan, et al., 2018) in order to remain interoperability between other file systems by utilizing the Multi Model approach by (Fuchs, et al., 2010).

## 3 ONTOLOGIES FOR DAMAGE RETRIEVAL

For the application of RDPs, a damage retrieval system has been conceived which outputs either damage based on given external influences on a structure or vice versa the causation of damage concluded from the inspected data. Therefore, two additional knowledge domains are developed and interlinked with the predefined RDP ontology through rules. The first domain describes information about the building structure. For this purpose, the component types are categorized in an ontology, classified by attributes such as material, geometry, function, age and other properties. In the other domain, the multiple influences which can cause damage are described in an ontology. By combining these two domains with each other, conclusion could be made by using corresponding rule engines. Therefore, in this research the rule engine Drools has been used.

### 3.1 Components ontology

The properties of the damaged structure have a significant influence on the identification of damage and its causes. Because of this, information about relevant components are stored in a corresponding ontology.

A fundamental distinction between different types of structures must first be made, such as buildings, bridges or tunnels, because although similar damage patterns can occur on each of them, they differ considerably in terms of construction, use and environment. Based on this differentiation, numerous other criteria exist, according to which the structural components can be classified. Two categories, the construction type and the material were used to classify the components. For these categories, two separate classes are used that allow the addition of attributes depending of its classification. Similar to RPDs, the components can be combined to a group which reduces the level of detail for the application of simpler reasoning mechanisms.

Additionally, each element possesses attributes that are essential for semantic reasoning, such as age, used construction techniques or region.

### 3.2 Ontology for damage causes

The factors that lead to the structural deterioration can be classified into four fundamental groups (Radomski, 2002):

– Inner Factors
– Traffic Load Factors
– Weather and environmental Factors
– Maintenance Factors

In this regard, inner factors refer to components in the structure itself that contain some factors of degradation, e.g. inadequacy of the structural design, material quality or age. These factors are already covered by the Components Ontology and therefore not represented in this domain. The same applies to maintenance factors that are related to the quality and intensity of preservation measures. However, traffic load factors as well as weather and environmental factors have been considered in the ontology, thus it does not focus on these factors entirely but rather tries to describe the damage causations. Therefore, a different classification scheme is necessary.

Despite its exclusion in the classification of damage, the ontology for describing damage causes is based on (DIN EN 1504-9:2008-11, 2008). Accordingly, the causes are subdivided into four types:

- Mechanical Cause
- Chemical Cause
- Physical Cause
- Biological Cause

Similar to Elementary Damage, these types can also be classified into further subtypes. In general these elements are damaging processes, which are already classified in common literature. For example, the alkali-silica reaction of concrete can be classified which occurs, if a humid environment wets the material. In this case, other factors such as rain or rising water could be defined as the original external influence. However, it is unclear, to which extend or how long the rain impinges as well as other parameters, e.g. the PH value of the water. Because of its unpredictability, rain and similar effects should be used only as an upper categorization class and initialized elements must be more detailed and unambiguous.

Biological causes share similarities with Biological Infestation elements from the RDP system. However, in this ontology the focus lies on the causation of damage and not its effects. Sometimes, this could be to some extent the same, e.g. when vermin or insects occur, they are not the damage itself, but cause it. However, when identifying damage, these vermin are often related as the damage itself, because their appearance in general leads to losses in the components structure.

### 3.3 *Ontology unification*

The two aforementioned ontologies must be connected with each other in order to process reasoning mechanisms for damage retrieval. The assertion components of both ontologies which refer to individual BIM models and their linked data for environment, age, loads, etc., are then connected via rules to the predefined RDP knowledge base through object properties. In order to manage the unified ontology structure, new superclasses have been created which are assigned to the domains, for a clearer assignment of damage data. As a result, elements from the damage domain, such as RDP or Elementary Damage are subclasses of Damage Element, while elements from the components ontology are classified as Structural Elements. The elements of the damage cause domain have no additional superclass, since its ontology already consists of only one superclass, which subdivides into the whole damage cause data structure. In Fig. 3 the framework of the unified ontology is

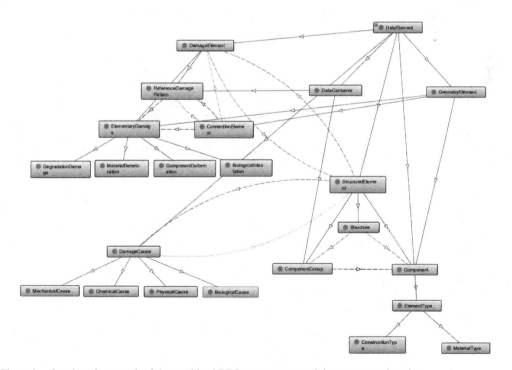

Figure 3. Ontology framework of the combined RDP, components and damage cause domains.

shown. Altough, its classes can be used for creating individuals, the current ontology serves only as a basic for further extensions that support specific fields, such as the construction of concrete buildings or steel composite bridges. For a detailed knowledge base, classification on a higher degree of differentation should be applied, so that classes that describe constructions from specific eras or regions would be implemented.

## 4 CONCEPT OF DAMAGE RETRIEVAL

Based on the ontology framework developed in this research, RDPs can be modeled as templates for recorded damage. However, in order to be applied or linked with a BIM model as damage patterns, the templates need to be instantiated as individuals. A similar problem has been solved by (Benevolenskiy, 2015) via using two separate ontologies for reference process patterns. Therefore, a pattern ontology is used for defining the templates as individuals. Afterwards an instance ontology is utilized for instantiating the predefined patterns by adding individual properties from the BIM model and other sources, such as an identificator, name or date. By using this method, it is possible to organize the templates in a knowledge base while retaining the ability to configure the instances of the patterns and linking them with other models. However, the use of two separate ontology systems requires an appropriate management framework and the retrieval process could become inefficient.

For this reason, another approach is used (see Fig. 5) which utilizes the Multi Model approach by (Fuchs, et al., 2010) and the linked Variation Model by (Luu, et al., 2018). Therefore, RDPs are instantiated in an ontology as predefined elements for multiple applications. For each damaged structure, specific ontologies for structural elements and causes are instantiated that are also

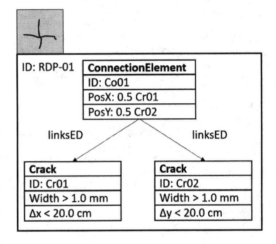

Figure 5. RDP consisting of two elementary damages.

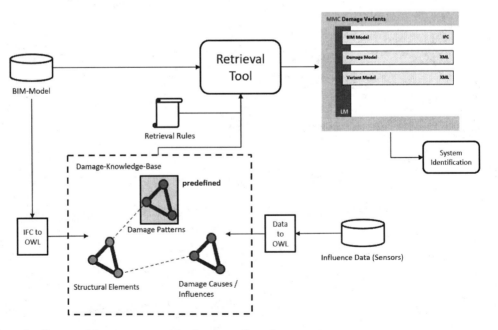

Figure 4. Concept of damage pattern retrieval and—configuration.

connected with each other through object properties. By using rules, the structural ontology is then connected with the RDP ontology and a damage model can be concluded together with a Variation Model which proposes possible damage and defects in the structure. Both of these models are connected with the BIM model through a Link Model. By performing a system identification, the multiple variants can be reduced to the most suitable results.

### 4.1 *Instantiation of reference damage patterns*

In order to create RDPs, the aforementioned concept of extending the developed ontology framework is utilized. The creation workflow of RDPs can be described in three steps:

1. Creation of Elementary Damages
2. Configuration of RDP
3. Connecting with structural elements and damage causes through rules

In Fig. 5 an example is shown that describes an RDP, consisting of two Elementary Damages and a Connection Element in the middle, so that a crack is modelled which forms a cross.

By combining multiple RDPs with the same configuration but a different position, a map-like structure is defined as crack pattern (see Fig. 6).

The resulting RDP can now be connected with structural elements and damage causes (see Fig.7). Since this is a reticulated crack pattern that could occur because of alkali-silica reactions, it can be concluded that it is on components that are exposed to moisture. For example, this is the case on bridge decks.

By using this approach, elements can be continuously added to ontology through importing knowledge and inspection results.

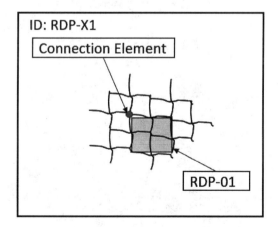

Figure 6. RDP consisting of multiple connected RDPs.

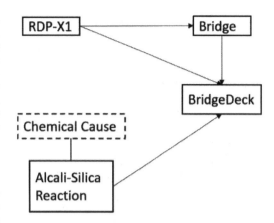

Figure 7. Connecting RDP with structural elements and damage causes.

### 4.2 *Retrieval of ontology elements*

The ontology is structured in such a way that at least two data elements from different domains are necessary, in order to conclude a third data element. Referring to the example in section 4.1, an alkali-silica reaction could occur on a bridge deck which concluded that RDP-X1 is present on this component. For this purpose, the rule engine Drools can be used to conclude unknown elements. By using Drools, it is also possible to modify existing values in the ontology, as well as creating new ones.

The elements in the pattern ontology have no definite values, but rather value ranges. Only in the instance ontology, concrete values are defined for the objects. Therefore, information from external models and data sources could be imported and adapted in the ontology during an additional implementation step.

## 5 CONCLUSION

In this paper an approach has been presented utilizing an OWL-ontology for identifying and classifying damage patterns based on the structure, in which they are positioned as well as deduce the damage causes.

Therefore, the ontology is subdivided in three domains which categorize damage-, structural— and damage cause-elements as well as linking three elements (one of each domain) with each other, so that unknown elements to two known elements can be queried by using the rule engine Drools. Reference Damage Patterns are used for describing complex combinations of Elementary Damages through Connection Elements. Since the templates in the TBox are not instantiated, two ontologies are utilized, one for initialized templates, and another one for the instances of the template-instances.

The concept of RDP has several advantages, but also some disadvantages, when identifying damage patterns. The major advantage is the modularization of damage for computational processing. Modelling damage is always difficult in terms of using an appropriate level of detail, since also many unknown factors remain after a structural inspection. However, by utilizing knowledge-based systems, multiple factors can be concluded. The approach of identifying damage causes by predefined RDP is also useful for the application of drones and other inspection gadgets that analyze structures and the damage on components. A possible scenario would be that a damage model which has been created from recorded drone data, could be evaluated by knowledge-based algorithms and critical areas as well as causations could be reasoned.

Nevertheless, the concept of RDP has not been tested to a sufficient extent. It is possible that several types of damage could not identified just by knowledge, but with the utilization of numerical approaches. This is often the case, when analyzing mechanical damages that occur through overstress or other forces. Additionally, different influences can lead to similar damage patterns, making an identification difficult. Since, the ontology need preexisting knowledge from previous inspections, another approach, such as artificial neural networks could perhaps be more suitable for this problem.

In conclusion, the concept of RDPs in an ontology for damage retrieval is in theory possible to utilize for damage assessment. Future tests will show the practical benefit of this approach.

ACKNOWLEDGEMENTS

This research work was enabled by the support of the Federal Ministry of Education and Research of Germany through the funding of the project wiSIB (project number 01|S16031C).

REFERENCES

Baradan, B. & Aydın, S. 2010. Betonun Dürabilitesi (Dayanıklılık, Kalıcılık). *Istanbul: THBB Yayınları, 2010*.

Becker, J. & Delfmann, P. & Knackstedt, R. 2007. Adaptive Reference Modeling: Integrating Configurative and Generic Adaption Techniques for Information Models. In J. Becker & P. Delfmann (eds), *Reference Modeling. Physica-Verlag Heidelberg, 2007, p. 27–58*.

Benevolenskiy, A. 2015. Ontology-based modeling and configuration of construction processes using process patterns.

Bertolini, L. & Elsener, B. & Pedeferri, P. & Polder, R.B. 2013. Corrosion of Steel in Concrete: Prevention, Diagnosis, Repair. *Wiley-VCH, 2013*.

Bundesministerium für Verkehr und digitale Infrastruktur. 2017. Richtlinien für die Erhaltung von Ingenieurbauten RI-ERH-ING. 2017.

DIN EN 1504-9:2008-11. 2008. Produkte und Systeme für den Schutz und die Instandsetzung von Betontragwerken- Definitionen, Anforderungen, Qualitätsüberwachung und Beurteilung der Konformität—Teil 9: Allgemeine Grundsätze für die Anwendung von Produkten und Systemen. *Deutsche Fassung EN 1504-9:2008. Berlin: Beuth Verlag GmbH, 2008*.

Espinosa, H.D. & Zavattieri, P.D. & Dwivedi, S.K. 1998. A finite deformation continuum\discrete model for the description of fragmentation and damage in brittle materials. *Journal of the Mechanics and Physics of Solids. 1. October 1998, p. 1909–1942*.

Fuchs, S. & Katranuschkov, P. & Scherer, R.J. 2010. A Framework for Multi-Model Collaboration and Visualisation. *eWork and eBusiness in Architecture, Engineering and Construction: ECCPPM 2010. 2010, p. 115–120*.

Hamdan, A. & Scherer, R.J. 2018. A Generic Model for the Digitalization of Structural Damage. *Ghent: IALCCE 2018*

Hammad, A. & Zhang, C. & Hu, Y. & Mozaffari, E. 2006. Mobile Model-Based Bridge Lifecycle Management System. *Computer-Aided Civil and Infrastructure Engineering 21, p. 530–547. 2006*.

Javor, T. 1991. Damage classification of concrete structures. The state of the art report of RILEM Technical Committee 104-DCC activity. *Materials and Structures. 1991, 24*.

Kalibatiene, D. & Vasilecas, O. 2011. Survey on Ontology Languages. *Lecture Notes in Business Information Processing, 2011*.

Luu, N.T. & Hamdan, A. & Polter, M. & Scherer, R.J. & Mansperger, T. 2018. A variation model method for real time system identification in bridge health monitoring. *Copenhagen: IABSE Conference 2018*.

Mongelli, M. & De Canio, G. & Roselli, I. & Malena, M. & Nacuzi, A & de Felice, G. 2017. 3D Photogrammetric Reconstruction by Drone Scanning for FE Analysis and Crack Pattern Mapping of the "Bridge of the Towers", Spoleto. *Key Engineering Materials. 2017, Bd. 747*.

Radomski, W. 2002. Bridge Rehabilitation. *London: Imperial College Press, 2002. 1-86094-122-2*.

Red Hat Inc. 2018. Drools. *https://www.drools.org/*. 2018.

Sacks, R. & Kedar, A. & Borrmann, A. & Ma, L. & Brilakis, I. & Hüthwohl, P. & Daum, S. & Kattel, U & Yosef, R. & Liebich, T. & Barutcu, B.E. & Muhic, S. 2018. SeeBridge as next generation bridge inspection: Overview, Information Delivery Manual and Model View Definition. *Automation in Construction. 2018, 90*.

W3C. 2012. OWL 2 Web Ontology Language (Second Edition). s.l.: W3C Recommendation 11 December 2012. *https://www.w3.org/TR/owl2-overview/*.

# A graph-based approach for management and linking of BIM models with further AEC domain models

A. Ismail & R.J. Scherer
*TU Dresden, Institute of Construction Informatics, Germany*

ABSTRACT: This paper presents a graph-based data management approach for automatic transformation and linking of BIM models with further domain models. The BIM transformation is based on IFC standard and covers both of IFC schema and IFC models. The aim of this research is to demonstrate the potential of graph theory concepts and applying a graph database framework for analyzing, managing and visualization the complex information and relationships inside BIM models and in particular to use this information to connect the BIM model with external models generating multimodels, i.e. a linked data set. It is demonstrated that graph-based approaches is an excellent alternative to current approaches which are based on traditional SQL databases and usually limited to query single BIM models. For validation of the approach and demonstration purpose, a set of advanced queries for data retrieval, compare, classification are shown. Also, a prototype web application with simple GUI is also developed for easy access of the connected graph model.

## 1 INTRODUCTION

The very fast development in the sector of information technology has been successfully exploited in construction and engineering field to adopt new digital methods such as Building Information Modelling (BIM) for construction project management. However, BIM models may contain a huge amount of information with complex relationships between the model entities. These information change with time and should be aligned with other project information like scheduling, security and cost models, which lead to a very huge complex data structure.

The BIM model information could remain inaccessible and cannot be linked with other information domains in several cases due to the use of closed property formats or the absent of suitable data management tools. A lot of data retrieval queries are hard to be accomplished using currently available software and most of them operate only on single BIM models even in the case of using IFC format. The rigid and complex hierarchical structure of the IFC schema prevents simple extraction of building information and requires deep understanding of the IFC object model itself. The BIM query languages introduced so far like EQL (Koonce, Huang, & Judd, 1998), Partial Model Query Langauge (PMQL) (Adachi, 2003), BIMQL (Mazairac & Beetz, 2013)have certain limitations, particularly with respect to the high level of knowledge about the IFC object model and about data mapping mechanisms required by the user.

Evidently, graphs have shown great capabilities in understanding and accessing and linking complex and rich datasets in many different domains. Graph models are extremely useful for representation and description of the complex relationships among building elements and data within BIMs (Isaac, Sadeghpour, & Navon, 2013). Therefore, converting BIM models based on the IFC standard into an effective information retrievable model based on graph databases could significantly facilitate the efforts of exploring and analyzing BIM highly connected data.

A graph-based schema, termed the graph data model (GDM) was presented by (Khalili & Chua, 2015). This schema can be used to employ semantic information, to extract, analyze and present the topological relationships among 3D objects in 3D space, and to perform topological queries faster. Another generic approach towards information retrieval using the IFC object model based on graph theory was presented by (Tauscher, Bargstädt, & Smarsly, 2016). In this approach a directed graph was generated that serve as semantic data pools facilitating generic queries. However, this approach is limited to apply queries on single IFC models.

This paper presents an approach for advanced analyzing and information management of BIM models based on IFC Standard using Graph Database. It presents a work flow for complete and automatic transformation of IFC models in the cloud into a labelled property graph-based model using the well-known graph database Neo4j (https://neo4j.com) as a graph database platform that is built to store, query, analyze and manage highly connected data efficiently. The paper explores also through examples the possibilities to extend the BIM graph database with further models.

In addition, this paper demonstrates the potential of using graph databases and concepts of graph theory in order to:

- Explore, check and analyze the complex relationships inside one or multiple BIM models
- Run complex queries for information retrieval
- Carry out advanced analysis of the building topology like escape route analysis and comparing of different IFC model versions

## 2 THE PROPERTY GRAPH DATA MODEL

A property graph is a directed, labelled, attributed multigraph. Both nodes and edges are labeled and they can have any number of properties and the edges are directed.

Each node of a property graph has a unique identifier and zero or more labels (object classes). Node labels could be associated to node typing in order to provide schema-based restrictions and a kind of basic classification mechanism. Similarly, each edge/relationship has a unique identifier, and one or more labels. In our research we use Neo4j graph database which implements the Property Graph Model and it provides full database characteristics including ACID transaction compliance (Robinson, Webber, & Eifrem, 2015). The Neo4j graph model can be accessed through a graph query language called Cypher that was introduced in the Neo4j. Cypher uses symbols to express patterns that correspond to a visual understanding of data, making it particularly well-suited to the challenges of querying connected data.

## 3 GRAPH REPRESENTATION OF BUILDINGS

In our approach the BIM model information is obtained from the IFC model and stored in a graph database. The property graph is built of connected entities (the nodes) which can hold any number of attributes (key-value-pairs). Nodes can also be tagged with labels representing their different roles (classes). In addition to contextualizing node and relationship properties, labels may also be used to attach metadata, index or constraint information to certain nodes. Relationships provide directed, named semantically meaningful connections between two nodes. A relationship in a property graph model always has a direction, a type, a start node, and an end node. Relationships can also have properties similar to those that can be attached to nodes. In most cases, relationships have quantitative properties, such as weights, costs, distances, sizes, positions, among others.

## 4 TRANSFORMATION OF IFC TO GRAPH DATABASE

The proposed approach targets two kinds of IFC to graph transformation:

- IFC EXPRESS schema into IFC Meta Graph (IMG)
- IFC models into IFC Objects Graph (IOG) based on STEP Physical File (SPF) format

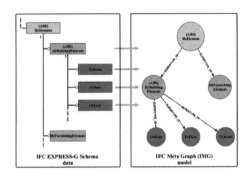

Figure 1. Mapping IFC EXPRESS into meta graph model.

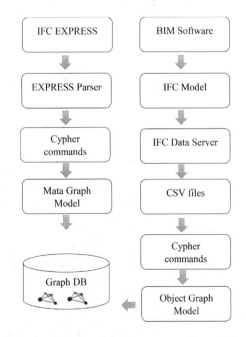

Figure 2. IFC to graph database conversation workflow.

The automatic workflow for the transformation (Figure 2) has been presented in details in the paper (Ismail, Nahar, & Scherer, 2017), where IFCWebServer (Ismail, 2011) is used as a tool for transforming building information into the graph model. Graph nodes correspond to building elements or their properties, while edges represent relations between these elements. Attributes assigned to element nodes describe the basic properties of building elements.

## 5 IMPLEMENTATION AND CASE STUDIES

After converting IFC models into a graph database it can be used as a data server to run data retrieval queries and advanced analysis of BIM models taking in account the advantages of applying graph theory concepts and algorithms like path finding and shortest path method.

### 5.1 Implementation

The implementation and validation of our approach is done based on the OpenBIM IFC Data Server and online viewer IFCWebServer (Ismail, 2011).

A special web application with simple GUI has been implemented in order to allow end-user to access and query the BIM graph without any experiences about IFC or graph databases. It can be accessed through http://ifcwebserver.org:3000. Through this application the user get a list of all building elements objects and their properties beside a general overview about the semantic information of the model (Figure 3). The web application is provided also with an API interface to access the BIM graph database through a RESTful HTTP API interface and get the response in JSON format.

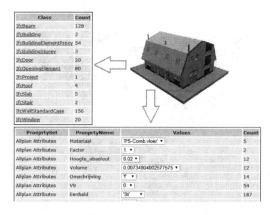

Figure 3. End-user web application to access BIM graph DB.

Another way to access the graph model is an interactive graph viewer, which can be accessed at http://ifcwebserver.org/bim_graph. The user can easily navigate through all BIM graph models and explore the relationships between graph nodes interactively (Figure 4).

In the following sections we present some applications of BIM Graph, which highlight the advantages of using graph databases.

### 5.2 Topology analysis of BIM models

In this example we used the graph database to do a simplified topology analysis of BIM models in order to analyze and generate emergency routes.

Information about the building, which is needed from the point of view of the problem considered in this example, includes information on topology of floor layouts, accessibility between spaces, stairs and doors types and sizes, if available. IFC entities, which store the data required by the proposed system, are of the types IfcSpace, IfcDoor, IfcWall and IfcStair.

The construction of possible routes is done in 2 steps:

– Extract the relationship "BoundedBy" between IfcSpace and IfcDoor entities
– Extract the relationship "RelatingSpace" between 2 IfcSpace neighbor objects

The relations between IFC entities required to compute the topological relationships of spaces are searched for. Two rooms are adjacent if two IfcSpace entities refer to the same IfcWall or to the same IfcWindowStandardCase using IfcRelSpaceBoundary relation. Two rooms are accessible if the wall between them has an opening or door. Therefore, IfcWall or IfcWindowStandardCase entity should refer to IfcOpeningElement by IfcRelVoidsElement relation, or additionally IfcDoor entity should refer to IfcOpeningElement by IfcRelFillsElement relation (the opening is filled with a door). The extracted information is then saved in the graph structure.

Space objects are represented as nodes while edges correspond to accessibility relations between these spaces through doors, virtual contact with other spaces, openings and accessibility between building floors through stairs/lifts. Labels assigned to graph nodes store names of spaces, while node attributes store other properties of spaces, for example their sizes or types.

The graph query in Cypher to get the space-door-space route:

MATCH (space:IfcSpace {model:'Office_A'})-[]-(storey:IfcBuildingStorey{ifcid:'1116'})

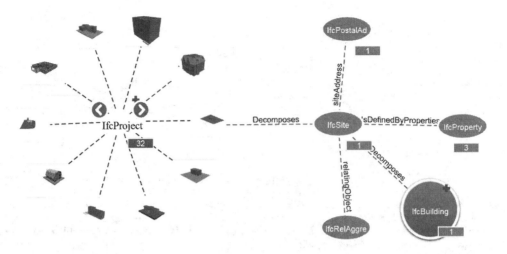

Figure 4. Interactively BIM graph explorer (http://ifcwebserver/bim graph).

MATCH p=(door:IfcDoor {ifcid:'807', model: 'Office_A'})-[:Bounded By]->(space)
RETURN p UNION
MATCH(sp1:IfcSpace{model:'Office_A'})-[]-(storey: IfcBuildingStorey {ifcid:'1116'})-[]- (sp2:IfcSpace)
MATCH  p=(space1{model:'Office_A'})<-[:B o u n d e d B y]-( d o o r : I f c D o o r )-[:Bounded By]->(sp2)
WHERE sp1.ifcid > sp2.ifcid RETURN p UNION
MATCH(sp1:IfcSpace{model:'Office_A'})-[]-(storey: IfcBuildingStorey{ifcid:'1116'})-[]-(sp2:IfcSpace)
MATCH
p = (sp1{model: 'Office_A'})-[:RelatingSpace]-(:IfcRelSpaceBoundary)-[:RelatingSpace]-(sp2)
WHERE space1.ifcid > sp2.ifcid RETURN p

Cypher reference card can be read at: https://neo4j.com/docs/cypher-refcard.

The results of the successive Cypher query represents the routes between spaces for a single floor in an office building (Figure 5).

Further examples for topology analysis of BIM models and the problem of searching for routes in building has been presented recently in (Ismail, Strug, & Slusarczyk, 2018).

### 5.3 Compare IFC models

Another important application of graph models is compare IFC model versions. There, two or more versions of the same IFC model can be compared to find out the new added building elements or the elements with geometrical changes. The result is visualized through colors inside the online BIM viewer (Figure 6).

The Cypher query of such a comparison looks like:

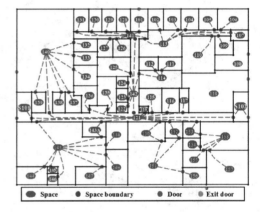

Figure 5. Space-door-space routes drawn on the floor layout of an office building.

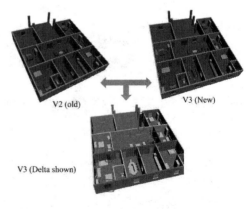

Figure 6. Deltas are shown on a standard BIM viewer using red frames.

MATCH(e2:IfcElement{model:'V2'}) with collect (distinct e2.globalId) as ids
MATCH (e3:IfcElement{model:'V3'})
WHERE NOT e3.globalId IN ids
return "new" as name, 'green' as color, collect('_' + e3.ifcid) as ids
UNION
MATCH (e3:IfcElement{model:'V3'})-[:IsDefinedByProperties]-()-[]-(p3{name:'"Gross Volume"'})
MATCH(e2:IfcElement{model:'V2'})-[:IsDefinedByProperties]-()-[]-(p2{name:'"GrossVolume"'})
WHERE e2.globalId = e3.globalId and to Float (p2.volumeValue) <> to Float (p3.volumeValue)
return "changed" as name, 'orange' as color, collect('_' + e1.ifcid) as ids

In this query, the new added elements in V3 have been detected through comparing the both lists of existing GUIDs of building elements in both versions. For detecting the elements with modified geometry without the need to process the low-level geometry information, the query uses the element property "Gross Volume" and return the IDs of elements, which have the same GUID but different "volume value". The graph query can be extended to find out deleted elements or changed elements regarding other semantic or geometrical criteria.

### 5.4 Classification of BIM objects

Using Graph database, BIM models can be classified in a flexible way. For each classification system a set of additional nodes will be added to the graph database and connected with objects through relationships, which are created based on predefined rules.

The following criteria have been applied to classify BIM objects according to: (1) Object class (2) Object material, (3) Object type, (4) Object layer, (5) Geometry type, (6) Object system.

For example, in case of classification of objects according to their materials, the classification rules should take in account the IFC specification which specify all possible ways to assign materials to objects. Namely through the IfcRel Associates Material relationship which connects one or more objects with one of the following material classes: IfcMaterial, IfcMaterialList, IfcMaterialLayerSetUsage, IfcMaterialLayerSet, IfcMaterialLayer.

The Cypher query to get a list of all objects which consist of one single material or contains this material partly (e.g. material layer) will look like:

MATCH(e:IfcObject{model:'MODELL_NAME'})-[]-(n:IfcMaterialLayerSetUsage {model:'MODELL_NAME'})-[:forLayerSet]-(IfcMaterialLayerSet{model:'MODELL_NAME'})-[:MaterialLayers]-(:IfcMaterialLayer{model:'MODELL_NAME'})-[:material]-(m:IfcMaterial {model:'MODELL_NAME',name: "MATERIAL_NAME"})
return collect('_' + e.ifcid) as IDs
UNION
MATCH(e:IfcObject{model:'MODELL_NAME'})-[]-(m:IfcMaterial {model:'MODELL_NAME', name: "MATERIAL_NAME"})
return collect('_' + e.ifcid) as IDs
UNION
MATCH(e:IfcObject{model:'MODELL_NAME'})-[]-(:IfcMaterialLayer{model:'MODELL_NAME'})-[:Material Layers]-(m:IfcMaterial{model:'MODELL_NAME',name: "MATERIAL_NAME"})
return collect('_' + e.ifcid) as IDs
UNION
MATCH(e:IfcObject{model:'MODELL_NAME'})-[]-(:IfcMaterialList{model:'MODELL_NAME'})-[:Materials]-(m:IfcMaterial{model:'MODELL_NAME', name:"MATERIAL_NAME"})
return collect('_' + e.ifcid) as IDs
UNION
MATCH(e:IfcObject{model:'MODELL_NAME'})-[:Has Associations]-(:Ifc Material Layer Set{model:'MODELL_NAME'})-[:MaterialLayers]-(m:Ifc MaterialLayer{model: 'MODELL_NAME'})-

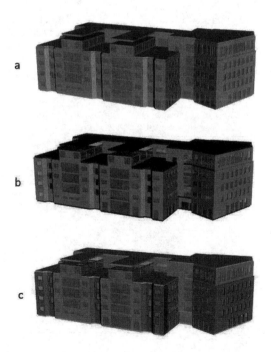

Figure 7. Classification of original model (a) by object class (b) and material (c).

[:material]-(m:IfcMaterial{model:'MODELL_NAME', name:"MATERIAL_NAME"})
return collect('_' + e.ifcid) as IDs

Such complex queries can be written once by IFC and graph experts and implemented within end-user application like an online BIM viewer, they can be applied on all BIM models simply by passing the model and material names as parameters (Figure 7).

### 5.5 Connecting BIM models with other domain models to create multimodels

The BIM Graph database can be extended with further external domain models in order to extend the possible applications and get more added value of the whole connected models. This extension can be done without the need to include all model information in the graph database. The BIM Graph database serves here as the link model, which contains additional link information connecting the BIM model with other external models, for example by linking BIM object with 2D drawing and documents on the Cloud through URLs or contains the cross relationships, when the cross topology model for a cross domain system is sought. BIM models can be linked also with project schedules resulting as 4D BIM models, or with cost and Facility Management models to get 5D/6D BIM models (Figure 8).

For each new external domain model, a set of graph nodes and relationships will be created and the new nodes can be linked with BIM object nodes through relationships, like for a usual multimodel, e.g. according to ISO 21597, ICDD.

The workflow of importing each domain model depends on the original format of the model and how to import them into the graph database e.g. by direct import from SQL database, import through CSV/Excel files, writing a special parser, etc.

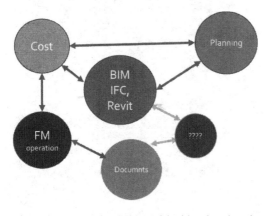

Figure 8. Connecting BIM model with other domain models.

In most cases the creation of relationships can be automated based on predefined rules or through direct importing of already linked models.

An application of this approach is developed by importing BIM-LV Multimodel containers (German Norm DIN SPEC 91350), which are based on the multimodel data exchange approach (Fuchs, Kadolsky, & Scherer, 2011). There, 3D objects in the BIM models are linked with the bill of quantities and work positions according to German data standard GAEB (www.gaeb.de). As a result, the graph database includes both of BIM and GAEB information and allows bidirectional queries to show related BIM object for a certain work position or the work position related to a certain BIM object. A demo of this example can be accessed at: http://ifcwebserver.org/BIM-LV.

Another approach to work with connected BIM models can be achieved by importing linked data models by applying sematic web technology like Resource Description Framework (RDF) & Web Ontology Language (OWL), for example by converting IFC models to RDF triples using the IfcOWL ontology (Beetz, Van Leeuwen, & De Vries, 2009), (Pauwels & Van Deursen, 2012) and import them into a graph databases, where triple stores are suitable for reasoning, graph databases are suitable for graph operations (shortest path, navigation-based queries) (Thakkar, Keswani, Dubey, Lehmann, & Auer, 2017).

## 6 CONCLUSIONS AND FUTURE WORK

This paper presented an approach for advanced analyzing and information management of BIM models based on IFC Standard using Graph Databases. The developed approach has been verified through applying graph queries for data retrieval and advanced topology analysis and compare queries of BIM models. In addition, a web application for accessing the BIM graph database has been also developed, which is available for free for teaching and research on http://ifcwebserver.org:3000.

Future research is needed in order to include the geometry information in the graph database by developing an interface between the graph database and an IFC geometry engine. As Another important future work is seen in the development of an IFC export interface for filtering sub-models or merging models together and export them as single IFC model.

Graph databases can be considered an efficient tool to manage and connect building information. The transformation of BIM models into a graph database and having the possibility to connect it with other related domain information opens the door for a lot of advanced application. However, writing complex queries without IFC and graph skills is a quite challenging task for end users. Such queries should be written by IFC and graph DB experts and provided as pre-defined function to the end-users. This research can be considered as an initial step in the direction of multi-connected graph models management for BIM and other information domains.

ACKNOWLEDGEMENT

We kindly acknowledge the support of the European Commission to the eeEmbedded project, Grant Agreement No. 609349, http://eeEmbedded.eu, where as a side effect this research was carried out.

REFERENCES

Adachi, Y. (2003). Overview of partial model query language. 10th ISPE International Conference on Concurrent Engineering. Madeira.

Beetz, J., Van Leeuwen, J., & De Vries, B. (2009). ifcowl: Acase of transforming express schemas into ontologies. Artificial Intelligence for Engineering Design, Analysis and Manufacturing (AI EDAM), 23, 89–101.

DIN SPEC 91350: Linked BIM data exchange of building models and bills of quantities. (2016, 11). Retrieved from https://www.beuth.de/de/technische-regel/din-spec-91350/263151122

Fuchs, S., Kadolsky, M., & Scherer, R. (2011). Formal Description of a Generic Multi-Model. 20th International Workshops on Enabling echnologies:Infrastructure for Collaborative Enterprises (WETICE 11), (pp. 205–210). Washington. DC.

Isaac, S., Sadeghpour, F., & Navon, R. (2013). Analyzing Building Information using Graph Theory. International Association for Automation and Robotics in Construction (IAARC), (pp. 1013–1020). Montreal.

Ismail, A. (2011). OpenBIM IFC Data Server and online viewer. Retrieved from http://ifcwebserver.org

Ismail, A., Nahar, A., & Scherer, R. (2017). Application of Graph Databases and Graph Theory Concepts for Advanced Analysing of BIM Models Based on IFC Standard. 24th International Workshop on Intelligent Computing in Engineering (EG-ICE 2017). Nottingham, UK.

Ismail, A., Strug, B., & Slusarczyk, G. (2018). Building Knowledge Extraction from BIM/IFC Data for Analysis in Graph Databases. International Conference on Artificial Intelligence and Soft Computing (ICAISC) (pp. 652–664). Springer International Publishing AG.

ISO 16739: Industry Foundation Classes (IFC) for data sharing in the construction and facility management industries. (2013, 04). Retrieved from https://www.iso.org/standard/51622.html

ISO 21597-Part1: Information container for data drop – Exchange specificaton – Part 1: Container. (n.d.). Retrieved from https://www.iso.org/standard/74389.html

ISO 21597-Part 2: Information container for data drop – Exchange specification – Part 2: Dynamic semantics. (n.d.). Retrieved from https://www.iso.org/standard/74390.html

Khalili, A., & Chua, D. (2015). IFC-Based Graph Data Model for Topological Queries on Building Elements. American Society of Civil Engineers, 29(3).

Koonce, D., Huang, L., & Judd, R. (1998). EQL an express query language. Computers & Industrial, 35(1), 271–274.

Mazairac, W., & Beetz, J. (2013). BIMQL—An open query language for building information models. Advanced Engineering Informatics, 27(4), 444–456.

Pauwels, P., & Van Deursen, D. (2012). IFC-to-RDF: Adaptation, Aggregation and Enrichment. 1st International Workshop on Linked Data in Architecture and Construction. Ghent, Belgium.

Robinson, I., Webber, J., & Eifrem, E. (2015). Graph Databases-New opportunities for conneced data. O'Reilly Media.

Tauscher, E., Bargstädt, H.-J., & Smarsly, K. (2016). Generic BIM queries based on the IFC object model using graph theory. The 16th International Conference on Computing in Civil and Building Engineering. Osaka, Japan.

Thakkar, H., Keswani, Y., Dubey, M., Lehmann, J., & Auer, S. (2017). Trying Not to Die Benchmarking—Orchestrating RDF and Graph Data Management Solution Benchmarks Using LITMUS. SEMANTiCS 2017. Amsterdam. DOI: 10.1145/3132218.3132232

# A building performance indicator ontology

A. Mahdavi & M. Taheri
*Department of Building Physics and Building Ecology, TU Wien, Vienna, Austria*

ABSTRACT: Building performance simulation results are commonly expressed in terms of numeric values of performance indicators. Such indicators involve multiple domains, different aspects, and varying degrees of resolution. They are used, amongst other things, to demonstrate compliance with building code requirements, generate building quality certificates, and rank building design alternatives. Despite their pervasiveness and familiarity, there have been few attempts to establish a robust and comprehensive ontology of building performance indicators. The present contribution is about an effort toward forming such an ontology. A number of motivations can be listed for an effort of this kind. Generally speaking, ontologies can help structure the conceptual and semantic constituents of a domain and thus improve the efficiency of communication processes and developmental work in that domain. More specifically, a performance indicator ontology can advance the specification process of building performance requirements, improve the understanding of performance procurement principles, and provide a structured basis for developing broadly deployable data visualization engines.

## 1 INTRODUCTION

Computational building performance assessment can be described as a process by which physical and operational data concerning buildings (i.e., building information models) are mapped onto building performance simulation results. Such results are expressed, not exclusively, but quite commonly in terms of numeric values of building performance indicators. Professionals engaged in building performance assessment operate with a large number of building performance indicators, involving multiple domains, different aspects, and varying degrees of resolution.

Building performance indicators are used, amongst other things, to demonstrate compliance with building code requirements, for certificate-type building quality specifications, and toward performance ranking of building design alternatives. Despite their pervasiveness and familiarity, there have been few attempts to establish a robust and comprehensive ontology of building performance indicators (Mahdavi et al. 2005).

The present contribution entails the approach and results of an effort toward forming such an ontology. A number of motivations can be listed for an effort of this kind. Generally speaking, ontologies can help structure the conceptual and semantic constituents of a domain and thus improve the efficiency of communication processes and developmental work in that domain. An augmented BIM (a core Building Information Model enriched with contextual information such as microclimate and user information) provides needed information for performance assessment applications. The resulting values of ontologically well-formed performance indicators can support visualization, optimization, and other decision support applications (Fig. 1). More specifically, a building performance indicator ontology can

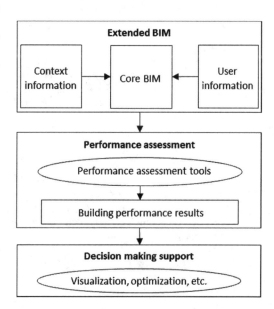

Figure 1. Core BIM, performance assessment, and decision support.

advance the specification process of building performance requirements, improve the understanding of building performance principles (e.g., in educational and training settings), and provide a highly structured basis for developing broadly deployable data visualization engines.

To establish the ontology, we utilized a comprehensive review of a large collection of building performance indicators in thermal, air quality, visual, and acoustical domains (Constantinou 2017). Moreover, we revisited a previous effort that resulted in an ontology for building monitoring data (Mahdavi et al. 2016, 2018, Mahdavi and Taheri 2017). Subsequently, we defined a conceptual space, whose dimensions provide a generic framework for the specification of building performance indicators and their values.

From the methodological standpoint, the robustness and coverage of the ontology can be demonstrated via fitness testing on the basis of the aforementioned collection of existing instances of building performance indicators.

## 2 ILLUSTRATIVE INSTANCES OF BUILDING PERFORMANCE INDICATORS

### 2.1 Overview

The list of building performance indicators is very long and is being continuously modified and extended, making a comprehensive review difficult. A recent effort compiled and commented on a number of such indicators in various domains (Constantinou 2017), including energy efficiency, hygro-thermal performance, thermal comfort, indoor air quality, indoor visual environment, and indoor acoustical environment. Whereas some indicators merely address technical systems performance issues, others may serve the evaluation of buildings regarding their suitability for the occupants, or in other words, its "habitability" (Mahdavi 1998). The latter aspect is typically captured in terms of criteria pertaining to human health, and safety, comfort, satisfaction, and productivity (Mahdavi 2011). Moreover, amongst indicators relevant to human occupancy, some may be phenomenally relevant (i.e., consciously experienced by inhabitants) and some not. To provide a general feel for the nature of building performance indicators, we briefly discuss, in the following (sections 2.2, 2.3 and 2.4.), three exemplary instances.

### 2.2 Annual energy demand

Buildings' energy demand (for heating, cooling, lighting, etc.) can be directly measured using energy meters, or computationally estimated. A commonly used instance of energy performance indicators is the annual energy demand. This indicator is typically normalized per reference floor area and aggregated over a period of a year, as reflected in the unit of the indicator's values [$kWh.m^{-2}.a^{-1}$] (ISO 2017). Energy demand belongs to the group of performance indicators that does not have a direct relevance to the habitability of a building or a space (i.e., thermal comfort conditions). Nor is it phenomenally relevant.

### 2.3 Indoor illuminance level

Performance indicators pertaining to indoor lighting conditions support evaluation of light availability, distribution over space, uniformity, etc. One of the key indicators for evaluation of the indoor lighting is the illuminance level. Illuminance is a measure of luminous flux incident on a given surface, hence the unit lx ($lm.m^{-2}$) (DiLaura et al. 2011). The illuminance at a point can be calculated or measured. Illuminance level is relevant to the quality of spaces in view of human occupancy requirements (health, comfort, productivity). However, it does not have a direct phenomenal correlate, i.e., people do not "see" illuminance.

### 2.4 Reverberation time

Room acoustics is concerned with the quality and effectiveness of sound propagation in spaces as relevant to criteria involving privacy, speech, communication, and music. Reverberation time is a long standing and common (but not necessarily the most important) indicator of rooms' acoustical performance (ISO 2008). It denotes the time (in seconds) during which the mean acoustical energy density in a space drops to a millionth of its initial level after the sound source is turned off. Reverberation time at different locations in the space may vary considerably, depending on the space characteristics, such as boundary materials and geometry. Nonetheless, it is often averaged over multiple locations in the space.

This indicator not only is directly relevant to human occupancy requirements, but has a phenomenal correlate. For instance, perceived levels of a room's impressions of liveliness as well as the intelligibility of speech and clarity of musical performance have been suggested to correlate with the reverberation time's values. Reverberation time is just of the many room acoustics indicators, which relate inhabitants' subjective impressions of the acoustic conditions to the physical properties of the sound field.

## 3 TOWARD AN ONTOLOGICAL SCHEMA FOR BUILDING PERFORMANCE INDICATORS

### 3.1 *Preliminary observations*

To approach the task at hand, a few preliminary observations could be useful:

- First, building performance has multiple and substantially varied facets, including but not limited to those related to indoor environment, energy and resources, building integrity, safety and security, structure, ecology, and economy. In the present treatment, we focus on indoor environmental factors (thermal, visual, air quality, acoustical) and energy-related building performance variables.
- Second, data underlying building performance indicators can be of different types. Here, we focus on two, namely quantitative (measurement) data and user-based evaluations (as long as they can be numerically expressed).
- Third, performance indicators can include both directly measureable parameters and derived variables (e.g., functions, rates). The ontology we seek to establish should cover both types.
- Fourth, some performance indicators are phenomenally relevant (e.g., temperature or glare rating), meaning they pertain to indoor environmental factors that can be perceived by inhabitants. Other indicators (e.g., heating and cooling demand) are perceptually irrelevant. Again, the ontological schema should cover both kinds.

Aside from the above considerations, we must address here a further non-trivial conceptual challenge in definition of building performance indicators. It can be argued that there is not always a sharp line between property specifications of buildings' constituent components/systems and building performance indicators. Consider, as a case in point, the thermal conductivity of building materials. We typically do not refer to this property as a building performance indicator, not because it does not refer to the whole building, but because it is not necessarily mapped into a clear and consistent "value system" that says what is desirable and what is not. For instance, low thermal conductivity may be considered an asset in case of thermal insulation elements, but many integral components of building constructions include materials of high thermal conductivity (such as structural steel) without any negative value connotations.

As a rule-of-thumb, building performance indicators can be mapped to a value system function (e.g., lower heating and cooling demands are generally preferable to higher ones). In contrast, general direct mappings between material properties and value systems are not meaningful by default. As already suggested, the demarcation between component properties and performance indicators is not always sharp. Consequently, the above criterion (correspondence of variable values with a clear preference scale) may not be universally applicable. For instance, thermal conductivities of multiple layers of a building enclosure components may be processed into an aggregate component property descriptor (thermal transmittance) and upon area-weighted averaging across multiple enclosure elements contribute to a kind of building thermal performance indicator (effective envelope U-value). Nonetheless, the multiple steps in this example (from thermal conductivity of material layers to weighted thermal transmittance of a whole building enclosure) suggest that it may be quite useful to distinguish between rather "neutral" property specifications on the one hand and performance indicators that can be readily mapped to a valuation scale.

One can of course argue that the association of the previously discussed reverberation time with a value scale is not universal, but is established based on specific functional requirements of a space. Nonetheless, the applicability to spaces and whole buildings as well as correspondence with value systems do imply a rough classification rule concerning building performance indicators on the one hand and specifications of materials, components, and systems on the other hand.

### 3.2 *A building performance indicator map*

There is probably no unique map of the building performance indicator categories. Nonetheless, efforts toward drawing such a map can help converging toward a common data schema underlying most—if not all—building performance indicators. Figure 2 includes such a map for a subset of building performance domains as alluded to in section 2. Relying on the aforementioned review (Constantinou 2017), this map includes five building performance domains. The map also includes sub-categories and illustrative instances of performance indicators and associated units in each sub-category.

### 3.3 *The fundamental schema*

Building upon the treatment in previous sections, we suggest that the schema shown in Table 1 captures the essential ontologically relevant characteristics of building performance indicators. Once the performance category and sub-categories are identified, features of the building performance variable may be specified. Aside from a name and assorted remarks (i.e., explanatory or documentary notes), the main content of a variable is its

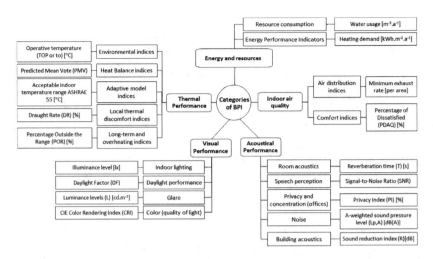

Figure 2. An overview map of five building performance domains with subcategories and illustrative performance indicator examples (based on the treatment in Constantinou 2017).

Table 1. General schema for building performance indicators.

| Category | | | |
|---|---|---|---|
| Sub-category | | | |
| | Name | | |
| | | Type | |
| | | Magnitude (size) | |
| | | Direction (in case of vectors) | |
| | | Unit | |
| Variable | Value | Spatial domain | Point |
| | | | Plane |
| | | | Volume |
| | | | Topological reference |
| | | | Aggregation method |
| | | | Grid size |
| | | Temporal domain | Time stamp |
| | | | Duration |
| | | | Time step |
| | | | Aggregation method |
| | | Frequency domain | Range |
| | | | Band (filter) |
| | | | Weighting |
| | | | Aggregation method |

value. To each value, a number of properties and attributes can be assigned as follows. The variable type denotes the relevant data type (e.g. quantitative versus ranking data).

In case of quantitative variables, a distinction can be made between scalar variables (such as temperature and reverberation time) and vector ones (such as air flow velocity and sound intensity). Variable values can have a magnitude (size) and—in case of vectors—also a direction. To meaningfully interpret variable magnitudes, they must be specified with relevant units.

The extent of the variables may be defined in spatial terms. These include, for instance, a specific point in a space (defined via x, y, z coordinates), a well-defined volume, or a room (possibly identified via a topological reference). If the value of a variable is derived base on integration over data across a spatial domain, aggregation method (e.g., arithmetic averaging) and respective grid size can be specified. Likewise, the temporal domain of a variable's value can be captured in the schema via a time stamp (e.g., in case of instantaneous spot measurements), duration (e.g., monthly or annual values). Analogous to the grid size in the spatial domain, discretization across the temporal domain can be specified by definition of the time step magnitude. Likewise, the aggregation method of data over the time domain must be declared in applicable cases.

In those cases, where the variable values pertain to phenomena with a wave character (sound, radiation, light), specifications concerning the frequency domain may be necessary. For instance, sound pressure levels or reverberation times may be integrated over the entire auditory range. Frequency analysis, on the other hand, involves the specification of discrete frequency bands (filters or intervals) such as octave bands or 1/3rd octave bands) or integrated over a certain frequency range (e.g., the entire audible frequency range).

Moreover, specific weighting schemes may be applied (e.g., A-weighted levels in acoustics) to

Table 2. Illustrative representation of three exemplary performance indicators following the structure of the proposed ontology.

| Category | | | | | Energy and resources | Thermal performance | Acoustical performance |
|---|---|---|---|---|---|---|---|
| Sub-category | | | | | Energy performance indicator | Environmental indices | Noise |
| | Name | | | | Heating Load | Air flow velocity | Sound pressure level |
| Variable | Value | Type | | | Quantitative | Quantitative | Quantitative |
| | | Magnitude | | | 50 | 0.25 | 60 |
| | | Direction | | | – | −51.17 ° downward from X axis | – |
| | | Unit | | | kWh.m$^{-2}$ | m.s$^{-1}$ | dB |
| | | Spatial domain | Point | | – | 1.50, 2.00, 1.10 | 1.00, 0.50, 1.20 |
| | | | Plane | | – | – | – |
| | | | Volume | | Building B1 | – | – |
| | | | Topological reference | | Building B_1 | Room R_1 | Room R_1 |
| | | | Aggregation method | | – | – | – |
| | | | Grid size | | – | 0.20 m | – |
| | | Temporal domain | Time stamp | | – | 08.05.2018 10:30:00 | 08.05.2018 10:30:00 |
| | | | Duration | | Annual | – | 45 min |
| | | | Time step | | 1 hour | – | 0.1 Sec |
| | | | Aggregation method | | Arithmetic summation | – | Continuous equivalent sound level |
| | | Frequency domain | Range | | – | – | 63–8000 Hz |
| | | | Band | | – | – | – |
| | | | Weighting | | – | – | A-Weighting |
| | | | Aggregation method | | – | – | – |

Notes:
Source: Simulation; no relevance for occupancy or phenomenal experience.
Source: Simulation; relevance for occupancy and phenomenal experience.
Source: Measurement; relevance for occupancy and phenomenal experience.

emulate human auditory system's characteristics via manipulation of values of physically relevant variables. Common photometric data (such as illuminance levels in rooms or luminance levels of rooms' surfaces) are typically broadband and weighted based on the standardized function of humans' spectral sensitivity to radiation in the visible range. However, photometric variables may also be specified in discrete frequency ranges (as common in analyses where color characteristics play a role).

3.4 *Application of the proposed schema*

A tight proof of the applicability of the proposed scheme to all kinds of building performance indicators cannot be provided here. Nonetheless, we conducted a test of its robustness via a randomly selected number of common performance indicators. Table 2 provides an illustrative instance of this test involving three different indicators.

4 CONCLUDING REMARKS

In this contribution, we introduced a building performance indicator ontology. Specifically, we described a general scheme to capture the essential features of building performance variables. The initial survey of existing instances of building performance indicators that accompanied the ontology formulation process was non-exhaustive. Likewise, the proposed schema could be tested only on the basis of a few instances from the population of building performance indicators. Hence, we are not suggesting that the robustness of the proposed ontology is strictly proven. Future in-depth studies may reveal the existence of building performance indicators that cannot be fully captured based on the proposed ontology. Specifically, the definition of specific performance domains and associated sub-category classifications include perspectival driving points and may be thus not fully objective. Moreover, as already alluded to in the paper, ambiguities in the definition of performance variables may arise from multiple and overlapping destinations of performance attribution (i.e., materials, elements, components, systems, spaces, whole buildings). These circumstances suggest that assuming there is a universally applicable building performance indicator ontology may be unfunded.

Nonetheless, experiences so far with the proposed ontology point to its considerable utility concerning the fields of its potential deployment, ranging from the code compliance checking, to building certification, performance contracting, and evidence-based design support. Future studies shall further address the applicability and robustness of the proposed ontology and explore its improvement and extension. Moreover, specific efforts are being undertaken to establish the ontology as the structural basis of a universal visualization engine for multi-domain building performance assessment applications.

## ACKNOWLEDGEMENT

The research presented in this paper is supported in part within the framework of the "Innovative Projekte" research-funding program of TU Wien.

## REFERENCES

Constantinou, N. 2017. *A comprehensive multi-domain building performance indicator catalogue*. Master thesis. Department pf Building Physics and Building Ecology, TUWien, Vienna, Austria.

DiLaura, D., Houser, W., Mistrick, G., Steffy, R. 2011. *The Lighting Handbook, 10th Edition*. Illuminating Engineering Society, New York, USA. ISBN-13: 978-0-87995-241-9.

ISO 2008. *ISO 3382-2. Acoustics – Measurement of room acoustic parameters – Part 2: Reverberation time in ordinary rooms*. International Organization for Standardization, Geneva, Switzerland.

ISO 2017. *ISO 52003-1. Energy performance of buildings. Indicators, requirements, ratings and certificates. Part1: General aspects and application to the overall energy performance*. International Organization for Standardization, Geneva, Switzerland.

Mahdavi A. 1998. *Steps to a General Theory of Habitability*. Human Ecology Review. Summer 1998, Volume 5, Number 1. pp. 23–30.

Mahdavi, A. 2011. *People in Building Performance Simulation*. In Building Performance Simulation for Design and Operation, edited by Hensen, J. and R. Lamberts, New York, Taylor & Francis Group. ISBN-13: 978-0415474146.

Mahdavi, A., Glawischnig, S., Schuss, M., Tahmasebi, F., and Heiderer, A. 2016. *Structured Building Monitoring: Ontologies and Platform*. Proceedings of ECPPM 2016: The 11th European Conference on Product and Process Modelling, Limassol, Cyprus.

Mahdavi, A., J. Bachinger, and G. Suter. 2005. *Toward a unified information space for the specification of building performance simulation results*. In: Building Simulation 2005, 9th International IBPSA Conference, Aug. 15–18, Montreal, Canada, I. Beausoleil-Morrison, M. Bernier (editors); International Conference: IBPSA, 2005, pp. 671–676.

Mahdavi, A., Taheri, M. 2017. *An Ontology for Building Monitoring*. Journal of Building Performance Simulation, 10:5–6, 499–508. DOI:10.1080/19401493.2016.1243730.

Mahdavi, A., Taheri, M., Schuss, M., Tahmasebi, F., and Glawischnig, S. 2018. *Structured Building Data Management: Ontologies, Queries, and Platforms*. In book: Exploring Occupant Behavior in Buildings. Publisher: Springer, Editors: Andreas Wagner, William O'Brien, Bing Dong. DOI: 10.1007/978-3-319-61464-9_10.

# From patterns to evidence: Enhancing sustainable building design with pattern recognition and information retrieval approaches

E. Petrova, K. Svidt & R.L. Jensen
*Aalborg University, Aalborg, Denmark*

P. Pauwels
*Ghent University, Ghent, Belgium*

ABSTRACT: Decision-making in design and engineering relies little on knowledge discovered in previous projects and embedded in digital data. Applying analytical computational techniques to available data and processes can be of significant influence for infusing decision-making with the evidence-based character that it is currently lacking. The design environment is where decisions are implemented, therefore, we aim to endow it with knowledge discovered in previous projects and existing buildings. We use an approach that combines data mining and semantic modelling for Case-Based Design (CBD). We investigate the character of the active design environment, what queries can be constructed automatically from the data available in that environment, and how they can be executed against a repository of design models and performance patterns obtained using Knowledge Discovery in Databases (KDD) and various machine learning approaches. We demonstrate this approach on a use case, highlighting its potential for evidence-based design decision support.

## 1 INTRODUCTION

The advancements in predictive analytics and simulations have led to the implementation of innovative performance assessment models in the building design domain. Yet, many of the decisions taken rely on design assumptions and previous experience, rather than documented evidence. The Architecture, Engineering and Construction (AEC) industry is more information-intensive than ever and that by itself unveils an unprecedented opportunity for discovery of hidden knowledge in the significant heterogeneous datasets generated during the design, construction, and operation of buildings (Soibelman & Kim, 2002; Bilal et al., 2016). Powerful cross-domain techniques such as machine learning and semantic query techniques have made prediction of performance outcomes and knowledge discovery not just possible, but much more accurate and reusable.

Being applied to available data, such approaches carry a powerful potential and can be of fundamental influence to the decision-making process by giving it an evidence-based character (Hamilton & Watkins, 2009). Relevant data sources may include operational building data from sensor networks, Building Information Models (BIM), design brief databases, performance targets relative to the sustainability criteria, etc. By employing the powerful potential of Knowledge Discovery in Databases (KDD) (Fayyad et al., 1996), data mining (Hand et al., 2001) and pattern recognition (Bishop, 2006), evidence can be found in patterns and potentially occurring links between patterns discovered in the data. And while traditional analytical and prescriptive approaches present issues when it comes to high-performance design, a combination of holistic performance-oriented approaches and computational technologies can more effectively contribute to achieving evidence-based decision-making. Besides the available data and the patterns discovered in the data, a decision support system is also essentially influenced by the design development environment, as it is the place that drives queries to any of the knowledge sources that are potentially available.

In this regard, we look specifically at the target data, and how discovered patterns in building operation can be retrieved and used to support the decision-making in new design processes. Therefore, this research effort focuses on enhancing sustainable building design through analytical computational approaches applied in the early design phase. We start from a design environment that is empowered by BIM tools. Furthermore, design brief requirements are considered to be an integral part of the design environment as well. Hence, a Common Data Environment (CDE) takes a prominent place in this research, as the CDE functions as the environment in which all design data is available. From this environment, knowledge is sought for in a pattern retrieval repository, which is based on an open repository of Industry Foundation Classes (IFC) models collected from previously executed building designs, for some of which motifs (frequent repetitive patterns) and association rules have been discovered.

In this article, we first look into related works (Section 2) aimed at informing building design with knowledge from existing buildings and/or similar designs. In Section 3, we explore the structure of design environments and propose the way in which such systems may be enhanced with evidence-based decision support. Section 4 documents the performed experiment, which consists of (1) a data repository containing building semantics and performance data, (2) a specifically considered building design, and (3) the tests conducted towards matching them. Section 5 discusses the results and future works, thereby leading to Section 6, which concludes this article.

## 2 RELATED WORKS

Using KDD and data mining approaches in AEC has gained momentum with regards to improvement of building performance. Promising advancements lie within the use of machine learning for model predictive control (Drgona, 2018), metamodelling for design space exploration (Geyer and Schlueter, 2014; Østergaard et al., 2018), use of data analytics for improvement of energy performance and building occupancy (Ahmed et al., 2011; Fan et al., 2015a), etc. Most prominently, research has shown great advancements related to use of data analytics for improvement of facility management and building operation. Included here are anomaly and fault detection diagnostics in systems operation, extraction of energy use and occupant behaviour patterns, improvement of occupant comfort, etc. (Fan et al., 2015b; Fan et al., 2018).

With regards to the use of KDD for design decision support, research efforts include pattern recognition in simulation data and extraction of information from BIM design log files (Yarmohammadi et al., 2016), extraction of 3D modelling patterns from unstructured temporal BIM log text data (Yarmohammadi et al., 2017), use of data-driven approaches to design energy-efficient buildings by mining of BIM data (Liu et al., 2015) and use of simulation data mining for energy efficient building design (Kim et al., 2011). Reuse of similarities in decision support has also been widely recognised in design practice. This is prominently present in case-based reasoning (CBR), which provides decision makers with a problem solving framework involving recalling and reusing previous knowledge and experience (Aamodt and Plaza, 1994). The use of CBR in design practice (case-based design (CBD)) differs with regards to the method of implementation. For instance, Dave et al. (1994) present a design system enabling case adaptation and combination for a more efficient generation of new design cases. Both Heylighen and Neuckermans (2000) and Richter et al. (2007) demonstrate the implementation of CBD in architecture to support knowledge renewal and exchange between designers. Eilouti (2009) further explores the possibility for recycling architectural design knowledge by reuse of design precedents.

In the context of sustainable building design, Xiao et al. (2017) develop an experience mining model for solving green building design problems by CBR, and thereby assist the decision maker in finding solutions. Shen at al. (2017) introduce an integrated system of text mining and CBR for retrieval of similar green building cases when producing new green building designs. In terms of energy efficiency, Abaza (2008) presents a model, where the computer evaluates design alternatives suggested by the designer and generates a matrix of design solutions. More recent approaches include that of Sabri et al. (2017) who apply CBR and graph matching techniques for retrieval of similar architectural floor plans in the early design stages. Ayzenshtadt et al. (2016) investigate the potential of rule-based and case-based retrieval coordination for architectural design search. Weber et al. (2010) propose a sketch-based retrieval system based on CBR and shape detection technologies, which gives access to a semantic floorplan repository. These approaches typically capture semantics in topology graphs, which is less complex and detailed compared to the rich semantics of BIM data.

However, despite coming a step closer to realizing the targeted process, these efforts rely mostly on design patterns for improvement of the design, or use performance patterns for improvement of building and system operation. Using knowledge discovered in performance data to influence design decision- making and improve future building design processes is an area that is rarely explored in detail. Furthermore, the combined use of semantics, KDD, and CBD is seldom achieved. Therefore, in this article, we aim to combine these three approaches for influencing design decisions using both design and operational building data.

## 3 PROPOSED TECHNICAL APPROACH

The way in which design professionals approach decision-making is characterized by iterative problem-solution cycles, in which solutions are widely based on tacit knowledge. Each design iteration explores a problem/solution space, which leads to a repetitive co-evolution of problems and solutions (Dorst and Cross, 2001). Figure 1 depicts that process, during which the design team aims to converge in the problem and solution spaces.

Convergence brings the team closer to a solution that fulfills the design brief and the performance targets, while avoiding widening of the cycles.

A typical design environment may include BIM authoring tools, parametric design tools, simulation tools, etc., by the use of which design professionals iterate through a number of proposals, both individually and in a collaborative manner. The generated design data is stored in the CDE. To be able to influence the above process, performance data and knowledge discovered in data need to be presented to the decision maker in the form of useful design alternatives matching the stated objectives. We therefore aim to connect the active design environment with a repository that collects data available from previous projects and the corresponding existing buildings. The data in the repository has various heterogeneous origins, representations, and purposes. Knowledge Discovery can be applied to this data, thereby following the KDD process defined by Fayyad et al. (1996), which consists of five steps. They include selection, cleansing, transformation, mining, and interpretation/evaluation of the data. It is important to note that a significant part of the workload is dedicated to data selection, cleansing and transformation. Furthermore, the evaluation step is critical to the interpretation of the meaning of the patterns found in the data. This study follows these five steps in creating the repository of design data with associated discovered patterns.

In this study, we aim to connect the outlined repository with the active design environment. This can be any BIM tool or the CDE itself. Recent initiatives aim at making the data available in an integrated manner using web technologies, both in the context of BIM tools and the CDE. In this regard, web technologies can enable a web-compliant and data-oriented information management approach. Such an approach is desirable as it (1) allows the integration of heterogeneous data sources, (2) enables federated query techniques over diverse data repositories for advanced information retrieval and (3) provides a well-defined formal data structure to capture building semantics. This results in a design environment as outlined in Fig. 2, with BIM tools on the left, and a web-based CDE on the right.

The adoption of web technologies for representing information in a design environment can be realized using a decentralized graph database approach. Promising in this regard are linked data and semantic web technologies (Berners-Lee et al., 2001; Pauwels et al., 2017a), which allow to build a decentralized web of semantic information, consisting of various repositories with relevant build-

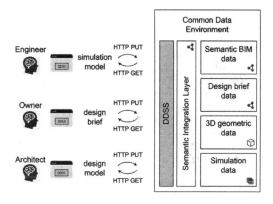

Figure 2. Integration of datasets in a web-based design environment.

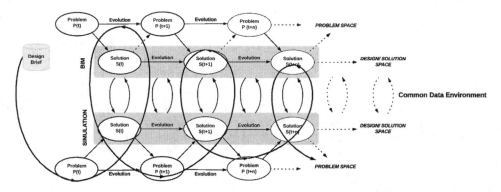

Figure 1. Problem-Solution iterations in collaborative design.

ing data. Such repositories can contain various kinds of data, including design brief data, user logs, BIM models, performance data, etc. For the purpose targeted in this paper, we therefore propose a semantic integration layer, which maintains the links between the individual datasets (Fig. 2). The semantic integration layer has a thin and modular structure which captures the semantics of available data, keeping the original data sources in their optimized structures.

## 4 USE CASE EXPERIMENT

In this section, we test the proposed approach and consider how the active design environment can be connected to a repository of design data that is enriched with patterns obtained using a KDD process, employing motif discovery and association rule mining algorithms (Fu, 2011; Patel et al. 2002). Each part of the experiment is documented here, including the repository (Section 4.1), the active design case (Section 4.2), and matching both (Section 4.3).

### 4.1 *Building the information retrieval repository*

A rich data repository should include heterogeneous data for multiple diverse buildings. That includes not only building models, but also design briefs, simulation and sensor data, and so forth. In this case, however, we limited to working with a collection of building models in the IFC data model, which was previously set up in the context of the performance benchmark in Pauwels et al. (2017b). At that time, the repository consisted of 369 building models of diverse size, origin and kind. The current version of the repository[1] consists of 531 IFC models.

As a first step, all models have been converted into linked data. This step makes the data easy to query, as linked data technologies come with an out-of-the-box query language (SPARQL), as opposed to the STEP and EXPRESS technology used by IFC. This conversion is done using an open source IFC to Linked Building Data converter[2]. The result is a set of RDF graphs in TTL format that are compliant with the BOT (Rasmussen et al., 2017), PRODUCT, and PROPS ontologies[3]. For this study, the conversion to LBD excludes geometry from the data, leaving only the semantic backbone and product data for the building models. Geometric data may be converted to linked data and made available, but is less useful for the purpose of the current semantic information retrieval effort. In order to be useful for information retrieval, the raw geometry should be processed first to contain semantically useful concepts (e.g. above, below, next to), which is out of scope for the current study.

The final result is a collection of two Stardog triple store databases, with in total 36 Million triples (24.951.647 triples and 11.425.589 triples). The data was spread over two databases, aiming to test and validate a decentralized information structure and a federated query approach. The data includes 372 *bot:Building* instances, 3,523 *bot:Zone* instances, 2,117 *bot:Space* instances, and 615,452 *bot:Element* instances. The *bot:Element* instances also have a more specific product type. For instance, one of the repositories includes 45 distinct product types, including *product:Wall*, *product:Fastener*, *mep:FlowTerminal*, *product:Pile*, etc. Each of these instances has a number of associated properties. Clearly, the majority of available triples consists of properties associated to building elements. At the moment, these properties come in various languages and notations, which makes it difficult to query them. Ideally, they should follow an ontology, which is the purpose of the PROPS ontology[4].

For some of the models in this repository, sensor data is available from the corresponding existing buildings. The sensor data is also modelled using linked data best practices[5]. More particularly, we used the SOSA ontology to describe the relationships between the spaces and the contained sensor nodes (data points), each of which has individual sensors, with observations and results. All data modelling is done according to the SOSA ontology, giving a semantic representation of the sensors and their observations and values in context of the spaces. The data values of the sensor data are not directly included in the semantic graph, in order not to make that graph too complex. Instead, links are maintained to the original locations where the sensor data is stored. This is done using a custom *gig:values* datatype property added to specific sensor nodes. These properties point to a web address that returns the data values as requested using the HTTP protocol. One is able to add attributes to an HTTP request, thereby setting query parameters such as time frame and refresh rate (e.g. from=now-30d&to=now&refresh=30s). The result includes the pointer to the data stream for a sosa:Result of a sosa:Observation. A short example snippet is provided in the Listing below:

---

1. http://smartlab1.elis.ugent.be:8889/IFC-repo/.
2. https://github.com/jyrkioraskari/IFCtoLBD.
3. https://www.w3.org/community/lbd/.
4. https://github.com/w3c-lbd-cg/props.
5. https://www.w3.org/TR/ld-bp/.

inst:room_16
  rdf:type bot:Space ;
  gig:hasSensorNode inst:sensorNode_0000014 ;
  gig:spaceType "Cafe" ;
  rdfs:label "Cafe" .

inst:sensorNode_00000014
  rdf:type gig:SensorNode ;
  rdfs:label "00000014" ;
  gig:observation "Indoor climate" ;
  gig:purpose "Thermal comfort in the lobby during big events when there is a gathering of a lot of people." ;
  sosa:hosts inst:sensor_00000014_1 ;
  sosa:hosts inst:sensor_00000014_2 ;
  sosa:hosts inst:sensor_00000014_3 ;
  sosa:hosts inst:sensor_00000014_4 ;
  sosa:hosts inst:sensor_00000014_5 ;
  sosa:hosts inst:sensor_00000014_6 ;
  gig:placement "Placed on a column in the cafe without direct sunlight." .

inst:sensor_00000014_1 ;
  rdf:type sosa:Sensor ;
  sosa:madeObservation inst:observation_1 ;
  sosa:observes inst:obsProperty_1 ;
  rdfs:label "00000014_1" .

inst:result_1 rdf:type sosa:Result ;
  rdfs:label "Result of observation of Relative Humidity" ;
  gig:values "https://gigantium.dk/Gigantium2018instances?orgId=1&datastream=true" .

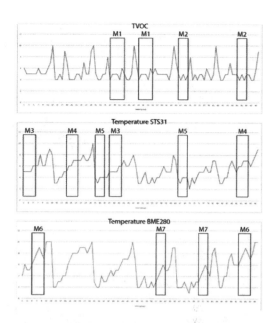

Figure 3. Obtained observation data and discovered patterns.

To make the use of collected sensor data more effective and based on the stated goal, multiple KDD techniques can be applied. We have specifically tested this approach in this study for some of the available sensor data. In this case, a combination of motif discovery and association rule mining has been applied to time series data. The detailed description and implementation of the KDD steps is performed in advance and is out of scope for the current paper. The resulting motifs and the related co-occurrence rules are added to the graph using a separate in-house developed pattern matching ontology. In more detail, a sensor node in the graph is directly linked to an instance of a *pattern:AssociationRule*, which furthermore links to a left hand side and right hand side in the rule. Both left hand and right hand side concepts furthermore link to *pattern:motif* concepts, such as M1 and M5 (Fig. 3).

In this example, these motifs occur in temperature and Total Volatile Organic Compounds (TVOC) observations for a cafeteria in a public building. These motifs are semantically described as well, eventually including the exact data sensor values for those observations.

### 4.2 The active design case

In addition to the repository of design models with sensor data and performance patterns, an active design model was selected, which forms the starting point for knowledge retrieval. We use a design model of a healthcare facility (Figs. 4 and 5), which is a part of the active design environment (in this case Autodesk Revit) and hence can be used to retrieve relevant knowledge from the repository (see Section 3).

The building design consists of two main parts (one wing with public access and another one accessible by medical professionals and patients only), connected with a connection spine. One part of the building contains the entrance, visitors' lobby, cafeteria and public spaces; the other part contains the patient wards, examination, operating and recovery rooms, staff rooms, etc. The basement area contains all necessary technical and equipment rooms.

In addition to the BIM design model in Revit, a number of design brief requirements are available. As they were unstructured in this case (textual document), we decided to not use them, in contrast to what is often done in related works with CBR and text mining techniques. Instead, we consider in our

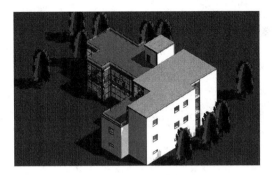

Figure 4. Revit design model of a healthcare facility.

Figure 5. Revit design model of a healthcare facility.

case a semantic building model directly linked to a semantic representation of design brief requirements. This semantic data can be used to perform case retrieval in the repository documented before, to then inform the designer of factual performance data in existing buildings.

4.3 *Information retrieval and pattern matching*

In order to obtain reference knowledge from the building data repository, a direct matching needs to be made between the new case and existing cases (cfr. CBR). Such matching can occur in a number of ways. As we have seen, most existing works perform geometric spatial layout matching using topology graphs. Even though many of these topology graphs have some semantics, the available semantics is in this case a lot more complex and rich. The semantics embedded in the design brief and the design model allows to perform semantically more specific queries and thus better matching. This does not rule out topology graph matching. Also user action log data can be useful for retrieving relevant cases. Depending on what actions the users take, their intentions may be tracked in a more intelligent way, thus improving the matches with the building knowledge repository.

As preliminary design decisions are made in an early planning stage that relies heavily on space types and configurations, for this case study we focus on matching cases based on space type.

Obviously, a full implementation can take into account a lot more of the available semantic data, aiming to match system configurations, material choices, expected usage patterns, and so on.

Matching the active design model is thus implemented using SPARQL queries, such as the one listed in Listing 2. This query shows how the repository is queried for buildings with spaces of type "cafeteria", aiming to retrieve not only those buildings, but also the corresponding performance data and patterns obtained using the data mining techniques briefly mentioned in Section 4.1. Querying is done through a federated query approach. The two repositories that are built for this use case are queried using the SERVICE keyword, as indicated in the Listing below:

```
SELECT ?b ?s ?o
WHERE {
SERVICE
<http://localhost:5820/BuildingRepo/query> {
    ?b rdf:type bot:Building .
    ?b bot:hasSpace ?s .
    ?s rdf:type bot:Space .
    ?s props:categoryDescription "cafeteria" }
SERVICE
<http://localhost:5820/BuildingRepo1/query> {
    ?b rdf:type bot:Building
    ?b bot:hasSpace ?s .
    ?s rdf:type bot:Space .
    ?s props:categoryDescription "cafeteria" }
}
```

These queries can be implemented in a plug-in for the corresponding design environment, or directly from the CDE, in which case more alternative data is available (briefs, logs, simulation data). The returned Unique Resource Identifiers (URIs) for spaces and buildings provide reference points for obtaining more data. These URIs can be used by plugins or CDE to subsequently query for building performance patterns that are available for the retrieved buildings and spaces.

In our case, the query in Listing 2 returns, among others, a cafe that is part of a visitors' lobby in a sports and cultural centre, for which operational data and performance patterns are available (Fig. 3). This data can be directly provided to the end user. Hence, users can be provided not only with a link to sample existing buildings of the kind they are developing (in this case, the bar in the hospital), they can also retrieve the knowledge about that place which is captured in patterns obtained from a KDD process. Being able to obtain this information during a design process is considered of utmost relevance in informing design decisions.

## 5 RESULTS AND DISCUSSION

The presented use case with the data repository consisting of 531 models and the healthcare facility design model provides a useful context to evaluate the proposal for decision support using a combination of CBR, KDD, and semantics from within a BIM environment. Results and discussion thus focus on those three main topics.

First and foremost, CBR provides a useful theoretical background for the given proposal around design decision support. In order to be fully effective, it would be useful to extend the amount and diversity of the data that is used, both to document the cases in the repository and to inform queries. In this regard, the availability of a CDE with user log data, design requirements, and performance data is potentially of tremendous relevance.

Second, the semantics provide effective and rich means to retrieve relevant cases. The semantic richness provides great opportunities to outperform case retrieval using topology graphs and text mining approaches. Nevertheless, there are also some boundaries. Namely, the effectiveness of the system relies a lot on the expressiveness and formal rigor of the ontologies used for capturing semantics. In this case, the *props:categoryDescription* predicate was used, for example, to retrieve spaces of a particular type; yet, very different predicates are used as well, making it difficult to cover an entire dataset. Also, the diversity of languages in a dataset is difficult to cope with. In this regard, a data dictionary that provides translations between terms is of important relevance.

Finally, one of the most important parts is the KDD process involved in retrieving the performance patterns and associations between them. The KDD process itself is out of scope in this article, yet, the results of that process are directly embedded in the knowledge graph. On-demand data mining is thus not performed. Such on-demand data mining, as well as the actual interpretation/evaluation of the discovered patterns is essential to turning the results into actionable knowledge. Therefore, user-driven KDD may be of relevance to be considered in future research.

## 6 CONCLUSION

In this article, we look into the ways in which knowledge about existing buildings and their performance patterns can be made accessible in an active design environment to give design processes a more evidence-based character. We particularly investigate how existing CBR approaches can be improved using a combination of BIM, KDD, and semantic data modelling, thereby aiming to enable BIM-based information retrieval in support of sustainable design. The article presents a technical approach, which indicates how decision support can be embedded in a BIM-based design environment and common data environment (CDE). The proposed technical approach is tested in a case study environment consisting of a building data repository and active design model of a healthcare facility. The building data repository consists of 531 building models, for some of which sensor data is available. All data is represented in semantic graphs and made available in a triple store using latest developments and techniques in linked data best practices. Data mining is performed over the sensor data using motif discovery and association rule mining. Finally, a number of semantic queries show how cases can be retrieved that match the active design model, including the retrieval of performance patterns. As such, the potential of the proposed technical approach is demonstrated for case retrieval in support of evidence-based sustainable design.

## REFERENCES

Aamodt, A., Plaza, E. (1994). Case-Based Reasoning: Foundational Issues, Methodological Variations, and System Approaches. AI Communications. IOS Press, Vol. 7: 1, 39–59.

Ahmed, A., Korres, N.E., Ploennigs, J., Elhadi, H., Menzel, K. (2011). Mining building performance data for energy efficient operation. Advanced Engineering Informatics 25, 341–354.

Ayzenshtadt, V., Langenhan, C., Roth, J., Bukhari, S.S., Althoff, K.-D., Petzold, F., and Dengel, A. (2016). Comparative evaluation of rule-based and case-based retrieval coordination for search of architectural building designs. 24th International Conference on Case Based Reasoning, Atlanta, GA, USA. Springer, Berlin, Heidelberg.

Berners-Lee, T., Hendler, J. & Lassila, O., 2001. The Semantic Web, Scientific American, pp. 29–37.

Bilal, M., Oyedele, L., Qadir, J., Munir, K., Ajayi, S., Akinade, O., Owolabi, H., Alaka, H., Pasha, M. (2016). Big Data in the construction industry: A review of present status, opportunities, and future trends. Advanced Engineering Informatics 30, 500–521.

Bishop, C.M. (2006). Pattern Recognition and Machine Learning. Springer.

Dave, B., Schmitt, G., Faltings, B., Smith, I. (1994). Case based design in architecture. Artificial Intelligence in Design- AID '94, J. Gero and F. Sudweeks (eds.), Kluwer Academic Publishers, Dordrecht, The Netherlands, 1994, 145–162.

Dorst, K. & Cross, N. (2001). Creativity in the design process: coevolution of problem-solution. Design Studies 22(5), 425–437.

Drgoňa, J., Picard, D., Kvasnica, M., Helsen, L. (2018). Approximate model predictive building control via machine learning. Applied Energy 218, 199–216.

Elouti, B.H. (2009). Design knowledge recycling using precedent-based analysis and synthesis models. Design Studies, 30, 340–368.

Fan, C., Xiao, F., Madsen, H. & Wang, D., (2015a). Temporal knowledge discovery in big BAS data for building energy management. Energy and Buildings, Vol 109, pp. 7589.

Fan, C., Xiao, F. & Yan, C., (2015b). A framework for knowledge discovery in massive building automation data and its application in building diagnostics. Automation in Construction, Vol 50, pp. 8190.

Fan, C., Xiao, F., Li, Z., Wang, J. (2018). Unsupervised data analytics in mining big building operational data for energy efficiency enhancement: A review. Energy and Buildings, 159, 296–308.

Fayyad, U., Piatetsky-Shapiro, G., Smyth, P. (1996). From Data Mining to Knowledge Discovery in Databases. AI Magazine 17(3), 37–54.

Fu, T.C. (2011). A review on time series data mining, Engineering Applications of Artificial Intelligence, 17, 164–181.

Geyer, P. and Schlueter, A. (2014). Automated meta-model generation for Design Space Exploration and decision-making–A novel method supporting performance-oriented building design and retrofitting. Applied Energy, 119, 537–556.

Hand D., Mannila H., Smyth P. (2001). Principles of Data Mining. MIT Press, Cambridge.

Hamilton, D.K. and Watkins, D. (2009). Evidence-Based Design for Multiple Building Types. John Wiley & Sons, New Jersey, USA.

Heylighen, A., Neuckermans, H. (2000). DYNAMO: A Dynamic Architectural Memory On-line. Educational Technology & Society (3)2.

Kim, H., Stumpf, A., Kim, W. (2011). Analysis of an energy efficient building design through data mining approach. Automation in Construction, 20, 37–43.

Liu, Y., Huang, Y.C., Stouffs, R. (2015). Using a data-driven approach to support the design of energy-efficient buildings. Journal of Information Technology in Construction, 20, 80–96.

Østergård, T., Jensen, R.L., Maagaard, S.E. (2018). A comparison of six metamodeling techniques applied to building performance simulations. Applied Energy, 211, 89–103.

Patel, P., Keogh, E., Lin, J., Lonardi, S. (2002). Mining Motifs in Massive Time Series Databases. In proceedings of the 2002 IEEE International Conference on Data Mining.

Pauwels, P., Zhang, S. & Lee, Y.C. (2017a). Semantic web technologies in AEC industry: A literature overview. Automation in Construction 73, 145–165.

Pauwels, P., de Farias, T.M., Zhang, C., Roxin, A, Beetz, J., De Roo, J., Nicolle, C. (2017b). A performance benchmark over semantic rule checking approaches in construction industry. Advanced Engineering Informatics 33, 68–88.

Rasmussen, M.H., Pauwels, P., Hviid, C.A. & Karlshøj, J. (2017). Proposing a central AEC ontology that allows for domain specific extensions. Proceedings of the Joint Conference on Computing in Construction (JC3), 237–244.

Richter, K., Heylighen, A., and Donath, D. (2007). Looking back to the future-an updated case base of case-based design tools for architecture. Knowledge Modelling eCAADe, 25, 285–292.

Sabri, Q.U., Bayer, J., Ayzenshtadt, V., Bukhari, S.S., Althoff, K.D., Dengel, A. (2017). Semantic Pattern-based Retrieval of Architectural Floor Plans with Case-based and Graph-based Searching Techniques and their Evaluation and Visualization. In Proceedings of the 6th International Conference on Pattern Recognition Applications and Methods, 50–60.

Shen, L., Yan, H., Fan, H., Wu, Y., Zhang, Y. (2017). An integrated system of text mining technique and case-based reasoning (TM-CBR) for supporting green building design. Building and Environment, 124, 388–401.

Soibelman, L. and Kim, H. (2002). Data preparation process for construction knowledge generation through knowledge discovery in databases. Journal of Computing in Civil Engineering, 16(1), 39–48.

Weber M., Langenhan C., Roth-Berghofer T., Liwicki M., Dengel A., Petzold F. (2010) a.SCatch: Semantic Structure for Architectural Floor Plan Retrieval. Case-Based Reasoning. Research and Development. Lecture Notes in Computer Science, vol 6176. Springer, Berlin, Heidelberg.

Xiao, X., Skitmore, M., Hu, X. (2017). Case-based reasoning and text mining for green building decision making. Energy Procedia, 111, 417–425.

Yarmohammadi, S., Pourabolghasem, R., Shirazi, A., Ashuri, B. (2016). A sequential pattern mining approach to extract information from BIM design log files. 33rd International Symposium on Automation and Robotics in Construction., 174–181.

Yarmohammadi, S., Pourabolghasem, R., Castro-Lacouture, D. (2017). Mining implicit 3D modeling patterns from unstructured temporal BIM log text data. Automation in Construction, 81, 17–24.

# Managing space requirements of new buildings using linked building data technologies

M.H. Rasmussen, C.A. Hviid & J. Karlshøj
*Department of Civil Engineering, Technical University of Denmark, Kgs. Lyngby, Denmark*

M. Bonduel
*Department of Civil Engineering, Technology Cluster Construction, KU Leuven, Ghent, Belgium*

ABSTRACT: Any stakeholder operating in the AEC industry knows that designing a building is a complex and highly iterative task. The project evolves over time and changes happen rapidly, meaning that design requirements, as well as solutions (often as a consequence), must undergo revision. Since building requirements are, however, documented and handled in a predominantly manual manner, the work processes are not aligned with the dynamic nature of the projects. Tracking and acting upon changes is a manual, and therefore an error-prone and labour intensive task. In this article, we suggest a generic method for working with the concept of spaces at different abstraction levels in order to compare requirements with actual properties in a non-static manner using semantic web technologies, primarily developed by the W3C Linked Building Data (LBD) Community Group. The generic modelling approach has the potential of also being applied to other concepts than building spaces.

## 1 INTRODUCTION

When buying a product you can rightly expect it to correspond to the technical specifications on which the purchase was originally based. When buying a building, however, the reality is unfortunately not always so (Kiviniemi 2005). Bertelsen (2003) describes construction as a complex system because of three main characteristics: (1) autonomous agents (2) undefined values and (3) non-linearity. Delivering a complete product specification in the form of a building program at day 1 is nearly impossible as everyone gains knowledge and insights as the design evolves, and as a result, the building program itself cannot be static during the design. The documentation and handling of it, therefore, needs to be dynamic, which is unfortunately typically not the case (Kiviniemi 2005). The majority of building design processes are today characterized by manual information extraction from static documents, and as the design progresses it becomes a cumbersome task for the project participants to keep track of, and meet the evolving client requirements. Because of the predominantly manual information handling, the quality of information exchange between project stakeholders is furthermore highly determined by the social capabilities and communicative skills of the individual practitioners (Bendixen 2007). This is a challenge that the methodology of Building Information Modelling (BIM) will hopefully remedy over time. However, unfortunately the BIM authoring tools of today are not delivering satisfactory interoperability, and data is therefore often trapped in data silos (Terkaj 2017).

In this article, we first provide a brief overview of existing software and data modelling approaches that focus on building requirements specification. We then argue why we believe semantic web technologies can possibly provide the means to overcome current challenges when dealing with the dynamic behaviour of building requirements. Based on knowledge manually deduced from existing document-based building programs and discussions with practitioners in the consulting engineering company, Niras, we have defined a set of competency questions. These were used as constraints for what the data model should be capable of. The model was developed accordingly, chiefly by using terminology defined in already existing and widely adopted ontologies. Lastly, we developed a set of tests to evaluate the modelling approach on the Common BIM Model "Duplex Apartment"[1]. The dataset was established partly by manually defining requirements as an RDF-graph (Resource Description Framework) following the suggested modelling approach, and partly by using a custom developed exporter for the BIM authoring tool, Revit[2]. The

---
[1] https://www.nibs.org/?page=bsa_commonbimfiles#project1.
[2] https://github.com/MadsHolten/revit-bot-exporter.

latter establishes an RDF-graph using ontologies provided by the World Wide Web Consortium Linked Building Data Community Group (W3C LBD-CG).

### 1.1 Open standards

The effort of storing knowledge in a construction project, including the information exchange between its stakeholders, has been addressed by the buildingSMART organisation. With standards such as Industry Foundation Classes (IFC) (Liebich and Wix 1999), Information Delivery Manuals (IDM) and Model View Definitions (MVD) they deliver a solid framework for information exchange and storage.

The W3C also has made efforts to standardize information exchange using semantic web technologies such as the Web Ontology Language (OWL) to construct formal vocabularies to describe a certain domain of interest. The scope of these technologies is not limited to the AEC industry alone, and therefore researchers and practitioners from a wide variety of domains are contributing to their continuous development.

One main difference between the above two methodologies is that OWL relies on an Open World Assumption (OWA), meaning that the schema can evolve over time to include concepts not initially thought of. This is quite different from typical database systems that depend on a Closed World Assumption (CWA) for defining schemas, such as IFC. Another benefit is that the full dataset does not need to be available at one location but can be combined with other datasets as needed, being both Linked Open Datasets (LOD) available online (material data, weather data, geographical data etc.) and private datasets, possibly hosted by other project stakeholders. Owners of such private datasets can restrict the access to specific partners.

The W3C Resource Description Framework (RDF) standard is used to describe Linked Data in a directed graph consisting of a collection of triples. A triple has three parts: a node (the subject), an edge (the predicate) and another node (the object) connected to the first node through the predicate-edge. All sub-elements of a triple are made globally unique by denoting them with a Uniform Resource Identifier (URI) except for objects that are literal values such as strings, integers, Booleans etc. The datatype of such literals are also described with a URI, and is often defined in an ontology version of the Extensible Markup Language (XML) Schema Definition (XSD). Both the terminology layer (TBox) – including semantics for classes and properties, and the data layer (ABox), covering individual instances and their interrelations, are described using RDF. The W3C encourages developers to make their ontologies publicly available so that useful ontology-related information can be retrieved from the URI. To continue, the W3C recommends that terms from widely adopted ontologies are used to explicitly describe the data layer.

An RDF graph is traversed using the SPARQL Protocol and RDF Query Language (SPARQL) and if it is described using widely adopted ontologies it is possible to structure generic, globally applicable queries to deduce knowledge. The semantics described in the TBox also allow reasoning engines to deduce implicit knowledge from what is explicitly defined in the ABox. A simple example: If chair is a sub-class of furniture (TBox), then all instances of chair are also instances of furniture (ABox).

### 1.2 Cloud-based BIM solutions

Although building programs are typically defined in static documents (Word, PDF) there are a few cloud-based BIM applications for building requirements management on the market. They typically consist of a user interface (UI) that enables the user to do create, read, update and delete (CRUD) operations on requirements stored in a central database along with a communication link to native BIM authoring tools. Since each internal database has a closed proprietary schema rather than a schema defined according to the previously described open standards, interlinking the requirements to information that exists outside the application is not easily accomplished. Additionally, migrating from one tool to another is seen as a cumbersome task. Some applications do offer a REST (representational state transfer) API (application programming interface) providing a machine-accessible interface to the internal data model. However, the design of this interface is also following a proprietary schema and therefore a deep understanding of this schema is a prerequisite for interpreting and using the data in other applications.

Onuma and dRofus are examples of BIM applications for requirements management that offer a REST API to interact with the data model[3,4], and they use XML and JavaScript Object Notation (JSON) respectively as data format. Both APIs offer only limited interaction with the data model and although accessible from outside, they are tightly coupled to their native data models.

The SPARQL Protocol (Feigenbaum et al. 2013) and SPARQL Graph Store HTTP protocol (Chimezie Ogbuji 2013) are W3C recommendations specifying how to make an RDF-graph

---

[3] http://www.onuma-bim.com/platform/api.
[4] https://wiki.drofus.com/display/DV/REST+API.

available through a REST architecture. Accessing the graph is achieved by sending a SPARQL query to a URI hosting a SPARQL endpoint, and this provides an interface for clients to do CRUD operations on the dataset. A cloud-based BIM tool using the W3C open standards to describe the schema could host a SPARQL endpoint in order to allow clients to access the data model using standardised SPARQL queries, but to our knowledge, no such tool currently exists.

## 1.3 Linked building data

Research has provided us with several examples of how semantic web technologies can be used to enhance data handling in the AEC industry. The typical research contribution is an ontology which describes a subset of the construction domain with a distinct scope such as smart homes and sensor data or even the construction domain as a whole. Pauwels & Terkaj (2016) proposed ifcOWL as the OWL-based counterpart for the IFC schema and probably the most widely adopted ontology in the AEC domain.

It has later been argued that this quite literal conversion of the IFC schema is not appropriate as it (1) contains artefacts from the EXPRESS schema from which it originates making queries less logic and (2) describes too wide a scope, thereby violating the W3C best practice of omitting redundancy and making it hard to get familiarized with (Pauwels & Roxin 2016; Rasmussen et al. 2017a).

Another, more modular approach for building-related ontologies is suggested by the W3C LBD-CG. A minimal ontology, the Building Topology Ontology (BOT) (Rasmussen et al. 2017a) describes the main concepts of a building and thereby serves as an extensible core for describing any concept in its context of a building. Another ontology, PROPS, describes building-related properties and is at the time of writing a conversion of the properties contained in the IFC4 schema[5]. The conversion approach is also used in the PRODUCT ontology which describes building-related products. Finally, the Ontology for Property Management (OPM) extends concepts from the Smart Energy-Aware Systems (SEAS) ontology to provide the means to describe property reliability as well as property changes over time using property states.

Both the IFCtoLBD-converter[6] (Bonduel et al. 2018) and an exporter for Revit[7] (Rasmussen et al. 2017b) generate LBD compliant RDF triples from conventional BIM models.

In this study we have used and extended a set of widely adopted web ontologies for property handling (schema.org/goodrelations), provenance data (PROV-O), literal units (Unified Code for Units of Measure (UCUM) (Lefrançois 2018)) along with the earlier mentioned LBD ontologies. Using these ontologies in combination with OWL description logics, we illustrate an approach for specifying project specific space classes that explicitly state the client's requirements. We further show how the architectural spaces can automatically inherit requirements based on the class they are assigned to using standard OWL reasoning engines. Queries to compare and evaluate requirements to actual properties of the space instances are further illustrated and a simple use case, is presented to simulate both requirement and property changes and the handling of these.

## 2 REQUIREMENTS MODELLING

In this section we illustrate how concepts defined in the BOT, OPM and schema.org ontologies can be used to model space requirements. Initially, various client requirements specifications for construction projects in which Danish consulting company Niras has been involved, were reviewed. In these specifications, it is common practice to specify space requirements at *type* level rather than at *instance* level.

IFC and various BIM authoring tools use the concept of *types* and include a mechanism for inheriting properties of a *type* to *instances* belonging to that *type*. *Instances* can further extend the set of properties at an individual level and properties can even be overridden (Borgo et al. 2014). It is clear that the instances belong at ABox level, but the concepts of space and object *types* are less obvious. In BIM tools, space and object *type instances* are defined at the data layer rather than the schema layer, but from an ontology engineering perspective, it would arguably be more correct to consider the *type instances* themselves at schema level.

In the following section, we will investigate a TBox modelling approach of *space types* that must be capable of providing answers to the following competency questions:

– *CQ1:* How to model a *space type*?
– *CQ2:* How to assign a quantitative requirement to a *space type*?
– *CQ3:* How to state that a *designed space* instance matches a *space type* of the client's requirements specification?
– *CQ4:* How to check if a property that also exists as a requirement is fulfilled by the architectural design?

---
[5]https://github.com/w3c-lbd-cg/props/blob/master/IFC4-output.ttl.
[6]https://github.com/jyrkioraskari/IFCtoLBD.
[7]https://github.com/MadsHolten/revit-bot-exporter.

- *CQ5:* How to check an adjacency or quantity requirement?
- *CQ6:* How to update a *space type* and its assigned requirements?

### 2.1 *CQ1: Modelling a space type*

Modelling a *space type* is achieved by defining a project-specific extension of <u>BOT</u>, in this case in the namespace of the building client. In Figure 1 the class `client:spacetype_bathroom1` is defined as a sub-class of `bot:Space` meaning that any instance of the class will be classified as a `bot:Space`. The `rdfs:label` and `rdfs:comment` are widely adopted predicates from the RDF Schema (RDFS) that provide a human-readable specification of the class. In this example, in Danish and English language.

### 2.2 *CQ2: Assigning a quantitative requirement*

In order to meet the demands for modelling a *space requirement*, it should be possible to capture the following information:

- Range, (minimum and maximum) or specific value to be matched
- Quantitative unit of the value
- Property changes over time (deleted, modified).

OWL includes logics to describe property restrictions for classes. For example, it is possible to describe that `that:BlueCars` is a sub-class of all cars that have a blue color, which entails that every instance of `that:BlueCars` class will consequently be blue. Figure 2 illustrates how an `owl:Restriction` can be used to describe that all instances of `client:spacetype_bathroom1` have a `props:area` with the value `client:property_001`. This objectified property belongs to the ABox of the client's dataset, which allows it to evolve over time.

Rasmussen et al. (2018) describe three levels of complexity for assigning properties to some feature of interest (FoI). Level 3, the most expressive form, satisfies the demand of allowing property changes over time and is therefore used to model *space requirements*. Figure 2 illustrates how the property has a property state (`client:state_p001_001`) assigned. This state is currently classified as the `opm:CurrentPropertyState`, which indicates that it is the most recent state of the property but this might change over time as the client requirements are revised. A new class opm:Required which we suggest to implement as an extension of <u>OPM</u> is used to specify that the state is a requirement rather than a designed property. A value range is specified using properties defined in <u>schema.org</u> and the generation time is captured using <u>PROV-O</u>. The unit is given as part of the value string using a custom datatype based on <u>UCUM</u>. Further metadata such as who created the property state for which reason can also be attached.

### 2.3 *CQ3: Mapping designed space instances to spaces requested by the client*

At one point, as the architectural design progresses, the architect's dataset will hold a number of *designed spaces* that should match the *space types* required by the client. At this point, the architectural spaces are geometrically defined, and therefore they have an actual area.

Mapping a *designed space* to a client *space type* is handled by stating that the *designed space* is an instance of the specific *space type* class. Figure 3 illustrates how properties of the client *space type* (`client:spacetype_bathroom1`) are inherited to all instances of this class. In this example, spaces `inst:room123` and `inst:room213` both

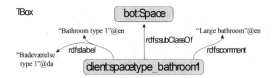

Figure 1. Modelling a *space type* with BOT.

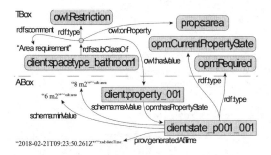

Figure 2. Assigning a requirement using (Rasmussen et al. 2018) Level 3.

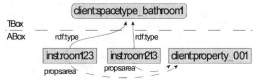

Figure 3. Two designed spaces are classified as client:spacetype_bathroom1. Therefore the properties (requirements) of the client *space type* are inherited by the *designed spaces*.

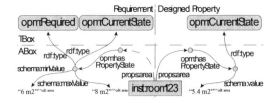

Figure 4. Requirement vs. property.

inherit `client:property_001` (and its property state) as the value for property `props:area`.

### 2.4 CQ4: Checking that a requirement is fulfilled

When the same space property exists both as a requirement and a designed property it is possible to do a comparison in order to check if the requirement is met. Figure 4 illustrates `inst:room123` which has the property `props:area` assigned twice. Explicitly as a result of its geometry and implicitly as a requirement inherited by the mechanism described in Figure 3. Performing the comparison is possible by traversing the graph using a SPARQL query.

Listing 1 shows a SPARQL query to retrieve all violations of the `props:area` requirement in the model, when both the requirements and designed properties are all in one database. The query is structured as a graph traversal which operates by matching the defined patterns. The first triple pattern maps anything that is an instance of `bot:Space` to the variable `?space`. The next pattern is a sub-query which is used to get data from the state of `props:area` that is classified as `opm:Required`. The variable `?space` is used to match the same space, and the URI of the property object is mapped to variable `?reqURI`. All states of the property are assigned to variable `?reqState` but the next two triples limit the result to only include the one state which is both classified as `opm:Required` and `opm:CurrentPropertyState`. Since a requirement can be specified either as an exact match or as a range, each of the `schema:value` patterns are optional.

A similar pattern is used to get the actual property and by using a filter it is ensured that the requirement is not assigned to variable `?propURI` (since both match the pattern). The value of `?propURI`'s latest state is assigned to variable `?val` and compared to the required range to check if it is violated. A result is returned only if the requirement is violated.

Replacing `?space` with `inst:room123` or the URI of any other space will return violated requirements for this particular space and this approach can be used to switch any variable with a constant.

Listing 1. SPARQL query to retrieve violated requirements.

```
SELECT *
WHERE {
  # Must be a space
  ?space rdf:type bot:Space .

  # Sub-query to get requirement
  {
    SELECT ?space ?reqURI ?reqVal ?reqMax ?reqMin
    WHERE {
      ?space      props:area            ?reqURI .
      ?reqURI     opm:hasPropertyState  ?reqState .
      ?reqState   rdf:type opm:Required .
      ?reqState   rdf:type opm:CurrentPropertyState .

      OPTIONAL {?reqState schema:value ?reqVal}
      OPTIONAL {?reqState schema:minValue ?reqMin}
      OPTIONAL {?reqState schema:maxValue ?reqMax}
    }
  }

  # Get property
  ?space props:area ?propURI .
  FILTER(?propURI != ?reqURI) # Disjoint from req
  ?propURI opm:hasPropertyState ?propState .
  ?propState rdf:type opm:CurrentPropertyState .
  ?propState schema:value ?val

  # Compare requirements to actual value
  BIND( ?value != ?reqVal AS ?matchViolated  )
  BIND( ?value <  ?reqMin AS ?minViolated    )
  BIND( ?value >  ?reqMax AS ?maxViolated    )

  # Show only results where a requirement is
  # violated
  FILTER( ?matchViolated || ?minViolated ||
                                    ?maxViolated )
```

### 2.5 CQ5: Adjacency and quantity requirements

Specifying adjacency or quantity requirements is not different from any other requirement. However, special queries must be used to check whether these are violated. The same is the case for other requirements such as zone or element containment.

Checking if the required quantity of spaces of a certain *space type* is met, is accomplished by the query shown in Listing 2. Accessing the requirement can be done in the main query since it is not necessary to distinguish between two properties of the same kind, but in order to count the number of *space type* occurences, a sub-query is necessary. Listing 2 shows the optional sub-query to count the number of *designed space* instances per client *space type*. Each *space type* is assigned a unique `?reqURI` for the `props:quantity` property requirement, so this can be used for the grouping. This query is executed before continuing to the next step where requirement `props:quantity` is compared to `?value`.

Listing 2. Sub-query to count number of *designed space* instances that have a specific quantity requirement assigned.

```
{
  SELECT ?reqURI (COUNT(?reqURI) AS ?qty)
  WHERE {
    ?space props:quantity ?reqURI .
  } GROUP BY ?reqURI
}
```

Finding violated adjacency requirements is likewise handled by first getting the requirement (like illustrated in Listing 1). Also in this case it can be done in the main query, and this time it is only necessary to get the `schema:value` and bind it to `?reqVal`. By using the MINUS clause a result is only returned if the space does not have an adjacency to a *designed space* defined as an instance of the required client *space type*.

Listing 3. SPARQL query to retrieve violated adjacency requirements.

```
# Return result if the space does not have an
# adjacent space of the required type
MINUS {
    ?space      bot:adjacentZone ?adjSpace .
    ?adjSpace rdf:type           ?reqVal .
}
```

### 2.6 CQ6: Performing updates

Changes to property requirements are according to Rasmussen et al. (2018) handled by creating a new current state and removing the `opm:CurrentPropertyState` from the evaluation that was previously defined as the current state. This can be achieved with an update query which can be generated using the OPM query generator JavaScript library[8]. Since all the queries explicitly look for the current state of both properties and requirements, the evaluations will automatically reflect the changes.

## 3 USE CASE

To illustrate a possible workflow for modelling, mapping and evaluating requirements, a simple use case was set up. The Common BIM Model "Duplex Apartment" was used as a reference, and the Revit BOT exporter[9] plugin (Rasmussen et al. 2017b) was extended to include the concept of *space types* and OPM *property states*. The exporter was used to export the architectural model in LBD format. The steps to establish the dataset were the following:

1. Define client requirements in RDF. (This step should preferably be accomplished through a UI)
2. Run BOT exporter in Revit to:
   - Create and assign a Revit URI parameter to spaces and elements
   - Create Revit SpaceTypeURI parameter
   - Export BOT relationships and properties to RDF
3. Specify *space type* URI corresponding to the URI used for the client *space type* and re-export triples (Figure 5).

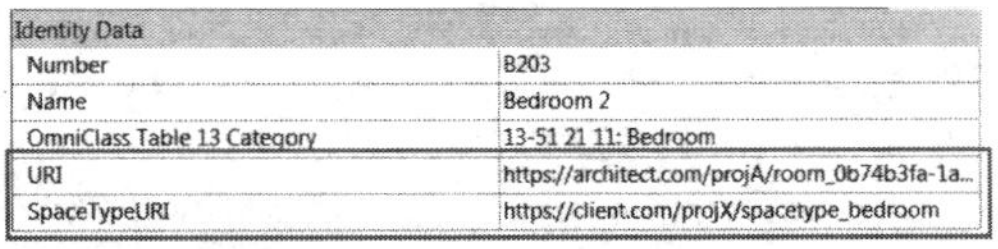

---
[8]https://www.npmjs.com/package/opm-qg.
[9]https://github.com/MadsHolten/revit-bot-exporter.

| Identity Data | |
|---|---|
| Number | B203 |
| Name | Bedroom 2 |
| OmniClass Table 13 Category | 13-51 21 11: Bedroom |
| URI | https://architect.com/projA/room_0b74b3fa-1a... |
| SpaceTypeURI | https://client.com/projX/spacetype_bedroom |

Figure 5. Revit shared parameters for URI and SpaceTypeURI.

4. Use Dynamo script to export zone adjacencies to RDF. This functionality will be implemented in the exporter plugin in the future.

Once the dataset was available it was loaded into a triplestore in order to do the checks described in the previous section. The checking is implemented in a JavaScript based testing tool that is available online[10], while the results are presented here.

### 3.1 Testing property requirements

All *space types* in the test have an area requirement specified. In general, the areas are fulfilled by the designed spaces, except for `inst:spacetype_bedroom` and `inst:spacetype_bathroom1`. Once the dataset is loaded into the triplestore, the tool performs the query from Listing 1 to find area requirement violations. Listing 4 shows the results.

Listing 4. Test tool output for violated property requirements. Numbers in parenthesis are (actual/range).

```
- 'Bathroom 2 B204' violates req. (5.44/(6-))
- 'Bathroom 2 A204' violates req. (5.42/(6-))
- 'Bedroom 1 B202'  violates req. (26.12/(20-25))
- 'Bedroom 2 B203'  violates req. (26.18/(20-25))
- 'Bedroom 2 A203'  violates req. (26.18/(20-25))
- 'Bedroom 1 A202'  violates req. (26.12/(20-25))
```

### 3.2 Checking quantity of spaces

Some *space types* have a requirement for quantity of *designed space instances*, and for `inst:spacetype_living_room` a requirement of seven occurrences is specified, which is not fulfilled in the case of the Duplex house model. A query to group the rooms into apartments based on the room numbers (which are suffixed with either A or B) was implemented in the test tool. The query from Listing 2 was modified slightly in order to accommodate this before counting the number of *designed space* occurrences for each *space type*. The result of this query was, correctly, that the requirement was not met as there is only one living room per apartment in the Duplex model.

Listing 5. Test tool output for violated quantity requirements. Numbers in parentheses are (actual/range).

```
- 'Living Room A102' (1/7)
- 'Living Room B102' (1/7)
```

---
[10]www.student.dtu.dk/~mhoras/ecppm2018/test.zip.

## 3.3 Testing adjacency requirements

Two adjacency requirements were given as a client requirement:

- `spacetype_living_room/spacetype_kitchen`
- `spacetype_bedroom/spacetype_bathroom1`.

The query from Listing 3 revealed that requirement 2 is only fulfilled by one of the bedrooms in each appartment, which is correct.

## 3.4 Changing requirements

By performing four SPARQL update queries, three client requirements were revised and a new one was added:

- Area requirement for `spacetype_bathroom1` relaxed from 6 m$^2$ to 5 m$^2$.
- `props:quantity` for `spacetype_living_room` relaxed from 7 to 1.
- New *space type* `spacetype_bedroom2` with `props:quantity` requirement of 1 and area requirement of minimum 9 m$^2$ added.
- Adjacency requirement between `spacetype_bedroom` and `spacetype_bathroom1` deleted by appending new `opm:PropertyState` of class `opm:Deleted`.

Re-running the tests from section 3.1 and 3.2 now concludes that the area requirements of 'Bathroom 2 A204' and 'Bathroom 2 B204', the `props:quantity` requirement for `spacetype_living_room` and the adjacency requirements for 'Bedroom 1 A202' and 'Bedroom 1 B202' are no longer violated.

The requirements for the new bedroom type cannot be evaluated with the queries presented in Section 2 since the class is not assigned to any spaces. In order to check for required spaces which have not been instantiated, one must do a query starting from the client *space type* itself, and even though this is less intuitive, it is possible. Listing 6 shows a query pattern to retrieve a *space type* which has a quantity requirement assigned, but is not instantiated.

Listing 6. Find *space types* with a quantity requirement but no instances.

```
# GET QUANTITY REQUIREMENT
?spaceType rdfs:subClassOf [
  rdf:type owl:Restriction ;
  owl:onProperty props:quantity ;
  owl:hasValue ?reqURI
] .
MINUS { ?space a ?spaceType }
```

Since the initial requirements are all available in the model, the architect is able to track the changes and relate a property compliance check to a certain state of a requirement.

## 4 CONCLUSIONS AND FUTURE WORK

The main outcome of this work is the illustration of how to use semantic web technologies and existing ontologies, BOT in particular, to establish a knowledge model of requirements for spaces of a new building. The model illustrates an approach to describe space requirements at *type level* in a way that utilizes OWL reasoning capabilities thereby providing best practice examples of how to extend BOT at project level.

In the use case presented in this work, *designed architectural spaces* inherit properties of the *space types* described by the client. The same approach could be used for (1) other features of interest such as building elements or the building as a whole or (2) other generalisations such as an automation control strategy. In the use case, the requirements were modelled manually, but it is obviously not practical for practitioners to do this, so some CRUD application with a user-friendly UI should be developed.

Another interesting use case to investigate is *derived requirements*. Specific requirements such as minimum and maximum temperature, fresh air supply etc. are a result of the more general requirement; the desired indoor climate class (according to EN15251) and can be deduced by taking into account properties of the users of the space (ie. activity level, clothing). The specific indoor climate requirements set the constraints for the technical systems to be designed by the HVAC engineer, and modelling these interdependencies could potentially provide a valuable tool for *design change consequence analysis*.

In the use case, all data was stored in the same triplestore, but in a real world implementation the client would probably make the project specific classes and associated requirements available to project participants as a SPARQL-endpoint hosted on a separate server or as part of a Common Data Environment (CDE). Further research in how such an implementation could be configured is a separate research topic.

The use of OPM enables documentation of design and requirement changes over time, and in the use case it was used to revise requirements. Inferring into the graph that a requirements check was made based on a specific state of a requirement could be used for documentation purposes, but this was out of the scope for this work. The legal aspects of being able to document design changes, potentially in combination with block chain technology could entail great benefits and composes a separate research topic.

In summary, this work illustrates a data modelling approach that provides all the means to overcome current challenges when dealing with evolving design data and requirements in the complex construction industry. It is our belief that future BIM tools can benefit from adopting these technologies and methodologies.

ACKNOWLEDGEMENTS

Special thanks to the NIRAS ALECTIA Foundation and Innovation Fund Denmark for funding.

REFERENCES

Bendixen, M. 2007 The challenges of consulting engineers. *PhD Thesis*. Kgs. Lyngby: Technical University of Denmark.

Bertelsen, S. 2003 Construction as a Complex System. *Proceedings of IGLC* 11(February):143–68.

Bonduel, M., Oraskari, J. & Pauwels, P. 2018 The IFC to Linked Building Data Converter—Current Status. *6th Linked Data in Architecture and Construction Workshop (LDAC)*.

Borgo, S. et al. 2014 Towards an Ontological Grounding of IFC *6th Workshop Formal Ontologies Meet Industry, Joint Ontology Workshops, CEUR*.

Chimezie O. 2013 SPARQL 1.1 Graph Store HTTP Protocol. Retrieved May 15, 2018 (https://www.w3.org/TR/sparql11-http-rdf-update/).

Feigenbaum, L. et al. 2013 SPARQL 1.1 Protocol. Retrieved May 15, 2018 (https://www.w3.org/TR/sparql11-protocol/).

Kiviniemi, A. 2005 Requirements Management Interface to Building Product Models *PhD Thesis*. Stanford University.

Lefrançois, M. & Zimmermann, A. 2018 The unified code for units of measure in RDF: cdt:ucum and other UCUM datatypes *Proceedings of the International Semantic Web Conference (ISWC)*, demonstration paper, submitted.

Liebich, T. & Wix, J. 1999 Highlights of the Development Process of Industry Foundation Classes. *Proceedings of the 1999 CIB W78 Conference*.

Pauwels, P. & Roxin, A. 2016 SimpleBIM : From Full IfcOWL Graphs to Simplified Building Graphs. *European Conference on Product and Process Modelling (ECPPM)*.

Pauwels, P. & Terkaj, W. 2016. EXPRESS to OWL for Construction Industry: Towards a Recommendable and Usable IfcOWL Ontology. *Automation in Construction* 63:100–133. doi: 10.1016/j.autcon.2015.12.003.

Rasmussen, M. H. et al. 2017a Proposing a Central AEC Ontology That Allows for Domain Specific Extensions. *Lean and Computing in Construction Congress—Volume 1: Proceedings of the Joint Conference on Computing in Construction* 237–44. doi: 10.24928/JC3–2017/0153.

Rasmussen, M. H., et al. 2017b Web—Based Topology Queries on a BIM Model. *5th Linked Data in Architecture and Construction (LDAC2017) Workshop*. doi: 10.13140/RG.2.2.22298.95685.

Rasmussen, M. H. et al. 2018 OPM: An Ontology for Describing Properties That Evolve over Time. *6th Linked Data in Architecture and Construction Workshop (LDAC), CEUR*.

Terkaj, W. et al. 2017 Reusing Domain Ontologies in Linked Building Data : The Case of Building Automation and Control. *8th Workshop Formal Ontologies Meet Industry, Joint Ontology Workshops, CEUR*.

# Linked building data for modular building information modelling of a smart home

G.F. Schneider
*Technische Hochschule Nürnberg and Fraunhofer Institute for Building Physics IBP, Nürnberg, Germany*

M.H. Rasmussen
*Technical University of Denmark, Copenhagen, Denmark*

P. Bonsma
*RDF Ltd., Bankya, Bulgaria*

J. Oraskari
*Aalto University, Helsinki, Finland*

P. Pauwels
*Ghent University, Ghent, Belgium*

ABSTRACT: Linked data and semantic web technologies offer promising characteristics in terms of meaningful integration of disparate information sources in the built environment and their exchange on a web scale. The current state of the art, however, focusses on large and monolithic schemas, which poses serious barriers for the effortless development of intelligent applications consuming this information. In this work, we propose a modular approach leveraging on the semantic integration capabilities of linked data to enable seamless information exchange among stakeholders in the exchange process as well as accessing it for web-based applications. We study the methodology by applying it to a use case related to the Open Smart Home Data Set; a data set obtained from a real-world smart home system.

## 1 INTRODUCTION

Tremendous amounts of data are generated throughout the design, commissioning and operation of the built environment. This includes data from various domains using heterogeneous formats (Curry et al. 2013). The use of linked data technologies originating from the Semantic Web (Berners-Lee et al. 2001) to address this heterogeneity in the buildings domain seems to be a promising path towards enabling seamless exchange of this information. An overview of the current state of the art of the use of Semantic Web Technologies (SWT) in the Architecture, Engineering and Construction (AEC) industry is provided in Pauwels et al. (2017).

The Industry Foundation Classes (IFC) model (ISO 16739:2013) constitutes a valuable asset by enabling the information exchange among various stakeholders of the AEC industry. It is an open standard and data model and is also available in a semantic web compliant version (ifcOWL, Pauwels & Terkaj (2016)). However, the IFC model is criticised for its monolithic approach (Pauwels & Roxin 2016) as it covers various domains related to construction industry in one singular model. Problems occur when using IFC in applications as its schema is comprised out of hundreds of classes and generating tools need to support the full schema including domains, which are not relevant to them (Pauwels & Roxin 2016; Terkaj & Pauwels 2017). Hence, developers have to face a significant overhead when implementing applications consuming data compliant to IFC. Moreover, the current practice to exchange this data is confined to single files, which are transferred between tools and stakeholders. Another downside is that by its technological basis in EXPRESS it does not provide a built in query functionality.

To address these issues, we propose a modular approach for modelling information arising from applying the Building Information Modelling (BIM) methodology. Existing domain ontologies are reused and linked to describe the full domain of interest. Through the use of SWT for the implementation the exchange and retrieval of this data is enabled on a web scale.

The use of the methodology is studied by applying it to a real world use case from home automation. Adapters are presented, which convert the information from its source format (IFC, Revit, CSV) to the respective instances compliant to Resource Description Framework (RDF) (Cyganiak et al. 2014).

We evaluate the proposed modular approach in a use case to describe monitoring data of a residential flat as well as the geometry and topology of the flat. This static information is linked to monitoring data obtained from a real smart home system. We describe the data set in detail and show a web application, which runs cross-domain queries to access the diverse data sets.

In the remainder of this paper we review existing work related to modular BIM approaches (Section 2). We then detail our modular BIM approach in Section 3. Finally, in Section 4 we study the proposed method by applying it to the Open Smart Home data set (OSH).

## 2 RELATED WORK

A number of works exists, which deal with a modular approach for BIM. Terkaj & Pauwels (2017) present a method to derive a modular IFC ontology. The resulting modules comprise different layers of the IFC model and are hierarchically structured. In the work by Pauwels & Roxin (2016) the SimpleBIM approach is presented. This approach investigates how the ifcOWL ontology (Pauwels & Terkaj 2016) can be simplified to be easier to handle in SWT applications. Four different methods for simplifying the ifcOWL ontology are studied and reduction of model sizes of up to 91 percent are achieved.

The Multimodel approach (Scherer & Schapke 2011) proposes to link domain-specific application models by linking models. The different application models can be integrated into an overarching ontology framework at project scale. The approach allows to track the evolvement of a project model over three dimensions (1) vertical in terms of level of detail, (2) horizontal in terms of domains and (3) longitudinal in terms of phases in the project's lifecycle. The Energy Systems Information Model (ESIM) (Kaiser & Stenzel 2015) constitutes an excellent application of the multimodel approach and covers several aspects of energy systems on a buildings' as well as a district scale.

The Building Topology Ontology (BOT) (Rasmussen et al. 2017) is proposed to act as a central domain ontology for the building domain capturing and standardising frequently reoccurring patterns in ontology-based domain descriptions of the built environment. The alignment of existing domain ontologies to BOT is studied in Schneider (2017) and in an initial attempt BOT is found to be capable of acting as a linking element between different domain descriptions.

An example of modular ontology-based modelling of building information is presented in relation to the European H2020 MOEEBIUS project (Schneider et al. 2018), where domain ontologies for data points in building managements systems, units and quantities are linked.

The presented approaches highlight the needs and interest of scholars in deriving modular BIM approaches. The works related to deriving simplified, modular ontologies from the IFC schema (Terkaj & Pauwels (2017), Pauwels & Roxin (2016)) are a potential option in defining modular domain ontologies. However, in the semantic web domain a number of well-defined ontologies already exist and, hence, would need to be aligned to the generated modules. The approach presented in this work follows the conceptual framework of the Multimodel approach (Scherer & Schapke 2011). However, a lack of the model is the proposed method of exchange, which stipulates to serialise models as files and does not reflect the needs for exchanging data on a web-scale.

Finally, the BOT ontology marks a starting point towards aligning domain ontologies from the built environment (Schneider 2017). However, the alignment task is conducted manually at schema level and is as challenging as ontology engineering to avoid misleading results when performing reasoning.

Hence, in the remainder of this paper, a modular BIM approach is proposed, which inherits from the multimodel approach while leveraging on the capabilities of SWT to allow the exchange of information on a web-scale.

## 3 MODULAR BIM APPROACH

### 3.1 *Modular domain ontologies*

This section presents the intended modular BIM approach pursued in this paper. In Figure 1 the utilised modular ontology framework is illustrated, where reused ontologies are displayed as solid ellipses and ontologies holding instances related to the studied use case are illustrated as dashed ellipses. The following modules reuse ontologies from the SWT domain.

The *ifc4osh* module allows to include data originating from IFC compliant to the ifcOWL (Pauwels & Terkaj 2016) ontology. We consider this beneficial as the IFC format has a strong support for representing geometry-related data and a number of free viewers exist for the visualisation.

In the *bot4osh* module, a number of ontologies are reused. First, the BOT ontology (Rasmussen

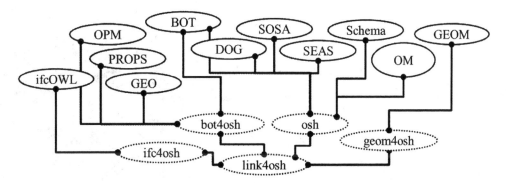

Figure 1. Modular ontology structure. Ellipses (solid line): reused domain ontology; Ellipses (dashed line): Instances of the OSH.

et al. 2017) is used to describe aspects of the building topology such as containment of building elements in locations of the building. Using GeoSPARQL (GEO) (Open Geospatial Consortium 2011) it is possible to describe geospatial aspects of some of the building components, such as their relation to their footprint. This is expressed using WKT[1] datatype strings. Finally, the PROPS[2] and OPM[3] ontologies are used to provide and describe additional information on properties related to building components as well as meta-information for the management of properties.

The reuse of ontologies in the *osh* module focusses solely on the formal definition of smart home systems. To achieve this, the DogOnt (DOG) ontology (Bonino & Corno 2008) is reused, which provides concepts and relationships beyond the basic nature of BOT on residential buildings equipped with a home automation system. The SEAS ontology (Lefrancois et al. 2017) is used to describe the logical binding of physical sensors and actuators in the smart home system. Finally, for the description of the sensors' measurements and actuator settings as well as the measurement values and units and quantities we use the SOSA (Haller et al. 2017), OM (Rijgersberg et al. 2011) and schema.org (Guha et al. 2016) ontologies.

In the *geom4osh* module the focus is solely on geometric aspects of tangible components of buildings and the GEOM ontology (McGlinn et al. 2017) is reused in this regard.

Finally, the linking model for establishing links between the different domains is covered by the *link4osh* module. The linking of different domains on instance level is discussed in the following section.

### 3.2 Linking modules at instance level

In the presented approach specialised ontologies are reused for describing particular domains of interest. This is beneficial when describing a certain domain without the need to take adjacent domains and their semantics immediately into account.

However, still, the instances should be linked to allow cross-domain information exchange and integration. In our approach, the link is established at instance level. In SWT, establishing links between individuals on instance level as well as formally specifying the implications of these links is discussed and researched in the past. Different possibilities are described to capture the different semantics of the relationships between two instances (Halpin et al. 2010). For example, owl:sameAs is a built-in property in the Web Ontology Language (OWL) (W3C OWL Working Group 2012), which allows to define that two different URIs refer to the same individual. Logically this implies that all properties applied to both URIs apply equally. In contrast, the property rdfs:seeAlso is a built-in property in RDFS (Brickley & Guha 2014) expressing that some "resource might provide additional information about the subject resource" (Brickley & Guha 2014). Its formal semantics are light-weight as it is defined as an annotation property in OWL ontologies and, hence, does not have implications when reasoning is applied.

In our modular approach, resources can be associated to each other. An example is a geometrical entity representing a space, which can be rendered in a visualisation tool. Additionally, there might exist an instance, which is related to

---

[1] http://www.opengis.net/doc/IS/wkt-crs/1.0, Last Accessed 15/05/2018.
[2] https://github.com/w3c-lbd-cg/props, Last Accessed 15/05/2018.
[3] https://w3id.org/opm#, Last Accessed 15/05/2018.

the space but represents this space in a different context, e.g. building topology. Then semantically it would be misleading to state that both instances are the same (owl:sameAs). Instead, both entities have a relation to each other (symmetry) but do not share exactly the same properties. These semantics are explicitly the definition of the skos:related object property (Miles & Bechhofer 2009), which we reuse in our modular approach to establish links between instances in the link model.

## 4 SMART HOME USE CASE

The described data set is hosted both in a public SPARQL endpoint[4] and the source files with an explanation how to access the SPARQL endpoint is hosted in a GitHub repository[5]. The data set is intended to be the basis to study smart home data with real data at hand. The data includes both the static data from a BIM authoring tool describing the building and its systems as well as the dynamic data from the sensor measurements and actuator set points.

### 4.1 Data set description

As a use case for evaluating the modular approach a flat in a residential building situated in Nürnberg, Germany is studied. The flat is placed on the ground floor and has three rooms, bathroom, kitchen, lobby and a toilet. A three-dimensional model of the flat is created from manual drawings using Autodesk Revit. The respective Revit model is provided as part of the data set as well as an IFC file exported using Revit's IFC exporting capabilities. In Figure 2 a screenshot of the IFC model is presented using the IFCViewer[6] tool.

The flat is equipped with a gas boiler for heating and hot water as well as space heaters in each room except the lobby. Heated water from the boiler circulates through the radiators to heat the flat. The windows can be shaded manually using outside shutters.

The flat is equipped with a smart home system. The system's capabilities include:

- Wall-mounted, remote sensors in each room where a space heater is placed (none in staircase and lobby), which measures the air temperature, illuminance and humidity at the place where the sensor is mounted to the wall;
- A remote-controlled thermostat valve at each space heater. The air temperature setpoint and the air temperature at the thermostat valve are logged;
- A base station, which performs the wireless communication with the sensors and actuators. The base station is connected to the internet to retrieve the current weather forecast and to allow connectivity if the occupant is not at home; A virtual outdoor air temperature for each room is calculated and part of the data set;
- A smartphone app to allow users to change setpoints, define a schedule and visualise current measured values.

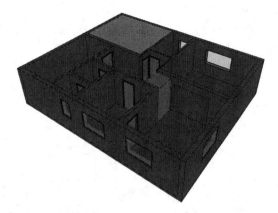

Figure 2. Screenshot of IFC model using the mentioned IFCViewer tool. Room (ID: Room 1) highlighted in green has IFC GUID 05i4VutGDCsQKCrT6CQvhu.

The described data set includes measurement data of this smart home system. Readings of the flat are obtained starting from 2017–03–09 00:56:17 and ending at 2017–06–06 06:06:22, local time. The sampling rate varies but does not exceed 15 minutes.

### 4.2 Conversion infrastructure and tool chain

One activity of the Linked Building Data Community Group (LBD) is related to the implementation of adapters, which allow to publish data created from BIM authoring tools as linked data.

The IFCtoRDF[7] converter is developed to batch process IFC-SPF files to RDF triples compliant to the ifcOWL ontology. The tool and its source code are freely available. The Revit-BOT-Exporter[8] tool is a plugin for the commercial Revit

---

[4]https://rdf.ontotext.com/4139541402/mydb/repositories/OpenSmartHomeDataSet, Last Accessed 15/05/2018.
[5]https://github.com/TechnicalBuildingSystems/OpenSmartHomeData, Last Accessed 15/05/2018.
[6]http://www.rdf.bg, Last Accessed 15/05/2018.
[7]https://github.com/pipauwel/IFCtoRDF, Last Accessed 15/05/2018.
[8]https://github.com/MadsHolten/revit-bot-exporter, Last Accessed 15/05/2018.

software. It is provided open source and allows to export aspects of a Revit model compliant to the BOT ontology as well as GEO, PROPS and OPM ontology modules. Also from the IFC-SPF file, a converter is available to generate instances compliant to the GEOM ontology (McGlinn et al. 2017). These instances solely focus on the high-fidelity visualisation of the geometry and are supported in a number of web-based geometry kernels provided by RDF Ltd. The measurement data of the smart home system is provided from the manufacturer of the smart home system as comma separated files and converted to RDF using custom code implemented in Python.

An overview on the described tools and data formats is given in Figure 3. The resulting files are hosted on the mentioned GitHub repository.

### 4.3 Ontology-based cross-domain data integration and access

To illustrate the cross-domain querying of the linked data sets, Listing 1 shows a query in SPARQL (Prud'hommeaux et al. 2017) syntax where the identifier (05i4VutGDCsQKCrT6C-Qvhu) retrieved from the IFCViewer tool as depicted in Figure 2 is used to retrieve the identifier of a temperature sensor of this entity. Here, the specified value is the GUID of the room highlighted in Figure 2. Note, the respective GUID is one reference to retrieve the respective information, of course other identifiers such as the Revit ID could be used as well.

The returned ID allows to retrieve measurement data from the dynamic data set. For instance, in Figure 4 the room air temperature of the respective sensor (ID: "Room1Temp") is plotted for a time period starting at 01 May and ending at 06 June 2017. Additionally, the set point of the thermostat and the outdoor air temperature are depicted.

Listing 1. Cross-domain query retrieving the sensor if of a room temperature of the space with the IFC GUID 05i4VutGDCsQKCrT6CQvhu.

```
PREFIX [...]
SELECT ?SensorIdent
WHERE{
?ifcglobID express:hasString
"05i4VutGDCsQKCrT6CQvhu" .
?roomIFC ifc:globalId_IfcRoot ?ifcglobID
.
?room skos:related ?roomIFC .
?room bot:containsElement ?TempSensor .
?TempSensor rdf:type
dog:TemperatureSensor .
?TempSensor
seas:connectsAt/dcterms:identifier ?SensorIdent .
}
```

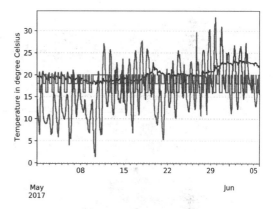

Figure 4. Plot of the outdoor air temperature (green), room air temperature (blue) and room air temperature setpoint (orange) of the smart home system from 01 May until 06 June 2017. Data resampled to 60s interval. Data is re-trieved from the identifier returned by query listed in Listing 1.

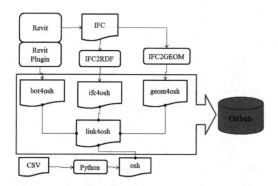

Figure 3. Conversion procedure and tool chain applied to generate modular ontology structure.

Other relevant measurements of the data set can be retrieved, for example, the illuminance and the relative humidity in a specified room.

We implemented a web application, which allows to access the data set in an intuitive way. Screenshots illustrating the use of the web application are presented in Figure 5 and Figure 6.

In Figure 5 the floor plan of the flat is plotted. The geometrical information needed by the web application to create this plan is stored in the database and has been previously generated by the mentioned Revit-BOT-Exporter tool.

By selecting one of the spaces an interaction window opens and plots the respective measurements of interest to the user. Again, the connec-

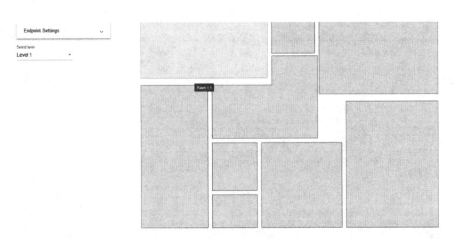

Figure 5. Automatically configured floor plan of the web application. The geometrical information for plotting the floor plan is retrieved from the server storing the OSH data set. By clicking on spaces measurements can be retrieved (see Figure 6).

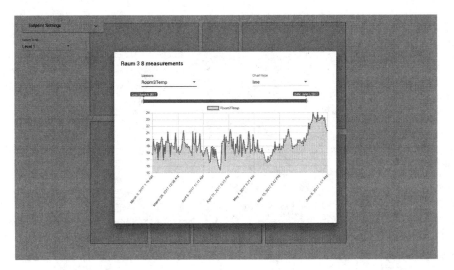

Figure 6. Interaction window displaying measurement values of the previously selected space. The time interval can be specified using sliders.

tion of the space plotted on the application canvas and the identifier needed to retrieve the respective measurements for plotting is established through the defined link model in RDF (SPARQL queries). Hence, the web application can run on any data set, which is provided in compliance to the described modular ontology structure.

## 5 DISCUSSION

During the course of this research and implementing the use cases of the proposed modular approach, some limitations and obstacles can be observed.

The proposed approach defines a link model, where relationships are established at the instance level using the semantically lightweight skos:related property. A benefit of this approach is that links can be established in a flexible manner without *semantic implications*, meaning that the reasoning is not effected on the respective individuals when linked. However, the links cannot be shared to other use cases as this would require their definition at schema level. Schema level alignments allow transferring the linking knowledge at the expense of the non-trivial effort needed for defining these alignments

In the described data set the actual control actions are removed from the author out of intellectual property reasons by the manufacturing company of the smart home system.

The actual observations of the smart home system in the OSH are published compliant to the SOSA ontology and serialised using the Turtle format. This creates an overhead as some information is provided redundantly (the unit will not change between the observations). As the data set is small, this procedure still can be applied, however, it is recommended for larger quantities of data to store the readings in a relational database and allow the ontology-based access via respective converters (Das et al. 2012).

## 6 CONCLUSION

In this work, a modular approach for implementing the Building Information Modelling (BIM) method is investigated. The approach uses Semantic Web Technologies for the integration of diverse domains. A multimodel paradigm is implemented as the respective domain models are linked at instance level in a link model. Some of the links are established manually and some by using permanent identifiers of the original sources, e.g. IFC GUID. The approach is intensively studied in a use case originating from the smart home domain. In particular, the Open Smart Home (OSH) data set is introduced, which is an ontology-based data set freely available on the web[9] containing static BIM data of a real world flat paired with dynamic readings obtained from a real smart home system of this flat.

The presented approach works particularly well in describing and linking the mentioned data and the retrieval of cross-domain data to be consumed by web applications is performed.

Future work is related to the addition of other data related to the data set such as HVAC equipment of the flat, meta data on the measurement capabilities of the smart home system and product data on all elements.

In addition it would be very interesting to apply the approach in further larger-scale use cases for thorough testing and evaluation. Related to this is also the further improvement and development of conversion tools and web applications, which can consume this data.

---

9. https://github.com/TechnicalBuildingSystems/OpenSmartHomeData, Last Accessed: 15/05/2018

## ACKNOWLEDGEMENTS

The authors would like to thank all members of the W3C Linked Building Data Community Group for their input through discussions and comments. We gratefully acknowledge the support by Ontotext (https://cloud.ontotext.com/) for providing a free instance of GraphDB Cloud. Georg F. Schneider gratefully acknowledges financial support from MOEEBIUS project, a Horizon 2020 research and innovation program under grant agreement No. 680517.

Mads H. Rasmussen thanks the NIRAS ALECTIA Foundation and Innovation Fund Denmark for funding.

## REFERENCES

Berners-Lee, T., Hendler, J. & Lassila, O. 2001. The Semantic Web *Scientific American*, 285(5): 28–37.

Bonino, D. & Corno, F. 2008. DogOnt - ontology modeling for intelligent domotic environments. In *Proc. of ISWC*, Karlsruhe, Germany, 26–30 October, pp. 790–803, Springer.

Brickley, D. & Guha, R.V. 2014. RDF Schema 1.1. Permanent URL: https://www.w3.org/TR/rdf-schema/, W3C, Cambridge, USA.

Curry, E., O'Donnell, J., Corry, E., Hasan, S., Keane, M. & O'Riain, S. 2013. Linking building data in the cloud: Integrating cross-domain building data using linked data. *Advanced Engineering Informatics*, 27(2): 206–219.

Cyganiak, R., Wood, D. & Lanthaler, M. 2014. Resource Description Framework (RDF): Concepts and Abstract Syntax. Permanent URL: http://www.w3.org/TR/rdf11-concepts/, W3C, Cambridge, USA.

Das, S., Sundara, S. & Cyganiak, R. 2012. R2RML: RDB to RDF Mapping Language. Permanent URL: http://www.w3.org/TR/r2rml, W3C, Cambridge, USA.

Guha, R.V., Brickley, D. & Macbeth, S. 2016. Schema.org: Evolution of structured data on the web. *Communications of the ACM*, 59(2): 44–51.

Haller, A., Krzysztof, J., Cox, S., Le Phuoc, S. & Taylor, K. 2017. Semantic Sensor Network Ontology. Permanent URL: https://www.w3.org/TR/vocab-ssn/, W3C, Cambridge, USA.

Halpin, H., Hayes, P.J., McCusker, J.P., McGuinness, D.L. & Thompson, H.S. 2010. When owl:sameAs isn't the same: An analysis of identity in linked data. In Proc. of ISWC, Shanghai, China, 7–11 November, pp. 305–320, Springer.

ISO 16739 2013. Industry Foundation Classes (IFC) for data sharing in the construction and facility management industries, Standard, ISO, Geneva, Switzerland.

Kaiser, J. & Stenzel, P. 2015. D4.2: Energy System Information Model – ESIM. eeEmbedded Consortium, Brussels, Belgium.

Lefrancois, M., Kalaoja, J., Ghariani, T. & Zimmermann, A. 2017. D2.2: The SEAS Knowledge Model. ITEA2 SEAS, Brussels, Belgium

McGlinn, K., Bonsma, P., Wagner, A. & Pauwels, P. 2017 Interlinking Building Geometry with Existing

and Developing Standards. In Proc. of LDAC, Dijon, France, 13–15 November.
Miles, A. and Bechhofer, S. 2009. SKOS Simple Knowledge Organization System Namespace Document. Permanent URL: https://www.w3.org/2009/08/skos-reference/skos.html, W3C, Cambridge, USA.
Open Geospatial Consortium 2011. GeoSPARQL - A geographic query language for RDF data. [Report], No. OGC 09-157r4.
Pauwels, P. & Roxin, A. 2016. SimpleBIM: From full ifcOWL graphs to simplified building graphs. In Proc. of ECPPM, Limassol, Cyprus, 7–9 September, pp. 11–18, CRC Press.
Pauwels, P. & Terkaj, W. 2016. EXPRESS to OWL for construction industry: Towards a recommendable and usable ifcOWL ontology, *Automation in Construction*, 63: 100–133.
Pauwels, P., Zhang, S. & Lee, Y.-C. 2017. Semantic web technologies in AEC industry: A literature overview, *Automation in Construction*, 73: 145–165.
Prud'hommeaux, E. & Seaborne, A. 2017. The SPARQL query language for RDF. Permanent URL: https://www.w3.org/TR/rdf-sparql-query/, W3C, Cambridge USA.
Rasmussen, M., Pauwels, P., Lefrançois, M., Schneider, G.F., Hviid, C. & Karlshøj, J. 2017. Recent changes in the building topology ontology. In Proc. of LDAC, Dijon, France, 11–13 November.
Rijgersberg, H., Wigham, M. & Top, J.L. 2011. How semantics can improve engineering processes: A case of units of measure & quantities, *Advanced Engineering Informatics*, 25(2): 276–287.
Scherer, R.J. & Schapke, S.-E. 2011. A distributed multi-model-based Management Information System for simulation and decision-making on construction projects, *Advanced Engineering Informatics*, 25(4): 582–599.
Schneider, G.F. 2017. Towards Aligning Domain Ontologies with the Building Topology Ontology. In Proc. of LDAC, Dijon, France, 13–15 November.
Schneider, G.F., Qiu, H. & Kontes, G.D. 2018. MOEEBIUS - An ontology for the MOEEBIUS demonstration sites. Permanent URL: https://w3id.org/moeebius/MOEEBIUSOntology.
Terkaj, W. & Pauwels, P. 2017. A Method to generate a Modular ifcOWL Ontology. In Proc. of JOWO, Bolzano, Italy, 21–23 September, CEUR-WS.org.
W3C OWL Working Group 2012. OWL 2 Web Ontology Language. Permanent URL: https://www.w3.org/TR/owl2-overview/, W3C, Cambridge, USA.

# APPENDIX

Table 1. Prefixes utilised in this paper.

| Prefix | URI |
| --- | --- |
| ifc: | http://www.buildingsmart-tech.org/ifcOWL/IFC4_ADD1# |
| bot: | https://w3id.org/bot# |
| express: | http://purl.org/voc/express# |
| rdf: | http://www.w3.org/1999/02/22-rdf-syntax-ns# |
| seas: | https://w3id.org/seas/ |
| dcterms: | http://purl.org/dc/terms/ |
| dog: | http://elite.polito.it/ontologies/dogont.owl# |
| skos: | http://www.w3.org/2004/02/skos/core# |
| geom: | http://rdf.bg/geometry.ttl# |

# Ontology and data formats for the structured exchange of occupancy related building information

M. Taheri & A. Mahdavi
*Department of Building Physics and Building Ecology, TU Wien, Vienna, Austria*

ABSTRACT: An important prerequisite for pervasive data sharing is the availability of a systematic ontology and standardized data formats. However, developers and users of building monitoring systems appear to operate without an explicitly documented, systematic, detailed, and comprehensive ontology. The authors have previously introduced an ontology for the representation and incorporation of multiple layers of occupancy-related monitored building data. Starting from a brief description of this ontology, the present contribution focuses on an effort towards the specifications of a process, by which various streams of monitored data can be mapped onto a unified data format. For this purpose, the potential of Hierarchical Data Format 5 (HDF5) is explored, which is open source and suitable for managing data collections of different sizes and complexity. This paper specifically illustrates the suitability of this data format for the structured representation of inhabitants in building performance simulation models.

## 1 INTRODUCTION

Buildings' energy and indoor environmental performance is influenced by people's presence, activities, and actions. Consequently, these influences must be considered while generating building models for performance assessments applications. Many efforts have been made to integrate occupants' presence and behavior models in building performance simulation. However, uncertainties associated with occupancy-related assumptions are still a key challenge in simulation-based performance predictions, given considerable gaps and limitations in our knowledge on this topic. Representation of inhabitants in building performance simulations would benefit from a good understanding of and comprehensive information on the inhabitants and their actions, internal and external environmental conditions, control systems and devices, equipment, and energy flows in buildings. In this regard, high quality and detailed input data is required to apply computational methods and to develop dependable models. As such, the relevant professional community is aware of the value and benefits of building monitoring and open access data. The related challenges concern scientists across multiple engineering disciplines and underline the importance of an infrastructure to ensure that research data is adequately structured and publicly accessible.

In some disciplines, such as life sciences, unlike building science, monitored data have been widely archived and shared.

Data sharing in general has various benefits, such as, facilitating new research, increasing efficiency concerning time and effort, collaborative and interdisciplinary data processing, assessing data quality, and effectively examining the reliability of research results (Tenopir et al. 2011). Scientific research in recent years has become increasingly collaborative and interdisciplinary. Even though the importance of open access data is undeniable, there is still a lack of uniform standardized platforms for open data sharing and processing (Gault and Koers 2015). Efforts have been made to provide such platforms for sharing scientific data and related material, for example, in Scientific Data (Springer Nature 2018) by Nature Publishing Group, Open Data pilot (Elsevier 2018a) and Data in Brief (Elsevier 2018b) by Elsevier.

To support research via data sharing, metadata must be systematically collected and stored, including all information regarding the collected data, relevant experimental setup, data cleaning, and pre-processing details (Grewe et al. 2011). The related experiences point to a further essential requirement of data sharing, namely the availability of a systematic ontology and standardized data formats. However, developers and users of building monitoring systems appear to operate without an explicitly documented, systematic, detailed, and comprehensive ontology. Well-developed schemes, such as IFC standards (ISO 2013), are available for representation of building fabric (e.g., walls, roofs, technical elements, etc.). However, there is a lack of explicit schemes for representation of buildings' sensory information. In this context, Mahdavi & Taheri (2017) introduced an ontology that specifically captures the multitude of dynamic state and performance data.

Standardized data formats represent another important requirement for pervasive data sharing. Different file formats are developed and commonly used to store data, including Comma-Separated Values (CSV), Plain Text (txt), JavaScript Object Notation (JSON), eXtensible Markup Language (XML), Hypertext Markup Language (HTML), and Hierarchical Data Format (HDF). For further data reuse and management, choosing the right data format is of importance. One of the most common structures used for data handling and exchange is "tabular", where data is organized into rows and columns, listing serial values as separated and not linked entries. One example of a tabular file structure is CSV. Some CSV features may be considered as unfavorable for certain applications. For instance, CSV appears to be not practical for highly complex datasets. In such cases, other formats such as XML, HDF, etc., are used. National Center for Supercomputing Applications (NCSA) introduced HDF as a scientific data format for high performance management of extensive and diverse scientific data (HDF 2017a). With a comprehensive set of libraries and management tools, HDF5 (HDF 2017b) is an open source data model and file format, which benefits from the support of large institutions, including NASA (National Aeronautics and Space Administration), NSF (National Science Foundation), and DOE (Department of Energy) (Castro et al. 2015).

The present study explores the two above-mentioned prerequisites for effective data sharing, i.e., a systematic ontology and a standardized data format. To address this issue at a fundamental level, we have introduced an ontology for the representation and incorporation of multiple layers of occupancy-related monitored building data (Mahdavi et al. 2016, Mahdavi & Taheri 2017). Subsequent to a brief description of this ontology, the present contribution focuses on an effort towards the specifications of a process, by which various streams of monitored data can be mapped onto a unified data format. For this purpose, we explored the potential of HDF5, which is open source and suitable for managing data collections of different sizes and complexity. In this paper, we specifically illustrate the suitability of this data format for the structured representation of inhabitants in building performance simulation models.

## 2 APPROACH

### 2.1 *Building monitored data ontology*

The identification of basic data categories does appear to be a necessary and fundamental step in any related effort. However, any proposed schema in this area should be independently tested in view of its applicability, robustness, and practicability. Note that, there may not exist a unique correct path to the construction of a well-formed schema for building monitoring.

Building upon previous efforts (Mahdavi et al. 2005, Mahdavi 2011a, 2011b, Zach et al. 2012, Mahdavi et al. 2016), Mahdavi & Taheri (2017) suggested six data categories as an effective classification to accommodate empirical information obtained from building monitoring systems. Such information pertains to occupants, indoor environmental conditions, external environmental conditions, control systems and devices, equipment, and energy flows. Note that the information in each main category of buildings' monitored data may be structured in terms of further sub-categories. Sensors, meters, simulation-powered virtual sensors, human agents, etc., generate streams of information in the above categories. For a clear definition of the nature of the monitored variables, we showed that all monitored data could be captured in terms of the profile shown in Figure 1 (Mahdavi & Taheri 2017).

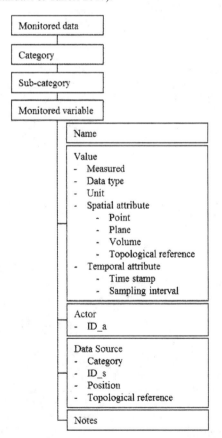

Figure 1. The structure of monitored data (Mahdavi & Taheri 2017).

## 2.2 HDF5 file format

The structure of HDF5 is similar to Windows folders that start with a root group to which other groups or data objects are linked. The linking system feature in HDF5 allows for a hierarchical structure and formation of complex relationships. Moreover, HDF file format is self-describing, which has significant implications for handling scientific data (NASA 2012). This means that a HDF file and its contents can be understood without the need for extraneous documents. HDF5 file can hold almost any collection of data in a single file, including scientific data arrays, tables, text annotations, graphical data, raster images, and documents. In addition, the structure of HDF5 facilitates the efficient storage of large arrays and high dimensional objects with an indexing option. This option enables, for instance, accessing a dataset without going through the entire file. In recent years, HDF5 has become the format of choice for organizing heterogeneous collections of complex and enormous datasets (NASA 2012). It has been suggested to guarantee the handling of long term high time resolution monitored data (Cali et al. 2013).

## 2.3 Case study

An existing office space in a university building (TU Wien) in Vienna, Austria, is selected for the present study. The office covers an area of around 200 m$^2$, and includes one single-occupancy office as well as an open space with multiple workstations. Figure 2 presents a schematic illustration of the office space.

For the purpose of the present treatment, the focus lies on the single-occupancy office marked in Figure 2. This office is equipped with a monitoring infrastructure that captures various streams of high-resolution sensor data, such as thermal, visual, air quality, and equipment states. Outdoor conditions are monitored via a weather station installed in close proximity of the building.

The installed sensors are integrated in the online monitoring system of the Department of Building Physics and Building Ecology, TU Wien. The data is automatically stored in real time in a central database (Schuss et al. 2017). The original data is stored with.csv format. For the purpose of sharing the comprehensive monitored data with the relevant professional community, R scripts have been developed that convert the data stored in the csv-files into a single structured HDF5 file.

For the purpose of the present study, monitored data from the year 2013 was used. Figure 3 demonstrates the monitored data points. This type of datasets represents examples of useful resources for the relevant research community, for instance, professionals interested in development and validation of occupancy-related models.

## 2.4 Implementation milestones

Ongoing research projects benefit from the Long-term, high resolution, and comprehensive monitored physical parameters in this case study. Monitored data supports, for instance, generation and calibration of simulation models, optimization of building operation, and systematic performance evaluation. Moreover, monitored data can be used to advance the state of knowledge and improve the quality of computational representations of occupants' presence and actions in buildings. Representing monitored data based on the ontology structure enables incorporation of data streams in computational applications. Figure 4 provides an example of the monitored variable specifications in selected sub-categories of the main category "occupants", according to the structure of the aforementioned ontology.

In order to convert the monitored data stored in the csv-files into a structured HDF5 file, an R script was developed. HDFView 2.9 was used, to read the created file. HDFView is a visual tool, which shows the file hierarchy in a tree structure format and enables exploring and editing of the files.

Figure 5 illustrates a screenshot of the viewer including the monitored data categories. The root group is called "Monitored_Data_ECPPM2018", to which six main data categories of monitored data are linked. To these categories, further sub-categories are linked. In the example of Figure 5, "Acoustical conditions", "Air quality", "Hygrothermal conditions", "Solar radiation", and "Visual conditions" constitute the sub-categories of the main category "External Conditions".

It is possible to access and conveniently navigate different components: When using HDF5, all data

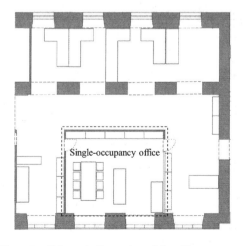

Figure 2. Schematic illustration of the office space.

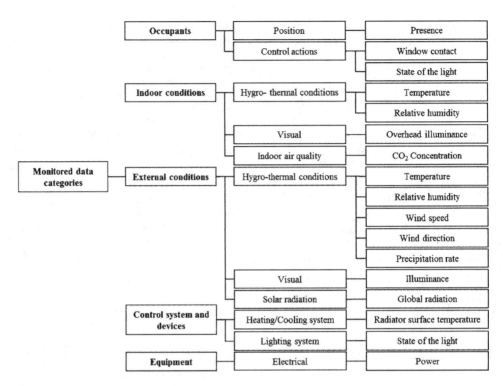

Figure 3. Monitored variables.

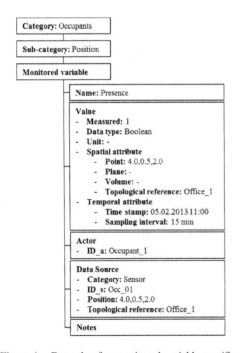

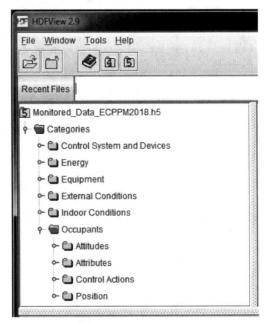

Figure 4. Example of a monitored variable specifications in category "occupants".

Figure 5. Screenshot from the data categories and examples of subcategories in HDFView 2.9.

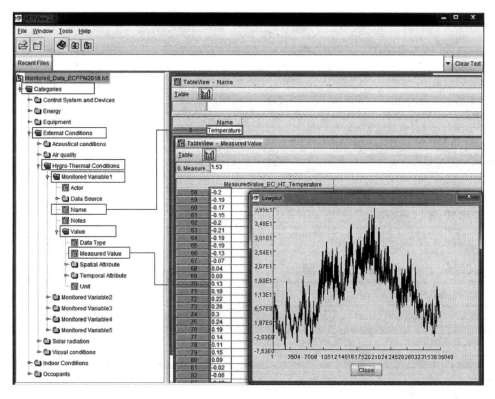

Figure 6. Screenshot from an example of the monitored data recorded on HDF5.

is passed along in one file. Figure 6 demonstrates an instance of a stored monitored variable. Here, the root group is "Monitored_Data_ECPPM2018". The respective sub-groups are linked to the main root group, including "Categories", "External Conditions", "Hygro-Thermal Conditions", and "Monitored Variable1". Following the structure of the ontology, each "Monitored Variable" further holds the information pertaining to the name of the variable, value, actor, data source, as well as essential remarks or notes. Figure 6 shows the measured values of the monitored variable "Temperature".

## 3 CONCLUSION

This contribution emphasized the importance of an infrastructure to ensure that research data is adequately structured and publicly accessible. The study addressed the challenges related to the availability of high quality and detailed monitored data. Such challenges concern scientists across multiple engineering disciplines. A brief description of an ontology for the representation and incorporation of occupancy-related monitored building data was provided. Moreover, the contribution included an effort towards the specifications of a process to map various streams of monitored data onto a unified data format. Towards this end, the potential of Hierarchical Data Format 5 (HDF5) and the suitability of this data format for the structured representation of monitored data was studied. Future research is required to develop standard evaluation methods and tools for visualization and analyses of the monitored data.

## ACKNOWLEDGEMENT

The research presented in this paper is supported in part within the framework of the "Innovative Projekte" research-funding program of TU Wien. This research also benefited from the authors' participation in the ongoing efforts of the IEA-EBC Annex 66 (Definition and Simulation of Occupant Behavior in Buildings) and the associated discussions. We also thank Dr. Farhang Tahmasebi for his contribution toward the preparation of the monitored data.

# REFERENCES

Cali, D., Streblow, R., Mueller, D., Osterhage, T. 2013. *Holistic Renovation and Monitoring of Residential Buildings*. ECEEE 2013 summer study. Toulon/Hyères, France.

Castro, A., Abadie, L., Makushok, Y., Ruizd, M., Sanz, D., Vega, J., Faig, J., Román-Pérez, G., Simrock, S., Makijarvi, P. 2015. *Data Archiving System Implementation in ITER's CODAC Core System*. Fusion Engineering and Design v. 96–97, pp. 751–755. DOI: 10.1016/j.fusengdes.2015.06.076.

Elsevier. 2018a. https://www.elsevier.com/authors/author-services/research-data/open-data.

Elsevier. 2018b. https://www.journals.elsevier.coc/data-in-brief.

Gault, B., Koers, H. 2015. *Supporting Data-sharing to Speed Up Innovation in Materials Science*. Available on: https://www.elsevier.com/connect/supporting-data-sharing-to-speed-up-innovation-in-materials-science.

Grewe, J., Wachtler, T., Benda, J. 2011. *A Bottom-up Approach to Data Annotation in Neurophysiology*. Frontiers in Neuroinformatics. Volume 5, Article 16. DOI: 10.3389/fninf.2011.00016.

HDF. 2017a. https://www.hdfgroup.org/about-us/.

HDF. 2017b. http://www.hdfgroup.org/HDF5/.

ISO 16739:2013. 2013. *Industry Foundation Classes (IFC) for Data Sharing in the Construction and Facility Management Industries*. International Organization for Standardization. Last accessed 15.02.2018.

Mahdavi, A. 2011. *People in Building Performance Simulation*. In Building Performance Simulation for Design and Operation, edited by Hensen, J. and R. Lamberts, New York, Taylor & Francis Group. ISBN-13: 978-0415474146.

Mahdavi, A., Glawischnig, S., Schuss, M., Tahmasebi, F., and Heiderer, A. 2016. *Structured Building Monitoring: Ontologies and Platform*. Proceedings of ECPPM 2016: The 11th European Conference on Product and Process Modelling, Limassol, Cyprus.

Mahdavi, A., Taheri, M. 2017. *An Ontology For Building Monitoring*. Journal of Building Performance Simulation, 10:5–6, 499–508. DOI:10.1080/19401493.2016.1243730.

Mahdavi, A., Taheri, M., Schuss, M., Tahmasebi, F., and Glawischnig, S. 2018. *Structured Building Data Management: Ontologies, Queries, and Platforms*. In book: Exploring Occupant Behavior in BuildingsPublisher: Springer, ChamEditors: Andreas Wagner, William O'Brien, Bing Dong. DOI: 10.1007/978-3-319-61464-9_10.

NASA. 2012. *Hierarchical Data Format*. https://eosweb.larc.nasa.gov/HBDOCS/hdf.html. Last accessed 24.02.2018.

Schuss, M., Glawischnig, S., Mahdavi, A. 2017. *Building Monitoring and Diagnostics: A Web-Based Approach*. Applied Mechanics and Materials, 861 (2017), 556–563.

Springer Nature. 2018. Macmillan Publishers Limited, part of Springer Nature. https://www.nature.com/sdata/.

Tenopir, C., Allard, A., Douglass, K., Aydinoglu, A., Wu, L., Read, E., Manoff, M., Frame, M. 2011. *Data Sharing by Scientists: Practices and Perceptions*. PLoS ONE, Volume 6, Issue 6. DOI: 10.1371/journal.pone.0021101.

Zach, R., Glawischnig, S., Hönisch, M., Appel, R., Mahdavi, A. 2012. MOST: *An Open-Source, Vendor and Technology Independent Toolkit for Building Monitoring, Data Preprocessing, and Visualization*. In: eWork and eBusiness in Architecture, Engineering and Construction. G. Gudnason, R. Scherer et al. (ed.), Taylor & Francis. ISBN: 978-0-415-62128-1.

Zehl, L., Jaillet, F., Stoewer, A., Grewe, J., Sobolev, A., Wachtler, T., Brochier, TG., Riehle, A., Denker, M., Grün, S. 2016. *Handling Metadata in a Neurophysiology Laboratory*. Front Neuroinform. 2016; 10: 26. DOI: 10.3389/fninf.2016.

# Graph representations and methods for querying, examination, and analysis of IFC data

H. Tauscher
*National University of Singapore, Singapore*

J. Crawford
*Ordnance Survey, UK*

ABSTRACT: This paper presents an approach to apply graph-based methods to IFC data. We discuss (1) the representation of object-oriented data and schemas as graphs including IFC-specific issues, (2) the specification of IFC queries as subgraph templates including their conversion to Model View Definition (MVD) fragments, BIMserver and Cypher queries, (3) graph grammars for IFC model transformation, and (4) visualization schemas for the different graph structures. These fundamentals provide the basis for future work in graph-based model transformation.

## 1 INTRODUCTION

We are considering graph-based transformation methods for the conversion of BIM data in IFC format into GIS data in CityGML format. Previous attempts at the conversion from IFC to CityGML have shown that no reliable solutions are possible without the employment of formal methods (e.g. Isikdag & Zlatanova 2009).

We specifically selected graph-based transformation methods since the sound mathematical model behind this approach allows to tackle questions of correctness, completeness, and consistency of transformations. Moreover, the rule-based nature of graph-based transformation methods works well with declarative specifications rather than imperative and allows thus an incremental development of the complex conversion process.

For the conversion of given IFC data into CityGML we have to consider three different graphs: the graph of the given IFC entities, the graph of CityGML entities to be created, and the graph of correspondence relations. In this paper we are focussing on the IFC-graph only.

### 1.1 Related work

Building information models lend themselves to graph methods due to their object-oriented nature.

Several other researchers have recently studied the application of graph-based methods to BIM data in general, e.g. Isaac et al. (2013) propose applications of graph theory for construction management. Vilgertshofer & Borrmann (2017) apply graph rewriting to generate parametric infrastructure models. Others targets IFC data in particular: Khalili & Chua (2015) trys to use graph methods for topological IFC queries. Tauscher et al. (2016) suggest to employ Dijstra's shortest path algorithm to simplify queries. Ismail et al. (2017) dump IFC data into neo4J for further processing.

### 1.2 Method and structure of the work

We investigate graph representations of IFC data and subsequently three different application scenarios for graph methods: queries with graph templates, decomposition with graph grammars, analysis with graph visualization. For each of these cases, we describe our approach and technology choices and compare alternatives.

First, in Section 2, we investigate how to represent object-oriented data, particularly the IFC schema and IFC object graphs, as typed directed graphs. For a reference implementation in a graph database we use Neo4J.

Second, in Section 3, we suggest subgraph templates as a generic way to express queries on IFC object graphs. We derive BIMserver queries, Neo4J cypher queries and MVD fragments from these templates, compare the three derivations regarding their expressiveness and evaluate the performance of queries on a BIMserver instance and a Neo4J database.

Subgraph templates are only capable of describing local fragments. With a graph grammar however, it is possible to capture the complete structure common to a set of graphs. To tap the grammar approach for IFC data, we derive a graph grammar from the IFC schema in Section 4. This grammar can then be used to generate valid instance graphs, but also to decompose a given graph, that is to parse the graph according to the grammar.

Finally, in Section 5, we aim to employ graph representations to foster visual inspection of IFC data sets, but also of operations carried out on them and analysis or validation results. To this end, we introduce a graphical representation schema for the different graph structures presented before: for schema and instance graphs, sub graph templates, as well as for transformation rules.

### 1.3 Technology choices and sample data

As part of Section 3 we present a sample implementation of the graph model and the subgraph template-based query process. We are using neo4J (Community edition, version 3.2.5) as a graph database and BIMserver 1.5.95 to store IFC data. As sample data set we exported the Revit advanced tutorial model with the default IFC4 (ISO 2013) Reference view settings. The resulting IFC file has 21 MB and contains 144521 IFC entities.

## 2 GRAPH REPRESENTATIONS OF OBJECT-ORIENTED DATA

For the graph representation of IFC schema and data, we identified key design issues: the representation of edges and attributes in the mathematical model, the treatment of objectified relations, the representation of lists, and the integration of the instance and metamodel graph. We are first focussing on the graph model of an IFC instance graph and then proceed integrating the graph model of the IFC schema.

### 2.1 Mathematical model of nodes and edges

To represent an IFC instance graph, we model entities as nodes and their entity data typed attributes as edges. For now we are excluding simple, enumeration, defined, and select data types[1]. A common mathematical definition of a graph $G$ is $G = (N, E)$ with a set of nodes $N$ and set of edges $E$. There are multiple ways to express the relation between edges and nodes.

The native model defines a directed edge $e \in E$ as the pair of its source node and target node: $(n_S, n_T) \in N \times N$. This node edge model is used by Isaac et al. (2013) and Tauscher et al. (2016). This model does not allow to express parallel edges between nodes, since those would be represented by the same tuple of nodes and could thus not be distinguished. However, investigating some IFC files using the Perl style regular expression(#\d+) (?=\D.*\1\D)revealed some cases where duplicate edges may exist: For example an IfcPolyline entity's Points attribute may contain the same IfcCartesionPoint entity as first and last element of the list for closed polylines. Another example is IfcSurfaceStyleRendering, where the attributes SurfaceColour and DiffuseColour may both refer to the same IfcColourRgb entity. We thus have to use a slightly more general model of directed edges. With source and target functions $s, t: E \rightarrow N$ assigning source or target nodes to edges we can describe parallel edges.

This can be mitigated by including edge labels in the tuples[2], but a more flexible way to model labelling of graph edges is through labelling functions which would not heal this shortcoming.

In Express, there are three types of attributes: explicit, derived, and inverse. We are currently excluding derived attributes. Inverse attributes are inferred from other explicit attributes relating to entities and are thus redundant. They are specified in order to name and constrain them. To facilitate navigation, we are adding inverse edges to our graph model when they are specified in the schema. An edge $e_1$ and its inverse edge $e_2$ fulfill $s(e_1) = t(e_2)$ and $s(e_2) = t(e_1)$.

### 2.2 Treatment of objectified relations

In the IFC schema, relations between entities of the core layer and above (shared and domain specific schemas), that are subtypes of IfcRoot, are modelled as entities of type IfcRelationship. Such a relationship entity references the relating and related entities through its attributes. In contrast, entities of the resource layer reference each other and are referenced by higher level entities directly via their attributes, without a mediating relationship entity.

We are representing the objectified relationships as nodes. That is, we have a dedicated set $N_R \subset N$ of relationship nodes and two sets $E_S, E_T \subseteq E$ of edges that connect to the relating and related nodes respectively. For each objectified relationship between two non-relationship nodes $n_S, n_T \in N \setminus N_R$, we then have a node $n_R \in N_R$ and two edges $e_S \in E_S$ and $e_T \in E_T$ such that $s(e_S) = s(e_T) = n_R$, $t(e_S) = n_S$, and $t(e_T) = n_T$.

Some graph approaches to IFC (e.g. Isaac et al. (2013), Ismail et al. (2017)) collapse objectified

---

[1]Select data types will pose an issue to node attribution, because they may resolve to both entity data types or defined data types which will be modelled differently.

[2]The resulting edge representation adheres more to the structure of a Resource Description Framework (RDF) triple.

relations into edges instead of modelling them as nodes. For the relationship between nodes $n_S$ and $n_T$ they would then just have an edge $e_R$ with $s(e_R) = n_S$ and $t(e_R) = n_T$.

In collapsing the edges, IFC information will be omitted from the graph representation for those relations that contain additional information. Examining IFC4 Add2 TC1 revealed a list of 15 subtypes of IfcRelationship that have at least one attribute in addition to the relating and related entities, which would be lost after collapsing. In total we found 27 such attributes, one of which is mandatory (IfcInterferesElements.ImpliedOrder). A prominent example is IfcRelSpaceBoundary.

### 2.3 Representation of collections: sets and lists

Although Express knows four kinds of aggregation data types—SET, BAG, LIST, ARRAY—(ISO 2004), in IFC only two of them are used. Thus we will limit our elaboration to ordinary sets and lists and ignore multisets (BAG)[3] and lists with unset elements (ARRAY).

Attributes of SET type (that is unordered aggregations) are represented as multiple edges of the same edge type with a common source node. Given an entity with an attribute having a set of $i$ entities as its value, we will then have one node $n_0$ for the initial entity, a set $N_A = \{n_1, \ldots, n_i\}$ of nodes for the attribute value set, and for each $n \in N_A$ we have an edge $e \in E$ with $s(e) = n_0$, $t(e) = n$.

For lists, we need to represent order of nodes, which is not a native concept in graph theory. Pauwels et al. (2015) discuss a similar problem for OWL. There are two fundamentally different ways to solve this. The first possibility is to construct a graph structure that connects successive elements, thus constructing something similar to the basic data type of a linked list. The second possibility is to enhance the node–edge construction for sets with an edge label denoting the position of the element, thus constructing something similar to the basic data type of an array with an index. The index is modelled as a labelling function $i : E \to \mathbb{N}$ over the range of the natural numbers.

Technically, the indexed representation would be more appropriate for an ARRAY type aggregate whereas the linked representation would better match the LIST type aggregate. However we have tentatively decided to use an indexed representation in our model for the sake of convenient direct element access[4].

---

[3]Note that the BAG type would result in duplicate edges between two nodes with the same label, which would definitely require to model edges as described in 2.1.

[4]Those familiar with Lisp data structures will know the beauty and pain of linked lists.

### 2.4 Integration of instance and metamodel graph

In the graph theoretic approach, the schema is also represented as a graph similar to the instance graph. Each entity type is represented as a node, while each entity data typed attribute is represented as a directed edge. These sets of entity type nodes $N_{IFC}$ and attribute edges $E_{IFC}$ with source and target functions $s_{IFC} : E_{IFC} \to N_{IFC}$ and $t_{IFC} : E_{IFC} \to N_{IFC}$ form the type graph $I_{FC} = (N_{IFC}, E_{IFC}, s_{IFC}, t_{IFC})$.

The relation between the instance and the type graph is modelled as a graph morphism, that is two functions mapping the nodes and edges of the graphs such that the graph structure is maintained. Given an instance graph $G = (N, E, s, t)$ and a type graph $IFC = (N_{IFC}, E_{IFC}, s_{IFC}, t_{IFC})$ this morphism is defined as $type = (G, IFC, m_N, m_E)$ with the two functions $m_N : N \to N_{IFC}$ and $m_E : E \to E_{IFC}$. To preserve the graph structure, it must hold that $s_{IFC} \circ m_E = m_N \circ s$ and $t_{IFC} \circ m_E = m_N \circ t$. Together, an instance graph $G$, a type graph $IFC$ and a morphism $type$ form the typed graph $G_{IFC} = (G, IFC, type)$. We are speaking of a graph $G$ typed over a type graph $IFC$ and write short type: $G \to IFC$ for the morphism. Figure 1 shows the morphism between type and instance graph.

Inheritance can be represented as dedicated graph $I = (N_{IFC}, E_I, s_I, t_I)$ which adds another set of directed edges $E_I$ to the nodes $N_{IFC}$ of the type graph with their target nodes being supertypes of their source nodes. This graph must be acyclic and in the case of IFC even forms a rooted in-tree, since multiple inheritance is not used in IFC (although theoretically possible in STEP). A subset $A \subseteq N_{IFC}$ is considered as abstract nodes which may not be instantiated, that is for any typed instance graph $G_{IFC} = (G, IFC, type)$ they shall not appear in the codomain of any type morphism's function $m_N$, that is $\forall n \in N \ \forall a \in A : m_N(n) \neq a$.

When inheritance is modelled like this, the inheritance tree has to be flattened in order to obtain a typing morphism as described above. Intuitively speaking this is done by duplicating all edges for

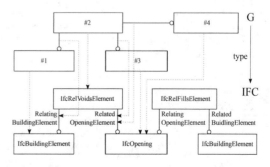

Figure 1. Type and instance graph with typing morphism.

all subtype nodes in the type graph. Alternatively, instead of the morphism to the flattened type graph, an equivalent mapping to higher levels of the inheritance tree called *clan-morphism* can be used. This model for typed graphs with inheritance was first described by Bardohl et al. (2004).

## 2.5 Implementation in neo4J

The implementation in neo4J is straight-forward according to the mathematical model described above. We created a BIMserver serializer to generate a Cypher query that creates the respective graph in a Neo4J database.

Currently, we are not creating the type graph in neo4J, but use labels to denote the concrete types of nodes and edges. This labelling is essentially a function from nodes and edges to the sets of entity type and attribute name labels respectively. Ehrig et al. (2006) has shown that labelling is equivalent to typing over a basic type graph (without inheritance) using a morphism. In order to address abstract supertypes in queries or other graph processing, we will either have to resort to the type graph with flattened inheritance or use multi labels in neo4J. We hypothesize that multi-labelling of supertypes can be shown to be equivalent to clan-morphism.

As described above, we are explicitly generating nodes for objectified relationships. The creation of edges for inverse relations is implemented as a serializer option. List order is persisted as an edge attribute $i$, that is essentially a named labelling function denoting an Integer index.

## 3 SUBGRAPH TEMPLATES FOR MODEL REDUCTION AND CHECKING

For model reduction we want to be able to identify specific parts of the graph that we want to retain or remove. For model checking we want to verify the existence or non-existence of specific parts of the graph. This section presents a method to specify such parts as *graph templates*, shows how these relate to query languages and MVDs, and finishes with a comparison of BIMserver queries with Neo4J cypher queries.

### 3.1 Mathematical model

To specify a part of an IFC typed graph, we use another IFC typed graph as a *template*. In order to identify a corresponding occurrence we are then looking for a morphism from the template graph to the instance graph. Such a morphism is called a *match*.

Given an IFC-typed graph $G_{IFC} = (G, IFC, type_G)$ and an IFC-typed template graph $T_{IFC} = (T,$

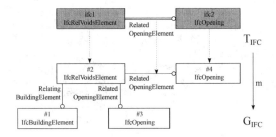

Figure 2. Template, instance graph and match morphism.

$IFC, type_T)$ a match is a morphism $m: T \rightarrow G$ such that $type_T \circ m = type_G$. Intuitively this means that the matched nodes and edges in the entity graph $G$ are of the same type as their counterparts in the template graph $T$. Figure 2 shows and example of a match morphism

### 3.2 Subgraph definition with a DSL

Despite previous attempts to standardize an exchange format for graphs (Taentzer 2001; Winter 2002), we are defining our own DSL in favour of a more compact notation than XML and more flexibility for the implementation of applications based on graph theory. Below the notation is described in Extended Backus-Naur form (EBNF).

```
lb       ::= NEWLINE | ';'
template ::= 'graph {' nodes edges '}'
nodes    ::= 'nodes {' nodeType { lb nodeType } '}' lb
edges    ::= 'edges {' edgeType { lb edgeType } '}' lb
nodeType ::= ifcType '{' node { lb node } '}' lb
edgeType ::= ifcType '= [' edge { ',' edge } ']' lb
node     ::= id, '{' [attr, { lb attr }] '}'
edge     ::= '[' id ',' id ']'
```

An example subgraph involving walls with their openings and filling doors reads as follows:

```
graph {
  nodes {
    IfcWall { wall {} }
    IfcRelVoidsElement { void {} }
    IfcOpeningElement { opening {} }
    IfcRelFillsElement { filling {} }
    IfcDoor { door {} }
  }
  edges {
    RelatingBuildingElement = [[void, wall]]
    RelatedOpeningElement = [[void, opening]]
    RelatingOpeningElement = [[filling, opening]]
    RelatedBuildingElement = [[filling, door]]
  }
}
```

Note that we are only using direct attribute names, not the inverse ones, but we could as well replace those with inverted edges.

## 3.3 Conversion

In this section we evaluate conversion of subgraph templates to MVD-XML fragments, BIMserver queries, and neo4J Cypher queries.

### 3.3.1 MVD-XML

This is an equivalent MVD-XML fragment:

```
<ConceptTemplate applicableEntity="IfcWall"><Rules>
  <AttributeRule AttributeName="HasOpenings"><EntityRules>
    <EntityRule EntityName="IfcRelVoidsElement"><AttributeRules>
      <AttributeRule AttributeName="RelatedOpeningElement"><EntityRules>
        <EntityRule EntityName="IfcOpeningElement"><AttributeRules>
          <AttributeRule AttributeName="HasFillings"><EntityRules>
            <EntityRule EntityName="IfcRelFillsElement"><AttributeRules>
              <AttributeRule AttributeName="RelatedBuildingElement"><EntityRules>
                <EntityRule EntityName="Door" />
              </EntityRules></AttributeRule>
            </AttributeRules></EntityRule>
          </EntityRules></AttributeRule>
        </AttributeRules></EntityRule>
      </EntityRules></AttributeRule>
    </AttributeRules></EntityRule>
  </EntityRules></AttributeRule>
</Rules></ConceptTemplate>
```

The reader is invited to compare this to the "Element Voiding" concept in the "General usage" MVD. Note that due to the tree-like structure of the specification, MVDs are confined to use inverse relation attributes. This applies to BIMserver queries as well.

### 3.3.2 BIMserver query

A similar BIMserver query would look like this:

```
{
  "type": {
    "name": "IfcWall"
  },
  "include": {
    "type": "IfcWall",
    "field": "HasOpenings",
    "include": {
      "type": "IfcRelVoidsElement",
      "field": "RelatedOpeningElement",
      "include": {
        "type": "IfcOpeningElement",
        "field": "HasFillings",
        "include": {
          "type": "IfcRelFillsElement",
          "field": "RelatedBuildingElement"
        }
      }
    }
  }
}
```

### 3.3.3 Neo4J Cypher query

With Neo4J we can use both queries with or without inverses. Here we show the version with inverses.

```
(:IfcWall)-[:HasOpenings]->(:IfcRelVoidsElement)
  -[:RelatedOpeningElement]->(:IfcOpeningElement)
  -[:HasFillings]->(:IfcRelFillsElement)
  -[:RelatedBuildingElement]->()
```

### 3.3.4 Discussion

While we can translate the graph template into a corresponding MVD fragment, BIMserver query and Neo4J Cypher query, there are subtle but essential differences in the meaning of these statements even though they appear similar. While our graph templates identify a set of matches, a BIMserver query does just restrict the whole object graph without identifying matches. The different purpose of defining what to include in an export is reflected in the term "filter language" instead of "query language". Neo4J can assemble result records similar to our graph template approach, but also create the restricted graph. For MVD definitions it depends on the use case: MVDs deliberately allow to be interpreted differently depending on whether they are used for filtering, validation, or documentation.

## 3.4 Comparison of BIMserver and neo4J

In the previous section we focussed on the expressiveness of different query DSLs, now we are going to compare the performance of IFC queries on BIMserver and neo4J.

### 3.4.1 Test setup

To execute the tests, queries are issued through the REST interfaces of BIMserver and neo4J running on local machine. To this end, cUrl is used with the option `-w "\n%{time_total}"` to measure response time. Two queries are issued this way:

1. all walls with their openings and building elements that fill those openings
2. all building elements that fill openings in a specific wall

To ensure comparability, we started tests with a fresh BIMserver install, a new home directory as well as new project and check-in. Similarly, a fresh neo4J database was setup with new project imported. We used the query without inverses.

### 3.4.2 BIMserver results

The import took 161.965 seconds as measured on the command line. Of those, BIMserver reported it needed 292 ms for writing and 2 min 36 s for geometry processing (triangulation). The database size after import was 61.3 MB.

The results of querying are listed in Table 1. We repeatedly issued the queries four times, clearing the cache after the second run. For both queries, there are three values listed: the time for initiating the query, the time for downloading the result, and the total. Query initiation and result download are separate calls to BIMserver to allow for clients to issue non-blocking (asynchronous) requests for long-running queries.

Table 1. Query profiling results for BIMserver.

|     | All openings | | | Specific opening | | |
|-----|------|--------|-------|-------|--------|-------|
| run | init | stream | total | init  | stream | total |
| 1   | 0.047 | 0.042 | 0.087 | 0.011 | 0.029 | 0.040 |
| 2   | 0.009 | 0.024 | 0.033 | 0.008 | 0.009 | 0.017 |
| 3   | 0.022 | 0.041 | 0.063 | 0.023 | 0.026 | 0.049 |
| 4   | 0.020 | 0.047 | 0.067 | 0.023 | 0.022 | 0.045 |

Table 2. Query profiling results for neo4J.

|           | All openings | Specific opening |
|-----------|--------------|------------------|
| first run | 0.158        | 0.035            |
| repeated  | 0.067        | 0.010            |

### 3.4.3 Neo4J results

The first attempt to import the IFC data into neo4J using the Cypher query exported from BIMserver used 6G of memory with a tendency to still claim more memory and did not finish in reasonable time. The import was issued as on big create statement, thus it was not possible to control progress. Following that the Cypher code was improved to use less memory due to separated statements. This might use more time, especially for the creation of edges, since start and end node have to be looked up from the database. Further improvements might be possible by joining the edge creation statements per entity. In total, the import took roughly 17 hours: 11am to 4am the next morning. The resulting DB size was 215.42 MiB.

The results of issueing the queries are listed in Table 2.

### 3.4.4 Comparison

BIMserver and Cypher queries perform equally well for small files, while for large data sets the import into Neo4J constitutes an upstream performance issue. The resulting database size is 4 times larger for neo4J compared to BIMserver.

## 4 TOWARDS GRAPH DECOMPOSITION WITH GRAMMARS

We are considering graph transformations following the double pushout (DPO) approach. In this approach, a graph production $p$ consists of a left- and a right-hand side typed template graph $L$ and $R$ as well as a glueing or interface graph $I$ and two graph morphisms $l: I \to L$ and $r: I \to R$. Figure 3 shows an example production. We restrict the morphisms to injective morphisms and thus

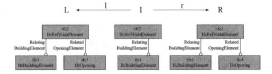

Figure 3. Transformation rule replacing an opening in a wall.

standard replacement.[5] From a production $p$ and a match $m: L \to G$ we can then construct a graph transformation $G \Rightarrow^{m,p} H$. We are saying that the production $p$ is applied to the graph $G$ via the match $m$. For details on the construction of these transformations, refer to König et al. (2018) and Ehrig et al. (2006).

Graph transformations can be concatenated, such that we obtain a transformation $G_0 \Rightarrow^* G_n$ from a sequence of direct transformations $G_i \Rightarrow G_{i+1}$. A typed graph transformation system consists of a type graph $IFC$ and a set of productions $P$. Such a system can describe potentially infinite amounts of transformations which are obtained by selecting sequences of $p \in P$ and successively applying these productions as appropriate matches to a typed start graph $S$. A graph transformation system together with a start graph is also called a grammar, because it implicitly describes the set of all graphs that can be obtained by application of sequences of productions. This language $L$ is denoted as $L = \{G \mid \exists S \Rightarrow^* G\}$.

We are now aiming to obtain a graph grammar from the IFC schema, that is an equivalent to the meta model in terms of graph transformations. Such a grammar could serve different use cases: First, it could foster the generation of instance graphs, e.g. for testing purposes. Second, it could aid "semantic analysis or transformation of model graphs" (Fürst et al. 2015), that is it can provide a framework to deconstruct instance graphs in order to either better understand or transform them in a piecewise manner.

Ehrig et al. (2006) and Fürst et al. (2015) describe different general approaches to convert metamodels into graph grammars. We intend to study their applicability in the context of IFC. We already identified key considerations for the IFC graph grammar: representation of mandatory and optional attributes, resources reuse, and inheritance.

For a graph grammar to be equivalent to the schema we need two conditions to be full-filled: completeness and consistency. For completeness we need all valid graphs to be contained in the language. For consistency, we need all graphs contained in the language to conform to the schema.

---

[5] A non-injective $l$ would result in splitting and a non-injective $r$ in merging of graph nodes or edges.

Depending on the use cases for such a grammar, we can relax the requirement on either consistency or completeness. E.g. to be able to analyse or transform all potentially valid IFC graphs, we need completeness, but can relax consistency, assuming that we only operate on valid instance graphs. In contrast, for instance graph creation, we need consistency, but can relax completeness, accepting that we will only retain a subset of potentially valid graphs. For validation purposes we would need both completeness and consistency, because lack of consistency would result in false positives whereas lack of completeness would result in false negatives.

## 5 EXPLORATION WITH GRAPH VISUALIZATION

In the first part of the paper we have used graphs with various connotations: to model the IFC schema, an IFC data set, query templates, and rules. Diagrammatic representations are easy to generate from these graphs, but in order to ease the understanding of the concepts behind them, it must be clear what each graphical element is denoting. That is why we use this section to develop unique graphical representations for the concepts introduced previously. The visualization schema we are contributing is inspired by Express-G, but extended and modified to accommodate the specifics of the different types of graphs.

First, to represent the schema in isolation, we are using a type graph. Express-G is a dedicated diagrammatic notation for Express schemas and also used in the IFC documentation. It essentially provides a direct representation of the type graph: entity type nodes are represented as rectangular boxes with the name of the type name written inside.

To represent the instance graph, we use a similar representation, but each node and edge is labeled with an object (and optionally relation identifier) instead of the type, e.g. numbered consecutively. We can represent the object and type graph explicitly and indicate the typing morphism with dotted lines as shown in Figure 1. A more compact notation integrates the object and type graph representations by adding type names to the node and edge identifiers as shown in Figure 4.

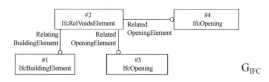

Figure 4. Integrated representation of type and instance graph (explicit version in Figure 1).

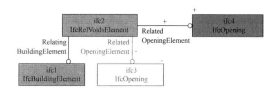

Figure 5. Integrated rule representation (explicit version in Figure 3).

To represent a template graph we use the same diagrammatic representation as for integrated object and type graphs, but add a filling to nodes and use a duplicate line for edges to distinguish them from ordinary type graphs. We can then represent the match morphism (between a template and an object graph) similar to the typing morphism with dotted lines as in Figure 2 or again in an integrated form where the match is hightlighted. While node IDs are following the IFC line numbers for object graphs, pure template graphs have freely chosen node IDs.

Transformation rules can be represented either as left-hand and right-hand side template graphs together with a gluing graph or a mapping (Taentzer 2001), see Figure 3. As Lambers (2005) points out, most graph transformation tools use an integrated, more compact notation, where the common parts of left- and right-hand graph templates and potentially the gluing graph are not repeated. Instead, deleted, retained and added graph parts receive a specific annotation or formatting. This representation is however restricted to specific forms of transformations where the left- and right-hand side morphisms are homomorphic. Since our current rules are not affected from this restriction, we will be using the integrated version as shown in Figure 5.

We can use the very same presentation to show rule application, that is we are representing a match in the source graph and the comatch in the target graph at the same time.

## 6 CONCLUSION AND OUTLOOK

We have shown our approach to the application of graph-based methods to IFC data. In future work, we will apply these methods to specific use cases, such as model-checking and requirements specification using subgraph templates, in-place transformation of IFC graphs, IFC schema conversion, or transformation from IFC to other formats. For the latter we will extend the methods described here to the triad of source, target and connection graph.

The methods still need further refinement, since we excluded some detailed investigation in favour of a comprehensive overview of the fundamentals and the different types of graphs that are to be considered in the context of graph-based model transformations. For instance we excluded data nodes or labels. When we consider them in a next step, we will arrive at the category of *attributed type graphs with inheritance (ATGI)* according to Ehrig et al. (2006). Also for the subgraph templates we just used a fairly simple query. More complex queries are work in progress. We only touched the subject of graph grammars, but we anticipate that they play an essential role in graph-based model analysis and transformation. Therefor it would be interesting to evaluate generic graph transformation tools against a custom IFC-specific implementation.

We acknowledge the funding support from the National Research Foundation (Singapore) under the Virtual Singapore programme grant NRF2015VSG-AA3DCM001-008. We thank Rudi Stouffs and Patrick Janssen for their comments on an early version of this manuscript.

## REFERENCES

Bardohl R., Ehrig H., Lara J. de, Taentzer G. (2004): Integrating Meta-Modelling Aspects with Graph Transformation for Efficient Visual Language Definition and Model Manipulation. In: M. Wermelinger & T. Margaria-Steffen (Eds.), *Proc. of Fundamental Approaches to Software Engineering (Fase)*, pp. 214–228, Springer, Berlin, Heidelberg, 2004.

Ehrig H., Ehrig K., Prange U., Taentzer G. (2006): *Fundamentals of Algebraic Graph Transformation*. Springer, Berlin.

Fürst L., Mernik M., Mahnič V. (2015): Converting Metamodels to Graph Grammars: Doing Without Advanced Graph Grammar Features. *Software & Systems Modeling*, Vol. 14, Nos. 3, pp. 1297–1317.

Isaac S., Sadeghpour F., Navon R. (2013): Analyzing Building Information Using Graph Theory. In:, *Proc. of the 30th International Symposium on Automation and Robotics in Construction (ISARC)*, pp. 1013–1020, Montreal, Canada, August 2013.

Isikdag U., Zlatanova S. (2009): Towards Defining a Framework for Automatic Generation of Buildings in CityGML Using Building Information Models. In: J. Lee & S. Zlatanova (Eds.), *3D Geo-Information Sciences*, pp. 79–96, Springer, Berlin, Heidelberg.

Ismail A., Nahar A., Scherer R. (2017): Application of Graph Databases and Graph Theory Concepts for Advanced Analysing of BIM Models Based on IFC Standard. In: C. Koch, W. Tizani, & J. Ninic (Eds.), *EG-ICE 2017: 24th International Workshop on Intelligent Computing in Engineering*, pp. 146–157, Nottingham, UK, July 2017.

ISO (2004): *Industrial Automation Systems and Integration. Product Data Representation and Exchange. Part 11: Description Methods: The EXPRESS Language Reference Manual*. International Organization for Standardization, Geneva, Switzerland.

ISO (2013): *Industry Foundation Classes (IFC) for Data Sharing in the Construction and Facility Management Industries*. International Organization for Standardization, Geneva, Switzerland.

Khalili A., Chua D.K.H. (2015): IFC-Based Graph Data Model for Topological Queries on Building Elements. *Journal of Computing in Civil Engineering*, Vol. 29, Nos. 3, p. 04014046.

König B., Nolte D., Padberg J., Rensink A. (2018): A Tutorial on Graph Transformation. In: R. Heckel & G. Taentzer (Eds.), *Graph Transformation, Specifications, and Nets: In Memory of Hartmut Ehrig*, pp. 83–104, Springer International Publishing, Cham.

Lambers L. (2005): A New Version of GTXL: An Exchange Format for Graph Transformation Systems. *Electronic Notes in Theoretical Computer Science*, Vol. 127, Nos. 1, pp. 51–63.

Pauwels P., Terkaj W., Krijnen T., Beetz J. (2015): Coping with Lists in the IfcOWL Ontology. In:, *Proceedings of the 22nd International Workshop on Intelligent Computing in Engineering (EG-ICE)*, pp. 111–120, Eindhoven, The Netherlands, June 2015.

Taentzer G. (2001): Towards Common Exchange Formats for Graphs and Graph Transformation Systems. *Electronic Notes in Theoretical Computer Science*, Vol. 44, Nos. 4, pp. 28–40.

Tauscher E., Bargstädt H.-J., Smarsly K. (2016): Generic BIM Queries Based on the IFC Object Model Using Graph Theory. In:, *Proc. of the 16th International Conference on Computing in Civil and Building Engineering (ICCCBE)*, pp. 905–912, Osaka, Japan, July 2016.

Vilgertshofer S., Borrmann A. (2017): Using Graph Rewriting Methods for the Semi-Automatic Generation of Parametric Infrastructure Models. *Advanced Engineering Informatics*, Vol. 33, pp. 502–515.

Winter A. (2002): Exchanging Graphs with GXL. In: P. Mutzel, M. Jünger, & S. Leipert (Eds.), *Graph Drawing*, pp. 485–500, Springer, Berlin, Heidelberg, 2002.

# Integration of an ontology with IFC for efficient knowledge discovery in the construction domain

## Z.S. Usman, J.H.M. Tah & F.H. Abanda
*Oxford Institute for Sustainable Development, School of the Built Environment, Faculty of Technology, Design and Environment, Oxford Brookes University, Oxford, UK*

## C. Nche
*School of Information, Technology and Computing, American University of Nigeria, Yola, Adamawa State, Nigeria*

ABSTRACT: In the most recent years, the Semantic Web applied to Building Information Modelling and Industry Foundation Classes are the most widely recognized efforts in the digitalization of products in the architectural, engineering and construction industry. This study performs knowledge discovery of construction products from domain experts to build an ontology. However, ontology developed becomes an isolated conceptualization. In order to get rich domain ontology, novel approach is to retrieve knowledge from domain-specific online product repositories such as the industry foundation classes. This study identifies similarities between the industry foundation classes and ontology likewise their differences with the hope that it would develop an enhanced knowledge conceptualization of the domain. Integration of IFC and ontology will mean that concepts represent well-established standard terms rich in domain knowledge. Other semantic web technologies can then be exploited on the resulting application to perform data querying, consistency checking, data reasoning, knowledge inference and so much more. Thus, the resulting ontology could be standardized and be easily applied for use in practice in the industry.

*Keywords*: Semantic Web (SW), Ontology, Web Ontology Language (OWL), Industry Foundation Classes (IFC), Photovoltaic Systems (PV)

## 1 INTRODUCTION

Recent applications in the construction domain rely on semantic web and its technologies to tackle issues concerning interoperability, knowledge extraction, representation, exchange, management and re-use. Knowledge identification, specification and refinement are key techniques used to capture basic theoretical and conceptual knowledge to develop domain ontology. Momentous effort is required and numerous complexities are faced when knowledge is obtained from books, journal articles, qualitative research through case study. Similarly, researchers face difficulties in communicating with domain experts and require a lot of work to resolve. This also means concepts will be manually entered into the ontology editing software. An ontology developed solely on knowledge from domain experts cannot serve its purpose for long, due to the fact that construction product terms often change. In order to get rich domain ontology, a novel approach is to retrieve knowledge from domain-specific online product repositories. Fortunately, the goals of the semantic web relatedly tend to overlap with Building Information Modelling (BIM) product classification schemas and taxonomies such as the Industry Foundation Classes (IFC). Being an international multi-platform standard for representing building products, IFC has drawn momentous attention in construction research and industry worldwide.

The aim of this study is to investigate the methodologies and benefits of integrating a component of the sematic web technology (ontology) with IFC and integrate them as a solution to a product problem of a domain of interest—the Photovoltaic (PV) systems technology. The research further reviews and analyzes numerous efforts that have been made in developing a Web Ontology Language (OWL) ontology from the EXPRESS schema that defines the IFC, the ifcOWL ontology. After making this analysis, we attempt to extend the World Wide Web Consortium (W3C) approved ifcOWL ontology to represent concepts essential to our problem domain and integrate it to the initial PVontology generated from domain knowledge to create an ifcOWL-PVontology.

## 2 SEMANTIC WEB

### 2.1 Overview of ontology

An ontology collects raw and unstructured data on a domain and transforms it to a formal format interpretable by human, agents and machines. It is the most powerful machine processable language for representing domain knowledge (Abanda et al., 2013). Ontologies share a collective structural pattern. It typically includes explicit representation of concepts (or classes), data and annotation properties (of concepts), restrictions (on properties), relations (between concepts), instances (of classes). An ontology together with its components establishes a domain knowledge base (Chen & Luo, 2016). From this established knowledge base, the rule engine of the ontology can create deduced components where the standard rule language of the semantic web such as the Semantic Web Rule Language (SWRL) can be easily applied (Chen & Luo, 2016). Nonetheless, an ontology development language must be used for representation of complex knowledge to be possible. The Web Ontology Language (OWL), an accepted ontology language standard is explained further in the ensuing section.

### 2.2 Web ontology language

The Web Ontology Language (OWL) is the standard language for the ontology layer of the semantic web. It is recommended by the World Wide Web Consortium (W3C) (W3C, 2004). OWL executes great machine interoperability of the contents of the web. It specifies a collection of operators to develop concept definitions and descriptions as well as reasoners to perform consistency checking of the ontology. Ontology developers usually adopt one of OWL and OWL 2 (OWL2 EL, QL and RL) sublanguages which best suits the needs of the application. The expressive power, computational completeness, reasoning capacity and limitations of the OWL sublanguages are amongst the characteristics that are analyzed during selection. The level of expressiveness of the OWL language determines what is represented in the OWL ontology (Pauwels & Terkaj, 2016). The reader is referred to the W3C OWL recommendation document for more details (W3C, 2004). OWL and SWRL together perform extensive semantic reasoning (Chen & Luo, 2016). Thus an ontology provides standard conceptualization and the semantic knowledge reasoning required on the selected domain. In this study, the Photovoltaic (PV) Systems is the selected domain of interest.

## 3 OVERVIEW OF PHOTOVOLTAIC SYSTEMS

### 3.1 PV-systems

The Photovoltaic (PV) system is a renewable and sustainable energy technology that gets its energy from the sun. It has been widely researched and used in practice as an alternative electricity source worldwide (Buckley & Nicholas, 2016; IRENA, 2016). Evaluations have shown that PV technology has the potential to provide adequate electricity supply in many parts of the world by 2021. A typical PV technology consists of a group of components. A group of solar cells (capture electric energy from the sun) are assembled together to form solar modules. A collection of modules in turn make up the solar array of huge electric output. Buildings use inverters to switch electricity form direct current (DC) to alternating current (AC) while batteries are used to store back-up energy. The size and design of the PV system is determined by the electric output required. This electric output is directly dependent on the size of the PV components installed.

PV components are made up of different materials and sizes with dissimilar electric capacities, efficiencies and prices. Because of this disparity, there are so many PV components available online that are manufactured by hundreds of manufacturers and sold by hundreds of suppliers. The exponential increase in the number of these online PV collections however makes it challenging to easily retrieve the right information that the user requires. Product information retrieval from these sources are often keyword based with very low precision. One product could be named using different terms by different manufacturers (synonyms). Relatedly, one term may have different general meanings (polysemy) (Liu et al., 2015). An ontology-based solution in the PV domain means dissimilar PV products from diverse sources could be easily unified and transformed into a common vocabulary for representation of the domain. An ontology developed solely on knowledge from domain human experts cannot serve its purpose for long, because construction product terms and descriptions evolve as construction product libraries continuously enlarge. The result is often an isolated private conceptualization of the domain. In order to develop a rich domain ontology that is a shared conceptualization of the PV-domain, novel approach is to acquire knowledge from domain-specific online product repositories.

### 3.2 Building domain-specific online product repositories

Product model libraries hold product terms with specific descriptions, meanings and relationships to one another. There are varieties of online product model libraries. BIMobject (bimobject, 2013) is one of the largest digital system for management of BIM objects. It allows professionals in the architecture, engineering and construction industry to easily retrieve BIM files with product specifications into their respective applications and projects. The contents of Autodesk Seek, which is accessible on BIMobject cloud holds about 23 million building products from thousands of manufacturers (bimobject, 2018), NBS National BIM Library (NBS, 2018), 3D Warehouse (Warehouse, 2018). Key information about construction products are stored in the documentation of these online repositories. These include detail descriptions of the product specifications (manufacturer, dimensions, durability, sustainability, performance, warranty) (Liu et al., 2015). They are perfect for domain-specific knowledge solutions.

## 4 BUILDING INFROMATION MODELLING (BIM) AND INDUSTRY FOUNDATION CLASSES (IFC)

### 4.1 Overview of BIM and IFC

BIM can be acclaimed as one of the most remarkable state-of-the-art platforms for building data management in recent years (Pauwels & Terkaj, 2016). BIM offers digital and geometric representation of building-related information in one model. It provides solutions to limitations faced in information integration, exchange and interoperability in different stages in the AEC industry. Likewise, the Industry Foundation Classes (IFC), a product classification library and taxonomy is a BIM multi-platform standard for representing building products' data and descriptions for effective exchange and re-use. IFC is developed by buildingSMART (buildingSMART, 2017). IFC provides data structures that support automated project modelling so BIM data can be shared across construction industry software applications (IFC, 2018). It is the state-of-the-art standard for information exchange in this domain. IFC data models are represented in EXPRESS data specification language of the International Organization for Standardization (ISO) (ISO, 2013). The EXPRESS language represents in detail entities, attributes, data types, properties, rules and restrictions that build up a schema (type)

```
ENTITY IfcSolarDevice
 SUBTYPE OF (IfcEnergyConversionDevice);
  PredefinedType : OPTIONAL IfcSolarDeviceTypeEnum;
 WHERE
  CorrectPredefinedType : NOT(EXISTS(PredefinedType)) OR
(PredefinedType <> IfcSolarDeviceTypeEnum.USERDEFINED) OR
((PredefinedType = IfcSolarDeviceTypeEnum.USERDEFINED) AND EXISTS
(SELF\IfcObject.ObjectType));
  CorrectTypeAssigned : (SIZEOF(IsTypedBy) = 0) OR
('IFCELECTRICALDOMAIN.IfcSolarDeviceType' IN
TYPEOF(SELF\IfcObject.IsTypedBy[1].RelatingType));
END_ENTITY;
```
EXPRESS-G diagram

Figure 1. Express specification of IFC entity IfcSolarDevice.

of EXPRESS in.exp. IFC has numerous of these schemas arranged in the order in which they were created such as the IFCS2X3.exp, IFC2X4_ADDI.exp (buildingSMART, 2018). In one of these schemas, a solar device (as an example in Figure 1) represented in an IFC format can be described in any BIM environment with details of its meaning and purpose to building and construction industry.

### 4.2 Similarities between Industry Foundation Classes (IFC) and Semantic Web (SW)

As explained in previous sections, the IFC represents and exchanges building-related data in the construction industry while ontology represents data on the web. While they are used within the scope of their domain, they both target the same specification and standardization goals of data exchange and interoperability. The IFC is of utmost importance to the construction industry comparable to ontology's significance in the semantic web. Similarly, researchers have identified similarities between the EXPRESS specification/modelling language of IFC and the OWL ontology language (Pauwels et al., 2015). Some scholars believe the semantic structures of the IFC and OWL model are somewhat equal (Pauwels & Terkaj, 2016), however, this study disagrees, if it were equal, semantic enhancement of the IFC's EXPRESS schema would not have been required.

### 4.3 Benefits of integration of SW, BIM, IFC

Some researchers believe more fundamental research on BIM's information representation, exchange and interoperability is relevant (Liu et al., 2015). While they boast of BIM international role as a standard for openBIM, it is confessed that its "…lack of semantic information greatly hinders its application" (Chen & Luo, 2016). They admit that added to the IFC EXPRESS schema, the logic-based nature of OWL would provide formal semantics to BIM data since its structure permits both geometric and semantic information of building models to be stored (Barbau et al.,; Liu et al., 2015). A combination of IFC and OWL ontology (namely ifcOWL) means that data querying tools (not applicable to IFC EXPRESS schemas) and semantic web tools could then be used on IFC data models to offer more meaning. In agreement, the semantic web also solves the limitation of EXPRESS' decreasing monotonic reasoning (Beetz et al., 2009). Scholars have added an integration of IFC and Semantic Web means enhanced reasoning and semantic searches would be available to experts in the domain of Architecture, Engineering and Construction (Beetz et al., 2009; Schevers & Drogemuller, 2006; Pauwels & Van Deusen, 2012 and Pauwels & Terkaj, 2016).

The integration of IFC and Semantic Web has led to many application developments and IT solutions in the construction industry. Table 1 gives a short review of latest applications in the construction industry that apply techniques from BIM technology, use building information represented by IFC and employ semantic web technologies for

Table 1. An appraisal of applications that have integrated SW, BIM and IFC.

| Semantic web technology/BIM | Description | Article/Journal sources |
|---|---|---|
| BIM, GIS, Ontology | Urban Information Modelling (UIM) is developed here as an extension to BIM. It offers several different forms of concept representation to create a unique ontology. | (Mignard & Nicolle, 2014) |
| Ontology, IFC, BIM | An ontology was developed that integrates information from IFC, building performance simulation and some sensors and executes performance checking on them. | (Pauwels et al., 2015) |
| Ontology, SWRL | A prototype application system named OntoSCS is developed here to enhance the design and selection of materials relating to sustainable concrete structures. | (Hou et al., 2015) |
| Semantic search engine, BIM | A semantic based search engine (called BIMSeek) is developed for easy retrieval of BIM-related data/knowledge/documents. | (Gao et al., 2015) |
| Ontology, Supply chain | An ontology is developed to address the challenges of intelligent maintenance systems and spare parts supply chain. The ontology provided a medium for conceptual representation and exchange. | (Saalmann et al., 2016) |
| Ontology, BIM | These researchers extracted information from BIM model using an "ontology-augmented model index". | (Zhang & Issa, 2012) |
| BIM, Ontology | Authors review state-of-the-art applications of BIM and ontology. They propose an intelligent software application that uses both BIM and ontology for "intelligent" information integration and reasoning. | (Chen & Luo, 2016) |
| Semantic web, BIM | This studied analyzed key role played by SW technologies in logic-based applications that require information from BIM, GIS, and energy in an attempt to solve challenges of interoperability, data connection of information from numerous domains and apply semantic web logical inferences. In addition, this study offers a robust groundwork for future research of applications of semantic web in the sectors. | Pauwels, Zhang and Lee, (2016) |
| BIM, Ontology | This paper uses an ontology-based BIM as a knowledge model to enable better understanding of building automation systems (BAS); their causes and consequence in order to improve error fixing and maintenance. | (Dibowski et al., 2017) |
| IFC, Semantic information retrieval | Researchers here use the **documentation** of the IFC schema as a domain knowledge repository (not the IFC schema itself) to perform effective searching through BIM product model libraries. | (Liu et al., 2017) |
| BIM, Ontology | A review of knowledge engineering technologies used in AEC is made. They focused on techniques that could be used to integrate these technologies to BIM and use an OWL ontology as a domain knowledge model for the development of the application. | (Perhavec & Kaucic, 2018) |

efficient information integration, exchange and semantic reasoning respectively.

## 5 INTEGRATION OF IFC EXPRESS AND OWL ONTOLOGY

### 5.1 *Benefits of integration of SW, BIM, IFC*

The OWL ontology developed from the EXPRESS schema that defines the IFC is named the ifcOWL ontology. Although IFC instances are represented in graph-like format similar to RDF, the EXPRESS schema contains a number of complex elements which are different from the OWL language elements. The two languages have similar and different features with distinctive expressive power. Therefore, a methodological technique is required to convert one instance from the EXPRESS schema of IFC into a corresponding OWL ontology instance.

In recent years, numerous efforts have been made to develop the right approach to convert the EXPRESS schema of IFC into an OWL ontology. Schevers & Drogemuller (2006), proposed the earliest approach of EXPRESS to OWL conversion. They provided cases studies where IFC EXPRESS entities were converted to OWL classes and attributes and outlined some of the central limitations (of the approach) so they could be addressed by future research. With the intention to increase the application of the IFC standard, Beetz et al. (2009) proposed a method of partial automatic conversion of IFC EXPRESS schema to OWL ontology. Subsequently, Barbau et al. (2012) created an automatic conversion of EXPRESS within the OntoSTEP research initiative. They designed a conversion tool which is currently available as a plugin in protégé. A similar conversion tool was developed by Pauwels and Van Deursen (2012). Both Barbau et al. (2012) and another conversion approach described in Krima et al. (2009) concentrated on creating a general conversion methodology for all forms of EXPRESS schema conversions to OWL ontology, and not just the EXPRESS schema that represents the IFC. However, in this conversion, the LIST cardinality restrictions are converted using EmptyList which is not implemented in the ontoSTEP tool (Pauwels & Terkaj, 2016).

Terkaj et al., (2012) suggested a partial IFC conversion approach with a focus on a single domain-specific data schema of IFC and its integration with other data models in the construction domain. Hoang and Torma (2014) and Hoang (2015) focused on two benchmarks: conversion of IFC to OWL between the three (3) OWL 2 sub-languages (OWL2 EL, QL and RL), and to create RDF graphs easily into linked data (Pauwels & Terkaj, 2016). One particular conversion approach was focused on Information Retrieval (Gao et al., 2015). Only concepts considered applicable to Information Retrieval were marked and retrieved from the IFC schema to create the IFC ontology. Terkaj and Sojic (2015) proposed ways to enhance previously developed ifcOWL ontologies by coming up with how rules in IFC schema's EXPRESS can also be represented in OWL.

The expressiveness of OWL is a fundamental part of the ontology development. Most of these conversion approaches were based on OWL 1 dialects and not OWL2. OWL 2 resolves limitations of OWL 1 and extends it with many useful features required to model complex domains that includes extended annotations, excellent cardinality constructors, extended property sets and so much more (SemWebTec, 2009). The use of OWL 2 means ontology developed is enriched with the semantics of SROIQ knowledge base. SROIQ is a part of first order logic that contains excellent computational properties (W3C, 2012). In the development of ifcOWL, only the approach presented by Hoang and Torma (2014) and Hoang (2015) recognized the relevance of the OWL2. Similarly, these researchers (Schevers & Drogemuller, 2006), (Beetz et al., 2009), (Pauwels & Van Deursen, 2012) have published in journal articles only the ideas behind their conversion procedures without providing the actual generated ontologies. Therefore, all further work adopted result in slightly different versions of the ifcOWL. Similarly, in most cases, large proportions of the semantic richness of the actual EXPRESS schema of the IFC is lost during conversion which limits the functionalities and applications of the approaches. Relatedly, because Barbau et al., and Krima et al.'s approaches created a converter for all forms of EXPRESS schema (not specific to IFC EXPRESS), it is not unique to the construction industry, this ontology does not suit the concepts within the IFC EXPRESS schema. Therefore, because recent studies attempt to extend IFC and apply semantic web technologies to provide solutions to certain product problem domains, creating private and different ifcOWL ontologies would not be practical in the long run. The need for a standardized and recommendable ifcOWL ontology brought the emergence of the current ifcOWL recognized by buildingSmart (openbimstandards, 2016).

### 5.2 *Standard ifcOWL*

The official standard ontology for IFC, ifcOWL was developed by Pauwels and Terkaj and a group of many contributors as part of the BuilingSMART Linked Data Working Group (W3C, 2018) and W3C Community Group on Linked

Building Data. The most recent version of the ifcOWL is available at openbimstandards (2016). This ifcOWL took as reference some conversion patterns from previous proposals. For example, the conversion of simple data types were adopted from class wrapping approach recommended by Schevers & Drogemuller (2006) and Beetz et al. (2009). The conversion procedure of the Onto-STEP initiative (2012) was equally relevant to this ifcOWL. From Hoang and Torma (2014) and Hoang (2015), ifcOWL adopts their use of OWL 2. In Pauwels et al. (2015) two suitable options were outlined for converting LIST (ifc:list) in EXPRESS to its OWL equivalent. Consequently, every other detail including naming of classes, properties and restrictions were comprehensively revised, removed, altered or improved. The most important difference between the ifcOWL and the previous proposals that are freely available (Onto-STEP (2012), Hoang and Torma (2014)) are its richness of axioms. Detailed comparison of the ifcOWL and the previous proposals can be found in (Pauwels & Terkaj, 2016). Ontology components such as equivalent classes, disjoint classes, functional and inverse properties as well as property ranges are all defined, making the standardized ifcOWL 'semantically' closer to the original EXPRESS schema of IFC (IFC, 2018).

## 6 PROPOSED APPROACH TO INTEGRATING SEMANTIC WEB AND IFC

### 6.1 *Description of the approach*

This study develops as a case study an ontology for photovoltaic systems called pvOntology. This ontology would then be integrated with ifcOWL ontology to develop the ifcOWL-pvOntology. Section 6.1.1 provides details of the design of a typical PV System and 6.1.2. outlines the implementation of the design in an ontology editing software. 6.1.3 then provides details of the integration of the IFC-semantic web ontology. The resulting ontology can be found at https://github.com/usmanzainab/IFC-Semantic-Web-pvOntology/tree/Code.

#### 6.1.1 *Design of PV-systems for development of pvOntology*

The design of the PV system solely depends on the size of the components to be installed. A comprehensive review of literature, analysis and recommendations have been made on the design of PV systems (Wade, 2008; Abanda et al., 2013; Leonics, 2013; Zeman, 2015; Bockliseh, 2016) that would not be repeated for the sake of brevity in this paper. Thorough calculations on how to obtain energy loads of buildings, determining of size of solar modules, arrays, inverters, batteries and length and width of cables and wires are available in Usman et al., (2018). Further details on data captured on PV-systems and PV process map for the complete design of a PV-system can be found at (Usman et al., 2018).

#### 6.1.2 *Implementation of pvOntology in Protégé*

The pvOntology was implemented by manually adding concepts and classes to an ontology editing software Protégé-OWL. Class hierarchies are created. Annotation properties, object properties and data properties provide comments on entities, defines relations between instances and define the relationship between class instances and their data types respectively. Instances of the pvOntology are given, each specific to its class with restrictions. Figure 2 shows a class hierarchy of the pvOntology.

Developing an ontology manually from scratch takes a lot of effort and time. It is painstakingly difficult and always results in errors and inconsistencies. Very often consistency errors cannot be fixed and one has to re-start the ontology all over again. Therefore, the BIM standardized and recommended ifcOWL ontology is extended to properly fit the scope of the photovoltaic system technology.

#### 6.1.3 *Integration of the IFC-semantic web pvOntology*

The ifcOWL ontology extended with photovoltaic attributes develops the ifcOWL-pvOntology. First and foremost, concepts and their corresponding properties that are similar in both ontologies are identified and the terms used in ifcOWL is used for representation. Concepts that are not common and only exist in pvOntology maintain their current terms. However, it is noticed that probably the most important classes and attributes of IFC relating to PV products are new and had just been released in the IFC2X4_ADDI.exp. Therefore, these new classes such as the *IfcSolarDevice, IfcSolarDeviceType* are not included in the ifcOWL ontology. For the purpose of this study, these new classes are directly extracted from IFC schema

Figure 2. Class hierarchy of pvOntology.

(through inherited definitions from supertypes shown in Figure 3) and inputted into the ontology editing software as seen in Figure 4. Notice *IfcSolarDevice* inherits the properties of its super classes. A visualization of the IfcSolarDevice entities are shown in Figure 5.

The ifcOWL-pvOntology is populated with classes and sub classes, annotation, object and data properties as well as individuals. The properties however are represented using the structure of the ifcPropertySets shown in Figure 6. This structure is obtained through levels of property data represented within the domain specific data schemas of the IFC. All entities obtained from IFC

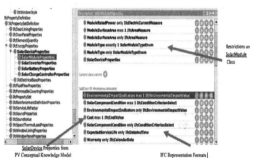

Figure 6. ifcOWL-pvOntology object properties definitions and restrictions.

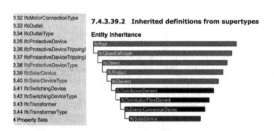

Figure 3. IfcSolarDevice IFC4 Add2 – Addendum http://www.buildingsmart-tech.org/ifc/IFC4/Add2/html/.

Figure 7. Hermit reasoner checking for Anomalies.

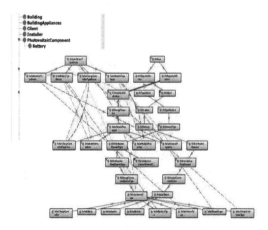

schema are represented in the ontology with a prefix *i8fc* while entities from PV conceptual model do not have the prefix. For example, in Figure 6, *EnvironmentImpactIndicator* is a solar device property while its value *IfcEnvironmentalImpactValue* is represented in IFC format.

### 6.2 Validation of the integrated (IFC-Semantic web) pvOntology

One of the many important stages of ontology development is evaluating the correctness and quality of the ontology developed. Throughout the development time of the ifcOWL-pvOntology, constant checking was done using ontology reasoners. Reasoners use first-order predicate logic to retrieve logical knowledge and check if that knowledge is clear and consistent. The reasoners used during the development of the ifcOWL-pvOntology are the Hermit, Fact++ and ELK reasoners. They have been used consistently to check for anomalies and inconsistencies. A typical case of ontology inconsistency could be incompatible properties, datatypes or range definitions. Figure 7 shows the Hermit reasoner in process. The ifcOWL-pvOntology consistency checking was successful without errors.

Semantic Web Rule Language (SWRL) and other semantic web technologies can now be

Figures 4. IfcSolarDevice in Protégé-OWL.

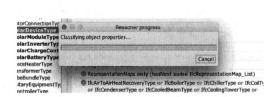

Figure 5: OntoGraf visualization of the entities of IfcSolarDevice.

435

exploited on the ifcOWL-pvOntology to perform data querying, semantic reasoning, knowledge inference and so much more. Thus, the resulting ontology could be standardized and could be integrated into other software applications for use in practice in the recommendation of products in the photovoltaic/solar/sustainable energy industry.

## 7 CONCLUSION

This paper presented a review of ontology and the advancement of the OWL language. This followed an overview of Building Information Modelling and Industry Foundation Classes elaborating their differences, similarities with semantic web and the utmost benefit of integrating all three (3) to develop software applications as solution to a problem domain; the Photovoltaic Systems domain. The domain has been reviewed and approaches towards its design have been outlined with thorough reference from literature. A successful review of several approaches to integrate IFC and OWL to develop the standard BIM ifcOWL ontology was undertaken. Because a stable and standardized application is required for the PV systems technology, the ifcOWL is extended to develop the ifcOWL-pvOntology. This ontology serves as a foundation for future ontological and semantic reasoning of the PV domain.

## REFERENCES

Abanda, F.H., Tah, J.H.M. & Duce, D., (2013), 'PV-Tons A Photovoltaic Technology Ontology System for the design of PV-Systems". pp. 1399–412.

Alatrish, E., (2013), 'Comparison Some of Ontology Editors", 8(2), pp. 018–24.

Barbau, R. et al., (2012), "OntoSTEP: Enriching Product Model Data using Ontologies", In *Computer Aided Design*.

Beetz, J., Van Leeuwen, B. & De Vries, (2009). IfcOWL: a Case of Transforming EXPRESS Schemas into Ontologies. *Artificial Intelligence for Engineering Design, Analysis and Manufacturing*, 23(1), pp. 89–101.

bimobject, 2013. bimobject. [Online] Available at: https://info.bimobject.com/ [Accessed 14 May1 2018].

bimobject, 2018. SEEK and BIMobject. [Online] Available at: https://info.bimobject.com/seek [Accessed 14 May 2018].

Bockliseh, T., (2016). Hybrid Energy Storage Approach for Renewable Energy Applications. *In Journal of Energy Storage*, 8, pp. 311–19.

Buckley, T. & Nicholas, S., (2016). *2016: Year in Review Three Trends Highlighting the Accelerating Global Energy Market Transformation*. Institute of Energy Economics and Financial Analysis. IEEFA.org.

buildingSMART, (2018). *Summary of IFC Releases*. [Online] Available at: HYPERLINK "http://www.buildingsmart-tech.org/specifications/ifc-releases/ summary" http://www.buildingsmart-tech.org/specifications/ifc-releases/summary [Accessed 2 May 2018].

Chen, G. & Luo, Y., (2016). A BIM and Ontology-Based Intelligent Application Framework. In *2016 IEEE Advanced Information Management, Communicates, Electronic and Automation Control Conference (IMCEC)*. Xi'an.China, 2016. IEEE.

Dibowski, H., Holub, O. & Rojicek, J., (2017). Knowledge-Based Fault Propagation in Building Automation Systems. In *International Conference on Systems Informatics, Modelling and Simulation (SIMS)*. Riga, Latvia, 2017. IEEE.

Gao, G. et al., (2015). A query expansion method for retrieving online BIM resources based on Industry Foundation Classes. *Automation in Construction*, 56, pp. 14–25.

Gesterkamp, L., Rebstadt, J. & Mertens, R., (2017). Tackling Complex Ontologies with AVOnEd — Aspect-Oriented Visual Ontology Editor. In *In 2017 IEEE 11th International Conference on Semantic Computing (ICSC).*, 2017.

Hartley, G., (2014). *Energy Efficiency is Big Business*. [Online] Available at: HYPERLINK "www.energysavingtrust.org.uk/" www.energysavingtrust.org.uk/ [Accessed 2 April 2018].

Hoang, N.V., (2015). IFC-to-Linked Data Conversion: Multilayering Approach. In *In The Third International Workshop on Linked Data in Architecture and Construction*. Eindhoven, Netherlands, 2015.

Hoang, N.V. & Torma, S., (2014). *Opening BIM to the Web - IFC-to-RDF Conversion Software*. [Online] Available at: HYPERLINK "http://rym.fi/results/opening-bim-to-the-web-ifc-to-rdf-con3003 version-software/" http://rym.fi/results/opening-bim-to-the-web-ifc-to-rdf-con3003 version-software/ [Accessed 2 May 2018].

Hou, S., Li, H. & Rezgui, Y., (2015). Ontology-Based Approach for Structural Design Considering Low Embodied Energy and Carbon. *Energy and Buildings*, 102, pp. 75–90.

IFC, (2018). *IFC Overview Summary*. [Online] Available at: HYPERLINK "http://www.buildingsmart-tech.org/specifications/ifc-overview" http://www.buildingsmart-tech.org/specifications/ifc-overview [Accessed 23 April 2018].

IRENA, (2016). *Solar PV in Africa: Costs and Markets*. Abu Dhabi, Dubai: International Renewable Energy Agency.

ISO, (2013). *International Organization for Standardization*. [Online] Available at: HYPERLINK "http://www.iso.org/iso/catalogue_detail.htm?csnumber=51622" http://www.iso.org/iso/catalogue_detail.htm? cs-number=51622 [Accessed 2 May 2018].

Kontopoulos, E., Martinopoulos, G., Lazarou, D. & Bassiliades, N., (2016). An ontology-based decision support tool for optimizing domestic solar hot water system selection. *Journal of Cleaner Production*, 112(5), pp. 4636–46.

Krima, S. et al., 2009. OntoSTEP: OWL-DL ontology for STEP. *National Institue of Standards and Technology, NISTIR*, 7561.

Leonics, (2013). *How to Design a Solar PV System*. [Online] Available at: HYPERLINK "http://www.leonics.com/support/article2_12j/

articles2_12 j_en.php" http://www.leonics.com/support/article2_12j/articles2_12 j_en.php [Accessed 2 May 2018].

Liu, Y.-S. et al., (2015). Recent Advances on Building Information Modeling. *The Scientific World Journal*, p.2.

Liu, H. et al., (2017). Enhanced Explicit Semantic Analysis for Product Model Retrieval in Construction Industry. *In IEEE Transactions on Industrial Informatics*, 13(6).

Mignard, C. & Nicolle, C., (2014). Merging BIM and GIS using Ontologies Application to Urban Facility Management in ACTIVe3D. *Computers in Industry*, 65, pp. 1276–90.

NeOn, (2016). *NeOn ToolKit*. [Online] Available at: HYPERLINK "http://www.neon-project.org/nw/Welcome_to_the_NeOn_Project" http://www.neon-project.org/nw/Welcome_to_the_NeOn_Project [Accessed 23 April 2018].

Nepal, M., Zhang, J., Pottinger, R. & Staub-French, S., (2012). Ontology-Based Feature Modeling for Construction Information Extraction from a Building Information Model. *Journal of Computing in Civil Engineering*, 27, pp. 555–69.

openbimstandards, 2016. *What is ifcOWL*. [Online] Available at: http://openbimstandards.org/standards/ifcowl/ [Accessed 14 May 2018].

Pauwels, P. et al., (2015). A performance assessment ontology for the environmental and energy management of buildings. *Automation in Construction*, 57, pp. 249–59.

Pauwels, P. & Terkaj, W., (2016). EXPRESS to OWL for Construction Industry: Towards a Recommendable and Usable ifcOWL Ontology. *In Automation in Construction*, 63, pp. 100–33.

Pauwels, P., Terkaj, W., Krijnen, T. & Beetz, J., (2015). Coping with Lists in the ifcOWL Ontology. In *Proceedings of the 22nd EG-ICE Workshop 2015*. Eindhoven, 2015.

Pauwels, P. & Van Deursen, D., (2012). IFC/RDF: Adaptation, Aggregation and Enrichment. In *in: Report of the First International Workshop on Linked Data in Architecture and Construction*. Ghent, Belgium, 2012.

Perhavec, D.D. & Kaucic, B., (2018). Knowledge Modelling of Buildings: Use Case for Heritage Buildings. In *2017 International Conference on Engineering, Technology and Innovation (ICE/ITMC)*. Funchal, Portugal, 2018. IEEE.

Protege, (2016). *Protege*. [Online] Available at: HYPERLINK "https://protege.stanford.edu/" https://protege.stanford.edu/ [Accessed 23 April 2018].

Saalmann, P. et al., (2016). Application Potentials for an Ontology-Based Integration of Intelligent Maintenance Systems and Spare Parts Supply Chain Planning. *Procedia CIRP*, 41, pp. 270–75.

Schevers, H. & Drogemuller, R., (2006). Converting the Industry Founda2964 tion Classes to the Web Ontology Language. In *In Proceedings of the First International Conference on Semantics, Knowledge and Grid, IEEE Computer Society*. Washington, 2006.

Terkaj, W., Pedrielli, G. & Sacco, M., (2012). Virtual Factory Data Model. *In CEURWorkshop Proceedings, Workshop on Ontology and Semantic Web for Manufacturing OSEMA 2012*, 886, pp. 29–43.

Terkaj, W. & Sojic, A., (2015). Ontology-Based Representation of IFC EXPRESS Rules: An Enhancement of the ifcOWL Ontology. *Automation in Construction*, 57, pp. 188–201.

W3C, (2004). *OWL Web Ontology Language*. [Online] Available at: HYPERLINK "https://www.w3.org/TR/2004/REC-owl-features-20040210/" [Accessed 23 April 2018].

Wade, H., (2008). PV System Sizing. In *Solar PV Design Implementation O & M*. Marshall Island.

Warehouse, 3., 2018. *3D Warehouse*. [Online] Available at: https://3dwarehouse.sketchup.com/ [Accessed 14 May 2018].

Zeman, M., (2015). Photovoltaic Systems. In *Solar Cells*.

Zhang, L. & Issa, R.R., (2012). Ontology-Based Partial Building Information Model Extraction. *Journal of Computing in Civil Engineering*, 27, pp. 576–84.

# The RIMcomb research project: Towards the application of building information modeling in Railway Equipment Engineering

S. Vilgertshofer, D. Stoitchkov, S. Esser & A. Borrmann
*Chair of Computational Modeling and Simulation, Leonhard Obermeyer Center, Technical University of Munich, Germany*

S. Muhič
*AEC3 Deutschland GmbH, Germany*

T. Winkelbauer
*SIGNON Deutschland GmbH, Germany*

ABSTRACT: This paper presents our research towards the utilization of Building Information Modeling (BIM) methods in Railway Equipment Engineering. While BIM is already applied in several railway infrastructure construction projects, so far, it is mainly used for construction engineering tasks, i.e. for modeling the alignment, bridges or tunnels. The RIMcomb research project targets the further application in scope of rail engineering subsections such as control and safety systems, train control systems, telecommunication systems, electric power systems, rail power supply or cable management. In this paper, we are going to give a general overview of the project and will discuss first findings of our approach to digitizing conventional technical drawings in order to translate their content into a machine-readable form.

## 1 INTRODUCTION

The planning and restoration of railway tracks including technical equipment is of high importance for maintaining this important backbone of the Europe's infrastructure. While Building Information Modeling (BIM) is widely established in the building domain, it is less often used for the planning and construction of infrastructure facilities. So far, it has not been applied as a general concept for railway infrastructure equipment engineering yet.

At the moment, the development of various proposals for extending BIM concepts and standards for the infrastructure domain is underway. buildingSMART International (bSI) is working on extending the Industry Foundation Classes (IFC) for representing infrastructure facilities. Currently, IFC is primarily focused on the building domain and comprises a wide range of classes, which enable the exchange of detailed BIM models. Although first infrastructure concepts (such as alignment and linear referencing), have been published as part of the latest version IFC4.1, more specific elements are still missing. To fill this gap, the Rail Room was established in 2017 as a subsection of bSI dedicated to the development of rail-specific extensions. In the past, several proposals had been published that provided the foundation for the respective standardization effort, Examples include IFC Tunnel (Vilgertshofer et al., 2016), IFC Rail (bSI SPEC) and IFC Road (bSI SPEC) (buildingSMART, 2016). Contributing to these efforts, is one of the main objectives of the RIMcomb project.

On a general level, the RIMcomb research project aims at introducing BIM methods into the sector of railway equipment engineering by analyzing current conventional workflows in order to identify beneficial BIM use cases. The main goal is the avoidance of data loss or inconsistent data during different planning stages as well as to identify labor-intensive tasks, which could benefit from automation. Especially, inconsistencies across the different disciplines involved in railway design is a severe problem that can be overcome by using BIM methods instead of conventional 2D drawings.

In this regard, we are looking into possibilities of using the existing version of the IFC data format to represent railway equipment.

Additionally, we analyzed preexisting data schemas that are currently used in this domain. As these are not necessarily compatible with one another we are developing a tool that can import various data formats and uses this information for creating an integrated BIM model, preferably in the IFC format. Furthermore, we use conventional 2D drawings as an additional data source for BIM model generation.

This paper is structured as follows: In Section 2 we give a general overview of the RIMcomb research project's scope in order to put the subsequently described approaches into context. Section 3 describes our approach of digitizing technical drawings. We give an overview of the methods applied and the conclusions of our testing process. Section 4 describes how we intend to further use the digitized plan data and other data sources for the creation of BIM models representing railway equipment. The paper ends with a summary and an outlook.

## 2 THE RIMCOMB PROJECT

The research project "RIMcomb: Railway Information Modeling for the Equipment of Railway Infrastructure" was initiated by SIGNON Deutschland GmbH in 2016 in cooperation with Technical University of Munich and AEC3 Deutschland GmbH. The project is funded by the Bavarian Research Foundation and started in early 2017.

The main focus of the research project is to develop and adapt new computer-supported methods for model-based collaboration between the different subsections of technical railway equipment in order to increase the efficiency of the planning process and the quality of the outcome.

During the design, planning and construction of railway infrastructure and technical equipment, a multitude of domains experts are involved. Therefore, data exchange between these participants is an issue that needs to be addressed, as there a various specialized software tools available for different tasks that do not implement any common data standard.

Also, most of railway construction projects involve the modernization or alteration of existing infrastructure and thus the industry has to rely on technical drawings from past decades that are not necessarily consistent with the real-world circumstances.

As described in Section 2 we developed a method that allows the automatic recognition of plan symbols in technical drawings of railway infrastructure. Besides the use of this data described in Section 3, we also aim at comparing the generated data with real-world data in order to identify discrepancies. This use case may create a significant benefit for the railway companies as the manual comparison of the as-built drawings with real-world stock data requires a considerable amount of effort but is nonetheless necessary.

Here, machine-learning and convolutional neural networks (as outlined in Section 2) will also be employed to process video files of railway tracks in order to identify objects in single frames for the mapping of objects such as signals, balises, switches or poles of overhead lines.

Another aspect in the scope of the research project is the development of a method that allows the automated checking of technical rules and regulations. As such approaches have already been applied outside of the infrastructure domain (Preidel and Borrmann, 2016), we see a huge benefit in introducing them into the domain of railway equipment, as the amount of rules and regulations in this domain requires a large amount of manual work.

## 3 DITIALIZATION OF PLAN DATA

One major topic in the research project is the digitalization of conventional drawings depicting railway equipment infrastructure. While most drawings are available digitally, the interpretation of these plans has to be undertaken manually. This is necessary when the accuracy of plans has to be compared to real-world circumstances or in case of stocktaking.

Our approach aims at supporting this process in order to reduce the manual effort by automating at least parts of this image interpretation process. The first step towards this goal is the automatic recognition and highlighting of plan symbols on a given drawing and the subsequent storing of their count and location.

### 3.1 *Theoretical background*

In a first step, three preexisting methods of image recognition are described. We evaluated those techniques in respect to the given problem. As none of those methods matched our requirements completely we also tested Convolutional Neural Networks, which are already widely used for image recognition, in respect of their ability to detect plan symbols.

#### 3.1.1 *Template matching*

Template Matching is a well-known method for the searching of a template image in a larger image. This is made by sliding the template image over the input (larger) image and comparing them at every position. The result of this method is a grayscale image with a size of *(W–w+1, H–h+1)*, where $W$ and $H$ are the width and the height of the input image, $w$ and $h$ are the width and the height of the template image. We investigated different comparison methods for each one of which there is a normalized version (Kaehler and Bradski, 2016).

In this work two of these methods are used:

– Normalized Square Difference Matching Method:

$$R(x,y) = \frac{\sum_{x',y'}(T(x',y')-I(x+x',y+x'))^2}{\sqrt{\sum_{x',y'}T(x'y')^2 \cdot \sum_{x',y'}I(x+x',y+y')^2}}$$

– Normalized Correlation Coefficient Matching Method:

$$R(x,y) = \frac{\sum_{x',y'}(T'(x',y')-I'(x+x',y+x'))^2}{\sqrt{\sum_{x',y'}T'(x'y')^2 \cdot \sum_{x',y'}I'(x+x',y+y')^2}}$$

Here $T$ is the template image, $I$ the input image, $R$ the result image and

$$T'(x'y') = T(x',y')-1/(w \cdot h) \cdot \sum_{x''y''} T(x'',y'')$$
$$I'(x+x',y+y') = I(x+x',y+y') - 1/(w \cdot h) \cdot \sum_{x''y''} T(x+x'',y+y'')$$

With the first comparison method a perfect match is 0 and a perfect mismatch is 1. For the second method 1 stands for a perfect match and –1 for a complete mismatch.

### 3.1.2 *Contours search*

This technique compares objects with their contours. A contour lies on the border between the black and white spaces. A contour tree contains the hierarchy between the contours or how they relate to one another. Contours can be compared with the help of image moments. An image moment is a characteristic of a given contour calculated by summing over the pixels of that contour. The Hu invariant moments (Hu, 1962) were used in our work to compare contours. The Hu moments are combinations of different normalized central moments and are scale, rotation and translation invariant. However, this method works only if the symbol is not connected to other lines or objects in the image, because then the contour around the symbol can't be defined properly.

### 3.1.3 *Cascade classifiers*

First proposed by Viola & Jones, 2001, it was originally used for face detection, but can be used for many types of objects. This method learns by searching for Haar-like features in an image. These features differentiate between dark and bright parts of an image by subtracting the sum of the pixels in the white parts from the sum of the pixels in the dark parts. The features are placed over the image in different locations and sizes to extract certain patterns.

A cascade classifier is a machine learning method in which the information computed in a given classifier is used for the next classifier and becomes more complex at each stage. "The Viola-Jones detector uses AdaBoost, but inside of a larger context called a "re-jection cascade". This "cascade" is a series of nodes, where each node is itself a distinct multi-tree Ada-Boosted classifier. The basic operation of the cascade is that sub-windows from an image are sequentially tested against all of the nodes, in a particular order, and those windows that "pass" every classifier are deemed to be members of the class being sought." (Kaehler and Bradski, 2016).

### 3.1.4 *Convolutional neural networks*

A neural network is a form of machine learning, in which the computer "learns" from given data. For the sake of this work, a convolutional neural network (CNN or ConvNets) is used, which speeds up the training process especially with images. One of the first CNN is called LeNet5 (LeCun et al., 1998), which has started an new era of state-of-the-art artificial intelligence. CNNs are so effective for image recognition because they use filters to detect patterns (or features) in an image. Different locations of the image are searched for these features and a value is saved, representing how well every pattern matches the image in a given location (Rohrer, 2016). This results in a map which represents where each feature occurs in the image. By matching the feature for every possible location in the image, a convolution is made. The features are then passed to the actual neural network (In Figure 2 from the left):

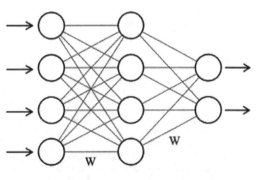

Figure 2. Simplified neural network with an input layer on the left, an output layer on the right and one hidden layer in the middle.

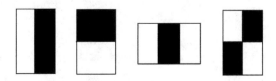

Figure 1. The Haar-like features.

Every layer has multiple neurons which are connected to one another between the layers (Figure 2). Each of the connections between the nodes is weighted (marked with *W*), which means that they are multiplied by a number, that is between −1 and 1. The neural network "learns" by adjusting the weight until the required output is obtained. An activation function is required to generate an output from given input in a processing unit (or neuron). The ReLU (Rectified Linear Unit) function was used in this work to normalize the values in the feature map that was calculated. This simplifies the calculation by setting all negative values to zero. The ReLU has the mathematical form: $f(x) = max(0, x)$. To make the calculations less complex the size of the feature map can be reduced with a max pooling layer. This is done by taking only the maximum value of a given window and saving it at the correct location in a new smaller feature map.

### 3.2 *Concept*

Even in today's modern world, in which almost everything is digitized, many technical drawings are still only available in paper form. The main reason for this is that most of the infrastructure that exists today was built before the widespread availability of computers. Many of these drawings are now being digitized to be used by modern computer programs. Much of the digitization process of a technical drawing consists of recognizing and locating different symbols which is often a difficult and time-consuming task. Therefore, this work aims to find a method of automating this process at least in parts and drastically reducing the processing time.

The different methods for symbol recognition were tested in different ways. The first two methods do not require training and can be implemented directly to find symbols. The Cascade Classifiers and the Convolutional Neural Networks require many images for training. These images are generated by cutting technical drawings into thousands of smaller images and then placing the searched symbol over half of the images in different sizes and locations. This generates two kinds of images—positive images with the symbol and negative images without it. After these methods have been trained, they must be tested to measure the accuracy of finding the symbol. This course of action differs from other approaches that use machine learning techniques for image detection. Normally, our way of generating the training data would result in a machine learning algorithm, that would only detect the exact image that is was trained with—which is not useful for e.g. detecting faces in a picture. However, since plan symbols are always of a similar shape or match one another exactly, the chosen training data work fine in our scope.

### 3.3 *Prototypical implementation*

The methods were tested on many different technical drawings to measure the accuracy. An example can be seen in Figure 4, which is a part of a telecommunication technical drawing.

Figure 3 shows the symbol for which the image was searched.

As a result, the corresponding symbols in the image are marked by red rectangles (Figure 5). The different methods show similarly results with this image, but the template matching method is not scale and rotation invariant and works only if the symbol always has the same size and rotation. The cascade classifiers method is scale and rotation invariant, but still fails if the shape of the symbol is simple and therefore does not have many distinct features. However, the convolutional neural networks were found to be very accurate, even when tested with many different types of symbols.

The CNN was not only tested with technical drawings, but also with artificially generated images to better test the accuracy. It showed above 95% of accuracy for detecting if a symbol is on the image or not and 80–85% of accuracy for finding the exact location of the symbol.

### 3.4 *Conclusion*

Both the template matching and the contour methods are very easy and fast to implement but work only under certain conditions. The cascade classifier is a more complex method, but still fails if the symbol has too few features. The last technique, the artificial neural networks, is the most "sophisticated" method and can be used in many situations with

Figure 3.  Test symbol.

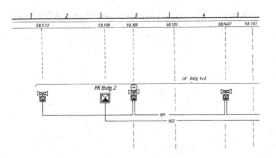

Figure 4.  Test image.

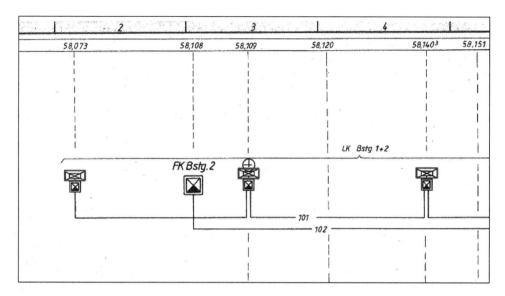

Figure 5. Exemplary results.

|  | Advantage | Disadvantage | Limitation |
|---|---|---|---|
| **Template matching** | Very easy and fast to implement | Not rotation and scale invariant, thereshold value | Works only if the symbol always occur with the same size and rotation |
| **Contours** | Easy and fast to implement, rotation and scale invariant | Connected or overlapped symbols, open shapes, threshold value | Not practical for big technical drawings in which the symbols are connected or crossed by lines |
| **Cascade classifiers** | Versatile, works under many conditions | Requires training, threshold values | Works only if the symbol has many features to differentiate it, requires big image data |
| **Convolutional neural networks** | Very versatile, works under almost any conditions | Requires training | Powerful GPU, requires big image data |

Figure 6. Overview of advantages and disadvantages of the different image recognition methods.

very few disadvantages and almost no limitations with today's powerful computers. The neural network, which was designed for this work, can be further modified for even better results. An overview of our findings is given in Figure 6 (Stoitchkov, 2018).

## 4 BIM MODELS FOR RAILWAY INFRASTRUCTURE

The previously described method of automatically detecting plan symbols is only one aspect of the RIMcomb research project's goals. To use the collected data for creating BIM models we need to investigate how such models can be created. Furthermore, the image recognition approach is only one of various sources of preexisting data that can be used for model creation. In the following section, an overview of data formats is given as well as an approach of how a model can be created from different data sources.

### 4.1 Data formats

One of the main aspects of Building Information Modeling is data consistency and the collaboration of different project partners using the same set of data. These two aspects require data formats, which can be imported and exported by different software applications.

An analysis of the software market offers many different software tools, which meet these requirements and can be used for creating building models or performing simulations on them. The vendor-neutral format "Industry Foundation Classes" (IFC) provides the possibility to exchange model data amongst an enormous range of different applications. Besides open formats, many software providers implement proprietary interfaces between their tools, which often lead to a higher data quality in the receiving application but is also limited to the use of a few tools.

In contrast to these developments for building constructions, no state-of-the-art application or exchange format does exist for modelling infrastructure projects, although projects such as IFC-Rail or the proposal of Allah Bukhsh et al. (2016) are under development. One reason is based in the geometrical project dimensions: Buildings normally have a ground area less than 100 meters per site and an elevation of a couple of meters. Therefore, a high information density occurs on a comparatively small volume (different layers of a wall, structural analysis, architectural properties, etc.). In contrast to this, infrastructural projects usually reach over many kilometers and therefore the data density can extremely vary within the project's alignment. Thinking of modelling a railway path between two stations makes this aspect a bit clearer: The nearer the station is, the more signals, switches or security systems are needed whereas the lines in the outer field only need rails, swells, railway signals and electric components.

This leads to the need of new storing approaches in (existing) data formats. There are already some data formats in use today, however, they are mostly limited to one discipline of railway engineering or can only be used by one specific software tool.

Modelling of infrastructural components is quite a challenge today, as most of the software products for infrastructure planning are based on the paradigm of drawing generation and thus are not capable to represent semantically rich 3D models.

At the same time, these tools are tailored and well-suited for the respective engineering task. On the other hand, the creation of BIM models for railway engineering is only partially supported. Creating model components in well-established BIM authoring tools, which are known from building modeling, can only be realized by either using building elements and append additional properties or creating new components based on generic templates. Both approaches can only offer the high data quality that is known from today's building models, when a lot of effort is put into the model creation.

Also, the data exchange of such models is a big challenge. Once they have been created in a BIM authoring tool, the export into a vendor-neutral format leads to another significant issue. Up to now, IFC has no preexisting classes for storing infrastructural components (they will be intro-

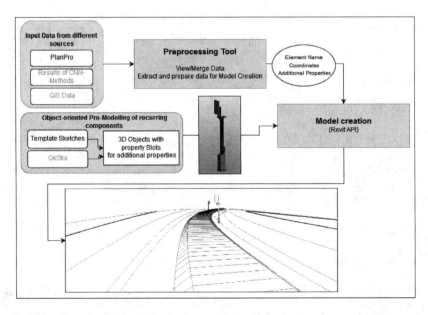

Figure 7. Workflow of creating BIM models of railway equipment infrastructure from various data sources by using predefined object-oriented model part templates.

duced in IFC 5). Besides IFC, there are some data formats such as *PlanPro* or *railML* providing schemata for storing and transferring information about infrastructural components (especially for railway infrastructure) but cannot represent the elements' geometry in a high quality.

4.2 *Model creation*

Due to the lack of a standardized data format and the need of the digitalization of existing drawings, a software-application is being developed in scope of the RIMcomb research project that combines data from different sources and automatically creates an integrated BIM model from this diverse information (Figure 7). To integrate as much import data formats as possible, a basic task is to define a set of the minimal necessary information to create a model, which is suitable for advanced modelling and simulation tasks (e.g. supplying the existing state of a railway path for redevelopment planning).

To start with a manageable set of data, the data schema *PlanPro* was used to set up the preprocessing tool. *PlanPro* is developed by DB Netz AG and it is used for gathering information needed in an interlocking system. Therefore, this format is powerful for storing information about different components that are necessary for security aspects of railway traffic. However, this schema holds only coarse information about the railway alignment, which causes to integrate additional GIS data. We plan to integrate this information in a later implementation step.

The link between the preprocessing tool and the BIM Editor is realized by an SQL database, which is controlled via the Entity Framework. This offers a lot of flexibility for the use of the preprocessed data and retains the possibility to use other programs for the modelling process as well.

The tool will serve as a testing platform in the research project to investigate how we can use and combine existing data in order to create a functioning BIM model.

## 5 SUMMARY AND OUTLOOK

This paper gives a general overview of the RIMcomb research project that aims at introducing BIM methods and technologies into the domain of railway equipment engineering. In addition to the main project goals, we discussed the ongoing research and first findings. In this scope, the automatic recognition of technical symbols in technical drawings by using different methods of image recognition is described in detail. Our results show that this method has large potential in automating a labor-intensive task and can also be employed to create semantic models of railway infrastructure. As a machine-learning based approach has the highest rates of correct recognitions we aim at improving this method in further research. This is especially important for the comparison of elements depicted in technical drawings with real-world video data gathered from railway track inspections.

## ACKNOWLEDGEMENTS

We gratefully acknowledge the support of the Bavarian Research Foundation for funding the project.

## REFERENCES

Allah Bukhsh, Z., Hartmann, T. & Stipanovic, I., 2016. From Analysis of Information Needs towards an Information Model of Railway Infrastructure. 23rd International Workshop of the European Group for Intelligent Computing in Engineering, Krakow, Poland.
buildingSMART, 2016. Standards Library, Tools and Services—buildingSMART.https://www.buildingsmart.org/standards/standards-tools-services/ (accessed 5.15.18).
Hu, M.-K., 1962. Visual pattern recognition by moment invariants. IRE transactions on information theory, 8(2):179–187. https://doi.org/10.1109/TIT.1962.1057692.
Kaehler, A. & Bradski, G., 2016. Learning OpenCV—computer vision in c++ with the opencv Library. O'Reilly Media.
LeCun, Y., Bottou, L., Bengio, Y. & Haffner, P., 1998. Gradient-based learning applied to document recognition, in: Proceedings of the IEEE. pp. 2278–2323. https://doi.org/10.1109/5.726791.
Preidel, C. & Borrmann, A., 2016. Towards code compliance checking on the basis of a visual programming language. ITcon 21, 402–421.
Rohrer, B., 2016. How do Convolutional Neural Networks work?https://brohrer.github.io/how_convolutional_neural_networks_work.html (accessed 5.15.18).
Stoitchkov, D., 2018. Analysis of Methods for Automated Symbol Recognition in Technical Drawings. Bachelor's thesis, Technical University of Munich.
Vilgertshofer, S., Jubierre, J.R. & Borrmann, A., 2016. IfcTunnel—A proposal for a multi-scale extension of the IFC data model for shield tunnels under consideration of downward compatibility aspects, in: Proc. of the European Conference on Product and Process Modeling (ECPPM). Limassol, Cyprus.
Viola, P. & Jones, M., 2001. Rapid object detection using a boosted cascade of simple features, in: Proceedings of the 2001 IEEE Computer Society Conference on Computer Vision and Pattern Recognition. IEEE Comput. Soc. https://doi.org/10.1109/CVPR.2001.990517.

# SolConPro: Describing multi-functional building products using semantic web technologies

A. Wagner, L.K. Möller, C. Leifgen & U. Rüppel
*Technische Universität Darmstadt, Darmstadt, Germany*

ABSTRACT: The research project Solar Construction Processes (SolConPro) focusses on developing new methods to facilitate planners in integrating solar active façade components, e.g. Building Integrated Photovoltaics (BIPV) and Building Integrated Solar Thermals (BIST), into the building processes. The decisive feature of such components is that they reunite product properties from multiple product categories and therefore cannot be assigned to a specific category, such as static component or technical building equipment. Well-established product exchange standards within the construction industry such as ICF, simple ifcXML or VDI 3805 are not suitable to cover the wide-ranging properties of multi-functional components. In order to provide a comprehensive digital representation of such components and to enable the integration within the methodology of Building Information Modelling, the following paper presents a data schema based on Semantic Web Technologies. The designed multi-layered ontology enables and supports interpretability the wider use of Semantic Web technologies in the construction industry.

## 1 INTRODUCTION

With regenerative energies becoming of more interest to reduce the usage of fossil energy, the application of technologies to harvest them becomes more popular, as well. To allow the utilisation of vertical urban surfaces as conversion areas, energy active façade components can be of significant value. However, the integration of such components into the building and planning processes is complex and not well supported, yet.

The research project Solar Construction Processes (SolConPro) focusses on developing methods to facilitate planners in working with energy active components (SolConPro, 2015). As results, methods for data pre-processing, including the reduction and transformation of data, and a concept for a distributed product catalogue were introduced by (Wagner et al., 2018). Still, a digital description of such components must be available in order to use the proposed methods. Such description must hold all available information relevant for any involved stakeholders while allowing manufacturers to describe their products freely. The latter is of particular interest, since energy active façade components are commonly designed as multi-functional components and use innovative technologies and approaches.

In this paper, we introduce a new data schema for describing multi-functional façade components. Next, we define the requirements that apply to such components. This is followed by a brief overview over existing data schemas that are analysed regarding their compliance towards the previously defined requirements. Based on this, we describe a multi-layered ontology as a solution to the defined problem in Section 3. Consecutively, we demonstrate the application of the introduced ontology in in Section 4. Finally, we conclude this paper with a discussion in Section 5.

### 1.1 Requirements to the product description schema

The requirements we are presenting in this section are mostly based on the multi-functionality as well as the innovative nature of energy active façade components. Besides those, there are also requirements that are valid for product description schemas in general.

While most building components can be clearly assigned to specific categories (e.g. doors), BIPV and BIST cannot always be assigned to one specific product category due to their multi-functional features. To allow the description of such components, the description schema must allow them to be described in high detail for every functionality they serve. This can result in a complex and extensive overall description that can in turn potentially cause big amounts of data. Also, it must be possible to define a component as instance of multiple categories and thereby to append properties to the component that are not necessarily defined for the same category. Based on this requirement, a template-driven approach is only applicable if the schema allows multi-classification and by that the assignment of multiple templates to a singular component.

Next to the multi-functionality, the innovative nature of energy active components results in the requirement of a flexible modelling approach. As the introduction of new technologies to harvest regenerative energies are introduced, manufacturers are eager to offer components that are exploiting their benefits. If the product description schema is developed too rigid, e.g. by using templates that predetermine required properties or constructions for components of certain product categories, the distribution of innovative products is hindered. Thus, a template-driven approach should be dismissed for describing innovative components, even though it could be used for multi-functional components.

Another aspect that arises requirements is the customisability of façade components. Since façades rarely consist only of components with equal geometries, façade components either must be described for any applicable geometry or the geometry needs to be parametric. For components with only the functionality of serving as a façade element, it can be assumed that the parametric description needs only to apply to geometry. In the case of energy active façade components, this assumption does not hold, since the component's properties might correlate to its geometry. For instance, the yield of a BIPV module is directly coherent to its size, as the geometry defines how many PV cells may be placed within the module. Conclusively, the product description schema must allow to define the component's geometry parametrically as well as to describe parametric coherences between geometric and semantic properties. Thereof, the component's semantic and geometric description should be given in the same data format to the susceptibility of errors by parsing or linking different data formats.

In order to ensure holistic integration into the building and planning processes, it is not only necessary to provide a comprehensive description of the data. It must also be easy to integrate it into the Building Information Modelling (BIM) processes. This includes the transfer of the data—semantic and geometric—into BIM models and the corresponding software used. Therefore, it must be possible to store the data completely and not on a proprietary basis thus separating data storage from application software.

## 2 RELATED WORK

The presented results are based on a profound analysis of existing data schemas used in various domains. These schemas can be separated into product description schemas in general and schemas developed based on Semantic Web Technologies. Results of the first part of the analysis were considered during the design of our proposed multi-layered product ontology. The second part is evaluated regarding their applicability for the proposed ontology in general, as the re-usage of ontologies is recommended by the World Wide Web Consortium (W3C) and widens the range of the developed ontology by providing a better information basis.

### 2.1 Existing product data schemas

To enable and support the exchange of product information within the Building Service sector the VDI (The Association of German Engineers) directive 3805 provides rules and guidelines. It defines a detailed schema for product data to provide the components' information in a machine-readable way. The schema combines technical, geometric and media data to describe the products as a whole. The directive implements separate definitions of the product data model for different product groups and defines a separate data structure for each of these groups including lead data, product main groups, properties, functions and BS numbers (Building Services numbers). However, these specifications are only intended for the exchange of data within Building Services (e.g. heat generators, solar collectors), so that products of other fields from the construction sector such as structural or multi-functional components cannot be modelled and exchanged using this guideline.

The STEP (standard for the exchange of product model data) standard, ISO 10303, introduces a computer-interpretable representation for the exchange of product manufacturing information including product and process data using Application Protocols (APs) and defining EXPRESS as a standard data modelling language. The schema is designed for and widely used in design and manufacturing within the Automotive sector.

Based on the physical file structure of STEP the open IFC (Industry Foundation Classes) standard (ISO 16739) and the corresponding IFC EXPRESS specification enable the data exchange with in the in the construction and facility management industries. (buildingSMART 2018a) Within the standard building components are modelled as according to certain element types with assigned characteristics. Therefore, a comprehensive description of multifunctional components including all their joint properties inherited from different element types is not feasible.

Current standards define specific schemas to represent and store product data classified by various product categories predefining specific properties of the component. This causes existing data schemes being not suitable for a comprehensive description of multi-functional façade, since they come together from different categories and aspects. In addition, the requirement of allowing parametric product descriptions with coherences between geometric properties and product attributes is not met by the IFC standard.

## 2.2 Existing ontologies

The ontologies presented in this section were designed in order to describe building data, geometries, or parametric products.

Various approaches exist for describing product data on the Semantic Web. One of these approaches is the ifcOWL ontology (Pauwels & Terkay, 2016), developed based on the IFC schema. This holds the advantage that a conversion from this popular schema can easily be performed. But it also holds disadvantages, as the rigid structure of the IFC schema remains and the complex and nested relations caused by the approach to define a concept for every object are not well suitable for Semantic Web Technologies as the data size is mostly dependent on the account of relations existent. In contrast, the ifcWoD (Mendes de Farias et al. 2015) is trying to overcome the complex relations by creating simplified relations. This is successful in the aspect of query-performance, but therefore causes information-loss.

A varying approach to define building data, is the BOT ontology (Rasmussen et al. 2017). This ontology is not designed to create one standard-ontology to describe entire building models but is kept simple and abstract in order to allow any building being described using it. If additional detail and information are needed, users are encouraged to use an ontology of their wishing and map it towards the BOT ontology. Thus, the BOT ontology acts as an abstract skeleton ontology aiming to allow interoperability between different ontology schemas for building models. As this approach seems feasible, the idea of creating a core-ontology for a specific domain (e.g. buildings or products) is applied in our studies.

Besides ontology focussing on the description of building models, other approaches also focus on product descriptions. The research project PROFICIENT, funded by the EU, presented the CMO with extensions ontology for this purpose (Bonsma et al. 2015). The ontology was designed in order to model data for an e-marketplace for buildings and can also depict parametric products. The ontology contains multiple domains, including geometry and parametric. Those domains are considered closely in this paper, as they are re-used directly or altered and re-used for the proposed approach. The geometry domain has also been revisited by the W3C Linked Building Data Community Group (LBDCG) and published under the name Geometry.

The W3C LBDCG is currently also working on defining an ontology for the description of building products (PROD, LDAC 2017). This development is just beginning, wherefore no results from this angle can be taken into account for developing this multi-layered product ontology.

Another re-used ontology originates from the QUDT Ontologies that provide a wide ranged collection of schema and vocabulary ontologies spanning various disciplines. As part of the NASA Exploration Initiatives Ontology Models (NExIOM) project the ontologies are free to the public. The specific ontology applied in this paper has its domain in the definition of units of measure, quantity kind dimensions, and data types (QUDT, 2018a).

## 3 MULTI-LAYERED PRODUCT-ONTOLOGY

Due to the gathered requirements and shown benefits of Semantic Web Technologies, we decided to design our product description schema based on these. One of the main benefits we are thereby exploiting, is the possibility of multi-classification. This allows to describe multi-functional products by assigning multiple product categories altogether with their corresponding properties and views to one product. Also, Semantic Web Technologies are a promising approach for flexible data schemas. Ontologies can easily be extended even after their initial launching without altering the already existent data designed using them. Another benefit is that ontologies are not just machine-readable but also machine-understandable, meaning machine can extract new knowledge from them based on the defined ontology-schema. Thereby, Semantic Web Technologies may be the basis for smart product descriptions.

With these benefits in mind, we decided to design a multi-layered product ontology consisting of three layers connected via multi-classification (bright grey arrows, dashed lines) and an ontology for parametric modelling (medium grey arrows, dotted lines) (see Fig. 1).

The upper, medium grey layer is a domain ontology used for defining components and their attributes semantically. Currently, this layer is realised using the buildingSMART Data Dictionary (bSDD, buildingSMART 2018b). Since the bSDD provides a REST API instead of a SPARQL endpoint, the connection could not be implemented

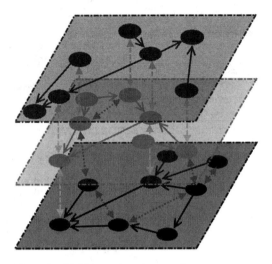

Figure 1. Multi-layered product ontology.

using multi-classification. Instead, the core ontology was adapted in order to allow the linkage via String-attributes, that are attached to nodes and contain the bSDD global unique identifier of the node's definition. However, the definition-layer could be exchanged for any suitable domain-ontology in respect of the considered use cases. Based on the knowledge stored within domain-ontologies, this layer might be of use for classifying products and their built-in components.

As core of the multi-layered ontology stands an abstract product ontology (light grey layer). This layer is introduced to describe the construction and properties of products. Since it does not contain any definitions—neither for concepts nor terms—this ontology is applicable not just to energy active façade components but potentially other components in the building sector. We designed the ontology in alignment to already existent schemas outside of the Semantic Web as these concepts have been proven viable. The abstract product ontology is further described in Section 3.1.

Finally, the ground, dark grey layer describes the product's geometry. This layer may be exchanged by any desirable geometry-ontology, as long as it can be mapped to the abstract product ontology (see Section 3.2). In the presented approach, we decided to use the Geometry-ontology (), which was developed as part of the CMO with extensions.

In order to describe parametric coherences between the geometry as well the construction layers, a part of the CMO with extensions ontology was re-used and adapted to our demands. This adaption is explained in more detail in Section 3.3.

Overall, the approach of a multi-layered ontology allows a high degree of freedom in modelling, as the singular layers can be exchanged and extended as needed (except of the core layer). Users could also define their own ontologies and map them to the abstract product ontology, similar to the BOT-ontology approach (Rasmussen et al. 2017). By interconnecting these layers via multi-classification, the knowledge defined in every layer can be exploited and applied to the other layers. E.g. if the existence of certain properties might imply the product being a photovoltaic element, this category could be added to the already defined categories of the considered product.

### 3.1 Abstract product-ontology

The abstract product-ontology was designed with the aim to allow its universal application for products in the building and construction industry. Therefore, it is stripped from any definitions of terms or concepts, as these are meant to be modelled within the definition layer. During the design phase, we focussed on multi-functional façade components as BIPV and BIST components. This does not imply the restriction to these components, however.

The design of the ontology is aligned towards established modelling concepts known from IFC and STEP based project models. Specifically, we decided upon an approach including collections of elements and the definition of concepts resp. classes and their instantiation within the data format. As these concepts are also the cause for disadvantages such as complex and nested relations between components and their attributes, we needed to overcome these disadvantages, as well. By using Semantic Web Technologies, we were able to simplify such relations while still allowing users to use nested relations in order to provide the highest level of detail possible. This was achieved using inferencing and thereby creating a smart product description.

The product construction can be defined using the ontology-excerpt shown in Fig. 2. The class *Component* builds an abstract super-class subsuming the classes *Assembly* and *Element*. The *Element*-class represents such components that cannot be split into its own parts. The level of detail defining which part should be defined as an *Element* resides with the users as long as they satisfy the constraint that an *Element*-object cannot consist of any other

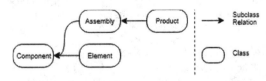

Figure 2. Modelling of the product construction.

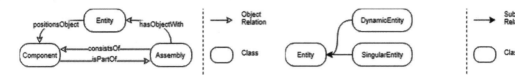

Figure 3. Modelling of component placements.

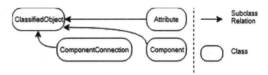

Figure 4. Modelling of entities.

components within the proposed ontology. For more complex components, the *Assembly*-class can be used. Objects of this class can consist of other components that can be placed within the object using the *Entity*-class (Fig. 3). The *Product*-class is a sub-class of the *Assembly*-class, designed to represent such components, that can be offered as products. Nonetheless, such products can still be used by other components using Linked Data methods.

For placing instances of the designed component concepts, we introduce the *Entity*-class, that—similar to the IFC-based approach—positions a *Component*-object (*Element* or *Assembly*) within an *Assembly*-object using the relations *hasObjectWith* and *positionsObject*. As previously mentioned, this holds the disadvantage of causing nested and complex relations. To reduce this problem, we defined the additional relations *consistsOf* and *isPartOf* that are inverse to each other. These relations are also transitive and enhanced with chain-axioms allowing the automatic generation of these connections (Eq.). This allows easier and more direct querying over all contained components and their attributes.

$\forall\, x,y,z(hasObjectWith(x,y)$
$\quad \wedge positionsObject(y,z))$
$\quad \rightarrow consistOf(x,z)$

$\forall\, x,y,z(consistOf(x,y) \leftrightarrow isPartOf(x,y))$

$\forall x,y,z(consistOf(x,y) \wedge consistOf(y,z)$
$\quad \rightarrow consistOf(x,z))$

Furthermore, we introduce sub-classes of the *Entity*-class to allow automatic placement of multiple instance of one component at once (*DynamicEntity*, Fig. 4). As the placement of multiple instances is not directly related to the placement of a singular instance, we also created the *SingularEntity*-class defined for the latter case.

To permit a holistic description of the product construction, the ontology should also be able to depict connections between components. Since various types of connections exist (electric, hydraulic, physical), the defined class to describe such connections is designed as an abstract concept. The definition of the connection-type can be appended

Figure 5. Connection between definition- and core-layer.

using the definition-layer. Figure 5 shows an overview of the *ComponentConnection*-class and its relation towards the *Entity*-class. An *Entity*-object can connect to multiple *ComponentConnection*-objects via the *incomingConnection*- and *outgoingConnection*-relation. By using these differentiated relations, the direction of the connection can also be defined. Similar to the product construction, these relations can be simplified by means of the *connectsTo*- and *connectsFrom*-relations. To allow easier access to data and bypass nested relations, these relations can be automatically generated based on a chain-axiom (similar Eq.). The simplified relations are also inverse and transitive in order to provide an easy and complete overview of all incoming and outgoing connections of a placed component. However, this reduction also causes an information loss regarding the type of the component connection.

Finally, the ontology provides means to describe the product's and its component's properties. Therefore, the *Attribute*-class is introduced (Fig. 6). The *Attribute*-class makes use of the NASA ontology for units (QUDT 2018b) to identify the used unit for the given property. Additionally, the class has data-relations for the specific value of its described property (*value*-relation). An *Attribute*-object can only use this relation once, as the value must be definite. In cases where properties do not have one singular value but a value range, interval, or list of values, the *DynamicAttribute*-class that inherits from the *Attribute*-class can be applied. This class is invested with multiple data-relations, defining the properties' minimal and maximal values, as well as step sizes in case it cannot attain any value between those boundary values. Furthermore, objects of this class can have object-relations to 2D Lists, Intervals, or Well Known Text (WKT). The meaning of this subclass is fully revealed considering the description

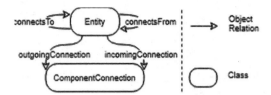

Figure 6. Modelling of component connections.

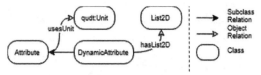

Figure 7. Modelling of attributes.

of parametric products (Section 3.3). To allow an uncomplicated extraction of the property values, all data-relations regarding values inherit from the originally defined *value*-relation of the *Attribute*-class. *Attribute*-objects can be connected to any *Component*-object using the *hasAttribute*-relation.

### 3.2 Multi-classification

The connection between the different layers is realised using multi-classification. However, since the bSDD only provides a REST API, the connection between the core- and definition-layer must be considered separately from the one between the geometry- and core-layer.

As the bSDD cannot be connected using multi-classification, an additional abstract super-class was introduced to the core-ontology. The *Classified Object*-class provides a data-relation to store the bSDD global unique identifier of the dedicated entry as String within the core-ontology. Every class, that may need to be classified inherits from this super-class and thereby implements this relation (Fig. 7). In case the bSDD will be replaced by a domain-ontology, every object, that needs to be defined in its terms and concepts, can be classified as its corresponding domain-ontology-class. Such connections would also allow the generation of suitable product categories based on the used attributes and built-in components of the product; e.g. a product being classified as PV component, since it has built-in PV cells, or a product being classified as façade component, because it has attributes for thermal transmittance values and hail resistance classes.

The connection between the core- and geometry-layer was implemented using multi-classification. In order to create suitable connections, we analysed the Geometry-ontology for matching classes. As a result, we decided to create three alignments; one on component-level and two on entity-level.

Components—disregarding whether they are classified as *Element*, *Assembly*, or *Product* – are mapped with the *Collection*-class of the Geometry-ontology. This allows users to freely model any geometry (primitives, Boolean operations, etc.) and connect it to elements. At the same time, *Assembly*-objects can consist of any entities and thereby sub-geometries. However, this approach holds the disadvantage of bloating the geometry-description unnecessary, as collections that are not needed to define the geometry must be created to align the geometry to the product construction.

Entities are differentiated between *Singular Entity*- and *DynamicEntity*-objects. The functionality of entities to place instances of components within another component, requires the associated Geometry-class to change the origin of a geometry (e.g. translate it) in a way that the altered geometry can be placed within a collection (and thereby assembly). For the placement of singular objects, the *Transformation*-class fulfils this requirement and is consecutively mapped to the *SingularEntity*-class. The *DynamicEntity*-class however is used to place multiple instances of one component automatically. This functionality is also given in the Geometry-ontology by the *Repetition*-class, which is why these classes are mapped accordingly.

### 3.3 Parametric coherences

The basis for the parametric-ontology is the parametric-part of the CMO with extensions ontology. However, since the modelling approach of the CMO with extensions ontology differs from the one we have chosen for the multi-layered ontology, some amendments must be made. The CMO with extension defines concepts and constructions within the T-box of an ontology by creating subclasses instead of instances in the A-box. We discarded this approach, as the specific instances of a parametric concept are stored within the A-box and thereby cannot be re-used in other T-box concepts.

The parametric-ontology is used for two purposes: Firstly, to define parametric products, and secondly, to create the connection between geometrical properties and attributes of the product construction.

For describing parametric products, the parametric-ontology was slightly adapted in order to operate with the different modelling approach. The *Variable*-class was extended by two object-relations: one to connect the variable to the instance that is hosting the property in question, and another one to define the data-relation, that connects the property-value to the previously defined instance. Furthermore, the connection to *Attribute*- and *DynamicAttribute*-objects allows a differentiation between parameters, that may be used as variables but cannot change its values (*Attribute*), and variables, that can freely attain any value within their boundaries (*DynamicAttribute*). Apart from this alteration, we added an optimisation problem (floor function) as sub-class of the *UnaryOperation*-class, since this function was needed to describe our use case being a parametric BIPV module. Already defined operations as well as the definition of equation have been adopted without further changes.

To create connections between geometric properties and attributes of the core-ontology, we created a sub-class of the *Equation*-class (*Constraint*). The *Constraint*-class defines as an equation that can only use variables on its left-hand- and right-hand-side. As only one of these variables must be named in order to re-use them in equations, we defined the variable pointing to the core-attribute to be named, while the variable pointing to the geometric property remains anonymous by using blank nodes. This, however, is a modelling suggestion and does not place any restriction upon the user.

## 4 APPLICATION

The introduced multi-layered product ontology was used within the research project SolConPro. It was used to model three different multi-functional façade components—two BIPV and one BIST module—in different levels of detail. These products were modelled in three different modelling approaches each in order to validate the freedom of modelling of the proposed ontology. Additionally, one BIPV module was modelled parametrically to verify our amendments to the parametric ontology. The parametric product also links to a static PV cell modelled according to our proposed schema to demonstrate the benefit of using Linked Data methods. Within the research project, other tools to store, distribute and pre-process data were demonstrated on the described products (Wagner et al. 2018).

### 4.1 *Different modelling approaches*

The BIPV and BIST modules were first analysed regarding the available level of detail and unique features. Based on this analysis, we decided individually how the modelling approaches should vary to achieve a maximum of variation within one product without changing its properties or appearance.

The variation mainly affected the grouping of elements within assemblies. This was achieved in three different ways: 1) existing elements were pooled in different assemblies (e.g. two glasses, the cavity between them and the gas within the cavity were pooled within one assembly for a double glazing in one approach while they were pooled in assemblies for glazing, cavities and gases in another approach); 2) assemblies were simplified to elements (e.g. double glazing being modelled as an element, consisting of one complex geometry); 3) additional assemblies were introduced to produce more modelling steps (e.g. the two glasses and the cavity between them were pooled within one assembly for glazing constructions, which then again would be pooled with the gas that is contained in the cavity within the double glazing assembly).

While querying the different modelling variations, it became clear that, since the bSDD REST API only provides a limited amount of data connected to the considered entry, it also limits the freedom of modelling. Therefore, we must specify the modelling restriction that if assemblies are simplified, the simplification can only concern their geometry while their element-specific properties must still be stored in elements that are connected to the respective assembly.

If this modelling constraint is followed, unified SPARQL queries filtering for product or built-in component properties or categories return the same results for every modelling variation.

### 4.2 *Parametric evaluation*

Modelling products with varying attribute values and geometry is part of the complex description coming with the facts multi-functional façade components bring part of the building's envelope. As described in section 3.3 the designed multi-layered ontology is capable of representing parametric features of the components. The ontology itself only holds the parametric relationships in the form of mathematical equations. These must be stored in such a way that solvable equation systems may be formed. For the determination of concrete products from the parametric representation, an evaluation of the equations or the systems as a whole is required. The ontology cannot evaluate this itself. Therefore, an external tool parses the queried parametric data from the ontology to form the respective equations systems. In contrast to the approach of the CMO with extensions, in this case no internal 'math-kernel' is used for the evaluation of the parametric equations. The generated equation system taken from the ontology's data is sent to and evaluated by the solvers of the free open-source mathematics software system Sage-Math (SageMath, 2018) which may also function as a webservice. This gives the flexibility to use any Math kernel capable of solving the stored parametric.

## 5 DISCUSSION AND CONCLUSION

Concluding, we will give a short discussion of the presented ontology and a brief outlook. The discussion will be split into the three topics of the multi-layered product ontology, the parametric evaluation, and the overall results.

The multi-layered ontology fulfils its defined purpose, as we demonstrated the free modelling of three different multi-functional façade components. It has proven that the approach of multi-classification is viable for connecting different layers of an ontology, and points of intersection can be found for geometry representations and product constructions. Because of the abstract nature of the core-ontology, the multi-layered ontology shows a promising outlook on a wide field of application, as for now any component and property defined in the bSDD can be modelled using it. Additionally, the introduction of the BOT ontology, which bases on the same approach of designing an abstract core ontology, shows that such a concept is needed, also for different domains as products.

However, the ontology does arise shortcomings. The biggest problem, that occurred during the application, was the insufficient quality of data retrieved via the bSDD REST API. Because we were not able to link to the underlying ontology, we were not able to exploit the entire potential of connecting a definition-layer containing a domain-ontology. Thus, we needed to specify modelling restrictions and thereby violate the requirement of free and flexible modelling. This also forestalled the analysis on whether it would be possible to assign product categories based on the product's and its built-in component's properties. Apart from the bSDD issue, we only covered a small field of application, being BIPV and BIST modules. Furthermore, even though we validated the possibility to model one product in different ways while still being able to receive the same results of unified queries, we did not analyse the impact this modelling variation bears upon the querying performance. Another un-validated feature of the multi-layered ontology is the assumption of an exchangeability of the definition- and geometry-layer, as neither have been tested with exchanged ontologies. This leads to a missing comparison to existing ontologies regarding the size of the data and querying-performance.

The parametric evaluation was validated in two use cases containing multiplication, division, subtraction, addition, and floor functions. It can be evaluated using an open-source kernel after the equations and operations are translated. Nonetheless, a kernel and interpreter are needed, and it must be tested using more use cases to fully validate its functionality.

Overall, the collected requirements are mostly met, and the approach is partly validated. For a complete validation, a broader field of application and more use cases are needed and small amendments regarding the definition- and geometry-layer must be conducted.

The next steps to address are—besides the aforementioned expansion of the field of application, exchange of definition- and geometry-layer, and further testing—a comparison of the multi-layered ontology to select (product) ontologies regarding several aspects: querying-performance, freedom in modelling, potential fields of application, and data size. Furthermore, the applicability of SPARQL extensions in order to evaluate the parametric ontology should be examined. After exchanging the definition-layer with a domain-ontology, inquiries regarding the possibility to use the domain-ontology for automated product classifications should be conducted.

[NOTE: In case of acceptance, the ontology, modelled products and parametric evaluation tool will be published online and linked in the final paper for the application's reproducibility.]

## ACKNOWLEDGEMENTS

This work is part of the research project Holistic Integration of Energy Active Façade Components in Building Processes (SolConPro, 2015) founded by the German Federal Ministry for Economic Affairs and Energy (BMWi).

## REFERENCES

Bonsma, P., Bonsma, I., Zayakova, T., van Delft, A., Sebastian, R., Böhms, M. 2015. Open standard CMO for parametric modelling based on semantic web, in Proceedings of the 10th ECPPM European Conference on Product and Process Modelling, 923–928.

building SMART 2018a. http://www.buildingsmart-tech.org/ifc/IFC4/Add2/html/, accessed May 15, 2018.

building SMART 2018b. http://bsdd.buildingsmart.org/, accessed May 15, 2018.

ISO 10303-1:1994, Industrial automation systems and integration – Product data representation and exchange – Part 1: Overview and fundamental principles, International Standard, ISO/TC 184/SC 4, 1994. https://www.iso.org/standard/20579.html, accessed May 15, 2018.

ISO 16739:2013, Industry Foundation Classes (IFC) for data sharing in the construction and facility management industries, International Standard, ISO/TC 184/SC 4, 2013. https://www.iso.org/standard/51622.html, accessed May 15, 2018.

LDAC 2017. Workshop Report of the 5th Workshop on Linked Data in Architecture and Construction, Nov 2017, Dijon, France.

Mendes de Farias, T., Roxin, A. & Nicolle, C. 2015. IfcWoD, Semantically Adapting IFC Model Relations into OWL Properties. In Proceedings of the 32nd CIB W78 Conference on Information Technology in Construction, 175–185.

Pauwels, P. & Terkaj, W. 2016. EXPRESS to OWL for construction industry: Towards a recommendable and usable ifcOWL ontology. Automation in Construction, 63:100–133. https://doi.org/10.1016/j.autcon.2015.12.003.

QUDT 2018a. http://www.qudt.org/, accessed May 15, 2018.

QUDT 2018b. http://qudt.org/1.1/vocab/unit, accessed May 15, 2018.

Rasmussen, M., Pauwels, P., Lefrançois, M., Schneider, G.F., Hviid, C., Karlshøj, J. 2017, Recent changes in the Building Topology Ontology, LDAC2017 – 5th Linked Data in Architecture and Construction Workshop, Nov. 2017, Dijon, France.

SageMath 2018. http://www.sagemath.org/, accessed May 15, 2018.

SolConPro 2015. Solar Construction Process—Holistic Integration of Energy Active Façade Components in Building Processes. http://www.solconpro.de/, accessed on January 12, 2018.

VDI 3805 Blatt 1; 2011-10 Produktdatenaustausch in der Technischen Gebäudeausrüstung; Grundlagen (Product data exchange in the Building Services; Fundamentals). Berlin, Beuth Verlag. https://www.vdi.de/technik/fachthemen/bauen-und-gebaeudetechnik/fachbereiche/technische-gebaeudeausruestung/richtlinienarbeit/richtlinienreihe-vdi-3805-produktdatenaustausch-in-der-tga, accessed May 15, 2018.

Wagner, A., Möller, L.K., Eller, C., Leifgen, C., Rüppel, U. 2018. SolConPro: An Approach for the Holistic Integration of Multi-Functional Façade Components into Buildings' Lifecycles. Accepted for oral presentation at ICCCBE 2018, 17th International Conference on Computing in Civil and Building Engineering, 5th–7th June 2018, Tampere, Finland.

*Regulatory and legal aspects*

# Semantic topological querying for compliance checking

N. Bus, F. Muhammad & B. Fies
*Centre Scientifique et Technique de Bâtiment, Sophia-Antipolis, France*

A. Roxin
*Université de Bourgogne, Dijon, France*

ABSTRACT: Architects and construction engineers need services to check their designs against specific standards, regulations and policies. Many works have been done during the last years to develop checking software. French National project "Digitizing building regulations" aims at formalizing regulation for automatic compliance checking purposes. Our approach is cloud ready, opened, extensible and based on international standard. In this paper, we focus on semantic topological aspects to show gains, limits and perspectives of this approach.

## 1 INTRODUCTION

In a world where industrial operations are increasingly complex and globalized, there is a growing need for delivering services informing the users about the levels of "safety" and compliance of a given building. While technical experts and audit specialists are moving towards BIM (Building Information Modeling), model-checking software allows to control building operations and processes, starting from the first lifecycle phase—building design. This improves project standard compliance while ensuring related costs remain affordable.

In this study, we will focus on how geometrical and semantic constraints can simultaneously be computed in a compliance checking context. This work is a part of the French national project "Digitizing building regulations" («Numeration des règles») which aims at converting existing AEC laws and regulations into logic rules.

## 2 RELATED WORK

### 2.1 *Model checking solutions*

The traditional approach of verification using MVDXML [1] [14] is very limited. Major limitations have been identified such as restricted scope of applying conditions and constraints on several branches of an IFC model; poor geometric analysis of an IFC model; lack of mathematical calculations; support of only static verification of a model, etc. Several works introduced a semantic rule engine oriented [2] approach as a viable alternative.

### 2.2 *Semantic approach*

In the context of BIM (Building Information Modelling), modeling building elements as resources has been identified as an interesting approach for achieving information interoperability (Pauwels, et al., 2011 and Farias, et al., 2014). Existing IFC-related ontologies were conceived as direct syntax mappings between EXPRESS and OWL languages (Beetz, et al., 2009 and Pauwels, et al., 2011). One of the latest and the most solid implementations of an IFC ontology is IfcOWL proposed in (W3C Linked Building Data Community Group, 2014). This version of IfcOWL is also a Candidate Standard (buildingSMART, 2017) for buildingSMART (meaning it is considered as an activity in the process of acquiring international consensus before being submitted to the Standards Committee for a final vote).

In the last years, several initiatives were proposed by the research community, most of them relying on Semantic Web technologies for addressing the issues that appear when relying solely on IFC models. Based on these, buildingSMART International accepted ifcOWL (OWL serialization of the IFC standard) as a candidate standard. Indeed, ifcOWL and its sub-related graph structure allow solving some issues regarding data partitioning, querying and reasoning. By doing so, buildingSMART International (notably its Linked Data Working Group – LDWG) have opened the road to expert model-checking solutions. Still, at this stage, several questions remain unanswered. Most requirements address simultaneously geometrical and semantics aspects.

Taking IFCOWL and IFC to RDF converter as a starting point, we have built transformation services and a methodology that is described in detail in the following subsections.

### 2.3 Collaborative BIM platform

In March 2018, the KROQI platform has been officially launched. It is a collaborative cloud platform for construction SMEs available free of charge to all construction stakeholders including SMEs and VSEs.

It provides document management functionalities, collaborative services, and innovative business services all along the building life cycle.

The MVDXML technology is deployed on the KROQI platform to check the quality of the IFC files that are uploaded by the users. It checks if the structure of the IFC files is compliant with project policies.

The semantic checker to be developed in the frame of the «Numérisation des règles» project will complement this first checker. It will be integrated to the KROQI platform to provide extended rule checking capabilities such as checking topology.

## 3 METHOD

### 3.1 Validating modeling options

The Building Executive Plan (BEP) describes how construction project stakeholders agree to model and share information. This document describes data exchange processes, modeling best practices, exports parameters, nomenclatures and classifications to be used. Most of the time the document references international classification frameworks like Uniformat or Omniclass. Having all actors of the project share the same modeling rules drastically decrease the number of models' variants. The BEP is also used to extend IFC properties with ad-hoc property sets dedicated to specific topics, such as certification level, equipment performances or environmental impact of product. CSTB aims at providing a national modeling charter core to be used as a common starting point when engineering the modeling plan of a project. We are working together with architects, building owners, engineers, audit experts, building quality control agencies and government to reach this goal.

### 3.2 Enriching model with semantic inferences

The semantic approach allows to describe a building as well as a requirement by using the same atomic fact formalism called triplet. The IFC and so ifcOWL standard vocabulary includes hundreds of classes and properties representing the building physical layer. This vocabulary is not natively designed to deal with national considerations, building performances, requirements and building functionalities.

During the first phase of the "Numérisation des règles" project, a first regulation vocabulary has been extracted from a set of regulatory documents. While trying to align IFC vocabulary with regulatory vocabulary we showed that this alignment cannot be based on a simple bijective relation. Indeed, some regulatory concepts can only be defined with a logical statement aggregating various IFC classes. For instance, in our regulation ontology, a simple regulatory concept like "highest storey" is inferred by comparing the "elevation" (ifcOWL property) values of the all "IfcBuildingStorey" (ifcOWL class) for a given building.

The Regulatory ontology is composed by complex regulatory concepts defined on the top of ifcOWL. Regulatory concepts are organized in layers. The ifcOWL vocabulary is the ground layer of a vocabulary pyramid. At the very top of the pyramid (see Figure 1), we find the very regulatory-specific vocabulary. A term of a specific layer is defined by using terms belonging to the lower layer. This layered approach is flexible as it keeps high-level definitions simple by increasing the number of terms.

The whole process that provides a semantic regulatory view from the IFC source model works as follows:

Firstly, we convert an IFC model into an ifcOWL triplet database by using the IFC-to-RDF-converter [3] – provided by BuildingSmart–;

A custom class and relations filter is used to remove non-significant information (concepts not involved in any rule). This operation can be compared to a MVD – Model View Definition—dedicated to regulation. As some IFC terms are almost never addressed by regulatory texts—such as high-level geometry, element history or sensor states, they can be filtered from the model to reduce the amount of data.

Then a "geometrical preprocessor" renders and computes additional geometry aspects. This preprocessor is detailed in the followingsubsection.

Eventually a "semantic preprocessor" infers the model according to the regulatory ontology.

As a major advantage, this last transformation step offers a simplified graph from an ifcOWL model [11].

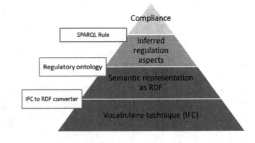

Figure 1.  Regulation vocabulary layers.

For instance, referencing to classification (e.g. Uniformat) takes at least ten triplets with ifcOWL vocabulary whereas one triplet is sufficient with the regulatory ontology.

### 3.3 Geometrical preprocessor

The regulation describes constraints that involve topological relations between objects such as: is below, is inside, is adjacent to. IFC allows to describe elements with various representation models: extrusion, BREP, boolean operations. To compute topological relations a low-level geometry such as bounding boxes is sufficient.

The building boxes are described by using only 6 semantic relations from the regulation ontology: Xmin, Xmax, Ymin, Ymax, Zmin and Zmax (see Figure 2).

The geometric preprocessor infers geometry data to compute triplets materializing bounding box representation of building objects (e.g. furniture20 xMin 50.12). According to ours tests, reasoning on the bounding box representation instead of BREP or geometrical operations representation are reliable and low CPU consuming.

### 3.4 Formalizing regulatory requirements

This section provides details on the methodology used to transform the regulatory constraints, expressed in natural language, into processable queries.

Firstly, regulatory texts are prepared and interpreted by working groups. Experts use a text editor with auto-completion and syntax highlighting capabilities to re-write regulatory texts into semi-formal rules [18]. Each semi-formal rule is supposed to detect non-compliant building elements concerning a very specific aspect. Depending on its complexity, a regulatory rule, in texts, can be divided into several atomic semi-formal rules dealing with complementary aspects. The idea is to keep each semi-formal rule as simple as possible. As a guideline, we suggested that each semi-formal rule begin with "IF" followed by a condition on specific elements and ends with "THEN NON-COMPLIANT".

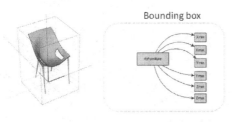

Figure 2. Bounding box corresponding to geometry of a chair.

The second step of this transformation brings the regulatory constraints from the semi-formal to the formal stage. This step is performed by computer scientists familiar with the BIM and with semantic technologies. The semi-formal stage makes it convenient to build formal rules. Each semi-formal rule is translated to a SPARQL query using the same level of vocabulary coming from the regulatory ontology. The direct use of the regulatory controlled vocabulary within the constraint-queries keeps them understandable for construction experts. Geometrical constraints are described by using the geoSPARQL (OGC standard) relations covering all the possible topological relations between two geometries as introduced by the Trinity College of Dublin [5]. Geometry simplification makes it easier to deal with geometry coordinates. This approach could be extended if needed by using functional extensions [6].

### 3.5 Checking regulatory requirements as a KROQI service

Automating code-compliance checking consists of chaining conversion algorithm with geometrical and semantic preprocessors introduced in §3.3. Each time new IFC model is submitted to the semantic checker this chain is triggered so that we obtain a triplet database with the right level of details, aligned with the regulatory ontology. The geometrical preprocessor consists of a java code that returns geometrical triplets from an IFC. The code executes a sequence of semantic forward chaining operations that filters useless information or enhances model with the high level vocabulary. Ontology alignment and extensions to IFCOWL are declared in the regulation ontology. Constraint-queries are organized by regulatory topics so that the KROQI user can select a set of constraints corresponding to his specific needs (fire safety, accessibility, ventilation, acoustics...).

## 4 SYSTEM ARCHITECTURE

Technically, the checking service is hosted by a micro-service architecture among other BIM service like account manager, document sharing, annotations, instant messaging, BIM viewer. Each service is programmatically independent but can communicate with others through the synchronous HTTP/REST queries or through an asynchronous messages queue.

The checking service is developed according to the SaaS pattern. This service can be executed locally or integrated to any collaborative BIM platform. The service (see Figure 3) can be executed through its own user interface that provides a list of rule set to check or silently.

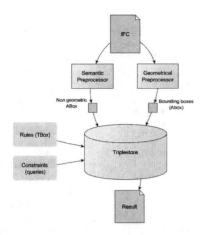

Figure 3. Checking service architecture.

The REST API published by the service is simple. It consists of a unique method "check" and two parameters. The first parameter defines the path to the IFC or IFCZIP model. Constraint-queries are packed as SPARQL files in a ZIP archive. The archive also contains the metadata for end-users' information purpose. The second parameter defines the path to the ZIP file corresponding to the Constraint-queries set.

Next step of the automation algorithm consists of executing, one by one, each formal constraint-query corresponding to the chosen topic.

The result can be returned as XML, BCF, PDF or JSON format according to the accept request-header defined by the REST specification. PDF is used for user display, JSON allows to chain services and BCF can interoperate with BIM-native tools. Whatever the format asked, the result contains for each constraint-query a list of non-compliant building elements identified by their GUID.

The BCF format [15] is simple and easy to implement [16]. The basic content is an issue with comments and references to affected objects (using IFC mechanisms for Global Unique ID's). The format also supports status-information, since issues may be discussed by different users during the workflow. The BCF file makes it convenient to display within a standard IFC viewer, on the top of a graphical representation of a building element the results of the evaluation. The BCF format is independent of the IFC schema version, so it can be used with IFC2 × 3 and IFC4.

## 5 RESULTS

We applied this approach on several constraints of the French regulation on both fire safety and accessibility domains. A first version of a French regulation ontology based on previous CSTB research works has been adapted [7]. A group of domain experts had analyzed, interpreted and finally converted a dozen of regulation texts into about one hundred semi-formal constraints implementing concepts of the regulatory ontology vocabulary. The regulation ontology was then extended with new concepts suggested by experts'. The following paragraphs illustrate in detail two specific use cases.

### 5.1 Checking fire safety requirements

The fire safety domain regulation put some constraint on the structure element performances. Indeed, from the fire safety perspective, the structure elements of the building must bear load during at least one hour in case of fire. This constraint differs according to the building height. On a fire safety point of view, the building height is measured from the ground up to the floor of the latest storey. Translated as a SPARQL query, this constraint implies various high-level concepts such as: highest level (storey with highest elevation), storey floor (lower slab of a storey), structure element (building element that bears load), load-bearing duration in case of fire. The following sequence is triggered by the constraint-query:

1. The building height according to the fire safety standard is calculated on the fly
2. The material performance threshold is determined according to the building height
3. All structure elements are retrieved and their performances are checked
4. Finally, the query returns all structure elements with fire resistance below than the threshold.

Fire safety regulation needs a circulation graph to be checked (see Figure 4).

This circulation graph can be inferred from the IFC model by using the *IfcRelSpaceBoundary* relation between spaces and openings. Sometime some expected relations are missing in the IFC model. To complete information, we also infer spaces adjacencies with a fuzzy intersection rule. The fuzzy intersection is a clash detection rule with tolerance on boundary boxes.

This architecture has been implemented on the KROQI platform providing a full automated checking process on various fire safety and accessibility topics.

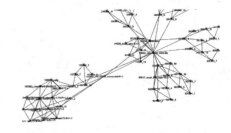

Figure 4. Circulation graph inferred for an IFC model.

## 5.2 Checking accessibility requirements

According to French building regulation, an accessible WC seat (water closet seat) must have a free space of $0.8 \times 1$ meter on its left OR on its right side. The *WC* is a high level regulatory concept corresponding to an *IfcFlowTerminal* [8] with a predefined type equal to *WCSEAT* [8]. Each free space is classified by using the high level regulatory concept *FreeSpace* (see Figure 5). It represents a virtual object (not physical) with three dimensions (bounding box). According to the regulation, this object must not intersect with any other physical element of the building.

The checking process of this constraint is detailed by the following sequence:

1. For each *WC* (water closet seat), two *FreeSpace* elements—and their bounding boxes—are created on-the-fly. One on the left and one on the right.
2. For each *FreeSpace*, the query retrieves all building elements that intersect.
3. Finally, the query returns all *WC* with at least one *FreeSpace* intersecting at least one physical building element.

In this use case, a bounding box representation of a building element is accurate enough to detect clashes with the *WC*. Yet, the bounding box must be oriented to be able to locate left and right sides. This orientation can either be explicit or inferred be comparing dimensions.

In addition to the list of elements that break rules, an explanation query displays the list of elements that intersects with a *FreeSpace*. During our test we manage to detect a handrail that was settled in a wrong location.

## 5.3 Test set

This approach has been validated on various IFC models (see Table 1) from the simplest to the heaviest by using a widely used triplestore implemented in Java with forward and backward chaining capabilities.

While the whole process takes 7 minutes for the Liberty model (2.1 M triplets) on a quad core i5

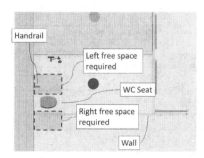

Figure 5. Materialized free space required around WC seat.

Table 1. Models tested.

| Building name | IFC size Mb | Spaces | RDF size M triplets |
|---|---|---|---|
| Chanteloup | 17 | 166 | 0.8 |
| AC9 Haus | 4 | 26 | 1.2 |
| CSTB Building | 15 | 59 | 2.7 |
| EM Building | 2 | 17 | 0.4 |
| Model 23 | 29 | 17 | 1.9 |
| Hitos | 62 | 243 | 13.6 |
| Liberty | 12 | 210 | 2.1 |

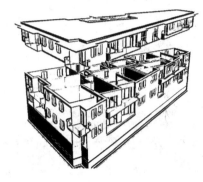

Figure 6. The "Liberty" model displayed in an IFC viewer.

CPU at 2,5Ghz, it takes 10 minutes for the HITOS model (13.6 M triplets). IFC conversion to RDF is the most consuming phase and should be optimized.

## 6 DISCUSSION AND FUTURE WORK

### 6.1 Enriching the regulatory ontology

Web semantic technologies bring a powerful formalism to express terminology and constraints. Yet many BIM platform practitioners do not have full understanding of web semantic technics which limit the number of contributors to a regulation ontology. We believe that it is possible to design tools that support visual query building and enable quick, easy and reusable presentation of terminologies and constraints.

### 6.2 Explaining results

Assigning an object to a rule to point out non-compliance can be insufficient to explain complex cases. The EXPLAIN capabilities of SPARQL could be used to give information on contexts and causes, other elements, properties and relation involved in the computation. Ideally the system should give advices on how to solve compliance issues. Topological requirement could be visualized within the model to point out obstructions, clashing or wrong dimensions. We could materialize those requirement as extra elements and then use the BIM-Snippet [15] brought by BCF 2.0 to return those small partial models in a standard way.

In addition, as each BCF-issue can carry a camera/viewport and even a screenshot of the situation those elements could be generated by the checker itself to illustrate the issue.

### 6.3 Expressing compliance on partial elements

For complex elements, we note that compliance should be expressed on a single part of the element. For instance, a specific aisle of a complex corridor could be less large than required. As the corridor is generally described as a single space we need a new formalism to assign the issue to a specific part of the space.

### 6.4 Improving topological treatments

Next, we should improve the geometrical preprocessor to compute and reason on the topological quantities such as intersection volume, distance between elements or projection surfaces. A full integration of geoSPARQL using WKT [17] description of geometric could help.

### 6.5 Standardizing regulation ontology

To reinforce domain interoperability and build solutions on international and acknowledged standards, we envision to improve the regulation ontology by considering the latest works done in the Linked Building Data Community group such as the Building Topology Ontology (BOT) [9], the ontology defining the core concepts of a building and OntoBREP [10] – ontology for CAD Data and Geometric Constraints.

## 7 CONCLUSION

We showed that our semantic topological approach allows addressing significant issues raised by the computation of building regulation requirements. We've built, explained and executed queries validating the approach by using portable SPARQL syntax and a triple store implementing W3C recommendations.

Obviously, this approach does not address all constraints defined in the regulations. For instance, when considering a complex corridor with multiples aisles—and their different related widths—the bounding box approximation will output erroneous results while computing intersection between spaces.

## REFERENCES

[1] MVDXML specification and support implementation—https://github.com/BuildingSMART/mvdXML.

[2] Fahad, M., BUS, N., Andrieux, F.: Towards Validation of IFC Models with IfcDoc and SWRL A Comparative Study, In the Twelfth International Conference on Internet and Web Applications and Services, pp. 7–13 (2017).

[3] Pauwels P., and Oraskari J.: IFC-to-RDF Converter, https://github.com/IDLabResearch/IFC-to-RDF-converter.

[4] Preidel, C., Bus, N., Borrmann, A., Fies, B.: Pre-Processing IFC Building Models for Code Compliance Checking based on Visual Programming, to appear in the proceedings of ICCCBE (2018).

[5] McGlinn, K., Debruyne,C., McNerney, L., O'Sullivan, D., Integrating Ireland's Geospatial Information to Provide Authoritative Building Information Models, In proc. of 13th conf. on semantic systems, pp. 57–64 (2017).

[6] Chi Zhang, Jakob Beetz: Querying Linked Building Data Using SPARQL with Functional Extensions.

[7] Bouzidi, K.R., Zucker, C.F., Fies, B., Corby, O., Thanh, N-Le.: Towards a Semantic-based Approach for Modeling Regulatory Documents in Building Industry, 9th European Conf. on Product & Process Modelling, ECPPM (2012).

[8] IFC 2 × 3, Specification, http://www.buildingsmart-tech.org/ifc/IFC2 × 3/TC1/html/ifcsharedbldgserviceelements/lexical/ifcflowterminal.htm.

[9] Rasmussen, M.H., Schneider, G.F., Pauwels, P.: Building Topology Ontology (BOT), https://github.com/w3c-lbd-cg/bot.

[10] Perzylo, A.C., Somani, N., Rickert, M., Knoll, A.: An ontology for CAD data and geometric constraints as a link between product models and semantic robot task descriptions. In proceedings of IROS 2015: 4197–4203.

[11] Pauwels P., Roxin, A.: SimpleBIM: from full ifcOWL graphs to simplified building graphs, Ework and ebusiness in architecture, engineering and construction. WORK AND EBUSINESS IN ARCHITECTURE, ENGINEERING AND CONSTRUCTION. pp.11–18 (2016).

[12] Daum, S., Borrmann, A.: Processing of Topological BIM Queries using Boundary Representation Based Methods, Advanced Engineering Informatics, vol. 28(4), pp. 272–286 (2014).

[13] Maïssa, S., Frachet, JP., Lombardo, JC., Bourdeau, M., Soubra, S.: Regulation checking in a Virtual Building, 2002.

[14] Zhang, C., Beetz, J., Weise, M.: Model view checking: automated validation for IFC building models. In Mahdavi, ed. eWork and eBusiness in Architecture, Engineering and Construction: ECPPM (2014).

[15] BuidlingSMART, BCF specification, http://www.buildingsmart-tech.org/specifications/bcf-releases.

[16] Mahdavi, A., Martens, B., Scherer, R., eWork and eBusiness in Architecture, Engineering and Construction: ECPPM 2014.

[17] Open Geospatial Consortium (OGC), geoSPARQL specification, http://www.opengeospatial.org/standards/geosparql.

[18] Wieringa, R., Dubois, E., Huyts, S. Integrating semi-formal and formal requirements, 2005.

# BIM-based compliance audit requirements for building consent processing

J. Dimyadi & R. Amor
*University of Auckland, Auckland, New Zealand*

ABSTRACT: The progressive uptake of ISO-standard BIM (Building Information Modelling) in the Architecture, Engineering, Construction, and Facilities Management (AEC/FM) industry in recent years has started to make common building information accessible to all project stakeholders and processes. Building Consent Authorities (BCA) are exploring how to take advantage of the benefits provided by BIM for automating some of the laborious and repetitive compliance auditing tasks that are part of the core consenting process. This paper describes a proof-of-concept project undertaken in conjunction with a BCA in New Zealand to identify requirements for implementing and to assess the industry's readiness for the BIM-based consenting process. An automated compliance audit tool, ACABIM, was used in the project. The findings of the project suggests that there is a clear indication of productivity, efficiency and analytical benefits gain. The main outcome of the project is a document specifying the minimum data requirements for BIM-based consent submission.

## 1 INTRODUCTION

### 1.1 Background

Compliance audit is the core function of the consenting process in the AEC/FM (Architecture, Engineering, Construction, and Facilities Management) domain. It typically involves a laborious review of every aspect of a submitted paper-based design proposal for compliance with voluminous normative requirements before it can be approved for construction. This is currently a manual undertaking that is inefficient, costly, and error-prone.

The industry's quest for a practical and sustainable automated compliance audit solution has been pursued by researchers for almost half of a century with little success. The emergence of BIM (Building Information Modelling) has provided a new incentive for more active research in this area with some promising results. One practical approach for BIM-based compliance auditing focuses on automating current repetitive and procedural auditing tasks that can easily be executed by machines and leaving those that require human expertise to be solved by human experts (Dimyadi & Amor 2017). A prototype compliance audit tool, ACABIM, has been implemented based on this approach. The philosophy behind the ACABIM approach is human-guided automation where machines are given the tasks of executing human-designed procedures, which treat the building model and compliance criteria as external data sources that can be queried (Dimyadi et al. 2016).

### 1.2 Objectives

This paper describes a proof-of-concept research project for BIM-based consenting undertaken by the University of Auckland (UoA) in conjunction with the Building Consent Authority (BCA) in the city of Christchurch, New Zealand. The first phase of the project was for 6 months and set out to identify requirements for implementing the BIM-based consenting process and to assess the industry's readiness for making BIM-based consent application submissions.

### 1.3 Conventional consenting practice

In the conventional consenting practice, each submission must contain a full set of a documented design with a reasonable level of detail that would enable the building consent processing officer (BCO) to understand the intent. The extent of the documentation required depends on the type of work being proposed. For a new building, there may be several design disciplines involved, e.g. architectural, structural engineering, electrical services, etc. Each discipline would contribute a separate set of documentation to the overall submission. These design disciplines may further be grouped as part of the process into different areas of consenting such as fire safety, natural ventilation, sanitary facilities, etc., and each is subject to compliance assessment with a different set of normative criteria.

If any information in the submission is vague or incomplete, the BCO would issue a Request for

Information (RFI) and put the consent processing on-hold until the requested information is provided.

A BCO typically follows a set of procedures that also acts as a checklist. Any aspect of the submission that does not fit within the scope of the procedures or is considered out of the ordinary would be referred to the specialist BCO for specific assessment and consideration.

### 1.4 New Zealand legal framework

All building construction projects in New Zealand are governed by two primary legislations, namely the Resource Management Act 1991 (RMA), and the Building Act 2004 (the Building Act).

The RMA controls the effects of buildings and the impact of their uses on the environment throughout their life-cycle. The Building Act is part of a framework that controls the actual design and construction of buildings. Another element of the framework is the Building Regulations, which incorporates the New Zealand Building Code (NZBC) containing normative standards on various technical aspects of the building construction, such as stability, fire safety, services and facilities, and energy efficiency. Each technical clause of the Building Code has a corresponding set of normative documents containing the compliance criteria against which each aspect of every building consent submission is audited.

Like many countries in the world, the New Zealand legal framework allows two means of compliance with the law, namely the deemed-to-comply prescriptive method, and the alternative method. In the consenting process, the latter is generally only managed by a specialist BCO and also often calls for a peer-review assessment by third party experts in the specific discipline from the industry.

## 2 PROOF-OF-CONCEPT BIM-BASED CONSENTING PROJECT

### 2.1 Project description

The BIM-based consenting proof-of-concept project was undertaken in conjunction with the BCA in the city of Christchurch, New Zealand. The main motivation behind the project was the general desire to reap the benefits of BIM. More importantly, however, the BCA would like to be able to provide a more efficient and cost-effective consenting service to their customers. Furthermore, having the capability to process BIM-based consent submissions is in-line with the industry's expectation as well as government's guiding principles and strategic priorities.

At the outset of the project, priority consenting requirements were identified based on the initial feedback from an industry consultation. The priorities related to three compliance areas that were controlled by three prescriptive documents from the NZBC, namely G1/AS1 for the provision of sanitary facilities, G4/AS1 for the provision of natural ventilation, and H1/AS1 for energy efficiency in buildings.

The BCA working group for this project consisted of three BCO in the commercial building consenting unit, each was assigned one of the priority compliance areas. All three members of the working group were set up as users on the ACABIM platform and were given two days hands-on training on the system and approach.

### 2.2 The ACABIM Approach

The ACABIM approach relies on three main input components. Firstly, the human-designed executable Compliance Audit Procedures (CAP) that incorporate instructions for accessing various information from two other input components, namely the BIM-based building model in IFC (Industry Foundation Classes, ISO16739), and the official and independently maintained Legal Knowledge Model (LKM) containing normative requirements encoded in the emerging open standard LegalDocML and LegalRuleML computable form (Dimyadi, Governatori, & Amor 2017). Additionally, the approach also supports supplementary human input and interfacing with external computation and simulation tools.

### 2.3 BIM-based building model (BIM)

Three BIM-based building models from current consenting projects, all based in the city of Christchurch, were selected and set up in ACABIM, namely the new public library building, an existing hostel building, and a private residential dwelling. Three other sample building projects

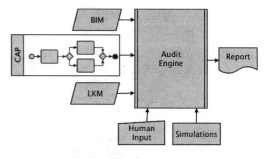

Figure 1. The ACABIM framework (Dimyadi et al. 2016).

located elsewhere in New Zealand were also added for comparative analysis.

The IFC file for each selected building was obtained from respective owners and made available for the project (Figure 2, Figure 3, Figure 4).

### 2.4 Compliance Audit Procedures (CAP)

In the current consenting practice, each BCO follows a set of audit procedures documented and managed using spreadsheets. For the ACABIM approach, these manual procedures are translated into a set of executable CAP workflows using a subset of the open standard BPMN 2.0, which natively supports XML serialisation. ACABIM supports a dedicated high-level domain-specific scripting language that can query BIM and LKM data from each BPMN scriptTask.

Figure 5 shows a CAP that gathers information from BIM and LKM and performs calculations prescribed by the G1/AS1 document to determine the required number of Unisex sanitary facilities. These calculations are based on the intended activity of each space, which also relates to the occupancy load of the building. For the purposes of G1/AS1, the scope of audit has been limited to commercial office buildings due to time constraints. For G4/AS1 and H1/AS1, the auditing process applies to both commercial and residential building types.

### 2.5 Legal Knowledge Model (LKM)

At the outset of the auditing process using ACABIM, the selected priority compliance documents (i.e. G1/AS1, G4/AS1, H1/AS1) were encoded into their computable representations using an interim XML schema with rules representing selected normative provisions for the audit process.

These digital LKM documents were preloaded into a repository for access by ACABIM. In the real situation, it is anticipated that the digital content of LKM will be served from a central repository managed by a central government agency, ideally the same agency that manages the authoring and maintenance of the paper-based source documents. One key feature of the ACABIM approach is process transparency, where the user has an assurance that the content of the LKM is current and reflects the latest amendment of the

Figure 2. The new Christchurch library building.

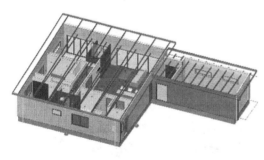

Figure 3. A private residential dwelling in Christchurch.

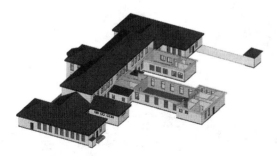

Figure 4. An admin office building in Auckland.

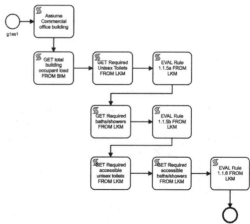

Figure 5. CAP for G1/AS1 (Unisex sanitary facilities).

source document. For this reason, the approach advocates that any amendment to the source document is managed concurrently with the digital computable LKM version.

### 2.6 Supplementary human input

The ACABIM approach recognises the need to support supplementary human input. One of the challenges with sharing BIM data currently is the different level of detail at different stages of a building project. Furthermore, there are variations in the BIM modelling standard as well as the interpretation of various aspects of the IFC standard. Consequently, certain objects and their attribute information may be represented slightly differently within the scope of the specification. More importantly, there are inevitably some interoperability issues pertaining to sharing information using such a highly complex data model.

For example, space activity information is essential for the evaluation of several compliance criteria. However, this information may be presented in the building model rather arbitrarily as part of the object's name instead of using a more specific IFC property set and standard labeling or category code. ACABIM has implemented a feature to translate the space name into a standard space function code in accordance with OmniClass or Uniclass, depending on the user's preference as specified in the CAP.

Another source of supplementary human input relates to information that is not expected to be present in a BIM-based building model, such as consent specific information (e.g. specific location details, zoning classification and code, construction staging details, RMA related information, etc.) as well as consent processing related information such as the legal name of the applicant, project status, etc.

## 3 BIM-BASED CONSENTING EXPERIENCE

### 3.1 BIM-based consenting process preparation

Following the training on using the ACABIM system and understanding the ACABIM approach, each BCO in the working group was provided with an online tool to develop a BPMN-compliant CAP workflow representing their spreadsheet-based audit procedures related to their assigned priority compliance area.

In parallel, a set of LKM representing the selected priority compliance documents were encoded and uploaded into a repository for access by ACABIM.

Each BIM-based building model obtained for the project was uploaded into ACABIM and analysed. The preliminary analysis includes simple checks to see if the building model contains any space objects, i.e. *IfcSpace*, or if the building model contains any window objects, i.e. *IfcWindow*, in the case of an audit against the requirements of G4/AS1, for example.

### 3.2 Compliance audit with G1/AS1

For this exercise, two CAP workflows were developed, one for calculating the number of gendered sanitary facilities (i.e. including WC pans, urinals, showers, hand-wash basins, and accessible facilities) and the other one for unisex sanitary facilities (including showers and accessible facilities).

The new library building model (Figure 2) was initially chosen for this audit. However, it was discovered that the model did not contain any sanitary fixtures or *IfcSanitaryTerminal* objects. Therefore, the audit using both CAP workflows failed prematurely. Subsequently, the admin office building model (Figure 4) was used instead and the audit could proceed to produce an outcome. In this case, the audit using the CAP for the Unisex option produced a FAIL outcome due to inadequate number of toilets provided, but a PASS for the provision of showers and accessible facilities (Figure 6).

### 3.3 Compliance audit with G4/AS1

For this exercise a simple CAP workflow was developed as shown in Figure 7 below. In this CAP, the ratio between the natural ventilation opening area and the floor area is used to evaluate a prescribed rule (i.e. Rule ID 1.2.2.1).

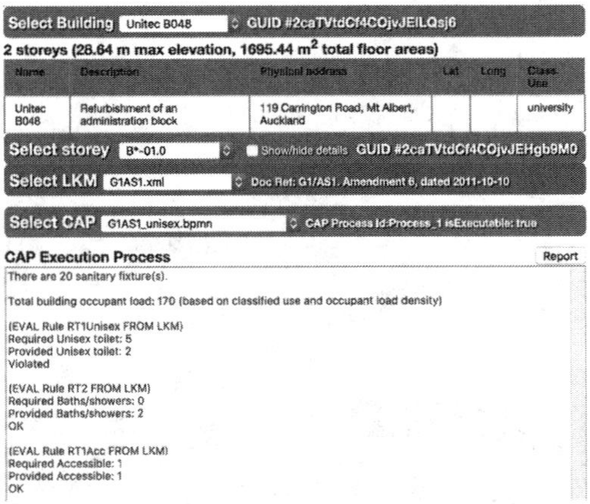

Figure 6. Exemplar ACABIM compliance audit result.

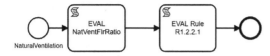

Figure 7. CAP for G4/AS1.

When executed in ACABIM, the process would first go through each space object (*IfcSpace*) and look within its property set *Pset_SpaceThermalRequirements* for the property *NaturalVentilation*. A space intended to be naturally ventilated must have this property set to TRUE. Otherwise, it would be set to FALSE indicating that the space is to be mechanically ventilated.

Once all spaces intended to be naturally ventilated have been identified, the process would then proceed to identify any openable vertical and horizontal vents in each space. The first step in the process is to eliminate all internal windows that have no connection to the outside, i.e. the property *IsExternal* is equal to FALSE. If any external vent is found, then the process looks at the property of each to quantify its openable area. The properties *IfcWindowLiningProperties* and *IfcWindowPanelProperties* are both defined by *IfcWindowStyle*, which is related to *IfcWindow* via *IfcRelDefinesByType*. *IfcWindow* also has *IfcWindowType* property, which defines the shapes, partitioning and the operation of its panels, which in turned is defined by *IfcWindowPanelProperties*, a set of predefined attributes, such as *OperationType*, *PanelPosition*, *etc*. The *OperationType* attribute is an enumerated list of 14 items including *FixedCasement*, which is the only operation mode that is not openable and can be excluded from the selection. However, the complexity increases when not all the panels are fixed. For example, a window with 4 panels, but only the top two panels are openable. So, the procedure adopts a principle that "all panels must be fixed" to be considered "non-openable".

Once all openable areas are identified, their sum is then divided by the floor area to get the required natural ventilation ratio of the space concerned. This complex procedure was described by a large sub-task CAP workflow, which is then called upon by the main CAP shown in Figure 7.

The residential dwelling model (Figure 3) was used for this audit exercise. This model had been developed with a great level of detail and incorporated all the information necessary for the audit. The audit found 7 spaces intended to be naturally ventilated. All of these spaces, except two, complied with the criteria specified in the LKM, i.e. the ratio of opening area to the floor area is greater than 0.05. The two spaces that failed the audit were internal spaces intended to be naturally ventilated but which do not have any openable vent.

An interesting issue was raised as the result of this audit exercise on what level of detail can be expected of a building model submitted at the consenting stage for compliance audit against the G4/AS1 criteria. There must be a minimum level of detail necessary to process a given design for compliance with certain criteria. This minimum level of detail may vary between different compliance aspects. A question was also raised as to whether it was considered reasonable to specify where certain information should be incorporated within the IFC model or to let the designer specify where the information could be found. For example, the designer may incorporate the design intent into a separate window schedule by specifying the *Reference* attribute in the *Pset_Window-Common*. So, a reference code could be assigned to a window object with a user-defined *OperationType* attribute, which specifies whether it is a fully *FixedPanel* window assembly. Otherwise, the total opening area for that window assembly may be specified instead.

### 3.4 Compliance audit with H1/AS1

For this exercise, three rules from the H1/AS1 document (fourth edition, amendment 3) were captured in the LKM, namely paragraphs 1.0.1, 2.1.6, and 3.0.1 of H1/AS1. There are four methods of compliance with H1/AS1, namely the schedule method, calculation method, modelling method, and BPI (Building Performance Index) method. Some of these methods involve using calculation and simulation tools, which ACABIM can support via its interfacing with external tools. It was identified that generating a gbXML (Green Building XML Schema) from ACABIM is a good solution for sharing common information with industry standard energy calculation tools.

The compliance auditing procedures prescribed by the H1/AS1 document were initially translated into one BPMN-compliant CAP, which resulted in a rather complex process model involving four branches of compliance options each with a different set of calculations and potentially interfacing with external tools. This large CAP was subsequently reduced to four smaller CAP workflows for different compliance options. The CAP to determine if the calculation method of compliance is acceptable is shown in Figure 8. The result of internal calculations performed in the process is then used to evaluate Rule 2.1.6 specified in the LKM representing the provision of paragraph 2.1.6 of

Figure 8. CAP for the H1/AS1 calculation method.

H1/AS1, which is a PASS if the ratio of glazing areas to the area of external walls is less than 0.5.

Information related to the energy analysis can be incorporated into the IFC model as part of the building modelling process. The IFC schema supports the *U-Value* property (i.e. thermal transmittance coefficient) for several entity types including the wall, door and window. For wall objects, this property is named *ThermalTransmittance* and can be found in the wall object's property set *Pset_WallCommon*. Similarly, the property Thermal-Transmittance can also be found in the property set *Pset_RoofCommon* for roof objects (e.g. *IfcRoof*) or Pset_SlabCommon for slab objects (*IfcSlab*).

Calculation procedures often use the thermal resistance property or R-Value, which is the inverse of the thermal transmittance of the U-Value. So, there needs to be an internal calculation as part of the procedure to convert the U-Value obtained from the IFC model into the R-Value.

To identify an external wall, window, door, roof, or slab, the property name *IsExternal* can be used, as described before.

## 4 MINIMUM DATA SPECIFICATION FOR BIM-BASED CONSENTING

The main deliverable of the proof-of-concept project is a document for industry feedback specifying the minimum data requirements for BIM-based consenting process. The scope of the specification is limited to three areas of compliance identified as the current priority, namely G1/AS1, G4/AS1, and H1/AS1.

The main reasoning behind the minimum data requirements is that an adequate level of detail must be provided on a set of documentation submitted for the current paper-based consenting process. Any lack of clarity or inadequate level of detail would make compliance audit processing difficult and would result in one or more RFI, which may cause costly delays to the project. The BIM-based consenting process should adopt the same principle that an adequate level of details must be provided in the building model submitted. Otherwise, the same consequence would apply. The need to provide an adequate level of detail in the building model to enable automated compliance audit has been identified as a current problem (Solihin et al. 2017).

The BCA concluded from this project that satisfying the minimum data requirements means having a complete data set in a building model submitted for consent, which would provide the immediate benefit of overall time and cost reduction in the consenting process.

The minimum data specification document specifies the type of information and specific details of each aspect of the compliance area, as relevant, as follows:

Table 1. Exemplar minimum data specification.

| | |
|---|---|
| Format & Standard | The building model must be supplied in the ISO16739 IFC format version IFC2X3 and above with the Coordination View 2.0 |
| Metadata | Project location including GIS coordinates, and full physical address; building name and description; *OccupancyType* property in *IfcBuildingCommon*; Building Importance Level (as the *FireProtectionClass* property in *Pset_BuildingCommon*) |
| Objects or entities | Basic building entities such as *IfcSite*, *IfcBuilding*, *IfcBuildingStorey*, *IfcSpace*, *IfcSpaceBoundaries*, *IfcWall* or *IfcWallStandardCase*, *IfcSlab*, *IfcRoof*, *IfcWindow*, *IfcDoor*, etc., as well as objects related to specific compliance areas such as *IfcSanitaryTerminals*, *IfcFlowTerminals* |
| Properties | Space function/activity; windows/doors with complete details of paneling and glazing areas, all properties in *Pset_WindowCommon* and *Pset_DoorCommon*, as appropriate to the compliance areas; all common quantities such as areas, height, etc.; accessible stairs and routes to be clearly identified; sanitary spaces to be clearly classified into male, female, unisex; all properties in the *Pset_SpaceOccupancyRequirements*; all properties in *Pset_spaceThermalRequirements*, *Pset_spaceThermalDesign*, and *Pset_SpaceThermalLoadDesignCriteria*; all properties in *Pset_SpaceFireSafetyRequirements*; all properties in *Pset_SpaceLightingRequirements*. |

## 5 DISCUSSIONS AND SUMMARY

This paper has presented a proof-of-concept BIM-based consenting project carried out in conjunction with a BCA in New Zealand. A prototype automated compliance audit system, ACABIM, was used to deliver the project successfully. Using the ACABIM approach, the ability to process a BIM-based submission for consent is very similar to the conventional practice. Every building model submitted must contain the right type of information and the level of detail appropriate for the type of consent to be processed. For example, a window in a room cannot be checked for compliance with the provision of adequate natural ventilation if it is unclear if the window is openable. There is currently no guidance available as to what constitutes the minimum level of detail or information in a BIM-based building model to enable consent processing. So, this project has provided preliminary guidance to the industry.

Any feedback received from the industry on the specification will contribute towards a more practical guidance to enable a more effective BIM-based consenting process in the future.

## ACKNOWLEDGEMENT

The work described in this paper is a collaboration between the University of Auckland and Christchurch City Council, New Zealand.

## REFERENCES

Dimyadi, J., & Amor, R. 2017. Automating Conventional Compliance Audit Processes. In *14th IFIP International Conference on Product Lifecycle Management (PLM2017)* (Vol. 14, pp. 209–219).

Dimyadi, J., Clifton, C., Spearpoint, M., & Amor, R. 2016. Computerizing Regulatory Knowledge for Building Engineering Design. *Journal of Computing in Civil Engineering*, *30*(5), pp. 1–13. doi:10.1061/(ASCE)CP.1943–5487.0000572.

Dimyadi, J., Governatori, G., & Amor, R. 2017. Evaluating legaldocml and legalruleml as a standard for sharing normative information in the aec/fm domain. In F. Bosche, I. Brilakis, & R. Sacks (Eds.), *Proceedings of the Joint Conference on Computing in Construction (JC3)* (Vol. I, pp. 637–644). Heraklion, Greece: Heriot-Watt University, Edinburgh, UK. doi:10.24928/JC3–2017/0012.

Solihin, W., Dimyadi, J., Lee, Y.-C., Eastman, C., & Amor, R. 2017. The Critical Role of the Accessible Data for BIM Based Automated Rule Checking System. In *Proceedings of the Joint Conference on Computing in Construction (JC3)* (Vol. I, pp. 53–60). doi:10.24928/JC3–2017/0161.

# Contract obligations and award criteria in public tenders for the case study of ANAS BIM implementation

F. Semeraro, N. Rapetti & A. Osello
*Politecnico di Torino, Turin, Italy*

ABSTRACT: Over the last decades, Building Information Modelling (BIM) methodology has flourished all over the world, heavily influencing processes and procedures of the construction sector. Significant reports outlined BIM as a strategic tool to improve infrastructure quality and better environmental performance. ANAS SpA is a public industrial company, leader into the Italian market of design, construction and maintenance of transportation infrastructures and it also represents the main case study of the research. The research describes ANAS efforts in BIM implementation, focusing on obligations in contract documents related to the information management and award criteria in public tenders. Data collected from ANAS tenders have been analysed in order to compare relative strictness of obligations and award criteria with market responses. In conclusion, main limitations and possible future developments were discussed, related to advantages deriving from the mandatory adoption of BIM methodology, lacks and misinterpretation of the Italian decree effectiveness with BIM methods.

## 1 INTRODUCTION

BIM is an acronym for Building Information Modelling. In different times, different definitions were given by various authors and organizations, each of these pointing out slightly different aspects of it. According to Succar (2009), Building Information Modeling (BIM) can be defined as a set of interacting policies, processes and technologies generating a methodology to manage project information throughout the overall life-cycle. In the horizontal world of infrastructure construction, BIM is contemplated from few years compared to buildings where it represents a methodology in use from several years. But the same BIM features for vertical construction hold equally strong benefits to horizontal infrastructure construction, and the industry has begun to take notice (McGHC, 2012).

Three-dimensional (3D) modelling methods have been applied in government-procured infrastructure projects for many years. However, BIM goes beyond the production of 3D models generally in its philosophy and applications, not entirely limited to the visualization of a facility. It refers to the realization and use of digital parametric objects real-time related to cost and time information during the entire project, as well as in operations & maintenance (O&M) phase. This method becomes relevant in particular for highly complex infrastructure projects, allowing a better management of expenses incurred during major road/highway construction. In particular, it involves satisfying multiple stakeholders and also coordinating the complexity determined by services providers, existing activities and environmental implications.

The European Union issued the 2014/24/EU directive on public procurement, recommending to its member states the adoption of BIM in public projects. In fact, art.22, c.4 the directive states: "For public works contracts and design contests, Member State may require the use of specific electronic tools, such as building information modelling tools or similar".

Italian Government did not defined a clear vision for a BIM program of adoption, since the publication of the EU Directive 24 in 2014. Before a solid public interest, single firms and groups of associations from the AEC sector such as OICE (Engineering and Economical Consulting Organizations) or ANCE (National Association of Building Constructors), recognized values and benefits of the BIM method and started working with existing software and tools, without any particular existing policy, procurement and legal framework. Most of them using foreign standards and protocols. Anafyo reports (Anafyo, 2016) point out in 2015 a business value of €1 billion among private and public sector works related to BIM, and a growing value of €2.6 billion has been reported in 2016 (Anafyo, 2017). A first standing position came from the Minister of Transportations and Infrastructures with the D.Lgs. 18th of April 2016, n.50 which states: "Public contracting authorities may require [...] the use of specific electronic methods and tools, such as building and infrastructure information modelling tools", along with the national strategy of digitization of the public sector. During the same period, a set of technical national standards and contract protocols have been produced by the UNI (Italian Standards Institution), in order to develop a compatible regulatory framework to encourage BIM implementation by professionals of the sector. A final program of BIM adoption for the public sector was defined with the D.M. 1st of December 2017, n.560. The plan consists in six steps of progressive mandatory requirement of BIM methods and tools for public works, starting from 2019 with works over €100 million cost, until 2025 for works under €1 million cost.

ANAS SpA is a public industrial company, leader into the Italian market of design, construction and maintenance of transportation infrastructures and opened to the international market thanks to the subsidiary company ANAS International Enterprise (AIE), 100% held by the Italian Government. A relevant business plan project started in 2016, related to the implementation of BIM methodology into current practices of design and management of ANAS transportation assets. In fact, according to the D.M. 560/2017, the BIM mandatory requirement will affect ANAS' planned investments of the Design and Works Unit for a 32% of the total in 2019, and for a 89% in 2020.

The purpose of this paper is to describe preliminary steps of BIM implementation taken by ANAS, in partnership with Politecnico di Torino, focusing on the ability to require BIM models through the use of specific contract documents and award criteria in public tenders.

## 2 STATE OF THE ART

As much as BIM adoption is often market driven, regulatory frameworks and in particular contract procurement methods have a major impact on the success of BIM-use on medium-to-large construction projects. Contractual frameworks applied to govern design and construction of projects typically predate the use of life-cycle BIM for the delivery of projects [9]. These frameworks may at times rather obstruct than support BIM use. Stakeholders from both project delivery as well as construction law [10] have started to review the contract methods in the light of BIM.

Even if the focus on BIM was put on technological advance, the legal and procurement aspect soon followed as an area of interest. Since 2000, researchers observed that the potential of information sharing and data interoperability offered via new technology can get constraint in practice by contractual frameworks or other legal considerations. Holzer (2007) describes how the use of BIM affects the distribution of roles and responsibilities of individual stakeholders and possible implications for planners who need to work towards specific BIM requirements. The US Associated General Contractors of America released a ConsensusDOCS 301 BIM Addendum (2008) for construction contracts with standardised BIM terminology for stakeholder and model type definition. Klimt (2011) highlights the need to resolve copyright and liability issues related to BIM models and Olatunji & Sher (2010) discuss model ownership and sharing of model data. In order to identify legal problems posed by the adoption of BIM, McAdam (2010) scrutinises UK contract procurement solutions.

The construction sector has a long experience in using traditional procurement methods such as Design-Bid-Build (DBB), Design-Build (DB), and Construction Management (CM). Starting from 1990s a growing interest in Public-Private Partnership (PPP) has emerged, especially after 2007–2008 global financial crises (Osei-Kyei and Chan, 2015). In addition, in order to reduce risks and costs associated to procurement, new procedures are spreading in the AEC industry. Some of them are Cost Led Procurement (CLP), Integrated Project Insurance (IPI), Two Stage Open Book, Integrated Project Delivery (IPD) and Project Alliancing (PA) (Bolpagni, 2013).

Whereas IPD was initially hailed as the ideal procurement method to allow teams to achieve 'full BIM collaboration' the industry is now viewing the idea of 'full BIM' more cautiously (Cleves & Dal Gallo, 2012). IPD as a delivery method is developing globally. It represent the best contractual option one could aspire in the context of BIM. At the same time IPD in its form may well be too idealistic for common adoption throughout construction projects globally. IPD currently gets applied under Integrated Forms of Agreement (IFOA) on a number of infrastructure projects, or by selected clients who have learned to adapt its use in order to manage the supply chain associated to procuring their projects (Alarcon et. al., 2011). It will require markets to mature in terms of BIM knowledge, and stakeholders to become more comfortable with novel ways of procurement.

## 3 METHODOLOGY

The ANAS' Design Coordination unit efforts in BIM implementation are mainly driven by the mandatory terms of requirement set by the DM 560/2017 which established the 1st January 2019 as the first step of BIM mandatory adoption for tenders over €100 million. Greater issues for ANAS related to the fulfilment of the mandate concern a) the internal reorganization of design processes, in particular the new information workflow resulting from Specialty fields involved into the infrastructure design process; b) the capability to require BIM models through the use of specific contracts for public tenders; and c) the ability to control and verify products delivered by external providers of services, especially engineering and architecture services.

Actions taken by ANAS in order to be able to subcontract BIM-compliant works and engineering and architectural services of infrastructure projects in public tenders are aimed at operating in the legal framework set by the D. Lgs. 50/2016, despite of the actual research trend which promotes the use of the collaborative agreement framework as a contract between parties. This is due to the broad idea of the legislator to separate parties and activities, which denies, for example, to constructors with very few exceptions, the possibility to design the last level of project development, according to the Italian three-level system of design development prior the construction phase. These conditions determine a time for adoption of the collaborative framework much longer than the term fixed by the DM 560/2017 for BIM mandatory requirement.

ANAS approach was to develop i) technical documents related to information management of BIM models (EIRs), and to define ii) standard award criteria in public tenders related to the BIM knowledge and expertise (AC), in a "design-bid-build" scenario, where contractors are selected by the Most Economically Advantageous Tender (MEAT) criterion.

### 3.1 Employer's information requirements characterisation

EIRs are produced as part of a wider set of documentation for use during project procurement and shall typically be issued as part of the tender documentation. The development of the EIR shall start either with the assessment of an existing asset, leading to the development of the employer's need, or directly with the employer's need if no existing asset or asset information model is to be considered (Succar, 2009). ANAS priority, as a contracting authority, is to be able to subcontract works, engineering and architecture services and support services. For this reason three different types of technical addendum to contract documents were developed, related to information management of BIM models. The first type is specifically designed for support services (SS), the second one concerns engineering and architectural services (EAS) and the last one is fitted for construction works (W). In Table 1 are synthetized EIR contents for each type.

Contents of the EIR can be grouped in three main areas: a) technical section, b) management section, and c) commercial section. Section a) is dedicated to technical specifications essential for the realization of BIM models, such as hardware and software infrastructure, data exchange formats or company standards. Section b) includes requirements concerning information modelling and management processes, such as BIM uses and objectives, level of developments (LODs), or roles and responsibilities. Section c) is related to specific provisions of the service/work, such as model delivery plan, payments or intellectual property.

Table 1. EIR contents per type.

| | CONTENTS | SS | SEA | W |
|---|---|---|---|---|
| Technical section | hardware & software | ● | ● | ● |
| | IT solutions (provided or available) | ● | ● | ● |
| | exchange formats | ● | ● | ● |
| | ANAS standards | ● | ● | ● |
| | objectives across phases | | ● | ● |
| | BIM objectives and uses | ● | ● | ● |
| | Attended information deliverables | ● | ● | ● |
| | LODs | ● | ● | ● |
| Management section | Roles & Responsibility | ● | ● | ● |
| | Management of subcontractors | ● | ● | ● |
| | Validation procedures | ● | ● | ● |
| | Clash and code detection | | ● | ● |
| | 4D | ● | ● | ● |
| | 5D | ● | ● | ● |
| | 6D | | ● | ● |
| | 7D | | ● | ● |
| Commercial section | Intellectual property | ● | ● | ● |
| | Communications | ● | ● | ● |
| | BEP dev and approv | ● | ● | ● |
| | Specific provisions | ● | ● | ● |

## 3.2 Award criteria characterization

The award criteria are the criteria that constitute the basis on which a contracting authority chooses the best tender and awards a contract. These criteria must be established in advance by the contracting authority and must not be prejudicial to fair competition [2]. In particular, award criteria had to ensure an equal treatment and non-discrimination, which means that must be non-discriminatory and must not be prejudicial to fair competition; and also the transparency, which means that the award criteria must be set in advance and duly disclosed to tenderers.

Award criteria are characterized by the type of criterion and its weight. ANAS developed a non-cost related BIM criterion, subdivided in sub-criteria, with a specific weight. Even if the weight of the criterion is rapidly increasing over the time, the BIM criterion is mainly composed by sub-criteria which tries to evaluate three main aspects of tenderers' offers: a) willingness to deliver BIM models as required by the contracting authority, even if is not yet compulsory; b) technical capability, related to the BIM system and managerial aspects proposed via BEP pre-contract document; and c) previous BIM modelling experiences of linear infrastructure projects, which represent the principal typology of ANAS projects.

In Table 2 were synthetized sub-criteria composing the BIM criterion for different type of tenders (EAS, W).

## 3.3 Testing EIRs and AC

Over the total amount of ended public telematic tenders published by ANAS in 2017, a sample of 19 tenders with open procedure have been selected for testing EIRs and AC, since the beginning of the ANAS' "BIM initiative" in July 2016. The sample is represented by 5 tenders of engineering and architecture services of different costs and for different stages of the project development, 2 tenders of construction works, 10 tenders of extraordinary maintenance works and 2 tenders of design support services. In each case, different strategies of EIRs and AC adoption were tested, in order to

Table 2. Synthesis of tenders with EIR and AC used.

| NUMBER | CODE | TYPE | DESCRIPTION | COST | EIR | AC | AC WEIGHT |
|---|---|---|---|---|---|---|---|
| 1 | DG 25-17 | SERVICES | DETAILED DESIGN | ≤ €30 MILLIONS | EAS | ● | 2 |
| 2 | DG 26-17 | SERVICES | DETAILED DESIGN | €30 < x < €50 MILLIONS | EAS | ● | 2 |
| 3 | DG 27-17 | SERVICES | DETAILED DESIGN | €50 < x < €100 MILLIONS | EAS | ● | 2 |
| 4 | DG 28-17 | SERVICES | DETAILED DESIGN | > €100 MILLIONS | EAS | ● | 2 |
| 5 | DG 29-17 | SERVICES | CONCEPT DESIGN | – | EAS | ● | 2 |
| 6 | BO 02-17 | WORKS | CONSTRUCTION | €5 MILLIONS | W | | – |
| 7 | CA 09-17 | WORKS | CONSTRUCTION | – | W | ● | 5 |
| 8 | DG 21-17 | WORKS | EXTRAORDINARY MAINTENANCE | – | W | ● | 7 |
| 9 | DG 20-17 | WORKS | EXTRAORDINARY MAINTENANCE | – | W | ● | 3 |
| 10 | DG 36-17 | WORKS | EXTRAORDINARY MAINTENANCE | – | W | ● | 3 |
| 11 | DG 37-17 | WORKS | EXTRAORDINARY MAINTENANCE | – | W | ● | 3 |
| 12 | DG 38-17 | WORKS | EXTRAORDINARY MAINTENANCE | – | W | ● | 3 |
| 13 | DG 45-17 | WORKS | EXTRAORDINARY MAINTENANCE | – | W | ● | 3 |
| 14 | DG 46-17 | WORKS | EXTRAORDINARY MAINTENANCE | – | W | ● | 7 |
| 15 | CB 19-17 | WORKS | EXTRAORDINARY MAINTENANCE | €5 MILLIONS | W | ● | 4 |
| 16 | CB 34-17 | WORKS | EXTRAORDINARY MAINTENANCE | €4 MILLIONS | W | ● | 4 |
| 17 | DG 31-17 | WORKS | EXTRAORDINARY MAINTENANCE | €10 MILLIONS | W | ● | 4 |
| 18 | DG UP 86 | SUPPORT | BIM MODELING | €20.000 | SS | | – |
| 19 | DG UP 89 | SUPPORT | BIM MODELING | €20.000 | SS | | – |

evaluate strictness of requirements and market responses.

The first test performed was the tender BO 02-17, a tender for construction works where the contractor had to update the existent BIM model to an as-built Level of Development (LOD). There were no BIM-oriented award criteria in the call for bid, but a specific EIR, W type, as a technical addendum to the contract which identified activities and duties about information management of the BIM model. Furthermore, the choice of not considering award criteria for the first test was made in order to avoid limitations in the number of participants.

Following tests were performed in combination between a BIM criterion in the call for bid, varying its weight according to the total cost of the service/work and to the level of confidence gained over the time, and a EIR as a contract document. In fact, both EIRs and AC were modified and implemented during tests, according to tenders outcomes. In Table 3 there is a detailed description of tenders and the use of EIRs and AC during tests.

Generally, in a standard ANAS MEAT, the maximum possible score is 100 points, subdivided in 70 points for the technical offer and 30 points for the price. The BIM criterion was involved in the technical area with a weight from 0 to 3 points, during first tests performed, according to lower cost of the tender, which was usually lower than € 5 million. For greater costs, the BIM criterion weight increased from a maximum of 3 to 7 points. In the majority of cases, a specific EIR was provided as addendum to the contract.

Table 3. BIM criterion.

| SUB CRITERIA | SEA | W |
|---|---|---|
| To develop the as-built model at LOD F, according to the standard UNI 11337-4 and the EIR, in open and proprietary exchange formats. | ● | ● |
| To provide a CDE for the BIM model, accessible by ANAS | ● | ● |
| To guarantee the interoperability with ANAS software platform | ● | ● |
| To implement the BIM model with operations and maintenance information | | ● |
| To deliver a BEP pre-contract | ● | ● |
| To report previous BIM experience (particularly in linear infrastructure projects) | ● | ● |

## 4 RESULTS AND DISCUSSION

ANAS developed a total of three different types of EIR, one for support services (SS), one for engineering and architecture services (EAS) and one for works (W). Thanks to tests performed in real public tenders, several versions of the EIR have been produced, which implemented specific areas of the technical document.

In the SS case, the BIM model delivered to ANAS was not aligned with expectations pointed out in the EIR. For this reason, a second version of the EIR for SS has been developed, implementing and making more restrictive some specific areas of the technical document. In particular, contents from the management section, such as LODs were detailed with specific information tables containing BIM uses and objectives. Furthermore, a BEP post-contract requirement was added, in order to better clarify roles and responsibilities of the contractor related to BIM activities.

In the EAS case, tests results are not completely available. This is due to the particular typology of the tender, which is generally a Frame Agreement (FA) of 4 years duration and which comprehend a series of design services until the realization of the total cost of the contract. In this context, a general EIR has been developed as a contract document during the tender phase; in conjunction with each single service, a specific addendum will be provided to the contractor in order to manage BIM activities of the particular project. Despite of that, preliminary considerations can be done. In fact, also in this case, evidences from tests demonstrate a useless contractual value of LODs if not combined with BIM objectives and uses. Furthermore, clear definitions of attended deliverables and validation procedures result in better project outcomes.

In the W case, tests performed can be grouped in two main type of works: construction and extraordinary maintenance. Furthermore, according to the typology of the contract, ANAS requirements were mainly oriented in managing 4D, 5D and 7D information related to the BIM model. Results from tests revealed that fields of implementation were mainly related to validation procedures and clash and code detection.

Areas of EIR implementation per type of contract are shown in Figure 1.

Lessons learnt from EIR development were useful to create an organized standard version of BIM AC for different type of tenders and contracts. The hypothesis of using AC in ANAS public tenders was confirmed as a positive action, not prejudicial for ANAS providers, only if criteria were not too strict. This is possible introducing a criteria which requires the delivery of a BEP pre-contract, in which the tenderer explains the proposed structure of the

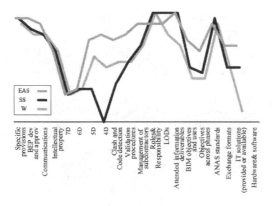

Figure 1. Areas of EIR implementation across versions.

BIM model and the managerial aspects of roles and responsibilities. In the Italian context, the BIM maturity level of the construction sector is dramatically low. For this reason, ANAS decided to assign 4 points over a total of 7 points to the sub-criterion of previous experiences in BIM projects, especially in linear infrastructure projects, as a synonym of a professional warranty, awarding BIM expertise.

## 5 CONCLUSIONS

The ANAS' Design Coordination unit efforts in BIM implementation presented in this work clearly defined a preliminary framework of BIM adoption for a contracting authority of transportation infrastructure projects. Detailed contract requirements concerning information modelling and management and organized standard versions of BIM award criteria are crucial for the contracting authority in order to subcontract services and works concerning BIM methods.

In terms of EIR, further investigations will analyse data deriving from tenderers, in order to implement specific requirements and new technical solutions for an always more flexible and useful contractual form. Moreover, further research will investigate on award criteria, in order to increase the weight of the BIM criterion, which is a 10% of the total score of the technical part, and more specific sub-criteria will be developed.

## REFERENCES

Alarcon, I. Digby, C., and Tommelein, I.D. (2011) Collaborating with a permitting agency to deliver a Healthcare Project: Case Study of the Sutter Medical Center Castro Valley (SMCCV), Proceedings of the 19th Annual Conference of the International Group for Lean Construction IGLC 19, Lima, Peru, July 13–15.

Anafyo (2016). Il BIM in Italia: un quadro della situazione. Technical report, Anafyo.

Anafyo (2017). Italian Bim Report 2016. Technical report, Anafyo.

Bolpagni, M. (2013) The implementation of BIM within the public procurement. A model-based approach for the construction industry. VTT Technology.

British Standard Institute. 2013. *PAS 1192-2:2013*. BSI Standard Limited.

ConsensusDOCS (2008) ConsensusDOCS 301: Building Information Modeling (BIM) Addendum.

Construction, M.H. (2012). The business value of BIM for infrastructure: Addressing America's infrastructure challenges with collaboration and technology. *Smart Market Report*.

Cleves, J. & Dal Gallo, L. (2012) Integrated Project Delivery: The Game Changer, Proceedings of the American Bar's Construction Law Forum 2012, Las Vegas (USA).

Ente Italiano di Unificazione (2017). UNI 11337, Gestione digitale dei processi informativi delle costruzioni.

European Parliament. (2014). Legislation. Official Journal of the European Union, 57(L94), 396.

Holzer, D. (2007) Are you talking to me? Why BIM alone is not the answer, Proceedings of AASA 2007 Conference, Sydney (Australia).

Klimt, M. (2011) Legalese: The Problem with BIM, Architects Journal (AJ), UK.

Kuiper, I. and Holzer, D. (2013) Rethinking the contractual context for Building Information Modelling (BIM) in the Australian built environment industry, Australasian Journal of Construction Economics and Building, Vol.13, No.4, UTS Press, Sydney (Australia).

McAdam, B. (2010) Building information modelling: the UK legal context, International Journal of Law in the Built Environment, Vol. 2 Iss: 3, pp.246–259.

Olatunji, O.A. & Sher, W.D. (2010) Legal implications of BIM: model ownership and other matters arising, CIB World Congress 2010 Proceedings, Salford Quays, UK.

Osei-Kyei, R. and Chan, A.P.C. (2015) Review of studies on the Critical Success Factors for Public-Private Partnership (PPP) Projects from 1990 to 2013. International Journal of Project Management 33 (2015) pp. 1335–1346.

Repubblica Italiana (2016). D.Lgs. 50/2016, Attuazione delle direttive 2014/23/UE, 2014/24/UE e 2014/25/UE sull'aggiudicazione dei contratti di concessione, sugli appalti pubblici e sulle procedure d'appalto degli enti erogatori nei settori dell'acqua, dell'energia, dei trasporti e dei servizi postali, nonche' per il riordino della disciplina vigente in materia di contratti pubblici relativi a lavori, servizi e forniture.

Repubblica Italiana (2017). Decreto Ministeriale n° 560 del 01/12/2017.

SIGMA. 2011. *Setting the Award Criteria, Public Procurement Brief 8*. Retrieved from: http://www.sigmaweb.org/publications/SettingtheAwardCriteria_Brief8_2011.pdf

Succar, B. (2009). Building information modelling framework: A research and delivery foundation for industry stakeholders. *Automation in construction*, *18*(3), 357–375.

# Author index

Abanda, F.H. 429
Abualdenien, J. 187
Aguinaga, S. 137
Alhamami, A. 31
Alhava, O. 127
Álvarez, S. 73, 223
Amor, R. 465
Amrollahibuki, R. 243
Andrés, M. 73
Arambarri, J. 169
Arenaza, X. 281
Arroyo, A. 169
Asgari, D. 39
Azhar, S. 97

Barbero, A. 291
Beltrani, L. 43
Björnsson, I. 297
Bonduel, M. 347, 399
Bonelli, S. 13, 21
Bonsma, P. 407
Borrmann, A. 147, 187, 231, 265, 439
Brachet, A. 137
Braun, A. 265
Brizzolari, A. 13, 21
Bus, N. 361, 459

Carré, S. 137
Castronovo, F. 113
Cerè, G. 51
Chauhan, K. 273
Crawford, J. 421

Dejaco, M.C. 21
Delval, T. 137
Deom, S. 137
Dimyadi, J. 465

Eftekharirad, R. 355
Eickeler, F. 147
Ekholm, A. 297
Esser, S. 439

Fahad, M. 361
Fakhimi, A.H. 97

Farhang, T. 65
Fies, B. 361, 459
Fioravanti, A. 123
Fisher, A. 329

Gade, P.N. 177, 305
Gasparella, A. 65
Ghoreishi, S.R. 97
Giangiacomi, S. 251
Giannakis, G. 57
Giudice, M.D. 291
Giuliani, L. 43

Haavisto, A. 127
Hamdan, A. 369
Hammad, A. 243, 355
Hansen, R. 305
Haronian, E. 257
Hjelseth, E. 3, 127
Hviid, C.A. 399

Ismail, A. 377
Izkara, J.L. 169

Jahr, K. 265
Jensen, R.L. 391
Jolibois, A. 137
Jones, S. 313
Jones, S.W. 127

Kamari, A. 197
Karlshøj, J. 43, 399
Katsigarakis, K. 57
Kazakov, K. 207
Kirkegaard, P.H. 197
Kiviniemi, A. 13, 21, 127, 313
Kjems, E. 177
Klein, R. 347
König, M. 337
Kouhestani, S. 103

Lavikka, R.H. 273
Leifgen, C. 447
Leon, M. 321
Lilis, G.N. 57
Liu, C. 87

Liu, C.H. 79
Liu, L. 79
Lodewijks, J. 177

Mahamadu, A.M. 321
Mahdavi, A. 65, 385, 415
Mailhac, A. 137
Manzone, F. 291
Mastrolembo Ventura, S. 113
Mêda, P. 215
Mediavilla, A. 169, 281
Mirarchi, C. 13, 21
Mistre, A. 147
Möller, L.K. 447
Molnár, M. 297
Moreira, J. 215
Morozov, S. 207
Muhammad, F. 459
Muhič, S. 439
Munir, M. 13, 313
Mwiya, B. 127

Namork, S.A. 155
Nche, C. 429
Nik-Bakht, M. 103, 355
Nordahl-Rolfsen, C. 155
Novembri, G. 123

Oraskari, J. 407
Osello, A. 473

Papadonikolaki, E. 321
Pauwels, P. 347, 391, 407
Peltokorpi, A. 273
Pernigotto, G. 65
Petri, I. 31
Petrova, E. 391
Philipps, P. 281
Pilati, G. 65
Pinti, L. 13, 21
Poinet, P. 329
Polter, M. 163
Prieto, I. 169

Rapetti, N. 473
Rasmussen, M.H. 347, 399, 407

Regidor, M. 73
Rezgui, Y. 31, 51
Rossini, F.L. 123
Rovas, D.V. 57, 223
Roxin, A. 459
Rüppel, U. 447

Sacks, R. 257
Sánchez, V. 281
Sanz, R. 223
Sardroud, J.M. 97
Scherer, R.J. 163, 369, 377
Scheurer, F. 329
Schneider Jakobsen, L. 177
Schneider, G.F. 407
Schultz, C. 197
Sebesi, Y. 281

Seferi, R. 251
Semenov, V. 207
Semeraro, F. 473
Seppänen, O. 273
Sigalov, K. 337
Soula, J. 137
Sousa, H. 215
Stoitchkov, D. 439
Sujan, S.F. 127
Svidt, K. 305, 391

Tah, J.H.M. 429
Taheri, M. 385, 415
Tamke, M. 329
Tauscher, H. 421
Thomsen, M.R. 329
Trzeciak, M. 231

Usman, Z.S. 429

Valmaseda, C. 223
Vergauwen, M. 347
Vilgertshofer, S. 439

Wagner, A. 447
Wang, J.L. 79
Wang, N. 79
Wheatcroft, J.M. 127
Winkelbauer, T. 439

Yang, H. 87

Zhao, W. 51
Zolotov, V. 207